Process Control INSTRUMENTATION TECHNOLOGY

Curtis D. Johnson

CUSTOM EDITION

Taken from:

Process Control Instrumentation Technology, Eighth Edition
by Curtis D. Johnson

Custom Publishing

New York Boston San Francisco
London Toronto Sydney Tokyo Singapore Madrid
Mexico City Munich Paris Cape Town Hong Kong Montreal

Cover Art: Courtesy of Photodisc / Getty Images

Taken from:

Process Control Instrumentation Technology, Eighth Edition
by Curtis D. Johnson

Published by Prentice Hall
Upper Saddle River, New Jersey 07458

This special edition published in cooperation with Pearson Custom Publishing.

Printed in the United States of America

7 8 V0356 10 11 12 13 14

2008200213

CF

Pearson
Custom Publishing
is a division of

PEARSON

www.pearsonhighered.com

ISBN 10: 0-555-00962-9
ISBN 13: 978-0-536-00962-8

PREFACE

This edition of *Process Control Instrumentation Technology* is the result of a very intense and hectic several months spent incorporating some of the major changes in the way control systems and process control are implemented. Many changes have been dictated by observations of the changing technology of control systems. Following is a brief summary of some of the most important revisions.

The section on Industrial Electronics is renamed Power Electronics and rewritten to include more current devices such as GTOs, MOSFETs, and IGBTs. The coverage of ac motors is reworked to bring it up to date. Although this is not an electrical machines and electronics text, the brief summary of these topics should at least be current.

The greatest revolution in control systems is the growing employment of computers and computer networks in process control. Chapter 11 is modified significantly to improve coverage of computer-based controllers and the rapidly developing field of distributed control and control system networks. The discussion and treatment of the fieldbus networks now in common use is expanded. Finally, every example and problem in the book is reworked and checked for accuracy, as are those in the solutions manual.

This book is intended for a reader who has at least a working knowledge of algebra and trigonometry, but familiarity with calculus would also enhance understanding. In addition, a solid understanding of basic electricity along with some working knowledge of analog and digital electronics is essential to good comprehension of the topics.

The text is based upon a two-semester sequence. The first semester is expected to cover the first six chapters. This starts with a general review of control systems and process control in Chapter 1, which also includes the very important topics of first- and second-order time response. Chapters 2 through 6 cover the general topics of measurement from analog and digital signal conditioning to the most common sensors used in the control industry.

Chapters 7 through 12 constitute the second semester and cover the final control element and PLCs, as well as the principles of controller action, analog and digital controllers, process-control networks, and an overview of control-loop performance and tuning. Many users simply pick those chapters that best suit the educational objectives of their program.

I used the World Wide Web extensively to research topics I wanted to include in this revision, and was amazed at the extent of information about every little aspect of control systems and process control. I found excellent material—from elegant tutorials on loop tuning to specifications associated with fieldbus standards. I was tempted to include the URLs of some of these sites, but decided against that because of the volatility of such sites. Instead, I urge the reader to use available search engines to seek out tutorials, informational sites, and application notes associated with process control and instrumentation. The Website maintained for this book (*www.uh.edu/~tech13v/pcit*) can be used for posting identified errors, corrections, and other information about process control.

The author can also be contacted directly via e-mail at: *cjohnson@uh.edu*.

Acknowledgments

I would like to thank my wife, Helene Blake, for her continued support through these many years and many revisions. She is helping me now, even as I type, with some of the necessary editing of the manuscript.

A number of reviewers provided excellent suggestions for improving the content and presentation in the book. Nearly all these suggestions were incorporated (I regret that not all of them could be made). The reviewers were Joe E. Ashby, Indiana State University; Gerald W. Cockrell, Indiana State University; Sri R. Kulla, Bowling Green State University; and Paul Ricketts, New Mexico State University.

Ancillaries

An online Instructor's Manual with PowerPoint™ slides is also available to instructors through the Johnson catalog page at **www.prenhall.com**. Instructors can search for a text by author, title, ISBN, or by selecting the appropriate discipline from the pull down menu at the top of the catalog home page. To access supplementary materials online, instructors need to request an instructor access code. Go to **www.prenhall.com**, click the **Instructor Resource Center** link, and then click Register Today for an instructor access code. Within 48 hours of registering you will receive a confirming e-mail including an instructor access code. Once you have received your code, go to the site and log on for full instructions on downloading the materials that you wish to use.

CONTENTS

CHAPTER 1

Introduction to Process Control

INSTRUCTIONAL OBJECTIVES

This chapter presents an introduction to process-control concepts and the elements of a process-control system. After you read this chapter and work through the example problems and chapter problems you will be able to:

- Draw a block diagram of a simple process-control loop and identify each element.
- List three typical controlled variables and one controlling variable.
- Describe three criteria to evaluate the performance of a process-control loop.
- Explain the difference between analog and digital control systems.
- Define supervisory control.
- Explain the concept behind process-control networks.
- Define accuracy, hysteresis, and sensitivity.
- List the SI units for length, time, mass, and electric current.
- Recognize the common P&ID symbols.
- Draw a typical first-order time response curve.
- Determine the average and standard deviation of a set of data samples.

1.1 INTRODUCTION

Human progress from a primitive state to our present complex, technological world has been marked by learning new and improved methods to control the environment. Simply stated, the term *control* means methods to force parameters in the environment to have specific values. This can be as simple as making the temperature in a room stay at 21°C or as complex as manufacturing an integrated circuit or guiding a spacecraft to Jupiter. In general, all the elements necessary to accomplish the control objective are described by the term *control system.*

The purpose of this book is to examine the elements and methods of control system operation used in industry to control industrial processes (hence the term *process control*). This chapter will present an overall view of process-control technology and its elements, including important definitions. Later chapters will study the elements of process control in more detail.

1.2 CONTROL SYSTEMS

The basic strategy by which a control system operates is logical and natural. In fact, the same strategy is employed in living organisms to maintain temperature, fluid flow rate, and a host of other biological functions. This is natural process control.

The technology of artificial control was first developed using a human as an integral part of the control action. When we learned how to use machines, electronics, and computers to replace the human function, the term *automatic control* came into use.

1.2.1 Process-Control Principles

In process control, the basic objective is to regulate the value of some quantity. To regulate means to maintain that quantity at some desired value regardless of external influences. The desired value is called the *reference value* or *setpoint.*

In this section, a specific system will be used to introduce terms and concepts employed to describe process control.

The Process Figure 1.1 shows the process to be used for this discussion. Liquid is flowing into a tank at some rate, Q_{in}, and out of the tank at some rate, Q_{out}. The liquid in the tank has some height or level, h. It is known that the output flow rate varies as the square root of the height, $Q_{out} = K\sqrt{h}$, so the higher the level, the faster the liquid flows out. If the output flow rate is not exactly equal to the input flow rate, the level will drop, if $Q_{out} > Q_{in}$, or rise, if $Q_{out} < Q_{in}$.

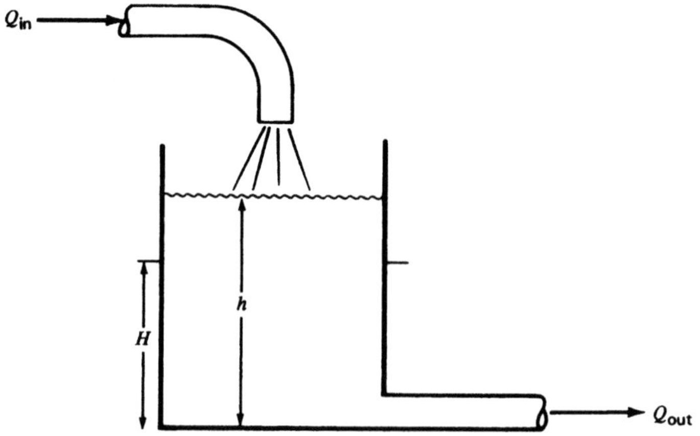

FIGURE 1.1
The objective is to regulate the level of liquid in the tank, h, to the value H.

This process has a property called *self-regulation.* This means that for some input flow rate, the liquid height will rise until it reaches a height for which the output flow rate matches the input flow rate. A self-regulating system does not provide regulation of a variable to any particular reference value. In this example, the liquid level will adopt some value for which input and output flow rates are the same, and there it will stay. But if the input flow rate changed, then the level would change also, so it is not regulated to a reference value.

EXAMPLE 1.1

The tank in Figure 1.1 has a relationship between flow and level given by $Q_{\text{out}} = K\sqrt{h}$ where h is in feet and $K = 1.156\ (\text{gal/min})/\text{ft}^{1/2}$. Suppose the input flow rate is 2 gal/min. At what value of h will the level stabilize from self-regulation?

Solution

The level will stabilize from self-regulation when $Q_{\text{out}} = Q_{\text{in}}$. Thus, we solve for h,

$$h = \left(\frac{Q_{\text{out}}}{K}\right)^2 = \left(\frac{2\ \text{gal/min}}{1.156\ (\text{gal/min})/\text{ft}^{1/2}}\right)^2 = 3\ \text{ft}$$

Suppose we want to maintain the level at some particular value, H, in Figure 1.1, regardless of the input flow rate. Then something more than self-regulation is needed.

Human-Aided Control Figure 1.2 shows a modification of the tank system to allow artificial regulation of the level by a human. To regulate the level so that it maintains the value H, it will be necessary to employ a sensor to measure the level. This has been provided via a "sight tube," S, as shown in Figure 1.2. The actual liquid level or height is called the *controlled variable.* In addition, a valve has been added so that the output flow rate can be changed by the human. The output flow rate is called the *manipulated variable* or *controlling variable.*

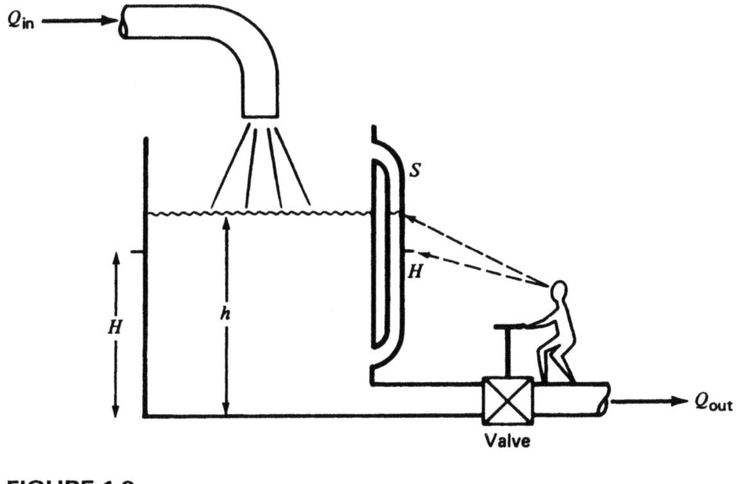

FIGURE 1.2
A human can regulate the level using a sight tube, S, to compare the level, h, to the objective, H, and adjust a valve to change the level.

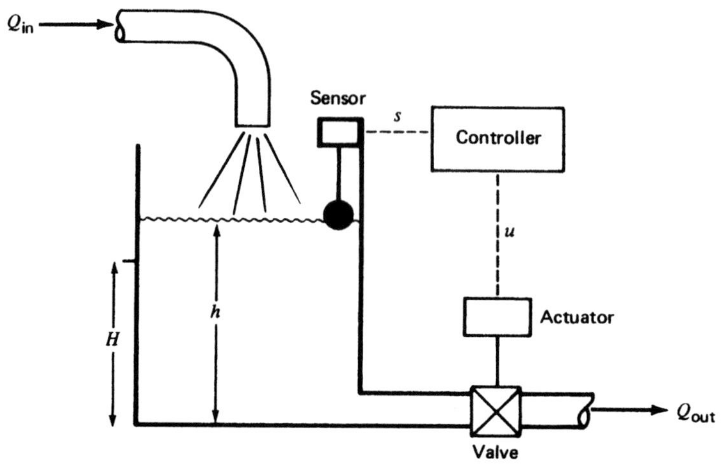

FIGURE 1.3
An automatic level-control system replaces the human with a controller and uses a sensor to measure the level.

Now the height can be regulated apart from the input flow rate using the following strategy: The human measures the height in the sight tube and compares the value to the setpoint. If the measured value is larger, the human opens the valve a little to let the flow out increase, and thus the level lowers toward the setpoint. If the measured value is smaller than the setpoint, the human closes the valve a little to decrease the flow out and allow the level to rise toward the setpoint.

By a succession of incremental opening and closing of the valve, the human can bring the level to the setpoint value, H, and maintain it there by continuous monitoring of the sight tube and adjustment of the valve. The height is regulated.

Automatic Control To provide automatic control, the system is modified as shown in Figure 1.3 so that machines, electronics, or computers replace the operations of the human. An instrument called a *sensor* is added that is able to measure the value of the level and convert it into a proportional signal, s. This signal is provided as input to a machine, electronic circuit, or computer called the *controller.* The controller performs the function of the human in evaluating the measurement and providing an output signal, u, to change the valve setting via an *actuator* connected to the valve by a mechanical linkage.

When automatic control is applied to systems like the one in Figure 1.3, which are designed to regulate the value of some variable to a setpoint, it is called *process control.*

1.2.2 Servomechanisms

Another commonly used type of control system, which has a slightly different objective from process control, is called a *servomechanism.* In this case, the objective is to force some parameter to vary in a specific manner. This may be called a tracking control system. In-

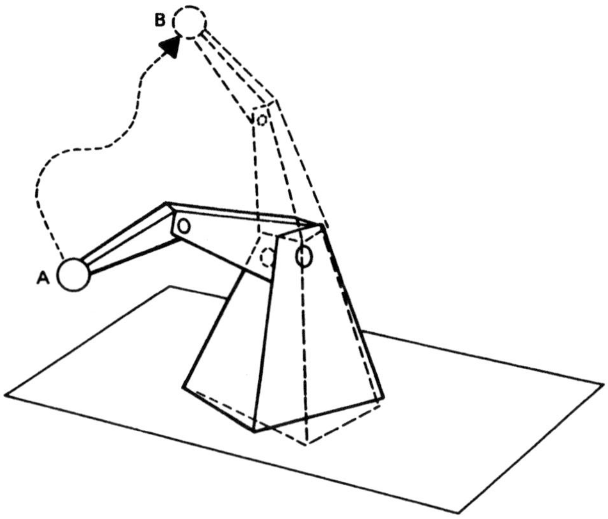

FIGURE 1.4
Servomechanism-type control systems are used to move a robot arm from point *A* to point *B* in a controlled fashion.

stead of regulating a variable value to a setpoint, the servomechanism forces the controlled variable value to follow variation of the reference value.

For example, in an industrial robot arm like the one shown in Figure 1.4, servomechanisms force the robot arm to follow a path from point *A* to point *B*. This is done by controlling the speed of motors driving the arm and the angles of the arm parts.

The strategy for servomechanisms is similar to that for process-control systems, but the dynamic differences between regulation and tracking result in differences in design and operation of the control system. This book is directed toward process-control technology.

1.2.3 Discrete-State Control Systems

This is a type of control system concerned with controlling a *sequence of events* rather than regulation or variation of individual variables. For example, the manufacture of paint might involve the regulation of many variables, such as mixing temperature, flow rate of liquids into mixing tanks, speed of mixing, and so on. Each of these might be expected to be regulated by process-control loops. But there is also a sequence of events that must occur in the overall process of manufacturing the paint. This sequence is described in terms of events that are timed to be started and stopped on a specified schedule. Referring to the paint example, the mixture needs to be heated with a regulated temperature for a certain length of time and then perhaps pumped into a different tank and stirred for another period.

The starting and stopping of events is a discrete-based system because the event is either *true* or *false,* (i.e., started or stopped, open or closed, on or off). This type of control system can also be made automatic and is perfectly suited to computer-based controllers.

These discrete-state control systems are often implemented using specialized computer-based equipment called programmable logic controllers (PLCs). Chapter 8 will present an in-depth consideration of discrete-state control systems and PLCs.

1.3 PROCESS-CONTROL BLOCK DIAGRAM

To provide a practical, working description of process control, it is useful to describe the elements and operations involved in more generic terms. Such a description should be independent of a particular application (such as the example presented in the previous section) and thus be applicable to *all* control situations. A model may be constructed using blocks to represent each distinctive element. The characteristics of control operation then may be developed from a consideration of the properties and interfacing of these elements. Numerous models have been employed in the history of process-control description; we will use one that seems most appropriate for a description of modern and developing technology of process control.

1.3.1 Identification of Elements

The elements of a process-control system are defined in terms of separate functional parts of the system. The following paragraphs define the basic elements of a process-control system and relate them to the example presented in Section 1.2.

Process In the previous example, the flow of liquid in and out of the tank, the tank itself, and the liquid all constitute a process to be placed under control with respect to the fluid level. In general, a process can consist of a complex assembly of phenomena that relate to some manufacturing sequence. Many variables may be involved in such a process, and it may be desirable to control all these variables at the same time. There are *single-variable* processes, in which only one variable is to be controlled, as well as *multivariable* processes, in which many variables, perhaps interrelated, may require regulation. The process is often also called the *plant.*

Measurement Clearly, to effect control of a variable in a process, we must have information about the variable itself. Such information is found by measuring the variable. In general, a *measurement* refers to the conversion of the variable into some corresponding *analog* of the variable, such as a pneumatic pressure, an electrical voltage or current, or a digitally encoded signal. A sensor is a device that performs the initial measurement and energy conversion of a variable into analogous digital, electrical, or pneumatic information. Further transformation or *signal conditioning* may be required to complete the measurement function. The result of the measurement is a representation of the variable value in some form required by the other elements in the process-control operation.

In the system shown in Figure 1.3, the controlled variable is the level of liquid in the tank. The measurement is performed by some sensor, which provides a signal, *s,* to the controller. In the case of Figure 1.2, the sensor is the sight tube showing the level to the human operator as an actual level in the tank.

The sensor is also called a *transducer.* However, the word *sensor* is preferred for the initial measurement device because "transducer" represents a device that converts any signal from one form to another. Thus, for example, a device that converts a voltage into a proportional current would be a transducer. In other words, all sensors are transducers, but not all transducers are sensors.

Error Detector In Figure 1.2, the human looked at the difference between the actual level, h, and the setpoint level, H, and deduced an error. This error has both a magnitude and polarity. For the automatic control system in Figure 1.3, this same kind of error determination must be made before any control action can be taken by the controller. Although the error detector is often a physical part of the controller device, it is important to keep a clear distinction between the two.

Controller The next step in the process-control sequence is to examine the error and determine what action, if any, should be taken. This part of the control system has many names, such as *compensator* or *filter,* but *controller* is the most common. The evaluation may be performed by an operator (as in the previous example), by electronic signal processing, by pneumatic signal processing, or by a computer. In modern control systems, the operations of the controller are typically performed by microprocessor-based computers. The controller requires an input of both a *measured indication* of the controlled variable and a representation of the *reference value* of the variable, expressed in the same terms as the measured value. The reference value of the variable, you will recall, is referred to as the setpoint. Evaluation consists of determining the action required to drive the controlled variable to the setpoint value.

Control Element The final element in the process-control operation is the device that exerts a direct influence on the process; that is, it provides those required changes in the controlled variable to bring it to the setpoint. This element accepts an input from the controller, which is then transformed into some proportional operation performed on the process. In our previous example, the control element is the valve that adjusts the outflow of fluid from the tank. This element is also referred to as the *final control element.*

Often an intermediate operation is required between the controller output and the final control element. This operation is referred to as an *actuator* because it uses the controller signal to actuate the final control element. The actuator translates the small energy signal of the controller into a larger energy action on the process.

1.3.2 Block Diagram

Figure 1.5 shows a general block diagram constructed from the elements defined previously. The controlled variable in the process is denoted by c in this diagram, and the measured representation of the controlled variable is labeled b. The controlled variable setpoint is labeled r, for reference. The controller uses the error input to determine an appropriate output signal, p, which is provided as input to the control element. The control element operates on the process by changing the value of the controlling process variable, u.

The error detector is a *subtracting-summing point* that outputs an *error signal,* $e = r - b$, to the controller for comparison and action.

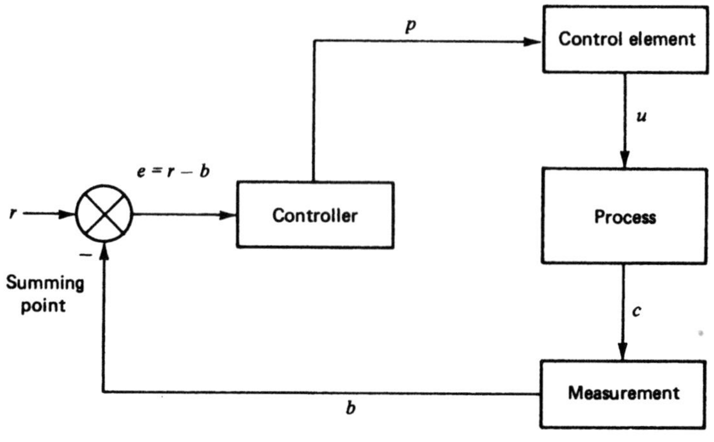

FIGURE 1.5
This block diagram of a control loop defines all the basic elements and signals involved.

Figure 1.6 shows how a physical control system is represented as a block diagram. The physical system for control of flow through a pipe is shown in Figure 1.6. Variation of flow through an obstruction (the orifice plate) produces a pressure-difference variation across the obstruction. This variation is converted to the standard signal range of 3 to 15 psi. The P/I converter changes the pressure to a 4- to 20-mA electric current, which is sent to the controller. The controller outputs a 4- to 20-mA control signal to signify the correct valve setting to provide the correct flow. This current is converted to a 3- to 15-psi pressure signal by the I/P converter and applied to a pneumatic actuator. The actuator then adjusts the valve setting.

Figure 1.6 shows how all the control system operations are condensed to the standard block diagram operations of measurement, error detection, controller, and final control element.

The purpose of a block diagram approach is to allow the process-control system to be analyzed as the interaction of smaller and simpler subsystems. If the characteristics of each element of the system can be determined, then the characteristics of the assembled system can be established by an analytical marriage of these subsystems. The historical development of the system approach in technology was dictated by this practical aspect: first, to specify the characteristics desired of a total system, and then, to delegate the development of subsystems that provide the overall criteria.

It becomes evident that the specification of a process-control system to regulate a variable, *c*, within specified limits and with specified time responses, determines the characteristics the measurement system must possess. This same set of system specifications is reflected in the design of the controller and control element.

From this concept, we conclude that the analysis of a process-control system requires an understanding of the overall system behavior and the reflection of this behavior in the properties of the system elements. Most people find that an understanding of

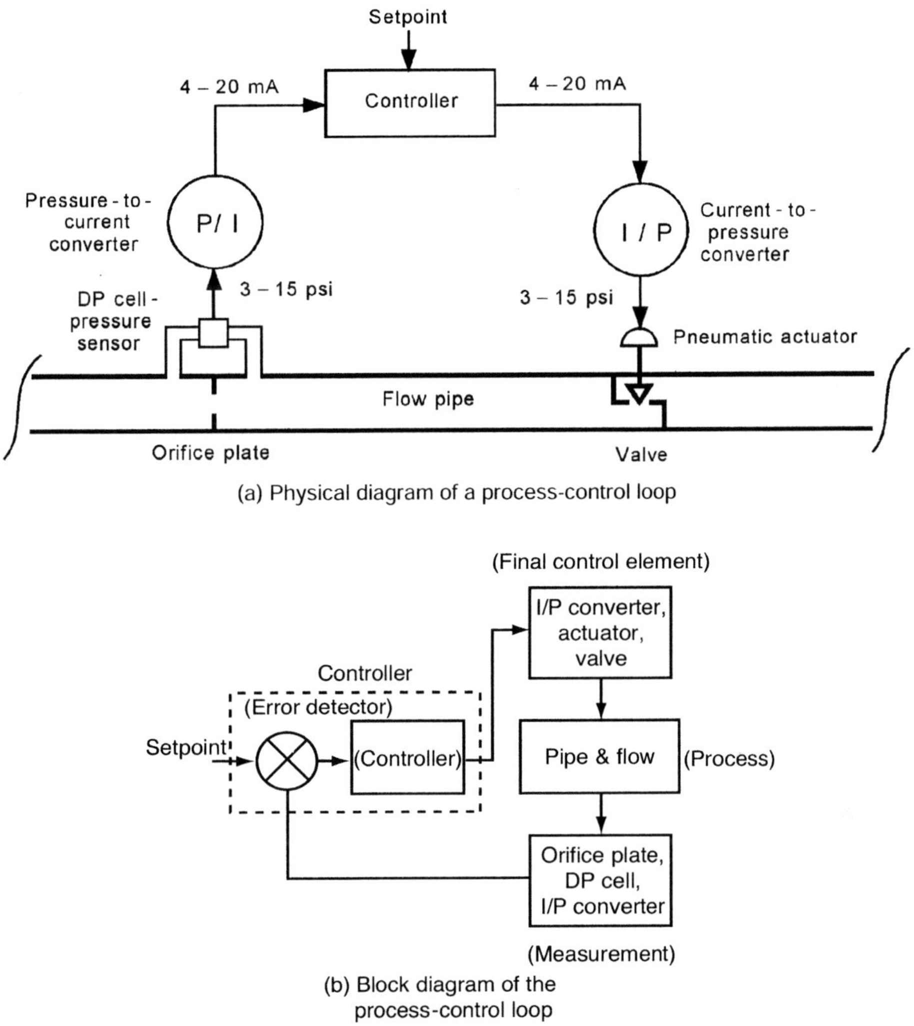

FIGURE 1.6
The physical diagram of a control loop and its corresponding block diagram look similar. Note the use of current- and pressure-transmission signals.

the parts leads to a better understanding of the whole. We will proceed with this assumption as a guiding concept.

The Loop Notice in Figure 1.5 that the signal flow forms a complete circuit from process through measurement, error detector, controller, and final control element. This is called a *loop,* and in general we speak of a process-control loop. In most cases, it is called a *feedback loop,* because we determine an error and feed back a correction to the process.

1.4 CONTROL SYSTEM EVALUATION

A process-control system is used to regulate the value of some process variable. When such a system is in use, it is natural to ask, How well is it working? This is not an easy question to answer, because it is possible to adjust a control system to provide different kinds of response to errors. This section discusses some methods for evaluating how well the system is working.

The variable used to measure the performance of the control system is the error, $e(t)$, which is the difference between the constant setpoint or reference value, r, and the controlled variable, $c(t)$.

$$e(t) = r - c(t) \tag{1.1}$$

Since the value of the controlled variable may vary in time, so may the error. (Note that in a servomechanism, the value of r may be forced to vary in time also.)

Control System Objective In principle, the objective of a control system is to make the error in Equation (1.1) exactly zero, but the control system responds only to errors (i.e., when an error occurs, the control system takes action to drive it to zero). Conversely, if the error were zero and stayed zero, the control system would not be doing anything and would not be needed in the first place. Therefore, this objective can never be perfectly achieved, and there will always be some error. The question of evaluation becomes one of how large the error is and how it varies in time.

A practical statement of control system objective is best represented by three requirements:

1. The system should be stable.
2. The system should provide the best possible steady-state regulation.
3. The system should provide the best possible transient regulation.

1.4.1 Stability

The purpose of the control system is to regulate the value of some variable. This requires that action be taken on the process itself in response to a measurement of the variable. If this is not done correctly, the control system can cause the process to become unstable. In fact, the more tightly we try to control the variable, the greater the possibility of instability.

Figure 1.7 shows that, prior to turning on a control system, the controlled variable drifts in a random fashion and is not regulated. After the control system is turned on, the variable is forced to adopt the setpoint value, and all is well for awhile. Notice that some time later, however, the variable begins to exhibit growing oscillations of value—that is, an instability. This occurs even though the control system is still connected and operational; in fact, it occurs *because* the system is connected and operational.

The first objective, then, simply means that the control system must be designed and adjusted so that the system is stable. Typically, as the control system is adjusted to give better control, the likelihood of instability also increases.

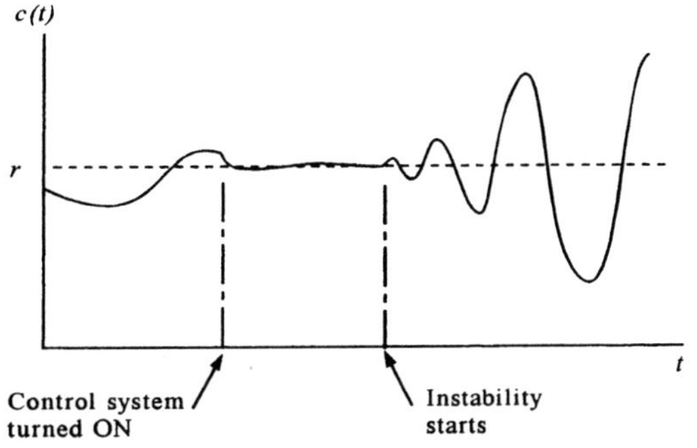

FIGURE 1.7
A control system can actually cause a system to become unstable.

1.4.2 Steady-State Regulation

The objective of the best possible steady-state regulation simply means that the steady-state error should be a minimum. Generally, when a control system is specified, there will be some allowable deviation, $\pm\Delta c$, about the setpoint. This means that variations of the variable within this band are expected and acceptable. External influences that tend to cause drifts of the value beyond the allowable deviation are corrected by the control system.

For example, a process-control technologist might be asked to design and implement a control system to regulate temperature at 150°C within ±2°C. This means the setpoint is to be 150°C, but the temperature may be allowed to vary within the range of 148° to 152°C.

1.4.3 Transient Regulation

What happens to the value of the controlled variable when some sudden transient event occurs that would otherwise cause a large variation? For example, the setpoint could change. Suppose the setpoint in the aforementioned temperature case were suddenly changed to 160°C. Transient regulation specifies how the control system reacts to bring the temperature to this new setpoint.

Another type of transient influence is a sudden change of some other process variable. The controlled variable depends on other process variables. If one of them suddenly changes value, the controlled variable may be driven to change also, so the control system acts to minimize the effect. This is called *transient response.*

1.4.4 Evaluation Criteria

The question of how well the control system is working is thus answered by (1) ensuring stability, (2) evaluating steady-state response, and (3) evaluating the response to setpoint changes and transient effects. There are many criteria for gauging the response. In general, the term *tuning* is used to indicate how a process-control loop is adjusted to provide the best control. This topic is covered in more detail in Chapter 12.

FIGURE 1.8
One of the measures of control system performance is how the system responds to changes of setpoint or a transient disturbance.

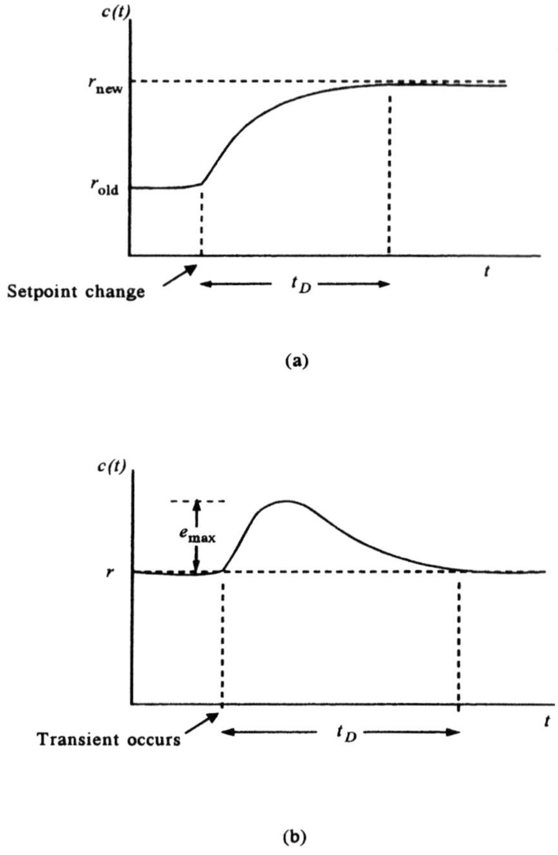

Damped Response One type of criterion requires that the controlled variable exhibit a response such as that shown in Figure 1.8 for excitations of both setpoint changes and transient effects. Note that the error is of only one polarity (i.e., it never oscillates about the setpoint). For this case, measures of quality are the duration, t_D, of the excursion and, for the transient, the maximum error, e_{max}, for a given input. The duration is usually defined as the time taken for the controlled variable to go from 10% of the change to 90% of the change following a setpoint change. In the case of a transient, the duration is often defined as the time from the start of the disturbance until the controlled variable is again within 4% of the reference.

Different tuning will provide different values of e_{max} and t_D for the same excitation. It is up to the process designers to decide whether the best control is larger duration with smaller peak error, or vice versa, or something in between.

Cyclic Response Another type of criterion applies to those cases in which the response to a setpoint change or transient is as shown in Figure 1.9. Note that the controlled variable oscillates about the setpoint. In this case, the parameters of interest are the maximum error,

FIGURE 1.9
In cyclic or underdamped response, the variable will exhibit oscillations about the reference value.

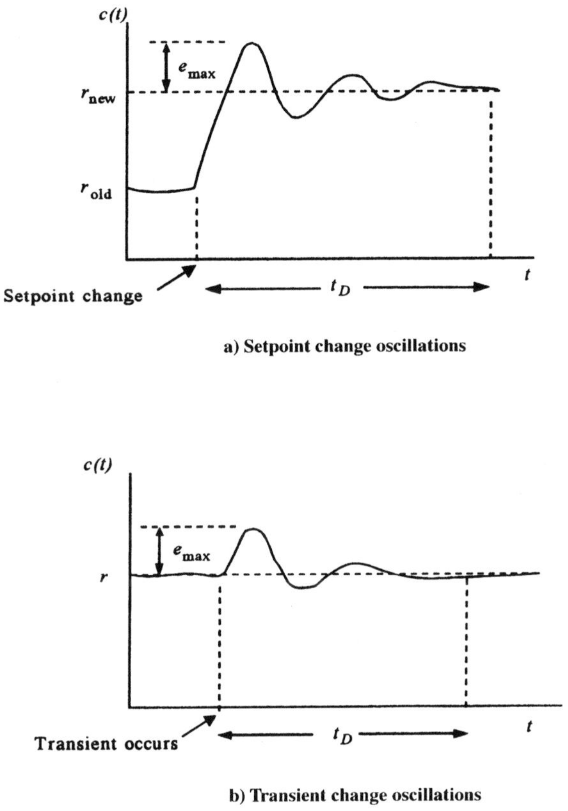

e_{max}, and the duration, t_D, also called the settling time. The duration is measured from the time when the allowable error is first exceeded to the time when it falls within the allowable error and stays.

The nature of the response is modified by adjusting the control loop parameters, which is called *tuning.* There may be large maximum error but short duration or long duration with small maximum error, and everything in between.

A number of standard cyclic tuning criteria are used. Two common types are minimum area and quarter amplitude. In *minimum area,* the tuning is adjusted until the net area under the error-time curve is a minimum, for the same degree of excitation (setpoint change or transient). Figure 1.10 shows the area as a shaded part of the curve. Analytically, this is given by

$$A = \int |e(t)| dt = \text{minimum} \qquad \textbf{(1.2)}$$

The *quarter-amplitude* criterion, shown in Figure 1.10, specifies that the amplitude of each peak of the cyclic response be a quarter of the preceding peak. Thus, $a_2 = a_1/4, a_3 = a_2/4$, and so on.

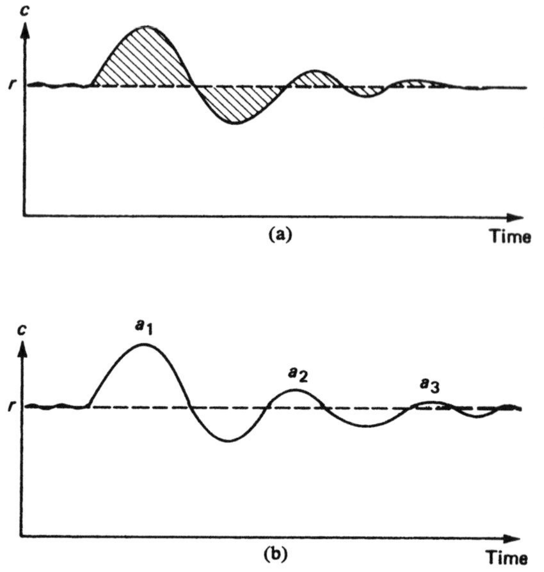

FIGURE 1.10
Two criteria for judging the quality of control-system response are the minimum area and quarter amplitude.

1.5 ANALOG AND DIGITAL PROCESSING

In the past, the functions of the controller in a control system were performed by sophisticated electronic circuits. Data were represented by the magnitude of voltages and currents in such systems. This is referred to as *analog processing.* Most modern control systems now employ digital computers to perform controller operations. In computers, data are represented as binary numbers consisting of a specific number of bits. This is referred to as *digital processing.* The paragraphs that follow contrast the analog and digital approaches to control system operation. These topics are covered in greater detail in later chapters.

1.5.1 Data Representation

The representation of data refers to how the magnitude of some physical variable is represented in the control loop. For example, if a sensor outputs a voltage whose magnitude varies with temperature, then the voltage represents the temperature. Analog and digital systems represent data in very different fashions.

Analog Data An analog representation of data means that there is a smooth and continuous variation between a representation of a variable value and the value itself. Figure 1.11 shows an analog relationship between some variable, c, and its representation, b. Notice that, for every value of c within the range covered, there is a unique value of b. If c changes by some small amount, δc, then b will change by a proportional amount, δb.

The relationship in Figure 1.11a is called *nonlinear* because the same δc does not result in the same δb, as shown. This is described in more detail later in this chapter.

FIGURE 1.11
Graph (a) shows how output variable *b* changes as an analog of variable *c*. Graph (b) shows how a digital output variable, *n*, would change with variable *c*.

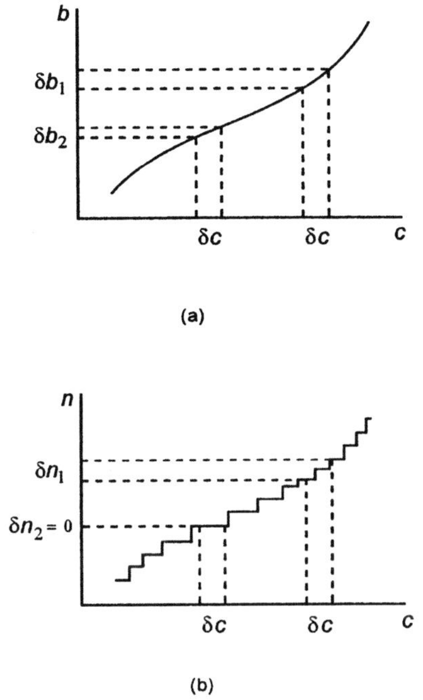

Digital Data Digital data means that numbers are represented in terms of binary digits, also called bits, which take on values of one **(1)** or zero **(0)**. When data are represented digitally, some range of analog numbers is encoded by a fixed number of binary digits. The consequence is a loss of information because a fixed number of binary digits has a limited resolution. For example, Table 1.1 shows how voltage from 0 to 15 volts could be encoded by four binary digits. A change of one volt produces a change of the least significant bit (LSB). You can see that, if the voltage changed by less than one volt, the digital representation would not change. So the representation cannot distinguish between 4.25 V and 4.75 V because both would be represented by $\mathbf{0100_2}$.

Table 1.1 also shows the hexadecimal (hex) representation of binary numbers in the digital representation. Often we use the hex representation when presenting digital data because it is a more compact and human-friendly way of writing numbers Appendix 2 reviews digital data representations.

The consequence of digital representation of data is that the smooth and continuous relation between the representation and the variable data value is lost. Instead, the digital representation can take on only discrete values. This can be seen in Figure 1.11, where a variable, *c*, is represented by a digital quantity, *n*. Notice that variations of *c*, such as δc, may not result in any change in *n*. The variable must change by more than some minimum amount, depending on where in the curve the change occurs, before a change in representation is assured.

Data Conversions Special devices are employed to convert analog voltages into a digital representation. These are called *analog-to-digital converters* (ADCs). In a control system, the sensor often produces an analog output such as a voltage. Then an ADC is used

TABLE 1.1
Decimal-binary-hex encoding

Voltage	Binary	Hex
0	0000	0h
1	0001	1h
2	0010	2h
3	0011	3h
4	0100	4h
5	0101	5h
6	0110	6h
7	0111	7h
8	1000	8h
9	1001	9h
10	1010	Ah
11	1011	Bh
12	1100	Ch
13	1101	Dh
14	1110	Eh
15	1111	Fh

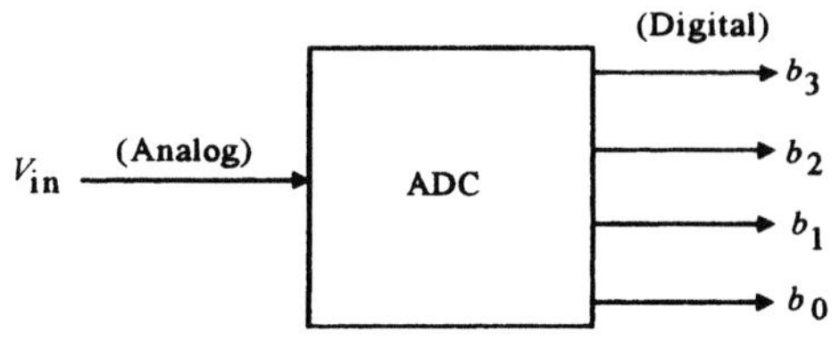

FIGURE 1.12
An ADC converts analog data, such as voltage, into a digital representation, in this case 4 bits.

to convert that voltage into a digital representation for input to the computer. Figure 1.12 shows how an ADC might be used to convert voltage into a 4-bit digital signal as illustrated in Table 1.1.

Digital-to-analog converters (DACs) convert a digital signal into an analog voltage. These devices are used to convert the control output of the computer into a form suitable for the final control element.

Both ADCs and DACs are covered in greater detail in Chapter 3.

1.5.2 ON/OFF Control

One of the most elementary types of digital processing has been in use for many years, long before the advent of computers, in fact. This is called ON/OFF control because the final control element has only two states, on and off. Thus, the controller output need have only these two states as well. It can be said that the controller output is a digital representation of a single binary digit, **0** or **1.**

Figure 1.13 shows a diagram of an elementary ON/OFF control system whose objective is to maintain the temperature in a system at some reference value, T_{ref}. A sensor

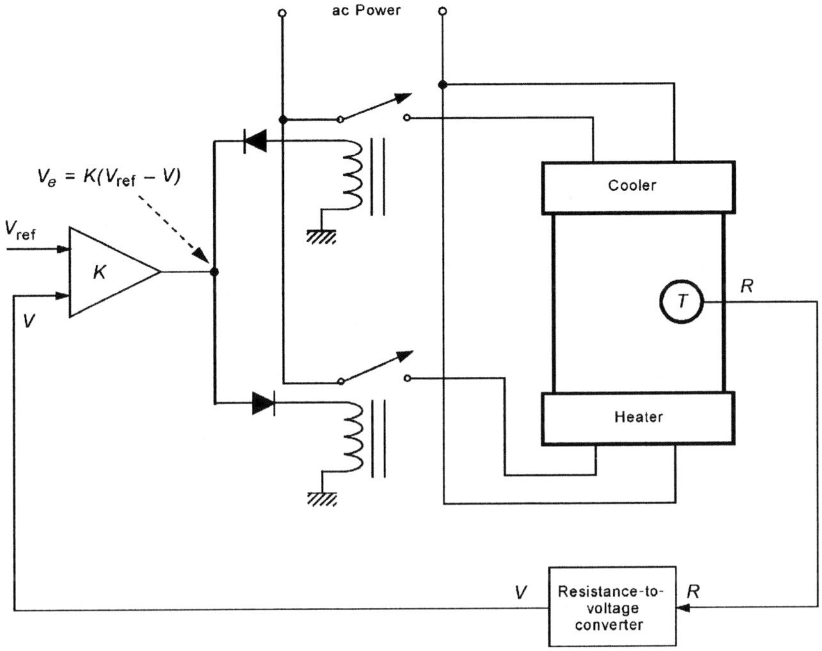

FIGURE 1.13
This ON/OFF control system can either heat or cool or do neither. No variation of the degree of heating or cooling is possible.

converts temperature values into a resistance in an analog fashion; that is, R varies smoothly and continuously with T. Signal conditioning converts the variable R into an analog voltage, V. Thus, V is an analog of T as well. The differential amplifier multiplies the difference between V and a reference voltage, V_{ref}, by a gain K to produce an error voltage, V_e.

$$V_e = K(V_{ref} - V)$$

V_{ref} is simply defined as the voltage from the converter that would be produced by T_{ref}.

At this point, the system becomes digital, because the relays will either be open or closed so that the heater or cooler will either be on or off. The diodes direct the current to the appropriate relay to produce heating or cooling, based on polarity. This system also exhibits a deadband and a hysteresis, since there is a difference between a relay "pull-in" voltage and the "release" voltage. A *deadband* is a range, of temperature in this case, wherein no action will occur. *Hysteresis* means that the behavior of the system is different at the same value of temperature, depending on whether the temperature is increasing or decreasing. [These topics are covered in more detail in this and later chapters.]

Our home and auto heaters and air conditioners, home water heaters, and a host of other basic control systems work according to the same ON/OFF mode.

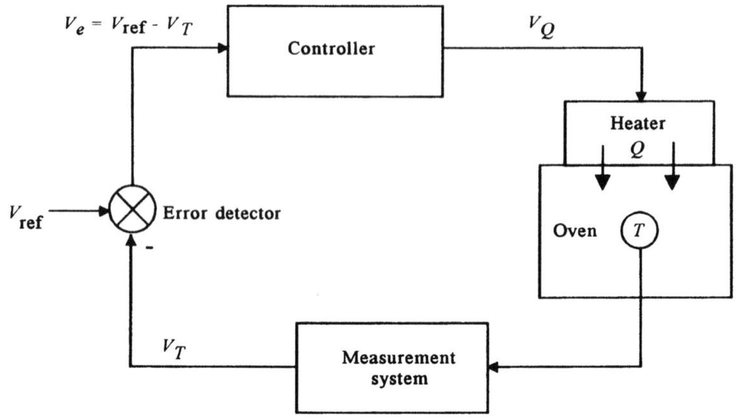

FIGURE 1.14
An analog control system such as this allows continuous variation of some parameter, such as heat input, as a function of error.

1.5.3 Analog Control

True analog control exists when all variables in the system are analog representations of another variable. Figure 1.14 shows a process in which a heater is used to control temperature in an oven. In this case, however, the heater output, Q, is an analog of the excitation voltage, V_Q, and thus heat can be varied continuously. Notice that every signal is an analog: V_T is an analog of T; the error E is an analog of the difference between the reference, V_{ref}, and the temperature voltage, V_T. The reference voltage is simply the voltage that would result from measurement of the specified reference temperature, T_{ref}.

1.5.4 Digital Control

True digital control involves the use of a computer in modern applications, although in the past, digital logic circuits were also used. There are two approaches to using computers for control.

Supervisory Control When computers were first considered for applications in control systems, they did not have a good reliability; they suffered frequent failures and breakdown. The necessity for continuous operation of control systems precluded the use of computers to perform the actual control operations. Supervisory control emerged as an intermediate step wherein the computer was used to monitor the operation of analog control loops and to determine appropriate setpoints. A single computer could monitor many control loops and use appropriate software to optimize the setpoints for the best overall plant operation. If the computer failed, the analog loops kept the process running using the last setpoints until the computer came back on-line.

Figure 1.15 shows how a supervisory computer would be connected to the analog heater control system of Figure 1.14. Notice how the ADC and DAC provide interface between the analog signals and the computer.

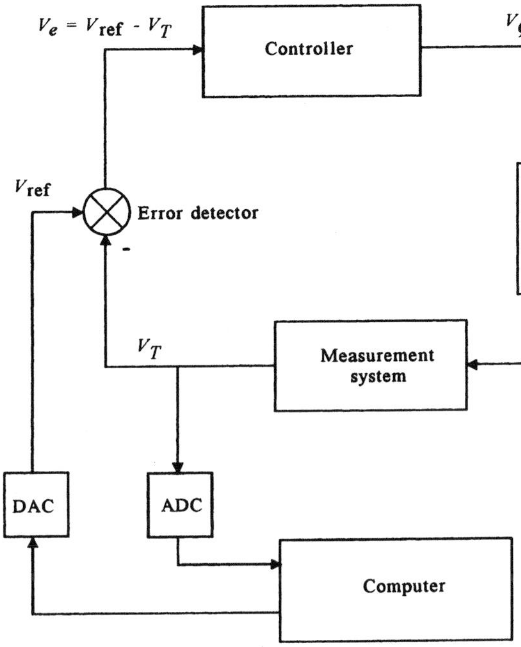

FIGURE 1.15
In supervisory control, the computer monitors measurements and updates setpoints, but the loops are still analog in nature.

Direct Digital Control (DDC) As computers have become more reliable and miniaturized, they have taken over the controller function. Thus, the analog processing loop is discarded. Figure 1.16 shows how, in a full computer control system, the operations of the controller have been replaced by software in the computer. The ADC and DAC provide interface with the process measurement and control action. The computer inputs a digital representation of the temperature, N_T, as an analog-to-digital conversion of the voltage, V_T. Error detection and controller action are determined by software. The computer then provides output directly to the heater via digital representation, N_Q, which is converted to the excitation voltage, V_Q, by the DAC.

Smart Sensor Along with the dramatic advances in computer technology has come an equally dramatic advance in the applications of computers to control systems. One of the most remarkable developments has been the integration of a microprocessor-based controller computer directly into the sensor assembly. Using modern integrated circuit technology, the sensor, signal conditioning, ADC, and computer controller are all contained within the sensor housing. In one form, the unit also contains a DAC with a 4- to 20-mA output to be fed to the final control element. The setpoint is programmed by connecting another computer to the unit using a serial interface line.

The most current technology, however, is to interface these smart sensors to a local area network, or field bus, to be described next, which allows the sensor to be connected to other computers and to the final control element over a common digital serial line.

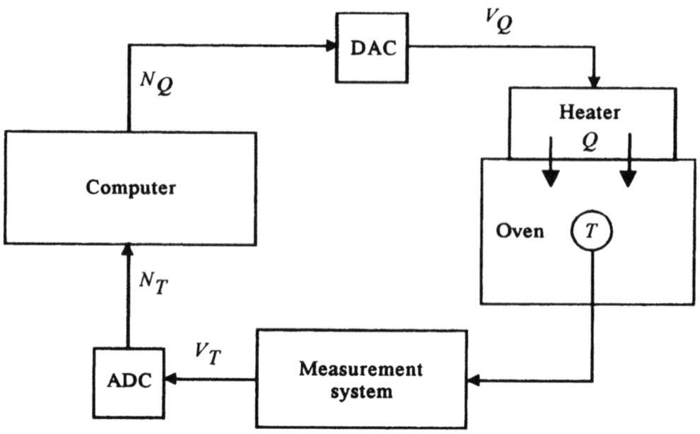

FIGURE 1.16
This direct digital control system lets the computer perform the error detection and controller functions.

Networked Control Systems When a plant uses DDC, it becomes possible to place the computer-based controller directly at the site of the plant where the control is needed. This is done by using smart sensors or by placing the computer controller in hermetically sealed instrument cases around the plant. In order to have coordinated control of the whole plant, all these DDC units are placed on a local area network (LAN). The LAN commonly provides communication as a serial stream of digital data over a variety of carriers such as wires and fiber optics. The LAN also connects to computers exercising master control of plant operations, fiscal computers for accounting and production control, and engineering computers for monitoring and modifying plant operations as needed. In control systems, these LANs are referred to as a *field bus.*

Figure 1.17 shows how the LAN or field bus connects the computers in a plant together. Each of the process-control computers operates one or more DDC loops like the one shown in Figure 1.16. Bus users can monitor the operations of any of the plant process-control loops, and those with authorization can modify control characteristics such as setpoints and gains.

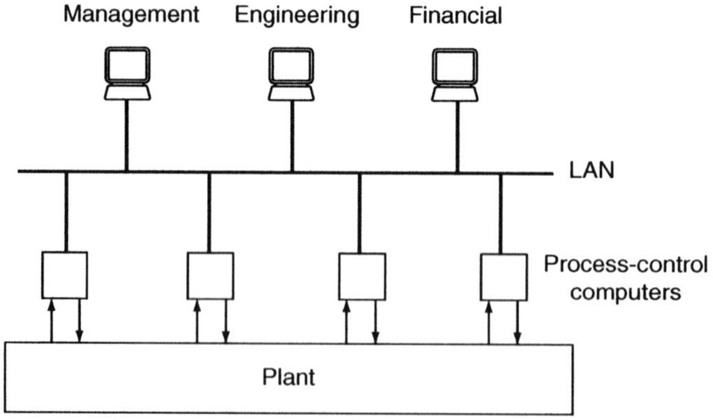

FIGURE 1.17
Local area networks (LANs) play an important role in modern process-control plants.

Special process-control bus standards have been developed for how data and information are represented and transmitted in these networks. The two most commonly implemented standards are the Foundation Fieldbus and the Profibus (Process Field Bus). The idea behind these standards is to have universal agreement among process equipment manufacturers on how data are represented on the bus line and how data are transmitted and received. This is an extension of the "plug and play" concept used for computer hardware. With standardization, successful interconnection and interfacing of equipment from a variety of manufacturers into a control loop is assured. Fieldbus (primarily in the United States) and Profibus (primarily in Europe) are vying to become the universal standard.

Chapter 10 presents more information about these bus standards.

1.5.5 Programmable Logic Controllers

Many manufacturing operations are ON/OFF in nature; that is, a conveyor or heater is either on or off, a valve is either open or closed, and so on. In the past, these types of discrete control functions were often provided by a system of electrical relays wired according to a complex diagram into what was called a relay logic controller.

In recent years, computers have also taken over the operation of such relay logic controllers, known as *programmable logic controllers* (PLCs). Even though originally designed to control discrete-state (ON/OFF) systems, they are now also used to implement DDC. Chapter 8 covers these systems in more detail.

Figure 1.18 shows how the problem of Figure 1.13 would be implemented using a PLC. Note that thermal-limit switches are used instead of a sensor to indicate when the

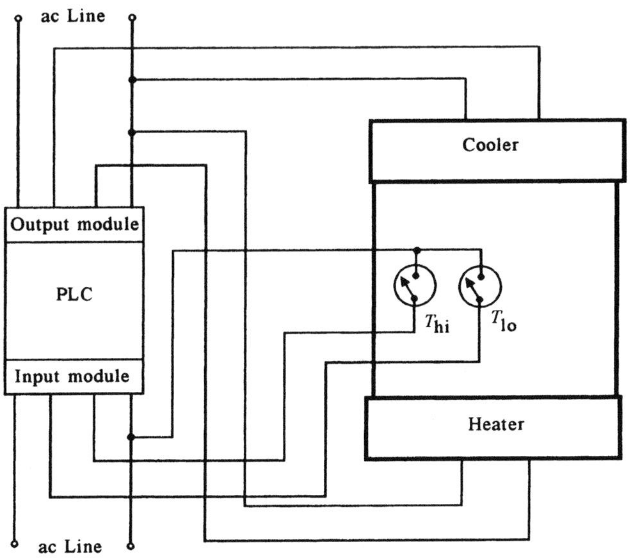

FIGURE 1.18
A programmable logic controller (PLC) is an outgrowth of ON/OFF-type control environments. In this case the heater and cooler are either ON or OFF.

temperature has risen above or fallen below the limit temperatures. These are simply switches designed to open (or close) when the temperature reaches certain preset limits.

1.6 UNITS, STANDARDS, AND DEFINITIONS

As in any other technological discipline, the field of process control has many sets of units, standards, and definitions to describe its characteristics. Some of these are a result of historical use, some are for convenience, and some are just confusing. As the discipline grew, there were efforts to standardize terms so that professional workers in process control could effectively communicate among themselves and with specialists in other disciplines. In this section, we summarize the present state of affairs relative to the common units, standards, and definitions.

1.6.1 Units

To ensure precise technical communication among individuals employed in technological disciplines, it is essential to use a well-defined set of units of measurement. The metric system of units provides such communication and has been adopted by most technical disciplines. In process control, a particular set of metric units is used (which was developed by an international conference) called the International System (SI, Système International D'Unités).

In the United States, the English system of units is still in common use, although use of SI units is gradually occurring more often. It is important for the process-control specialist to know English units and to be able to perform conversions with the SI system.

International System of Units The international system of units is maintained by an international agreement for worldwide standardization. The system is based on seven well-defined base units and two supplementary, dimensionless units. Everything else falls into the category of defined units, which are defined in terms of the seven base and two supplementary units.

Quantity	Unit	Symbol
BASE		
Length	Meter	m
Mass	Kilogram	kg
Time	Second	s
Electric current	Ampere	A
Temperature	Kelvin	K
Amount of substance	Mole	mol
Luminous intensity	Candela	cd
SUPPLEMENTARY		
Plane angle	Radian	rad
Solid angle	Steradian	sr

All other SI units can be derived from these nine units, although in many cases a special name is assigned to the derived quantity. Thus, a force is measured by the newton (N), where $1\ \text{N} = 1\ \text{kg}\cdot\text{m/s}^2$; energy is measured by the joule (J) or watt-second (W·s), given by $1\ \text{J} = 1\ \text{kg}\cdot\text{m}^2/\text{s}^2$; and so on, as shown in Appendix 1.

Other Units Although the SI system is used in this text, other units remain in common use in some technical areas. The reader, therefore, should be able to identify and translate between the SI system and other systems. The centimeter-gram-second system (CGS) and the English system are also given in Appendix 1. The following examples illustrate some typical translations of units.

EXAMPLE 1.2 Express a pressure of $p = 2.1 \times 10^3\ \text{dyne/cm}^2$ in pascals. $1\ \text{Pa} = 1\ \text{N/m}^2$.

Solution

From Appendix 1, we find $10^2\ \text{cm} = 1\ \text{m}$ and $10^5\ \text{dyne} = 1$ newton; thus,

$$p = (2.1 \times 10^3\ \text{dyne/cm}^2)\left(10^2 \frac{\text{cm}}{\text{m}}\right)^2\left(\frac{1\ \text{N}}{10^5\ \text{dyne}}\right)$$

$$p = \mathbf{210\ N/m^2 = 210\ pascals\ (Pa)}$$

EXAMPLE 1.3 Find the number of feet in 5.7 m.

Solution

Reference to the table of conversions in Appendix 1 shows that 1 m = 39.37 in.; therefore,

$$(5.7\ \text{m})\left(39.37 \frac{\text{in.}}{\text{m}}\right)\left(\frac{1\ \text{ft}}{12\ \text{in.}}\right) = \mathbf{18.7\ ft}$$

EXAMPLE 1.4 Express 6.00 ft in meters.

Solution

Using 39.37 in./m and 12 in./ft,

$$(6.00\ \text{ft})\ (12\ \text{in./ft})\left(\frac{1\ \text{m}}{39.37\ \text{in.}}\right) = \mathbf{1.83\ m}$$

EXAMPLE 1.5 Find the mass in kilograms of a 2-lb object.

Solution

The conversion factor between mass in kilograms and pounds is found from Appendix 1 to be 0.454 kg/lb. Therefore, we have

$$m = (2\ \text{lb})(0.454\ \text{kg/lb})$$

$$m = \mathbf{0.908\ kg}$$

Metric Prefixes With the wide variation of variable magnitudes that occurs in industry, there is a need to abbreviate very large and small numbers. Scientific notation allows the expression of such numbers through powers of 10. A set of standard metric prefixes has been adopted by the SI to express these powers of 10, which are employed to simplify the expression of very large or very small numbers. These prefixes are given in Appendix 1.

EXAMPLE 1.6 Express 0.0000215 s and 3,781,000,000 W using decimal prefixes.

Solution

We first express the quantities in scientific notation and then find the appropriate decimal prefix from Appendix 1.

$$0.0000215 \text{ s} = 21.5 \times 10^{-6}\text{s} = \mathbf{21.5\ \mu s}$$

and

$$3,781,000,000 \text{ W} = 3.781 \times 10^{9} \text{ W} = \mathbf{3.781\ GW}$$

1.6.2 Analog Data Representation

For measurement systems or control systems, part of the specification is the range of the variables involved. Thus, if a system is to measure temperature, there will be a range of temperature specified, for example, 20° to 120°C. Similarly, if the controller is to output a signal to a continuous valve, this signal will be designed to cover the range from fully closed to fully open, with all the various valve settings in between.

Two analog standards are in common use as a means of representing the range of variables in control systems. For electrical systems, we use a range of electric current carried in wires, and for pneumatic systems we use a range of gas pressure carried in pipes. These signals are used primarily to transmit variable information over some distance, such as to and from the control room and the plant. Figure 1.19 shows a diagram of a process-control

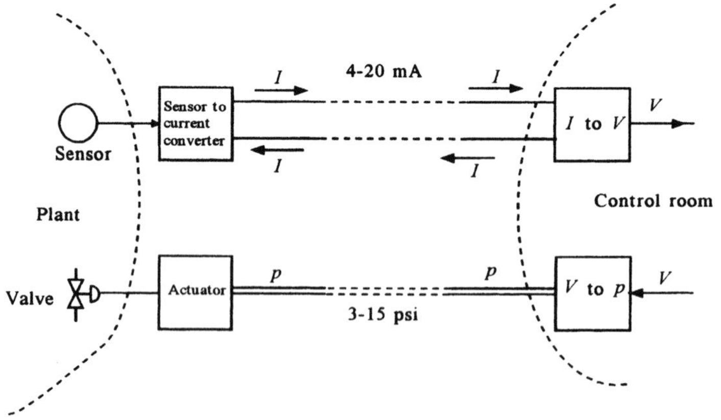

FIGURE 1.19
Electric current and pneumatic pressures are the most common means of information transmission in the industrial environment.

installation in which current is used to transmit measurement data about the controlled variable to the control room, and gas pressure in pipes is used to transmit a feedback signal to a valve to change flow as the controlling variable.

Current Signal The most common current transmission signal is 4 to 20 mA. Thus, in the following temperature example, 20°C might be represented by 4 mA and 120°C by 20 mA, with all temperatures in between represented by a proportional current.

EXAMPLE 1.7 Suppose the temperature range 20° to 120°C is linearly converted to the standard current range of 4 to 20 mA. What current will result from 66°C? What temperature does 6.5 mA represent?

Solution

The easiest way to solve this kind of problem is to develop a linear equation between temperature and current. We can write this equation as $I = mT + I_0$, and we know from the given data that $I = 4$ mA when $T = 20^\circ$C and that $I = 20$ mA when $T = 120^\circ$C. Thus, we have two equations in two unknowns:

$$4 \text{ mA} = (20^\circ\text{C})m + I_0$$

$$20 \text{ mA} = (120^\circ\text{C})m + I_0$$

Subtracting the first from the second gives

$$16 \text{ mA} = (100^\circ\text{C})m$$

so that $m = 0.16 \text{ mA}/^\circ\text{C}$. Then we find I_0:

$$I_0 = 4 \text{ mA} - (20^\circ\text{C})(0.16 \text{ mA}/^\circ\text{C})$$

$$I_0 = 0.8 \text{ mA}$$

Thus, the equation relating current and temperature is

$$I = (0.16 \text{ mA}/^\circ\text{C})T + 0.8 \text{ mA}$$

Now answering the questions is easy. For 66°C, we have

$$I = (0.16 \text{ mA}/^\circ\text{C})66^\circ\text{C} + 0.8 \text{ mA} = \mathbf{11.36 \text{ mA}}$$

For 6.5 mA, we solve for T:

$$6.5 \text{ mA} = (0.16 \text{ mA}/^\circ\text{C})T + 0.8 \text{ mA}$$

for which $T =$ **35.6°C**.

Current is used instead of voltage because the system is then less dependent on load. The sensor-to-current converter in Figure 1.19, also called a transmitter, is designed to launch a current into the line regardless of load, to a degree. In Figure 1.20, a resistor, R, has been added to the lines connecting the plant to the control room. In the control room, the incoming current has been converted to a voltage using resistor R_L. Note that if the short around resistor R is cut

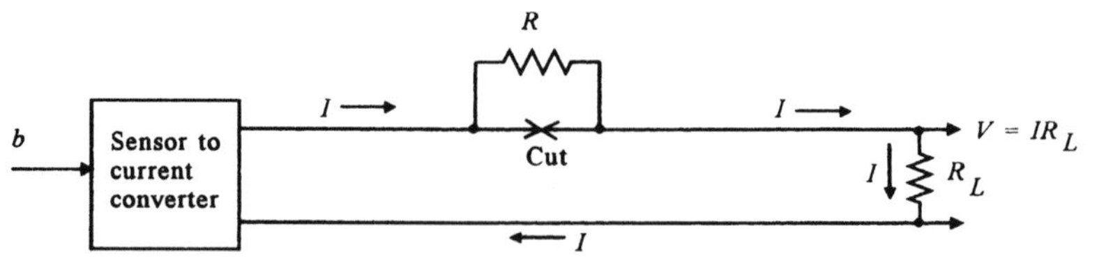

FIGURE 1.20
One of the advantages of current as a transmission signal is that it is nearly independent of line resistance.

so that R is now in the circuit, no change in current will occur. The transmitter is designed to adjust conditions (in this case, output voltage) so that the current is held constant. Practically speaking, most current transmitters can work into any load, from 0 to about 1000 Ω.

Voltage is not used for transmission because of its susceptibility to changes of resistance in the line.

Pneumatic Signals In the United States, the most common standard for pneumatic signal transmission is 3 to 15 psi. In this case, when a sensor measures some variable in a range, it is converted into a proportional pressure of gas in a pipe. The gas is usually dry air. The pipe may be many hundreds of meters long, but as long as there is no leak in the system, the pressure will be propagated down the pipe. This English system standard is still widely used in the United States, despite the move to the SI system of units. The equivalent SI range that will eventually be adopted is 20 to 100 kPa.

1.6.3 Definitions

This section presents definitions of some of the common terms and expressions used to describe process-control elements.

Error The most important quantity in control systems is the error. When used to describe the results of a measurement, error is the difference between the actual value of a variable and the measured indication of its value. In that case, the *accuracy* of the measurement system places bounds on the possible error.

When used for a controlled variable in a control system, error is the difference between the measured value of the variable and the desired value—that is, the reference or setpoint value.

Block Definitions As noted in Section 1.3.2, control systems are often described in terms of blocks. One block represents the measurement, another the controller, and so on. In order to work effectively in control systems, one must understand the terms and expressions used to describe the characteristics of a block. Figure 1.21 shows a block that has an input of some variable, $x(t)$, and an output of another variable, $y(t)$. This model will be used in the following paragraphs to define the characteristics of the block.

Transfer Function The transfer function, $T(x, y, t)$ in Figure 1.21, describes the relationship between the input and output for the block. The transfer function is often described in two parts, the static part and the dynamic part. The static transfer function describes the

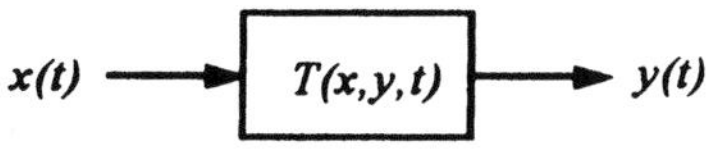

FIGURE 1.21
A transfer function shows how a system-block output variable varies in response to an input variable, as a function of both static input value and time.

input/output relationship when the input is not changing in time. The dynamic transfer function describes the input/output relationship when there is time variation of the input.

Static transfer functions may be presented in the form of equations, tables, or graphs. For example, a flow meter may relate flow, Q, in gallons per minute, to a differential pressure, Δp, in psi, via an equation such as

$$Q = 119.5\sqrt{\Delta p}$$

However, an RTD temperature sensor is usually represented by a table of resistance versus temperature. Graphs are often used to visually display how input and output vary. Frequently, the transfer is valid only over a certain range of variable values.

The dynamic transfer function is often represented by a differential equation in time. Common examples of simple dynamic transfer functions are presented in Section 1.7.

Accuracy This term is used to specify the maximum overall error to be expected from a device, such as measurement of a variable. Accuracy is usually expressed as the *inaccuracy* and can appear in several forms:

1. Measured variable; the accuracy is $\pm$ 2°C in some temperature measurement. Thus, there would be an uncertainty of $\pm$2°C in any value of temperature measured.
2. Percentage of the instrument full-scale (FS) reading. Thus, an accuracy of $\pm$0.5% FS in a 5-V full-scale range meter would mean the inaccuracy or uncertainty in any measurement is $\pm$0.025 V.
3. Percentage of instrument span—that is, percentage of the range of instrument measurement capability. Thus, for a device measuring $\pm$3% of span for a 20 to 50 psi range of pressure, the accuracy would be $(\pm 0.03)(50 - 20) = \pm 0.9$ psi.
4. Percentage of the actual reading. Thus, for a $\pm$2% of reading voltmeter, we would have an inaccuracy of $\pm$0.04 V for a reading of 2 V.

EXAMPLE 1.8

A temperature sensor has a span of 20°–250°C. A measurement results in a value of 55°C for the temperature. Specify the error if the accuracy is **(a)** $\pm$0.5% FS, **(b)** $\pm$0.75% of span, and **(c)** $\pm$0.8% of reading. What is the possible temperature in each case?

Solution

Using the given definitions, we find

a. Error $= (\pm 0.005)(250°C) = \pm$**1.25°C**. Thus, the actual temperature is in the range of 53.75° to 56.25°C.

b. Error $= (\pm 0.0075)(250 - 20)°C = \pm$**1.725°C.** Thus, the actual temperature is in the range of 53.275° to 56.725°C.

c. Error $= (\pm 0.008)(55°C) = \pm$**0.44°C.** Thus, the temperature is in the range of 54.56° to 55.44°C.

EXAMPLE 1.9 A temperature sensor has a transfer function of 5 mV/°C with an accuracy of $\pm 1\%$. Find the possible range of the transfer function.

Solution

The transfer function range will be $(\pm 0.01)(5\ \text{mV}/°\text{C}) = \pm 0.05\ \text{mV}/°\text{C}$. Thus, the range is 4.95 to 5.05 mV/°C.

EXAMPLE 1.10 Suppose a reading of 27.5 mV results from the sensor used in Example 1.9. Find the temperature that could provide this reading.

Solution

Because the range of transfer function is 4.95 to 5.05 mV/°C, the possible temperature values that could be inferred from a reading of 27.5 mV are

$$(27.5\ \text{mV})\left(\frac{1}{4.95\ \text{mV}/°\text{C}}\right) = \mathbf{5.56°C}$$

$$(27.5\ \text{mV})\left(\frac{1}{5.05\ \text{mV}/°\text{C}}\right) = \mathbf{5.45°C}$$

Thus, we can be certain only that the temperature is between 5.45°C and 5.56°C.

The application of digital processing has necessitated an accuracy definition compatible with digital signals. In this regard, we are most concerned with the error involved in the digital representation of analog information. Thus, the accuracy is quoted as the percentage deviation of the analog variable per bit of the digital signal. As an example, an A/D converter may be specified as 0.635 volts per bit $\pm$ 1%. This means that a bit will be set for an input voltage change of 0.635 $\pm$ 0.006 V, or 0.629 to 0.641 V.

System Accuracy Often, one must consider the overall accuracy of *many* elements in a process-control loop to represent a process variable. Generally, the best way to do this is to express the accuracy of each element in terms of the transfer functions. For example, suppose we have a process with two transfer functions that act on the dynamic variable to produce an output voltage as shown in Figure 1.22. We can describe the output as

$$V \pm \Delta V = (K \pm \Delta K)(G \pm \Delta G)C \tag{1.3}$$

where
V = output voltage
$\pm\Delta V$ = uncertainty in output voltage
K, G = nominal transfer functions
$\Delta K, \Delta G$ = uncertainties in transfer functions
C = dynamic variable

From Equation (1.3), we can find the output uncertainty to be

$$\Delta V = \pm GC\ \Delta K \pm KC\ \Delta G \pm \Delta K\ \Delta GC \tag{1.4}$$

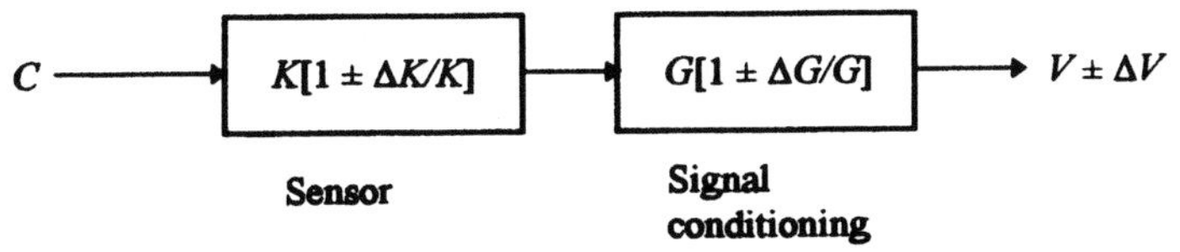

FIGURE 1.22
Uncertainties in block transfer functions build up as more blocks are involved in the transformation.

Equation (1.4) can be written in terms of fractional uncertainties by factoring out V. In this case we get,

$$\frac{\Delta V}{V} = \pm\frac{\Delta K}{K}\pm\frac{\Delta G}{G}\pm\left(\frac{\Delta V}{V}\right)\left(\frac{\Delta K}{K}\right)$$

Of course the fractional uncertainties will be small (<< 1) so the last term will be the product of two small numbers and will thus be really small. Therefore, we usually ignore the last term and write the uncertainty as:

$$\frac{\Delta V}{V} = \pm\frac{\Delta K}{K}\pm\frac{\Delta G}{G} \tag{1.5}$$

Thus Equation (1.5) shows that the worst-case uncertainty would be the sum of the individual uncertainties.

Statistical analysis teaches us that it is more realistic to use the root-mean-square (rms) representation of system uncertainty. This will give an overall uncertainty somewhat less than worst-case but more likely to reflect the actual value. This is found from the relation

$$\left[\frac{\Delta V}{V}\right]_{\text{rms}} = \pm\sqrt{\left(\frac{\Delta K}{K}\right)^2 + \left(\frac{\Delta G}{G}\right)^2}$$

EXAMPLE 1.11 Find the system accuracy of a flow process if the transducer transfer function is $10\ \text{mV}/(\text{m}^3/\text{s}) \pm 1.5\%$ and the signal-conditioning system-transfer function is $2\ \text{mA}/\text{mV} \pm 0.5\%$.

Solution

Here we have a direct application of

$$\frac{\Delta V}{V} = \pm\left[\frac{\Delta K}{K} + \frac{\Delta G}{G}\right]$$

$$\frac{\Delta V}{V} = \pm[0.015 + 0.005]$$

$$\frac{\Delta V}{V} = \pm 0.02 = \pm 2\%$$

so that the net transfer function is 20 mA/(m^3/s) ± 2%. If we use the more statistically appropriate rms approach, the system accuracy would be

$$\left[\frac{\Delta V}{V}\right]_{\text{rms}} = \pm\sqrt{(0.015)^2 + (0.005)^2} = \pm 0.0158$$

So the accuracy is about ±1.6%.

Sensitivity Sensitivity is a measure of the change in output of an instrument for a change in input. Generally speaking, high sensitivity is desirable in an instrument because a large change in output for a small change in input implies that a measurement may be taken easily. Sensitivity must be evaluated together with other parameters, such as *linearity* of output to input, *range,* and *accuracy.* The value of the sensitivity is generally indicated by the transfer function. Thus, when a temperature transducer outputs 5 mV per degree Celsius, the sensitivity is 5 mV/°C.

Hysteresis and Reproducibility Frequently, an instrument will not have the same output value for a given input in repeated trials. Such variation can be due to inherent uncertainties that imply a limit on the *reproducibility* of the device. This variation is random from measurement to measurement and is not predictable.

A similar effect is related to the history of a particular measurement taken with an instrument. In this case, a different reading results for a specific input, depending on whether the input value is approached from higher or lower values. This effect, called *hysteresis,* is shown in Figure 1.23, where the output of an instrument has been plotted against input. We see that if the input parameter is varied from low to high, curve *A* gives values of the output. If the input parameter is decreasing, curve *B* relates input to output. Hysteresis is usually specified as a percentage of full-scale maximum deviation between the two curves. This effect is predictable if measurement values are always approached from one direction, because hysteresis will not cause measurement errors.

Resolution Inherent in many measurement devices is a minimum measurable value of the input variable. Such a specification is called the *resolution* of the device. This

FIGURE 1.23
Hysteresis is a predictable error resulting from differences in the transfer function as the input variable increases or decreases.

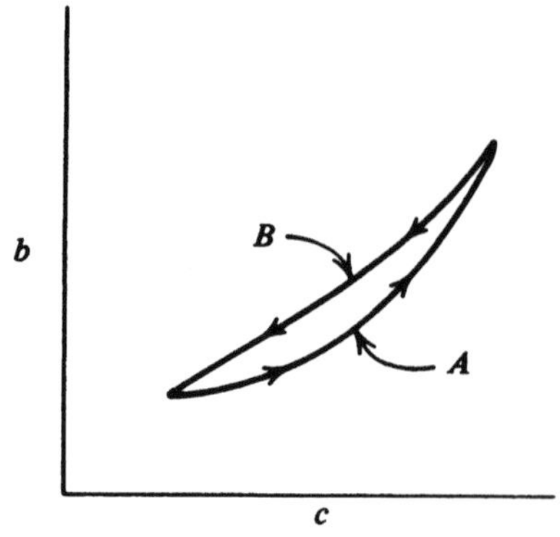

characteristic of the instrument can be changed only by redesign. A good example is a wire-wound potentiometer in which the slider moves across windings to vary resistance. If one turn of the winding represents a change of ΔR ohms, then the potentiometer cannot provide a resistance change *less* than ΔR. We say that the potentiometer resolution is ΔR. This is often expressed as a percentage of the full-scale range.

EXAMPLE 1.12 A force sensor measures a range of 0 to 150 N with a resolution of 0.1% FS. Find the smallest change in force that can be measured.

Solution

Because the resolution is 0.1% FS, we have a resolution of (0.001)(150 N) = 0.15 N, which is the smallest measurable change in force.

In some cases, the resolution of a measurement system is limited by the sensitivity of associated signal conditioning. When this occurs, the resolution can be improved by employing better conditioning.

EXAMPLE 1.13 A sensor has a transfer function of 5 mV/°C. Find the required voltage resolution of the signal conditioning if a temperature resolution of 0.2°C is required.

Solution

A temperature change of 0.2°C will result in a voltage change of

$$\left(5\frac{\text{mV}}{{}^\circ\text{C}}\right)(0.2^\circ\text{C}) = \mathbf{1.0\ mV}$$

Thus, the voltage system must be able to resolve 1.0 mV.

In analog systems, the resolution of the system is usually determined by the smallest measurable change in the analog output signal of the measurement system. In digital systems, the resolution is a well-defined quantity that is simply the change in dynamic variable represented by a *1-bit change* in the binary word output. In these cases, resolution can be improved only by a different coding of the analog information or adding more bits to the word. (This will be discussed further in Section 3.4.)

Linearity In both sensor and signal conditioning, output is represented in some functional relationship to the input. The only stipulation is that this relationship be unique; that is, for each value of the input variable, there exists one unique value of the output variable. For simplicity of design, a linear relationship between input and output is highly desirable. When a linear relationship exists, a straight-line equation can be used to relate the measured variable and measurement output

$$c_m = mc + c_0 \tag{1.6}$$

where c = variable to be measured
m = slope of straight line
c_0 = offset or intercept of straight line
c_m = output of measurement

No simple relationship such as Equation (1.6) can usually be found for the nonlinear cases, although in some cases approximations of a linear or quadratic nature are fitted to portions of these curves, as will be shown in Chapter 4.

EXAMPLE 1.14

A sensor resistance changes linearly from 100 to 180 Ω as temperature changes from 20° to 120°C. Find a linear equation relating resistance and temperature.

Solution

Using Equation (1.6) as a guide, the desired equation would be of the form

$$R = mT + R_0$$

To find the two constants, m and R_0, we form two equations and two unknowns from the facts given:

$$100\ \Omega = (20^\circ\text{C}\ m + R_0)$$
$$180\ \Omega = (120^\circ\text{C}\ m + R_0)$$

Subtracting the first equation from the second gives

$$80\ \Omega = (100^\circ\text{C})m \quad \text{or} \quad m = 0.8\ \Omega/^\circ\text{C}$$

Then, from the first equation, we find

$$100\ \Omega = (20^\circ\text{C})(0.8\ \Omega/^\circ\text{C}) + R_0$$

from which

$$R_0 = 84\ \Omega$$

The equation relating resistance and temperature is

$$R = 0.8T + 84$$

One of the specifications of sensor output is the degree to which it is linear with the measured variable and the span over which this occurs. A measure of sensor linearity is to determine the deviation of the sensor output from a best-fit straight line over a particular range. A common specification of linearity is the maximum deviation from a straight line expressed as percent of FS.

Consider a sensor that outputs a voltage as a function of pressure from 0 to 100 psi with a linearity of 5% FS. This means that, at some point on the curve of voltage versus pressure, the deviation between actual pressure and linearly indicated pressure for a given voltage deviates by 5% of 100 psi, or 5 psi.

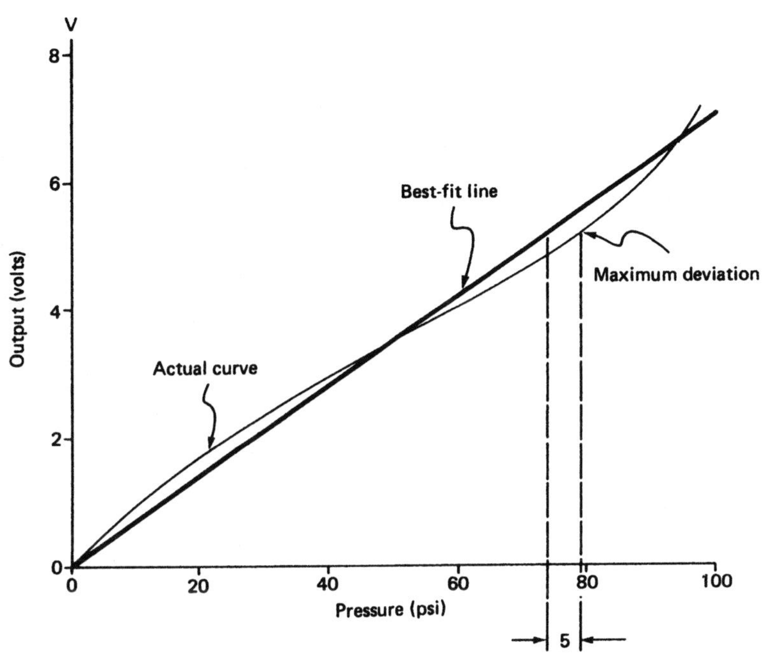

FIGURE 1.24
Comparison of an actual curve and its best-fit straight line, where the maximum deviation is 5% FS.

Figure 1.24 indicates this graphically. A straight line has been fitted to the slightly nonlinear sensor curve. One can either specify that, for a given voltage, there is a deviation between actual and linearly predicted pressure or that, for a given pressure, there is a deviation between actual and linearly predicted voltage.

1.6.4 Process-Control Drawings

An electrical schematic is a drawing that employs a standard set of symbols and definitions so that anyone who knows the standards can understand the operation of the circuit. In just the same way, process control employs a standard set of symbols and definitions to represent a plant and its associated control systems. This standard was developed and approved by a collaboration between the American National Standards Institute (ANSI) and the Instrumentation, Systems, and Automation (ISA) society. The standard is designated ANSI/ISA S5.1-1984 (R1992) Instrumentation Symbols and Identification.

This standard is used for the early design phases of a plant control system to construct simplified process-control diagrams and then to render the final, detailed plant design as a piping and instrumentation diagram (P&ID). Appendix 5 presents some of the more common definitions of the standard and how they are used in a P&ID.

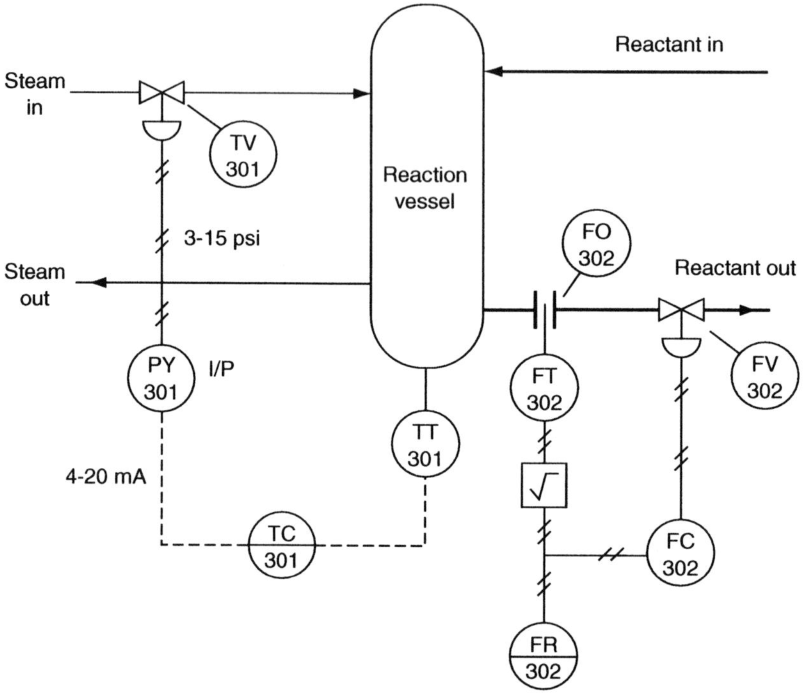

FIGURE 1.25
A P&ID uses special symbols and lines to show the devices and interconnections in a process-control system.

Essential Elements The P&ID depicts the entire plant and associated control systems. This includes plant operating units, product flow lines, measurement and control signal lines, sensors, controllers, final control elements, computers, and programmable logic controllers (PLCs). Also included are letter and number designations to identify the function of an element and notes to further explain features of the P&ID. Let's use the P&IDs shown in Figures 1.25 and 1.26 to summarize some of these features. Please refer to Appendix 5 for more information.

Figure 1.25 shows a section of a plant with a reaction vessel into and out of which a reactant (heavy line) flows. The vessel is heated by steam input. Temperature within the vessel is controlled by controlling the steam input, and a flow control system regulates the reactant flow out of the vessel.

Figure 1.26 shows a system in which the level in a tank is controlled using a cascade control system. In a cascade control system, the setpoint of one loop is the controller output of another loop. The system in Figure 1.26 uses computers to provide the controller operations. Again, the primary product flow is shown as a heavy line.

Instrument Lines Symbols Figure 1.25 shows that the standard 4- to 20-mA current signal is represented as a dashed line in the P&ID, while the pneumatic signal

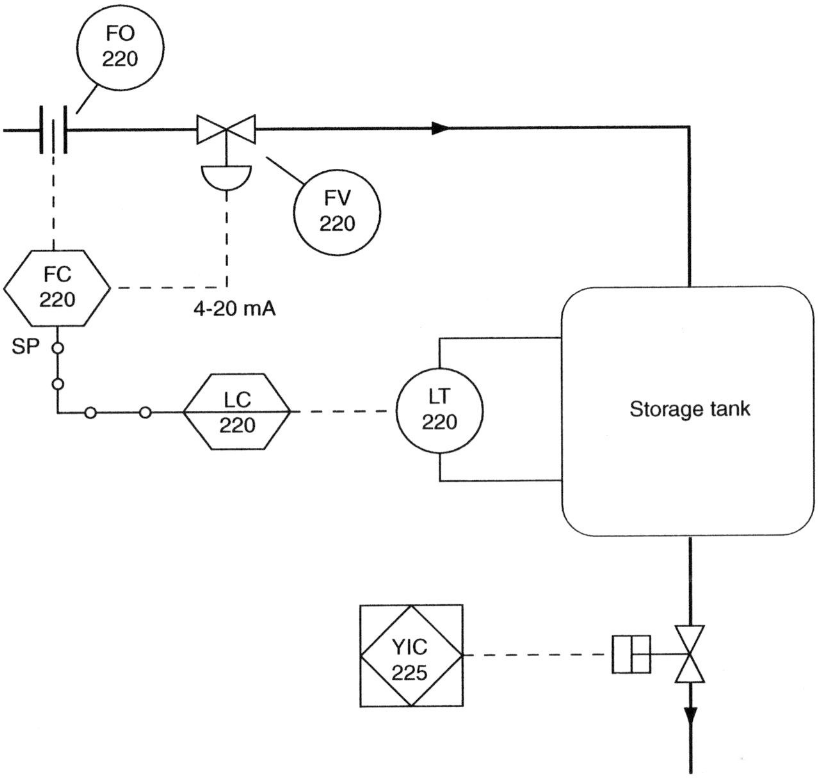

FIGURE 1.26
Computers and programmable logic controllers are included in the P&ID.

(e.g., 3 to 15 psi) is presented as a line with crosshatches. Figure 1.26 shows electrical (current) lines and also the digital data feed from a computer as a solid line with small bubbles. Appendix 5 shows the standards for other signal lines.

Instrument Symbols The instrumentation associated with control systems varies from sensors and transmitters to controllers, computers, and PLCs. These are drawn as bubbles with or without rectangles. Each has its own special symbol and designation as further described in Appendix 5. In general, the instrument symbol will be identified by a letter code, which denotes its function, and by a number code assigned by the designers, which may identify the loop or some region of the plant.

Figure 1.25 shows that the temperature control loop has a temperature sensor and transmitter designated by TT/301, which is connected to a temperature controller, TC/301. The solid line through the controller bubble means it is accessible by an operator, as in a control room panel. When the setpoint is not indicated, that means it is manually set. The controller is connected to an I/P converter designated by PY/301. The flow control loop is all pneumatic and has a flow transmitter (FT/302), a flow controller (FC/302), and a flow recorder (FR/302), which is located in the control room. Note that the sensor actually measures pressure and that the flow is proportional to the square root of pressure. Therefore, a square-root extractor is indicated following the flow transmitter. The flow recorder charts variations of flow rate for an operator.

Figure 1.26 shows a flow loop under computer control. This flow control computer, FC 220, is located in the field (i.e., at the site of control). This is an all-electrical loop that inputs a 4- to 20-mA signal for flow measurement and outputs a 4- to 20-mA signal to the control valve, FV/220. A level transmitter, LT/220, provides input to a computer, LC/220, which is located in the control room. The output of this computer is the setpoint of the flow loop. Finally, a PLC, shown as a diamond within a box, YIC/225, can open or close the tank drain valve.

Other Symbols Figures 1.25 and 1.26 also show the symbols for several other elements associated with the control systems and plants. For example, the control valves are often identified in terms of what they control. In Figure 1.25, the steam valve is labeled TV/301 to indicate that it is for temperature control even though it is actually controlling steam flow. Both figures show an orifice plate used to measure flow through a pipe. The valve actuator shown in Figure 1.26 with the PLC indicates it is spring-loaded and electrically actuated.

Appendix 5 shows more P&ID symbols defined by the S5.1 standard and some expanded information about its application.

1.7 SENSOR TIME RESPONSE

The static transfer function of a process-control loop element specifies how the output is related to the input if the input is constant. An element also has a time dependence that specifies how the output changes in time when the input is changing in time. It is independent from the static transfer function. This dynamic transfer function, which is independent from the static transfer function, is often simply called the *time response.* It is particularly important for sensors because they are the primary element for providing knowledge of the controlled variable value. This section will discuss the two most common types of sensor time responses.

Figure 1.27 shows a sensor that produces an output, $b(t)$, as a function of the input, $c(t)$. The static transfer function determines the output when the input is not changing in time. To specify the time response, the nature of the time variation in output, $b(t)$, is given

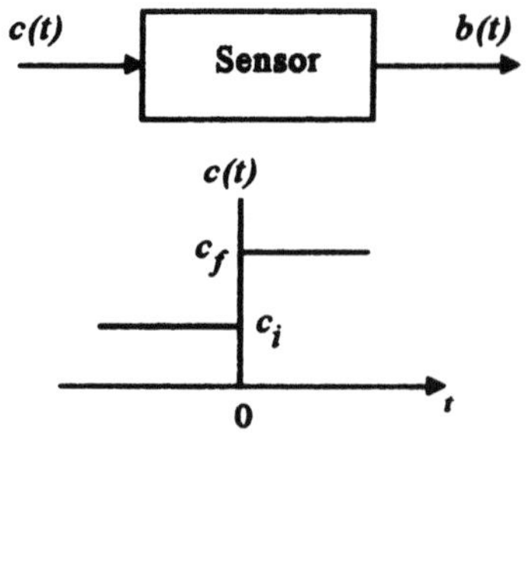

FIGURE 1.27
The dynamic transfer function specifies how a sensor output varies when the input changes instantaneously in time (i.e., a step change).

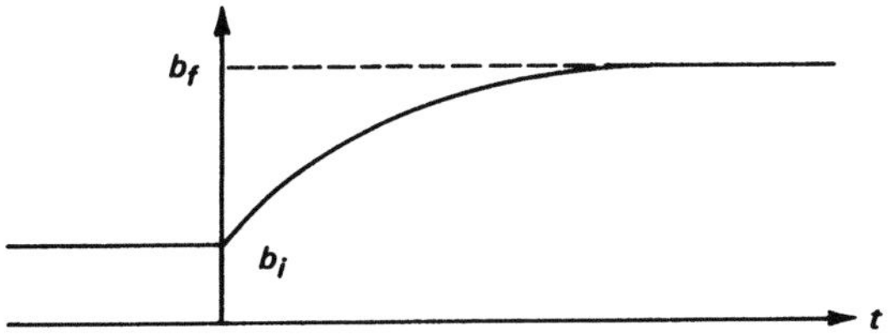

FIGURE 1.28
Characteristic first-order exponential time response of a sensor to a step change of input.

when the input exhibits a step change as shown. Note that at $t = 0$ the input to the sensor is suddenly changed from an initial value, c_i, to a final value, c_f. If the sensor were perfect, its output would be determined by the static transfer function to be b_i before $t = 0$ and b_f after $t = 0$. However, all sensors exhibit some lag between the output and the input and some characteristic variation in time before settling on the final value.

1.7.1 First-Order Response

The simplest time response is shown in Figure 1.28 as the output change in time following a step input as in Figure 1.27. This is called *first order* because, for all sensors of this type, the time response is determined by the solution of a first-order differential equation.

A general equation can be written for this response independent of the sensor, the variable being measured, or the static transfer function. The equation gives the sensor output as a function of time following the step input (i.e., it traces the curve of Figure 1.28 in time):

$$b(t) = b_i + (b_f - b_i)[1 - e^{-t/\tau}] \quad \textbf{(1.7)}$$

where b_i = initial sensor output from static transfer function and initial input
b_f = final sensor output from static transfer function and final input
τ = sensor time constant

The sensor output is in error during the transition time of the output from b_i to b_f. The actual variable value was changed instantaneously to a new value at $t = 0$. Equation (1.7) describes transducer output very well except during the initial time period—that is, at the *start* of the response near $t = 0$. In particular, the actual transducer response generally starts the change with a *zero* slope, and Equation (1.7) predicts a finite starting slope.

Time Constant The time constant, τ, is part of the specification of the sensor. Its significance can be seen by writing Equation (1.7) as follows

$$b(t) - b_i = (b_f - b_i)[1 - e^{-t/\tau}] \quad \textbf{(1.8)}$$

In this equation, the quantity on the left is the *change* in output as a function of time, whereas $(b_f - b_i)$ is the total change that will occur. Thus, the square-bracketed term in Equation (1.8) is the fraction of total change as a function of time.

Suppose we wish to find the change that has occurred at a time numerically equal to τ. Then we set $t = \tau$ in Equation (1.8) and find

$$b(\tau) - b_i = (b_f - b_i)[1 - e^{-1}] \tag{1.9}$$
$$b(\tau) - b_i = 0.6321(b_f - b_i)$$

Thus, we see that one time constant represents the time at which the output value has changed by approximately 63% of the total change.

The time constant τ is sometimes referred to as the 63% time, the *response time,* or the e-folding time. For a step change, the output response has approximately reached its final value after five time constants, since from Equation (1.8) we find

$$b(5\tau) - b_i = 0.993(b_f - b_i)$$

EXAMPLE 1.15

A sensor measures temperature linearly with a static transfer function of 33 mV/°C and has a 1.5-s time constant. Find the output 0.75 s after the input changes from 20° to 41°C. Find the error in temperature this represents.

Solution

We first find the initial and final values of the sensor output:

$$\begin{aligned} b_i &= (33\,\text{mV}/^\circ\text{C})(20^\circ\text{C}) \\ b_i &= 660\,\text{mV} \\ b_f &= (33\,\text{mV}/^\circ\text{C})(41^\circ\text{C}) \\ b_f &= 1353\,\text{mV} \end{aligned}$$

Now,

$$\begin{aligned} b(t) &= b_i + (b_f - b_i)[1 - e^{-t/\tau}] \\ b(0.75) &= 660 + (1353 - 660)[1 - e^{-0.75/1.5}] \\ b(0.75) &= \mathbf{932.7\ mV} \end{aligned}$$

This corresponds to a temperature of

$$\begin{aligned} T &= \frac{932.7}{33\ \text{mV}/^\circ\text{C}} \\ T &= \mathbf{28.3^\circ C} \end{aligned}$$

Thus the indicated temperature differs from the actual temperature by −12.7°C because of the lag in sensor output. After a time of about five times constants (~7.5 s) the sensor output would be about 1.353 V, correctly indicating the actual temperature of 41°C.

In many cases, the transducer output may be inversely related to the input. Equation (1.7) still describes the time response of the element where the final output is less than the initial output.

Note that the time response analysis is *always* applied to the output of the sensor, never the input. In Example 1.15 the temperature changed suddenly from 20° to 41°C, so it would be wrong to write equations for the temperature in terms of first-order time response. It is only the *output* of the sensor that lagged. Particularly if the sensor output varies nonlinearly with its input, terrible errors will occur if the time response equation is applied to the input.

Real-Time Effects The concept of the exponential time response and associated time constant is based on a sudden discontinuous change of the input value. In the real world, such instantaneous changes occur rarely, if ever, and thus we have presented a worst-case situation in the time response. In general, a sensor should be able to track any changes in the physical dynamic variable in a time less than one time constant.

1.7.2 Second-Order Response

In some sensors, a step change in the input causes the output to oscillate for a short period of time before settling down to a value that corresponds to the new input. Such oscillation (and the decay of the oscillation itself) is a function of the sensor. This output transient generated by the transducer is an error and must be accounted for in any measurement involving a transducer with this behavior.

This is called a *second-order response* because, for this type of sensor, the time behavior is described by a second-order differential equation. It is not possible to develop a universal solution, as it is for the first-order time response. Instead, we simply describe the general nature of the response.

Figure 1.29 shows a typical output curve that might be expected from a transducer having a second-order response for a discontinuous change in the input. It is impossible to describe this behavior by a simple analytic expression, as it is with the first-order response. However, the general behavior can be described in time as

$$R(t) \propto R_0 e^{-at} \sin(2\pi f_n t) \tag{1.10}$$

where
$R(t)$ = the transducer output
a = output damping constant
f_n = natural frequency of the oscillation
R_0 = amplitude

This equation shows the basic damped oscillation output of the device. The damping constant, a, and natural frequency, f_n, are characteristics of the transducer itself and must be considered in many applications.

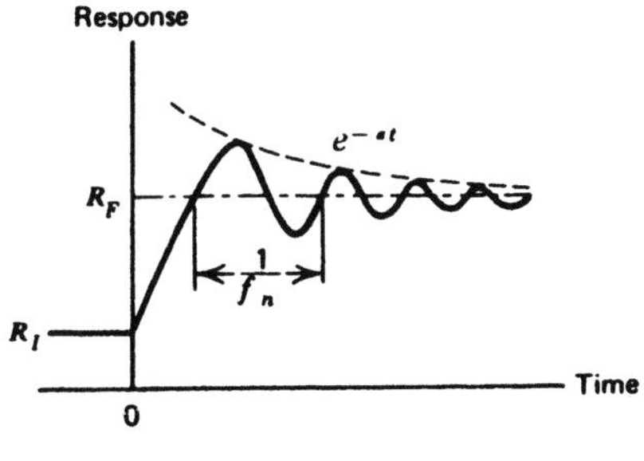

FIGURE 1.29
Characteristic second-order oscillatory time response of a sensor.

In general, such a transducer can be said to track the input when the input changes in a time that is *greater* than the period represented by the natural frequency. The damping constant defines the time one must wait after a disturbance at $t = 0$ for the transducer output to be a true indication of the transducer input. Thus, we see that in a time of $(1/a)$, the amplitude of the oscillations would be down to e^{-1}, or approximately 37%. More will be said about the effects of natural frequency and damping in the treatment of specific transducers that exhibit this behavior.

1.8 SIGNIFICANCE AND STATISTICS

Process control is vitally concerned with the value of variables, as the stated objective is to regulate the value of selected variables. It is therefore very important that the true significance of some measured value be understood. We have already seen that inherent errors and accuracy may lend uncertainty to the value indicated by a measurement. In this section we need to consider another feature of measurement that may be misleading about the actual value of a variable, as well as a method to help interpret the significance of measurements.

1.8.1 Significant Figures

In any measurement, we must be careful not to attach more significance to a variable value than the instrument can support. This is particularly true with the growing use of digital reading instruments and calculators with 8 to 12-digit readouts. Suppose, for example, that a digital instrument measures a resistance as 125 kΩ. Even if we ignore the instrument accuracy, this does not mean that the resistance is 125,000 Ω. Rather, it means that the resistance is closer to 125,000 than it is to 124,000 or to 126,000 Ω. We can use the 125 kΩ number in subsequent calculations, but we cannot draw conclusions about results having more than three numbers—that is, three significant figures. The significant figures are the digits (places) actually read or known from a measurement or calculation.

Significance in Measurement When using a measuring instrument, the number of significant figures is indicated either by readability, in the case of analog instruments, or by the number of digits, in a digital instrument. This is not to be confused with accuracy, which supplies an uncertainty to the reading itself. The following example illustrates how significant figures in measurement and accuracy are treated in the same problem.

EXAMPLE 1.16 A digital multimeter measures the current through a 12.5-kΩ resistor as 2.21 mA, using the 10-mA scale. The instrument accuracy is ±0.2% FS. Find the voltage across the resistor and the uncertainty in the value obtained.

Solution

First, we note that current is given to three significant figures, so no result we find can be significant to more than three digits. Then we see that the given accuracy becomes an uncertainty in the current of ±0.02 mA. From Ohm's law, we find the voltage as

$$V = IR = (2.21 \text{ mA})(12.5 \text{ k}\Omega) = 27.625 \text{ V}$$

But in terms of significant figures, we give this as $V = 27.6$ V. The accuracy means the current could vary from 2.19 to 2.23 mA, which introduces an uncertainty of ± 0.25 V. Thus, the complete answer is 27.6 ± 0.3 V, because we must express the uncertainty so that our significance is not changed.

Significance in Calculations In calculations, one must be careful not to obtain a result that has more significance than the numbers employed in the calculation. The answer can have no more significance than the least of the numbers used in the calculation.

EXAMPLE 1.17 A transducer has a specified transfer function of 22.4 mV/°C for temperature measurement. The measured voltage is 412 mV. What is the temperature?

Solution

Using the values given, we find

$$T = (412\,\text{mV})/(22.4\,\text{mV}/°\text{C}) = 18.392857°\text{C}$$

This was found using an 8-digit calculator, but the two given values are significant to only three places. Thus, our result can be significant to only three places; the answer is **18.4°C**.

Significance in Design The reader should be aware of [the difference in significant figures associated with measurement and conclusions drawn from measurement and significant figures associated with design.] A design is a hypothetical development that makes implicit assumptions about selected values in the design. If the designer specifies a 1.1-kΩ resistor, the assumption is that it is exactly 1100 Ω. If the designer specifies that there are 4.7 V across the resistor, then there are exactly 4.7 V and the current can be calculated as 4.2727272 mA. Now, suppose we measure the resistor when the design is built and find it to be 1.1 kΩ (two significant figures) and measure the voltage and find it to be 4.7 V (two significant figures). In this case, we report the calculated current of 4.3 mA, because we are dealing with two significant figures.

In the examples and problems in this book, we try to maintain the distinction between design values and measurement values. Whenever a problem or example involves design, perfect values are assumed. Thus, if a design specifies a 4.7-kΩ resistor, it is assumed that the value is exactly 4.7 kΩ (i.e., 4700.00 Ω). Whenever measurement is suggested, the figures given are assumed to be the significant figures. If a problem specifies that the measured voltage is 5.0, it is assumed that it is known to only two significant figures.

1.8.2 Statistics

Often, confidence in the value of a variable can be improved by elementary statistical analysis of measurements. This is particularly true where random errors in measurement cause a distribution of readings of the value of some variable.

Arithmetic Mean If many measurements of a particular variable are taken, the arithmetic mean is calculated to obtain an average value for the variable. There are many instances in process control when such an average value is of interest. For example, one may wish to control the average temperature in a process. The temperature might be measured in 10 locations and averaged to give a controlled variable value for use in the control loop.

Another common application is the calibration of transducers and other process instruments. In such cases, the average gives information about the transfer function. In digital or computer process control, it is often easier to use the average value of process variables. The arithmetic mean of a set of n values, given by $x_1, x_2, x_3, \ldots, x_n$ is defined by the equation

$$\bar{x} = \frac{x_1 + x_2 + x_3 + \cdots + x_n}{n} \tag{1.11}$$

where
$\bar{x}$ = arithmetic mean
n = number of values to be averaged
$x_1, x_2, \ldots, x_n$ = individual values

We often use the symbol Σ to represent a sum of numbers such as that used in Equation (1.11). Here we would write the equation as

$$\bar{x} = \frac{\Sigma x_i}{n} \tag{1.12}$$

where Σx_i = symbol for a sum of the values $x_1, x_2, \ldots, x_n$

Standard Deviation Often it is insufficient to know the value of the arithmetic mean of a set of measurements. To interpret the measurements properly, it may be necessary to know something about how the individual values are spread out about the mean. Thus, although the mean of the set (50, 40, 30, 70) is 47.5 and the mean of the set (5, 150, 21, 14) is also 47.5, the second group of numbers is obviously far more spread out. The standard deviation is a measure of this spread. Given a set of n values $x_1, x_2, \ldots, x_n$, we first define a set of deviations by the difference between the individual values and the arithmetic mean of the values, $\bar{x}$. The deviations are

$$d_1 = x_1 - \bar{x}$$
$$d_2 = x_2 - \bar{x}$$

and so on, until

$$d_n = x_n - \bar{x}$$

The set of these n deviations is now used to define the standard deviation according to the equation

$$\sigma = \sqrt{\frac{d_1^2 + d_2^2 + d_3^2 + \cdots + d_n^2}{n - 1}} \tag{1.13}$$

or, using the summation symbol,

$$\sigma = \sqrt{\frac{\Sigma d_i^2}{n - 1}} \tag{1.14}$$

Of course, the larger the standard deviation, the more spread out the numbers from which it is calculated.

EXAMPLE 1.18

Temperature was measured in eight locations in a room, and the values obtained were 21.2°, 25.0°, 18.5°, 22.1°, 19.7°, 27.1°, 19.0°, and 20.0°C. Find the arithmetic mean of the temperature and the standard deviation.

Solution

Using Equation (1.11), we have

$$\overline{T} = \frac{21.2 + 25 + 18.5 + 22.1 + 19.7 + 27.1 + 19 + 20}{8}$$

$$\overline{T} = \mathbf{21.6°C}$$

The standard deviation is found from Equation (1.13):

$$\sigma = \sqrt{\frac{(21.2 - 21.6)^2 + (25 - 21.6)^2 + \cdots + (20 - 21.6)^2}{(8 - 1)}}$$

$$\sigma = \mathbf{3.04°C}$$

Interpretation of Standard Deviation Figure 1.30 shows two curves constructed from many samples of two sensors measuring some variable with a fixed value, *V*. Due to sensor uncertainty the sampled values provided by the sensors exhibit variation about the average; however, both sensors are providing the correct value as the average. Notice that the distribution of readings for sensor *A* is much more narrowly distributed around the average than sensor *B*. This means that any single reading from sensor *A* is more likely to give the actual value of the measured variable. The standard deviation of the readings from sensor *A* would be much smaller than the standard deviation of sensor *B*.

A more quantitative evaluation of spreading can be made if we make certain assumptions about the set of data values used. In particular, we assume that the errors are truly

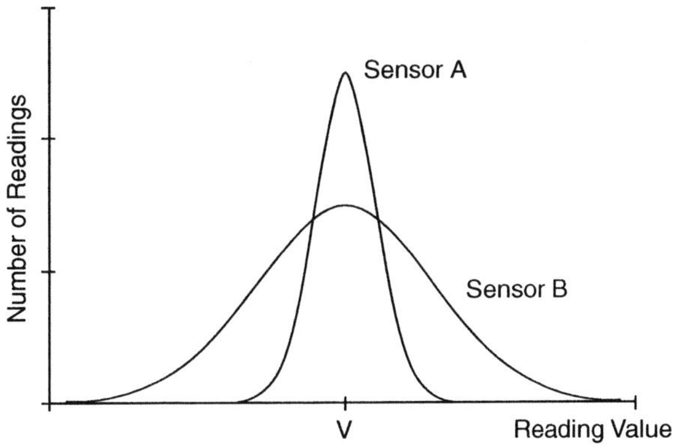

FIGURE 1.30
Multiple readings are taken of some variable with an actual value, *V*. The distributions show that sensor A has a smaller standard deviation than sensor B.

random and that we have taken a large sample of readings. We then can claim that the standard deviation and data are related to a special curve called the *normal probability curve,* or *bell curve.* If this is true, then

1. 68% of all readings lie within $\pm 1\sigma$ of the mean.
2. 95.5% of all readings lie within $\pm 2\sigma$ of the mean.
3. 99.7% of all readings lie within $\pm 3\sigma$ of the mean.

This gives us the added ability to make quantitative statements about how the data are spread about the mean. Thus, if one set of pressure readings has a mean of 44 psi with a standard deviation of 14 psi and another a mean of 44 psi with a standard deviation of 3 psi, we know the latter is much more peaked about the mean. In fact, 68% of all the readings in the second case lie from 41 to 47 psi, whereas in the first case, 68% of readings lie from 30 to 58 psi.

EXAMPLE 1.19

A control system was installed to regulate the weight of potato chips dumped into bags in a packaging operation. Given samples of 15 bags drawn from the operation before and after the control system was installed, evaluate the success of the system. Do this by comparing the arithmetic mean and standard deviations before and after. The bags should be 200 g.

Samples before: 201, 205, 197, 185, 202, 207, 215, 220, 179, 201, 197, 221, 202, 200, 195

Samples after: 197, 202, 193, 210, 207, 195, 199, 202, 193, 195, 201, 201, 200, 189, 197

Solution

In the before case, we use Equations (1.12) and (1.14) to find the mean and standard deviation, and get

$$\overline{W_b} = 202\,\text{g}$$
$$\sigma_b = 11\,\text{g}$$

Now the mean and standard deviations are found for the after case:

$$\overline{W_a} = 199\,\text{g}$$
$$\sigma_a = 5\,\text{g}$$

Thus, we see that the control system has brought the average bag weight closer to the ideal of 200 g and that it has cut the spread by a factor of 2. In the before case, 99% of the bags weighed 202 ± 33 g, but with the control system, 99% of the bags weighed in the range of 199 ± 15 g.

SUMMARY

This chapter presents an overview of process control and its elements. Subsequent chapters will examine these topics in more detail and provide a more quantitative understanding.

The following list will help the reader master the key points of the chapter.

1. Process control itself has been described as suitable for application to any situation in which a variable is regulated to some desired value or range of values. Figure 1.5 shows a block diagram in which the elements of measurement, error detector, controller, and control element are connected to provide the required regulation.
2. Numerous criteria have been discussed that allow the evaluation of process-control loop performance, of which the settling time, peak error, and minimum area are the most indicative of loop characteristics.
3. Both analog and digital processing are used in process-control applications. The current trend is to make analog measurements of the controlled variable, digitize them, and use a digital controller for evaluation. The basic technique of digital encoding allows each bit of a binary word to correspond to a certain quantity of the measured variable. The arrangement of "0" and "1" states in the word then serves as the encoding.
4. The SI system of units forms the basis of computations in this book as well as in the process-control industry in general. However, it is still necessary to understand conversions to other systems, notably the English system (see Appendix 1).
5. A standard adopted for analog process-control signals is the 4- to 20-mA current range to represent the span of measurements of the dynamic variable.
6. The definitions of accuracy, resolution, and other terms used in process control are necessary and are similar to those in related fields.
7. The concept of transducer time response was introduced. The time constant becomes part of the dynamic properties of a transducer.
8. The use of significant figures is important to properly interpret measurements and conclusions drawn from measurements.
9. Statistics can help interpret the validity of measurements through the use of the arithmetic mean and the standard deviation.
10. P&ID drawings and symbols are the typical representation used to display process-control systems.

PROBLEMS

Section 1.2

1.1 Explain how the basic strategy of control is employed in a room air-conditioning system. What is the controlled variable? What is the manipulated variable? Is the system self-regulating?

1.2 Is driving an automobile best described as a servomechanism or a process-control system? Why?

Section 1.3

1.3 Construct a block diagram of a refrigerator control system. Define each block in terms of the refrigerator components. (If you do not know the components, look them up in an encyclopedia or the Internet.)

FIGURE 1.31
Figure for Problem 1.4.

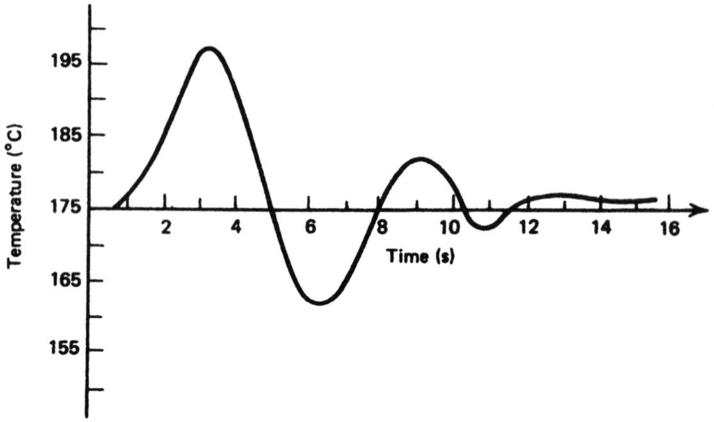

Section 1.4

1.4 A process-control loop has a setpoint of 175°C and an allowable deviation of ±5°C. A transient causes the response shown in Figure 1.31. Specify the maximum error and settling time.

1.5 Two different tunings of a process-control loop result in the transient responses shown in Figure 1.32. Estimate which would be preferred to satisfy the minimum area criteria.

1.6 The second cyclic transient error peak of a response test measures 4.4%. For the quarter-amplitude criteria, what error should be the third peak value?

1.7 Does the response of Figure 1.31 satisfy the quarter-amplitude criterion?

Section 1.5

1.8 An analog sensor converts flow linearly so that flow from 0 to 300 m^3/h becomes a current from 0 to 50 mA. Calculate the current for a flow of 225 m^3/h.

1.9 What binary word would represent the decimal number 16 if Table 1.1 were continued?

1.10 Suppose each bit change in a 4-bit ADC represents a level of 0.15 m.
a. What would the 4 bits be for a level of 1.7 m?
b. Suppose the 4 bits were $\mathbf{1000_2}$. What is the range of possible levels?

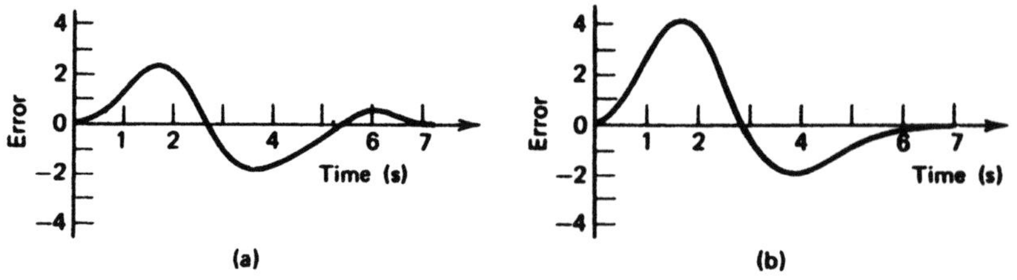

FIGURE 1.32
Figure for Problem 1.5.

1.11 For the process-control system in Figure 1.13, suppose that the relays close at |1.5| V and open at |1.1| V. This means that as the voltage on the relay reaches ± 1.5 V, it closes, and does not open again until the voltage drops to 1.1 V (i.e., there is a deadband). The amplifier has a gain of 10, the reference is 3 V, and the sensor outputs 150 mV/°C. Calculate the temperatures at which the heater turns on and off and at which the cooler turns on and off.

1.12 Show how the control system in Figure 1.6 would be modified to use **(a)** supervisory computer control and **(b)** DDC or computer control.

1.13 Think about how you adjust the water temperature coming from a single nozzle using the hot and cold hand valves in a kitchen faucet. Construct the block diagram of an automatic system as follows: The desired water temperature is selected by the user, perhaps by a knob and LCD readout. One hand valve turns on the cold water. The hot water valve is automatically set to keep the temperature at the selected value. Describe what elements would be necessary to do this using **(a)** analog control and **(b)** computer control.

Section 1.6

1.14 What is your mass in kilograms? What is your height in meters?

1.15 Atmospheric pressure is about 14.7 lb/in.^{2} (psi). What is this pressure in pascals?

1.16 An accelerometer is used to measure the constant acceleration of a race car that covers a quarter mile in 7.2 s.

a. Using $x = at^2/2$ to relate distance, x, acceleration, a, and time, t, find the acceleration in ft/s^2.

b. Express this acceleration in m/s^2.

c. Find the car speed, v, in m/s at the end of the quarter mile using the relation $v^2 = 2ax$.

d. Find the car energy in joules at the end of the quarter mile if it weighs 2000 lb, where the energy $W = mv^2/2$.

1.17 Suppose a liquid level ranging from 5.5 to 8.6 m is linearly converted to pneumatic pressure ranging from 3 to 15 psi. What pressure will result from a level of 7.2 m? What level does a pressure of 4.7 psi represent?

1.18 A controller output is a 4- to 20-mA signal that drives a valve to control flow. The relation between current and flow is $Q = 45[I - 2\text{ mA}]^{1/2}$ gal/min. What is the flow for 12 mA? What current produces a flow of 162 gal/min?

1.19 An instrument has an accuracy of $\pm 0.5\%$ FS and measures resistance from 0 to 1500 Ω. What is the uncertainty in an indicated measurement of 397 Ω?

1.20 A sensor has a transfer function of 0.5 mV/°C and an accuracy of $\pm 1\%$. If the temperature is known to be 60°C, what can be said with absolute certainty about the output voltage?

1.21 The sensor in Problem 1.20 is used with an amplifier with a gain of 15 ± 0.25 and displayed on a meter with a range of 0 to 2 V at $\pm 1.5\%$ FS. What is the worst-case and rms uncertainty for the total measurement?

1.22 Using the nominal transfer function values, what is the maximum measurable temperature of the system in Problems 1.20 and 1.21?

1.23 A temperature sensor transfer function is 44.5 mV/°C. The output voltage is measured at 8.86 V on a 3-digit voltmeter. What can you say about the value of the temperature?

1.24 A level sensor inputs a range from 4.50 to 10.6 ft and outputs a pressure range from 3 to 15 psi. Find an equation such as Equation (1.6) between level and pressure. What is the pressure for the level of 9.2 ft?

1.25 Draw Figure 1.6 in the standard P&ID symbols.

Section 1.7

1.26 A temperature sensor has a static transfer function of 0.15 mV/°C and a time constant of 3.3 s. If a step change of 22° to 50°C is applied at $t = 0$, find the output voltages at 0.5 s, 2.0 s, 3.3 s, and 9 s. What is the *indicated* temperature at these times?

1.27 A pressure sensor measures 44 psi just before a sudden change to 70 psi. The sensor measures 52 psi at a time 4.5 s after the change. What is the sensor time constant?

1.28 A photocell with a 35-ms time constant is used to measure light flashes. How long after a sudden dark-to-light flash before the cell output is 80% of the final value?

1.29 An alarm light goes ON when a pressure sensor voltage rises above 4.00 V. The pressure sensor outputs 20 mV/kPa and has a time constant of 4.9 s. How long after the pressure rises suddenly from 100 kPa to 400 kPa does the light go ON?

1.30 A pressure sensor has a resistance that changes with pressure according to $R = (0.15\ \text{k}\Omega/\text{psi})p + 2.5\ \text{k}\Omega$. This resistance is then converted to a voltage with the transfer function

$$V = \frac{10R}{R + 10k} \text{ volts}$$

The sensor time constant is 350 ms. At $t = 0$, the pressure changes suddenly from 40 psi to 150 psi.

a. What is the voltage output at 0.5 s? What is the indicated pressure at this time?

b. At what time does the output reach 5.0 V?

1.31 At $t = 0$, a temperature sensor was suddenly changed from 25° to 100°C. The sensor outputs voltage given by the expression $V = (0.06\ \text{V/°C})\,[T - 20°\text{C}]$. The following table gives the voltages measured and the times. Determine the average time constant of the sensor.

t (seconds)	0	0.1	0.2	0.3	0.4	0.5
V (volts)	0.3	1.8	2.8	3.4	3.9	4.2

Section 1.8

1.32 A circuit design calls for a 1.5-kΩ resistor to have 4.7 V across its terminals. What would be the expected current? The circuit is built, and the resistance is measured at 1500 Ω and the voltage at 4.7 V. What is the current through the resistor?

1.33 Flow rate was monitored for a week, and the following values were recorded as gal/min: 10.1, 12.2, 9.7, 8.8, 11.4, 12.9, 10.2, 10.5, 9.8, 11.5, 10.3, 9.3, 7.7, 10.2, 10.0, and 11.3. Find the mean and the standard deviation for these data.

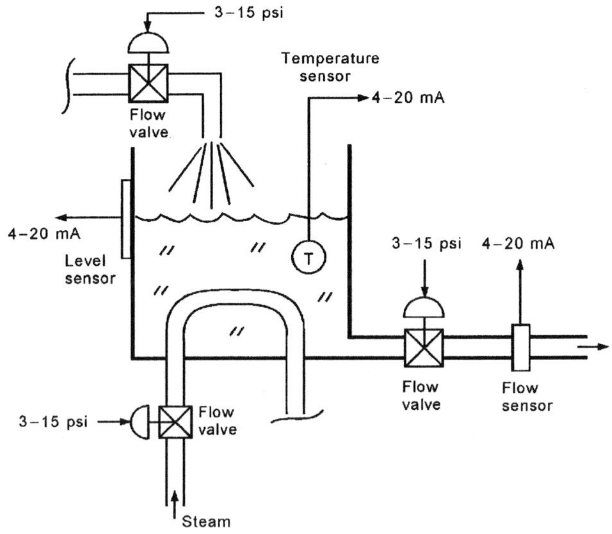

FIGURE 1.33
Figure for Problem S1.1.

1.34 A manufacturer specification sheet lists the transfer function of a pressure sensor as 45 ± 5% mV/kPa with a time constant of 4 ± 10% s. A highly accurate test system applies a step change of pressure from 20 kPa to 100 kPa.

a. What is the range of sensor voltage outputs initially and finally?

b. What range of voltages would be expected to be measured 2 s after the step change is applied?

SUPPLEMENTARY PROBLEMS

S1.1 Figure 1.33 shows a manufacturing process diagram. In this process, the following independent control requirements must be satisfied:

a. Control the level at L_{sp}.

b. Control the temperature at T_{sp}.

c. Control the output flow rate at Q_{sp}.

Complete the diagram showing the control loops by using the block diagram error-detector symbols and controller blocks. Include blocks for necessary signal converters.

S1.2 Prepare a process-control drawing of the system shown in Figure 1.33 similar to that shown in Figure 1.25. Be sure to include signal conversions.

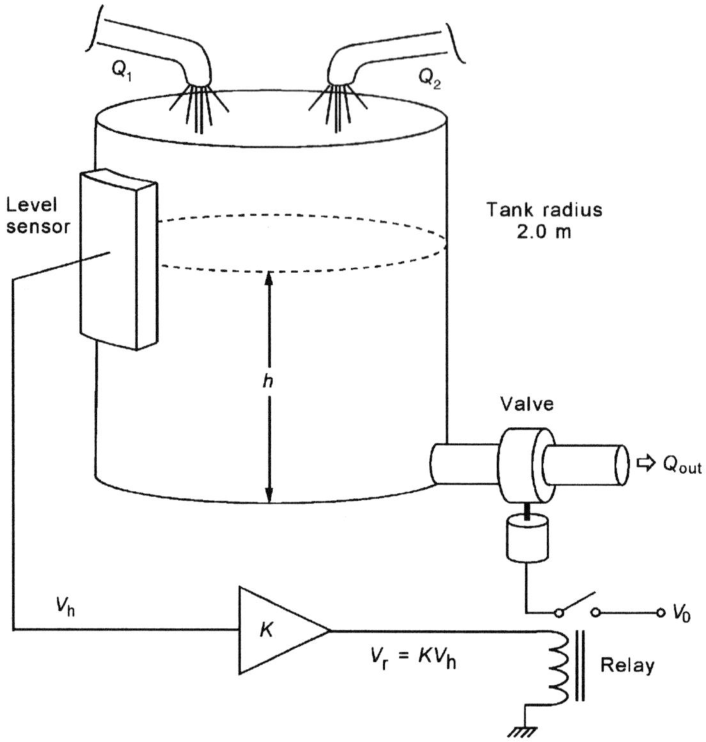

FIGURE 1.34
Figure for Problem S1.5.

S1.3 Explain which of the control loops in Figure 1.33 are self-regulating; give reasons why and why not.

S1.4 Assume control systems are in place to provide the requirements of Problem S1.1. Now, suppose the output-control valve suddenly sticks in the closed state. Explain why the tank will not overflow.

S1.5 Figure 1.34 shows a simple level-control system in which a closed relay opens the valve and an open relay closes the valve. Input flow is not controlled. The relay closes at 6.0 V and opens again at 4.8 V. The level sensor has a transfer function of $V_h = 0.8h + 0.4$ V.

a. Find the value of amplifier gain, *K*, required to open the valve when the level reaches 1.5 m.

b. At what level does the valve close?

c. Suppose $Q_1 = 5\ \text{m}^3/\text{min}$, $Q_2 = 2\ \text{m}^3/\text{min}$, and $Q_{out} = 9\ \text{m}^3/\text{min}$ (when open). What is the period of the level oscillation?

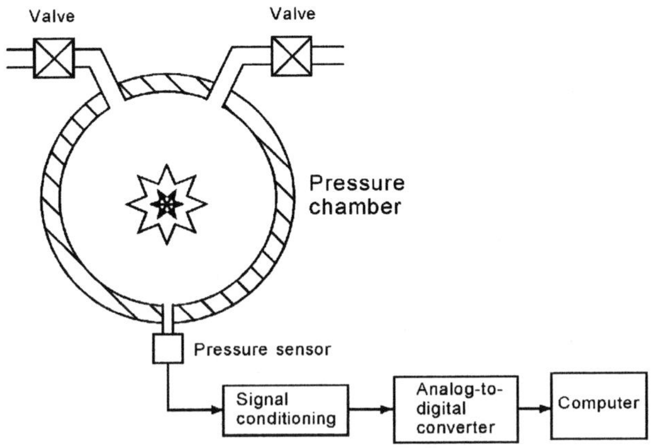

FIGURE 1.35
Figure for Problem S1.7.

S1.6 A pressure-measurement system uses a sensor that converts pressure into voltage according to the transfer functions, $V_p = 0.5\sqrt{p}$. This voltage is then converted into a current. As the pressure varies from 0 to 100 psi, the current varies from 4 to 20 mA.

a. Find the transfer function equation for the conversion of voltage to current.

b. What pressure change, Δp, will cause the current to change by 1 mA from 19 mA to 20 mA?

c. What pressure change, Δp, will cause the current to change by 1 mA from 4 mA to 5 mA? Why is the pressure change not the same as in the previous case, even though the current changed by 1 mA in both cases?

d. Prepare a graph of current versus pressure. Is it linear or nonlinear?

S1.7 Figure 1.35 shows a system for measuring the pressure of exploding gases inside a steel chamber. A computer is used to measure the pressure. The pressure sensor has a transfer function of $V_p = 0.05\sqrt{p + 500}$ and a first-order time constant of $\tau = 2.0$ s. When an explosion occurs, the pressure rises virtually instantaneously from 0 to some maximum, p_{max}. At $t = 0$, the explosion occurs, and the computer *must* take a reading at $t = 1$ s and determine the pressure, p_{max}. This is before the sensor signal has stabilized.

a. Explain how p_{max} can be determined from a measurement taken at $t = 1.0$ s.

b. Suppose the sensor signal at $t = 1.0$ s is 1.45 V. What is the value of p_{max}?

c. Suppose $p_{max} = 2500$ psi; what value will the sensor voltage have at 1.0 s?

d. What equations will the computer be programmed to use in order to find p_{max} from the sensor voltage taken at 1.0 s?

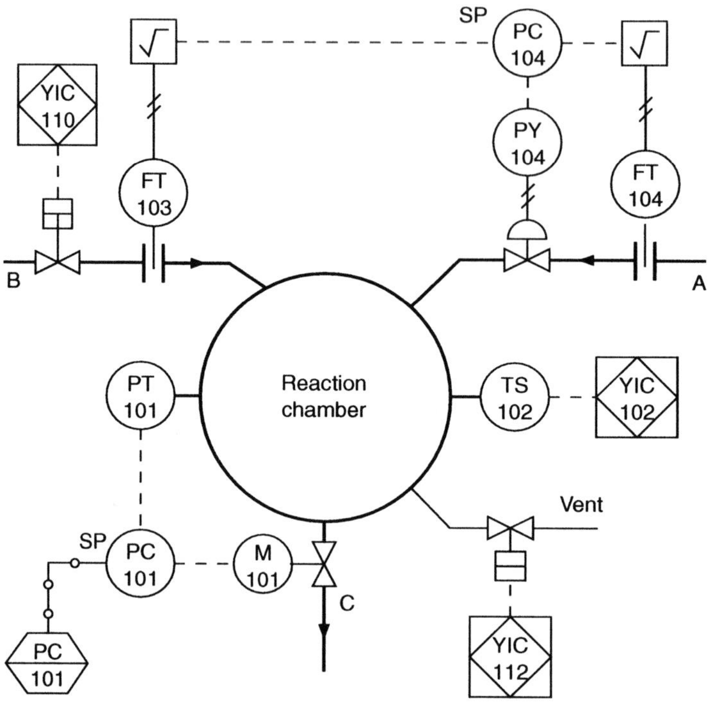

FIGURE 1.36
Figure for Problem S1.8.

S1.8 Figure 1.36 shows the P&ID for a process wherein materials A and B react in a chamber to create product C. The reaction generates heat and pressure within the chamber.

a. Provide a description of each element in the diagram and the signals that connect the element.

b. Describe the nature of the control loops shown and an overall view of how the process operates.

c. Explain the purpose of the PLC units and the computer.

CHAPTER 2

Analog Signal Conditioning

INSTRUCTIONAL OBJECTIVES

The purpose of this chapter is to introduce the reader to a variety of analog signal-conditioning methods used in process-control systems. Both passive methods and active methods, based upon the use of op amps, are defined. After reading this chapter and working through the examples and problems you will be able to:

- Explain the purpose of analog signal conditioning.
- Design a Wheatstone bridge circuit to convert resistance change to voltage change.
- Design RC low-pass and high-pass filter circuits to eliminate unwanted signals.
- Draw the schematics of four common op amp circuits and provide the transfer functions.
- Explain the operation of an instrumentation amplifier and draw its schematic.
- Design an analog signal-conditioning system to convert an input range of voltages to some desired output range of voltage.
- Design analog signal conditioning so that some range of resistance variation is converted into a desired range of voltage variation.

2.1 INTRODUCTION

Signal conditioning refers to operations performed on signals to convert them to a form suitable for interfacing with other elements in the process-control loop. In this chapter, we are concerned only with analog conversions, where the conditioned output is still an analog representation of the variable. Even in applications involving digital processing, some type of analog conditioning is usually required before analog-to-digital conversion is made. Specifics of digital signal conditioning are considered in Chapter 3.

2.2 PRINCIPLES OF ANALOG SIGNAL CONDITIONING

A sensor measures a variable by converting information about that variable into a dependent signal of either electrical or pneumatic nature. To develop such transducers, we take advantage of fortuitous circumstances in nature where a dynamic variable influences some characteristic of a material. Consequently, there is little choice of the type or extent of such proportionality. For example, once we have researched nature and found that cadmium sulfide resistance varies inversely and nonlinearly with light intensity, we must then learn to employ this device for light measurement within the confines of that dependence. Analog signal conditioning provides the operations necessary to transform a sensor output into a form necessary to interface with other elements of the process-control loop. We will confine our attention to electrical transformations.

We often describe the effect of the signal conditioning by the term *transfer function.* By this term we mean the effect of the signal conditioning on the input signal. Thus, a simple voltage amplifier has a transfer function of some constant that, when multiplied by the input voltage, gives the output voltage.

It is possible to categorize signal conditioning into several general types.

2.2.1 Signal-Level and Bias Changes

One of the most common types of signal conditioning involves adjusting the level (magnitude) and bias (zero value) of some voltage representing a process variable. For example, some sensor output voltage may vary from 0.2 to 0.6 V as a process variable changes over a measurement range. However, equipment to which this sensor output must be connected perhaps requires a voltage that varies from 0 to 5 V for the *same* variation of the process variable.

We perform the required signal conditioning by first changing the zero to occur when the sensor output is 0.2 V. This can be done by simply subtracting 0.2 from the sensor output, which is called a *zero shift,* or a bias adjustment.

Now we have a voltage that varies from 0 to 0.4 V, so we need to make the voltage larger. If we multiply the voltage by 12.5, the new output will vary from 0 to 5 V as required. This is called *amplification,* and 12.5 is called the *gain.* In some cases, we need to make a sensor output smaller, which is called *attenuation.* You should note that the circuit that does either chore is called an *amplifier.* We distinguish between amplification and attenuation by noting whether the gain of the amplifier is greater than or less than unity.

In designing bias and amplifier circuits, we must be concerned with issues such as the frequency response, output impedance, and input impedance.

2.2.2 Linearization

As pointed out at the beginning of this section, the process-control designer has little choice of the characteristics of a sensor output versus a process variable. Often, the dependence that exists between input and output is nonlinear. Even those devices that are approximately linear may present problems when precise measurements of the variable are required.

FIGURE 2.1
The purpose of linearization is to provide an output that varies linearly with some variable even if the sensor output does not.

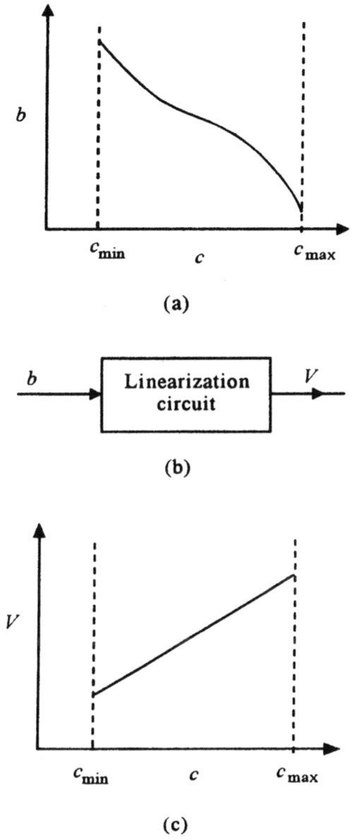

Historically, specialized analog circuits were devised to linearize signals. For example, suppose a sensor output varied nonlinearly with a process variable, as shown in Figure 2.1a. A linearization circuit, indicated symbolically in Figure 2.1b, would ideally be one that conditioned the sensor output so that a voltage was produced which was linear with the process variable, as shown in Figure 2.1c. Such circuits are difficult to design and usually operate only within narrow limits.

The modern approach to this problem is to provide the nonlinear signal as input to a computer and perform the linearization using software. Virtually any nonlinearity can be handled in this manner and, with the speed of modern computers, in nearly real time.

2.2.3 Conversions

Often, signal conditioning is used to convert one type of electrical variation into another. Thus, a large class of sensors exhibit changes of resistance with changes in a dynamic variable. In these cases, it is necessary to provide a circuit to convert this resistance change either to a voltage or a current signal. This is generally accomplished by bridges when the fractional resistance change is small and/or by amplifiers whose gain varies with resistance.

Signal Transmission An important type of conversion is associated with the process-control standard of transmitting signals as 4- to 20-mA current levels in wire. This gives rise to the need for converting resistance and voltage levels to an appropriate current level at the transmitting end and for converting the current back to voltage at the receiving end. Of course, current transmission is used because such a signal is independent of load variations other than accidental shunt conditions that may draw off some current. Thus, voltage-to-current and current-to-voltage converters are often required.

Digital Interface The use of computers in process control requires conversion of analog data into a digital format by integrated circuit devices called analog-to-digital converters (ADCs). Analog signal conversion is usually required to adjust the analog measurement signal to match the input requirements of the ADC. For example, the ADC may need a voltage that varies between 0 and 5 V, but the sensor provides a signal that varies from 30 to 80 mV. Signal conversion circuits can be developed to interface the output to the required ADC input.

2.2.4 Filtering and Impedance Matching

Two other common signal-conditioning requirements are filtering and matching impedance.

Often, spurious signals of considerable strength are present in the industrial environment, such as the 60-Hz line frequency signals. Motor start transients may also cause pulses and other unwanted signals in the process-control loop. In many cases, it is necessary to use high-pass, low-pass, or notch *filters* to eliminate unwanted signals from the loop. Such filtering can be accomplished by *passive* filters, using only resistors, capacitors, and inductors, or *active* filters, using gain and feedback.

Impedance matching is an important element of signal conditioning when transducer internal impedance or line impedance can cause errors in measurement of a dynamic variable. Both active and passive networks are employed to provide such matching.

2.2.5 Concept of Loading

One of the most important concerns in analog signal conditioning is the loading of one circuit by another. This introduces uncertainty in the amplitude of a voltage as it is passed through the measurement process. If this voltage represents some process variable, then we have uncertainty in the value of the variable.

Qualitatively, loading can be described as follows. Suppose the open-circuit output of some element is a voltage, say V_x, when the element input is some variable of value x. This element could be a sensor or some other part of the signal-conditioning circuit, such as a bridge circuit or amplifier. *Open circuit* means that nothing is connected to the output. Loading occurs when we do connect something, a load, across the output, and the output voltage of the element drops to some value, $V_y < V_x$. Different loads result in different drops.

Quantitatively, we can evaluate loading as follows. Thévenin's theorem tells us that the output terminals of any two terminal elements can be defined as a voltage source in se-

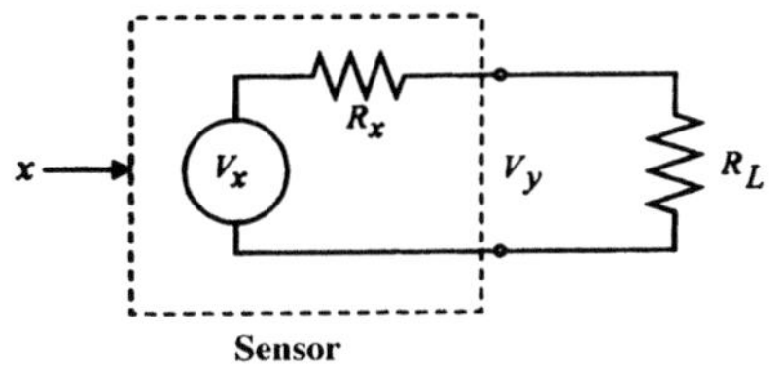

FIGURE 2.2
The Thévenin equivalent circuit of a sensor allows easy visualization of how loading occurs.

ries with an output impedance. Let's assume this is a resistance (the output resistance) to make the description easier to follow. This is called the Thévenin equivalent circuit model of the element.

Figure 2.2 shows such an element modeled as a voltage V_x and a resistance R_x. Now suppose a load, R_L, is connected across the output of the element as shown in Figure 2.2. This could be the input resistance of an amplifier, for example. A current will flow, and voltage will be dropped across R_x. It is easy to calculate that the loaded output voltage will thus be given by

$$V_y = V_x\left(1 - \frac{R_x}{R_L + R_x}\right) \tag{2.1}$$

The voltage that appears across the load is reduced by the voltage dropped across the internal resistance.

This equation shows how the effects of loading can be reduced. Clearly, the objective will be to make R_L much larger than R_x—that is, $R_L \gg R_x$. The following example shows how the effects of loading can compromise our measurements.

EXAMPLE 2.1

An amplifier outputs a voltage that is 10 times the voltage on its input terminals. It has an input resistance of 10 kΩ. A sensor outputs a voltage proportional to temperature with a transfer function of 20 mV/°C. The sensor has an output resistance of 5.0 kΩ. If the temperature is 50°C, find the amplifier output.

Solution

The naive solution is represented by Figure 2.3a. The unloaded output of the sensor is simply $V_T = (20\ \text{mV}/^\circ\text{C})50^\circ\text{C} = 1.0\ \text{V}$. Since the amplifier has a gain of 10, the output of the amplifier appears to be $V_{\text{out}} = 10\ V_{\text{in}} = (10)1.0\ \text{V} = 10\ \text{V}$. But this is wrong, because of loading!

Figure 2.3b shows the correct analysis. Here we see that there will be a voltage dropped across the output resistance of the sensor. The actual amplifier input voltage will be given by Equation (2.1),

$$V_{\text{in}} = V_T\left(1 - \frac{5.0\ \text{k}\Omega}{5.0\ \text{k}\Omega + 10\ \text{k}\Omega}\right)$$

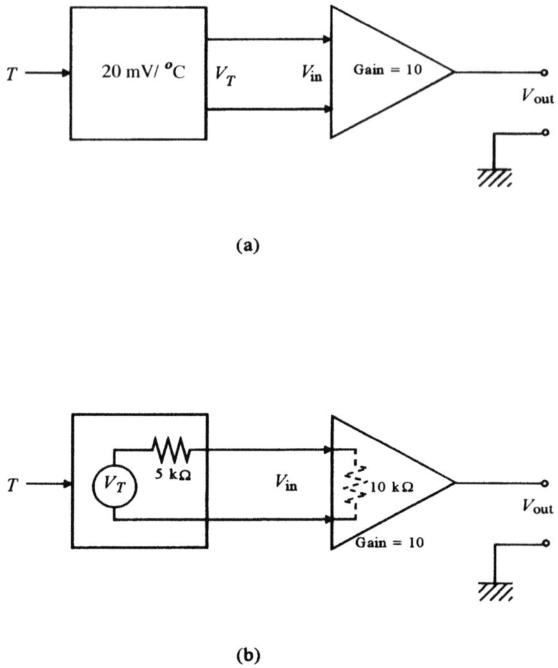

FIGURE 2.3
If loading is ignored, serious errors can occur in expected outputs of circuits and gains of amplifiers.

where $V_T = 1.0$ V, so that $V_{in} = 0.67$ V. Thus, the output of the amplifier is actually $V_{out} = 10(0.67\text{ V}) = 6.7$ V.

This concept plays an important role in analog signal conditioning and is referred to many times in this and later chapters.

If the electrical quantity of interest is frequency or a digital signal, then loading is not such a problem. That is, if there is enough signal left after loading to measure the frequency or to distinguish ones from zeros, there will be no error. Loading is important mostly when signal amplitudes are important.

2.3 PASSIVE CIRCUITS

Bridge and divider circuits are two passive techniques that have been extensively used for signal conditioning for many years. Although modern active circuits often replace these techniques, there are still many applications where their particular advantages make them useful.

Bridge circuits are used primarily as an accurate means of measuring changes in impedance. Such circuits are particularly useful when the fractional changes in impedance are very small.

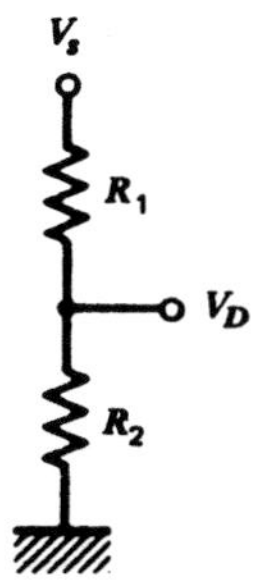

FIGURE 2.4
The simple voltage divider can often be used to convert resistance variation into voltage variation.

Another common type of passive circuit involved in signal conditioning is for filtering unwanted frequencies from the measurement signal. It is quite common in the industrial environment to find signals that possess high- and/or low-frequency noise as well as the desired measurement data. For example, a transducer may convert temperature information into a dc voltage, proportional to temperature. Because of the ever-present ac power lines, however, there may be a 60-Hz noise voltage impressed on the output that makes determination of the temperature difficult. A passive circuit consisting of a resistor and capacitor often can be used to eliminate both high- and low-frequency noise without changing the desired signal information.

2.3.1 Divider Circuits

The elementary voltage divider shown in Figure 2.4 often can be used to provide conversion of resistance variation into a voltage variation. The voltage of such a divider is given by the well-known relationship

$$V_D = \frac{R_2 V_s}{R_1 + R_2} \tag{2.2}$$

where V_s = supply voltage
R_1, R_2 = divider resistors

Either R_1 or R_2 can be the sensor whose resistance varies with some measured variable.

It is important to consider the following issues when using a divider for conversion of resistance to voltage variation:

1. The variation of V_D with either R_1 or R_2 is nonlinear; that is, even if the resistance varies linearly with the measured variable, the divider voltage will not vary linearly.
2. The effective output impedance of the divider is the parallel combination of R_1 and R_2. This may not necessarily be high, so loading effects must be considered.
3. In a divider circuit, current flows through both resistors; that is, power will be dissipated by both, including the sensor. The power rating of both the resistor and sensor must be considered.

EXAMPLE 2.2

The divider of Figure 2.4 has $R_1 = 10.0\ \text{k}\Omega$ and $V_s = 5.00\ \text{V}$. Suppose R_2 is a sensor whose resistance varies from 4.00 to 12.0 kΩ as some dynamic variable varies over a range. Then find **(a)** the minimum and maximum of V_D, **(b)** the range of output impedance, and **(c)** the range of power dissipated by R_2.

Solution

a. The solution is given by Equation (2.2). For $R_2 = 4\ \text{k}\Omega$, we have

$$V_D = \frac{(5\ \text{V})(4\ \text{k}\Omega)}{10\ \text{k}\Omega + 4\ \text{k}\Omega} = 1.43\ \text{V}$$

For $R_2 = 12\ \text{k}\Omega$, the voltage is

$$V_D = \frac{(5\ \text{V})(12\ \text{k}\Omega)}{10\ \text{k}\Omega + 12\ \text{k}\Omega} = 2.73\ \text{V}$$

b. Thus, the voltage varies from 1.43 to 2.73 V.

c. The range of output impedance is found from the parallel combination of R_1 and R_2 for the minimum and maximum of R_2. Simple parallel resistance computation shows that this will be from 2.86 to 5.45 kΩ.

d. The power dissipated by the sensor can be determined most easily from V^2/R_2, as the voltage across R_2 has been calculated. The power dissipated varies from 0.51 to 0.62 mW.

2.3.2 Bridge Circuits

Bridge circuits are used to convert impedance variations into voltage variations. One of the advantages of the bridge for this task is that it can be designed so the voltage produced varies around zero. This means that amplification can be used to increase the voltage level for increased sensitivity to variation of impedance.

Another application of bridge circuits is in the precise static measurement of an impedance.

Wheatstone Bridge The simplest and most common bridge circuit is the dc Wheatstone bridge, as shown in Figure 2.5. This network is used in signal-conditioning applications where a sensor changes resistance with process variable changes. Many modifications of this basic bridge are employed for other specific applications. In Figure 2.5, the object labeled D is a *voltage detector* used to compare the potentials of points a and b of the network. In most modern applications, the detector is a very high-input impedance differential amplifier. In some cases, a highly sensitive galvanometer with a relatively low impedance may be used, especially for calibration purposes and spot measurement instruments.

For our initial analysis, assume the detector impedance is infinite—that is, an open circuit.

In this case, the potential difference, ΔV, between points a and b is simply

$$\Delta V = V_a - V_b \tag{2.3}$$

FIGURE 2.5
The basic dc Wheatstone bridge.

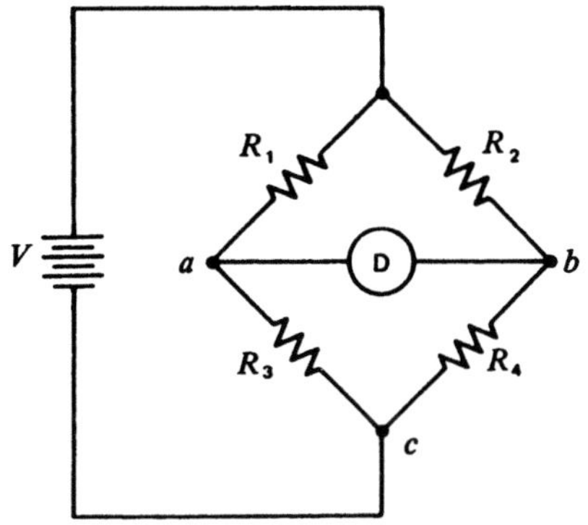

where V_a = potential of point a with respect to c

V_b = potential of point b with respect to c

The values of V_a and V_b now can be found by noting that V_a is just the supply voltage, V, divided between R_1 and R_3.

$$V_a = \frac{VR_3}{R_1 + R_3} \tag{2.4}$$

In a similar fashion, V_b is a divided voltage given by

$$V_b = \frac{VR_4}{R_2 + R_4} \tag{2.5}$$

where V = bridge supply voltage

If we now combine Equations (2.3), (2.4), and (2.5), the voltage difference or voltage offset can be written

$$\Delta V = \frac{VR_3}{R_1 + R_3} - \frac{VR_4}{R_2 + R_4} \tag{2.6}$$

Using algebra, the reader can show that this equation reduces to

$$\Delta V = V\frac{R_3R_2 - R_1R_4}{(R_1 + R_3)\cdot(R_2 + R_4)} \tag{2.7}$$

Equation (2.7) shows how the difference in potential across the detector is a function of the supply voltage and the values of the resistors. Because a difference appears in the numerator of Equation (2.7), it is clear that a particular combination of resistors can be found that will result in zero difference and zero voltage across the detector—that is, a *null.* Obviously, this combination, from examination of Equation (2.7), is

$$R_3R_2 = R_1R_4 \tag{2.8}$$

Equation (2.8) indicates that whenever a Wheatstone bridge is assembled and resistors are adjusted for a detector null, the resistor values must satisfy the indicated equality. It does not matter if the supply voltage drifts or changes; the null is maintained. Equations (2.7) and (2.8) underlie the application of Wheatstone bridges to process-control applications using high-input impedance detectors.

EXAMPLE 2.3 If a Wheatstone bridge, as shown in Figure 2.5, nulls with $R_1 = 1000\ \Omega$, $R_2 = 842\ \Omega$, and $R_3 = 500\ \Omega$, find the value of R_4.

Solution

Because the bridge is nulled, find R_4 using Equation (2.8):

$$R_1R_4 = R_3R_2$$

$$R_4 = \frac{R_3R_2}{R_1} = \frac{(500\ \Omega)\,(842\ \Omega)}{1000\ \Omega}$$

$$R_4 = \mathbf{421\ \Omega}$$

EXAMPLE 2.4 The resistors in a bridge are given by $R_1 = R_2 = R_3 = 120\ \Omega$ and $R_4 = 121\ \Omega$. If the supply is 10.0 V, find the voltage offset.

Solution

Assuming the detector impedance to be very high, we find the offset from

$$\Delta V = V\frac{R_3R_2 - R_1R_4}{(R_1 + R_3)\cdot(R_2 + R_4)}$$

$$\Delta V = 10\ \text{V}\ \frac{(120\ \Omega)(120\ \Omega) - (120\ \Omega)(121\ \Omega)}{(120\ \Omega + 120\ \Omega)\cdot(120\ \Omega + 121\ \Omega)}$$

$$\Delta V = \mathbf{-20.7\ mV}$$

Notice that the result in Example 2.4 is a negative voltage. Remember that ΔV is simply the difference between V_a and V_b in Figure 2.5, as given by Equation (2.3). So the fact that ΔV is negative in this example simply means that V_b is larger than V_a.

Galvanometer Detector The use of a galvanometer as a null detector in the bridge circuit introduces some differences in our calculations because the detector resistance may be low and because we must determine the bridge offset as current offset. When the bridge is nulled, Equation (2.8) still defines the relationship between the resistors in the bridge arms. Equation (2.7) must be modified to allow determination of current drawn by the galvanometer when a null condition is not present. Perhaps the easiest way to determine this offset current is first to find the Thévenin equivalent circuit between points *a* and *b* of the bridge (as drawn in Figure 2.5 with the detector removed). The Thévenin voltage is simply the open-circuit voltage difference between points *a* and *b* of the circuit. But wait! Equation (2.7) is the open-circuit voltage, so

$$V_{\text{Th}} = V\frac{R_3R_2 - R_1R_4}{(R_1 + R_3)(R_2 + R_4)} \tag{2.9}$$

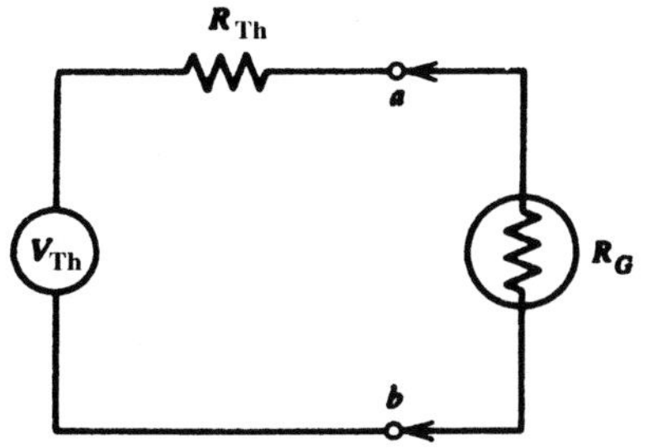

FIGURE 2.6
When a galvanometer is used for a null detector, it is convenient to use the Thévenin equivalent circuit of the bridge.

The Thévenin resistance is found by replacing the supply voltage by its internal resistance and calculating the resistance between terminals a and b of the network. We may assume that the internal resistance of the supply is negligible compared to the bridge arm resistances. It is left as an exercise for the reader to show that the Thévenin resistance seen at points a and b of the bridge is

$$R_{Th} = \frac{R_1 R_3}{R_1 + R_3} + \frac{R_2 R_4}{R_2 + R_4} \qquad \textbf{(2.10)}$$

The Thévenin equivalent circuit for the bridge enables us to easily determine the current through any galvanometer with internal resistance, R_G, as shown in Figure 2.6. In particular, the offset current is

$$I_G = \frac{V_{Th}}{R_{Th} + R_G} \qquad \textbf{(2.11)}$$

Using this equation in conjunction with Equation (2.8) defines the Wheatstone bridge response whenever a galvanometer null detector is used.

EXAMPLE 2.5

A bridge circuit has resistance of $R_1 = R_2 = R_3 = 2.00\ \text{k}\Omega$ and $R_4 = 2.05\ \text{k}\Omega$ and a 5.00-V supply. If a galvanometer with a 50.0-Ω internal resistance is used for a detector, find the offset current.

Solution

From Equation (2.9), the offset voltage is V_{Th}.

$$V_{Th} = 5\ \text{V}\frac{(2\ \text{k}\Omega)(2\ \text{k}\Omega) - (2\ \text{k}\Omega)(2.05\ \text{k}\Omega)}{(2\ \text{k}\Omega + 2\ \text{k}\Omega)(2\ \text{k}\Omega + 2.05\ \text{k}\Omega)}$$

$$V_{Th} = \mathbf{-30.9\ mV}$$

We next find the bridge Thévenin resistance from Equation (2.10):

$$R_{Th} = \frac{(2\ \text{k}\Omega)(2\ \text{k}\Omega)}{(2\ \text{k}\Omega + 2\ \text{k}\Omega)} + \frac{(2\ \text{k}\Omega)(2.05\ \text{k}\Omega)}{(2\ \text{k}\Omega + 2.05\ \text{k}\Omega)}$$

$$R_{Th} = \mathbf{2.01\ k\Omega}$$

Finally, the current is given by Equation (2.11):

$$I_G = \frac{-30.9 \text{ mV}}{2.01 \text{ k}\Omega + 0.05 \text{ k}\Omega}$$

$$I_G = \mathbf{-15.0\,\mu A}$$

The negative sign on the current simply means that the current flows through the galvanometer from right to left (i.e., from point *b* to point *a* in the circuit of Figure 2.5).

Bridge Resolution The resolution of the bridge circuit is a function of the resolution of the detector used to determine the bridge offset. Thus, referring to the case where a voltage offset occurs, we define the resolution in resistance as that resistance change in one arm of the bridge that causes an offset voltage that is equal to the resolution of the detector. If a detector can measure a change of 100 μV, this sets a limit on the minimum measurable resistance change in a bridge using this detector. In general, once given the detector resolution, we may use Equation (2.7) to find the change in resistance that causes this offset.

EXAMPLE 2.6 A bridge circuit has $R_1 = R_2 = R_3 = R_4 = 120.0\text{-}\Omega$ resistances and a 10.0-V supply. Clearly, the bridge is nulled, as Equation (2.8) shows. Suppose a $3\frac{1}{2}$-digit DVM on a 200-mV scale will be used for the null detector. Find the resistance resolution for measurements of R_4.

Solution

On a 200-mV scale, the DVM measures from 000.0 to 199.9 mV, so the smallest measurable change is 0.1 mV, or 100 μV. So, we need to find out how much R_4 has changed from 120 Ω to create this much off null voltage.

We simply use Equation (2.6), with R_4 changed to some unknown value so that 100 μV results:

$$100\,\mu\text{V} = \frac{(120\,\Omega)(10\text{ V})}{120\,\Omega + 120\,\Omega} - \frac{R_4(10\text{ V})}{120\,\Omega + R_4}$$

This equation is a bit of an algebraic challenge to solve, but eventually we find

$$R_4 = 119.9952\,\Omega$$

So the smallest change in resistance that can be measured is

$$\Delta R_4 = 120\,\Omega - 119.9952\,\Omega = 0.0048\,\Omega$$

A bridge offset of $+100\,\mu$V is caused by a reduction of R_4. It follows that a bridge offset of $-100\,\mu$V would be caused by an increase of R_4.

In Example 2.6, we see that a minimum resistance change of 0.0048 Ω must occur before the detector indicates a change in offset voltage.

One may also view this as an overall *accuracy* of the instrument, because it can also be said that ΔR represents the uncertainty in any determination of resistance using the given bridge and detector.

The same arguments can be applied to a galvanometer measurement, where the resolution is limited by the minimum measurable current.

Lead Compensation In many process-control applications, a bridge circuit may be located at considerable distance from the sensor whose resistance changes are to be measured. In such cases, the remaining fixed bridge resistors can be chosen to account for the resistance of leads required to connect the bridge to the sensor. Furthermore, any measurement of resistance can be adjusted for lead resistance to determine the actual resistance. Another problem exists that is not so easily handled, however. There are many effects that can change the resistance of the long lead wires on a transient basis, such as frequency, temperature, stress, and chemical vapors. Such changes will show up as a bridge offset and be interpreted as changes in the sensor output. This problem is reduced using *lead compensation,* where any changes in lead resistance are introduced equally into two (both) arms of the bridge circuit, thus causing no effective change in bridge offset. Lead compensation is shown in Figure 2.7. Here we see that R_4, which is assumed to be the sensor, has been removed to a remote location with lead wires (1), (2), and (3). Wire (3) is the power lead and has no influence on the bridge balance condition. If wire (2) changes in resistance because of spurious influences, it introduces this change into the R_4 leg of the bridge. Wire (1) is exposed to the same environment and changes by the same amount, but is in the R_3 leg of the bridge. Effectively, both R_3 and R_4 are identically changed, and thus Equation (2.8) shows that no change in the bridge null occurs. This type of compensation is often employed where bridge circuits must be used with long leads to the active element of the bridge.

Current Balance Bridge One disadvantage of the simple Wheatstone bridge is the need to obtain a null by variation of resistors in bridge arms. In the past, many process-control applications used a feedback system in which the bridge offset voltage was amplified and used to drive a motor whose shaft altered a variable resistor to renull the bridge. Such a system does not suit the modern technology of electronic processing because it is not very fast, is subject to wear, and generates electronic noise. A technique that provides for an electronic nulling of the bridge and that uses only fixed resistors (except as may be required for calibration) can be

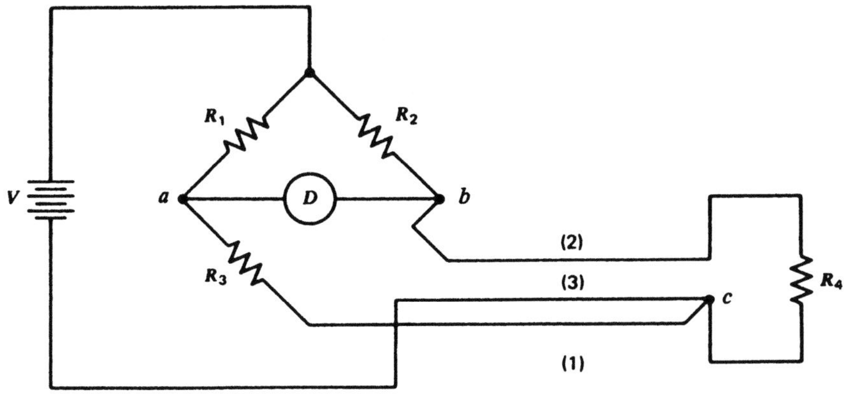

FIGURE 2.7
For remote sensor applications, this compensation system is used to avoid errors from lead resistance.

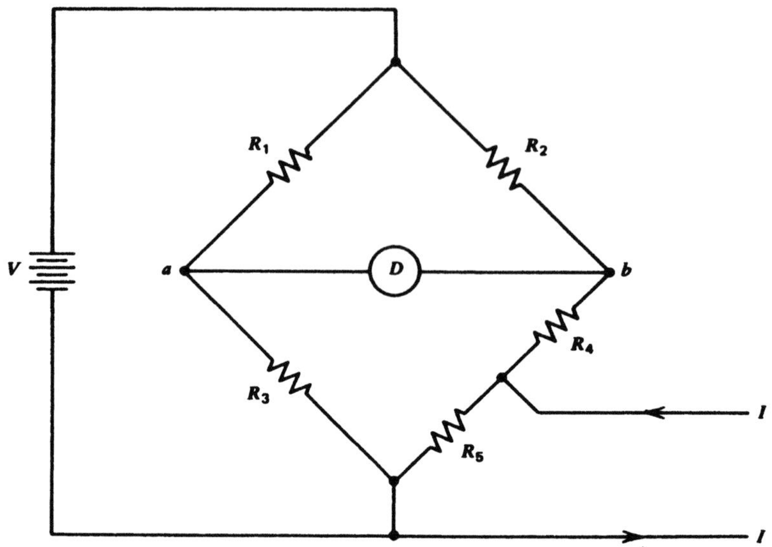

FIGURE 2.8
The current balance bridge.

used with the bridge. This method uses a *current* to null the bridge. A closed-loop system can even be constructed that provides the bridge with a *self-nulling* ability.

The basic principle of the current balance bridge is shown in Figure 2.8. The standard Wheatstone bridge is modified by splitting one arm resistor into two, R_4 and R_5. A current, I, is fed into the bridge through the junction of R_4 and R_5 as shown. We now stipulate that the size of the bridge resistors is such that the current flows predominantly through R_5. This can be provided for by any of several requirements. The least restrictive is to require

$$R_4 \gg R_5 \tag{2.12}$$

Often, if a high-impedance null detector is used, the restriction of Equation (2.12) becomes

$$(R_2 + R_4) \gg R_5 \tag{2.13}$$

Assuming that either conditions of Equations (2.12) or (2.13) are satisfied, the voltage at point b is the sum of the divided supply voltage plus the voltage dropped across R_5 from the current, I.

$$V_b = \frac{V(R_4 + R_5)}{R_2 + R_4 + R_5} + IR_5 \tag{2.14}$$

The voltage of point a is still given by Equation (2.4). Thus, the bridge offset voltage is given by $\Delta V = V_a - V_b$, or

$$\Delta V = \frac{VR_3}{R_1 + R_3} - \frac{V(R_4 + R_5)}{R_2 + R_4 + R_5} - IR_5 \tag{2.15}$$

This equation shows that a null is reached by adjusting the magnitude and polarity of the current I until IR_5 equals the voltage difference of the first two terms. If one of the bridge resistors changes, the bridge can be renulled by changing current I. In this manner, the bridge is electronically nulled from any convenient current source. In most applications, the bridge is nulled at some nominal set of resistances with zero current. Changes of a bridge resistor are detected as a bridge offset signal that is used to provide the renulling current. The action is explained in Example 2.7.

EXAMPLE 2.7 A current balance bridge, as shown in Figure 2.8, has resistors $R_1 = R_2 = 10\ \text{k}\Omega$, $R_4 = 950\ \Omega$, $R_3 = 1\ \text{k}\Omega$, $R_5 = 50\ \Omega$ and a high-impedance null detector. Find the current required to null the bridge if R_3 changes by $1\ \Omega$. The supply voltage is 10 V.

Solution

First, for the nominal resistance values given, the bridge is at a null with $I = 0$, because

$$V_a = \frac{(10\ \text{V})(1\ \text{k}\Omega)}{10\ \text{k}\Omega + 1\ \text{k}\Omega}$$

$$V_a = 0.9091\ \text{V}$$

With $I = 0$, Equation (2.14) gives

$$V_b = \frac{(10\ \text{V})(950\ \Omega + 50\ \Omega)}{10\ \text{k}\Omega + 950\ \Omega + 50\ \Omega}$$

$$V_b = 0.9091\ \text{V}$$

When R_3 increases by $1\ \Omega$ to $1001\ \Omega$, V_a becomes

$$V_a = \frac{(10\ \text{V})(1001)}{10\ \text{k}\Omega + 1001}$$

$$V_a = 0.9099\ \text{V}$$

which shows that the voltage at b must increase by 0.0008 V or 0.8 mV to renull the bridge. This can be provided by a current, from Equation (2.15) and $\Delta V = 0$, found from $50I = 0.8\ \text{mV}$.

$$I = \mathbf{16.0\ \mu A}$$

Potential Measurements Using Bridges A bridge circuit is also useful to measure small potentials at a very high impedance, using either a conventional Wheatstone bridge or a current balance bridge. This type of measurement is performed by placing the potential to be measured in series with the detector, as shown in Figure 2.9. The null detector responds to the potential between points c and b. In this case, V_b is given by Equation (2.14) and V_c by

$$V_c = V_x + V_a \tag{2.16}$$

where V_a is given by Equation (2.4), and V_x is the potential to be measured. The voltage appearing across the null detector is

$$\Delta V = V_c - V_b = V_x + V_a - V_b$$

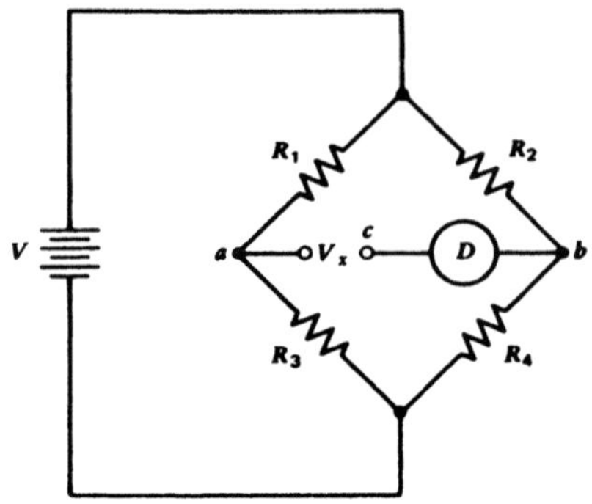

FIGURE 2.9
Using the basic Wheatstone bridge for potential measurement.

A null condition is established when $\Delta V = 0$; furthermore, no current flows through the unknown potential when such a null is found. Thus, V_x can be measured by varying the bridge resistors to provide a null with V_x in the circuit and solving for V_x using the null condition

$$V_x + \frac{R_3 V}{R_1 + R_3} - \frac{V R_4}{R_2 + R_4} = 0 \qquad \textbf{(2.17)}$$

The current balance bridge of Figure 2.8 can also be used for potential measurement with no current through the potential to be measured. Here again, the potential is placed in series with the detector, and ΔV is defined exactly as before. Now, however, V_b is given by Equation (2.14), so the null condition becomes

$$V_x + \frac{R_3 V}{R_1 + R_3} - \frac{V(R_4 + R_5)}{R_2 + R_4 + R_5} - IR_5 = 0 \qquad \textbf{(2.18)}$$

If the fixed resistors are chosen to null the bridge with $I = 0$ when $V_x = 0$, then the two middle terms in Equation (2.18) cancel, leaving a very simple relationship between V_x and the nulling current:

$$V_x - IR_5 = 0 \qquad \textbf{(2.19)}$$

EXAMPLE 2.8

A bridge circuit for potential measurement nulls when $R_1 = R_2 = 1\ \text{k}\Omega$, $R_3 = 605\ \Omega$, and $R_4 = 500\ \Omega$ with a 10-V supply. Find the unknown potential.

Solution

Here we simply use Equation (2.17) to solve for V_x.

$$V_x + \frac{(605)(10)}{605 + 1000} - \frac{(10)(500)}{1000 + 500} = 0$$

$$V_x + 3.769 - 3.333 = 0$$

$$V_x = \mathbf{-0.436\ V}$$

The negative sign tells us the polarity of the unknown voltage, V_x. Since V_x numerically subtracts from V_a, we see that its positive terminal must be connected to point a in Figure 2.9.

EXAMPLE 2.9

A current balance bridge is used for potential measurement. The fixed resistors are $R_1 = R_2 = 5\ \text{k}\Omega$, $R_3 = 1\ \text{k}\Omega$, $R_4 = 990\ \Omega$, and $R_5 = 10\ \Omega$ with a 10-V supply. Find the current necessary to null the bridge if the potential is 12 mV.

Solution

First, an examination of the resistances shows that the bridge is nulled when $I = 0$ and $V_x = 0$ because, from Equation (2.18),

$$\frac{VR_3}{R_1 + R_3} = \frac{10(1\ \text{k})}{1\ \text{k} + 5\ \text{k}} = 1.667\ \text{V}$$

and

$$\frac{V(R_4 + R_5)}{R_2 + R_4 + R_5} = \frac{10(990 + 10)}{5\ \text{k} + 990 + 10} = 1.667\ \text{V}$$

Thus, we can use Equation (2.19):

$$12\ \text{mV} - 10\,I = 0$$
$$I = \mathbf{1.2\ mA}$$

ac Bridges The bridge concept described in this section can be applied to the matching of impedances in general, as well as to resistances. In this case, the bridge is represented as in Figure 2.10 and employs an ac excitation, usually a sine wave voltage signal. The analysis of bridge behavior is basically the same as in the previous treatment, but impedances replace resistances. The bridge offset voltage then is represented as

$$\Delta E = E\frac{Z_3Z_2 - Z_1Z_4}{(Z_1 + Z_3)(Z_2 + Z_4)}$$
$$\Delta E = \text{ac offset voltage} \qquad \textbf{(2.20)}$$

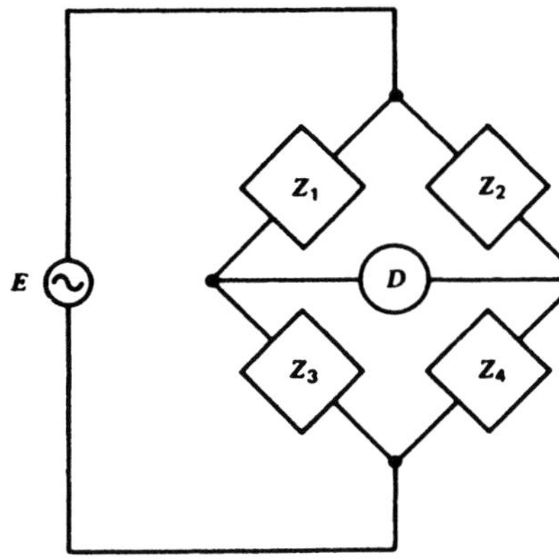

FIGURE 2.10
A general ac bridge circuit.

where E = sine wave excitation voltage
Z_1, Z_2, Z_3, Z_4 = bridge impedances

A null condition is defined as before by a zero offset voltage $\Delta E = 0$. From Equation (2.20), this condition is met if the impedances satisfy the relation

$$Z_3 Z_2 = Z_1 Z_4 \tag{2.21}$$

This condition is analogous to Equation (2.8) for resistive bridges.

A special note is necessary concerning the achievement of a null in ac bridges. In some cases, the null detection system is phase sensitive with respect to the bridge excitation signal. In these instances, it is necessary to provide a null of both the in-phase and quadrature (90° out-of-phase) signals before Equation (2.21) applies.

EXAMPLE 2.10 An ac bridge employs impedances as shown in Figure 2.11. Find the value of R_x and C_x when the bridge is nulled.

Solution

Because the bridge is at null, we have

$$Z_2 Z_3 = Z_1 Z_x$$

or

$$R_2\left(R_3 - \frac{j}{\omega C}\right) = R_1\left(R_x - \frac{j}{\omega C_x}\right)$$

$$R_2 R_3 - j\frac{R_2}{\omega C} = R_1 R_x - \frac{jR_1}{\omega C_x}$$

The real and imaginary parts must be independently equal, so that

$$R_x - \frac{R_2 R_3}{R_1} = 0$$

$$R_x = \frac{(2\ \text{k}\Omega)(1\ \text{k}\Omega)}{1\ \text{k}\Omega}$$

$$R_x = \mathbf{2\ k\Omega}$$

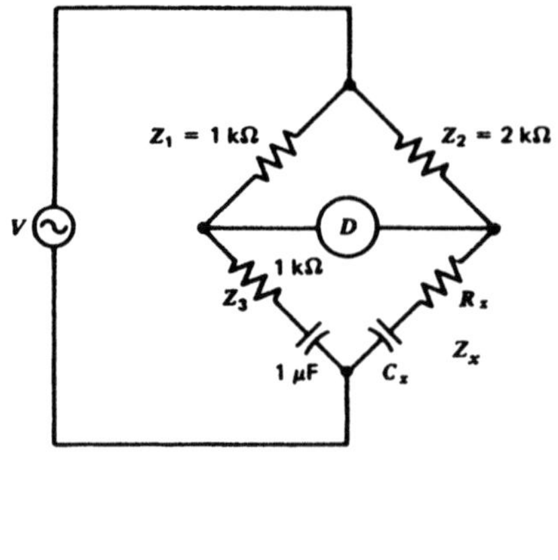

FIGURE 2.11
The ac bridge circuit and components for Example 2.10.

and

$$C_x = C\frac{R_1}{R_2}$$

$$C_x = (1\ \mu\text{F})\frac{1\ \text{k}\Omega}{2\ \text{k}\Omega}$$

$$C_x = \mathbf{0.5\mu F}$$

Bridge Applications The primary application of bridge circuits in modern process-control signal conditioning is to convert variations of resistance into variations of voltage. This voltage variation is then further conditioned for interface to an ADC or other system. It is thus important to note that the variation of bridge offset as given by Equation (2.7) is nonlinear with respect to any of the resistors. The same nonlinearity is present in ac bridge offset as given by Equation (2.20). Thus, if a sensor has an impedance that is linear with respect to the variable being measured, such linearity is lost when a bridge is used to convert this to a voltage variation. Figure 2.12a shows how ΔV varies with R_4 for a bridge with $R_1 = R_2 = R_3 = 100\ \Omega$ and $V = 10$ V. Note the nonlinearity of ΔV with R_4 as it varies from 0 to 500 Ω.

If the range of resistance variation is small and centered about the null value, then the nonlinearity of voltage versus resistance is small. Figure 2.12b shows that if the range of variation of R_4 is small (90 to 110 Ω), then the variation of ΔV with R_4, on an expanded scale, is relatively linear. Amplifiers can be used to amplify this voltage variation, since it is centered about zero, to a useful range.

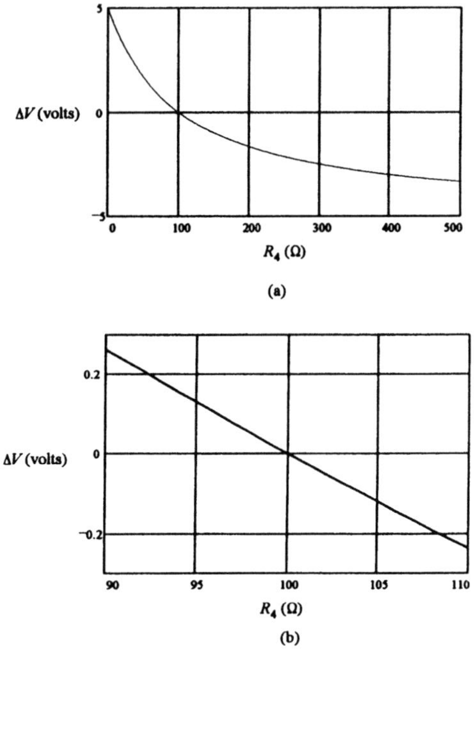

FIGURE 2.12
(a) Bridge off-null voltage is clearly nonlinear for large-scale changes in resistance. (b) However, for small ranges of resistance change, the off-null voltage is nearly linear.

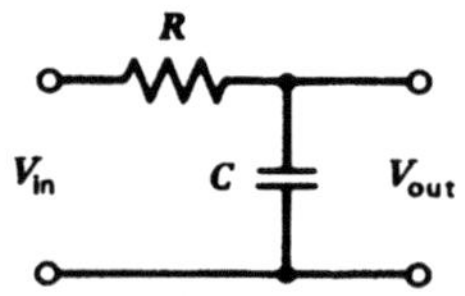

FIGURE 2.13
Circuit for the low-pass *RC* filter.

2.3.3 *RC* Filters

To eliminate unwanted noise signals from measurements, it is often necessary to use circuits that block certain frequencies or bands of frequencies. These circuits are called *filters.* A simple filter can be constructed from a single resistor and a single capacitor.

Low-pass *RC* Filter The simple circuit shown in Figure 2.13 is called a *low-pass RC* filter. It is called low-pass because it blocks high frequencies and passes low frequencies. It would be most desirable if a low-pass filter had a characteristic such that all signals with frequency above some critical value are simply rejected. Practical filter circuits approach that ideal with varying degrees of success.

In the case of the low-pass *RC* filter, the variation of rejection with frequency is shown in Figure 2.14. In this graph, the vertical is the ratio of output voltage to input voltage without regard to phase. When this ratio is one, the signal is passed without effect; when it is very small or zero, the signal is effectively blocked.

The horizontal is actually the logarithm of the ratio of the input signal frequency to a *critical frequency.* This critical frequency is that frequency for which the ratio of the output to the input voltage is approximately 0.707. In terms of the resistor and capacitor, the critical frequency is given by

$$f_c = 1/(2\pi RC) \tag{2.22}$$

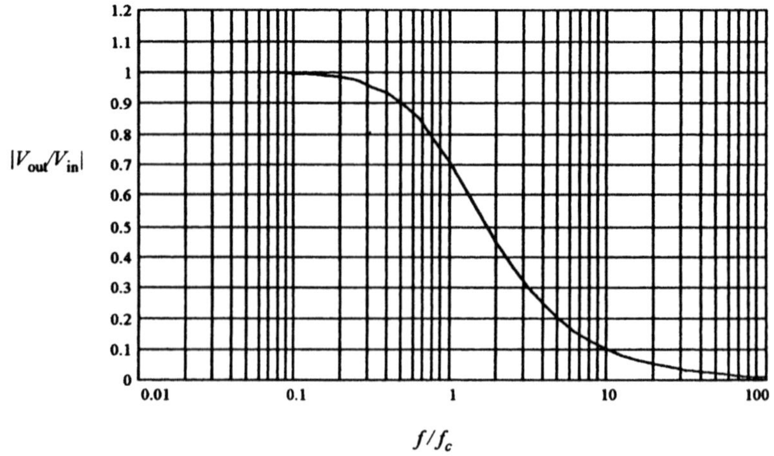

FIGURE 2.14
Response of the low-pass *RC* filter as a function of the frequency ratio.

The output-to-input voltage ratio for any signal frequency can be determined graphically from Figure 2.14 or can be computed by

$$\left|\frac{V_{out}}{V_{in}}\right| = \frac{1}{[1 + (f/f_c)^2]^{1/2}} \tag{2.23}$$

Design Methods A typical filter design is accomplished by finding the critical frequency, f_c, that will satisfy the design criteria. The resistor and capacitor are then determined to provide the required critical frequency from Equation (2.22). Because there are two unknowns (R and C) but only one equation, we can often just select one of the component values and compute the other. Often the capacitor is selected and the required resistor value is computed from Equation (2.22). The following practical guidelines are offered on this process:

1. Select a standard capacitor value in the μF to pF range.
2. Calculate the required resistance value. If it is below 1 kΩ or above 1 MΩ, try a different value of capacitor so that the required resistance falls within this range, which will avoid noise and loading problems.
3. If design flexibility allows, use the nearest standard value of resistance to that calculated.
4. Always remember that components such as resistors and capacitors have a tolerance in their indicated values. This must be considered in your design. Quite often, capacitors have a tolerance as high as ±20%.
5. If exact values are necessary, it is usually easiest to select a capacitor, measure its value, and then calculate the value of the required resistance. Then a trimmer (variable) resistor can be used to obtain the required value.

EXAMPLE 2.11 A measurement signal has a frequency <1 kHz, but there is unwanted noise at about 1 MHz. Design a low-pass filter that attenuates the noise to 1%. What is the effect on the measurement signal at its maximum of 1 kHz?

Solution

Use Equation (2.23) to determine what critical frequency will give $(V_{out}/V_{in}) = 0.01$ at 1 MHz. To do this, we have the relationship

$$0.01 = \frac{1}{[1 + (1\,\text{MHz}/f_c)^2]^{1/2}}$$

which gives us

$$(1\,\text{MHz}/f_c)^2 = 9999$$

or

$$f_c = 10\,\text{kHz}$$

Let's try a capacitor of 0.47 μF. Then Equation (2.22) can be used to solve for the required resistance,

$$R = \frac{1}{(2\pi)(10^4\,\text{Hz})(0.47 \times 10^{-6}\,\text{F})} = 33.9\ \Omega$$

This is a very small resistance, which could lead to excessive current and loading. So we try a smaller value of capacity, which will result in a larger value of resistance. Let's try 0.01 μF:

$$R = \frac{1}{(2\pi)(10^4\,\text{Hz})(0.01 \times 10^{-6}\,\text{F})} = 1591\ \Omega$$

This value is much more practical. Now suppose we try a standard value of resistance of 1.5 kΩ. This will give an actual critical frequency of

$$f_c = \frac{1}{(2\pi)(1500\ \Omega)(10^{-2}\,\text{F})} = 10{,}610\ \text{Hz}$$

which is a 6% difference. So the noise at 1 MHz would be down by

$$\left|\frac{V_{out}}{V_{in}}\right| = \frac{1}{[1 + (1\,\text{MHz}/10{,}610\,\text{Hz})^2]^{1/2}} = 0.0099995$$

instead of 0.01 as specified.

To check the effect of our filter on the signal at 1 kHz, we have

$$\left|\frac{V_{out}}{V_{in}}\right| = \frac{1}{[1 + (1\,\text{kHz}/10.61\,\text{kHz})^2]^{1/2}} = 0.996$$

Thus, the data have been reduced by about 0.4%.

High-Pass *RC* Filter A high-pass filter passes high frequencies (no rejection) and blocks (rejects) low frequencies. A filter of this type can be constructed using a resistor and a capacitor, as shown in the schematic of Figure 2.15. Similar to the low-pass filter, the rejection is not sharp in frequency but distributed over a range around a critical frequency. This critical frequency is defined by the same value—Equation (2.22)—as for the low-pass filter.

The graph of voltage output to input versus logarithm of frequency to critical frequency is shown in Figure 2.16. Note that the magnitude of $V_{out}/V_{in} = 0.707$ when the frequency is equal to the critical frequency.

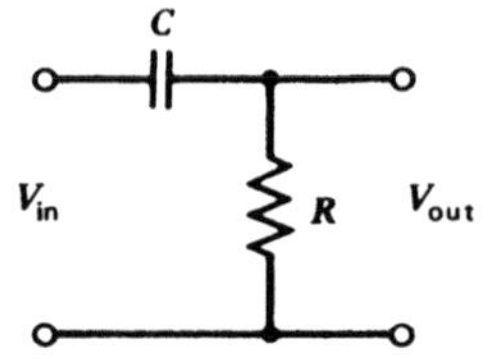

FIGURE 2.15
Circuit for the high-pass *RC* filter.

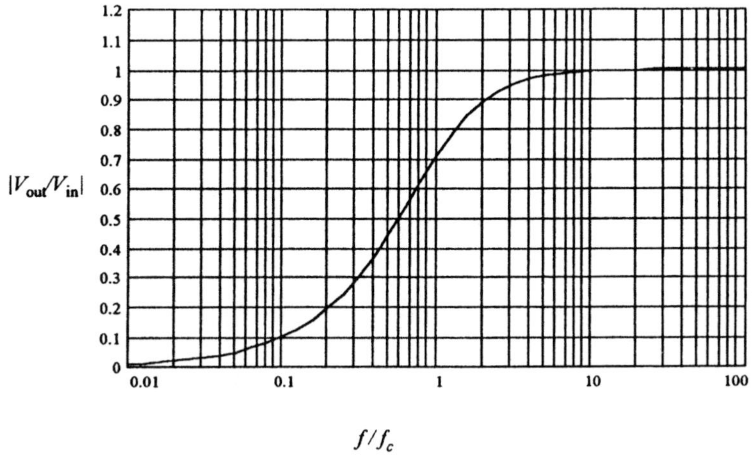

FIGURE 2.16
Response of the high-pass *RC* filter as a function of frequency ratio.

An equation for the ratio of output voltage to input voltage as a function of the frequency for the high-pass filter is found to be

$$|V_{out}/V_{in}| = \frac{(f/f_c)}{[1 + (f/f_c)^2]^{1/2}} \tag{2.24}$$

EXAMPLE 2.12 Pulses for a stepping motor are being transmitted at 2000 Hz. Design a filter to reduce 60-Hz noise but reduce the pulses by no more than 3 dB.

Solution

Let us first find what voltage ratio corresponds to a 3-dB reduction. We remember that

$$P(\text{dB}) = 20\log(V_{out}/V_{in})$$

so "down by 3 dB" means that $P = -3$. Therefore,

$$(V_{out}/V_{in}) = 10^{-3/20} = 0.707$$

You probably saw that coming. **The critical frequency is that frequency for which the output is attenuated by 3 dB.** Thus, in this case, $f_c = 2\,\text{kHz}$. The effect on 60-Hz noise is found using Equation (2.24), with $f = 60\,\text{Hz}$.

$$V_{out}/V_{in} = \frac{(60/2000)}{[1 + (60/2000)^2]^{1/2}}$$

$$V_{out}/V_{in} = 0.03$$

Thus, we see that only 3% of the 60-Hz noise remains; that is, it has been reduced by 97%.

Let's try a capacitor of 0.01 μF. Then Equation (2.22) shows that the resistor should have a value of 7.96 kΩ. Perhaps design criteria would allow use of either a 7.5-kΩ or 8.2-kΩ resistor, because these are standard values.

Practical Considerations There are a number of issues that arise when designing a system which will use the high-pass or low-pass *RC* filters. Consider the following items:

1. After the critical frequency is determined, the values of *R* and *C* are selected to satisfy Equation (2.22). In principle, any values of *R* and *C* can be employed that will satisfy this critical frequency equation. A number of practical issues limit the selection, however. Some of these are:
   a. Very small values of resistance are to be avoided because they can lead to large currents, and thus large loading effects. Similarly, very large capacitance should be avoided. In general, we try to keep the resistance in the kΩ and above range and capacitors in the μF or less range.
   b. Often, the exact critical frequency is not important, so that fixed resistors and capacitors of approximately the computed values can be employed. If exact values are necessary, it is usually easier to select and measure a capacitor, then compute and obtain the appropriate resistance using a trimmer resistance.
2. The effective input impedance and output impedance of the *RC* filter may have an effect on the circuit in which it is used because of loading effects. If the input impedance of the circuit being fed by the filter is low, you may want to place a voltage follower (see Section 2.5) between the filter output and the next stage. Similarly, if the output impedance of the feeding stage to the filter is high, you may want to isolate the input of the filter with a voltage follower.
3. It is possible to cascade *RC* filters in series to obtain improved sharpness of the filter cutoff frequency. However, it is important to consider the loading of one *RC* stage by another. The output impedance of the first-stage filter must be much less than the input impedance of the next stage to avoid loading.

Apart from loading, cascading filters raises the filter equation by a power of the number cascaded. Thus, two high-pass filters of the same critical frequency would respond by Equation (2.24) squared.

EXAMPLE 2.13 A 2-kHz data signal is contaminated by 60 Hz of noise. Compare a single-stage and a two-stage high-pass *RC* filter for reducing the noise by 60 dB. What effect does each have on the data signal? (Ignore loading.)

Solution

An attenuation of 60 dB means that the output-to-input voltage ratio at 60 Hz will be

$$20\log_{10}(V_{out}/V_{in}) = -60$$
$$V_{out}/V_{in} = 10^{-3}$$

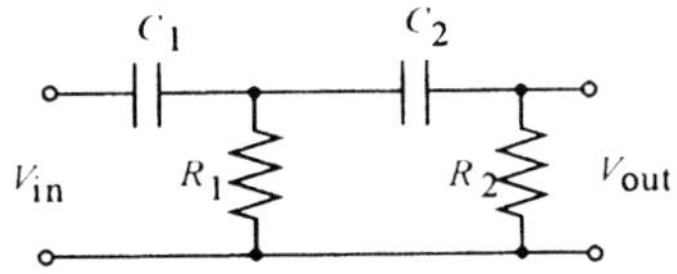

FIGURE 2.17
Cascaded high-pass *RC* filter for Example 2.13.

For the single stage, we use Equation (2.24) to find the critical frequency that will give this attenuation:

$$0.001 = \frac{(60/f_c)}{\sqrt{1 + (60/f_c)^2}}$$

Solving this for the critical frequency gives $f_c = 60\,\text{kHz}$. To see what effect this has on the data signal, we evaluate Equation (2.24) at 2 kHz,

$$\frac{V_{out}}{V_{in}} = \frac{(2\text{ kHz}/60\text{ kHz})}{\sqrt{1 + (2\text{ kHz}/60\text{ kHz})^2}} = 0.033$$

Thus, only about 3.3% of the data signal is left!

Now, for the cascaded *RC* filter we have two high-pass *RC* filters in series, each with the same critical frequency. If the filters are independent—that is, if the output impedance of the first is much less than the input impedance of the second—the total response will be Equation (2.24) squared. Therefore, to get 60-dB reduction, we have

$$0.001 = \frac{(60/f_c)^2}{1 + (60/f_c)^2}$$

Solving this for the critical frequency gives $f_c = 1896\,\text{Hz}$. The effect on the data signal is found by evaluating Equation (2.24) squared at 2 kHz with this critical frequency:

$$\frac{V_{out}}{V_{in}} = \frac{(2\text{ kHz}/1896\text{ Hz})^2}{1 + (2\text{ kHz}/1896\text{ Hz})^2} = 0.53$$

Thus, about 53% of the data signal is retained, which is a great improvement over the single stage. Each stage contributes an attenuation of $(0.53)^{1/2} = 0.73$. If the impedance condition is not satisfied, there will be loading of the first stage by the second and greater overall attenuation of the data signal. Figure 2.17 shows the circuit.

EXAMPLE 2.14 Suppose we require the first stage of Example 2.13 to use a capacitor of $C = 0.001\,\mu\text{F}$. Find the appropriate value of resistance, *R*. Suppose these same values are used for the second stage. How much further attenuation occurs at 2 kHz because of loading? What output impedance does the series filter present? Assume the V_{in} source resistance is very small.

Solution

The value of *R* is found from the equation for critical frequency, Equation (2.22).

$$R = (2\pi f_c C)^{-1} = [(2\pi)(1896\,\text{Hz})(0.001\,\mu\text{F})]^{-1}$$

$$R = 83.9\ k\Omega$$

To study the loading effect, we compute the reactance of the capacitor at 2 kHz.

$$X = (2\pi f C)^{-1} = [(2\pi)(2000\,\text{Hz})(0.001\,\mu\text{F})]^{-1}$$
$$X = 79.6\ \text{k}\Omega$$

Figure 2.18a shows the filter drawn with the elements replaced by their complex impedances, in polar form, at 2 kHz. In Example 2.13, we saw that the first stage alone had an output at 2 kHz of $V_{out} = 0.73\ V_{in}$. Now we will calculate how much this first-stage output voltage will be loaded by the second filter, which has an input impedance of $83.9\ \text{k}\Omega - j\,79.6\ \text{k}\Omega = 115.65\ \text{k}\Omega\ \angle{-43.5°}$. The first filter and source can be replaced by its Thévenin equivalent circuit where the Thévenin voltage is simply $0.73\ V_{in}$ and the Thévenin impedance is the parallel combination of the capacitor and resistor (source impedance is very small). Using the parallel rule for impedances, we find $Z_{th} = 57.5\,\text{k}\Omega\ \angle{-46.5°}$. Figure 2.18b shows the effective circuit for the cascaded filters to determine loading. The loaded input to the second filter at points *a-b* is the Thévenin voltage divided between the two impedances. Thus, we use the ac divider analysis to find

$$V_{ab} = \frac{(115.65\ \angle{-43.5°})}{57.5\ \angle{-46.5°} + 115.65\ \angle{-43.5°}}\,0.73\ V_{in}$$

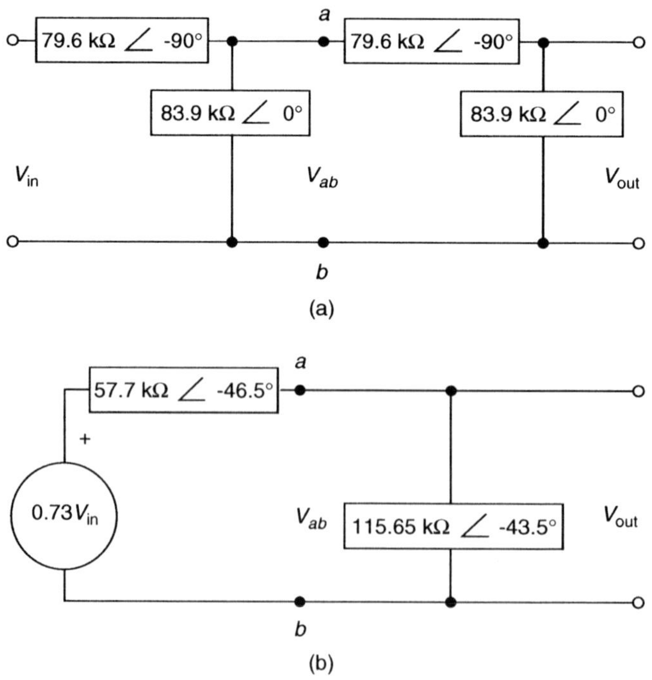

FIGURE 2.18
Analysis of loading for a high-pass *RC* filter in Example 2.14.

Since we are only interested in the amplitude of the voltages, we find the magnitudes of the voltage so the result is $|V_{ab}| = |0.49\ V_{in}|$. This is the voltage amplitude that is presented to the second filter stage. We have already seen that this stage has an output that is 0.73 of the input voltage. Therefore, the overall filter output voltage is

$$V_{out} = 0.73(0.49\ V_{in}) = 0.36\ V_{in}$$

Thus, loading has reduced the data signal from 53% to 36%. Of course this is still far better than a single-stage filter, which left only 3.3% of the signal at 2 kHz.

Band-Pass *RC* Filter It is possible to construct a filter that blocks frequencies below a low limit and above a high limit while passing frequencies between the limits. These are called *band-pass* filters. Passive band-pass filters can be designed with resistors and capacitors, but more efficient versions use inductors and/or capacitors. You should refer to specialized books on filtering to learn these circuits. In general, the band-pass filter is defined by the response shown in Figure 2.19. The lower critical frequency, f_L, defines the frequency below which the ratio of output voltage to input voltage is down by at least 3 dB, or 0.707. The higher critical frequency, f_H, defines the frequency above which the ratio of output voltage to input voltage is down by at least 3 dB, or 0.707. The frequency range between f_L and f_H is called the *passband*.

The band-pass *RC* filter is shown in Figure 2.20. Notice that it is simply a low-pass filter followed by a high-pass filter. Care must be taken that the second filter does not load the first. The lower critical frequency is that of the high-pass filter, whereas the high critical frequency is that of the low-pass filter. If the low and high critical frequencies are too close together, the passband region never reaches unity (i.e., the output is attenuated for all frequencies).

Equation (2.25) gives the ratio of the magnitude of output voltage to input voltage for this filter as a function of frequency. This equation includes the effects of loading by a constant, *r*, which is the ratio of the high-pass filter resistance to the low-pass filter resistance, $r = R_H/R_L$:

$$\left|\frac{V_{out}}{V_{in}}\right| = \frac{f_H f}{\sqrt{(f^2 - f_H f_L)^2 + [f_L + (1 + r)f_H]^2 f^2}} \tag{2.25}$$

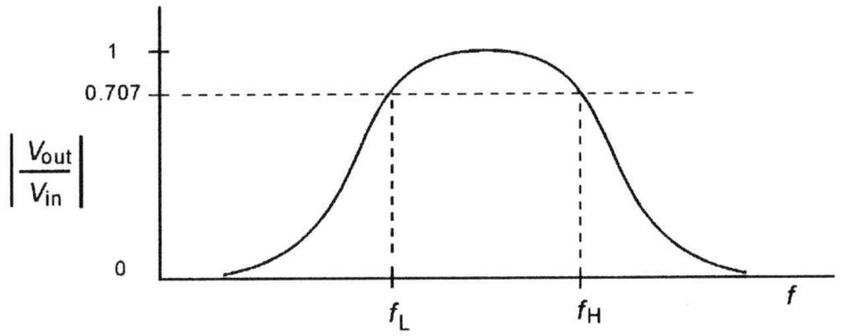

FIGURE 2.19
The response of a band-pass filter shows that high and low frequencies are rejected.

FIGURE 2.20
A band-pass *RC* filter can be made from cascaded high-pass and low-pass *RC* filters.

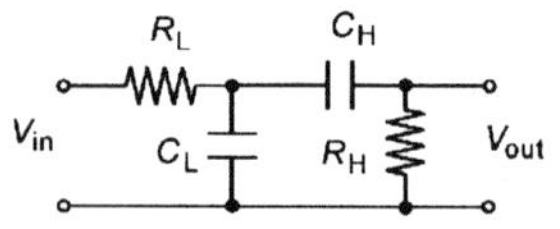

where f is the frequency in hertz, and the critical frequencies are defined in terms of the circuit components as follows:

$$f_H = \frac{1}{2\pi R_L C_L} \quad f_L = \frac{1}{2\pi R_H C_H} \tag{2.26}$$

To provide a good passband, it is essential that the critical frequencies be as far apart as possible *and* that the resistor ratio be kept below 0.01. Example 2.15 illustrates a typical design application.

EXAMPLE 2.15 A signal-conditioning system uses a frequency variation from 6 kHz to 60 kHz to carry measurement information. There is considerable noise at 120 Hz and at 1 MHz. Design a band-pass filter to reduce the noise by 90%. What is the effect on the desired passband frequencies?

Solution

First we ask what critical high-pass frequency will reduce signals at 120 Hz by a factor of 0.1 using Equation (2.24),

$$0.1 = \frac{120/f_L}{\sqrt{(1 + (120/f_L)^2)}} \qquad \text{which gives } f_L = 1200 \text{ Hz}$$

Then we ask what critical low-pass frequency will reduce 1-MHz signals by a factor of 0.1 using Equation (2.22),

$$0.1 = \frac{1}{\sqrt{(1 + (10^6/f_H)^2)}} \qquad \text{which gives } f_H = 100 \text{ kHz}$$

Equation (2.25) has been plotted in Figure 2.21 for the two cases $r = 0.1$ and $r = 0.01$. You can see that the overall passband response is reduced (because of loading) with the larger value of r. The effect on the signal can be found by evaluating Equation (2.25) at 6 kHz and 60 kHz, respectively. The results are as follows: The voltage ratio at 6 kHz is 0.969, for a 3% reduction, and the voltage ratio at 60 kHz is 0.851, for a 15% reduction.

To pick component values, we must have $r = 0.01$, so if we pick $R_L = 100$ kΩ, then $R_H = rR_L = 1$ kΩ. The capacitor values are then found from Equation (2.26) to be $C_H = 0.133\ \mu$F and $C_L = 15.9$ pF.

Band-Reject Filter Another kind of filter of some importance is one that blocks a specific range of frequencies. Often such a filter is used to reject a particular frequency or a small range of frequencies that are interfering with a data signal. In general, the defi-

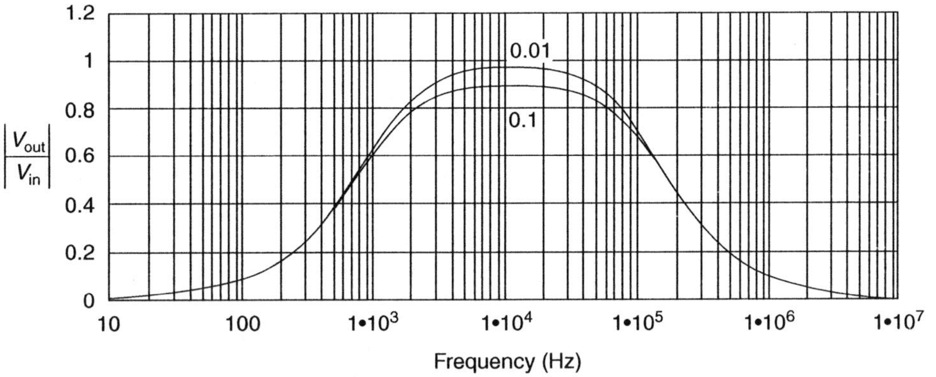

FIGURE 2.21
Band-pass response for the filter in Example 2.15.

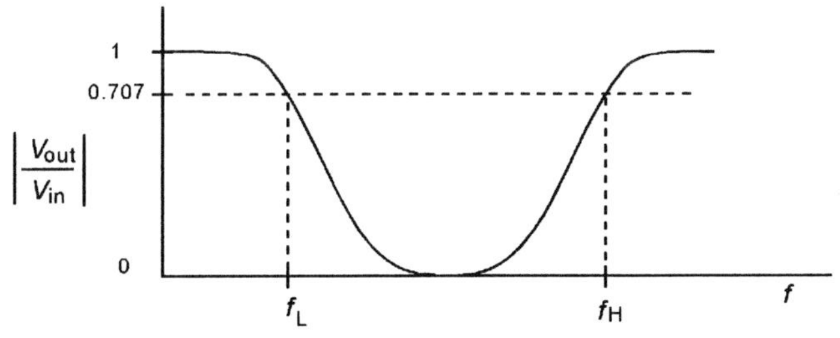

FIGURE 2.22
Response of a band-reject, or notch, filter shows that a middle band of frequencies are rejected.

nition of such a filter is given in Figure 2.22. The definitions are much the same as the band-pass filter in that f_L is a critical frequency above which the signal is attenuated by 3 dB, or about 0.707, whereas f_H is a critical frequency below which signals are attenuated by 3 dB, or about 0.707.

It is difficult to realize such filters with passive *RC* combinations. It is possible to construct band-reject frequencies using inductors and capacitors, but the most success is obtained using active circuits. You should look these up in a reference on filters.

One very special band-reject filter, which can be realized with *RC* combinations, is called a *notch filter* because it blocks a very narrow range of frequencies. This circuit, called a *twin-T filter,* is shown in Figure 2.23. The characteristics of this filter are determined strongly by the value of the grounding resistor and capacitor labeled R_1 and C_1 in Figure 2.23.

For the particular combination of $R_1 = \pi R/10$ and $C_1 = 10C/\pi$, the filter response versus frequency is shown in Figure 2.24. The critical "notch" frequency occurs at a frequency given by

$$f_n = 0.785 f_c \quad \text{where} \quad f_c = 1/(2\pi RC) \tag{2.27}$$

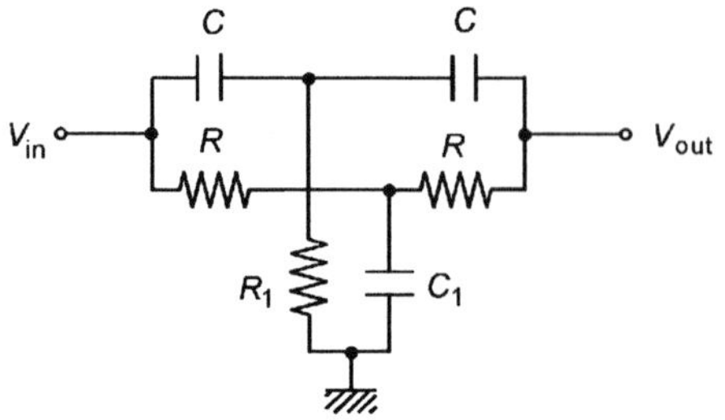

FIGURE 2.23
One form of a band-reject *RC* filter is the twin-T.

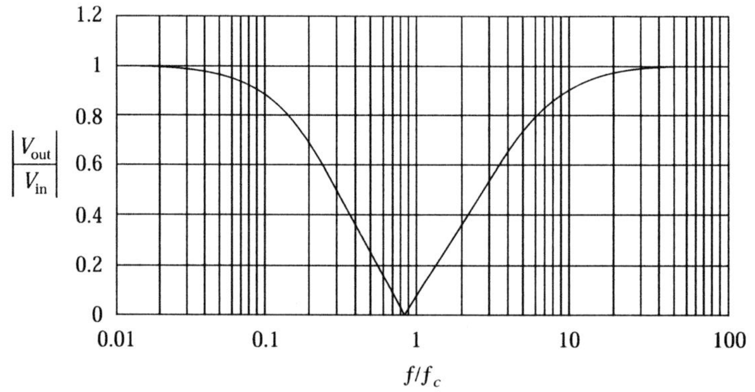

FIGURE 2.24
The twin-T rejection notch is very sharp for one set of components.

Note that at the notch frequency, the output computes to about 0.002, so the rejection is not complete. The frequencies for which the output is down 3 dB (about 0.707) from the passband are given by

$$f_L = 0.187 f_c \quad \text{and} \quad f_H = 4.57 f_c \tag{2.28}$$

EXAMPLE 2.16

A frequency of 400 Hz prevails aboard an aircraft. Design a twin-T notch filter to reduce the 400-Hz signal. What effect would this have on voice signals at 10 to 300 Hz? At what higher frequency is the output down by 3 dB?

Solution

If we are to have a notch frequency of 400 Hz, the critical frequency can be found from Equation (2.27):

$$400 = 0.785 f_c \quad \text{so} \quad f_c = \frac{400}{0.785} = 510 \text{ Hz}$$

If we pick $C = 0.01\ \mu\text{F}$, then the resistance is given by

$$R = \frac{1}{2\pi(510)(0.01 \times 10^{-6})} = 31{,}206\ \Omega$$

Now the values of the grounding components are found as

$$R_1 = \frac{\pi(31{,}206)}{10} = 9800\ \Omega$$

$$C_1 = \frac{10(0.01\ \mu\text{F})}{\pi} = 0.03\ \mu\text{F}$$

The effect can be estimated from Figure 2.24 by noting that $10/400 = 0.025$ and $300/400 = 0.75$. From the graph, this means the output/input ratios are about 0.99 and 0.03, respectively. So, the higher low frequencies are attenuated significantly. Higher frequencies will be down 3 dB for frequencies below $4.57f_c$, or about 2300 Hz.

It must be noted that much more improved band-reject and notch filters can be realized using active circuits, particularly using op amps.

2.4 OPERATIONAL AMPLIFIERS

As discussed in Section 2.2, there are many diverse requirements for signal conditioning in process control. In Section 2.3 we considered common, passive circuits that can provide some of the required signal operations, the divider, bridge, and *RC* filters. Historically, the detectors used in bridge circuits consisted of tube and transistor circuits. In many other cases where impedance transformations, amplification, and other operations were required, a circuit was designed that depended on discrete electronic components. With the remarkable advances in electronics and integrated circuits (ICs), the requirement to implement designs from discrete components has given way to easier and more reliable methods of signal conditioning. Many special circuits and general-purpose amplifiers are now contained in integrated circuit (IC) packages, producing a quick solution to signal-conditioning problems together with small size, low power consumption, and low cost.

In general, the application of ICs requires familiarity with an available line of such devices and their specifications and limitations, before they can be applied to a specific problem. Apart from these specialized ICs, there is also a type of amplifier that finds wide application as the building block of signal-conditioning applications. This device, called an operational amplifier (op amp), has been in existence for many years. It was first constructed from tubes, then from discrete transistors, and now as integrated circuits. Although many lines of op amps with diverse specifications exist from many manufacturers, they all have common characteristics of operation that can be employed in basic designs relating to any general op amp.

2.4.1 Op Amp Characteristics

An op amp is a circuit composed of resistors, transistors, diodes, and capacitors. It typically requires connection of *bipolar* power supplies—that is, both $+V_s$ and $-V_s$ with respect to ground. When considered as a functional element of some larger circuit, however, all we are concerned with are its input and output signals. For that reason, the op amp is usually shown in a larger circuit using its own schematic symbol as in Figure 2.25. Notice that the power supply connections are not shown—only two input terminals and an output terminal.

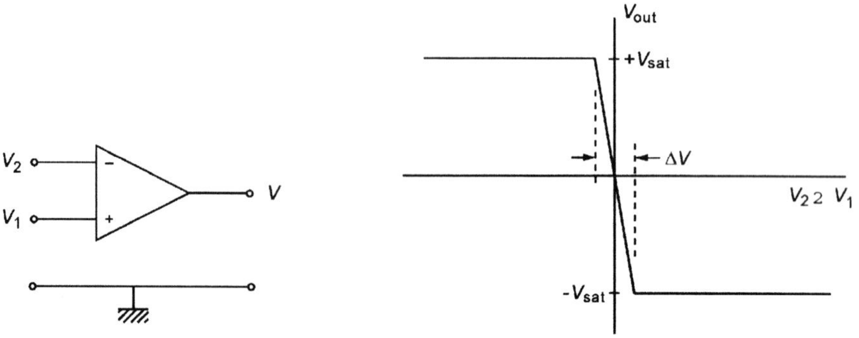

a) Operational amplifier schematic symbol   b) Relationship between input and output voltage

FIGURE 2.25
The schematic symbol and response of an op amp.

In the schematic symbol of Figure 2.25a, one input is labeled with a minus $(-)$ sign and is called the *inverting* input. The other is labeled with a plus $(+)$ sign and is called the *noninverting* input. The sign labels are part of the symbol and must always be included.

Transfer Function Figure 2.25b shows the relationship between voltages applied to the two input terminals and the resulting output voltage. The output voltage, V_{out}, is plotted versus the difference between the two input voltages, $(V_2 - V_1)$. This input is called the *differential input voltage.* Figure 2.25a shows that there is no ground connection directly to the op amp. However, the voltages shown are with respect to ground. Notice we have applied V_1 to the noninverting input and V_2 to the inverting terminal.

The interpretation of Figure 2.25b is made as follows. When V_2 is much larger than V_1, so that $(V_2 - V_1)$ is positive, the output is *saturated* at some negative voltage, $-V_{sat}$. Conversely, when V_1 is much larger than V_2, the output is saturated at some positive voltage, $+V_{sat}$. This is why the terminals are called inverting and noninverting, respectively. When the voltage on the $(-)$ terminal is more positive, the output is negative (i.e., the sign is inverted).

There is a narrow range of differential input voltage, labeled ΔV in Figure 2.25b, within which the output changes from $+$ *saturation* to $-$ *saturation.* For most op amps, this input voltage range is less than a millivolt, whereas the saturation voltages are typically on the order of 10 V. Thus, the slope of the transition between saturation levels is very large, typically exceeding 100,000 V/V.

Other characteristics of the op amp are that the input impedances are very high, typically exceeding 1 MΩ, whereas the output impedance is very low, typically less than 100 Ω.

Other characteristics of ideal op amps are (1) an infinite impedance between inputs and (2) a zero output impedance. In practice, the device is always used with feedback of output to input. Such feedback permits implementation of many special relationships between input and output voltage.

Ideal Inverting Amplifier To see how the op amp is used, let us consider the circuit in Figure 2.26. Resistor R_2 is used to feed back the output to the inverting input of the op amp, and R_1 connects the input voltage V_{in} to this same point. The common connection

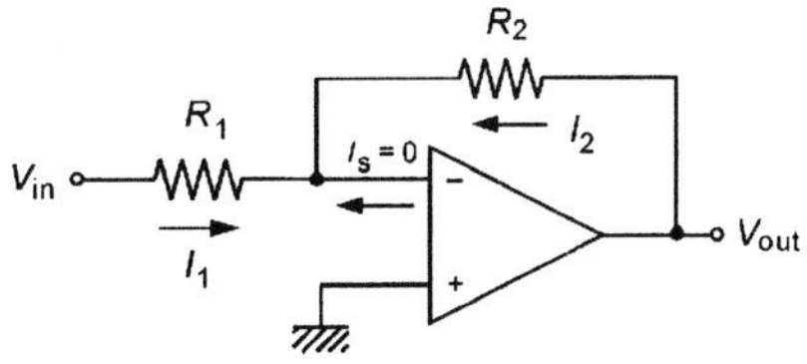

FIGURE 2.26
The op amp inverting amplifier.

is called the summing point. We can see that with no feedback and the (+) grounded, $V_{in} > 0$ saturates the output negative and $V_{in} < 0$ saturates the output positive. With feedback, the output adjusts to a voltage such that

1. The summing point voltage is equal to the (+) op amp input level, zero in this case.
2. No current flows through the op amp input terminals because of the assumed infinite impedance.

In this case, the sum of currents at the summing point must be zero.

$$I_1 + I_2 = 0 \tag{2.29}$$

where I_1 = current through R_1
I_2 = current through R_2

Because the summing point potential is assumed to be zero, by Ohm's law we have

$$\frac{V_{in}}{R_1} + \frac{V_{out}}{R_2} = 0 \tag{2.30}$$

From Equation (2.29), we can write the circuit response as

$$V_{out} = -\frac{R_2}{R_1} V_{in} \tag{2.31}$$

Thus, the circuit of Figure 2.26 is an inverting *amplifier* with gain R_2/R_1 that is shifted 180° in phase (inverted) from the input. This device is also an *attenuator* by virtue of making $R_2 < R_1$.

This example suggests two rules that can be applied to analyze the ideal operation of any op amp circuit. In most cases, such an analysis will provide the circuit transfer function with little error. The design rules are:

Rule 1 Assume that no current flows through the op amp input terminals—that is, the inverting and noninverting terminals.

Rule 2 Assume that there is no voltage difference between the op amp input terminals—that is, $V_+ = V_-$.

Nonideal Effects Analysis of op amp circuits with nonideal response is performed by considering the following parameters:

1. *Finite open-loop gain* The design rules presented here assume the op amp has infinite open-loop gain. Of course, a real op amp does not, as shown in

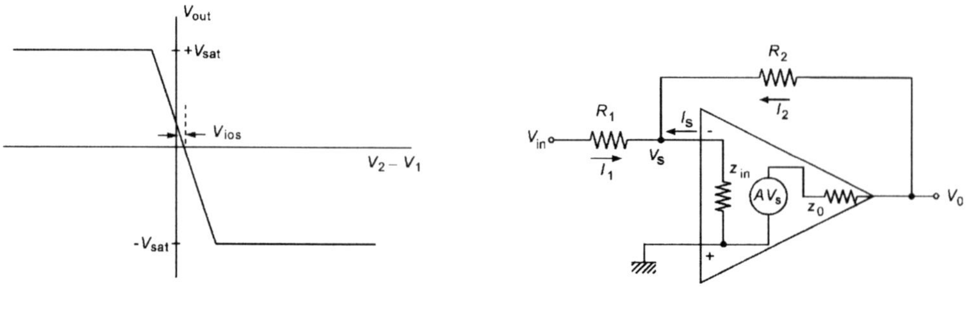

a) Nonideal input/output characteristic

b) Inverting amplifier with nonideal op amp properties

FIGURE 2.27
Nonideal characteristics of an op amp include finite gain, finite impedance, and offsets.

Figure 2.25a and again in Figure 2.27a. The gain is defined as the slope of the voltage-transfer function,

$$A = \left|\frac{\Delta V_{out}}{\Delta(V_2 - V_1)}\right| \approx \left|\frac{2V_{sat}}{\Delta V}\right|$$

For a typical op amp, $V_{sat} \sim 10$ V and $\Delta V \sim 100\,\mu$V, so $A \sim 200{,}000$!

2. *Finite input impedance* A real op amp has a finite input impedance and, consequently, a finite voltage across and current through its input terminals.
3. *Nonzero output impedance* A real op amp has a nonzero output impedance, although this low output impedance is typically only a few ohms.

In most modern applications, these nonideal effects can be ignored in designing op amp circuits. For example, consider the circuit in Figure 2.27b, where the finite impedances and gain of the op amp have been included. We can employ standard circuit analysis to find the relationship between input and output voltage for this circuit. Summing the currents at the summing point gives

$$I_1 + I_2 + I_3 = 0$$

Then, each current can be identified in terms of the circuit parameters to give

$$\frac{V_{in} - V_s}{R_1} + \frac{V_0 - V_s}{R_2} - \frac{V_s}{z_{in}} = 0$$

Finally, V_0 can be related to the op amp gain as

$$V_0 = AV_s - \left(\frac{V_0 - V_s}{R_2}\right) z_0$$

Now, combining the equations, we find

$$V_0 = -\frac{R_2}{R_1}\left(\frac{1}{1-\mu}\right) V_{in} \tag{2.32}$$

where

$$\mu = \frac{\left(1 + \frac{z_0}{R_2}\right)\left(1 + \frac{R_2}{R_1} + \frac{R_2}{z_{in}}\right)}{\left(A + \frac{z_0}{R_2}\right)} \tag{2.33}$$

If we assume that μ is very small compared with unity, then Equation (2.32) reduces to the ideal case given by Equation (2.31). Indeed, if typical values for an IC op amp are chosen for a case where $R_2/R_1 = 100$, we can show that $\mu \ll 1$. For example, a common, general-purpose IC op amp shows

$$\begin{aligned} A &= 200{,}000 \\ z_0 &= 75\ \Omega \\ z_{in} &= 2\ \text{M}\Omega \end{aligned}$$

If we use a feedback resistance R_2 of 100 kΩ and substitute the aforementioned values into Equation (2.33), we find $\mu \simeq 0.0005$, which shows that the gain from Equation (2.32) differs from the ideal by only 0.05%. This is, of course, only one example of the many op amp circuits that are employed, but in most cases a similar analysis shows that the ideal characteristics may be assumed.

2.4.2 Op Amp Specifications

There are characteristics of op amps other than those given in the previous section that enter into design applications. These characteristics are given in the specifications for particular op amps together with the open-loop gain and input and output impedance previously defined. Some of these characteristics are as follows:

- *Input offset voltage* In most cases, the op amp output voltage may not be zero when the voltage across the input is zero. The voltage that must be applied across the input terminals to drive the output to zero is the *input offset voltage,* V_{ios}. This is shown in Figure 2.27a.
- *Input offset current* Just as a voltage offset may be required across the input to zero the output voltage, so a net current may be required between the inputs to zero the output voltage. Such a current is referred to as an input offset current. This is taken as the difference of the two input currents.
- *Input bias current* This is the average of the two input currents required to drive the output voltage to zero.
- *Slew rate* If a voltage is suddenly applied to the input of an op amp, the output will saturate to the maximum. For a step input, the slew rate is the rate at which the output changes to the saturation value. This typically is expressed as volts per microsecond ($V/\mu s$).
- *Unity gain frequency bandwidth* The frequency response of an op amp is typically defined by a Bode plot of open-loop voltage gain versus frequency. Such a plot is very important for the design of circuits that deal with ac signals. It is beyond the scope of this book to consider the details of designs employing Bode plots.

FIGURE 2.28
Some op amps provide connections for an input offset compensation trimmer resistor.

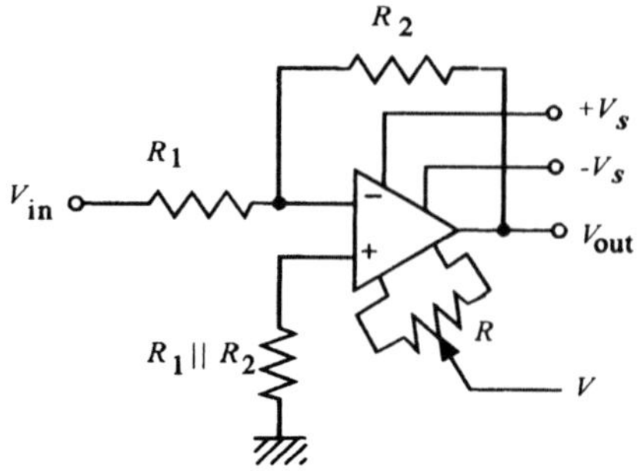

Instead, the overall frequency behavior can be seen by determination of the frequency at which the open-loop gain of the op amp has become unity, thus defining the unity gain frequency bandwidth.

Practical Issues There are several practical issues associated with op amp applications that appear as extra components in op amp circuits but which do not contribute to the circuit transfer function. The following paragraphs summarize the issues.

In general, op amps require bipolar power supplies, $+V_s$ and $-V_s$, of equal magnitude, which are connected to designated pins of the IC. Typically, the value of these supply voltages is in the range of 9 to 15 volts, although op amps are available with many other supply requirements. Figure 2.28 shows the inverting amplifier with the power supply connections.

Approximate input offset current compensation can be provided by making the resistance feeding both input terminals approximately the same. In Figure 2.28 for the inverting amplifier, this has been provided by a resistor on the noninverting terminal whose value is the same as R_1 and R_2 in parallel, since that is the effective resistance seen by the inverting terminal.

Compensation for input offset voltage can be provided in one of two ways. Many modern IC op amps provide terminals to allow input offset compensation. This has been shown in Figure 2.28 as a variable resistor connected to two terminals of the op amp. The wiper of the variable resistor is connected to the supply voltage, either $+V_s$ or $-V_s$, according to the specifications of the op amp. This resistor need be adjusted only one time and not again, unless the particular op amp used is changed.

The concept here is that the transfer function for the op amp circuit, Equation (2.31) in this case, will be incorrect for dc signals if there is input offset voltage. In this case, if the input is set to zero, $V_{in} = 0$, then the output by Equation (2.31) should be zero, but it will not be because of input offset voltage. Thus, the variable resistor is adjusted until the output is zero, thereby compensating for input offset voltage. Now the transfer function of Equation (2.31) will be correct. As long as the op amp is not replaced or the circuit components are not changed, further adjustments of the resistor are not required.

Some op amps do not provide terminals for input offset compensation in the manner described. In these cases, a small bias voltage must be placed on the input to provide the required compensation. Figure 2.29 shows one way to do this in the case of the inverting amplifier.

Another practical issue in the application of op amps involves the current drive capability. General-purpose IC op amps can source, or sink, no more than about 20 mA, which

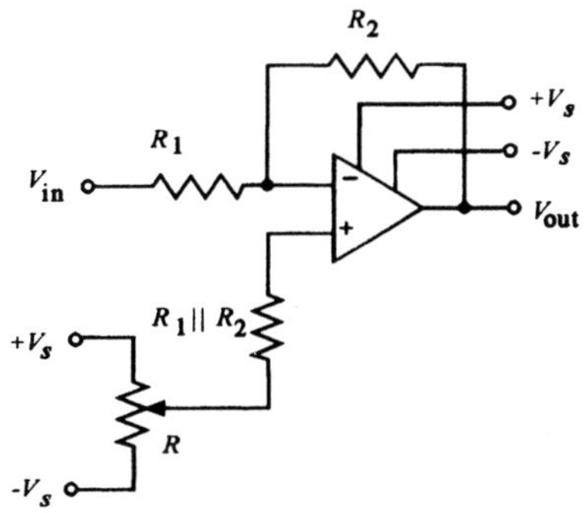

FIGURE 2.29
Input offset can also be compensated using external connections and trimmer resistors.

includes the current in the feedback circuit. This leads to a general design criterion to be applied to design with op amps: *Think mA and kΩ when designing circuits that use op amps.* The following example illustrates this point.

EXAMPLE 2.17

Specify the circuit and components for an op amp circuit with a gain of −4.5.

Solution

Since the gain is negative, we can use an inverting amplifier, as in Figure 2.26, to provide the solution. From Equation (2.31),

$$\text{gain} = -4.5 = -\frac{R_2}{R_1}$$

Thus, *any* two resistors with a ratio of 4.5 can be used. Here is where the practical issue comes into play. Suppose we picked $R_1 = 1\ \Omega$ and $R_2 = 4.5\ \Omega$. This satisfies the design equation; however, it is impractical for the following reason. Suppose the input voltage were 2.0 V. Then the output would be expected to be $V_{out} = -4.5(2.0) = -9.0$ V. However, this would mean that the feedback current would have to be $I_2 = (-9\text{ V}/4.5\ \Omega) = -2$ A! The poor op amp can provide only about 20 mA, so the circuit would not work. Thus, we think mA and kΩ and select, for example, $R_1 = 1$ kΩ and $R_2 = 4.5$ kΩ, or any combination with a ratio of 4.5, but in kΩ.

Most of the op amp circuits shown in this book will not include power supply connections or compensation components. This is done to simplify the circuits so the essential working principles can be understood. You should realize, however, that a practical, working circuit will usually need these compensation elements.

2.5 OP AMP CIRCUITS IN INSTRUMENTATION

As the op amp became familiar to the individuals working in process-control and instrumentation technology, a large variety of circuits were developed with direct application to this field. In general, it is much easier to develop a circuit for a specific service using op

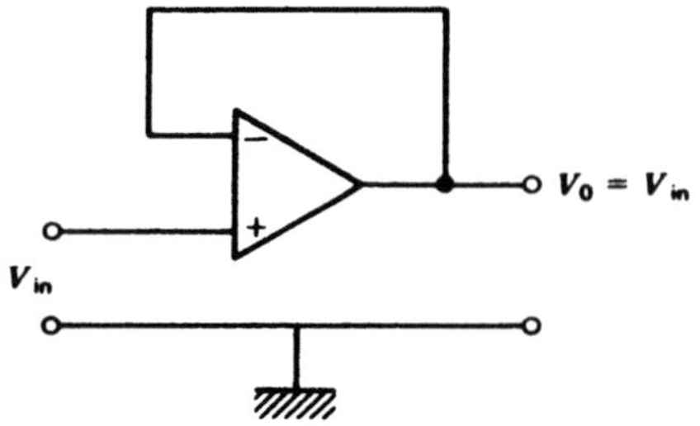

FIGURE 2.30
The op amp voltage follower. This circuit has unity gain but very high input impedance.

amps than discrete components; with the development of low-cost IC op amps, it is also practical. Perhaps one of the greatest disadvantages is the requirement of a bipolar power supply for the op amp. This section presents a number of typical circuits and their basic characteristics, together with a derivation of the circuit response assuming an ideal op amp.

2.5.1 Voltage Follower

Figure 2.30 shows an op amp circuit with unity gain and very high input impedance. The input impedance is essentially the input impedance of the op amp itself, which can be greater than 100 MΩ. The voltage output tracks the input over a range defined by the plus and minus saturation voltage outputs. Current output is limited to the short circuit current of the op amp, and output impedance is typically much less than 100 Ω. In many cases, a manufacturer will market an op amp voltage follower whose feedback is provided internally. Such a unit is usually specifically designed for very high input impedance. The unity gain voltage follower is essentially an impedance transformer in the sense of converting a voltage at high impedance to the same voltage at low impedance.

2.5.2 Inverting Amplifier

The inverting amplifier has already been discussed in connection with our treatment of op amp characteristics. Equation (2.31) shows that this circuit inverts the input signal and may have either attenuation or gain, depending on the ratio of input resistance, R_1, and feedback resistance, R_2. The circuit for this amplifier is shown in Figure 2.25. It is important to note that the input impedance of this circuit is essentially equal to R_1, the input resistance. In general, this resistance is not large, and hence the input impedance is not large. The output impedance is low.

Summing Amplifier A common modification of the inverting amplifier is an amplifier that sums or adds two or more applied voltages. This circuit is shown in Figure 2.31 for the case of summing two input voltages. The transfer function of this amplifier is given by

$$V_{out} = -\left[\frac{R_2}{R_1}V_1 + \frac{R_2}{R_3}V_2\right] \tag{2.34}$$

The sum can be scaled by proper selection of resistors. For example, if we make $R_1 = R_2 = R_3$, then the output is simply the (inverted) sum of V_1 and V_2. The average can be found by making $R_1 = R_3$ and $R_2 = R_1/2$.

FIGURE 2.31
The op amp summing amplifier.

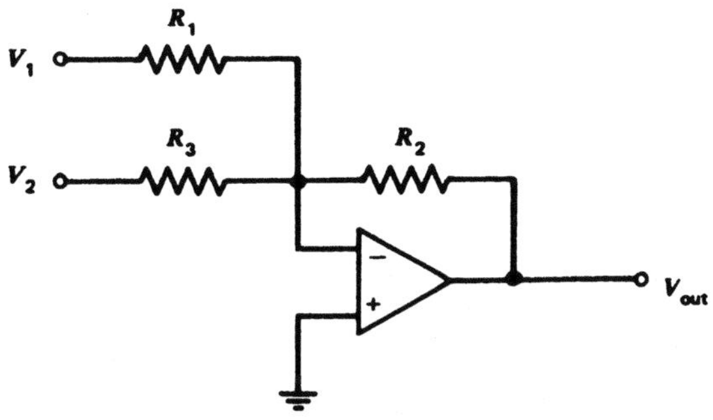

EXAMPLE 2.18

Develop an op amp circuit that can provide an output voltage related to the input voltage by

$$V_{out} = 3.4\,V_{in} + 5$$

Solution

There are many ways to do this. One way is to use a summing amplifier with V_{in} on one input and 5 V on the other. The gains will be selected to be 3.4 and 1.0, respectively. The summing amplifier of Figure 2.31 is also an inverter, however, so the sign will be wrong. Thus, a second amplifier will be used with a gain of -1 to make the sign correct. The result is shown in Figure 2.32. Selection of the values of the resistors is based on the general notion of keeping the currents in milliamperes.

2.5.3 Noninverting Amplifier

A noninverting amplifier may be constructed from an op amp, as shown in Figure 2.33. The gain of this circuit is found by summing the currents at the summing point, *S*, and using the fact that the summing point voltage is V_{in} so that no voltage difference appears across the input terminals.

$$I_1 + I_2 = 0$$

where I_1 = current through R_1
I_2 = current through R_2

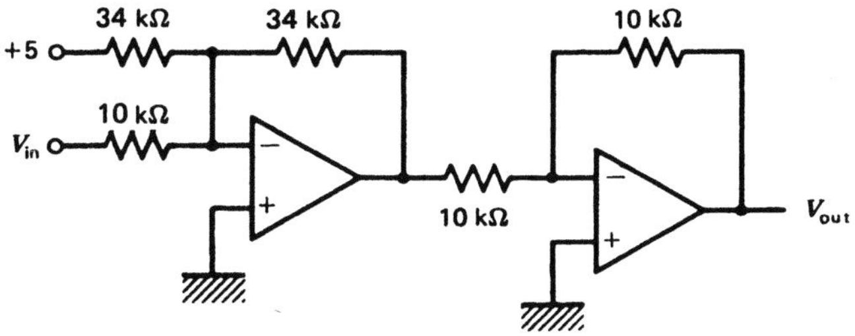

FIGURE 2.32
The op amp circuit for Example 2.18.

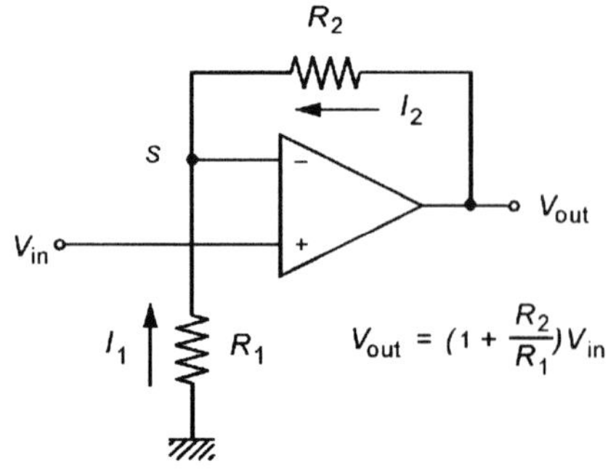

FIGURE 2.33
A noninverting amplifier.

But these currents can be found from Ohm's law such that this equation becomes

$$\frac{V_{in}}{R_1} + \frac{V_{in} - V_{out}}{R_2} = 0$$

Solving this equation for V_{out}, we find

$$V_{out} = \left[1 + \frac{R_2}{R_1}\right]V_{in} \tag{2.35}$$

Equation (2.35) shows that the noninverting amplifier has a gain that depends on the ratio of feedback resistor R_2 and ground resistor R_1, but this gain can never be used for voltage attenuation because the ratio is added to 1. Because the input is taken directly into the noninverting input of the op amp, the input impedance is very high, since it is effectively equal to the op amp input impedance. The output impedance is very low.

EXAMPLE 2.19 Design a high-impedance amplifier with a voltage gain of 42.

Solution

We use the noninverting circuit of Figure 2.33 with resistors selected from

$$V_{out} = \left[1 + \frac{R_2}{R_1}\right] V_{in}$$

$$42 = 1 + \frac{R_2}{R_1}$$

$$R_2 = 41\ R_1$$

so we could choose $R_1 = \mathbf{1\ k\Omega}$, which requires $R_2 = \mathbf{41\ k\Omega}$.

2.5.4 Differential Instrumentation Amplifier

There are many instances in measurement and control systems in which the difference between two voltages needs to be conditioned. A good example is the Wheatstone bridge, where the offset voltage, $\Delta V = V_a - V_b$, is the quantity of interest.

An ideal differential amplifier provides an output voltage with respect to ground that is some gain times the difference between two input voltages:

$$V_{out} = A(V_a - V_b) \tag{2.36}$$

where A is the differential gain and both V_a and V_b are voltages with respect to ground.

Such an amplifier plays an important role in instrumentation and measurement.

Common Mode Rejection Notice that the output voltage as given by Equation (2.36) does not depend on the values or polarity of either input voltage, but only on their difference. Thus, if the gain were 10 and $V_a = 0.3$ V and $V_b = 0.2$ V, the output would be given by $V_{out} = 10(0.3 - 0.2) = 1.0$ V. But if the inputs were $V_a = 7.8$ V and $V_b = 7.7$ V, the output would still be 1.0 V, since $V_{out} = 10(7.8 - 7.7) = 1.0$. Even if $V_a = -2.4$ V and $V_b = -2.5$ V, the output would still be given by 1.0 V since $V_{out} = 10(-2.4 - [-2.5]) = 1.0$ V. Real differential amplifiers can only approach this ideal.

To define the degree to which a differential amplifier approaches the ideal, we use the following definitions. The *common-mode input voltage* is the average of voltage applied to the two input terminals,

$$V_{cm} = \frac{V_a + V_b}{2} \tag{2.37}$$

An ideal differential amplifier will not have any output that depends on the value of the common-mode voltage; that is, the circuit gain for common-mode voltage, A_{cm}, will be zero.

The common-mode rejection ratio (CMRR) of a differential amplifier is defined as the ratio of the differential gain to the common-mode gain. The common-mode rejection (CMR) is the CMRR expressed in dB,

$$\text{CMRR} = \frac{A}{A_{cm}} \tag{2.38}$$

$$\text{CMR} = 20\log_{10}(\text{CMRR}) \tag{2.39}$$

Clearly, the larger these numbers, the better the differential amplifier. Typical values of CMR range from 60 to 100 dB.

Differential Amplifier There are a number of op amp circuits for differential amplifiers. The most common circuit for this amplifier is shown in Figure 2.34. Notice that the circuit uses two pairs of matched resistors, R_1 and R_2. When the matching is perfect and the op amp is ideal, the transfer function for this amplifier is given by (see Appendix 6)

$$V_{out} = \frac{R_2}{R_1}(\text{V}_2 - \text{V}_1) \tag{2.40}$$

If the resistors are not well matched, the CMR will be poor. The circuit of Figure 2.33 has a disadvantage in that its input impedance is not very high and, further, is not the same for the two inputs. For this reason, voltage followers are often used on the input to provide high input impedance. The result is called an *instrumentation amplifier.*

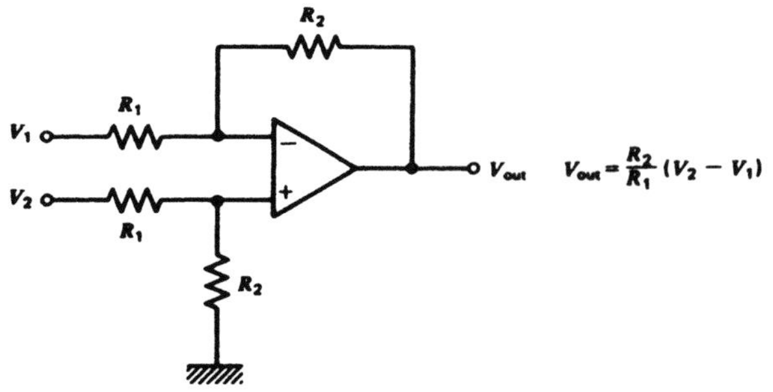

FIGURE 2.34
The basic differential amplifier configuration.

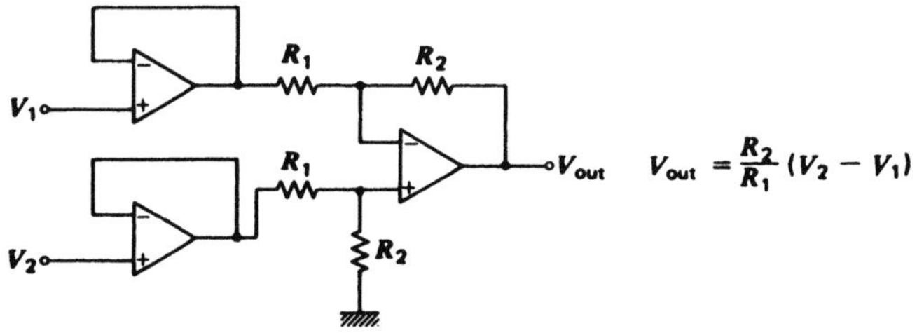

FIGURE 2.35
An instrumentation amplifier includes voltage followers for input isolation.

Instrumentation Amplifier Differential amplifiers with high input impedance and low output impedance are given the special name of instrumentation amplifier. They find a host of applications in process-measurement systems, principally as the initial stage of amplification for bridge circuits.

Figure 2.35 shows one type of instrumentation amplifier in common use. Voltage followers are simply placed on each input line. The transfer function is still given by Equation (2.40). One disadvantage of this circuit is that changing gain requires changing two resistors and having them carefully matched in value. Input offset compensation can be provided using only the differential amplifier op amp to compensate for overall offsets of all three op amps.

EXAMPLE 2.20 A sensor outputs a range of 20.0 to 250 mV as a variable varies over its range. Develop signal conditioning so that this becomes 0 to 5 V. The circuit must have very high input impedance.

Solution

A logical way to approach problems of this sort is to develop an equation for the output in terms of the input, such as that shown in Example 2.18. A circuit can then be developed to provide the variation of the equation. The equation is that of a straight line; we can then write

$$V_{out} = mV_{in} + V_0$$

where m is the slope of the line and represents the gain $(m > 1)$ or attenuation $(m < 1)$ required, and V_0 is the intercept; that is, the value V_{out} would be V_0 if $V_{in} = 0$.

For the two conditions in this problem, form two equations to solve for m and V_0.

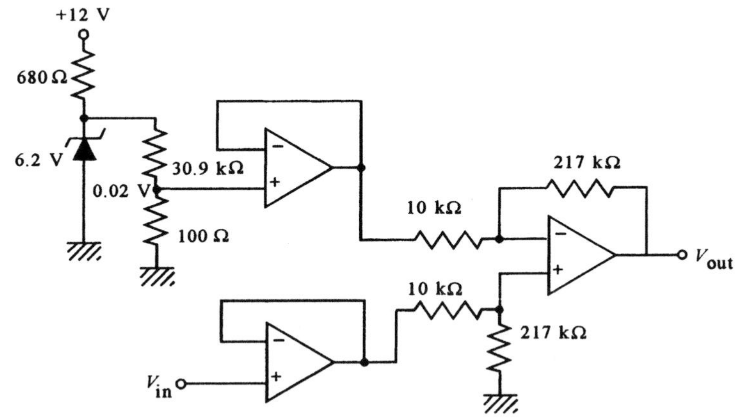

FIGURE 2.36
Solution for Example 2.20.

$$0 = m(0.02) + V_0$$
$$5 = m(0.25) + V_0$$

We get $m = 21.7$ and $V_0 = -0.434$ V using standard algebra. The equation is

$$V_{out} = 21.7\, V_{in} - 0.434$$

In this form, you can see that a summing amplifier could be used, such as the one shown in Figure 2.31.

This also can be written in the form

$$V_{out} = 21.7\,(V_{in} - 0.02)$$

This looks like a differential amplifier with a gain of 21.7 and a fixed input of 0.02 volts to the inverting side. Thus, the schematic in Figure 2.36 shows how this could be done with an instrumentation amplifier. Note the voltage divider, which is used to provide the 0.02-V bias. The zener diode is used to keep the bias voltage constant against changes of the supply voltage.

This example illustrates an important point—that more than one signal-conditioning circuit can often be used to satisfy the requirements. The choice of which one to use often comes down to the number of parts or cost.

In general, when designing a signal-conditioning circuit, any required bias voltage should be provided by a resistance divider and a regulated power supply. Otherwise, any variations of the supply voltage will cause the signal conditioning to be in error since the bias voltage would be wrong. Often a simple way to provide some comfort about the stability of the source to be divided is to use a zener diode as was done in Example 2.20.

Many integrated circuit manufacturers package instrumentation amplifiers in single ICs, since they are in such common use. Many use the circuit system in Figure 2.35, in some cases with fixed gain and in others providing for user selection of resistors so gain can be changed.

A more common configuration of instrumentation amplifier, however, is the circuit shown in Figure 2.37. This circuit allows for selection of gain, within certain limits, by adjustment of a single resistor, R_G. It can be shown that the CMR of this circuit, although still

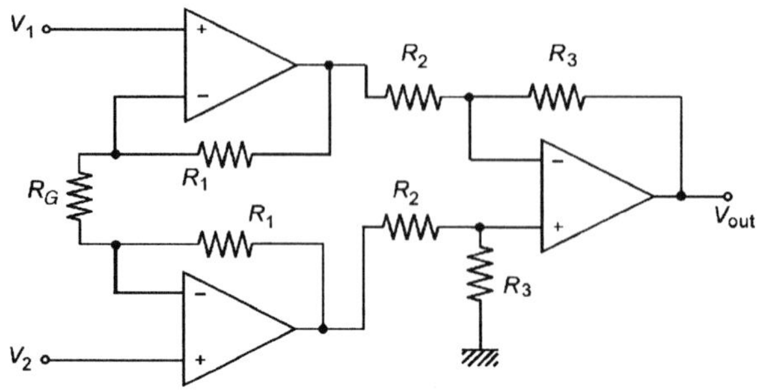

FIGURE 2.37
This instrumentation amplifier allows the gain to be changed using a single resistor.

dependent on careful matching of the differential amplifier resistors, does not depend on matching of the two R_1s. The transfer function of this amplifier is given by

$$V_{out} = \left(1 + \frac{2R_1}{R_G}\right)\left(\frac{R_3}{R_2}\right)(V_2 - V_1) \quad \textbf{(2.41)}$$

The input impedance is very high, and the output impedance very low. Many IC manufacturers provide this circuit with fixed differential gain and R_1, but allow the user to insert external R_G so the desired gain can be selected. They can thus ensure a high CMR.

EXAMPLE 2.21

Figure 2.38 shows a bridge circuit for which R_4 varies from 100 Ω to 102 Ω. Show how an instrumentation amplifier like that in Figure 2.37 could be used to provide an output of 0 to 2.5 V. Assume that $R_2 = R_3 = 1$ kΩ and that $R_1 = 100$ kΩ.

Solution

Clearly, the bridge is at null when $R_4 = 100$ Ω. When $R_4 = 102$ Ω, the bridge offset voltage is found from Equation (2.7):

$$\Delta V = (V_a - V_b) = 5\left[\frac{100}{100 + 100} - \frac{102}{100 + 102}\right]$$

$$\Delta V = -24.75 \text{ mV}$$

FIGURE 2.38
Bridge for Example 2.21.

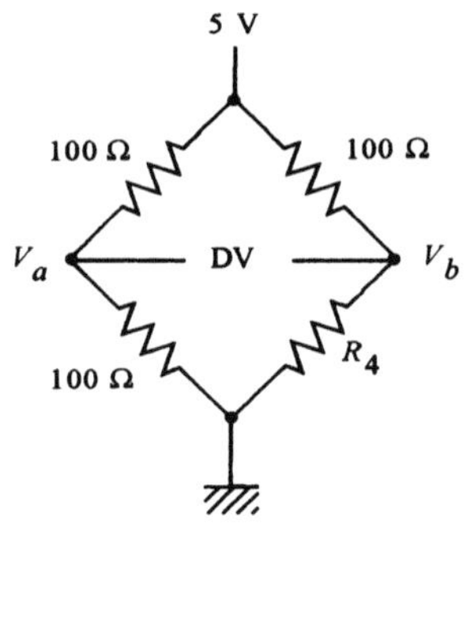

The negative sign shows that $V_b > V_a$. To get an output of 2.5 V at 102 Ω means that we need a differential gain of $A = (2.5\ \text{V}/24.75\ \text{mV}) = 101$. From Equation (2.41), we have

$$101 = \left(1 + \frac{200,000}{R_G}\right)\left(\frac{1000}{1000}\right)$$

Solving this, we find $R_G = 2000\ \Omega$. Note also that the input of the amplifier must be connected to the bridge with V_a connected to V_1 and V_b connected to V_2 so that the polarity comes out correctly.

2.5.5 Voltage-to-Current Converter

Because signals in process control are most often transmitted as a current, specifically 4 to 20 mA, it is often necessary to employ a linear voltage-to-current converter. Such a circuit must be capable of sinking a current into a number of different loads without changing the voltage-to-current transfer characteristics. An op amp circuit that provides this function is shown in Figure 2.39. An analysis of this circuit shows that the relationship between current and voltage is given by (see Appendix 6)

$$I = -\frac{R_2}{R_1 R_3} V_{\text{in}} \tag{2.42}$$

provided that the resistances are selected so that

$$R_1(R_3 + R_5) = R_2 R_4 \tag{2.43}$$

The circuit can deliver current in either direction, as required by a particular application.

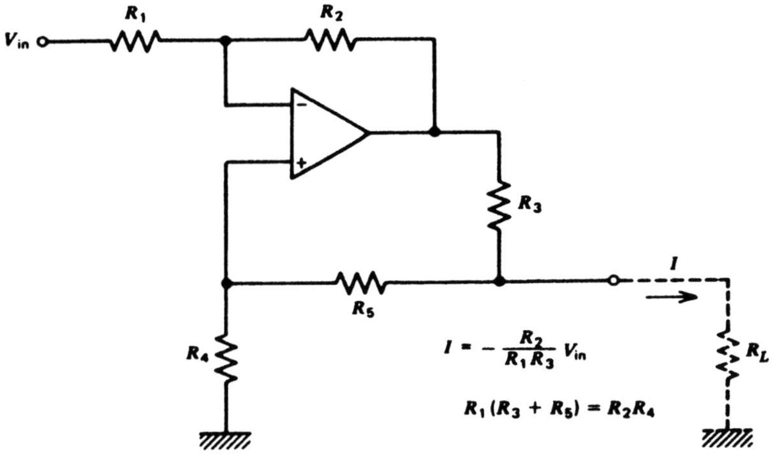

FIGURE 2.39
A voltage-to-current converter using an op amp.

The maximum load resistance and maximum current are related and determined by the condition that the amplifier output saturates in voltage. Analysis of the circuit shows that when the op amp output voltage saturates, the maximum load resistance and maximum current are related by

$$R_{ml} = \frac{(R_4 + R_5)\left[\frac{V_{sat}}{I_m} - R_3\right]}{R_3 + R_4 + R_5} \quad \textbf{(2.44)}$$

where R_{ml} = maximum load resistance
V_{sat} = op amp saturation on voltage
I_m = maximum current

A study of Equation (2.44) shows that the *maximum* load resistance is always less than V_{sat}/I_m. The *minimum* load resistance is zero.

EXAMPLE 2.22 A sensor outputs 0 to 1 V. Develop a voltage-to-current converter so that this becomes 0 to 10 mA. Specify the maximum load resistance if the op amp saturates at ± 10 V.

Solution

The voltage-to-current converter of Figure 2.39 inverts since there is a negative sign in Equation (2.42). Thus, we will need to use an inverting amplifier in the first stage so that the input to the voltage-to-current converter varies from 0 to -1 V to produce a current of 0 to 10 mA. If we make $R_1 = R_2$ in Figure 2.39, then Equation (2.42) reduces to $I = V_{in}/R_3$. To satisfy 10 mA at 1 V, we must have

$$R_3 = 1\ \text{V}/10\ \text{mA} = 100\ \Omega$$

Let us take $R_5 = 0$ (which is allowed) so that Equation (2.43) also specifies

$$R_3 = R_4 = 100\ \Omega$$

This completes the voltage-to-current converter. The maximum load resistance is found from Equation (2.44):

$$R_{ml} = 100\,[10\ \text{V}/10\ \text{mA} - 100]/200$$
$$R_{ml} = 450\ \Omega$$

2.5.6 Current-to-Voltage Converter

At the receiving end of the process-control signal transmission system, we often need to convert the current back into a voltage. This can be done most easily with the circuit shown in Figure 2.40. This circuit provides an output voltage given by

$$V_{out} = -IR \quad \textbf{(2.45)}$$

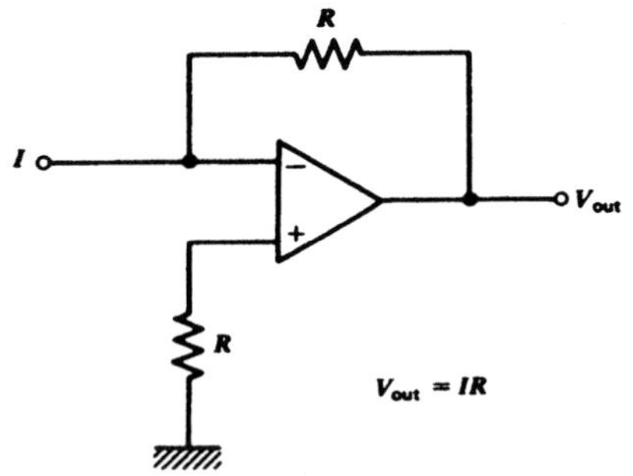

FIGURE 2.40
A current-to-voltage converter using an op amp. Care must be taken that the current output capability of the op amp is not exceeded.

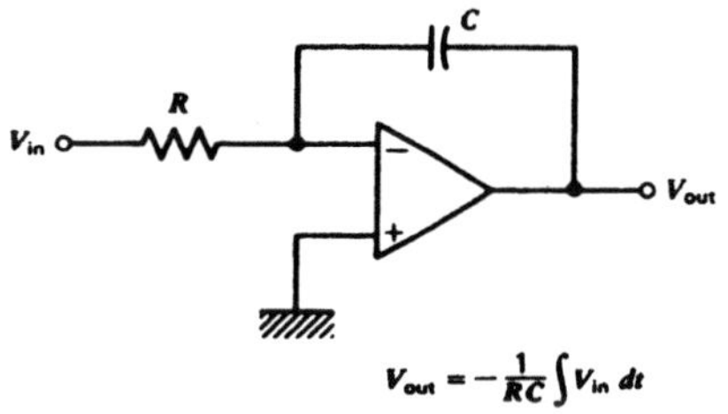

FIGURE 2.41
An integrator circuit using an op amp.

provided the op amp saturation voltage has not been reached. The resistor, R, in the noninverting terminal is employed to provide temperature stability to the configuration.

2.5.7 Integrator

Another op amp circuit to be considered is the *integrator.* This configuration, shown in Figure 2.41, consists of an input resistor and a feedback capacitor. Using the ideal analysis, we can sum the currents at the summing point as

$$\frac{V_{in}}{R} + C\frac{dV_{out}}{dt} = 0 \tag{2.46}$$

which can be solved by integrating both terms so that the circuit response is

$$V_{out} = -\frac{1}{RC}\int V_{in}dt \tag{2.47}$$

This result shows that the output voltage varies as an integral of the input voltage with a scale factor of $-1/RC$. This circuit is employed in many cases where integration of a transducer output is desired.

Other functions also can be implemented, such as a highly linear ramp voltage. If the input voltage is constant, $V_{in} = K$, Equation (2.47) reduces to

$$V_{out} = -\frac{K}{RC}t \tag{2.48}$$

which is a linear ramp, a negative slope of K/RC. Some mechanism of reset through discharge of the capacitor must be provided, because otherwise V_{out} will rise to the output saturation value and remain fixed there in time.

EXAMPLE 2.23

Use an integrator to produce a linear ramp voltage rising at 10 V per ms.

Solution

An integrator circuit, as shown in Figure 2.41, produces a ramp of

$$V_{out} = -\frac{V_{in}}{RC}t$$

when the input voltage is constant. If we make $RC = 1$ ms and $V_{in} = -10$ V, then we have

$$V_{out} = (10 \cdot 10^{+3})t$$

which is a ramp rising at 10 V/ms. A choice of $R = 1\ \text{k}\Omega$ and $C = 1\ \mu\text{F}$ will provide the required RC product.

2.5.8 Differentiator

It is also possible to construct an op amp circuit with an output proportional to the derivative of the input voltage. This circuit, which is shown in Figure 2.42, is realized with only a single capacitor and a single resistor, as in the case of the integrator. Using ideal analysis to sum currents at the summing point gives the equation

$$C\frac{dV_{in}}{dt} + \frac{V_{out}}{R} = 0 \tag{2.49}$$

Solving for the output voltage shows that the circuit response is

$$V_{out} = -RC\frac{dV_{in}}{dt} \tag{2.50}$$

Therefore, the output voltage varies as the derivative of the input voltage.

Practically speaking, this circuit exhibits erratic or even unstable response and can be used only in combination with other circuitry to depress this instability. Section 10.3.2 demonstrates how this circuit can be modified for use in controller implementation.

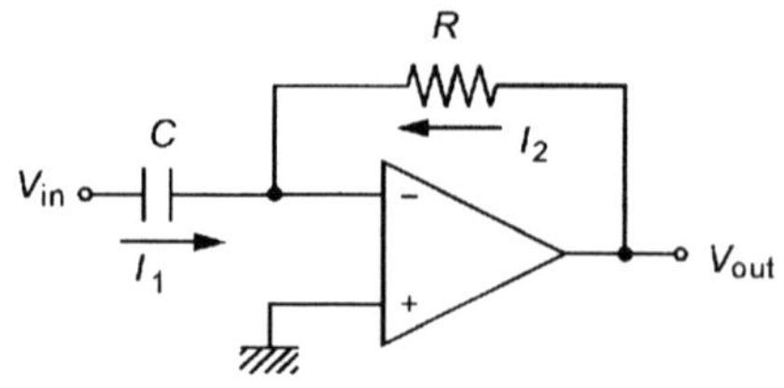

FIGURE 2.42
This circuit takes the time derivative of the input voltage.

2.5.9 Linearization

The op amp can also implement linearization. Generally, this is achieved by placing a *nonlinear* element in the feedback loop of the op amp, as shown in Figure 2.43. The summation of currents provides

$$\frac{V_{\text{in}}}{R} + I(V_{\text{out}}) = 0 \tag{2.51}$$

where V_{in} = input voltage
R = input resistance
$I(V_{\text{out}})$ = nonlinear variation of current with voltage

If Equation (2.51) is solved (in principle) for V_{out}, we get

$$V_{\text{out}} = G\left(\frac{V_{\text{in}}}{R}\right) \tag{2.52}$$

where V_{out} = output voltage
$G\left(\frac{V_{\text{in}}}{R}\right)$ = a nonlinear function of the input voltage [actually the inverse function of $I(V_{\text{out}})$]

Thus, as an example, if a diode is placed in the feedback as shown in Figure 2.44, the function $I(V_{\text{out}})$ is an exponential

$$I(V_{\text{out}}) = I_0 \exp(\alpha V_{\text{out}}) \tag{2.53}$$

where I_0 = amplitude constant
α = exponential constant

FIGURE 2.43
A nonlinear amplifier uses a nonlinear feedback element.

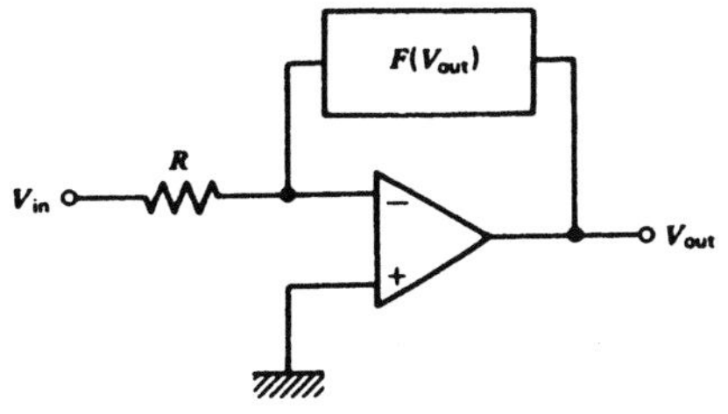

FIGURE 2.44
A diode in the feedback as a nonlinear element produces a logarithmic amplifier.

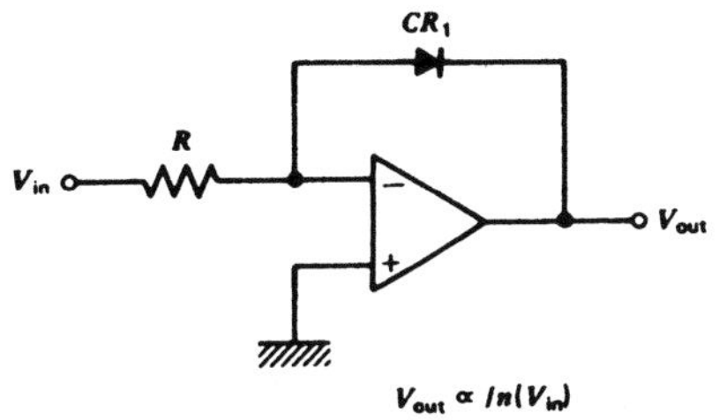

The inverse of this is a logarithm, and thus Equation (2.53) becomes

$$V_{out} = \frac{1}{\alpha}\log_c(V_{in}) - \frac{1}{\alpha}\log_e(I_0 R) \tag{2.54}$$

which thus constitutes a logarithmic amplifier.

Different feedback devices can produce amplifiers that only smooth out nonlinear variations or provide specified operations, such as the logarithmic amplifier.

2.6 DESIGN GUIDELINES

This section discusses typical issues that should be considered when designing an analog signal-conditioning system. The examples show how the guidelines can be used to develop a design. The guidelines ensure that the problem is clearly understood and that the important issues are included.

Not every guideline will be important in every design, so some will not be applicable. In many cases, not enough information will be available to address an issue properly; then the designer must exercise good technical judgment in accounting for that part of the design.

Figure 2.45 shows the measurement and signal-conditioning model. In some cases, the entire system is to be developed, from selecting the sensor to designing the signal conditioning. In other cases, only the signal conditioning will be developed. The guidelines are generalized. Since the sensor is selected from what is available, the actual design is really for the signal conditioning.

Guidelines for Analog Signal-Conditioning Design

1. *Define the measurement objective.*
   a. *Parameter* What is the nature of the measured variable: pressure, temperature, flow, level, voltage, current, resistance, and so forth?

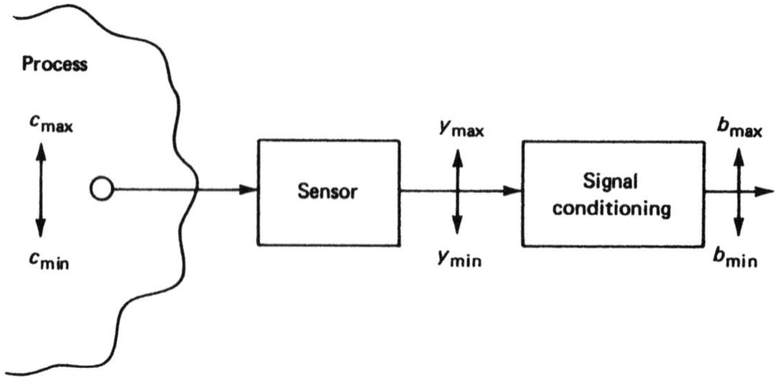

FIGURE 2.45
Model for measurement and signal-conditioning objectives.

   b. *Range* What is the range of the measurement: 100° to 200°C, 45 to 85 psi, 2 to 4 V, and so forth?
   c. *Accuracy* What is the required accuracy: 5% FS, 3% of reading, and so forth?
   d. *Linearity* Must the measurement output be linear?
   e. *Noise* What is the noise level and frequency spectrum of the measurement environment?
2. *Select a sensor (if applicable).*
   a. *Parameter* What is the nature of the sensor output: resistance, voltage, and so forth?
   b. *Transfer function* What is the relationship between the sensor output and the measured variable: linear, graphical, equation, accuracy, and so forth?
   c. *Time response* What is the time response of the sensor: first-order time constant, second-order damping, and frequency?
   d. *Range* What is the range of sensor parameter output for the given measurement range?
   e. *Power* What is the power specification of the sensor: resistive dissipation maximum, current draw, and so forth?
3. *Design the analog signal conditioning (S/C).*
   a. *Parameter* What is the nature of the desired output? The most common is voltage, but current and frequency are sometimes specified. In the latter cases, conversion to voltage is still often a first step.
   b. *Range* What is the desired range of the output parameter (e.g., 0 to 5 V, 4 to 20 mA, 5 to 10 kHz)?
   c. *Input impedance* What input impedance should the S/C present to the input signal source? This is very important in preventing loading of a voltage signal input.
   d. *Output impedance* What output impedance should the S/C offer to the output load circuit?
4. *Notes on analog signal-conditioning design.*
   a. If the input is a resistance change and a bridge or divider must be used, be sure to consider both the effect of output voltage nonlinearity with resistance and the effect of current through the resistive sensor.
   b. For the op amp portion of the design, the easiest design approach is to develop an equation for output versus input. From this equation, it will be clear what types of circuits may be used. This equation represents the static transfer function of the signal conditioning.
   c. Always consider any possible loading of voltage sources by the signal conditioning. Such loading is a direct error in the measurement system.

The following examples apply these guidelines to measurement signal-conditioning problems. In later chapters (4, 5, and 6) on sensors, many other examples will be presented.

EXAMPLE 2.24 A sensor outputs a voltage ranging from −2.4 to −1.1 V. For interface to an analog-to-digital converter, this needs to be 0 to 2.5 V. Develop the required signal conditioning.

Solution

For this type of problem, no information is provided about the measured variable, the measurement environment, or the sensor. We are simply asked to provide a voltage-to-voltage conversion. Since the source impedance is not known, it is good design practice to assume it is high and to design a high-input-impedance system to avoid loading. Most ADCs have input impedances of at least tens of kilohms, and the output impedance of op amp circuits is quite low, so there is no real concern for the output impedance of the S/C system.

For this type of problem, it is easiest to develop an equation for the output in terms of the input. From this, circuits can be envisioned.

$$V_{out} = mV_{in} + V_0$$

Using the specified information, we form two equations for the unknown slope (gain), m, and offset (bias), V_0.

$$0 = -2.4m + V_0$$
$$2.5 = -1.1m + V_0$$

Clearly, from the first equation we have $V_0 = 2.4m$, and when this is substituted into the second equation, we get

$$2.5 = -1.1m + 2.4m$$

Then, solving for m,

$$m = 2.5/(2.4 - 1.1) = 1.923$$

The transfer function equation is thus

$$V_{out} = 1.923\,V_{in} + 4.615$$

There are many ways to satisfy this equation. A summing amplifier could be used, but it does not have high input impedance, so a voltage follower would be needed at the input. Also, the summing amplifier inverts, so an inverter would be required to get the correct sign. The circuit is shown in Figure 2.46. Note that the bias has been provided by a divider. A 15-V supply has been assumed for the divider resistance calculations. The 100-Ω resistor was selected to keep loading by the op amp circuit small. A trimmer (variable) resistor has been used, so both loading of the divider by the op amp circuit and variation of the supply from exactly 15 V can be compensated for by adjusting until the bias is exactly 4.615 V.

The design could also be accomplished by a differential amplifier. If the 1.923 in the transfer equation is factored, we get

$$V_{out} = 1.923(V_{in} + 2.4)$$

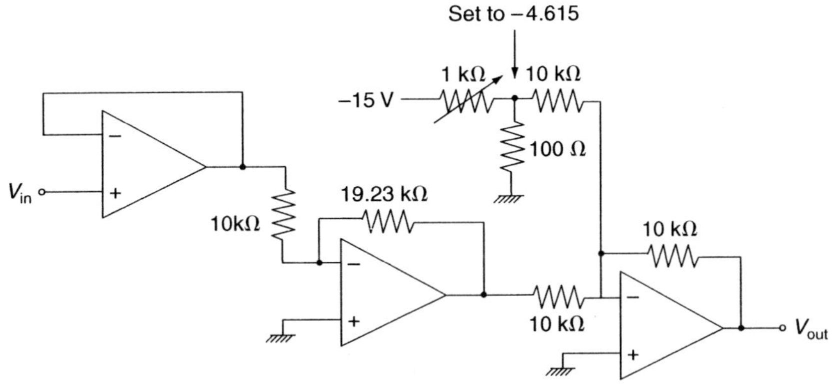

FIGURE 2.46
One possible solution to Example 2.24.

So this is the equation of a differential amplifier with a gain of 1.923 and one input fixed at 2.4 V. A voltage follower would still be required on the input. (The reader should complete this design.)

EXAMPLE 2.25 Temperature is to be measured in the range of 250°C to 450°C with an accuracy of ±2°C. The sensor is a resistance that varies linearly from 280 Ω to 1060 Ω for this temperature range. Power dissipated in the sensor must be kept below 5 mW. Develop analog signal conditioning that provides a voltage varying linearly from −5 to +5 V for this temperature range. The load is a high-impedance recorder.

Solution

Following the guidelines, let us first identify all the elements of the problem.

Measured Variable Parameter: Temperature

Range: 250° to 450°C

Accuracy: ±2°C

Noise: unspecified

Sensor Signal

Parameter: resistance

Transfer function: linear

Time response: unspecified

Range: 280 Ω to 1060 Ω, linear

Power: maximum 5 mW dissipated in sensor

Signal Conditioning

Parameter: voltage, linear

Range: −5 to +5 V

Input impedance: keep power in sensor below 5 mW

Output impedance: no problem, high-impedance recorder

The accuracy is ±0.8% at the low end and ±0.44% at the high end. Therefore, we will keep three significant figures to provide 0.1% on values selected.

The 5-mW maximum sensor dissipation means the current must be limited. To find the maximum current, we note that

$$P = I^2R$$
$$0.005 = I^2R$$
$$I = \sqrt{0.005/R}$$

The minimum current will thus occur at the maximum resistance,

$$I_{max} = \sqrt{0.005/1060} = 2.17\,\text{mA}$$

Thus, the design must always keep the sensor current below 2 mA.

Since the system must be linear, we should set up a linear equation between the sensor resistance and the output voltage. Then it is a matter of determining what circuits will implement the equation.

$$V_{out} = mR_s + V_0$$

We solve for m and V_0 by using the given information,

$$-5 = 280m + V_0$$
$$+5 = 1060m + V_0$$

Subtracting the first equation from the second gives

$$10 = 780m \quad \text{or} \quad m = 0.0128$$

Then, using this in the first equation,

$$-5 = 280(0.0128) + V_0$$
$$V_0 = -8.58$$

So the transfer function equation is

$$V_{out} = 0.0128R_s - 8.58$$

This can be provided by an inverting amplifier with the sensor resistor in the feedback, followed by an inverting summer to get the signs correct. Figure 2.47 shows one possible so-

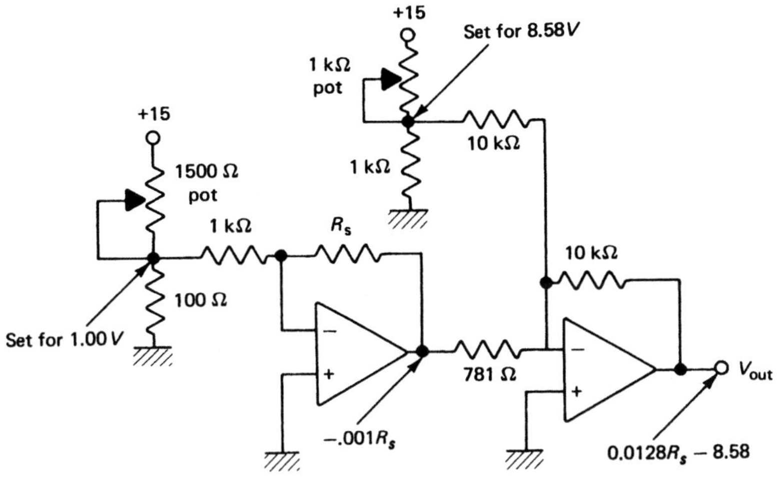

FIGURE 2.47
One possible solution for Example 2.25.

lution. The fixed input voltage and input resistor of the first op amp have been selected to satisfy the 5-mW maximum power dissipation. This has been done by noting that the current through the sensor is just equal to the current through the input circuit. Thus, by using 1.00 kΩ and 1.00 V, the current will always be 1 mA and thus less than 2 mA, as required.

As in Example 2.24, trimmers are used in dividers so the fixed voltages can be adjusted to 1.00 and 8.58 V and thus account for supply voltage differences. The alternative would be to use a zener diode as the source.

SUMMARY

The signal conditioning discussed in this chapter relates to the standard techniques employed for providing signal compatibility and measurement in analog systems. The reader is introduced to the basic concepts that form the foundation of such analog conditioning.

To present a complete picture of analog signal conditioning, the following points were considered:

1. The need for analog signal conditioning was reviewed and resolved into the requirements of signal-level changes, linearization, signal conversions, and filtering and impedance matching.
2. Bridge circuits are a common example of a conversion process where a changing resistance is measured either by a current or by a voltage signal. Many modifications of the bridge are used, including *electronic balancing* and techniques of *lead compensation.*

3. The high- and low-pass *RC* filters are passive circuits used to block undesired frequencies from data signals.
4. Operational amplifiers (op amps) are a special signal-conditioning building block around which many special-function circuits can be developed. The device was demonstrated in applications involving amplifiers, converters, linearization circuits, integrators, and several other functions.

PROBLEMS

Section 2.2

2.1 Derive Equation (2.1) for general circuit loading.

2.2 The unloaded output of a sensor is a sinusoid at 200 Hz and 5 V rms amplitude. Its output impedance is $2000 + 600j$. If a 0.22- μF capacitor is placed across the output as a load, what is the sensor output rms voltage amplitude?

Section 2.3

2.3 A sensor resistance varies from 520 to 2500 Ω. This is used for R_1 in the divider of Figure 2.4, along with $R_2 = 500\ \Omega$ and $V_s = 10.0$ V. Find (a) the range of the divider voltage, V_D, and (b) the range of power dissipation by the sensor.

2.4 Prepare graphs of the divider voltage versus transducer resistance for Example 2.2 and Problem 2.3. Does the voltage vary linearly with resistance? Does the voltage increase or decrease with resistance?

2.5 Show how the bridge offset equation given as Equation (2.7) can be derived from Equation (2.6).

2.6 Derive Equation (2.10) for the bridge circuit Thévenin resistance.

2.7 A Wheatstone bridge, as shown in Figure 2.5, nulls with $R_1 = 227\ \Omega$, $R_2 = 448\ \Omega$, and $R_3 = 1414\ \Omega$. Find R_4.

2.8 A sensor with a nominal resistance of 50 Ω is used in a bridge with $R_1 = R_2 = 100\ \Omega$, $V = 10.0$ V, and $R_3 = 100$-Ω potentiometer. It is necessary to resolve 0.1-Ω changes of the sensor resistance.

a. At what value of R_3 will the bridge null?

b. What voltage resolution must the null detector possess?

2.9 A bridge circuit is used with a sensor located 100 m away. The bridge is not lead compensated, and the cable to the sensor has a resistance of 0.45 Ω/ft. The bridge nulls with $R_1 = 3400\ \Omega$, $R_2 = 3445\ \Omega$, and $R_3 = 1560\ \Omega$. What is the sensor resistance?

2.10 The bridge in Figure 2.5 has $R_1 = 250\ \Omega$, $R_3 = 500\ \Omega$, $R_4 = 340\ \Omega$, and $V = 1.5$ V. The detector is a galvanometer with $R_G = 150\ \Omega$.

a. Find the value of R_2 that will null the bridge.

b. Find the offset current that will result if $R_2 = 190\ \Omega$.

2.11 A current balance bridge, shown in Figure 2.8, has resistances of $R_1 = R_2 = 1\text{ k}\Omega$, $R_4 = 590\ \Omega$, $R_5 = 10\ \Omega$, and $V = 10.0$ V.

a. Find the value of R_3 that nulls the bridge with no current.
b. Find the value of R_3 that balances the bridge with a current of 0.25 mA.

2.12 A potential measurement bridge, such as that in Figure 2.9, has $V = 10.0$ V, $R_1 = R_2 = R_3 = 10$ kΩ. Find the unknown potential if the bridge nulls with $R_4 = 9.73$ kΩ.

2.13 An ac Wheatstone bridge with all arms as capacitors nulls when $C_1 = 0.4\ \mu$F, $C_2 = 0.31\ \mu$F, and $C_3 = 0.27\ \mu$F. Find C_4.

2.14 The ac bridge of Figure 2.48 nulls with $R_1 = 1$ kΩ, $R_2 = 2$ kΩ, $R_3 = 100$ Ω, and $L_3 = 250$ mH.
a. Find the values of R_4 and L_4.
b. If the circuit is excited by a 5-V rms, 1-kHz oscillator, find the offset voltage for $L_4 = 510$ mH.
c. What are the amplitudes of the in-phase and quadrature (90°) components of the offset voltage?

2.15 Develop a low-pass *RC* filter to attenuate 0.5 MHz noise by 97%. Specify the critical frequency, values of *R* and *C*, and the attenuation of a 400-Hz input signal.

2.16 A low-pass *RC* filter has $f_c = 3.5$ kHz. Find the attenuation of a 1-kHz signal.

2.17 A high-pass *RC* filter must drive 120 Hz noise down to 1%. Specify the filter critical frequency, values of *R* and *C*, and the attenuation of a 30-kHz signal.

2.18 A high-pass filter is found to attenuate a 1-kHz signal by 20 dB. What is the critical frequency?

2.19 Design a band-pass filter with critical frequencies of 100 Hz and 10 kHz, respectively. Use a resistance ratio of 0.05. Draw a semilog graph like that in Figure 2.21 showing voltage output to input from 10 Hz to 100 kHz.

2.20 A sensor output needs to feed an amplifier with a 10-kΩ input impedance. There is significant noise in the range of 4 to 5 kHz. The data spectrum lies below 200 Hz.

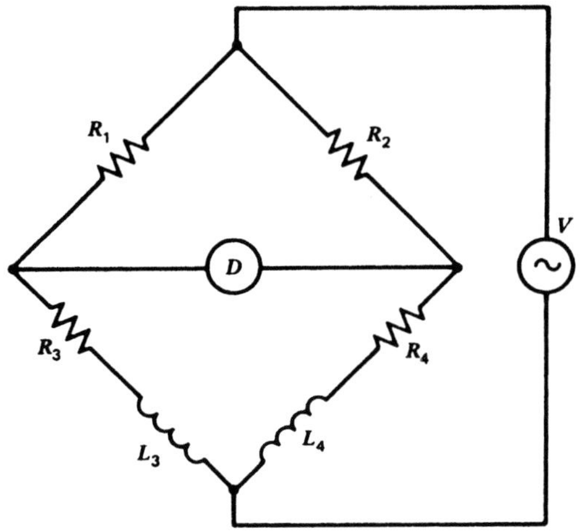

FIGURE 2.48
ac bridge for Problem 2.14.

Design a low-pass filter for use between the sensor and the amplifier that reduces the data by not more than 1%. By how much is the noise reduced?

2.21 A telephone line will be used to carry measurement data as a frequency-modulated signal from 5 to 6 kHz. The line is shared with voice data below 500 Hz, and switching noise occurs above 500 kHz. Design a band-pass *RC* filter that reduces the voice by 80% and the switching by 90%. Use a resistance ratio of $r = 0.02$. What is the effect on the passband frequencies?

2.22 A single line is multiplexed to carry sensor signals in a frequency range below 1 kHz and communication signals ranging from 10 to 50 kHz. There is a large noise component at 4.5 kHz from a turbine in the plant. Design a twin-T notch filter for the 4.5-kHz noise. Evaluate the effect on sensor and communication signals.

Section 2.5

2.23 Show how op amps can be used to provide an amplifier with a gain of +100 and an input impedance of 1.5 kΩ. Show how this can be done using both inverting and noninverting configurations.

2.24 Specify the components of a differential amplifier with a gain of 22.

2.25 Using an integrator with $RC = 10$ s and any other required amplifiers, develop a voltage ramp generator with 0.5 V/s.

2.26 Signal-conditioning analysis shows that the following equation must relate output voltage to input voltage:

$$V_{out} = 3.35\ V_{in} - 2.68$$

Design circuits to do this using (a) a summing amplifier and (b) a differential amplifier.

2.27 A differential amplifier has $R_2 = 470$ kΩ and $R_1 = 2.7$ kΩ. When $V_a = V_b = 2.5$ V the output is 87 mV. Find the CMR and CMRR.

2.28 Derive Equation (2.41) for the instrumentation amplifier of Figure 2.37.

2.29 Design an instrumentation amplifier like that of Figure 2.37 with switch-selectable gains of 1, 10, 100, and 1000. Show the complete circuit using 741 op amps, pin connections, and input offset adjustment.

2.30 A control system needs the average of temperature from three locations. Sensors make the temperature information available as voltages, V_1, V_2, and V_3. Develop an op amp circuit that outputs the average of these voltages.

2.31 Use an inverting amplifier, an integrator, and a summing amplifier to develop an output voltage given by

$$V_{out} = 10\ V_{in} + 4 \int V_{in} dt$$

2.32 Develop a voltage-to-current converter that satisfies the requirement $I = 0.0021\ V_{in}$. If the op amp saturation voltage is ±12 V and the maximum current delivery is 5 mA, find the maximum load resistance.

Section 2.6

2.33 A bridge circuit has $R_1 = R_2 = R_4 = 120\ \Omega$ and $V = 10.0$ V. Design a signal-conditioning system that provides an output of 0.0 to 5.0 V as R_3 varies from 120 to 140 Ω. Plot V_{out} versus R_3. Evaluate the linearity.

2.34 Develop signal conditioning for Example 2.2 so an output voltage varies from 0 to 5 V as the resistance varies from 4 to 12 kΩ.

2.35 Develop signal conditioning for Problem 2.3 so the output voltage varies from 0 to 5 V as the resistance varies from 520 to 2500 Ω, where 0 V corresponds to 520 Ω.

2.36 A sensor varies from 1 to 5 kΩ. Use this in an op amp circuit to provide a voltage varying from 0 to 5 V as the resistance changes.

2.37 A process signal varies from 4 to 20 mA. The setpoint is 9.5 mA. Use a current-to-voltage converter and a summing amplifier to get a voltage error signal with a scale factor of 0.5 V/mA.

2.38 Sensor resistance varies from 25 to 1.5 kΩ as a variable changes from c_{min} to c_{max}. Design a signal-conditioning system that provides an output voltage varying from −2 to +2 V as the variable changes from min to max. Power dissipation in the sensor must be kept below 2.5 mW.

2.39 A pressure sensor outputs a voltage varying as 100 mV/psi and has a 2.5- kΩ output impedance. Develop signal conditioning to provide 0 to 2.5 V as the pressure varies from 50 to 150 psi.

2.40 A system is needed to measure flow, which continuously cycles between 20 and 30 gal/min with a period of 30 s. The required output is a voltage varying from −2.5 to +2.5 V for the cycling flow range. The sensor to be used has a transfer function of $\sqrt{Q}$ volts, where Q is in gal/min, and an output impedance of 2.0 kΩ. Tests show that the output of the sensor has 60 Hz noise of 0.8 V rms. Design a signal-conditioning system, including noise filtering, and evaluate your design as follows.

a. Plot output voltage versus flow, and comment on the linearity.

b. Determine the noise on the output as percent FS.

SUPPLEMENTARY PROBLEMS

Figure 2.49 shows a system proposed as a scale for weighing. The basic sensor is a resistor, R_w, that *linearly* converts weight to resistance; for 0.00 lb, it nominally has a resistance of 119 Ω, and at 299 lb it has a resistance of 127 Ω. The bridge offset voltage is amplified by a differential amplifier and sent to a DVM whose voltage, by design, will equal the weight (i.e., a weight of 134 lb should result in a voltage of 1.34 V, so the DVM will read 134). Neat, huh?

The purpose of the resistor combination in the bridge is to allow *resetting* the bridge to zero using the variable resistor, R_z. This will allow compensation for changes of R_W or any of the other resistors for that matter. The next three questions are related to this system.

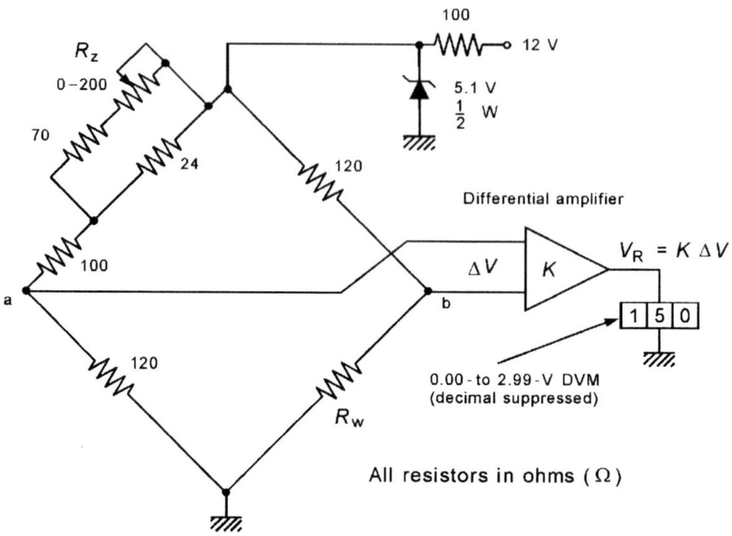

FIGURE 2.49
Circuit for supplementary problems.

S2.1 Consider first only the bridge circuit in Figure 2.49.

a. What value of R_z will be required to null the bridge at 0.00 lb?

b. For what minimum and maximum values of R_W can the bridge be nulled using R_z?

c. What offset voltage, ΔV, results when the weight is 299 lb (assuming the bridge is nulled at 0.0 lb)?

S2.2 Let us consider the amplifier of Figure 2.49 next.

a. Find the gain, *K,* so that the DVM indicates the weight but in volts (i.e., when the weight is 299 lb, the voltmeter reads 2.99 Vs).

b. Provide the circuit for a differential amplifier that can provide this gain. Specify the resistors in the amplifier circuit. Show how the amplifier inputs are connected to bridge points *a* and *b* to give the right output polarity.

S2.3 Let's evaluate how well the system in Figure 2.49 works and propose a change.

a. First suppose the weight is 150 lb. What will the DVM read? Therefore, what is the error in lb? Suppose the scale is exact at 0.00 lb and 299 lb but has errors at 150 lb. Why is there an error?

b. Change the gain so the reading is exact at 150 lb. Now specify the error at 0 lb (which no one weighs) and 299 lb (which few people weigh).

c. Prepare a plot of voltage reading versus weight.

S2.4 Figure 2.50 shows how a single wire can be used to carry measurement data at the same time. This is done by modulating two widely different carrier frequencies with data and using filters to extract the data at the receiving end. Suppose one data channel is a modulated signal of 1.0 to 1.5 kHz and the other is a modulated signal of 50 to 55 kHz. Design the extraction filters using simple *RC* filters such that the data loss

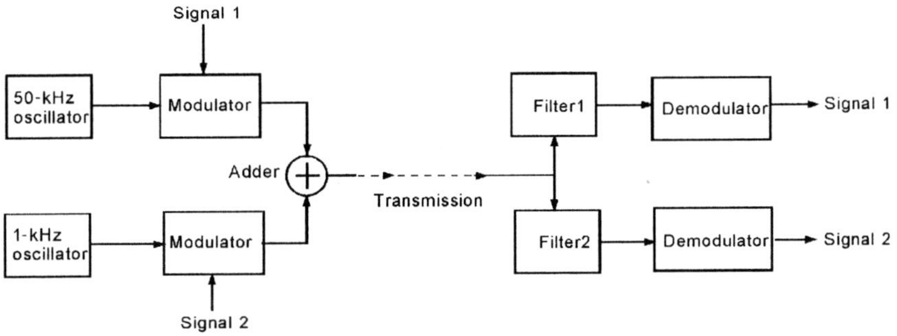

FIGURE 2.50
System for Problem S2.4.

is restricted to 0.707 (3 dB, or 50% power) in the data signals. How much amplitude crossover results? (What is the amplitude ratio of the 50–55 kHz in the 1–1.5-kHz channel and vice versa?)

S2.5 A humidity sensor resistance varies linearly from 250 kΩ to 120 kΩ as humidity varies from 0% to 100%. Power dissipation in the sensor must be kept below 100 μW. Design analog signal conditioning to provide a voltage of 0.00 to 1.00 V as the humidity varies from 0% to 100%.

S2.6 In some cases, we need amplifiers that have high gain when the input voltage is low and decreasing gain as the input voltage increases. So this is an amplifier whose *gain* depends upon the input voltage! An op amp circuit such as that in Figure 2.51 can provide this response. Here diodes are used to isolate feedback resistors via their forward voltage drop until the output voltage rises above predefined levels. For the circuit in Figure 2.51, assume the forward voltage drop of the diodes is 1.4 V (i.e., they do not begin to conduct until the voltage across them is above 1.4 V). Prepare a plot of output voltage versus input voltage for this circuit. (*Hint:* When diode D_1 begins to conduct, the output must be −1.4 V, and this effectively just puts the 100-kΩ resistors in parallel.)

FIGURE 2.51
Nonlinear amplifier using diodes for Problems S2.6 and S2.7.

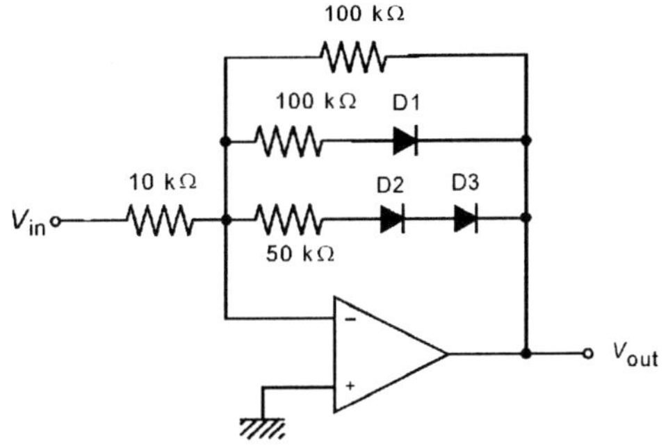

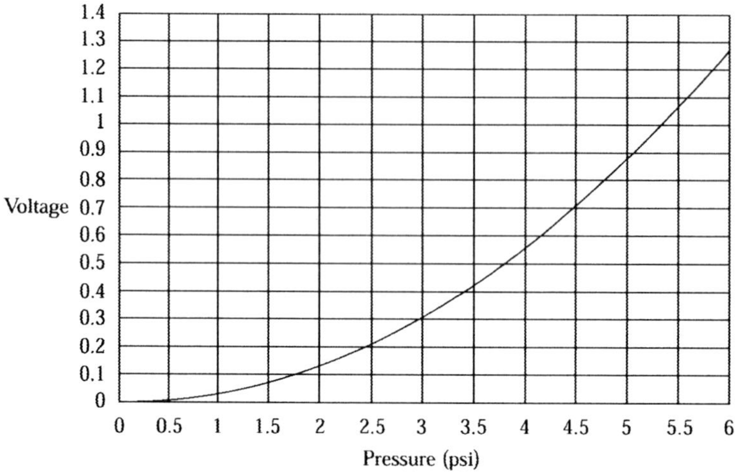

FIGURE 2.52
Voltage versus pressure for Problem S2.7.

S2.7 Circuits such as Figure 2.51 can also be used to provide some degree of linearization. To see this, consider a sensor whose output voltage varies nonlinearly with input pressure by the equation $V(p) = 0.035p^2$. This response is shown in Figure 2.52. Now, assume the sensor voltage is provided as input to the circuit in Figure 2.51. Determine the output voltage over the pressure range, and plot V_{out} versus p. You will see that the resulting voltage is more nearly linear.

CHAPTER 3

Digital Signal Conditioning

INSTRUCTIONAL OBJECTIVES

In this chapter the basic principles of digital signal processing will be studied. This includes digital-to-analog converters (DAC) and analog-to-digital converters (ADC), as well as the characteristics of digital data. After reading this chapter and working through the examples and problems at the end of the chapter you will be able to:

- Develop Boolean equations for multivariable alarms.
- Design an application using a comparator with hysteresis.
- Calculate the expected output of a biopolar DAC for a given input.
- Explain how a successive approximation ADC operates.
- Describe how a sample-and-hold circuit operates.
- Explain the operation of a frequency-based ADC.
- Describe the consequence of sampling rate on data acquisition.

3.1 INTRODUCTION

Why *digital signal conditioning?* And what is digital signal conditioning, anyway?

The answer to the first question is found in the recognition that digital electronics and digital computers have taken a major role in nearly every aspect of life in our modern world. Of course digital electronics is at the heart of computers, but there are lots of direct applications of digital electronics in our world. Everyday things like automatic door openers in stores, motion sensors in security systems, and seat-belt warning systems are implemented with digital electronics. All these digital electronic systems require data to be presented to them in a digital format (i.e., the data have to be *digitally conditioned*).

When most people think of a computer, they visualize the common home and office *personal computer*, or PC. In fact, however, one of the most common applications of computers is for controlling something, and we seldom ever even see the computer. When we learn that our microwave oven has a microprocessor-based computer inside, it is there for the purpose of controlling the operation of the oven. So it is with the use of computers in automobiles, washing machines, airplanes, and a vast host of other examples.

Computers are digital electronic devices and so all the information they work with has to be digitally formatted. Therefore, if they are used to control a variable such as temperature, then the temperature has to be represented digitally. So that's why we need digital signal conditioning—to condition process-control signals to be in an appropriate digital format.

The second question is the topic of this chapter. Digital signal conditioning in process control means finding a way to represent analog process information in a digital format. We will review some common digital electronics principles and then study how analog signals are converted into a digital format.

You should realize that there is no greater accuracy in using digital techniques to represent data; in fact, accuracy is usually lost. But digital data are much more immune from spurious influences that would cause subsequent inaccuracy, such as noise, amplifier gain changes, power supply drifts, and so on.

Use of computers in control systems is particularly valuable for a number of other reasons, however:

1. A computer can control multivariable process-control systems.
2. Nonlinearities in sensor output can be linearized by the computer.
3. Complicated control equations can be solved quickly and modified as needed.
4. Networking of control computers allows a large process-control complex to operate in a fully integrated fashion.

3.2 REVIEW OF DIGITAL FUNDAMENTALS

A working understanding of the application of digital techniques to process control requires a foundation in basic digital electronics. The design and implementation of control logic systems and microcomputer control systems require a depth of understanding that can be obtained only by taking several courses devoted to the subject. In this text, we assume a sufficient background that the reader can appreciate the essential features of digital electronic design and its application to process control. A summary of basic digital electronics concepts is presented in Appendix 2.

3.2.1 Digital Information

The use of digital techniques in process control requires that process variable measurements and control information be encoded into a digital form. Digital signals themselves are simply two-state (binary) levels. These levels may be represented in many ways (e.g., two voltages, two currents, two frequencies, or two phases). We speak, then, of the digital information as a high state (H, or **1**) or a low state (L, or **0**) on a wire that carries the digital signal.

Digital Words Given the simple binary information that is carried by a single digital signal, it is clear that multiple signals must be used to describe analog information. Generally, this is done by using an assemblage of digital levels to construct a binary number, often called a *word*. The individual digital levels are referred to as *bits* of the word. Thus, for example, a 6-bit word consists of six independent digital levels, such as $\mathbf{101011_2}$, which can be thought of as a six-digit base 2 number. An important consideration, then, is how the analog information is encoded into this digital word.

Decimal Whole Numbers One of the most common schemes for encoding analog data into a digital word is to use the straight counting of decimal (or base 10) and binary number representations. The principles of this process are reviewed in Appendix 2, together with octal and hexadecimal representations.

EXAMPLE 3.1 Find the base 10 equivalent of the binary whole number $\mathbf{00100111_2}$.

Solution
As in the base 10 system, zeros preceding the first significant digit do not contribute. Thus, the binary number is actually $\mathbf{100111_2}$, and $n = 5$. To find the decimal equivalent, we use Appendix 2 and compute as follows:

$$N_{10} = a_5 2^5 + a_4 2^4 + \cdots + a_1 2^1 + a_0 2^0$$
$$N_{10} = (1)2^5 + (0)2^4 + (0)2^3 + (1)2^2 + (1)2^1 + (1)2^0 \qquad \textbf{(A.2)}$$
$$N_{10} = 32 + 4 + 2 + 1$$
$$N_{10} = 39$$

EXAMPLE 3.2 Find the binary equivalent of the base 10 number 47.

Solution
Starting the successive division, we get

$$\frac{47}{2} = 23 \text{ with a remainder of } \frac{1}{2} \text{ so that } a_0 = 1$$

then

$$\frac{23}{2} = 11 \text{ with a remainder of } \frac{1}{2} \text{ so that } a_1 = 1$$

then

$$\frac{11}{2} = 5 + \frac{1}{2} \quad \therefore\ a_2 = 1$$

$$\frac{5}{2} = 2 + \frac{1}{2} \quad \therefore\ a_3 = 1$$

$$\frac{2}{2} = 1 + 0 \quad \therefore\ a_4 = 0$$

$$\frac{1}{2} = 0 + \frac{1}{2} \quad \therefore\ a_5 = 1$$

We find that base 10 number 47 becomes binary number $\mathbf{101111_2}$.

The representation of negative numbers in binary format takes on several forms, as discussed in Appendix 2.

Octal and Hex Numbers It is cumbersome for humans to work with digital words expressed as numbers in the binary representation. For this reason, it has become common to use either the octal (base 8) or hexadecimal (base 16, called hex) representations, which are reviewed in Appendix 2. Octal numbers are conveniently formed from groupings of three binary digits; that is, $\mathbf{000_2}$ is 0_8 and $\mathbf{111_2}$ is 7_8. Thus, a binary number like $\mathbf{101011_2}$ is equivalent to 53_8. Hex numbers are formed easily from groupings of four binary digits; that is, $\mathbf{0000_2}$ is 0H and $\mathbf{1111_2}$ is FH. The letter H (or h) is used to designate a hex number instead of a subscript 16. Also recall that the hex counting sequence is 0, 1, 2, 3, 4, 5, 6, 7, 8, 9, A, B, C, D, E, and F to cover the possible states. Because microcomputers must frequently use either 4-bit, 8-bit, or 16-bit words, the hex notation is commonly used with these machines. In hex, a binary number like $\mathbf{10110110_2}$ would be written B6H.

3.2.2 Fractional Binary Numbers

Although not as commonly used, it is possible to define a fractional binary number in the same manner as whole numbers, using only the **1** and **0** of this counting system. Such numbers, just as in the decimal framework, represent divisions of the counting system to values less than unity. A correlation can be made to decimal numbers in a similar fashion to Equation (A.2.1), as

$$N_{10} = b_1 2^{-1} + b_2 2^{-2} + \cdots + b_m 2^{-m} \tag{3.1}$$

where

N_{10} = base 10 number less than 1

$b_1 b_2 \ldots b_{m-1} b_m$ = base 2 number less than 1

m = number of digits in base 2 number

EXAMPLE 3.3 Find the base 10 equivalent of the binary number $\mathbf{0.11010_2}$.

Solution

This can be found most easily by using

$$N_{10} = b_1 2^{-1} + b_2 2^{-2} + \cdots + b_m 2^{-m} \tag{3.1}$$

with

$$m = 5$$
$$N_{10} = (1)2^{-1} + (1)2^{-2} + (0)2^{-3} + (1)2^{-4} + (0)2^{-5}$$
$$N_{10} = \frac{1}{2} + \frac{1}{4} + \frac{1}{16}$$
$$N_{10} = 0.8125_{10}$$

Converting a base 10 number that is less than 1 to a binary equivalent requires repeated multiplication by 2. The result of each multiplication is a fractional part and either a 0 or 1 whole-number part, which determines whether that digit is a **0** or a **1**. The first multiplication gives the most significant bit, b_1, and the last gives either a 0 or a 1 for the least significant bit, b_m.

EXAMPLE 3.4 Find the binary, octal, and hex equivalents of 0.3125_{10}.

Solution

Using successive multiplication, we find

$$\begin{aligned} 2(0.3125) &= 0.6250 & \text{so } b_1 &= 0 \\ 2(0.625) &= 1.250 & \text{so } b_2 &= 1 \\ 2(0.25) &= 0.5 & \text{so } b_3 &= 0 \\ 2(0.5) &= 1.0 & \text{so } b_4 &= 1 \end{aligned}$$

Thus, we find that 0.3125_{10} is equivalent to $\mathbf{0.0101_2}$. It can be represented as $\mathbf{0.010100_2}$, because trailing zeros are not significant in a number less than 1, and thus as 0.24 octal, because $\mathbf{010_2} = 2_8$ and $\mathbf{100_2} = 4_8$. Similarly, this is 0.50H.

3.2.3 Boolean Algebra

In process control, as well as in many other technical disciplines, action is taken on the basis of an evaluation of observations made in the environment. In driving an automobile, for example, we are constantly observing such external factors as traffic, lights, speed limits, pedestrians, street conditions, low-flying aircraft, and such internal factors as how fast we wish to go, where we are going, and many others. We evaluate these factors and take actions predicated on the evaluations. We may see that a light is green, streets are dry, speed is low, there are no pedestrians or aircraft, we are late, and thus conclude that an action of pressing on the accelerator is required. Then we may observe a parked police radar unit with all other factors the same, negate the aforementioned conclusion, and apply the brake. Many of these parameters can be represented by a *true* or *not true* observation; in fact, with enough definition, all the observations could be reduced to simple true or false conditions. When we learn to drive, we are actually setting up internal responses to a set of such true/false observations in the environment.

In the industrial world, an analogous condition exists relative to the external and internal influences on a manufacturing process, and when we control a process, we are in effect teaching a control system response to a set of true/false observations. This teaching may consist of designing electronic circuits that can logically evaluate the set of true/false conditions and initiate some appropriate action. To design such an electronic system, we must first be able to mathematically express the inputs, the logical evaluation, and the corresponding outputs. *Boolean algebra* is a mathematical procedure that allows the combinations of true/false conditions in various logical operations by equations so that conclusions can be drawn. For the purposes of this text, we do not require expertise in Boolean technique, but only an operational familiarity with it that can be applied to a process-control environment.

Before a particular problem in industry can be addressed using digital electronics, it must be analyzed in terms that are amenable to the binary nature of digital techniques. Generally, this is accomplished by stating the problem in the form of a set of true/false-type conditions that must be applied to derive some desired result. These sets of conditions are then stated in the form of one or more Boolean equations. We will see in Section 3.2.4 that a Boolean equation is in a form that is readily implemented with existing digital circuits. The mathematical approach of Boolean algebra allows us to write an analytical expression to represent these stipulations. The fundamentals of Boolean algebra are summarized in Appendix 2.

Let us consider a simple example of how a Boolean equation may result from a practical problem. Consider a mixing tank for which there are three variables of interest: liquid level, pressure, and temperature. The problem is that we must signal an alarm when certain combinations of conditions occur among these variables. Referring to Figure 3.1, we denote level by A, pressure by B, and temperature by C, and assume that setpoint values have been assigned for each variable so that the Boolean variables are either **1** or **0** as the physical quantities are above or below the setpoint values. The alarm will be triggered when the Boolean variable D goes to the logic true state. The alarm conditions are

1. Low level with high pressure
2. High level with high temperature
3. High level with low temperature and high pressure

We now define a Boolean expression with AND operations that will give a $D = \mathbf{1}$ for each condition:

1. $D = \overline{A} \cdot B$ will give $D = \mathbf{1}$ for condition 1.
2. $D = A \cdot C$ will give $D = \mathbf{1}$ for condition 2.
3. $D = A \cdot \overline{C} \cdot B$ will give $D = \mathbf{1}$ for condition 3.

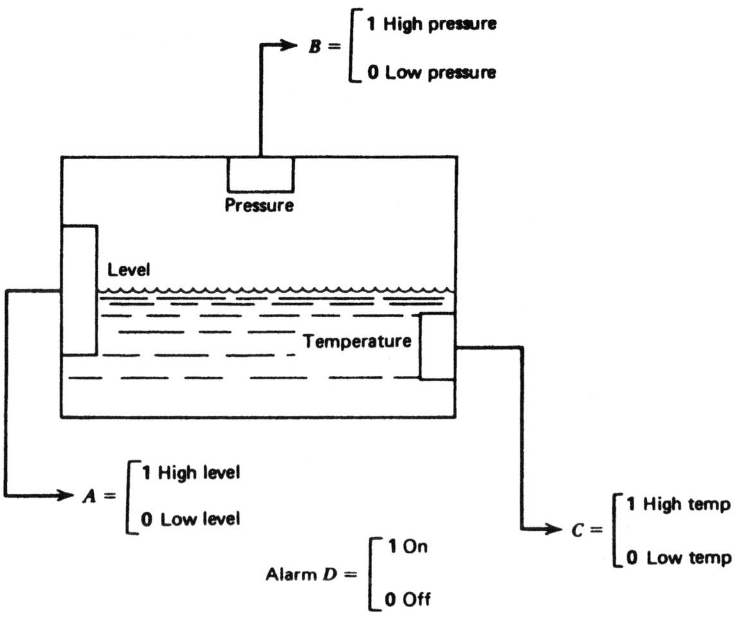

FIGURE 3.1
System for illustrating Boolean applications to control.

The final logic equation results from combining all three conditions so that if any is true, the alarm will sound ($D = \mathbf{1}$). This is accomplished with the OR operation

$$D = \overline{A} \cdot B + A \cdot C + A \cdot \overline{C} \cdot B \tag{3.2}$$

This equation would now form the starting point for a design of electronic digital circuitry that would perform the indicated operations.

3.2.4 Digital Electronics

The electronic building blocks of digital electronics are designed to operate on the binary levels present on digital signal lines. These building blocks are based on families of types of electronic circuits, as discussed in Appendix 2, that have their specific stipulations of power supplies and voltage levels of the **1** and **0** states. The basic structure involves the use of AND/OR logic and NAND/NOR logic to implement Boolean equations.

EXAMPLE 3.5 Develop a digital circuit using AND/OR gates that implements Equation (3.2).

Solution

The problem posed in Section 3.2.3 (with Figure 3.1) has a Boolean equation solution of

$$D = \overline{A} \cdot B + A \cdot C + A \cdot \overline{C} \cdot B \tag{3.2}$$

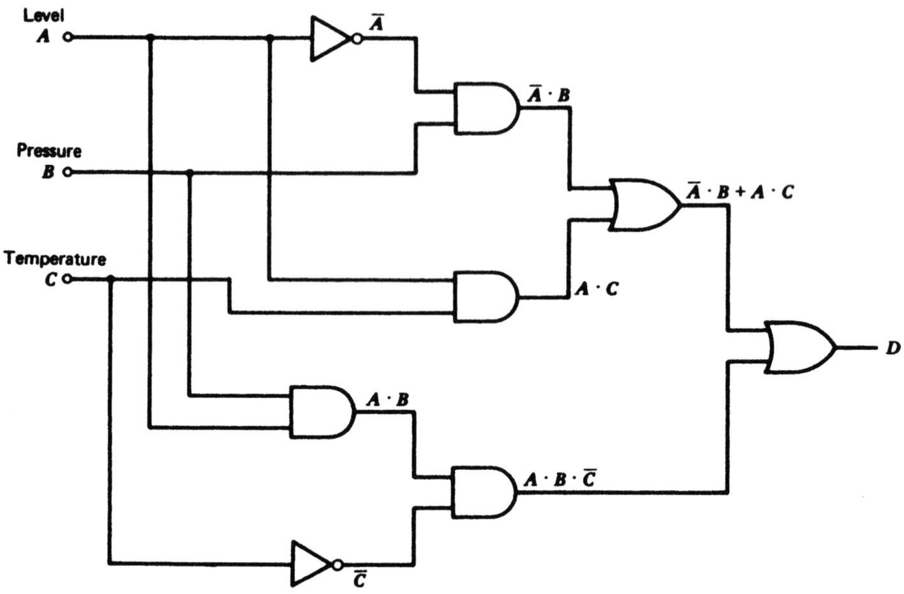

FIGURE 3.2
Solution for Example 3.5.

The implementation of this equation using AND/OR gates is shown in Figure 3.2. The AND, OR, and inverter are used in a straightforward implementation of the equation. It should be noted that Equation (3.2) can be greatly reduced by someone skilled in the art of digital logic to: $D = A \cdot C + B$. The reader can show that this can be implemented by one AND gate and one OR gate!

EXAMPLE 3.6 Repeat Example 3.5 using NAND/NOR gates.

Solution

One way to implement the equation in NAND/NOR would be to provide inverters after every gate, in effect, to convert the devices back to AND/OR gates. In this case, the circuit developed would look like Figure 3.2 but with an inverter after every gate. A second approach is to use the Boolean theorems to reformulate the equation for better implementation using NAND/NOR logic. For example, if we are to get the desired equation for D as output from a NAND gate, the inputs must have been

$$\overline{\overline{A} \cdot B + A \cdot C}$$

and

$$\overline{A \cdot B \cdot \overline{C}}$$

because NAND between these produces

$$\overline{\overline{(\overline{A} \cdot B + A \cdot C)} \cdot \overline{(A \cdot B \cdot \overline{C})}}$$

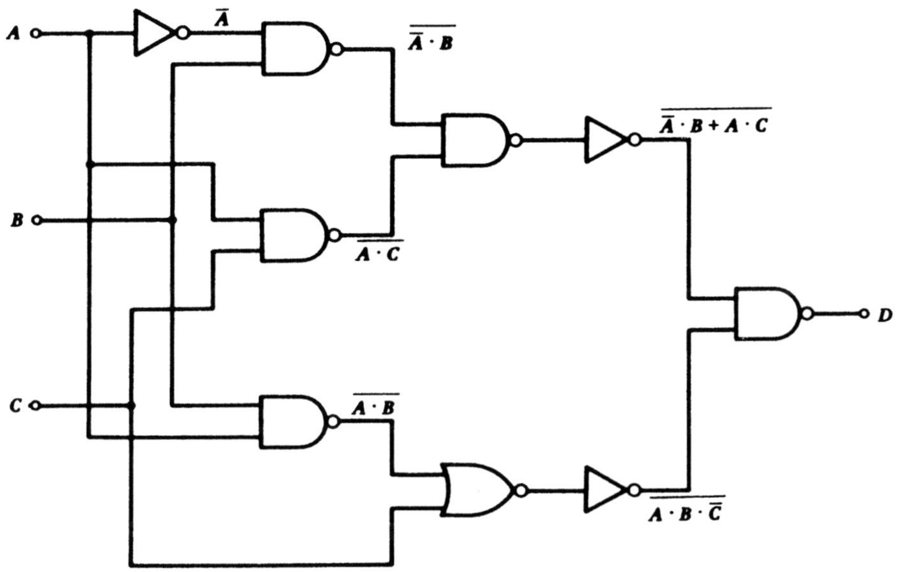

FIGURE 3.3
Solution for Example 3.6.

which, by DeMorgan's theorem, becomes

$$\bar{A} \cdot B + A \cdot C + A \cdot B \cdot \bar{C}$$

that is, the desired output. Working backward from this result allows the circuit to be realized, as shown in Figure 3.3. However, using the simplified version given in Example 3.5, we can use DeMorgan's theorem to write

$$D = A \cdot C + B = \overline{\overline{(A \cdot C)} \cdot \bar{B}}$$

The reader can now show that this can be provided by two NAND gates and one inverter.

3.2.5 Programmable Logic Controllers

The move toward digital logic techniques and computers in industrial control paralleled the development of special controllers called *programmable logic controllers* (PLCs), or simply *programmable controllers* (PCs). These devices are particularly suited to the solution control problems associated with Boolean equations and binary logic problems in general. They are a computer-based outgrowth of relay sequence controllers. Detailed treatment of this type of control system is given in Chapter 8.

3.2.6 Computer Interface

Figure 3.4 shows a simple model of a computer system. The processor is connected to external equipment via three parallel sets of digital lines. The *data lines* carry data to and from the processor. The *address lines* allow the computer to select external locations for input

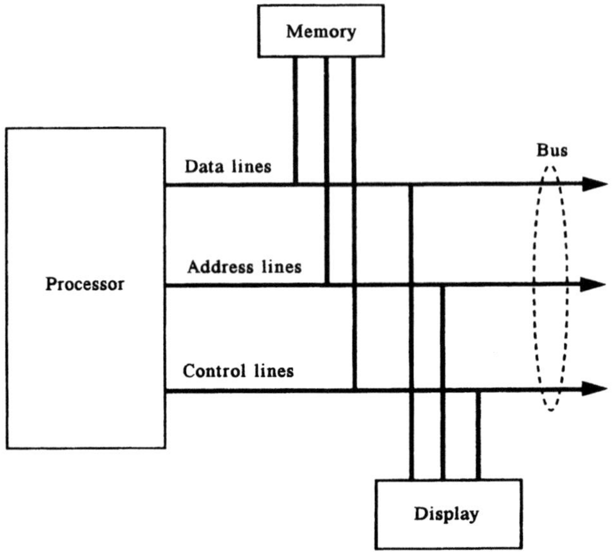

FIGURE 3.4
Generic model of a computer bus system.

and output. The *control lines* carry information to and from the computer related to operations, such as reading, writing, interrupts, and so on. This collection of lines is called the *bus* of the computer.

The term *interface* refers to the hardware connections and software operations necessary to input and output data using connections to the bus. All of the equipment connected to the computer must share the bus lines.

It is an important consideration for interface hardware that a bus line not be compromised by some external connection. This means that the external equipment must not hold a bus line in a logic state when that equipment is not using the bus. If a data line is held at **0** by some equipment even when it is not performing data transfer, then no other equipment could raise that line to the **1** state during its data transfer operations. This problem is prevented by the use of tri-state buffers.

Tri-State Buffers Isolation of a bus line is accomplished by making all connections via a special digital device called a *tri-state buffer.* This device acts like a simple switch. When the switch is closed, the logic level on its input is impressed upon the output. When open, the output is placed in a high-impedance state—that is, an open circuit.

Figure 3.5 shows how two digital signals can both be connected to a single data line through tri-state buffers. Normally, both tri-states are disabled—that is, in the high-impedance state. When the computer needs to input signal A, an enable signal, E_1, is sent to tri-state 1 so that the state of A is placed on the data line. After the computer reads the line, tri-state 1 is disabled again. Similarly, when the computer needs the state of signal B, an enable, E_2, is sent to tri-state 2 to place B on the line.

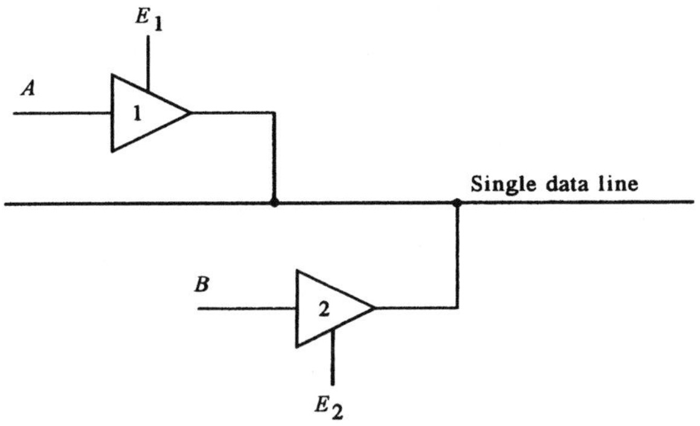

FIGURE 3.5
Tri-state buffers allow multiple signals to share a single digital line in the bus.

3.3 CONVERTERS

The most important digital tool for the process-control technologist is one that translates digital information to analog and vice versa. Most measurements of process variables are performed by devices that translate information about the variable to an analog electrical signal. To interface this signal with a computer or digital logic circuit, it is necessary first to perform an analog-to-digital (A/D) conversion. The specifics of this conversion must be well known so that a unique, known relationship exists between the analog and digital signals. Often, the reverse situation occurs, where a digital signal is required to drive an analog device. In this case, a digital-to-analog (D/A) converter is required.

3.3.1 Comparators

The most elementary form of communication between the analog and digital is a device (usually an IC) called a *comparator*. This device, which is shown schematically in Figure 3.6, simply compares two analog voltages on its input terminals. Depending on which voltage is larger, the output will be a **1** (high) or a **0** (low) digital signal. The comparator is extensively used for alarm signals to computers or digital processing systems. This element is also an integral part of the analog-to-digital and digital-to-analog converter, to be discussed in Section 3.3.2.

One of the voltages on the comparator inputs, V_a or V_b in Figure 3.6, will be the variable input, and the other a fixed value called a trip, trigger, or reference voltage. The reference

FIGURE 3.6
A basic comparator compares voltages and produces a digital output.

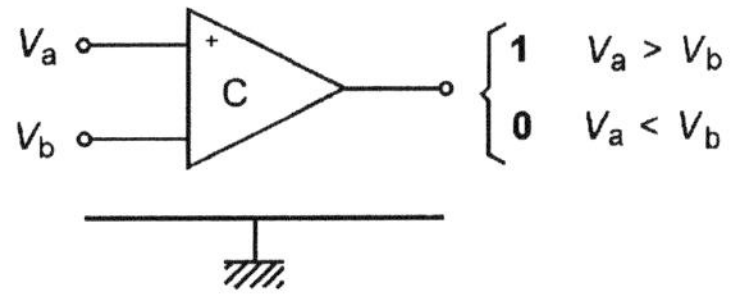

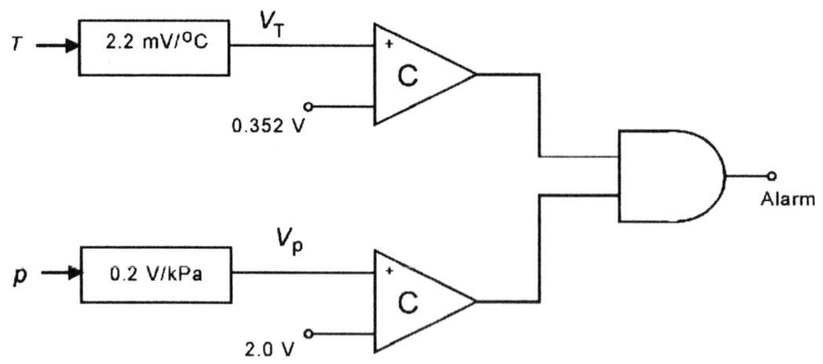

FIGURE 3.7
Diagram of a solution to Example 3.7.

value is computed from the specifications of the problem and then applied to the appropriate comparator input terminal, as illustrated in Example 3.7. The reference voltage may be provided from a divider using available power supplies.

EXAMPLE 3.7

A process-control system specifies that temperature should never exceed 160°C if the pressure also exceeds 10 kPa. Design an alarm system to detect this condition, using temperature and pressure transducers with transfer functions of 2.2 mV/°C and 0.2 V/kPa, respectively.

Solution

The alarm conditions will be a temperature signal of $(2.2\ \text{mV}/°\text{C})(160°\text{C}) = 0.352$ V coincident with a pressure signal of $(0.2\ \text{V}/\text{kPa})(10\ \text{kPa}) = 2$ V. The circuit in Figure 3.7 shows how this alarm can be implemented with comparators and one AND gate. The reference voltages could be provided from dividers.

Open-Collector Comparators Some comparator models have a special method of providing the digital output signal. Figure 3.8a shows that the output terminal of the comparator is connected internally to the collector of a transistor in the comparator and nowhere else! This is called an *open-collector output* because it is just that. Of course, even if there is base-emitter current in the transistor, no voltage will show up on the collector until it is connected to a supply through some collector resistor. In fact, this is exactly what is done in an application. Figure 3.8b shows that an external resistor is connected from the output to an appropriate power supply. This is called a collector *pull-up* resistor. Now the output terminal will show either a **0** (0 V) if the internal transistor is ON or **1** (V_s) if the internal transistor is OFF.

There are a number of advantages to using the open-collector output:

1. It is possible to use a different power source for the output. For example, suppose you want to activate a +12-V relay with the output of a comparator that operates on +5 V. By using an open-collector model, you can connect the pull-up resistor to a +12-V supply and power the relay directly from the output.

FIGURE 3.8
Many comparators use an open-collector output.

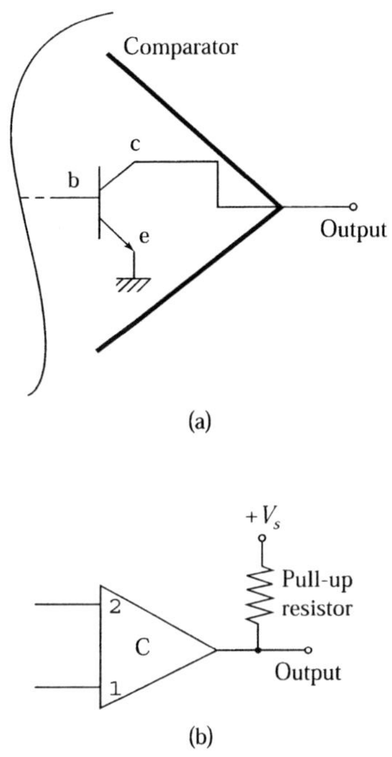

2. It is possible to OR together several comparators' outputs by connecting all open-collector outputs together and then using a common pull-up resistor. If any one of the comparator's output transistors is turned ON, the common output will go low.

Hysteresis Comparator When using comparators, there is often a problem if the signal voltage has noise or approaches the reference value too slowly. The comparator output may "jiggle" back and forth between high and low as the reference level is reached. This effect is shown in Figure 3.9. Such fluctuation of output may cause problems with the equipment designed to interpret the comparator output signal.

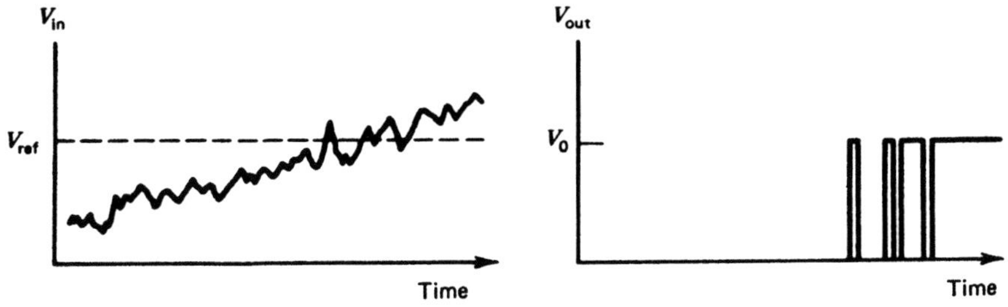

FIGURE 3.9
A comparator output will "jiggle" when a noisy signal passes through the reference voltage level.

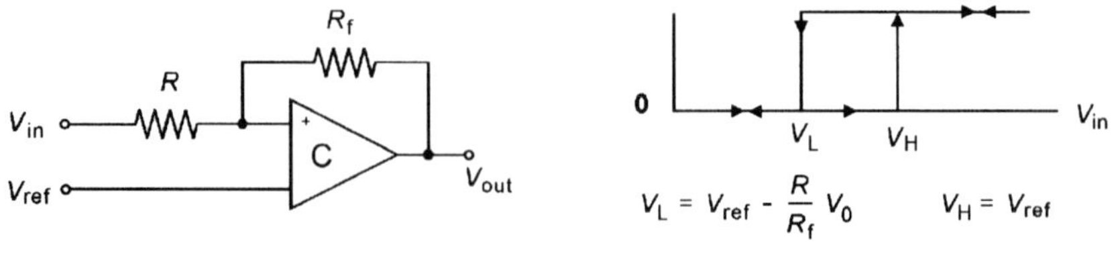

a) Hysteresis comparator circuit

b) Hysteresis comparator input-output relationship and equations

FIGURE 3.10
A generic DAC diagram, showing typical input and output signals.

This problem can often be solved by providing a *deadband* or *hysteresis* window to the reference level around which output changes occur. Once the comparator has been triggered high, the reference level is automatically reduced so that the signal must fall to some value below the old reference before the comparator goes to the low state.

There are many ways this hysteresis can be provided, but Figure 3.10a shows one common technique. Feedback resistor R_f is provided between the output and one of the inputs of the comparator, and that input is separated from the signal by another resistor, R. Under the condition that $R_f \gg R$, the response of the comparator is shown in Figure 3.10b.

The condition for which the output will go high (V_0) is defined by the condition

$$V_{in} \geq V_{ref} \tag{3.3}$$

Once having been driven high, the condition for the output to drop back to the low (0 V) state is given by the relation

$$V_{in} \leq V_{ref} - (R/R_f)V_0 \tag{3.4}$$

The deadband or hysteresis is given by $(R/R_f)V_0$, and is thus selectable by choice of the resistors, as long as this relation is satisfied. The response of this comparator is shown by the graph in Figure 3.10b. The arrows indicate increasing or decreasing input voltage.

EXAMPLE 3.8 A sensor converts the liquid level in a tank to voltage according to the transfer function (20 mV/cm). A comparator is supposed to go high (5 V) whenever the level becomes 50 cm. Splashing causes the level to fluctuate by ± 3 cm. Develop a hysteresis comparator to protect against the effects of splashing.

Solution

The nominal reference for the comparator occurs at 50 cm, which is $V_{ref} = (20\ \text{mV/cm})\,(50\ \text{cm}) = 1\ \text{V}$. The splashing, however, causes a "noise" of $(20\ \text{mV/cm}) \cdot (\pm 3\ \text{cm}) = \pm 60\ \text{mV}$. This is a total range of 120 mV. We need a deadband of at least 120 mV, but let us make it 150 mV for security. Thus, we have

$$(R/R_f)\,(5\ \text{V}) = 150\ \text{mV}$$
$$(R/R_f) = 0.03$$

If we make $R_f = 100\ \text{k}\Omega$, then $R = 3\ \text{k}\Omega$. Thus, use of these resistors, as shown in Figure 3.10, with a reference of 1 V will meet the requirement.

3.3.2 Digital-to-Analog Converters (DACs)

A DAC accepts digital information and transforms it into an analog voltage. The digital information is in the form of a binary number with some fixed number of digits. Especially when used in connection with a computer, this binary number is called a binary *word* or *computer word*. The digits are called *bits* of the word. Thus, an 8-bit word would be a binary number having eight digits, such as $\mathbf{10110110_2}$. A unipolar DAC converts a digital word into an analog voltage by scaling the analog output to be zero when all bits are zero and some maximum value when all bits are one. This can be mathematically represented by treating the binary number that the word represents as a *fractional* number. In this context, the output of the DAC can be defined using Equation (3.1) as a *scaling* of some reference voltage:

$$V_{\text{out}} = V_R[b_1 2^{-1} + b_2 2^{-2} + \cdots + b_n 2^{-n}] \tag{3.5}$$

where
V_{out} = analog voltage output
V_R = reference voltage
$b_1 b_2 \ldots b_n$ = n-bit binary word

The minimum V_{out} is zero, and the maximum is determined by the size of the binary word because, with all bits set to one, the decimal equivalent approaches V_R as the number of bits increases. Thus, a 4-bit word has a maximum of

$$V_{\text{max}} = V_R[2^{-1} + 2^{-2} + 2^{-3} + 2^{-4}] = 0.9375 V_R$$

and an 8-bit word has a maximum of

$$V_{\text{max}} = V_R[2^{-1} + 2^{-2} + 2^{-3} + 2^{-4} + 2^{-5} + 2^{-6} + 2^{-7} + 2^{-8}] = 0.9961 V_R$$

An alternative equation to Equation (3.5) is often easier to use. This is based on noting that the expression in brackets in Equation (3.5) is really just the fraction of total counting states possible with the n bits being used. With this recognition, we can write

$$V_{\text{out}} = \frac{N}{2^n} V_R \tag{3.6}$$

where N = base 10 whole-number equivalent of DAC input

Suppose an 8-bit converter with a 5.0-V reference has an input of $\mathbf{10100111_2}$, or A7H. If this input is converted to base 10, we get $N = 167_{10}$ and $2^8 = 256$. From Equation (3.6), the output of the ADC will be

$$V_{\text{out}} = \frac{167}{256} 5.0 = 3.2617 \text{ volts}$$

EXAMPLE 3.9 What is the output voltage of a 10-bit DAC with a 10.0-V reference if the input is **(a)** $\mathbf{0010110101_2}$ = 0B5H, **(b)** 20FH? What input is needed to get a 6.5-V output?

Solution

Let's use Equation (3.5) for part **(a)** and Equation (3.6) for part **(b)**. Thus, for the 0B5H input, we have

$$V_{out} = 10.0[2^{-3} + 2^{-5} + 2^{-6} + 2^{-8} + 2^{-10}]$$
$$V_{out} = 10.0[0.1767578]$$
$$V_{out} = 1.767578 \text{ V}$$

For **(b)**, we have 20FH $= 527_{10}$ and $2^{10} = 1024$, so

$$V_{out} = (527/1024)10.0$$
$$V_{out} = (0.514648)10.0$$
$$V_{out} = 5.14648 \text{ V}$$

We can use Equation (3.6) to determine the input needed to get a 6.5-V output by solving for N,

$$N = 2^n(V_{out}/V_R)$$
$$N = 1024(6.5/10)$$
$$N = 665.6$$

The fact that there is a fractional remainder tells us that we cannot get exactly 6.5 V from the converter. The best we can do is get an output for $N = 665 = 299\text{H}$ or $666 = 29\text{AH}$. The outputs for these two inputs are 6.494 V and 6.504 V, respectively. The only way to get exactly 6.5 V of output would be to change the value of the reference slightly.

Bipolar DAC Some DACs are designed to output a voltage that ranges from plus to minus some maximum when the input binary ranges over the counting states. Although computers frequently use 2s complement to represent negative numbers, this is not common with DACs. Instead, a simple *offset-binary* is frequently used, wherein the output is simply biased by half the reference voltage of Equation (3.6). The bipolar DAC relationship is then given by

$$V_{out} = \frac{N}{2^n} V_R - \frac{1}{2} V_R \tag{3.7}$$

Notice that if $N = 0$, the output voltage will be given by the minimum value, $V_{out}(\min) = -V_R/2$. However, the maximum value for N is equal to $(2^n - 1)$, so that the maximum value of output voltage will be

$$V_{out}(\max) = \frac{(2^n - 1)}{2^n} V_R - \frac{1}{2} V_R = \frac{1}{2} V_R - \frac{V_R}{2^n}$$

EXAMPLE 3.10 A bipolar DAC has 10 bits and a reference of 5 V. What outputs will result from inputs of 04FH and 2A4H? What digital input gives a zero output voltage?

Solution

The inputs of 04FH and 2A4H can easily be converted to base 10 numbers 79_{10} and 676_{10}. Then, from Equation (3.7), we find

$$V_{out} = \frac{79}{1024}(5) - \frac{(5)}{2} = -2.1142578\text{ V}$$

$$V_{out} = \frac{676}{1024}(5) - \frac{(5)}{2} = 0.80078\text{ V}$$

The zero occurs when Equation (3.7) equals zero. Solving for N gives

$$0 = \frac{N}{1024}(5) - \frac{(5)}{2}$$

or $N = 512_{10} = 200\text{H} = \mathbf{1000000000_2}$.

Conversion Resolution The conversion resolution is a function of the reference voltage and the number of bits in the word. The more bits, the smaller the change in analog output for a 1-bit change in binary word, and hence the better the resolution. The smallest possible change is simply given by

$$\Delta V_{out} = V_R 2^{-n} \qquad \textbf{(3.8)}$$

where ΔV_{out} = smallest output change
V_R = reference voltage
n = number of bits in the word

Thus, a 5-bit word D/A converter with a 10-V reference will provide changes of $\Delta V_{out} = (10)(2^{-5}) = 0.3125$ V per bit.

EXAMPLE 3.11 Determine how many bits a D/A converter must have to provide output increments of 0.04 V or less. The reference is 10 V.

Solution

One way to find the solution is to continually try word sizes until the resolution falls below 0.04 V per bit. A more analytical procedure is to use Equation (3.8):

$$\Delta V = 0.04 = (10)(2^{-y})$$

Any n larger than the integer part of the exponent of two in this equation will satisfy the requirement. Taking logarithms

$$\log(0.04) = \log[(10)(2^{-y})]$$

$$\log(0.04) = \log(10) - y\log 2$$

$$y = \frac{\log(10) - \log(0.04)}{\log 2}$$

$$y = 7.966$$

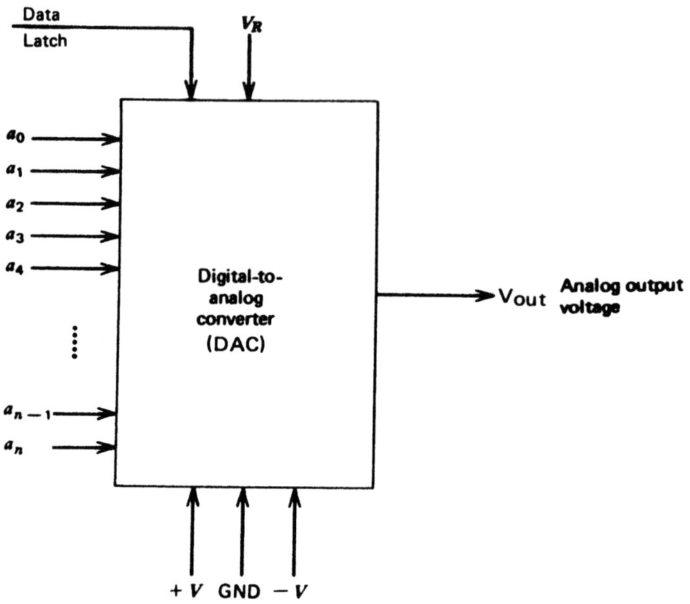

FIGURE 3.11
A generic DAC diagram, showing typical input and output signals.

Thus, a $n = 8$ will be satisfactory. This can be proved by Equation (3.8):

$$\Delta V_{out} = (10)(2^{-8})$$
$$\Delta V_{out} = \mathbf{0.0390625\ V}$$

DAC Characteristics For modern applications, most DACs are integrated circuit (IC) assemblies, viewed as a black box having certain input and output characteristics. In Figure 3.11, we see the essential elements of the DAC in terms of required input and output. The associated characteristics can be summarized as follows by referring to this figure:

1. *Digital input* Typically, digital input is a parallel binary word composed of a number of bits specified by the device specification sheet. TTL logic levels are usually required, unless otherwise noted.
2. *Power supply* The power supply is bipolar at a level of ±12 to ±18 V as required for internal amplifiers. Some DACs operate from a single supply.
3. *Reference supply* A reference supply is required to establish the range of output voltage and resolution of the converter. This must be a stable, low-ripple source. In some units, an internal reference is provided.
4. *Output* The output is a voltage representing the digital input. This voltage changes in steps as the digital input changes by bits, with the step determined by Equation (3.8). The actual output may be bipolar if the converter is designed to interpret negative digital inputs.

5. *Offset* Because the DAC is usually implemented with op amps, there may be the typical output offset voltage with a zero input (see Section 2.4.2). Typically, connections will be provided to facilitate a zeroing of the DAC output with a zero word input.
6. *Data latch* Many DACs have a data latch built into their inputs. When a logic command is given to latch data, whatever data are on the input bus will be latched into the DAC, and the analog output will be updated for that input data. The output will stay at that value until new digital data are latched into the input. In this way, the input of the DAC can be connected directly onto the data bus of a computer, but it will be updated only when a latch command is given by the computer.
7. *Conversion time* A DAC performs the conversion of digital input to analog output virtually instantaneously. From the moment that the digital signal is placed on the inputs to the presence of the analog output voltage is simply the propagation time of the signal through internal amplifiers. Typically, settling time of the internal amplifiers will be a few microseconds.

DAC Structure Generally speaking, a DAC is used as a black box, and no knowledge of the internal workings is required. There is some value, however, in briefly showing how such conversion can be implemented. The simplest conversion uses a series of op amps for input for which the gains have been selected to provide an output as given by Equation (3.5). The most common variety, however, uses a *resistive ladder network* to provide the transfer function. This is shown in Figure 3.12 for the case of a 4-bit converter. With the *R-2R* choice of resistors, it can be shown through network analysis that the output voltage is given by Equations (3.5) or (3.6). The switches are analog electronic switches.

EXAMPLE 3.12 A control valve has a linear variation of opening as the input voltage varies from 0 to 10 V. A microcomputer outputs an 8-bit word to control the valve opening using an 8-bit DAC to generate the valve voltage.

a. Find the reference voltage required to obtain a full open valve (10 V).
b. Find the percentage of valve opening for a 1-bit change in the input word.

Solution

a. The full open-valve condition occurs with a 10-V input. If a 10-V reference is used, a full digital word $\mathbf{11111111_2}$ will not quite give 10 V, so we use a larger reference. Thus, we have

$$V_{\text{out}} = V_R(b_1 2^{-1} + b_2 2^{-2} + \cdots + b_8 2^{-8})$$

$$10 = V_R\left(\frac{1}{2} + \frac{1}{4} + \cdots + \frac{1}{256}\right)$$

$$V_R = \frac{10}{0.9961} = 10.039 \text{ V}$$

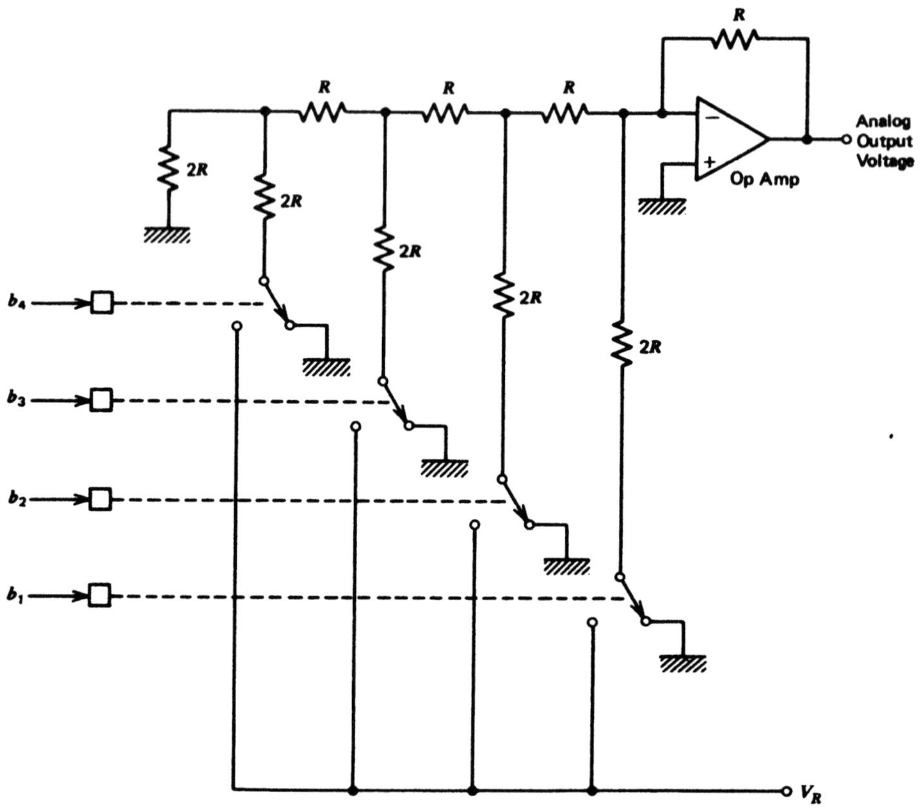

FIGURE 3.12
A typical DAC is often implemented using a ladder network of resistors.

b. The percentage of valve change per step is found first from

$$\Delta V_{\text{out}} = V_R 2^{-8}$$

$$\Delta V_{\text{out}} = (10.039)\frac{1}{256}$$

$$\Delta V_{\text{out}} = 0.0392 \text{ V}$$

Thus,

$$\text{percent} = \frac{(0.0392)(100)}{10} = \mathbf{0.392\%}$$

Data Output Boards It is now common and convenient to obtain a printed circuit board that plugs into a personal computer expansion slot and is a complete data output system. The board has all necessary DACs, address decoding, and bus interface. In most cases, the supplier of the board also provides elementary software—often written in C, BASIC, or assembly language—as necessary to use the board for data output.

3.3.3 Analog-to-Digital Converters (ADCs)

Although there are sensors that provide a direct digital signal output and more are being developed, most still convert the measured variable into an analog electrical signal. With the growing use of digital logic and computers in process control, it is necessary to employ an ADC to provide a digitally encoded signal for the computer. The transfer function of the ADC can be expressed in a similar way to that of the DAC as given in Equation (3.5). In this case, however, the interpretation is reversed. The ADC will find a fractional binary number that gives the closest approximation to the fraction formed by the input voltage and reference.

$$b_1 2^{-1} + b_2 2^{-2} + \cdots + b_n 2^{-n} \leq \frac{V_{in}}{V_R} \qquad \textbf{(3.9)}$$

where
$b_1 b_2 \ldots b_n$ = n-bit digital output
V_{in} = analog input voltage
V_R = analog reference voltage

We use an *inequality* in this equation because the fraction on the right can change continuously over all values, but the fraction derived from the binary number on the left can change only in fixed increments of $\Delta N = 2^{-n}$. In other words, the only way the left side can change is if the LSB changes from **1** to **0** or from **0** to **1**. In either case, the fraction changes by only 2^{-n} and nothing in between. Therefore, there is an inherent uncertainty in the input voltage producing a given ADC output, and that uncertainty is given by

$$\Delta V = V_R 2^{-n} \qquad \textbf{(3.10)}$$

Minimum and Maximum Voltages Equation (3.9) shows that if the ratio of input voltage to reference is less than ΔV, then the digital output will be all **0**s, (i.e., $\mathbf{0000 \ldots 000_2}$). The LSB will not change until the input voltage becomes at least equal to ΔV, and then the output will be $\mathbf{0000 \ldots 001_2}$. Therefore, if the ADC output is all zeros, you know *only* that V_{in} is less than $V_R 2^{-n}$, so it could even be a negative voltage, for example.

Now notice that the MSB changes from **0** to **1** when the input voltage becomes equal to or greater than $V_R - \Delta V$. Therefore, if the ADC output is all **1**s (i.e., $\mathbf{1111 \ldots 1111_2}$) then you know *only* that V_{in} is greater than $V_R(1 - 2^{-n})$.

This uncertainty must be taken into account in design applications. If the problem under consideration specifies a certain resolution in analog voltage, then the word size and reference must be selected to provide this in the converted digital number.

EXAMPLE 3.13 Temperature is measured by a sensor with an output of 0.02 V/°C. Determine the required ADC reference and word size to measure 0° to 100°C with 0.1°C resolution.

Solution

At the maximum temperature of 100°C, the voltage output is

$$(0.02\ \text{V/°C})\,(100\text{°C}) = 2\ \text{V}$$

so a 2-V reference is used.

A change of 0.1°C results in a voltage change of

$$(0.1°\text{C})\,(0.02\ \text{V}/°\text{C}) = 2\ \text{mV}$$

so we need a word size where

$$0.002\ \text{V} = (2)\,(2^{-y})$$

Choose a size n that is the integer part of y plus 1. Thus, solving with logarithms, we find

$$y = \frac{\log(2) - \log(0.002)}{\log 2}$$

$$y = 9.996 \approx 10$$

so a **10**-bit word is required for this resolution. A 10-bit word has a resolution of

$$V = (2)\,(2^{-10})$$
$$V = 0.00195\ \text{V}$$

which is better than the minimum required resolution of 2 mV.

Notice that the output actually changes from $\mathbf{1111111110_2}$ to $\mathbf{1111111111_2}$ at a voltage of

$$V_R(1 - 2^{-n}) = (2\ \text{V})\,(1 - 2^{-10}) = 1.9980\ \text{V}$$

which corresponds to a temperature of $(1.9980\ \text{V})/(0.02\ \text{V}/°\text{C}) = 99.90°\text{C}$. This means we are actually measuring temperature between 0.1°C and 99.9°C.

EXAMPLE 3.14 Find the digital word that results from a 3.127-V input to a 5-bit ADC with a 5-V reference.

Solution

The relationship between input and output is given by Equation (3.9). Thus, we are to encode a fractional number of V_{in}/V_R, or

$$b_1 2^{-1} + b_2 2^{-2} + \cdots + b_5 2^{-5} = \frac{3.127}{5} = 0.6254$$

Using the method of successive multiplication defined in Section 3.2.2, we find

$$\begin{aligned}
0.6254(2) &= 1.2508 &\quad \therefore b_1 &= 1\\
0.2508(2) &= 0.5016 &\quad \therefore b_2 &= 0\\
0.5016(2) &= 1.0032 &\quad \therefore b_3 &= 1\\
0.0032(2) &= 0.0064 &\quad \therefore b_4 &= 0\\
0.0064(2) &= 0.0128 &\quad \therefore b_5 &= 0
\end{aligned}$$

so that the output is $\mathbf{10100_2}$.

Equation (3.9) can be written in a simpler fashion by expressing the fractional binary number as the fraction of counting states, as was done for the DAC. In this case, the base 10 value of the digital output can be expressed as

$$N = \mathrm{INT}\left(\frac{V_{\mathrm{in}}}{V_R}2^n\right) \tag{3.11}$$

where INT( ) means to take the integer part of the quantity in brackets. This is not a round-off, but rather a truncation, so that INT(3.3) = 3 and INT(3.99) = 3 also. The value of N is then converted to hex and/or binary to demonstrate the ADC output. In the previous example, we would have

$$N = \mathrm{INT}\left(\frac{3.127}{5}2^5\right) = \mathrm{INT}(20.0128) = 20_{10}$$

or 14 H = $\mathbf{10100_2}$, as already found.

EXAMPLE 3.15

The input to a 10-bit ADC with a 2.500-V reference is 1.45 V. What is the hex output? Suppose the output was found to be 1B4H. What is the voltage input?

Solution

We will use Equation (3.11) to find the solution to these questions. For the first part, we can form the expression

$$N = \mathrm{INT}((1.45/2.5)2^{10})$$
$$N = \mathrm{INT}(593.92)$$
$$N = 593$$
$$N = 251\,\mathrm{H}$$

So the output of the ADC is 251H for a 1.45-V input. To get the voltage input for a 1B4H-output, we solve Equation (3.11) for the voltage:

$$V_{\mathrm{in}} = \frac{N}{2^n}V_R$$

A conversion yields 1B4H = 436_{10}.

$$V_{\mathrm{in}} = (436/1024)2.50$$
$$V_{\mathrm{in}} = 1.06445\ \mathrm{V}$$

However, it is important to realize that any voltage from this to 1.06445 + 2.5/1024 = 1.06689 will give the same output, 1B4H. So the correct answer to the question is that the input voltage lies in the range 1.06445 to 1.06689 V.

Bipolar Operation A bipolar ADC is one that accepts bipolar input voltage for conversion into an appropriate digital output. The most common bipolar ADCs provide an output called *offset-binary*. This simply means that the normal output is shifted by half the scale so that all-zeros corresponds to the negative maximum input voltage instead of zero. In equation form, the relation would be written as

$$N = \text{INT}\left[\left(\frac{V_{\text{in}}}{V_R} + \frac{1}{2}\right)2^n\right] \tag{3.12}$$

From this equation, you can see that if $V_{\text{in}} = -V_R/2$, the output is zero, $N = 0$. If $V_{\text{in}} = 0$, the output is half of 2^n. The output will be the maximum count when the input is $V_R/2 - V_R 2^n$. For example, for 8 bits with a 10.0-V reference, the step size is $\Delta V_{\text{in}} = (10)2^8 \approx 0.039\ V$. Looking at the possible states, we would have

$$\begin{array}{ll}
V_{\text{in}} = -5.000 & N = \mathbf{00000000_2} \\
V_{\text{in}} = -4.961 & N = \mathbf{00000001_2} \\
\quad\text{etc.} & \\
V_{\text{in}} = -0.039 & N = \mathbf{01111111_2} \\
V_{\text{in}} = 0.000 & N = \mathbf{10000000_2} \\
V_{\text{in}} = +\ 0.039 & N = \mathbf{10000001_2} \\
\quad\text{etc.} & \\
V_{\text{in}} = +\ 4.961 & N = \mathbf{11111111_2}
\end{array}$$

There is an asymmetry to the result so that the converter cannot represent the full range from minus to plus $V_R/2$.

EXAMPLE 3.16 What are the hex and binary output of a bipolar 8-bit ADC with a 5.00-V reference for inputs of −0.85 V and +1.5 V? What input voltage would cause an output of 72H?

Solution

Using Equation (3.12), we get

$$N = \text{INT}((1/5.00)[-0.85 + 2.50]2^8)$$
$$N = \text{INT}(84.48)$$
$$N = 84_{10}$$
$$N = 54\text{H} = \mathbf{01010100_2}$$

and

$$N = \text{INT}((1/5.00)[1.5 + 2.50]256)$$
$$N = \text{INT}(204.8)$$
$$N = 204_{10}$$
$$N = \text{CCH} = \mathbf{11001100_2}$$

To get an output of 72H, we solve Equation (3.12) for V_{in}:

$$V_{\text{in}} = (N/2^n)V_R - V_R/2$$
$$V_{\text{in}} = (114/256)5.00 - 2.50$$
$$V_{\text{in}} = -0.2734\ \text{V}$$

But of course the actual answer is any voltage between −0.2734 V and $(-0.2734 + 5/256) = -0.2539$ V.

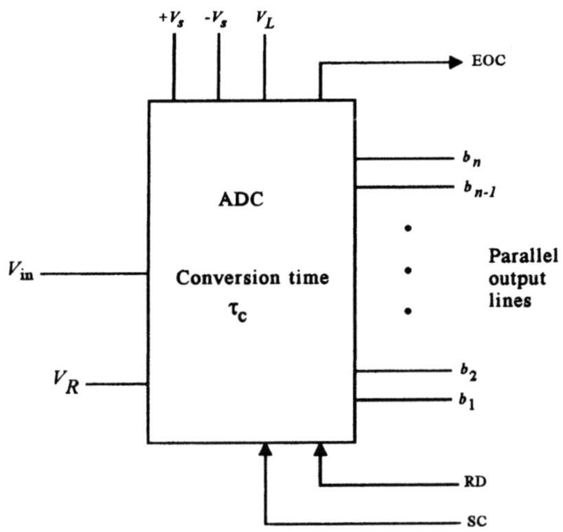

FIGURE 3.13
A generic ADC diagram, showing typical input and output signals and noting the conversion time.

ADC Characteristics Figure 3.13 shows a generic ADC with all the typical connections. It is quite possible, and even appropriate in many cases, to regard the ADC as simply a black box with certain input and output characteristics. The following list summarizes the important characteristics of the ADC:

1. *Analog voltage input* This is for connection of the voltage to be converted. As will be explained later, it is important that this voltage be constant during the conversion process.
2. *Power supplies* Generally, an ADC requires bipolar supply voltages for internal op amps and a digital logic supply connection.
3. *Reference voltage* The reference voltage must be from a stable, well-regulated source. Special, integrated circuit reference-source voltages are available for this purpose.
4. *Digital outputs* The converter will have n output lines for connection to digital interface circuitry. Generally, the levels are typical TTL values for definition of the high and low states. It is common for the output lines to be tri-state outputs so that the ADC can be connected directly to a bus.
5. *Control lines* The ADC has a number of control lines that are single-bit digital inputs and outputs designed to control operation of the ADC and allow for interface to a computer. The most common lines are:
   a. **SC** (Start-convert) This is a digital input to the ADC that starts the converter on the process of finding the correct digital outputs for the given analog voltage input. Typically, conversion starts on a falling edge.
   b. **EOC** (End-of-convert) This is a digital output from the ADC to receiving equipment, such as a computer. Typically, this line will be high during the conversion process. When the conversion is complete, the line will go low. Thus, the falling edge indicates that the conversion is complete.
   c. **RD** (Read) Since the output is typically buffered with tri-states, even though the conversion is complete, the correct digital results do not appear on the

output lines. The receiving equipment must take the RD line low to enable the tri-states and place the data on the output lines.

6. *Conversion time* This is not an input or an output, but a very important characteristic of ADCs. A typical ADC does not produce the digital output instantaneously when the analog voltage is applied to its input terminal. The ADC must sequence through a process to find the appropriate digital output, and this process takes time. This is one of the reasons that handshaking lines are required. Figure 3.14 shows a typical timing diagram for taking a sample of data via an ADC.

The existence of a finite conversion time complicates the use of ADCs in data acquisition. The computer cannot have a data input at any time; rather, it must request an input, wait for the ADC to perform a conversion, and then input the data.

ADC Structure Most ADCs are available in the form of integrated circuit (IC) assemblies that can be used as a black box in applications. To fully appreciate the characteristics of these devices, however, it is valuable to examine the standard techniques employed to perform the conversions. There are two methods in use that represent very different approaches to the conversion problem.

Parallel-Feedback ADC The parallel-feedback A/D converter employs a feedback system to perform the conversion, as shown in Figure 3.15. Essentially, a *comparator* is used to compare the input voltage, V_x, to a feedback voltage, V_F, that comes from a DAC as shown. The comparator output signal drives a logic network that steps the digital output (and hence DAC input) until the comparator indicates the two signals are the same within the resolution of the converter. The most popular parallel-feedback converter is the *successive approximation* device. The logic circuitry is such that it successively sets and tests each bit, starting with the most significant bit of the word. We start with all bits zero. Thus, the first operation will be to set $b_1 = \mathbf{1}$ and test $V_F = V_R 2^{-1}$ against V_x through the comparator.

If V_x is greater, then b_1 will be **1**, b_2 is set to **1**, and a test is made of V_x versus $V_V = V_R(2^{-1} + 2^{-2})$, and so on.

If V_x is less than $V_R 2^{-1}$, then b_1 is reset to zero, b_2 is set to **1**, and a test is made for V_x versus $V_R 2^{-2}$. This process is repeated to the least significant bit of the word. The operation can be illustrated best through an example.

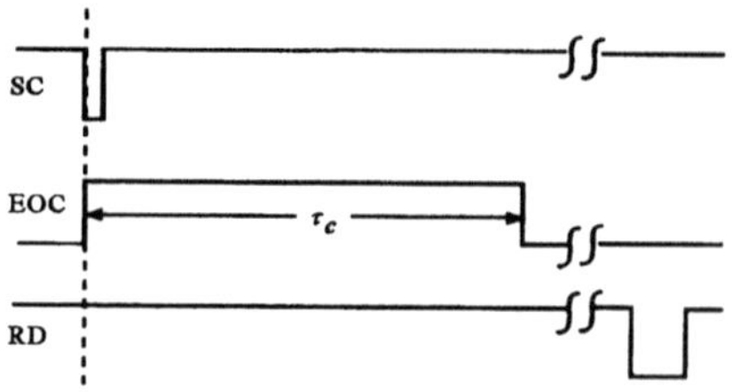

FIGURE 3.14
A typical data-acquisition timing diagram using an ADC. The read operation may occur at any time after the end-of-convert has been issued by the ADC.

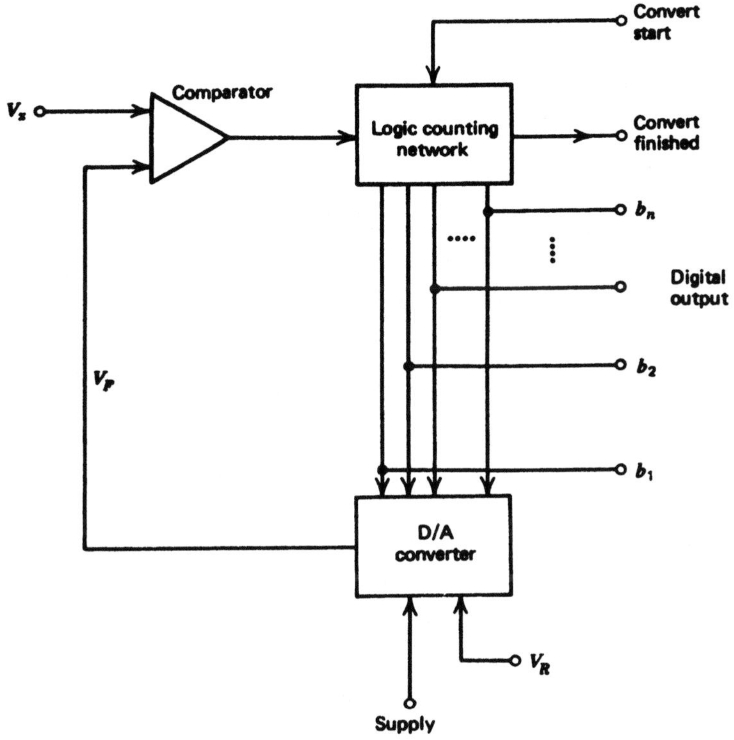

FIGURE 3.15
One common method of implementing an ADC is the successive approximation of parallel-feedback system using an internal DAC.

EXAMPLE 3.17 Find the successive approximation ADC output for a 4-bit converter to a 3.217-V input if the reference is 5 V.

Solution

Following the procedure outlined, we have the following operations: Let $V_x = 3.217$; then

(1) Set $b_1 = \mathbf{1}$	$V_F = 5(2^{-1}) = 2.5$ V
$V_x > 2.5$	leave $b_1 = \mathbf{1}$
(2) Set $b_2 = \mathbf{1}$	$V_F = 2.5 + 5(2^{-2}) = 3.75$
$V_x < 3.75$	reset $b_2 = \mathbf{0}$
(3) Set $b_3 = \mathbf{1}$	$V_F = 2.5 + 5(2^{-3}) = 3.125$
$V_x > 3.125$	leave $b_3 = \mathbf{1}$
(4) Set $b_4 = \mathbf{1}$	$V_F = 3.125 + 5(2^{-4})$
$V_x < 3.4375$	reset $b_4 = \mathbf{0}$

By this procedure, we find the output is a binary word of $\mathbf{1010}_2$.

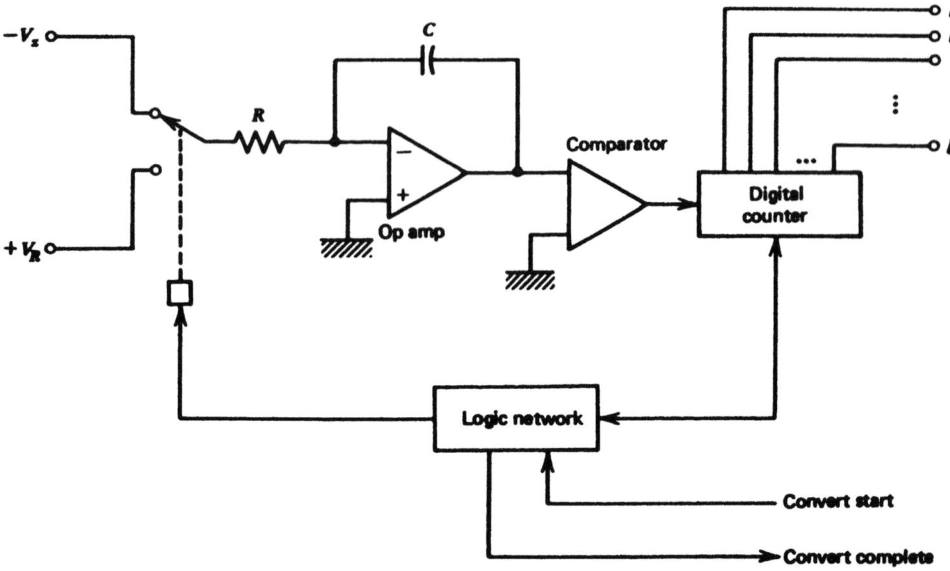

FIGURE 3.16
The dual-slope ADC uses an op amp integrator, comparator, and counter. This is commonly used in digital voltmeters.

The conversion time of successive approximation-type ADCs is on the order of 1 to 5 μs per bit. Thus, a low-priced 8-bit ADC might require 5 μs/bit for a total conversion time of about 40 μs. A higher-quality (and price) 12-bit might be able to perform the full conversion in only 15 μs.

These conversion times depend on a clock that is internal to the ADC and not crystal controlled. Thus, there will be variation of the conversion time from unit to unit.

Ramp ADC The ramp-type A/D converters essentially compare the input voltage against a linearly increasing ramp voltage. A binary counter is activated that counts ramp steps until the ramp voltage equals the input. The output of the counter is then the digital word representing conversion of the analog input. The ramp itself is typically generated by an op amp integrator circuit, discussed in Section 2.5.8.

Dual-Slope Ramp ADC This ADC is the most common type of ramp converter. A simplified diagram of this device is shown in Figure 3.16. The principle of operation is based on allowing the input signal to drive the integrator for a fixed time, T_1, thus generating an output of

$$V_1 = \frac{1}{RC}\int V_x dt \tag{3.13}$$

or, because V_x is constant,

$$V_1 = \frac{1}{RC} = T_1 V_x \tag{3.14}$$

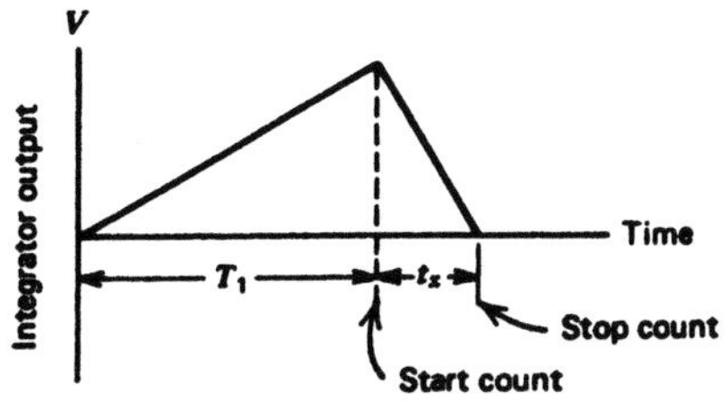

FIGURE 3.17
A typical timing diagram of a dual-slope ADC. Since both slopes depend upon *R* and *C*, the ADC output is independent of the values of these components.

After time T_1, the input of the integrator is electronically switched to the reference supply. The comparator then sees an input voltage that decreases from V_1 as

$$V_2 = V_1 - \frac{1}{RC}\int V_R dt \tag{3.15}$$

or, because V_R is constant and V_1 is given from Equation (3.14),

$$V_2 = \frac{1}{RC}T_1 V_x - \frac{1}{RC}tV_R \tag{3.16}$$

A counter is activated at time T_1 and counts until the comparator indicates $V_2 = 0$, at which time t_x [Equation (3.16)] indicates that V_x will be

$$V_x = \frac{t_x}{T_1}V_R \tag{3.17}$$

Thus, the counter time, t_x, is linearly related to V_x and is independent of the integrator characteristics—that is, *R* and *C*. This procedure is shown in the timing diagram in Figure 3.17. Conversion *start* and *stop* digital signals are also used in these devices, and (in many cases) internal or external references may be used.

EXAMPLE 3.18 A dual-slope ADC as shown in Figure 3.16 has $R = 100\ \text{k}\Omega$ and $C = 0.01\ \mu\text{F}$. The reference is 10 V, and the fixed integration time is 10 ms. Find the conversion time for a 6.8-V input.

Solution

We find the voltage after an integration time of 10 ms as

$$V_1 = \frac{1}{RC}T_1 V_x$$

$$V_1 = \frac{(10\ \text{ms})(6.8\ \text{V})}{(100\ \text{k}\Omega)(0.1\mu\text{F})}$$

$$V_1 = 6.8\ \text{V}$$

Then we find the time required to integrate this to zero as $V_2 = 0$ in

$$V_2 = \frac{T_1 V_x}{RC} - \frac{t_x}{RC}V_R$$

Thus,

$$t_x = \frac{T_1 V_x}{V_R}$$

$$t_x = \frac{(10\text{ ms})(6.8\text{ V})}{10\text{ V}}$$

$$t_x = 6.8\text{ ms}$$

The total conversion time is then 10 ms + 6.8 ms = **16.8 ms**.

Example 3.18 illustrates an important characteristic of the dual-slope ADC. It has a much longer conversion time than a successive approximation type. In fact, the conversion time for the dual-slope is frequently from tens to several hundreds of milliseconds.

One of the most common applications of the dual-slope ADCs is in digital multimeters. Here, input circuitry converts the input voltage into an appropriate range for the ADC. The ADC performs conversions continuously; that is, when one conversion is finished, the output is latched into a display register and another conversion is started. In applications such as this, a few hundred milliseconds' conversion time is plenty fast and allows for the display to be updated several times per second.

EXAMPLE 3.19

A measurement of temperature using a sensor that outputs 6.5 mV/°C must measure to 100°C. A 6-bit ADC with a 10-V reference is used. **(a)** Develop a circuit to interface the sensor and the ADC. **(b)** Find the temperature resolution.

Solution

To measure to 100°C means the sensor output at 100°C will be

$$(6.5\text{ mV/°C})(100\text{°C}) = 0.65\text{ V}$$

a. The interface circuit must provide a gain so that at 100°C the ADC output is **111111**. The input voltage that will provide this output is found from

$$V_x = V_R(a_1 2^{-1} + a_2 2^{-2} + \cdots + a_6 2^{-6})$$

$$V_x = 10\left(\frac{1}{2} + \frac{1}{4} + \cdots + \frac{1}{64}\right)$$

$$V_x = \mathbf{9.84375\ V}$$

Thus, the required gain must provide this voltage when the temperature is 100°C.

$$\text{gain} = \frac{9.84375}{0.65}$$

$$\text{gain} = \mathbf{15.14}$$

The op amp circuit of Figure 3.18 will provide this gain.

b. The temperature resolution can be found by working backward from the least significant bit (LSB) voltage change of the ADC:

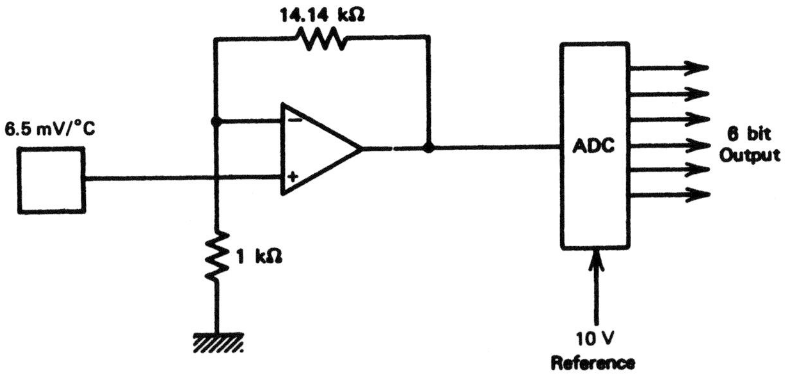

FIGURE 3.18
Analog circuit for Example 3.19.

$$\Delta V = V_R 2^{-n}$$
$$\Delta V = (10)(2^{-6}) = 0.15625 \text{ V}$$

Working back through the amplifier, this corresponds to a sensor change of

$$\Delta V_T = \frac{0.15625}{15.14} = 0.01032 \text{ V}$$

or a temperature of

$$\Delta T = \frac{0.01032 \text{ V}}{0.0065 \text{ V/°C}} = \mathbf{1.59}\text{°C}$$

Conversion-Time Consequences The finite conversion time of the ADC has serious consequences on the rate of change of signals presented for conversion. An ADC performs the conversion process by referring back to the input signal while the conversion is taking place. Obviously, if the input is changing while this process is taking place, errors will occur.

Consequently, the ADC output will be in error if the magnitude of the input voltage changes by more than one LSB voltage, ΔV, during the time of conversion, τ_c. This is a serious limitation. Since the change in time of the input voltage is just the derivative, this condition can be written in the form

$$\frac{dV_{\text{in}}}{dt} \leq \frac{\Delta V}{\tau_c} = \frac{V_R}{2^n \tau_c} \quad \textbf{(3.18)}$$

Consider, for example, a 10-bit ADC with a 5.0-V reference and a 20-μs conversion time. According to Equation (3.18), the maximum rate of change of the input voltage for this converter would be

$$\frac{dV_{\text{in}}}{dt} \leq \frac{5.0}{2^{10}(20 \times 10^{-6})} \approx 244 \text{ V/s}$$

This result, 244 V/s, doesn't seem so bad. Let's put this in perspective by asking what frequency it would correspond to if the input were a sinusoidal voltage,

$$V_{in} = V_0 \sin(\omega t)$$

Taking the derivative and using Equation (3.18) gives the result

$$\omega V_0 \cos(\omega t) \leq \frac{V_R}{2^n \tau_c}$$

or, since the maximum value of the cosine function is unity, we get the condition

$$\omega \leq \frac{V_R}{2^n \tau_c V_0} \tag{3.19}$$

or, in terms of frequency where $\omega = 2\pi f$,

$$f \leq \frac{V_R}{2^{n+1} \pi \tau_c V_0} \tag{3.20}$$

Returning to the 244 V/s, let's assume that the full range is in use, $V_0 = V_R$. Then we find that the maximum angular frequency is

$$\omega \leq \frac{1}{2^{10}(20 \times 10^{-6})} \approx 48.8 \text{ rad/s}$$

or, in terms of frequency,

$$f \leq \frac{\omega}{2\pi} \approx 7.8 \text{ Hz}$$

This is a remarkable result! It says that this 20-μs converter cannot find a 10-bit representation of an oscillating signal greater than 7.8 Hz. Yet it is true, and means that if the frequency is greater than this, there will be errors in the lower-order bits; that is, it is no longer converting to 10 bits.

EXAMPLE 3.20 An 8-bit, bipolar ADC with a 5-V reference will be used to take samples of a triangular wave as shown in Figure 3.19. What is the maximum frequency of the wave if the ADC conversion time is 12 μs?

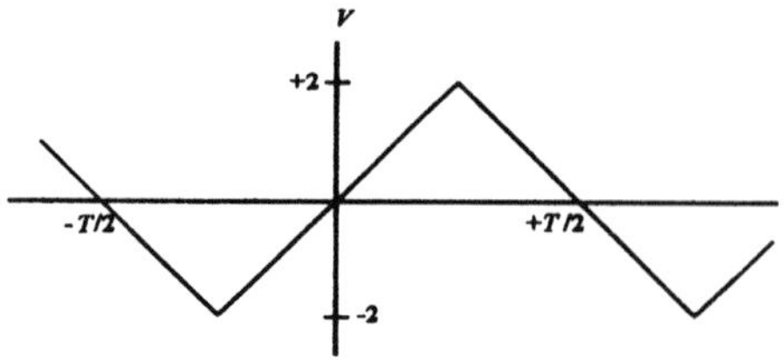

FIGURE 3.19
Input signal for Example 3.20.

Solution

The solution is found from the condition expressed by Equation (3.18). From Figure 3.19 it is clear that the derivative of the input signal is simply the slope of the triangular wave,

$$\frac{dV_{\text{in}}}{dt} = \frac{2}{T/4} = \frac{8}{T} = 8f$$

where T is the period and f is the frequency. Then, from Equation (3.18),

$$8f \leq \frac{5}{2^8(12 \times 10^{-6})} = 1627.6 \text{ Hz}$$

or $f \leq 203.5$ Hz.

Obviously, the limitations on frequency described previously severely limit the application of ADCs and computer data-acquisition systems. There is a solution, however. What is needed is simply that the signal not change during the conversion process. Therefore, the answer is to hold the value constant during that process. This is accomplished with a sample-and-hold (S/H) circuit.

Sample-and-Hold The basic concept of the sample-and-hold circuit is shown in Figure 3.20, where the S/H is connected to the input of an ADC. When the electronic switch is closed, the capacitor voltage will "track" the input voltage, $V_c(t) = V_{\text{in}}(t)$. At some time, t_s, when a conversion of the input voltage is desired, the electronic switch is opened, isolating the capacitor from the input. Thus, the capacitor will hold (stay charged) to the voltage when the switch opened, $V_c = V_{\text{in}}(t_s)$.

The voltage follower allows this voltage to be impressed upon the ADC input, but the capacitor does not discharge because of the very high input impedance of the follower. The start-convert is then issued, and the conversion proceeds with the input voltage remaining constant, so the problem of Equation (3.18) does not arise.

When the conversion is complete, the electronic switch is reclosed, and tracking continues until another conversion is needed. Figure 3.21 shows how $V_{\text{in}}(t)$ and $V_c(t)$ would appear during a sample collection sequence of a sinusoidal signal.

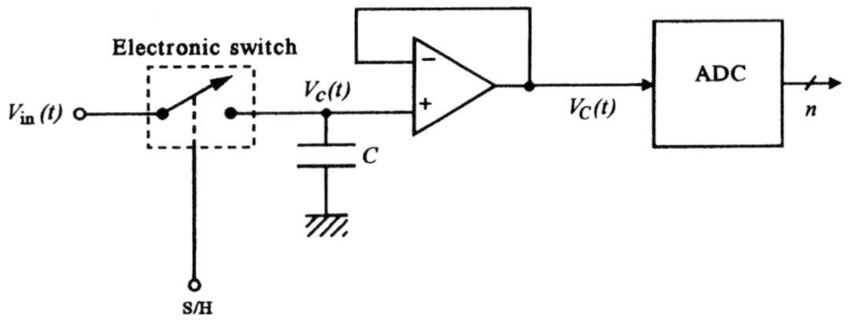

FIGURE 3.20
The basic concept of a sample-and-hold circuit for use with the ADC.

FIGURE 3.21
The sampled signal is literally "held" during the ADC conversion process.

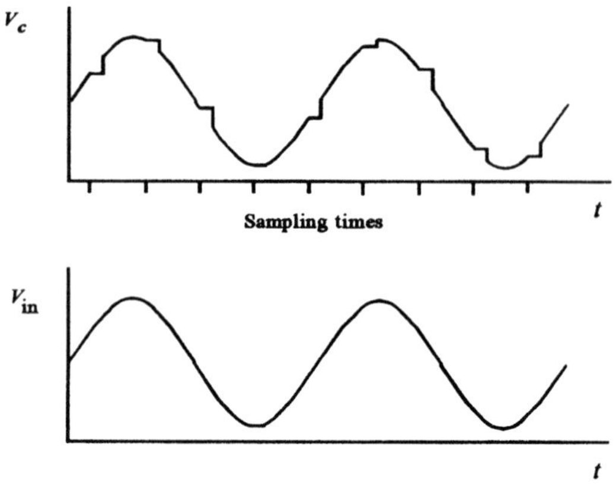

Practical S/H Issues Because of the severe limitations of Equation (3.18), ADC systems virtually always use S/H circuits on the input. Whenever an SC is issued to the ADC, a "hold" is also issued to the S/H circuit. When the EOC is issued by the ADC, this usually automatically switches the S/H back to the "sample" mode. Figure 3.22 shows how IC S/H circuits are often implemented using FETs as the electronic switches.

There are several practical issues associated with the nonideal electrical characteristics of the elements involved in S/H circuits.

- There is a nonzero resistance path from the input voltage to the capacitor. This resistance consists of the output resistance of the source of V_{in} and the finite "ON" resistance of the electronic switch (the FET, for example).

Figure 3.23a shows a model of the sample mode of the S/H. You can see that this constitutes a low-pass filter with $R = R_s + R_{ON}$ and the capacitor. Therefore, there will be a limitation on the frequency that the system can track. The capacitor voltage will be down 3 dB (0.707) at the critical frequency given by

$$f_c = \frac{1}{2\pi(R_s + R_{ON})C} \tag{3.21}$$

Most commercial S/H circuits reduce this limitation by using a voltage follower before the switch, since it has very low output resistance.

FIGURE 3.22
A S/H often uses a FET as an electronic switch.

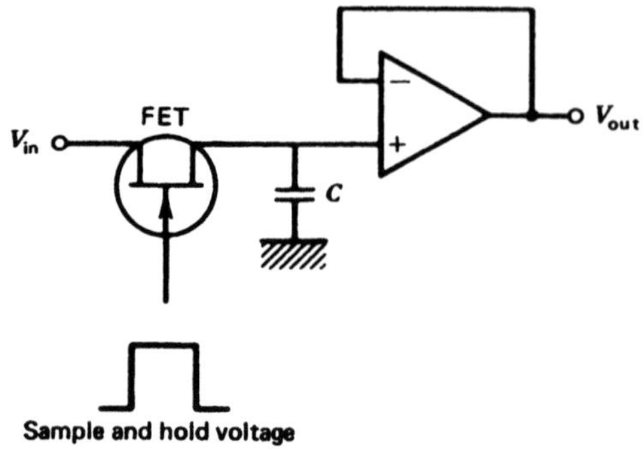

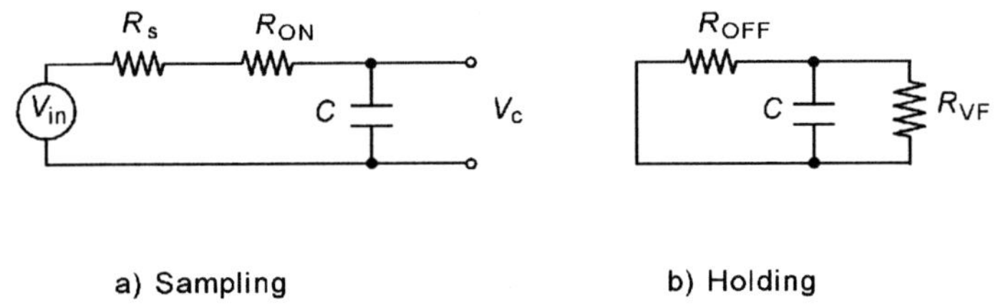

FIGURE 3.23
During (a) sampling and (b) holding, equivalent circuit resistance creates nonideal effects.

- During the hold mode, the capacitor is shunted by the parallel combination of the "OFF" resistance of the switch and the input resistance of the voltage follower, as shown in Figure 3.23b. In this case, there will be a gradual discharge, or "droop," of the capacitor voltage. The droop time constant for this discharge is simply

$$\tau_D = \frac{R_{\mathrm{OFF}} R_{VF}}{R_{\mathrm{OFF}} + R_{VF}} C \tag{3.22}$$

The condition here is that the droop of the capacitor voltage cannot have a slope larger than that given by Equation (3.18). For the exponential decay of the capacitor voltage, this leads to the condition

$$\frac{V_C}{\tau_D} \leq \frac{V_R}{2^n \tau_c} \tag{3.23}$$

or

$$\tau_D \geq 2^n \tau_c \frac{V_C}{V_R} \tag{3.24}$$

Equation (3.24) shows how the value of capacitor C can be selected to ensure that droop will not exceed the limitation imposed by Equation (3.18). Usually, the equation is evaluated for $V_C = V_R$ as a worst-case condition.

EXAMPLE 3.21 A S/H will be used with a 12-bit, unipolar ADC with a 30-μs conversion time. The S/H switch ON resistance is 10 Ω, and its OFF resistance is 10 MΩ. The voltage follower input resistance is also 10 MΩ, while the signal source output resistance is 50 Ω.

a. What value of capacitor should be used?
b. Determine the sampling cutoff frequency.

Solution

a. Equation (3.24), under the worst-case condition that $V_C = V_R$, will determine the minimum droop time, $\tau_D \geq 2^{12}(30 \times 10^{-6}) = 0.12288$ s. Now Equation (3.22) will allow determination of C,

$$\frac{(10^7)(10^7)}{(10^7 + 10^7)} C \geq 0.12288$$

which gives $C \geq 0.025\ \mu$F.

b. From this result and Equation (3.22), the critical frequency during sampling is found to be

$$f_c = \frac{1}{(2\pi)(50 + 10)(0.025 \times 10^{-6})} = 108 \text{ kHz}$$

- Other important characteristics of the S/H circuit are the acquisition time (τ_{acq}) and the aperture time (τ_{ap}). The acquisition time is the time required for the S/H to reacquire the signal when changing from the hold to the sample mode. Typical times are on the order of microseconds. This places limits on the frequency with which samples can be taken.

The aperture time is the time between when a command to hold is given and the actual signal level is held. This delay means that the value presented to the ADC is not exactly the value at the time the sample was requested. This time is typically less than 1 μs.

In general, the minimum time between samples taken from a S/H and ADC system is given by the sum

$$T = \tau_c + \tau_{acq} + \tau_{ap} \tag{3.25}$$

or, if expressed in terms of the maximum throughput frequency, $f_{max} = 1/T$.

EXAMPLE 3.22 A S/H has a 50-ns aperture time and a 4-μs acquisition time, and the ADC has a 40-μs conversion time. What is the maximum throughput frequency?

Solution

The frequency is given by the inverse of the time between samples, given by Equation (3.25):

$$T = 40\ \mu\text{s} + 0.05\ \mu\text{s} + 4\ \mu\text{s} = 44.05\ \mu\text{s}$$

Thus, the frequency is $f_{max} = 22.7$ kHz.

Microprocessor-Compatible ADCs Just as with DACs, a whole line of ADCs have been developed that interface easily with microprocessor-based computers. The ADCs have built-in tri-state outputs so that they can be connected directly to the data bus of the computer. Data from the ADC are placed on the data bus lines only when the computer issues an appropriate enable command (often called a READ). Figure 3.24 shows how the ADC appears when connected to the environment of the microprocessor-based computer. The ADC appears much the same as memory. In some cases, an ADC input is actually taken by the computer using a memory-read instruction.

The decoding circuitry is necessary to provide the start-convert command, to input the convert-complete response from the ADC, and to issue the tri-state enable back to the ADC.

3.3.4 Frequency-Based Converters

There is another important method by which an analog sensor signal can be converted into a digital signal. This is based upon converting the sensor signal into a variable frequency

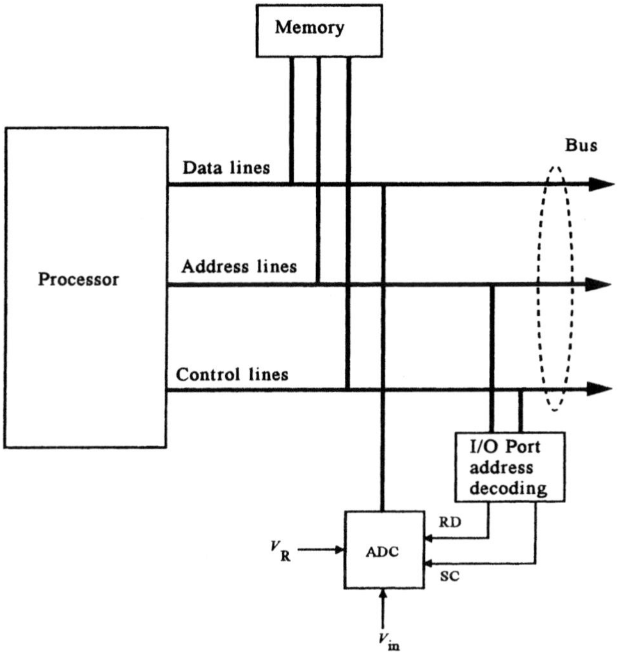

FIGURE 3.24
An ADC can be interfaced directly to the computer bus if it has tri-state outputs. Address decoding is required so the ADC can be operated by computer software.

and then using this frequency as input to a counter for a fixed interval of time. The output of the counter is then a measure of the frequency and thus the sensor signal.

Figure 3.25 shows a diagram that describes the essential elements of this type of analog-to-digital converter. An as yet not identified device converts the sensor signal into a proportional frequency, f_s. This frequency signal is typically a square wave, as suggested in the figure. The square wave is fed to an n-bit counter, which counts every rising (or falling) edge of the wave and hence every cycle. The counter often has a latch on the output that allows the counter to be accumulating a new count of input frequency while still maintaining the previous output.

A conversion cycle starts with a start-convert (SC) signal from the computer. This clears the counter and triggers a one-shot convert multivibrator (MV), which controls the operation. The MV is a simple digital IC that, when triggered, outputs a single pulse of some desired time duration. The count MV outputs a pulse of duration T_c when triggered by the SC. This pulse acts as a start/stop signal to the counter and so defines the time over which the frequency signal will be counted. If the input frequency is high, the count will be high; if the frequency is low, the count will be low. The latch MV issues a short pulse to latch the latest count into the output latches when the count time, T_c, is finished. The falling edge of T_c also signals the computer that a conversion is complete (EOC). The computer can then read the count by enabling the tri-state output of the counter latch with the RD signal taken low.

A typical design starts from the range of frequency of the converted sensor signal, f_{min} to f_{max}. For maximum resolution, we then make the count time, T_c, such that if the

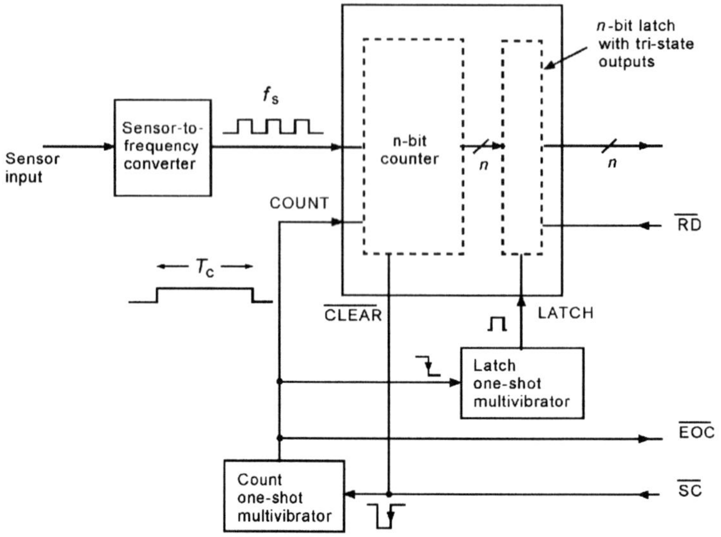

FIGURE 3.25
General diagram of a frequency-based analog-to-digital converter.

sensor signal produces the maximum frequency, the count will also be at its maximum. For an n-bit binary counter, the maximum count is 2^n-1, so the relation we need is

$$T_c = \frac{2^n - 1}{f_{max}} \tag{3.26}$$

The counter output for any other frequency is simply $N = fT_c$.

Example 3.23 illustrates this concept.

EXAMPLE 3.23 A sensor signal is converted to a frequency that varies from 2.0 to 20 kHz. This signal is to be converted into an 8-bit digital signal. Specify the count time, T_c. What is the range of count output for the sensor signal's frequency range?

Solution

An 8-bit counter has a maximum output of $255_{10}(\mathbf{11111111_2})$. Therefore, when the frequency is at its maximum, the count time must allow the counter to reach a full 255. From Equation (3.26),

$$T_c = \frac{2^8 - 1}{20{,}000 \text{ Hz}} = 0.01275 \text{ s}$$

So the one-shot multivibrator is configured to provide a 12.75-ms pulse. When the frequency is at the minimum of 2.0 kHz, the count will be

$$N = (2000 \text{ Hz})(0.01275 \text{ s}) = 25.5$$

or simply $25_{10} = \mathbf{00011001_2}$, since the counter can count only in integers.

Sensor-to-Frequency Conversion Of course, this technique of analog-to-digital conversion depends upon converting the measured variable information into a variable frequency. In fact, this is not so hard to do. Common ICs exist that readily convert voltage or current to frequency. An example is the LM331 voltage-to-frequency IC. Figure 3.26 shows a generic LM331 circuit for controlling output frequency with an input voltage. For this circuit, the output frequency is determined by the relation

$$f_{\text{out}} = \frac{R_S}{R_L} \frac{1}{R_t C_t} \frac{V_{\text{in}}}{2.09} \tag{3.27}$$

where the components are defined in the schematic. Actually, the frequency cannot really be zero even when V_{in} is zero. It is typically small, perhaps 10 Hz. Generally R_S is in the range of 10 to 20 kΩ and is adjustable to allow fine-tuning of the frequency-to-voltage scale factor. R_L is used to provide a discharge path for the 1-μF capacitor. It typically has a value of about 100 kΩ. The supply voltage can be up to 40 V. The output is open collector (OC), so a pull-up resistor is required to the logic supply voltage (for example, +5 V for TTL).

As an example, suppose we want an input voltage of 0 to 5.0 V to generate a frequency from $\approx$ 0 to 10 kHz. If we use typical values of $R_S = 15\ \text{k}\Omega$ and $R_L = 100\ \text{k}\Omega$, Equation (3.27) provides

$$10{,}000 = \frac{15\ \text{k}\Omega}{100\ \text{k}\Omega} \frac{5.0}{2.09} \frac{1}{R_t C_t}$$

so we get $R_t C_t = 3.59 \times 10^{-5}$ s. Thus, picking $C_t = 0.01\ \mu\text{F}$ means $R_t = 3.59\ \text{k}\Omega$.

The value of output frequency also depends upon resistors and a capacitor, so by fixing the control voltage, the frequency can also be made to depend upon these values. Therefore, if the sensor is a varying resistance or capacitance, it can often be used directly in the circuit to provide a frequency that varies with the sensed quantity. You must remember, however, that the variation of frequency with either resistance or capacitance is often nonlinear.

Another common IC that can be used for converting resistance or capacity to a frequency is the 555 timer. This IC has an output frequency that depends upon an external

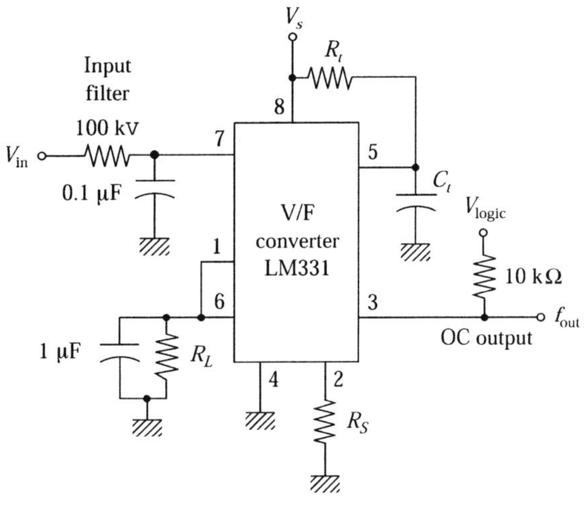

FIGURE 3.26
The LM331 is a common voltage-to-frequency converter useful in frequency-based ADCs.

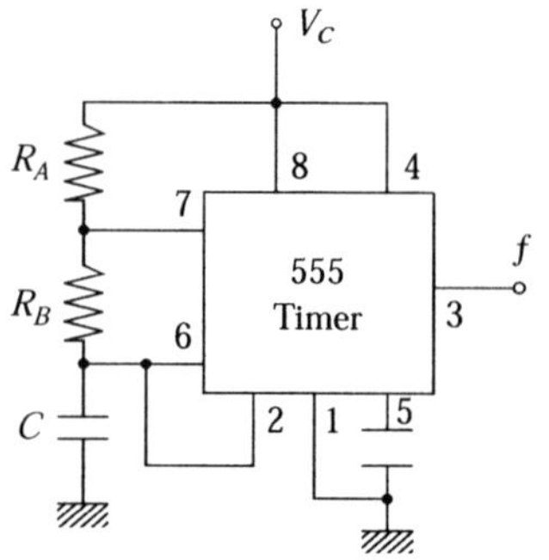

FIGURE 3.27
The 555 timer is useful for generation of a frequency that depends upon resistance or capacity.

resistor and capacitor. Figure 3.27 shows a standard 555 timer circuit that produces a steady, nonsymmetrical rectangular wave with frequency

$$f = \frac{1}{0.693(R_A + 2R_B)C} \tag{3.28}$$

The following example illustrates how the 555 can be used to produce a varying frequency and analog-to-digital conversion.

EXAMPLE 3.24

A resistor varies from 36 to 4 kΩ as light intensity varies from 1.5 to 10 W/m^2. Devise a frequency-based 10-bit ADC of this signal. What are the frequencies at 1.5 and 10 W/m^2? Plot the counter output versus light intensity. Notice the nonlinearity.

Solution

We want the maximum frequency to produce a count of $2^{10} - 1 = 1023$, but maximum frequency is unspecified, so we cannot determine the count time. We get to pick one, so let us pick the count time to be, say, 10 ms. Then we get the maximum frequency as $f_{max} = (1023/10\text{ ms}) = 102{,}300$ Hz. This frequency will occur when the sensor is at 4 kΩ:

$$102{,}300\text{ Hz} = \frac{1}{0.693(4\text{ k}\Omega + 2R_B)C}$$

This leaves another value we can pick. Let's make $R_B = 2$ kΩ; then we get

$$C = \frac{1}{0.693(4\text{ k}\Omega + 4\text{ k}\Omega)(102{,}300\text{ Hz})} = 0.0018\ \mu\text{F}$$

We get the frequency at 1.5 W/m^2 by using 36 kΩ for R_A.

$$f_{min} = \frac{1}{0.693(36\text{ k}\Omega + 4\text{ k}\Omega)(0.0018\ \mu\text{F})} = 20{,}042\text{ Hz}$$

So the count is $N = (20{,}042\text{ Hz})(0.01\text{ s}) = 200.42$, or simply 200. A plot of the count versus light intensity is shown in Figure 3.28. You can see the nonlinearity, which results from the resistor being in the denominator of the frequency equation.

In general, it is important for you to realize that there are simple IC circuits that allow conversion of the sensed quantity directly to frequency. If this frequency is counted using a

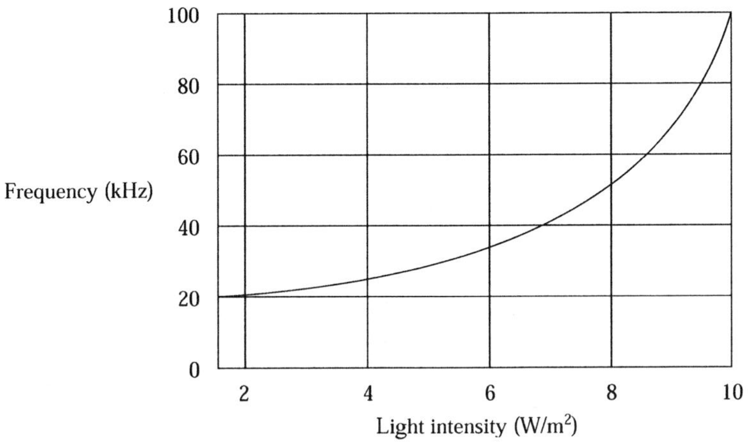

FIGURE 3.28
Response from Example 3.24.

simple system, as shown in Figure 3.26, then analog-to-digital conversion has resulted. The conversion time of these circuits is determined by T_c and the maximum sensor frequency. Most of the IC circuits can have a maximum frequency of around 100 kHz. This means, for example, that an 8-bit converter would need a count time of about 2.55 ms. So you can see that conversion times are long compared with successive approximation types of ADCs. For most process-control applications, the slower conversion time is not a limitation.

3.4 DATA-ACQUISITION SYSTEMS

Microprocessor-based personal computers (PCs) are used extensively to implement direct digital control in the process industries. These familiar desktop computers are designed much like the system shown in Figure 3.24, using a bus that consists of the data lines, address lines, and control lines. All communication with the processor is via these bus lines. This includes essential equipment such as RAM, ROM, disk, and CD-ROM.

The PC also connects the bus lines to a number of printed circuit board (PCB) sockets, using an industry standard configuration of how the bus lines are connected to the socket. These sockets are referred to as *expansion slots*. Many special types of peripheral equipment such as fax/modem boards, game boards, and network connection boards are designed on PCBs that plug into these expansion slots.

Special PCBs called data-acquisition systems (DASs) have been developed for the purpose of providing for input and output of analog data. These are used when the PC is to be used in a control system. The following paragraphs provide general information about the hardware and software of data-acquisition systems.

3.4.1 DAS Hardware

The hardware features of a general data-acquisition system are shown in Figure 3.29. Although there is variation from manufacturer to manufacturer, the system shown in this figure and described herein demonstrates the essential features of DASs.

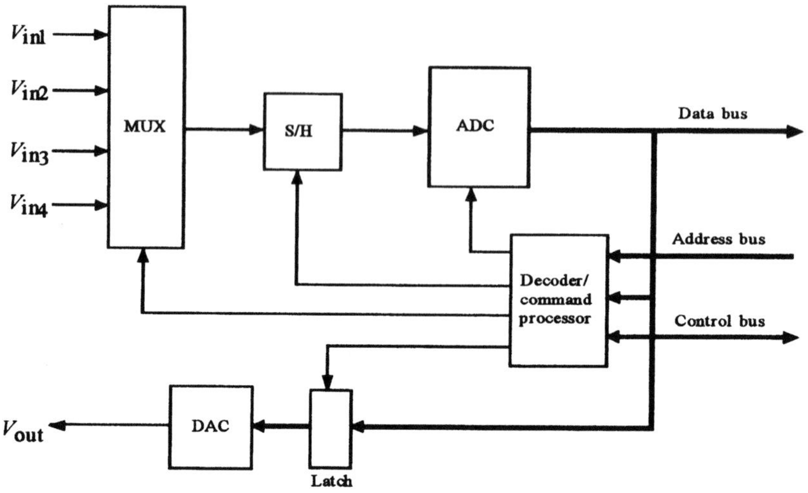

FIGURE 3.29
Typical layout of a data-acquisition board for use in a personal computer expansion slot.

ADC and S/H The DAS typically has a high-speed, successive approximation-type ADC and a fast S/H circuit. Whenever the DAS is requested to obtain a data sample, the S/H is automatically incorporated into the process. The ADC conversion time constitutes the major part of the data sample acquisition time, but the S/H acquisition time must also be considered to establish maximum throughput.

Analog Multiplexer The analog multiplexer (MUX) allows the DAS to select data from a number of different sources. The MUX has a number of input channels, each of which is connected to a different analog input voltage source. The MUX acts like a multiple set of switches, as illustrated in Figure 3.30, arranged in such a fashion that any one of the input channels can be selected to provide its voltage to the S/H and ADC. In some cases, the DAS can be programmed to take channel samples sequentially.

Address Decoder/Command Processor The computer can select to input a sample from a given channel by sending an appropriate selection on the address lines and control lines of the computer bus. These are decoded to initiate the proper sequence of commands to the MUX, ADC, and S/H. Another common feature is the ability to program the DAS to take a number of samples from a channel with a specified time between samples. In this case, the computer is notified by interrupt when a sample is ready for input.

DAC and Latch For output purposes, the DAS often includes a latch and DAC. The address decoder/command processor is used to latch data written to the DAS, which is then converted to an appropriate analog signal by the DAC.

3.4.2 DAS Software

The process of selecting a channel and initiating a data input from that channel involves some interface between the computer and the DAS. This interface is facilitated by software

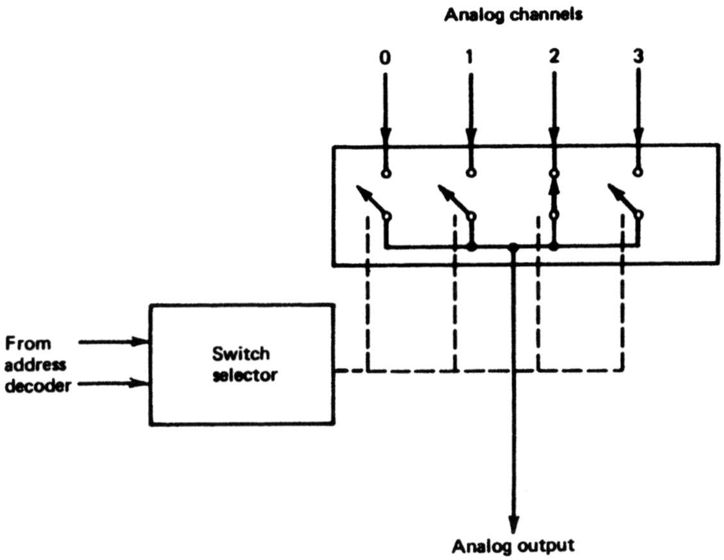

FIGURE 3.30
An analog multiplexer acts as a multiposition switch for selecting particular inputs to the ADC.

that the computer executes. The software can be written by the user, but is often also provided by the DAS manufacturer in the form of programs on disk.

Figure 3.31 is a flowchart of the basic sequence of operations that must occur when a sample is required from the DAS. The following paragraphs describe each element of the sequence.

Generally, the DAS is mapped into a base port address location in the PC system. In the PC, this address can be from 000H to FFFH, but many addresses are reserved for use by the processor and other peripherals. A common address for input/output (I/O) systems such as the DAS is port 300H.

The sequence starts with selection of a channel for input. This is accomplished by a write to the DAS decoder that identifies the required channel. The MUX then places that channel input voltage at the S/H input.

The software then issues a start-convert (SC) command according to the specifications of the DAS. This is often accomplished by a write to some base + offset address. The DAS internally activates the hold mode of the S/H and starts the converter.

The end-of-convert (EOC) is provided in a status register in the DAS. The contents of this status register can be read by the processor by a port input of a base + offset address. The appropriate bit is then tested by the software to deduce whether the EOC has been issued.

Once the EOC has been issued, the software can input the data itself by a read of an appropriate address, again a base + offset, which enables tri-states, placing the ADC output on the data bus.

There is one problem with the operations described, shown in Figure 3.31. If the DAS fails, the computer will be locked in the loop waiting for the EOC to be issued. One way to resolve this is to add an additional timer loop for a time greater than the conversion time of

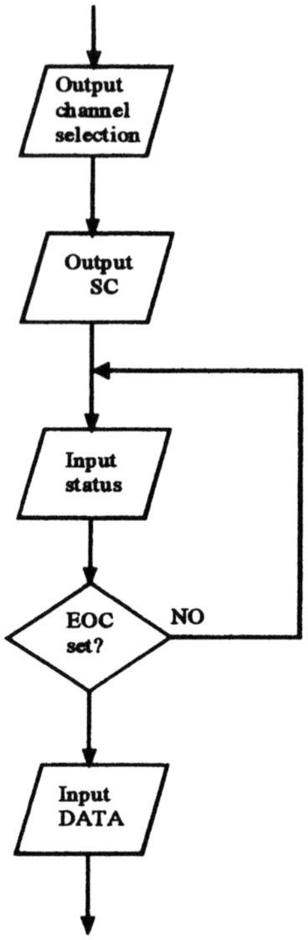

FIGURE 3.31
Software for data acquisition involves operations to start the ADC, test the EOC, and input the data.

the ADC. If the EOC is not detected prior to time-out, an error is announced, and the computer is returned to an error-handling routine.

In some cases, the EOC detection is handled by an interrupt service routine. In this way, the computer is free to execute other software until the interrupt occurs. Then the data is input. Again, there needs to be a system to detect that an EOC was not provided to protect against DAS failure.

EXAMPLE 3.25

A DAS has the following specifications:

1. 8 channels
2. 8-bit, bipolar ADC with a 5.0-V reference and a 25-μs conversion time
3. S/H with a 10-μs acquisition time
4. 8-bit unipolar DAC with a 10.0-V reference.

ADDRESSING: BASE = 000H to FFFH by switches

BASE + 0: READ inputs data sample
WRITE selects input channel:
b_0 set selects channel 0, and so forth, to
b_7 set selects channel 7

BASE + 1: READ inputs ADC status with EOC indicated by b_7 going low
WRITE initializes the DAS if b_7 is high, issues SC by taking b_0 low when b_7 is low

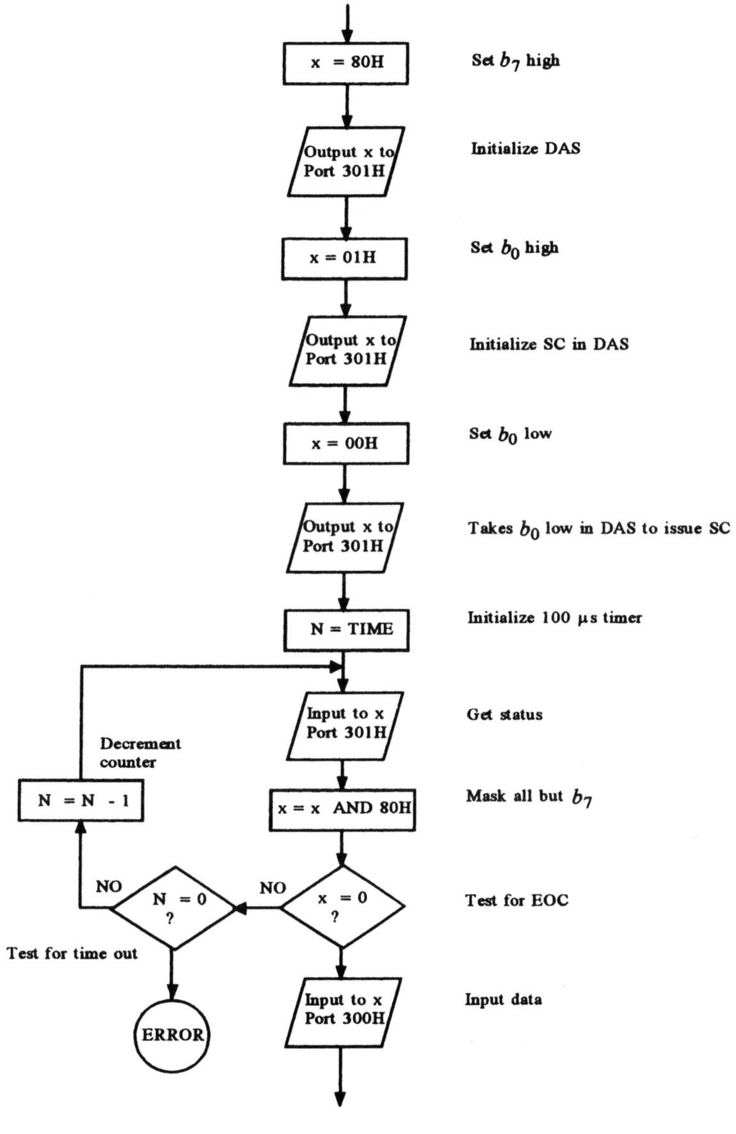

FIGURE 3.32
Solution to Example 3.25.

BASE + 2: READ has no action
WRITE sends data to DAC

Prepare a flowchart showing how a program would take a sample from channel 3. Include a time-out routine that jumps to ERROR (some unspecified routine) if the EOC is not issued after 100 μs. Use port 300H as the base address.

Solution

Figure 3.32 shows a flowchart indicating how the software would provide for an input from channel 3. It would be necessary to discover the value N to provide the 100-μs delay by determining the time to sequence through the EOC testing loop. Such a flowchart could be implemented in assembly language or a higher-level language such as C, BASIC, or FORTRAN.

3.5 CHARACTERISTICS OF DIGITAL DATA

There are many advantages to using computers for the controller function. For example, one computer can handle many loops, interaction between loops can be accounted for in the software, data are less susceptible to noise-induced errors, and linearization can be easily provided by software. Other advantages include self-tuning, error correction, and automatic failure recovery.

It seems there should be a price for these advantages, and indeed, there are some disadvantages to the use of computers in controller operations. A serious disadvantage is that conversion of analog data into digital data results in a loss of knowledge about the value of the variable. The nature and consequences of this will be considered.

3.5.1 Digitized Value

Consider first analog-to-digital conversion (ADC) of analog data into a digital format. The format of the ADC output is an n-bit binary representation of the data. With n-bits it is possible to represent 2^n values, including zero. There is a finite *resolution* of the physical data being represented of one part in 2^n, and that means we now are ignorant about the value of the variable after it has been converted into the binary representation.

In equation form, we can write the relation between a physical variable and its n-bit digital representation as

$$N = \frac{(V - V_{min})}{(V_{max} - V_{min})} 2^n \tag{3.29}$$

where
N = base 10 equivalent of binary representation
V = input value
V_{max} = maximum input value
V_{min} = minimum input value

Only the integer part of the right side of Equation (3.29) is used to determine N. Equation (3.29) assumes that the measurement system and ADC have been designed so that the binary output switches from the equivalent of $2^n - 1$ to 2^n just at V_{max}. The resolution of the measurement can be found by noting what change in V will produce a single-integer (bit) change of N. This is easily seen to be

$$\Delta V = (V_{max} - V_{min})/2^n \tag{3.30}$$

The following example illustrates some of the consequences of the digital conversion of data. A careful study of this example will help you understand the limitations of digital representation.

EXAMPLE 3.26 A temperature between 100°C and 300°C is converted into a 0- to 5.0-V signal. This signal is fed to an 8-bit ADC with a 5.0-V reference. What is the actual measurement range of the system?

a. What is the resolution?
b. What hex output results from 169°C?
c. What temperature does a hex output of C5H represent?

Solution

The nature of the ADC is that the output will change to FFH at a voltage of $5.0 - 5/256 = 4.98$ V (299.22°C) and would change to 100H at exactly 5.0 V (300°C). Thus, FFH would seem to mean any temperature between 299.22°C and 300°C. Because there is no 100H or higher (only 8 bits), FFH actually means any temperature *greater* than 299.22°C. Similarly, the output will be 00H for any voltage less than $5/256 = 0.0195$ V, which is a temperature of 100.78°C, so 00H is output for any temperature *less* than 100.78°C. Thus, the actual measurement range is from 100.78°C to 299.22°C.

a. The temperature span of $(300°\text{C} - 100°\text{C}) = 200°\text{C}$ is divided into $2^8 = 256$ values. Therefore, the resolution is given by Equation (3.30).

$$\Delta T = 200°\text{C}/256 = 0.78°\text{C/bit}$$

Let us make sure we understand what this means. If the temperature is 100°C, the output will be 00H. It will stay at this value until the temperature reaches 100.78°C; then it will change to 01H. Thus, the resolution of 0.78°C means that we are ignorant about the value of the temperature by this amount with any reading.

b. For every temperature within the range, there is one specific output value. Thus, for a temperature of 169°C, we can find that value by Equation (3.29). We find the fraction of the measurement range 169°C represented by

$$N = \frac{(169 - 100)}{(300 - 100)} 2^8 = \frac{69}{200} 256 = (0.345)(256)$$

$$N = 88.32$$

But only the integer part is used, so $88_{10} \rightarrow 58$ H and, therefore, 169°C $\rightarrow$ 58 H. Another way to get this result is to divide the quantity $(169 - 100)$ by the resolution to find the fraction of the binary number, $69/0.78 = 88.46 \rightarrow 58$ H (round-off error accounts for the difference).

c. Now our ignorance of value really shows up. What temperature does C5H represent? The procedure is quite straightforward. We simply solve Equation (3.29) for T knowing that C5H $\rightarrow 197_{10}$.

$$197 = \frac{(T - 100)}{(300 - 100)} 256$$

$$T = (197)(200)/256 + 100$$

$$T = 253.9°\text{C}$$

But wait! The fact is that the hex value will stay C5H until the temperature increases by the resolution, 0.78°C. So the actual answer can only be correctly stated as: the temperature is between 253.9°C and 254.68°C.

One of the consequences of the digitizing resolution is that we cannot be expected to control a value any more closely than this resolution. If we were supposed to control temperature to within ±0.2°C using the measurement system of Example 3.26, it would be impossible, because we do not know its value within that tolerance.

Problems of resolution are reduced by using more bits in the digital word. With 16 bits, for example, the resolution is one part in 65,536. With a 16-bit ADC using a 5.0-V reference, the least significant bit is toggled for voltage changes of only $5.0\ \text{V}/65{,}536 = 76.3\ \mu\text{V}$! Thus, noise becomes a severe problem in a typical industrial environment.

3.5.2 Sampled Data Systems

The previous section dealt with the consequences of having only discrete knowledge of the value of the physical variable. Consider also that we have only discrete knowledge of the value *in time.* That is, the computer control system takes only periodic samples of the variable value. Thus, we are ignorant of the value or variation of the variable between samples. For the control system to function correctly, certain conditions must be assumed about variations between samples. That is what sampled data systems are all about. In the following section, we will consider the nature and consequences of having only periodic samples of the physical variable.

Sampling Rate The key issue with respect to sampling in a computer-based controller is the rate at which samples must be taken. The sample rate is expressed either through t_s, the time between samples, or $f_s = 1/t_s$, the sampling frequency.

There is a maximum sampling rate in any system—that is, the time required to take a sample (ADC conversion time) plus the time required to solve the controller equations to determine the appropriate output (program execution time).

There is a minimum sampling rate in any system that depends on the nature of the time variation of the sampled variable. Simply put, samples must be taken at a high enough rate so that the signal can be reconstructed from the samples. There are serious consequences to sampling at too small a rate. For example, the control system will not be able to correct variations of the controlled variable that are missed because too few samples were taken.

Figure 3.33 illustrates the consequences of sampling rate on knowledge of signal variation. The actual signal is shown in Figure 3.33a. Figures 3.33b, 3.33c, and 3.33d illustrate knowledge about signal variation deduced from various sample rates. Reconstructions of the original signal are indicated by the dashed lines between samples.

The sampling rate of Figure 3.33b is much too slow, because little information about the actual signal variation is contained in the reconstruction from the samples.

For Figure 3.33c, the signal seems to possess a frequency of variation that is not in fact present in the actual signal. This is called *aliasing,* and it is one consequence of too small a sampling rate.

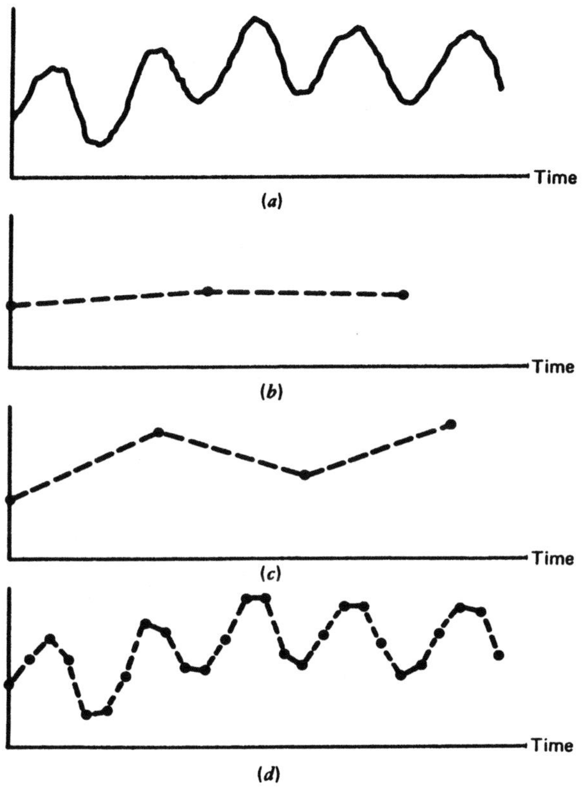

FIGURE 3.33
The sampling rate can disguise actual signal details.

The sampling rate of Figure 3.33d shows that the essential features of the signal can be reconstructed from the samples. A general rule for the minimum sampling rate can be deduced from the maximum frequency of the signal. The rule is that for adequate reconstruction of the signal from samples, the samples must be taken at a frequency that is about 10 times the maximum frequency of the signal.

$$f_s = 10 f_{\text{max}} \tag{3.31}$$

where f_s = sampling frequency
f_{max} = maximum signal frequency

This is, of course, equivalent to taking 10 samples within the shortest period of the signal.

To determine the minimum sampling rate, an estimate must be made of the highest possible frequency (shortest possible period) of the signal. The sampling frequency will be 10 times that value. For Figure 3.33d, 6 samples are taken instead of 10 in one basic period of the data. The reconstruction from samples is somewhat crude, but the basic structure is present.

EXAMPLE 3.27 The plot of Figure 3.34 shows typical data taken from pressure variations in a reaction vessel. Determine the maximum time between samples for a computer control system to be used with this system.

Solution

An examination of the signal of Figure 3.34 shows that the shortest time between any two peaks is 0.15 s. This gives a maximum signal frequency of $f_{max} = 1/0.15 = 6.7$ Hz. From Equation (3.31), the minimum sampling frequency is given by $f_s = 10f_{max} = (10)(6.7 \text{ Hz}) = 67 \text{ Hz}$. The maximum time between samples is $t_s = 1/f_s =$ **15 ms.**

It is important to understand the connection between the suggested sampling frequency of Equation (3.31) and that suggested by the Nyquist sampling frequency. The Nyquist sampling theorem presents a result that, if a signal is bandlimited to a frequency, f_{max}, then the signal can be reconstructed if sampled at a rate of twice this maximum—that is, $f_{sample} = 2f_{max}$. This is an apparent contradiction to Equation (3.31).

This contradiction is resolved by two observations. First, practical signals in industry are not bandlimited, which would mean the maximum frequency was infinite. This can be alleviated by filtering the signal to block frequencies beyond some value f_{max}, effectively providing an artificial band limit, but beyond that which presents real, practical data.

Second, the Nyquist criterion does not specify how many samples must be taken to reconstruct the signal. Indeed, in the limit of sampling at $2f_{max}$, an infinite number of samples must be taken. In the practical world, the signal is not regular and periodic. Equation (3.31) represents a compromise that has been found to provide sufficient samples for practical signal reconstruction.

The cycle time of a process-control fieldbus can also impact the issue of the maximum signal frequency. If the fieldbus is used to take measurement data from a sensor and send the response back to the final control element, the potential delays because of bus traffic limit the controllable maximum frequency.

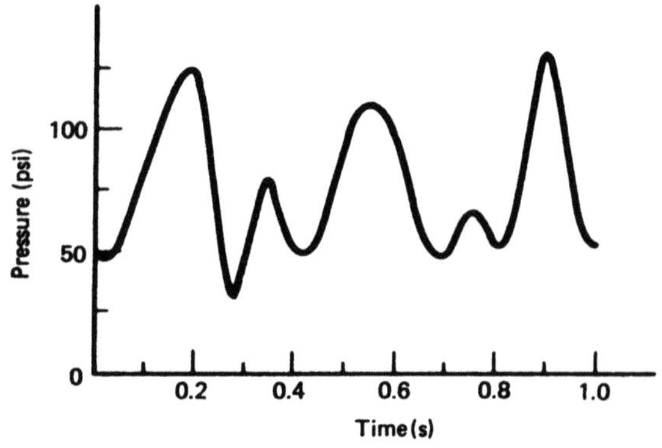

FIGURE 3.34
Pressure data for Example 3.27.

EXAMPLE 3.28 A network application is used with a control system that limits cycle time to 450 ms because of heavy bus usage. One of the control systems is to regulate periodic variations of temperature in a reaction chamber. Assuming the computer processing time is negligible, what is the maximum temperature variation frequency that can be controlled?

Solution

The 450 ms corresponds to the time for requesting a sample from the sensor and then receiving this sample. The total reaction time of the control system should include the time to get a signal back to the final control element. Thus, we add another 225 ms to the total processing time. In this case, the sampling frequency is given by

$$f_s = 1/(0.45 + 0.225) = 1.48 \text{ Hz}$$

The maximum frequency of the temperature variation is found from Equation (3.31):

$$f_{max} = f_s/10 = 0.148 \text{ Hz},$$

or a period of about 6.8 seconds.

3.5.3 Linearization

In many cases, the input binary number and the controlled variable are not linearly related. In such cases, it is necessary to execute a program that will linearize the binary number so that it is proportional to the controlled variable value. There are two common approaches: equation inversion and table look-up.

Linearization by Equation When an equation is known that relates the value of the controlled variable and the binary number in the computer, an equation can be developed to determine the linearized value of the variable. For example, suppose that a transducer outputs a voltage related to pressure by

$$V = K[p]^{1/2} \tag{3.32}$$

This voltage is converted to a binary number, *DV*, by an ADC. Then it is also true that the binary number and pressure are still related by the square root:

$$DV \text{ varies as } [p]^{1/2} \tag{3.33}$$

What we want is a binary number that is linearly related to pressure. The way to get this is to square *DV*:

$$DP = DV * DV \text{ varies as } p \tag{3.34}$$

Thus, the program would input a sample *DV* and multiply it by itself. The resulting number would be linearly related to the pressure. Of course, there may have to be scale shifts and offsets before we have a number equal to the pressure, as the following example shows.

EXAMPLE 3.29 Pressure from 50 to 400 psi is converted to voltage by the relation

$$V = 0.385[p]^{1/2} - 2.722$$

This is input to an ADC with a 5.0-V reference, which provides 00H to FFH over the pressure range. A program uses an instruction $DV = UDF(1)$ to input the data from the ADC as a base 10 number DV that varies from 0 to 255 over the pressure range. Develop a linearization equation to give a quantity, p, in the program that is equal to the actual pressure.

Solution

We have enough information to work backward through the ADC, signal conditioning, and measurement. Thus, we know the voltage is related to DV by the ADC transformation

$$DV = (V/V_{\text{ref}})256 = (V/5)256$$

or

$$V = (5/256)DV$$

Then, using the known relation for V in terms of p,

$$0.385[p]^{1/2} - 2.722 = (5/256)DV$$

and solving for p,

$$p = (0.0507\,DV + 7.071)^2$$

This number p in the program is equal to the actual pressure value, 50 to 400 psi.

Linearization by Table Look-up There are many measurement processes where it is impossible to find a simple equation such as Equation (3.32) to relate the controlled variable and the binary number. Also, when the program must be written in assembly language, it may be difficult to evaluate even simple equations such as that of Example 3.29. In these cases, it becomes much easier to use the look-up table approach.

This is really just what we humans do when we use thermocouple tables, for example. We measure the voltage, go to a table and look up the temperature, and sometimes interpolate. It is the exact same thing with the software approach. The table of input values and corresponding physical variable values are stored in a table in memory. Following a measurement, the input value is looked up in the table and the correct measured value found.

Figure 3.35 shows a software approach for table look-up in flowchart format. It is assumed that the input values are stored in ascending value in N memory locations and the corresponding physical variable values are stored in the following N locations. Thus, if the input is found at the Ith memory location from the start of the table, then the actual variable value is found at the $N + I$ location.

Of course, there are many other methods of table construction and search. In many cases, it is necessary to write interpolation routines by programming equations such as Equation (4.14) to refine the values between table values.

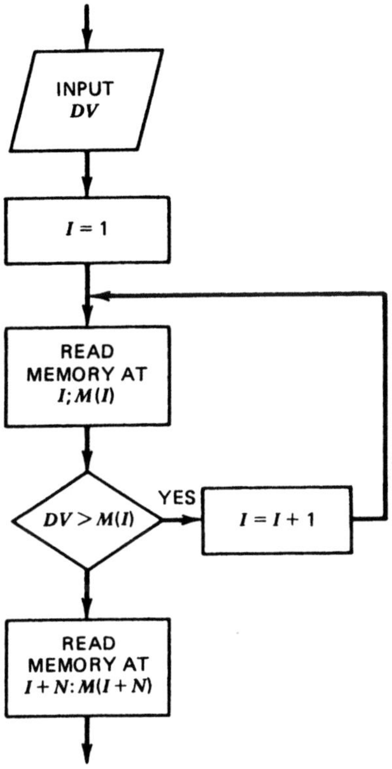

FIGURE 3.35
Linearization by table look-up can be accomplished by the operations in this flowchart.

SUMMARY

This chapter provides a digital electronics background to make the reader conversant with the elements of digital signal conditioning and able to perform simple analysis and design as associated with process control.

1. The use of digital words enables the encoding of analog information into a digital format.
2. It is possible to encode fractional decimal numbers as binary, and vice versa, using

$$N_{10} = b_1 2^{-1} + b_2 2^{-2} + \cdots + b_m 2^{-m} \qquad \textbf{(3.1)}$$

3. Boolean algebraic techniques can be applied to the development of process alarms and elementary control functions.
4. Digital electronic gates and comparators allow the implementation of process Boolean equations.
5. DACs are used to convert digital words into analog numbers using a fractional-number representative. The resolution is

$$\Delta V = V_R 2^{-n} \qquad \textbf{(3.8)}$$

6. An ADC of the successive approximations type determines an output digital word for an input analog voltage in as many steps as bits to the word.

7. The dual-slope ADC converts analog to digital information by a combination of integration and time counting.
8. The data-acquisition system (DAS) is a modular device that interfaces many analog signals to a computer. Signal address decoding, multiplexing, and ADC operations are included in the device.
9. The sampling rate of a signal must be high enough to assure the signal can be reconstructed from the samples. Generally we must sample about 10 times the maximum signal frequency.
10. One of the great advantages of digitizing data and feeding it into a computer is that nonlinearities can then be removed by software. This is done by either an equation or by a table look-up process.

PROBLEMS

Section 3.2

3.1 Convert the following binary numbers into decimal, octal, and hex:
- **a.** $\mathbf{1010_2}$
- **b.** $\mathbf{111011_2}$
- **c.** $\mathbf{010110_2}$

3.2 Convert the following binary numbers into decimal, octal, and hex:
- **a.** $\mathbf{1011010_2}$
- **b.** $\mathbf{0.1101_2}$
- **c.** $\mathbf{1011.0110_2}$

3.3 Convert the following decimal numbers into binary, octal, and hex:
- **a.** 21_{10}
- **b.** 630_{10}
- **c.** 427_{10}

3.4 Convert 27.156_{10} into a binary number with the fractional binary part expressed in 6 bits. What actual decimal does this binary fraction equal?

3.5 Find the 2s complement of
- **a.** $\mathbf{1011_2}$
- **b.** $\mathbf{10101100_2}$

3.6 Prove by a table of values that $\overline{A \cdot B} = \overline{A} + \overline{B}$ (DeMorgan's theorem).

3.7 Show that the Boolean equation $A \cdot B + A \cdot \overline{A \cdot B}$ reduces to A.

3.8 A process involves moving speed, load weight, and rate of loading in a conveyor system. The variables are provided as high (**1**) and low (**0**) levels for digital control. An alarm should be initiated whenever any of the following occur:
- **a.** Speed is low; both weight and loading rate are high.
- **b.** Speed is high; loading rate is low.

Find a Boolean equation describing the required alarm output. Let the variables be S for speed, W for weight, and R for loading rate.

3.9 Implement Problem 3.8 with
- **a.** AND/OR logic and
- **b.** NAND/NOR logic

FIGURE 3.36
System for Problems 3.10 and 3.14.

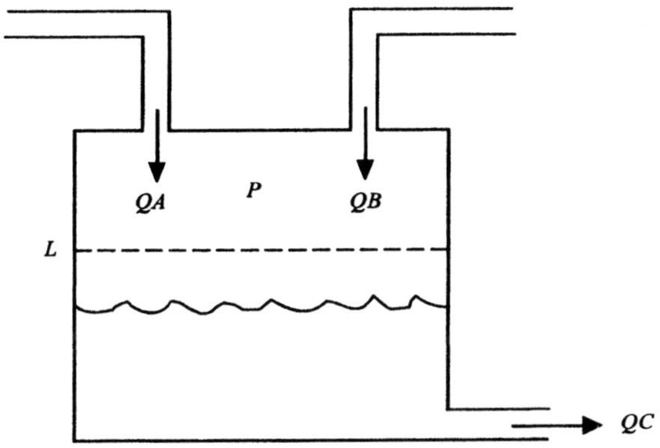

3.10 A tank shown in Figure 3.36 has the following Boolean variables: flow rates, QA, QB, and QC; pressure, P; and level, L. All are high if the variable is high and low otherwise. Devise Boolean equations for two alarm conditions as follows:

a. OV = overfill alarm
  1. If either input flow rate is high while the output flow rate is low, the pressure is low and the level is high.
  2. [If both input flow rates are high while the output flow rate is low and the pressure is low.]

b. EP = empty alarm
  1. If both input flow rates are low, the level is low and the output flow rate is high.
  2. If either input flow rate is low, the output flow rate is high and the pressure is high.

3.11 Devise logic circuits using NAND/NOR logic that will provide the two alarms of Problem 3.10.

Section 3.3

3.12 A sensor provides temperature data as 360 μV/°C. Develop a comparator circuit that goes high when the temperature reaches 530°C.

3.13 A light level is to trigger a comparator high (5 V) when the intensity reaches 30 W/m^2. Intensity is converted to voltage according to a transfer function of 0.04 V/(W/m^2). Noise is found to contribute $\pm$1.6 W/m^2 of intensity fluctuations. Develop a hysteresis comparator to provide the required output and immunity for noise.

3.14 Sensors are available for the system shown in Figure 3.36 that provide voltage outputs with the following transfer functions:

Flow rate: $V_Q = 0.15\sqrt{Q}$ (Q in gal/min)

Pressure: $V_P = \dfrac{20}{20 + p}$ (p in psi)

Level: $V_L = 0.05L$ (L in meters)

The critical values for the high/low conditions are $QA = 55$ gal/min, $QB = 30$ gal/min, $QC = 100$ gal/min, $L = 3.6$ m, and $P = 120$ psi. Design comparator circuits to provide the Boolean variable values. Include hysteresis of 1% of the reference.

3.15 Design hardware for a liquid delivery system alarm that provides a binary high output when the system temperature is between 40° and 50°C and flow is between 2.0 and 3.0 L/min. Sensors to be used have transfer functions of

$$R_T = 1000e^{-.05(T-25)} \text{ with } R_T \text{ in } \Omega \text{ and } T \text{ in } ^\circ\text{C}$$

$$V_Q = \frac{5}{Q + 5} \text{ with } V_Q \text{ in V and } Q \text{ in L/min}$$

3.16 A 6-bit DAC has an input of $\mathbf{100101_2}$ and uses a 10.0-V reference.
- **a.** Find the output voltage produced.
- **b.** Specify the conversion resolution.

3.17 A 4-bit DAC must have an 8.00-V output when all inputs are high. Find the required reference.

3.18 An 8-bit DAC with a 5.00-V reference connects to a light source with an intensity given by $I_L = 45V^{(3/2)}\text{W/m}^2$.
- **a.** What is the range of intensity that can be produced?
- **b.** What intensities are produced by digital inputs of 1BH, 7AH, 9FH, and E5H?
- **c.** Plot the intensity versus hex input, and comment on linearity.

3.19 A 12-bit bipolar DAC has a 10.00-V reference.
- **a.** What output voltage results from digital inputs of 4A6H, 02BH, and D5DH?
- **b.** An output of 4.740 V is needed. What digital input would come closest to this value? By what percentage is the actual output different?
- **c.** Suppose the output picks up a high-frequency noise of 50 mV rms. How many output bits are obscured by this noise?

3.20 A computer will be used to generate a ramp voltage signal in time (i.e., a triangular wave like Figure 3.19), with the following specifications: (1) −5 to +5 V, (2) period of 2.5 ms, (3) step size of 10 mV minimum.
- **a.** Specify the requirements of the DAC.
- **b.** Prepare flowcharts of the software that will generate the wave.

3.21 An 8-bit ADC with a 10.0-V reference has an input of 3.797 V. Find the digital output word. What range of input voltages would produce this same output? Suppose the output of the ADC is $\mathbf{10110111_2}$. What is the input voltage?

3.22 An ADC that will encode pressure data is required. The input signal is 666.6 mV/psi.
- **a.** If a resolution of 0.5 psi is required, find the number of bits necessary for the ADC. The reference is 10.0 V.
- **b.** Find the maximum measurable pressure.

3.23 A bipolar ADC has 10 bits and a 10.00-V reference. What output is produced by inputs of −4.3 V, −0.66 V, +2.4 V, and +4.8 V? What is the input voltage if the output is 30BH?

3.24 A 10-bit, unipolar ADC with a reference of 5.00 V and a 44-μs conversion time will be used to collect data on time constant measurement. Thus, the input will be of the form

$$V(t) = 4(1 - e^{t/\tau})$$

What is the minimum value of τ for which reliable data samples can be taken if no S/H circuit is used?

3.25 An 8-bit, 20-μs bipolar ADC with a 5.00-V reference will monitor a sinusoidal signal with a 3.00-V peak amplitude. What is the maximum frequency that can be tracked to 8-bit accuracy?

3.26 A sample-and-hold circuit like the one shown in Figure 3.22 has $C = 0.47\ \mu$F, and the ON resistance of the FET is 75 Ω. For what signal frequency is the sampling capacitor voltage down 3 dB from the signal voltage? How does this limit the application of the sample hold?

3.27 A S/H and 12-bit bipolar ADC combination has the following characteristics: ADC conversion time $= 35\ \mu$s, S/H aperture time $= 0.8\ \mu$s, acquisition time $= 5\ \mu$s, ADC reference $=$ 5.00 V, S/H ON resistance $= 75\ \Omega$, OFF resistance $= 20$ MΩ, and source resistance $= 1\ \Omega$. A 0.0075-μF capacitor is used for the S/H function, and the voltage follower has an input resistance of 7.5 MΩ. Determine:

a. the sampling cut-off frequency.
b. the effect of droop on the conversion process.
c. the effect of the S/H ON resistance on the sampling process.

3.28 A S/H and ADC combination has a throughput expressed as 100,000 samples per second. Explain the consequences of using this system to take samples every 5 ms.

3.29 A sensor signal is converted to a frequency that varies from 4.6 to 37 kHz. This is to be used with a counter-based ADC of 10 bits. Specify the count time, T_c. What is the count for the minimum frequency of 4.6 kHz?

3.30 A capacitor varies linearly from 0.04 to 0.26 μF as pressure varies from 0 to 100 kPa gauge. Devise a circuit using an LM331 that converts this to a frequency with a maximum of 20 kHz at 0.04 μF. Then determine the appropriate T_c of a counter-based ADC that will provide an 8-bit output of 255 at 20 kHz. What is the frequency at 0.26 μF? Plot the frequency versus pressure. Comment on the linearity.

Section 3.4

3.31 A DAS has an ADC of 8 bits with a 0- to 2.5-V range of input for 00H to FFH output (i.e., FEH to FFH occurs at 2.50 V). Inputs are temperature from 20° to 100°C scaled at 40 mV/°C, pressure from 1 to 100 psi scaled at 100 mV/psi, and flow from 30 to 90 gal/min scaled at 150 mV/(gal/min). Develop a block diagram of the required signal conditioning so that each of these can be connected to the DAS and so that the indicated variable range corresponds to 00H to FFH. Specify the resolution of each in terms of the change in each variable that corresponds to an LSB change in the ADC output.

3.32 Design op amp circuits that will provide the signal conditioning specified in Problem 3.31.

3.33 A data-acquisition system has eight input channels to be sampled continuously and sequentially. The multiplexer can select and settle on a channel in 3.1 μs, the ADC converts in 33 μs, and the computer processes a single channel of data in 450 μs. What is the minimum time between samples for a particular channel?

3.34 Flow is to be measured and input to a computer. For maximum resolution, we want the minimum flow, 30 m^3/h, to correspond to 00H, and the maximum, 60 m^3/h, to correspond to FFH from an 8-bit converter. The flow is measured as a voltage given by $V = 0.0022Q^2$, where Q is the flow in m^3/h. Develop the signal conditioning and ADC reference that will provide this specification. What is the resolution of flow when the flow is at the lower limit and at the upper limit? Explain why there is a difference.

3.35 An 8-bit ADC has an 8.00-V reference.

a. Find the output for inputs of 3.4 V and 6.7 V.

b. What range of inputs could have caused the output to become B7H?

3.36 Tests show that the output of a position sensor is 12 mV/mm, but there is 60-Hz noise on the output of a constant 5-mV rms. The sensor output impedance is 2.5 kΩ. This sensor is to measure work-piece motion, which oscillates between −10 mm and +10 mm with a period of 1.5 s. The position is to be interfaced to a 12-bit bipolar, offset binary ADC with a 5.000-V reference. Design the interface system such that −10 mm corresponds to 000H and +10 mm corresponds to FFFH.

a. With no filter, determine how many bits are being toggled by noise (i.e., are lost to any real data).

b. Introduce a filter, reevaluate the effect of noise on the ADC output, and determine how many bits represent real data.

3.37 A sensor linearly changes resistance from 2.35 to 3.57 kΩ over a range of some measured variable. The measurement must have a resolution of at least 1.25 Ω and be interfaced to a computer. Design the signal conditioning and specify the characteristics of the required ADC.

3.38 Using the DAS of Example 3.25, prepare a flowchart of a system that inputs a sample from channel 5, decrements it by one, and outputs to the DAC. Use port 300H for the base address.

3.39 What is the maximum rate that data can be taken from one channel of the DAS of Example 3.25, apart from software time? What is the maximum rate that data can be taken if all eight channels are sequentially sampled?

3.40 Prepare a flowchart of software that inputs data from channel 2 and then channel 6. These are multiplied, and the higher 8 bits of the 16-bit product are sent to the DAC for output. Develop an equation that relates the analog output voltage to the two analog input voltages.

Section 3.5

3.41 A digital control system is to provide regulation of pressure within 1.2 kPa in the range 30 to 780 kPa. How many bits must be used for the data acquisition?

3.42 A computer will be used to control flow through 10 pumping stations. The pumps exhibit a surging effect with a period of 2.2 s. What is the minimum sampling rate to ensure quality data? How much time can be spent processing each station's data? Data-acquisition hardware and software take 200 μ for a channel.

3.43 The output voltage of a silicon photovoltaic cell is given by $V = 0.11 \log_e(I_L)$, where I_L is the light intensity in W/m^2. Intensity is to be measured from 100 to 400 W/m^2, and input to a computer is via an 8-bit ADC with a 5.00-V reference.

a. Develop signal conditioning to interface the cell to the ADC so that 00H to 01H occurs at 100 W/m^2 and FEH to FFH occurs at 400 W/m^2.

b. Develop software in the language of your choice to linearize the input data so that a program variable is the intensity in W/m^2.

3.44 A type-K TC with a 20°C reference will be used to measure temperature from 200°C to 280°C in a computer controller application. The data-acquisition hardware and software make the actual measured voltage, in mV, available in software as a variable VTC. Devise a flowchart and/or program in the language of your choice that makes the reference correction and performs linearization by table look-up and interpolation. Use table values in 5°C increments for a type-K TC with a 0°C reference.

3.45 A gauge is used in a bridge circuit as in Figure 2.5 with $R_1 = R_2 = R_3 = 120\ \Omega$, $V = 10.0$ V. Assume the bridge offset voltage is provided as input to a computer and appears in a variable DELTAV. Prepare a flowchart and/or program that will obtain the value of R_4 from this voltage.

SUPPLEMENTARY PROBLEMS

S3.1 A monitoring station has a flow sensor that outputs a voltage proportional to the square root of high-pressure gas-flow rate, as given by $V_Q = 0.45\sqrt{(Q + 9)}$ volts with the flow rate in kg/h. Normal flow is 4.0 ± 1.0 kg/h, and the flow typically surges periodically within these limits. There are two alarm conditions: (1) if the flow surges above and below 6 kg/h twice without dropping below 4 kg/h in between, and (2) if a surge above 5.1 kg/h is followed by a drop below 2.9 kg/h within 2 s. Devise a system of comparators, *F/Fs*, one-shot multivibrators, and any other needed circuitry to provide the alarms.

S3.2 A pressure sensor outputs 2.5 mV/kPa, but there is 60-Hz noise of a steady peak 55 mV on the line. Pressure in the range 0.0 to 70 kPa must be converted by an ADC and presented to a computer. The resolution must be at least 0.2 kPa.

a. Design a system of filter, signal-conditioning, and ADC specifications to satisfy this requirement.

b. Suppose the pressure suddenly changed from 20 to 50 kPa. How long would it be before the ADC presented the correct digital signal for this pressure to the computer?

S3.3 A system is presented in Figure 3.37 to provide computer temperature control. The heater requires a voltage from 0.0 to 5.0 V for off (0%) to full on (100%). The following algorithm will be used by the computer to determine the heater setting:

1. If $T < 14$°C: heater at 100%
2. If 14°C $< T <$ 19°C: heater at 70%
3. If 19°C $< T <$ 22°C: heater at 38%
4. If $T > 22$°C: heater at 0%

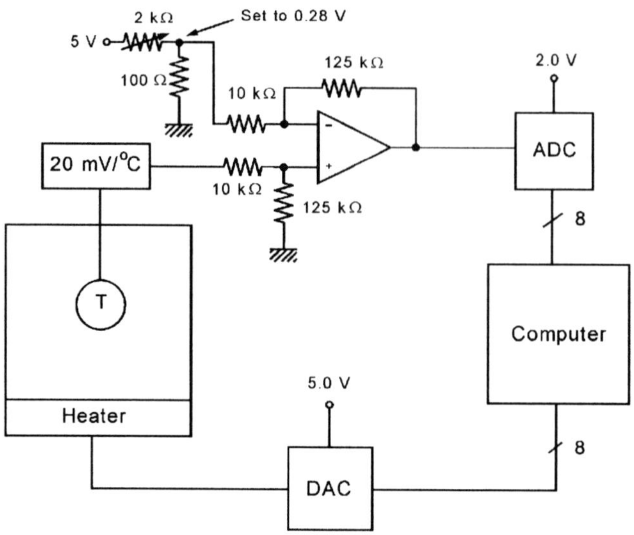

FIGURE 3.37
System for Problem S3.3.

a. Determine the ADC output for each critical temperature.
b. Determine the necessary heater voltages and the proper DAC input to produce those voltages.
c. Prepare a flowchart showing how a computer program would satisfy the algorithm requirements.

CHAPTER 4

Thermal Sensors

INSTRUCTIONAL OBJECTIVES

The objectives of this and the following two chapters stress the understanding required for application of measurement and instrumentation sensors. After you have read this chapter, you should be able to

- Define thermal energy, the relation of temperature scales to thermal energy, and temperature scale calibrations.
- Transform a temperature reading among the Kelvin, Rankine, Celsius, and Fahrenheit temperature scales.
- Design the application of an RTD temperature sensor to specific problems in temperature measurement.
- Design the application of a thermistor to specific temperature measurement problems.
- Design the application of a thermocouple to specific temperature measurement problems.
- Explain the operation of a bimetal strip for temperature measurement.
- Explain the operation of a gas thermometer and a vapor pressure thermometer.
- Develop the design of a system to measure temperature using a solid-state temperature sensor.

4.1 INTRODUCTION

Process control is a term used to describe any condition, natural or artificial, by which a physical quantity is regulated. There is no more widespread evidence of such control than that associated with temperature and other thermal phenomena. In our natural surroundings, some of the most remarkable techniques of temperature regulation are found in the bodily functions of living creatures. On the artificial side, humans have been vitally concerned with temperature control since the first fires were struck for warmth. Industrial temperature regulation has always been of paramount importance and becomes even more so with the advance of technology. In this chapter, we will be concerned first with developing an understanding of the principles of thermal energy and temperature, and then with developing a working knowledge of the various thermal sensors employed for temperature measurement.

4.2 DEFINITION OF TEMPERATURE

The materials that surround us and, indeed, of which we are constructed are composed of assemblages of atoms. Each of the 92 natural elements of nature is represented by a particular type of atom. The materials that surround us typically are not pure elements, but combinations of atoms of several elements that form molecules. Thus, helium is a natural element consisting of a particular type of atom; water, however, is composed of molecules, with each molecule consisting of a combination of two hydrogen atoms and one oxygen atom. In presenting a physical picture of thermal energy, we need to consider the physical relations or interaction of elements and molecules in a particular material as either solid, liquid, or gas. These statements actually refer to how the molecules of the material are interacting and to the thermal energy of the molecules.

4.2.1 Thermal Energy

Solid In any solid material, the individual atoms or molecules are strongly attracted and bonded to each other, so that no atom is able to move far from its particular location, or *equilibrium position.* Each atom, however, is capable of vibration about its particular location. We introduce the concept of thermal energy by considering the molecules' vibration.

Consider a particular solid material in which molecules are exhibiting no vibration; that is, the molecules are at rest. Such a material is said to have zero thermal energy, $W_{TH} = 0$. If we add energy to this material by placing it on a heater, for example, this energy starts the molecules vibrating about their equilibrium positions. We may say that the material now has some finite thermal energy, $W_{TH} > 0$.

Liquid If more and more energy is added to the material, the vibrations become more and more violent as the thermal energy increases. Finally, a condition is reached where the bonding attractions that hold the molecules in their equilibrium positions are overcome and the molecules "break away" and move about in the material. When this occurs, we say the material has *melted* and become a liquid. Now, even though the molecules are still attracted to one another, the thermal energy is sufficient to cause the molecules to move about and to no longer maintain the rigid structure of the solid. Instead of vibrating, one considers the molecules as randomly sliding about each other, and the average speed with which they move is a measure of the thermal energy imparted to the material.

Gas Further increases in thermal energy of the material intensify the velocity of the molecules until finally, the molecules gain sufficient energy to escape completely from the attraction of other molecules. Such a condition is manifested by *boiling* of the liquid. When the material consists of such unattached molecules moving randomly throughout a containing volume, we say the material has become a gas. The molecules still collide with each other and the walls of the container, but otherwise move freely throughout the container. The average speed of the molecules is again a measure of the thermal energy imparted to the molecules of the material.

Not all materials undergo these transitions at the same thermal energy, and indeed some not at all. Thus, nitrogen can be solid, liquid, and gas, but paper will experience a breakdown of its molecules before a liquid or gaseous state can occur. The whole subject of thermal sensors is associated with the measurement of the thermal energy of a material or an environment containing many different materials.

4.2.2 Temperature

If we are to measure thermal energy, we must have some sort of units by which to classify the measurement. The original units used were "hot" and "cold." These were satisfactory for their time but are inadequate for modern use. The proper unit for energy measurement is the *joules* of the sample in the SI system, but this would depend on the size of the material because it would indicate the total thermal energy contained. A measurement of the average thermal energy per molecule, expressed in joules, would be better to define thermal energy. We say "would be" because it is not traditionally used. Instead, special sets of units, whose origins are contained in the history of thermal energy measurements, are employed to define the *average energy per molecule* of a material. We will consider the four most common units. In each case, the name used to describe the thermal energy per molecule of a material is related by the statement that the material has a certain *degree of temperature;* the different sets of units are referred to as *temperature scales.*

Calibration To define the temperature scales, a set of *calibration points* is used; for each, the average thermal energy per molecule is well defined through equilibrium conditions existing between solid, liquid, or gaseous states of various pure materials. Thus, for example, a state of equilibrium exists between the solid and liquid phase of a pure substance when the *rate of phase change* is the same in either direction: liquid to solid, and solid to liquid. Some of the standard calibration points are

1. Oxygen: liquid/gas equilibrium
2. Water: solid/liquid equilibrium
3. Water: liquid/gas equilibrium
4. Gold: solid/liquid equilibrium

The various temperature scales are defined by the assignment of numerical values of temperatures to the list and additional calibration points. Essentially, the scales differ in two respects: (1) the location of the zero of temperature, and (2) the size of one unit of measure; that is, the average thermal energy per molecule represented by one unit of the scale.

The SI definition of the kelvin unit of temperature is in terms of the triple point of water. This is the state at which an equilibrium exists between the liquid, solid, and gaseous state of water maintained in a closed vessel. This system has a temperature of 273.16 K.

TABLE 4.1
Temperature scale calibration points

	Temperature			
Calibration Point	K	°R	°F	°C
Zero thermal energy	0	0	−459.6	−273.15
Oxygen: liquid/gas	90.18	162.3	−297.3	−182.97
Water: solid/liquid	273.15	491.6	32	0
Water: liquid/gas	373.15	671.6	212	100
Gold: solid/liquid	1336.15	2405	1945.5	1063

Absolute Temperature Scales An absolute temperature scale is one that assigns a zero temperature to a material that has no thermal energy, that is, no molecular vibration. There are two such scales in common use: the Kelvin scale in kelvin (K) and Rankine scale in degrees Rankine (°R). These temperature scales differ only by the quantity of energy represented by one unit of measure; hence, a simple proportionality relates the temperature in °R to the temperature in K. Table 4.1 shows the values of temperature in kelvin and degrees Rankine at the calibration points introduced earlier. From this table we can determine the transformation of temperature between the water liquid/solid point and water liquid/gas point is 100 K and 180°R, respectively. Because these two numbers represent the same difference of thermal energy, it is clear that 1 K must be larger than 1°R by the ratio of the two numbers:

$$(1\,\mathrm{K}) = \frac{180}{100}(1^{\circ}\mathrm{R}) = \frac{9}{5}(1^{\circ}\mathrm{R})$$

Thus, the transformation between scales is given by

$$T(\mathrm{K}) = \frac{5}{9}T(^{\circ}\mathrm{R}) \tag{4.1}$$

where $T(\mathrm{K})$ = temperature in K

$T(^{\circ}\mathrm{R})$ = temperature in °R

EXAMPLE 4.1 A material has a temperature of 335 K. Find the temperature in °R.

Solution

$$T(^{\circ}\mathrm{R}) = \frac{9}{5}T(\mathrm{K})$$

$$T(^{\circ}\mathrm{R}) = \frac{9}{5}(335\,\mathrm{K}) = \mathbf{603^{\circ}R}$$

Relative Temperature Scales The *relative* temperature scales differ from the absolute scales only in a shift of the zero axis. Thus, when these scales indicate a zero of temperature, the thermal energy of the sample is not zero. These two scales are the Celsius (related to the kelvin) and the Fahrenheit (related to the Rankine), with temperature indicated by °C and °F, respectively. Table 4.1 shows various calibration points of these scales. The quantity of energy represented by 1°C is the same as that indicated by 1 K, but the zero has been shifted in the Celsius scale, so that

$$T(^\circ\mathrm{C}) = T(\mathrm{K}) - 273.15 \tag{4.2}$$

Similarly, the size of 1°F is the same as the size of 1°R, but with a scale shift, so that

$$T(^\circ\mathrm{F}) = T(^\circ\mathrm{R}) - 459.6 \tag{4.3}$$

To transform from Celsius to Fahrenheit, we simply note that the two scales differ by the size of the degree, just as in K and °R, and a scale shift of 32 separates the two; thus,

$$T(^\circ\mathrm{F}) = \frac{9}{5}T(^\circ\mathrm{C}) + 32 \tag{4.4}$$

Relation to Thermal Energy It is possible to relate temperature to actual thermal energy in joules by using a constant called *Boltzmann's constant.* Although not true in all cases, it is a good approximation to state that the average thermal energy, W_{TH}, of a molecule can be found from the absolute temperature in K from

$$W_{\mathrm{TH}} = \frac{3}{2}kT \tag{4.5}$$

where $k = 1.38 \times 10^{-23}$ J/K is Boltzmann's constant. Thus, it is possible to determine the average thermal speed or velocity, v_{TH}, of a gas molecule by equating the kinetic energy of the molecule to its thermal energy

$$\frac{1}{2}mv_{\mathrm{TH}}^2 = W_{\mathrm{TH}} = \frac{3}{2}kT$$

and

$$v_{\mathrm{TH}} = \sqrt{\frac{3\,kT}{m}} \tag{4.6}$$

where m is the molecule mass in kilograms.

EXAMPLE 4.2 Given temperature of 144.5°C, express this temperature in **(a)** K and **(b)** °F.

Solution

a. $T(\mathrm{K}) = T(^\circ\mathrm{C}) + 273.15$

$T(\mathrm{K}) = 144.5 + 273.15$

$T(\mathrm{K}) = \mathbf{417.65\ K}$

$$\textbf{b.}\quad T(°\text{F}) = \frac{9}{5}T(°\text{C}) + 32$$

$$T(°\text{F}) = \frac{9}{5}(144.5°\text{C}) + 32$$

$$T(°\text{F}) = \mathbf{292.1°F}$$

EXAMPLE 4.3 A sample of oxygen gas has a temperature of 90°F. If its molecular mass is 5.3×10^{-26} kg, find the average thermal speed of a molecule.

Solution

We first convert 90°F to K and then use Equation (4.6) to find the speed.

$$T(°\text{R}) = T(°\text{F}) + 459.6$$

$$T(°\text{R}) = 90°\text{F} + 459.6$$

$$T(°\text{R}) = 549°\text{R}$$

$$T(\text{K}) = \frac{5}{9}T(°\text{R}) = \frac{5}{9}(549.6\ \text{K})$$

$$T(\text{K}) = 305.33\ \text{K}$$

Then the speed is

$$v_{\text{TH}} = \sqrt{\frac{3\,kT}{m}}$$

$$v_{\text{TH}} = \left[\frac{(3)(1.38 \times 10^{-23}\ \text{J/K})(305.33\ \text{K})}{5.3 \times 10^{-26}\ \text{kg}}\left(1\ \frac{\text{kg}\cdot\text{m}^2}{\text{s}^2\text{J}}\right)\right]^{1/2}$$

$$v_{\text{TH}} = \mathbf{488.37\ m/s}$$

4.3 METAL RESISTANCE VERSUS TEMPERATURE DEVICES

One of the primary methods for electrical measurement of temperature involves changes in the electrical resistance of certain materials. In this, as well as other cases, the principal measurement technique is to place the temperature-sensing device in contact with the environment whose temperature is to be measured. The sensing device then takes on the temperature of the environment. Thus, a measure of its resistance indicates the temperature of the device and the environment. Time response becomes very important in these cases because the measurement must wait until the device comes into thermal equilibrium with the environment. The two basic devices used are the *resistance-temperature detector* (RTD), based on the variation of metal resistance with temperature, and the *thermistor,* based on the variation of semiconductor resistance with temperature.

4.3.1 Metal Resistance versus Temperature

A metal is an assemblage of atoms in the solid state in which the individual atoms are in an equilibrium position with superimposed vibration induced by the thermal energy. The chief

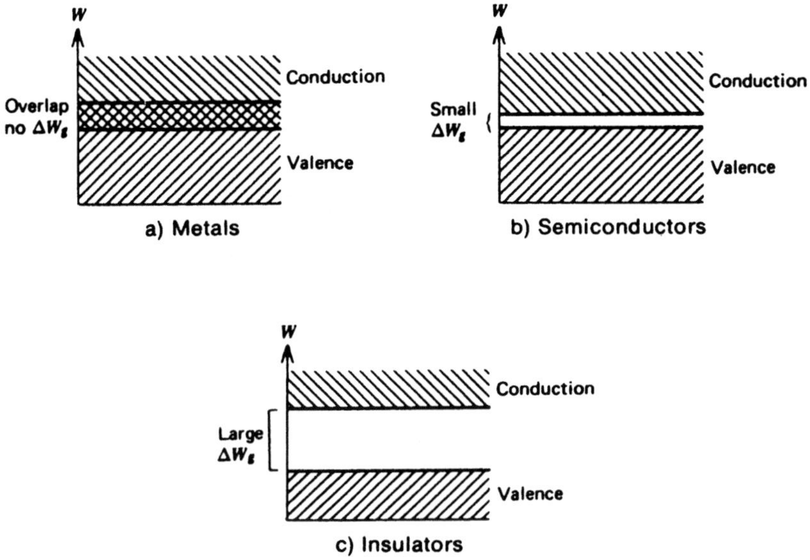

FIGURE 4.1
Energy bands for solids. Only conduction-band electrons are free to carry current.

characteristic of a metal is the fact that each atom gives up one electron, called its *valence electron,* that can move freely throughout the material; that is, it becomes a *conduction electron.* We say, then, for the whole material, that the valence band of electrons and the conduction band of electrons in the material overlap in energy, as shown in Figure 4.1a. Contrast this with a semiconductor, where a small gap exists between the top electron energy of the valence band and the bottom electron energy of the conduction band, as shown in Figure 4.1b. Similarly, Figure 4.1c shows that an insulator has a large gap between valence and conduction electrons. When a current is to be passed through a material, it is the conduction band electrons that carry the current.

As electrons move throughout the material, they collide with the stationary atoms or molecules of the material. When a thermal energy is present in the material and the atoms vibrate, the conduction electrons tend to collide even more with the vibrating atoms. This impedes the movement of electrons and absorbs some of their energy; that is, the material exhibits a *resistance* to electrical current flow. Thus, metallic resistance is a function of the vibration of the atoms and thus of the temperature. As the temperature is raised, the atoms vibrate with greater amplitude and frequency, which causes even more collisions with electrons, further impeding their flow and absorbing more energy. From this argument, we can see that metallic resistance should increase with temperature, and it does.

The graph in Figure 4.2 shows the effect of increasing resistance with temperature for several metals. To compare the different materials, the graph shows the relative resistance versus temperature. For a specific metal of high purity, the curve of relative resistance versus temperature is highly repeatable, and thus either tables or graphs can be used to determine the temperature from a resistance measurement using that material. It is possible to

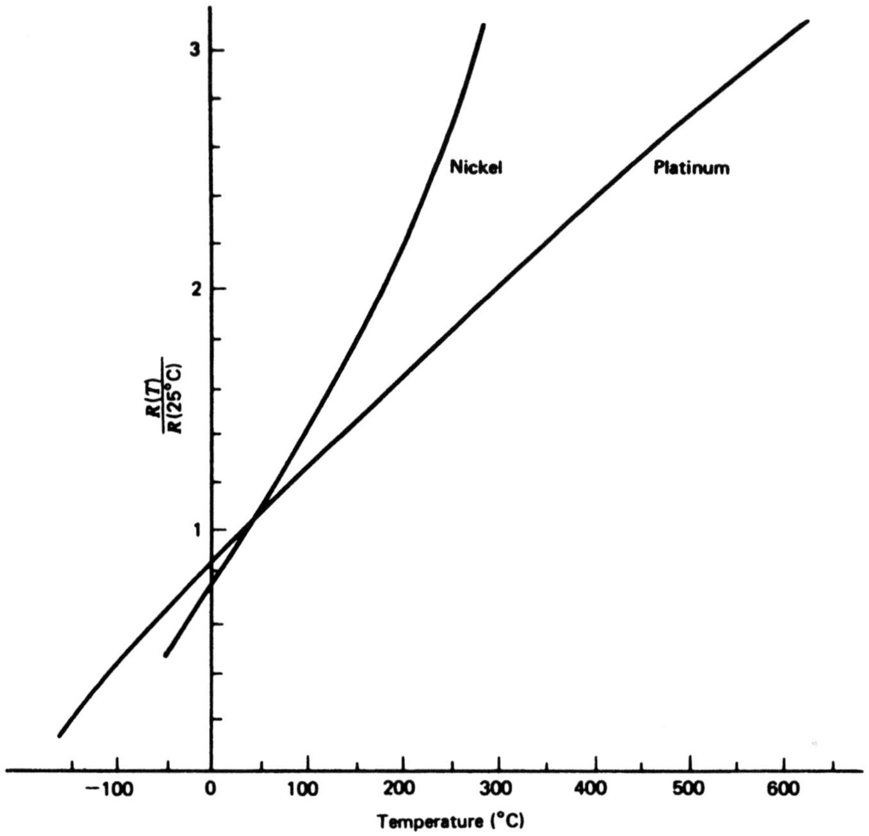

FIGURE 4.2
Metal resistance increases almost linearly with temperature.

express the resistance of a particular metal sample at a constant temperature (T) analytically using the equation

$$R = \rho \frac{l}{A} \ (T = \text{constant}) \tag{4.7}$$

where R = sample resistance(Ω)
l = length (m)
A = cross-sectional area(m^2)
ρ = resistivity($\Omega \cdot m$)

In Equation (4.7), the principal increase in resistance with temperature is due to changes in the resistivity (ρ) of the metal with temperature. If the resistivity of some metal is known as a function of temperature, then Equation (4.7) can be used to determine the resistance of any particular sample of that material at the same temperature. In fact, curves such as those in Figure 4.2 are curves of resistivity versus temperature because, for example,

$$\frac{R(T)}{R(25^\circ)} = \frac{\rho(T)l/A}{\rho(25^\circ)l/A} = \frac{\rho(T)}{\rho(25^\circ)} \tag{4.8}$$

The use of Equation (4.7), resistance versus temperature graphs, or resistance versus temperature tables is practical only when high accuracy is desired. For many applications, we can use an analytical approximation of the curves, for which we simply insert the temperature and quickly calculate the resistance, as described in Section 4.3.2.

4.3.2 Resistance versus Temperature Approximations

The curves of Figure 4.2 cover a large span of temperature, from about −150°C to about 600°C, or a nearly 750°C span. By examining the curves you can see that for smaller ranges of temperature, say the 100°C span between 100° and 200°C, the curves are nearly linear. This observation leads to development of a linear approximation of the sensor resistance versus temperature.

Linear Approximation A linear approximation means that we may develop an equation for a straight line that approximates the resistance versus temperature $(R - T)$ curve over some specified span. In Figure 4.3, we see a typical $R - T$ curve of some ma-

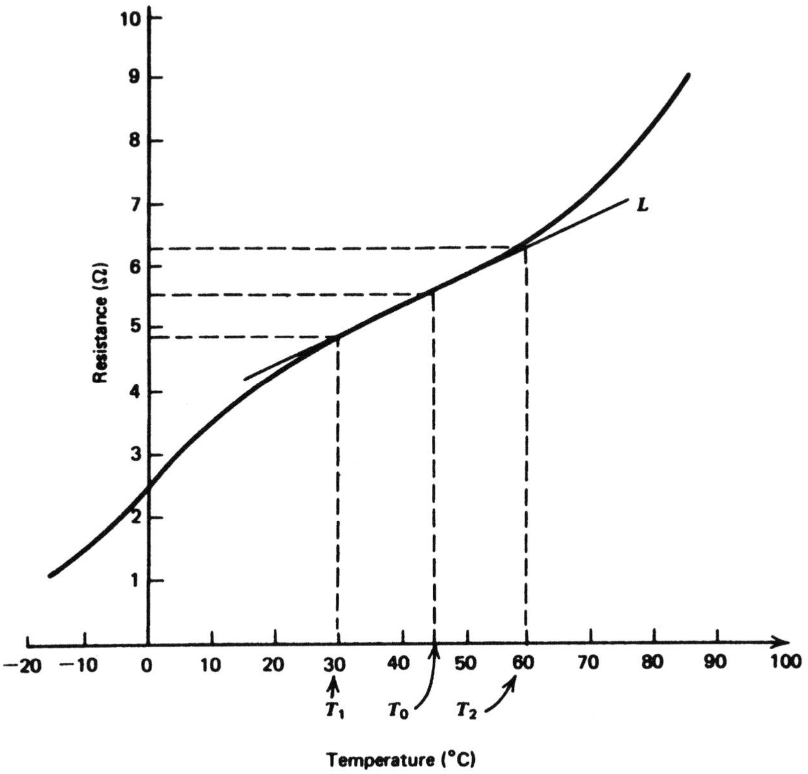

FIGURE 4.3
Line L represents a linear approximation of resistance versus temperature between T_1 and T_2.

terial. A straight line has been drawn between the points of the curve that represent temperature, T_1 and T_2 as shown, and T_0 represents the midpoint temperature. The equation of this straight line is the linear approximation to the curve over the span T_1 to T_2. The equation for this line is typically written as

$$R(T) = R(T_0)[1 + \alpha_0 \Delta T] \qquad T_1 < T < T_2 \tag{4.9}$$

where
$R(T)$ = approximation of resistance at temperature T
$R(T_0)$ = resistance at temperature T_0
$\Delta T = T - T_0$
α_0 = fractional change in resistance per degree of temperature at T_0

The reason for using α_0 as the fractional slope of the $R - T$ curve is that this same constant can be used for cases of other physical dimensions (length and cross-sectional area) of the same kind of wire. Note that α_0 depends on the midpoint temperature T_0, which simply says that a straight-line approximation over some other span of the curve would have a different slope.

The value of α_0 can be found from values of resistance and temperature taken either from a graph, as given in Figure 4.2, or from a table of resistance versus temperature, as given in Problem 4.9 (see end of chapter). In general, then,

$$\alpha_0 = \frac{1}{R(T_0)} \cdot (\text{slope at } T_0) \tag{4.10}$$

or, for example, from Figure 4.3,

$$\alpha_0 = \frac{1}{R(T_0)} \cdot \left(\frac{R_2 - R_1}{T_2 - T_1} \right) \tag{4.11}$$

where
R_2 = resistance at T_2
R_1 = resistance at T_1

The quantity α_0 has units of *inverse* temperature degrees and therefore depends on the temperature scale being used. Thus, the units of α_0 are typically 1/°C or 1/°F.

Quadratic Approximation A quadratic approximation to the $R - T$ curve is a more accurate representation of the $R - T$ curve over some span of temperatures. It includes both a linear term, as before, and a term that varies as the square of the temperature. Such an analytical approximation is usually written as

$$R(T) = R(T_0)[1 + \alpha_1 \Delta T + \alpha_2 (\Delta T)^2] \tag{4.12}$$

where
$R(T)$ = quadratic approximation of the resistance at T
$R(T_0)$ = resistance at T_0
α_1 = linear fractional change in resistance with temperature
$\Delta = T - T_0$
α_2 = quadratic fractional change in resistance with temperature

Values of α_1 and α_2 are found from tables or graphs, as indicated in the following examples, using values of resistance and temperature at three points. As before, both α_1 and α_2 depend on the temperature scale being used and have units of $1/°C$ and $(1/°C)^2$ if Celsius temperature is used, and $1/°F$ and $(1/°F)^2$ if the Fahrenheit scale is used.

The following examples show how these approximations are formed.

EXAMPLE 4.4 A sample of metal resistance versus temperature has the following measured values:

$T(°F)$	$R(\Omega)$
60	106.0
65	107.6
70	109.1
75	110.2
80	111.1
85	111.7
90	112.2

Find the linear approximation of resistance versus temperature between 60° and 90°F.

Solution

Since 75°F is the midpoint, this will be used for T_0 so that $R(T_0) = 110.2\ \Omega$. Then the slope can be found from Equation (4.11):

$$\alpha_0 = \frac{1}{110.2}\frac{(112.2 - 106.0)}{(90 - 60)} = 0.001875/°F$$

Thus, the linear approximation for resistance is

$$R(T) = 110.2[1 + 0.001875\,(T - 75)]\Omega$$

EXAMPLE 4.5 Find the quadratic approximation of resistance versus temperature for the data given in Example 4.4 between 60° and 90°F.

Solution

Again, since 75°F is the midpoint, we will use this for T_0, and therefore $R(T_0) = 110.2\ \Omega$. To find α_1 and α_2, two equations can be set up using the endpoints of the data, namely, $R(60°F)$ and $R(90°F)$.

$$112.2 = 110.2\,[1 + \alpha_1(60 - 75) + \alpha_2(60 - 75)^2]$$
$$106.0 = 110.2\,[1 + \alpha_1(90 - 75) + \alpha_2(90 - 75)^2]$$

Adding these two equations eliminates α_1 so that we can solve for α_2:

$$\alpha_2 = -44.36 \times 10^{-6}/(°F)^2$$

This can be used in either equation to find the value of $\alpha_1 = 0.001875/°F$. Thus, the quadratic approximation for the resistance versus temperature is

$$R(T) = 110.2\,[1 + 0.001875(T - 75) - 44.36 \times 10^{-6}(T - 75)^2]$$

EXAMPLE 4.6 By what percentage do the predictions of the linear and quadratic approximations vary from the actual values at 60°F and 85°F?

Solution

From Example 4.4, the predictions of the linear model can be found as

$$\begin{aligned} R(60°\text{F}) &= 110.2\,[1 + 0.001875(60 - 75)] \\ &= 107.1\ \Omega \end{aligned}$$

This is an error from the actual value of 106 Ω of +1%. At 85°F, we find $R(85°\text{F}) = 112.3\ \Omega$, which is in error by 0.54%.

The quadratic model from Example 4.5 finds the resistances to be

$$\begin{aligned} R(60°\text{F}) &= 110.2\,[1 + 0.001875(60 - 75) - 44.36 \times 10^{-6}(60 - 75)^2] \\ &= 106.0\ \Omega \end{aligned}$$

This is a zero error, because the endpoint was one of those used to determine the constants. For 85°F, we find that $R(85°\text{F}) = 111.8\ \Omega$, which is an error of +0.09%.

Clearly, the quadratic approximation provides a much better approximation of the resistance versus temperature.

4.3.3 Resistance-Temperature Detectors

A *resistance-temperature detector* (RTD) is a temperature sensor that is based on the principles discussed in the preceding sections; that is, metal resistance increasing with temperature. Metals used in these devices vary from platinum, which is very repeatable, quite sensitive, and very expensive, to nickel, which is not quite as repeatable, more sensitive, and less expensive.

Sensitivity An estimate of RTD *sensitivity* can be noted from typical values of α_0, the linear fractional change in resistance with temperature. For platinum, this number is typically on the order of 0.004/°C, and for nickel a typical value is 0.005/°C. Thus, with platinum, for example, a change of only 0.4 Ω would be expected for a 100-Ω RTD if the temperature is changed by 1°C. Usually, a specification will provide calibration information either as a graph of resistance versus temperature or as a table of values from which the sensitivity can be determined. For the same materials, however, this number is relatively constant because it is a function of resistivity.

Response Time In general, RTD has a response time of 0.5 to 5 s or more. The slowness of response is due principally to the slowness of thermal conductivity in bringing the device into thermal equilibrium with its environment. Generally, time constants are specified either for a "free air" condition (or its equivalent) or an "oil bath" condition (or its equivalent). In the former case, there is poor thermal contact and hence slow response, and in the latter, good thermal contact and fast response. These numbers yield a range of response times, depending on the application.

Construction An RTD, of course, is simply a length of wire whose resistance is to be monitored as a function of temperature. The construction is typically such that the wire is wound on a form (in a coil) to achieve small size and improve thermal conductivity to decrease response time. In many cases, the coil is protected from the environment by a *sheath* or protective tube that inevitably increases response time but may be necessary in hostile environments. A loosely applied standard sets the resistance at multiples of 100 Ω for a temperature of 0°C.

Signal Conditioning In view of the very small fractional changes of resistance with temperature (0.4%), the RTD is generally used in a *bridge* circuit. Figure 4.4 illustrates the essential features of such a system. The *compensation line* in the R_3 leg of the bridge is required when the lead lengths are so long that thermal gradients along the RTD leg may cause changes in line resistance. These changes show up as false information, suggesting changes in RTD resistance. By using the compensation line, the same resistance changes also appear on the R_3 side of the bridge and cause no net shift in the bridge null.

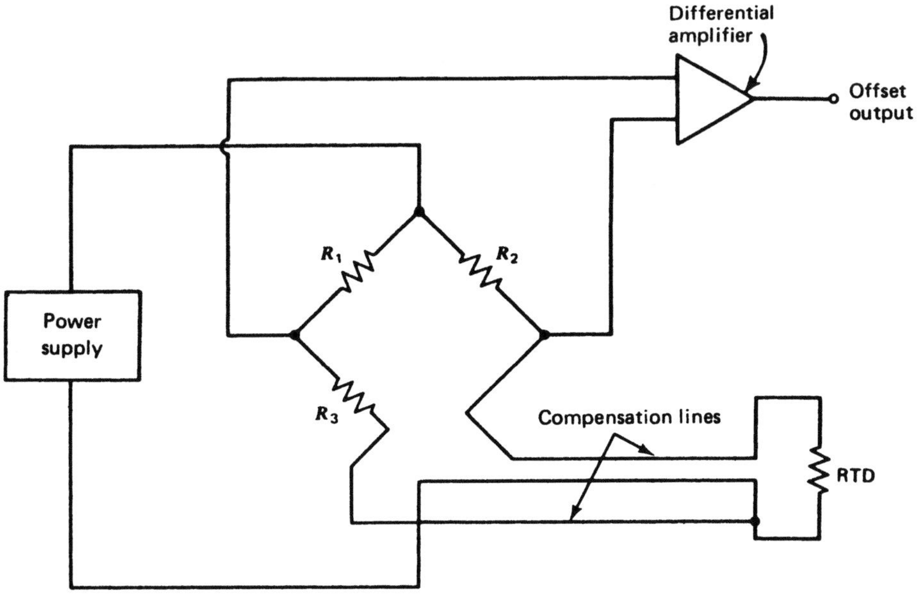

FIGURE 4.4
Note the compensation lines in this typical RTD signal-conditioning circuit.

Dissipation Constant Because the RTD is a resistance, there is an I^2R power dissipated by the device itself that causes a slight heating effect, a *self-heating*. This may also cause an erroneous reading or even upset the environment in delicate measurement conditions. Thus, the current through the RTD must be kept sufficiently low and constant to avoid self-heating. Typically, a *dissipation constant* is provided in RTD specifications. This number relates the power required to raise the RTD temperature by one degree of temperature. Thus, a 25-mW/°C dissipation constant shows that if I^2R power losses in the RTD equal 25 mW, the RTD will be heated by 1°C.

The dissipation constant is usually specified under two conditions: free air and a well-stirred oil bath. This is because of the difference in capacity of the medium to carry heat away from the device. The self-heating temperature rise can be found from the power dissipated by the RTD, and the dissipation constant from

$$\Delta T = \frac{P}{P_D} \tag{4.13}$$

where ΔT = temperature rise because of self-heating in°C
P = power dissipated in the RTD from the circuit in W
P_D = dissipation constant of the RTD in W/°C

EXAMPLE 4.7

An RTD has $\alpha_0 = 0.005/°\text{C}$, $R = 500\ \Omega$, and a dissipation constant of $P_D = 30\,\text{mW}/°\text{C}$ at 20°C. The RTD is used in a bridge circuit such as that in Figure 4.4, with $R_1 = R_2 = 500\ \Omega$ and R_3 a variable resistor used to null the bridge. If the supply is 10 V and the RTD is placed in a bath at 0°C, find the value of R_3 to null the bridge.

Solution

First we find the value of the RTD resistance at 0°C without including the effects of dissipation. From Equation (4.9), we get

$$R = 500[1 + 0.005(0 - 20)]\Omega$$
$$R = 450\ \Omega$$

Except for the effects of self-heating, we would expect the bridge to null with R_3 equal to 450 Ω also. Let's see what self-heating does to this problem. First, we find the power dissipated in the RTD from the circuit, assuming the resistance is still 450 Ω. The power is

$$P = I^2R$$

and the current I to three significant figures is found from

$$I = \frac{10}{500 + 450} = 0.011\ \text{A}$$

so that the power is

$$P = (0.011)^2(450) = 0.054\text{ W}$$

We get the temperature rise from Equation (4.13):

$$\Delta T = \frac{0.054}{0.030} = 1.8°\text{C}$$

Thus, the RTD is not actually at the bath temperature of 0°C, but at a temperature of 1.8°C. We must find the RTD resistance from Equation (4.9) as

$$R = 500[1 + 0.005(1.8 - 20)]\Omega$$
$$R = 454.5\ \Omega$$

Thus, the bridge will null with R_3 = **454.5 Ω**.

Range The effective range of RTDs principally depends on the type of wire used as the active element. Thus, a typical platinum RTD may have a range of −100° to 650°C, whereas an RTD constructed from nickel might typically have a specified range of −180° to 300°C.

4.4 THERMISTORS

The thermistor represents another class of temperature sensor that measures temperature through changes of material resistance. The characteristics of these devices are very different from those of RTDs and depend on the peculiar behavior of semiconductor resistance versus temperature.

4.4.1 Semiconductor Resistance versus Temperature

In contrast to metals, electrons in semiconductor materials are bound to each molecule with sufficient strength that no conduction electrons are contributed from the valence band to the conduction band. We say that a gap of energy, ΔW_g, exists between valence and conduction electrons, as shown in Figure 4.1b. Such a material behaves as an insulator because there are no conduction electrons to carry current through the material. This is true only when no thermal energy is present in the sample—that is, at a temperature of 0 K. When the temperature of the material is increased, the molecules begin to vibrate. In the case of a semiconductor, such vibration provides additional energy to the valence electrons. When such energy equals or exceeds the gap energy, ΔW_g, some of these electrons become free of the molecules. Thus, the electron is now in the conduction band and is free to carry current through the bulk of the material. As the temperature is further increased, more and more electrons gain sufficient energy to enter the conduction band. It is then clear that the semiconductor becomes a better conductor of

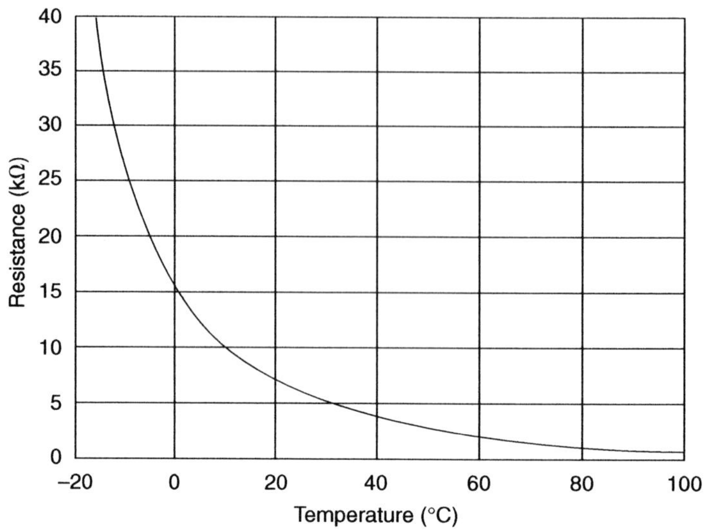

FIGURE 4.5
Thermistor resistance versus temperature is highly nonlinear and usually has a negative slope.

current as its temperature is increased—that is, as its resistance decreases. From this discussion, we form a picture of the resistance of a semiconductor material decreasing from very large values at low temperature to smaller resistance at high temperature. This is just the opposite of a metal. An important distinction, however, is that the change in semiconductor resistance is highly nonlinear, as shown in Figure 4.5. The reason semiconductors (but not insulators and other materials) behave this way is that the energy gap between conduction and valence bands is small enough to allow thermal excitation of electrons across the gap.

It is important to note that the effect just described requires that the thermal energy provide sufficient energy to overcome the band gap energy, ΔW_g. In general, a material is classified as a semiconductor when the gap energy is typically $0.01 - 4\,\text{eV}(1\,\text{eV} = 1.6 \times 10^{-19}\text{J})$. That this is true is exemplified by a consideration of silicon, a semiconductor that has a band gap of $\Delta W_g = 1.107\,\text{eV}$. When heated, this material passes from insulator to conductor. The corresponding thermal energies that bring this about can be found using Equation (4.5) and the joules-to-eV conversion, thus:

$$\begin{array}{ll} \text{For } T = 0\,\text{K} & W_{\text{TH}} = 0.0\ \text{eV} \\ \text{For } T = 100\,\text{K} & W_{\text{TH}} = 0.013\,\text{eV} \\ \text{For } T = 300\,\text{K} & W_{\text{TH}} = 0.039\ \text{eV} \end{array}$$

With average thermal energies as high as 0.039 eV, sufficient numbers of electrons are raised to the conduction level for the material to become a conductor. In *true* insulators, the gap energy is so large that temperatures less than destructive to the material cannot provide sufficient energy to overcome the gap energy.

4.4.2 Thermistor Characteristics

A thermistor is a temperature sensor that has been developed from the principles just discussed regarding semiconductor resistance change with temperature. The particular semiconductor material used varies widely to accommodate temperature ranges, sensitivity, resistance ranges, and other factors. The devices are usually mass-produced for a particular configuration, and tables or graphs of resistance versus temperature are provided for calibration. Variation of individual units from these nominal values is indicated as a net percentage deviation or a percentage deviation as a function of temperature.

Sensitivity The sensitivity of the thermistors is a significant factor in their application. Changes in resistance of 10% per °C are not uncommon. Thus, a thermistor with a nominal resistance of 10 kΩ at some temperature may change by 1 kΩ for a 1°C change in temperature. When used in null-detecting bridge circuits, sensitivity this large can provide for control, in principle, to less than 1°C in temperature.

Construction Because the thermistor is a bulk semiconductor, it can be fabricated in many forms. Thus, common forms include discs, beads, and rods, varying in size from a bead 1 mm in diameter to a disc several centimeters in diameter and several centimeters thick. By variation of doping and use of different semiconducting materials, a manufacturer can provide a wide range of resistance values at any particular temperature.

Range The temperature range of thermistors depends on the materials used to construct the sensor. In general, there are three range limitation effects: (1) melting or deterioration of the semiconductor, (2) deterioration of encapsulation material, and (3) insensitivity at higher temperatures.

The semiconductor material may melt or otherwise deteriorate as the temperature is raised. This condition generally limits the upper temperature to less than 300°C. At the low end, the principal limitation is that the thermistor resistance becomes very high, into the MΩs, making practical applications difficult. For the thermistor shown in Figure 4.5, if extended, the lower limit is about −80°C, where its resistance has risen to over 3 MΩ! Generally, the lower limit is −50 to −100°C.

In most cases, the thermistor is encapsulated in plastic, epoxy, Teflon, or some other inert material. This protects the thermistor itself from the environment. This material may place an upper limit on the temperature at which the sensor can be used.

At higher temperatures, the slope of the R-T curve of the thermistor goes to zero. The device then is unable to measure temperature effectively because very little change in resistance occurs. You can see this occurring for the thermistor resistance versus temperature curve of Figure 4.5.

Response Time The response time of a thermistor depends principally on the quantity of material present and the environment. Thus, for the smallest bead thermistors in an oil bath (good thermal contact), a response of 1/2 s is typical. The same thermistor in still air will respond with a typical response time of 10 s. When encapsulated, as in Teflon or other materials, for protection against a hostile environment, the time response is increased by the poor thermal contact with the environment. Large disc or rod thermistors may have response times of 10 s or more, even with good thermal contact.

Signal Conditioning Because a thermistor exhibits such a large change in resistance with temperature, there are many possible circuit applications. In many cases, however, a bridge circuit is used because the nonlinear features of the thermistor make its use difficult as an actual measurement device. Because these devices are resistances, care must be taken to ensure that power dissipation in the thermistor does not exceed the limits specified or even interfere with the environment for which the temperature is being measured. *Dissipation constants* are quoted for thermistors as the power in milliwatts required to raise a thermistor's temperature 1°C above its environment. Typical values vary from 1 mW/°C in free air to 10 mW/°C or more in an oil bath.

EXAMPLE 4.8

A thermistor is to monitor room temperature. It has a resistance of 3.5 kΩ at 20°C with a slope of −10%/°C. The dissipation constant is $P_D = 5\,\text{mW/°C}$. It is proposed to use the thermistor in the divider of Figure 4.6 to provide a voltage of 5.0 V at 20°C. Evaluate the effects of self-heating.

Solution

It is easy to see that the design seems to work. At 20°C, the thermistor resistance will be 3.5 kΩ, and the divider voltage will be

$$V_D = \frac{3.5\,\text{k}\Omega}{3.5\,\text{k}\Omega + 3.5\,\text{k}\Omega} 10 = 5\,\text{V}$$

Let us now consider the effect of self-heating. The power dissipation in the thermistor will be given by

$$P = \frac{V^2}{R_{\text{TH}}} = \frac{(5)^2}{3.5\,\text{k}\Omega} = 7.1\,\text{mW}$$

The temperature rise of the thermistor can be found from Equation (4.13):

$$\Delta T = \frac{P}{P_D} = \frac{7.1\text{ mW}}{5\text{ mW/°C}} = 1.42\text{°C}$$

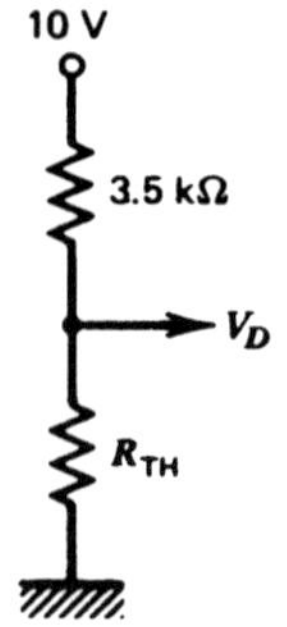

FIGURE 4.6
Divider circuit for Example 4.8.

But this means the thermistor resistance is really given by

$$R_{TH} = 3.5\,k\Omega - 1.42°C(0.1/°C)(3.5\,k\Omega)$$
$$= 3.0\,k\Omega$$

and so the divider voltage is actually $V_D = 4.6$ V. The actual temperature of the environment is 20°C, but the measurement indicates that this is not so. Clearly, the system is unsatisfactory.

This example shows the importance of including dissipation effects in resistive-temperature transducers. The real answer to this problem involves a new design that reduces the thermistor current to a value giving perhaps 0.1°C of self-heating.

4.5 THERMOCOUPLES

In previous sections, we considered the change in material resistance as a function of temperature. Such a resistance change is considered a variable parameter property in the sense that the measurement of resistance, and thereby temperature, requires external power sources. There exists another dependence of electrical behavior of materials on temperature that forms the basis of a large percentage of all temperature measurement. This effect is characterized by a voltage-generating sensor in which an electromotive force (emf) is produced that is proportional to temperature. Such an emf is found to be almost linear with temperature and very repeatable for constant materials. Devices that measure temperature on the basis of this thermoelectric principle are called *thermocouples* (TCs).

4.5.1 Thermoelectric Effects

The basic theory of the thermocouple effect is found from a consideration of the electrical and thermal transport properties of different metals. In particular, when a temperature differential is maintained across a given metal, the vibration of atoms and motion of electrons is affected so that a difference in potential exists across the material. This potential difference is related to the fact that electrons in the hotter end of the material have more thermal energy than those in the cooler end, and thus tend to drift toward the cooler end. This drift varies for different metals at the same temperature because of differences in their thermal conductivities. If a circuit is closed by connecting the ends through another conductor, a current is found to flow in the closed loop.

The proper description of such an effect is to say that an emf has been established in the circuit and is causing the current to flow. Figure 4.7a shows a pictorial representation of this effect, called the *Seebeck effect*, in which two different metals, *A* and *B*, are used to close the loop with the connecting junctions at temperatures T_1 and T_2. We could not close the loop with the same metal because the potential differences across each leg would be the same, and thus no net emf would be present. The emf produced is proportional to the difference in temperature between the two junctions. Theoretical treatments of this problem involve the thermal activities of the two metals.

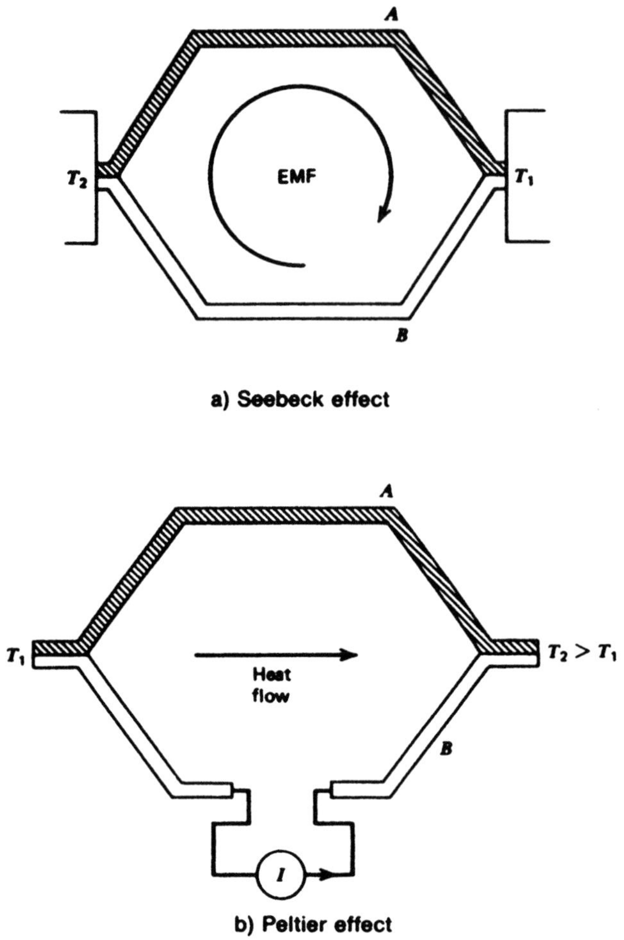

FIGURE 4.7
The Seebeck and Peltier effects refer to the relation between emf and temperature in a two-wire system.

Seebeck Effect Using solid-state theory, the aforementioned situation may be analyzed to show that its emf can be given by an integral over temperature

$$\varepsilon = \int_{T_1}^{T_2} (Q_A - Q_B)\, dT$$

where ε = emf produced in volts
T_1, T_2 = junction temperatures in K
Q_A, Q_B = thermal transport constants of the two metals

This equation, which describes the Seebeck effect, shows that the emf produced is proportional to the *difference* in temperature and, further, to the difference in the metallic thermal transport constants. Thus, if the metals are the same, the emf is zero, and if the temperatures are the same, the emf is also zero.

In practice, it is found that the two constants, Q_A and Q_B, are nearly independent of temperature and that an approximate linear relationship exists as

$$\varepsilon = \alpha(T_2 - T_1)$$

where α = constant in V/K
T_1, T_2 = junction temperatures in K

However, the small but finite temperature dependence of Q_A and Q_B is necessary for accurate considerations.

EXAMPLE 4.9 Find the Seebeck emf for a material with $\alpha = 50\,\mu\text{V}/^\circ\text{C}$ if the junction temperatures are 20°C and 100°C.

Solution

The emf can be found from

$$\varepsilon = \alpha(T_2 - T_1)$$
$$\varepsilon = (50\,\mu\text{V}/^\circ\text{C})(100^\circ\text{C} - 20^\circ\text{C})$$
$$\varepsilon = \mathbf{4\ mV}$$

Peltier Effect An interesting and sometimes useful extension of the same thermoelectric properties occurs when the reverse of the Seebeck effect is considered. In this case, we construct a closed loop of two different metals, *A* and *B*, as before. Now, however, an external voltage is applied to the system to cause a current to flow in the circuit, as shown in Figure 4.7b. Because of the different electrothermal transport properties of the metals, it is found that one of the junctions will be *heated* and the other *cooled;* that is, the device is a refrigerator! This process is referred to as the *Peltier effect.* Some practical applications of such a device, such as cooling small electronic parts, have been employed.

4.5.2 Thermocouple Characteristics

To use the Seebeck effect as the basis of a temperature sensor, we need to establish a definite relationship between the measured emf of the thermocouple and the unknown temperature. We see first that one temperature must already be known because the Seebeck voltage is proportional to the *difference* between junction temperatures. Furthermore, every connection of different metals made in the thermocouple loop for measuring devices, extension leads, and so on will contribute an emf, depending on the difference in metals and various junction temperatures. To provide an output that is definite with respect to the temperature to be measured, an arrangement such as shown in Figure 4.8a is used. This shows that the measurement junction, T_M, is exposed to the environment whose temperature is to be measured. This junction

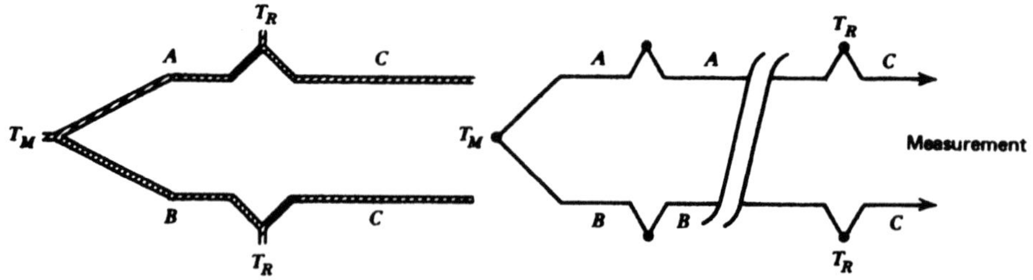

a) Three-wire thermocouple system

b) Use of extension wires in a thermocouple system

FIGURE 4.8
Practical measurements with a thermocouple system often employ extension wires to move the reference to a more secure location.

is formed of metals A and B as shown. Two other junctions are then formed to a common metal, C, which then connects to the measurement apparatus. The "reference" junctions are held at a common, known temperature T_R, the reference junction temperature. When an emf is measured, such problems as voltage drops across resistive elements in the loop must be considered. In this arrangement, an open-circuit voltage is measured (at high impedance) that is then a function of only the temperature difference $(T_M - T_R)$ and the type of metals A and B. The voltage produced has a magnitude dependent on the absolute magnitude of the temperature difference and a polarity dependent on which temperature is larger, reference or measurement junction. Thus, it is not necessary that the measurement junction have a higher temperature than the reference junctions, but both magnitude and sign of the measured voltage must be noted.

To use the thermocouple to measure a temperature, the reference temperature must be known, and the reference junctions must be held at the same temperature. The temperature should be constant, or at least not vary much. In most industrial environments, this would be difficult to achieve if the measurement junction and reference junction were close. It is possible to move the reference junctions to a remote location without upsetting the measurement process by the use of *extension wires,* as shown in Figure 4.8b. A junction is formed with the measurement system, but to wires of the same type as the thermocouple. These wires may be stranded and of different gauges, but they must be of the same type of metal as the thermocouple. The extension wires now can be run a significant distance to the actual reference junctions.

Thermocouple Types Certain standard configurations of thermocouples using specific metals (or alloys of metals) have been adopted and given letter designations; examples are shown in Table 4.2. Each type has its particular features, such as range, linearity, inertness to hostile environments, sensitivity, and so on, and is chosen for specific applications accordingly. In each type, various sizes of conductors may be employed for specific cases, such as oven measurements, highly localized measurements, and so on. The curves of voltage versus temperature in Figure 4.9 are shown for a reference temperature

TABLE 4.2
Standard thermocouples

Type	Materials[a]	Normal Range
J	Iron-constantan[b]	−190°C to 760°C
T	Copper-constantan	−200°C to 371°C
K	Chromel-alumel	−190°C to 1260°C
E	Chromel-constantan	−100°C to 1260°C
S	90% platinum + 10% rhodium-platinum	0°C to 1482°C
R	87% platinum + 13% rhodium-platinum	0°C to 1482°C

[a]First material is more positive when the measurement temperature is more than the reference temperature.
[b]Constantan, chromel, and alumel are registered trade names of alloys.

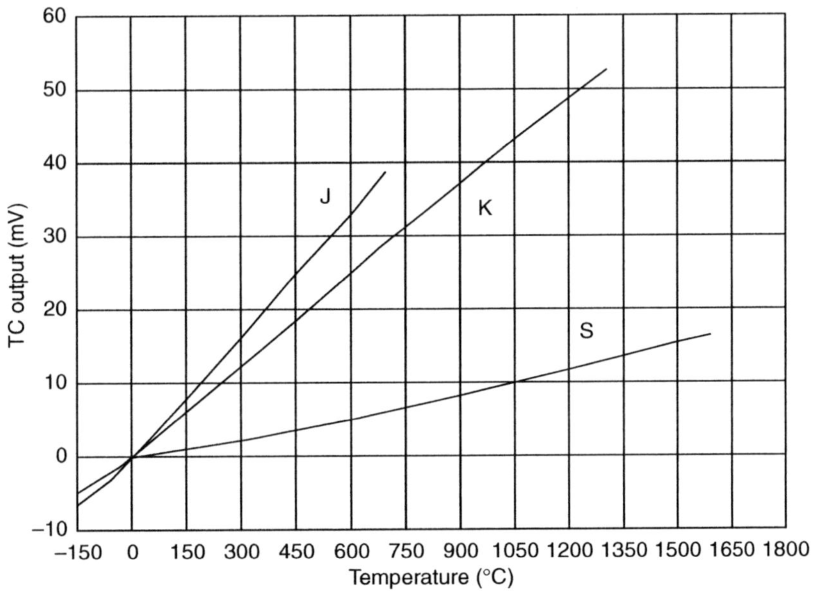

FIGURE 4.9
These curves of thermocouple voltage versus temperature for a 0°C reference show the different sensitivities and nonlinearities of three types.

of 0°C and for several types of thermocouples. We wish to note several important features from these curves.

First, we see that the type J and K thermocouples are noted for their rather large slope, that is, high sensitivity—making measurements easier for a given change in temperature. We note that the type S thermocouple has much less slope and is appropriately less sensitive. It has the significant advantages of a much larger possible range of measurement, including very high temperatures, and is highly inert material. Another important feature is that these curves are not exactly linear. To take advantage of the inherent accuracy possible with these devices, comprehensive tables of voltage versus temperature have been determined for many types of thermocouples. Such tables are found in Appendix 3.

Thermocouple Polarity The voltage produced by a TC is differential in the sense that it is measured between the two metal wires. As noted in the footnote to Table 4.2, by convention the description of a TC identifies how the polarity is interpreted. A type J thermocouple is called iron-constantan. This means that if the reference temperature is less than the measurement junction temperature, the iron will be more positive than the constantan. Thus, a type J with a 0°C reference will produce +5.27 mV for a measurement junction of 100°C, meaning that the iron is more positive than the constantan. For a measurement junction of −100°C, the polarity changes, and the voltage will be −4.63 mV, meaning that the iron is less positive than the constantan.

Thermocouple Tables The thermocouple tables simply give the voltage that results for a particular type of thermocouple when the reference junctions are at a particular reference temperature, and the measurement junction is at a temperature of interest. Referring to the tables, for example, we see that for a type J thermocouple at 210°C with a 0°C reference, the voltage is

$$V(210^\circ\text{C}) = 11.34\ \text{mV} \qquad (\text{type J, } 0^\circ\text{C ref})$$

Conversely, if we measure a voltage of 4.768 mV with a type S and a 0°C reference, we find from the table

$$T(4.768\ \text{mV}) = 555^\circ\text{C} \qquad (\text{type S, } 0^\circ\text{C ref})$$

In most cases, the measured voltage does not fall exactly on a table value. When this happens, it is necessary to *interpolate* between table values that bracket the desired value. In general, the value of temperature can be found using the following interpolation equation:

$$T_M = T_L + \left[\frac{T_H - T_L}{V_H - V_L}\right](V_M - V_L) \qquad \textbf{(4.14)}$$

The measured voltage, V_M, lies between a higher voltage, V_H, and a lower voltage, V_L, which are in the tables. The temperatures corresponding to these voltages are T_H and T_L, respectively, as shown in Example 4.10.

EXAMPLE 4.10 A voltage of 23.72 mV is measured with a type K thermocouple at a 0°C reference. Find the temperature of the measurement junction.

Solution

From the table, we find that $V_M = 23.72$ lies between $V_L = 23.63$ mV and $V_H = 23.84$ mV with corresponding temperatures of $T_L = 570$°C and $T_H = 575$°C, respectively. The junction temperature is found from Equation (4.14):

$$T_M = 570°\text{C} + \frac{(575°\text{C} - 570°\text{C})}{(23.84 - 23.63 \text{ mV})}(23.72 \text{ mV} - 23.63 \text{ mV})$$

$$T_M = 570°\text{C} + \frac{5°\text{C}}{0.21}(0.09 \text{ mV})$$

$$T_M = \mathbf{572.1°C}$$

The reverse situation occurs when the voltage for a particular temperature, T_M, which is not in the table, is desired. Again, an interpolation equation can be used, such as

$$V_M = V_L + \left[\frac{V_H - V_L}{T_H - T_L}\right](T_M - T_L) \qquad \textbf{(4.15)}$$

where all terms are as defined for Equation (4.14).

EXAMPLE 4.11 Find the voltage of a type J thermocouple with a 0°C reference if the junction temperature is −172°C.

Solution

We do not let the signs bother us but merely apply the interpolation relation directly. From the Appendix 3 tables, we see that the junction temperature lies between a high (algebraically) $T_H = -170°\text{C}$ and a low $T_L = -175°\text{C}$. The corresponding voltages are $V_H = -7.12$ mV and $V_L = -7.27$ mV. The TC voltage will be

$$V_M = -7.27 \text{ mV} + \frac{-7.12 + 7.27}{-170 + 175}(-172°\text{C} + 175°\text{C})$$

$$V_M = -7.27 \text{ mV} + \frac{0.15 \text{ mV}}{5°\text{C}}(3°\text{C})$$

$$V_M = \mathbf{-7.18 \text{ mV}}$$

Change of Table Reference It has already been pointed out that thermocouple tables are prepared for a particular junction temperature. It is possible to use these tables with a TC that has a different reference temperature by an appropriate shift in the table scale. The key point to remember is that the voltage is proportional to the difference between the reference and measurement junction temperature. Thus, if a new reference is greater than the table reference, all voltages of the table will be less for this TC. The amount less will be just the voltage of the new reference as found on the table.

Consider a type J thermocouple with a 30°C reference. The tables show that a type J thermocouple with a 0°C reference produces 1.54 mV at 30°C. This, then, is the correction factor that will be applied to any voltage expected when the reference is 30°C. Consider a temperature of 400°C.

$$V(400°\text{C}) = 21.85 \text{ mV} \qquad \text{(Type J, 0°C ref)}$$

and

$$V(30°\text{C}) = 1.54 \text{ mV} \qquad \text{(Type J, 0°C ref)}$$

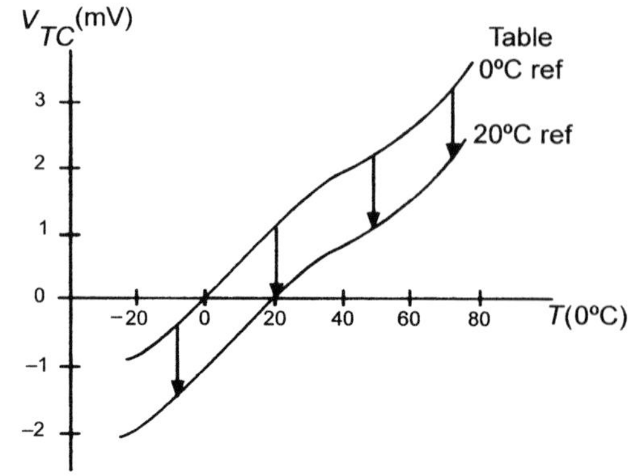

FIGURE 4.10
A change of reference from 0°C to 20°C is equivalent to sliding the TC curve down in voltage.

The correction factor is subtracted because the difference between 400° and 30°C is less than the difference between 400° and 0°C, and the voltage depends upon this difference. Therefore,

$$V(400°\text{C}) = 20.31 \text{ mV} \qquad \text{(Type J, 30°C ref)}$$

To avoid confusion on the reference, all TC voltages will henceforth be represented with a subscript of the type and reference. Thus, V_{J0} means a type J with a 0°C reference, and V_{J30} means a type J with a 30°C reference.

To consider a couple more temperatures, verify the following:

$$V_{J30}(150°\text{C}) = 8.00 - 1.54 = 6.46 \text{ mV}$$
$$V_{J30}(-90°\text{C}) = -4.21 - 1.54 = -5.75 \text{ mV}$$

In the last case, the magnitude of the voltage is larger because the difference between −90°C and 30°C is greater than the difference between −90°C and 0°C, and the voltage depends on this difference.

In summary, tables of TC voltage versus temperature are given for a specific reference temperature. These tables can be used to relate voltage and temperature for a different reference temperature by using a voltage correction factor. This correction factor is simply the voltage that the new reference would produce from the tables. The correction factor is algebraically subtracted from table values, if the new reference is less than the table reference, and added if the new reference is less than the table reference.

Figure 4.10 illustrates the correction process graphically. The curve for a 0°C reference is assumed to be the table values. You can see, then, that for a reference of 20°C, the curve is simply reduced everywhere by the correction voltage of 1.02 mV. In effect, the original curve slides down by 1.02 mV.

Note that you cannot simply add or subtract the new reference temperature as a correction factor. Correction is always applied to voltages.

EXAMPLE 4.12 A type J thermocouple with a 25°C reference is used to measure oven temperature from 300° to 400°C. What output voltages correspond to these temperatures?

Solution

Since the reference is not 0°C, a correction factor must be applied to the table voltages. That correction factor is simply the table voltage for a temperature of 25°C—that is, the reference. From the table, we find

$$V_{J0}(25°C) = 1.28 \text{ mV}$$

This, then, is the correction factor. From the tables, we now find the voltages for 300° and 400°C,

$$V_{J0}(300°C) = 16.33 \text{ mV}$$
$$V_{J0}(400°C) = 21.85 \text{ mV}$$

Now, since the reference is actually closer to these temperatures than to the table reference, we expect the voltages to be smaller. Therefore, the correction factor is subtracted to find the actual voltages:

$$V_{J25}(300°C) = 16.33 - 1.28 \text{ mV} = 15.05 \text{ mV}$$
$$V_{J25}(400°C) = 21.85 - 1.28 \text{ mV} = 20.57 \text{ mV}$$

EXAMPLE 4.13 A type K thermocouple with a 75°F reference produces a voltage of 35.56 mV. What is the temperature?

Solution

First, the reference temperature must be converted to Celsius, $T(°C) = 5(75 - 32)/9 = 23.9°C$. So we get $V_{K23.9} = 35.56$ mV and now need to determine the temperature. The voltage correction factor is determined using the 0°C tables and interpolation. The reference falls between 20° and 25°C, so

$$V_{K0}(23.9°C) = 0.8 + \frac{(1.00 - 0.80)}{(25 - 20)}(23.9 - 20) = 0.96 \text{ mV}$$

Since the reference is greater than the table reference, we would expect the table voltages to be larger for the same temperature. Thus, the correction is added to the measured voltage:

$$V_{K0}(T) = 35.56 + 0.96 = 36.52 \text{ mV}$$

From the tables, this voltage lies between 36.35 mV at 875°C and 36.55 mV at 880°C. Using interpolation,

$$T = 875 + \frac{(880 - 875)}{(36.55 - 36.35)}(36.52 - 36.35) = 879.3°C$$

The correction factors are not just minor adjustments. In this last example, if you ignored the correction factor and used 35.56 mV directly in the tables, the erroneously indicated temperature would be 850.5°C, or a 28.8°C error.

4.5.3 Thermocouple Sensors

The use of a thermocouple for a temperature sensor has evolved from an elementary process with crudely prepared thermocouple constituents into a precise and exacting technique.

Sensitivity A review of the tables shows that the range of thermocouple voltages is typically less than 100 mV. The actual sensitivity strongly depends on the type of signal conditioning employed and on the TC itself. We see from Figure 4.9 the following worst and best case of sensitivity:

- Type J: 0.05 mV/°C (typical)
- Type S: 0.006 mV/°C (typical)

Construction A thermocouple by itself is, of course, simply a welded or even twisted junction between two metals, and in many cases, that is the construction. There are cases, however, where the TC is sheathed in a protective covering or even sealed in glass to protect the unit from a hostile environment. The size of the TC wire is determined by the application and can range from #10 wire in rugged environments to fine #30 AWG wires or even 0.02-mm microwire in refined biological measurements of temperature.

Range The thermocouple temperature sensor has the greatest range of all the types considered. The tables in Appendix 3 show that the general-purpose, type J thermocouple is usable from −150° to 745°C. The type S is usable up to 1765°C. Other special types have ranges above and below these.

Time Response Thermocouple time response is simply related to the size of the wire and any protective material used with the sensor. The time response equates to how long it takes the TC system to reach thermal equilibrium with the environment.

Large, industrial TCs using thick wire or encased in stainless steel sheathing may have time constants as high as 10 to 20 s. However, a TC made from very small-gauge wire can have a time constant as small as 10 to 20 ms.

Often, the time constant is specified under conditions of good thermal contact and poor thermal contact as well, so that you can account for the environment.

Signal Conditioning The key element in the use of thermocouples is that the output voltage is very small, typically less than 50 mV. This means that considerable amplification will be necessary for practical application. In addition, the small signal levels make the devices susceptible to electrical noise. In most cases, the thermocouple is used with a high-gain differential amplifier.

Reference Compensation A problem with the practical use of thermocouples is the necessity of knowing the reference temperature. Because the TC voltage is proportional to the difference between the measurement and reference junction temperatures, variations of the reference temperature show up as direct errors in the measurement temperature determination. The following techniques are employed for reference junction compensation:

1. *Controlled temperature reference block* In some cases, particularly when many thermocouples are in use, extension wires bring all reference junctions to a temperature-controlled box in the control room. Then, a local control system maintains this box at a precisely controlled temperature so that the reference is regulated. Readouts of temperature from the TC voltage take into account this known reference temperature.
2. *Reference compensation circuits* The modern approach to reference correction is supplied by specialized integrated circuits (ICs) that add or subtract the correction factor directly to the TC output. These ICs, which are called *cold junction compensators* or *ice point compensators,* are actually temperature sensors themselves that measure the reference junction temperature. The ICs include circuitry that provides a scaled correction voltage, depending on the type of TC being used. Figure 4.11 shows a block diagram of how the compensator is used. The actual reference junctions are at the connection to the IC, so the IC temperature is the reference temperature.
3. *Software reference correction* In computer-based measurement systems, the reference junction temperature can be measured by a precision thermistor or another IC temperature sensor and provided as an input to the computer. Software routines then can provide necessary corrections to the thermocouple temperature signal that is also an input to the computer.

Noise Perhaps the biggest obstacle to the use of thermocouples for temperature measurement in industry is their susceptibility to electrical noise. First, the voltages generated generally are less than 50 mV and often are only 2 or 3 mV, and in the industrial environment it is common to have hundreds of millivolts of electrical noise generated by large electrical machines in any electrical system. Second, a thermocouple constitutes an excellent antenna for pickup of noise from electromagnetic radiation in the radio, TV, and

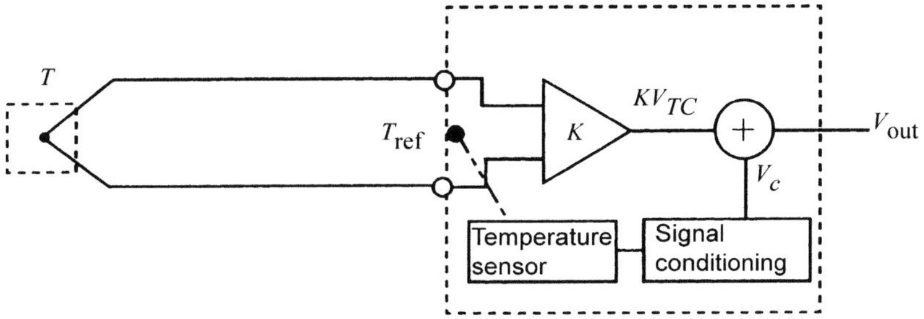

FIGURE 4.11
Automatic reference correction is now common in TC systems. Frequently thermistors or other solid-state temperature sensors are used for the reference measurement.

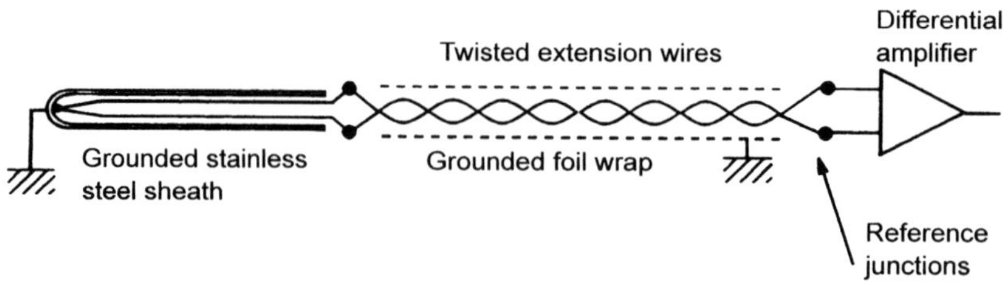

FIGURE 4.12
Since TC voltages are small, great care must be taken to protect against electrical noise by using shielding, twisting, and differential amplification.

microwave bands. In short, a bare thermocouple may have many times more noise than temperature signal at a given time.

To use thermocouples effectively in industry, a number of noise reduction techniques are employed. The following three are the most popular:

1. The extension or lead wires from the thermocouple to the reference junction or measurement system are twisted and then wrapped with a grounded foil sheath.
2. The measurement junction itself is grounded at the point of measurement. The grounding is typically to the inside of the stainless steel sheath that covers the actual thermocouple.
3. An instrumentation amplifier that has excellent common-mode rejection is employed for measurement.

Figure 4.12 shows a typical arrangement for measurement with a thermocouple. Note that the junction itself is grounded through the stainless steel sheath. The differential amplifier must have very good common-mode rejection to aid in the noise-rejection process.

The advantage to grounding the measurement junction is that the noise voltage will be distributed equally on each wire of the TC. Then the differential amplifier will, at least partially, cancel this noise because the voltage on these lines is subtracted.

Twisting is done to decouple the wires from induced voltages from varying electric and magnetic fields that permeate our environment. In principle, equal voltages are induced in each loop of the twisted wires but of opposite phase, so they cancel.

4.6 OTHER THERMAL SENSORS

The sensors discussed in the previous sections cover a large fraction of the temperature measurement techniques used in process control. There remain, however, numerous other devices or methods of temperature measurement that may be encountered. Pyrometric methods involve measurement of temperature by the electromagnetic radiation that is emitted in proportion to temperature. This technique is discussed in detail in Chapter 6.

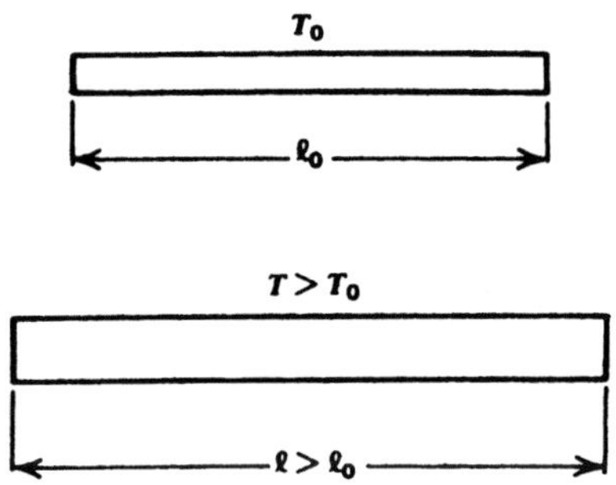

FIGURE 4.13
A solid object experiences a physical expansion in proportion to temperature. Here the effect is highly exaggerated.

4.6.1 Bimetal Strips

This type of temperature sensor has the characteristics of being relatively inaccurate, having hysteresis, having relatively slow time response, and being low in cost. Such devices are used in numerous applications, particularly where an ON/OFF cycle rather than smooth or continuous control is desired.

Thermal Expansion We have seen that greater thermal energy causes the molecules of a solid to execute greater-amplitude and higher-frequency vibrations about their average positions. It is natural to expect that an expansion of the volume of a solid would accompany this effect, as the molecules tend to occupy more volume on the average with their vibrations. This effect varies in degree from material to material because of many factors, including molecular size and weight, lattice structure, and others. Although one can speak of a volume expansion, as described, it is more common to consider a length expansion when dealing with solids, particularly in the configuration of a rod or beam. Thus, if we have a rod of length l_0 at temperature T_0 as shown in Figure 4.13, and the temperature is raised to a new value, T, the rod will be found to have a new length, l, given by

$$l = l_0[1 + \gamma \Delta T] \tag{4.16}$$

where $\Delta T = T - T_0$ and γ is the linear thermal expansion coefficient appropriate to the material of which the rod is produced. Several different expansion coefficients are given in Table 4.3.

TABLE 4.3
Thermal expansion coefficients

Material	Expansion Coefficient
Aluminum	25×10^{-6}/°C
Copper	16.6×10^{-6}/°C
Steel	6.7×10^{-6}/°C
Beryllium/copper	9.3×10^{-6}/°C

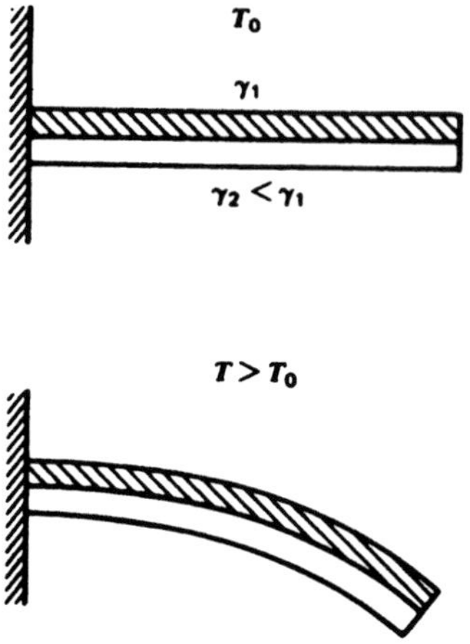

FIGURE 4.14
A bimetal strip will curve when exposed to a temperature change because of different thermal expansion coefficients. Metal thickness has been exaggerated in this view.

Bimetallic Sensor The thermal sensor exploiting the effect discussed previously occurs when two materials with grossly different thermal expansion coefficients are bonded together. Thus, when heated, the different expansion rates cause the assembly curve shown in Figure 4.14. This effect can be used to close switch contacts or to actuate an ON/OFF mechanism when the temperature increases to some appropriate setpoint. The effect also is used for temperature indicators, by means of assemblages, to convert the curvature into dial rotation.

EXAMPLE 4.14 How much will an aluminum rod of 10-m length at 20°C expand when the temperature is changed from 0° to 100°C?

Solution

First, find the length at 0°C and at 100°C; then find the difference. Using Equation (4.16) at 0°C, we get

$$l_1 = (10 \text{ m})[1 + (2.5 \times 10^{-5}/°\text{C})(0°\text{C} - 20°\text{C})]$$
$$l_1 = 9.995 \text{ m}$$

and at 100°C,

$$l_2 = (10 \text{ m})[1 + (2.5 \times 10^{-5}/°\text{C})(100°\text{C} - 20°\text{C})]$$
$$l_2 = 10.02 \text{ m}$$

Thus, the expansion is

$$l_2 - l_1 = 0.025 \text{ m} = \mathbf{25 \text{ mm}}$$

4.6.2 Gas Thermometers

The operational principle of the gas thermometer is based on a basic law of gases. In particular, if a gas is kept in a container at constant volume and the pressure and temperature vary, the ratio of gas pressure and temperature is a constant:

$$\frac{p_1}{T_1} = \frac{p_2}{T_2} \tag{4.17}$$

where p_1, T_1 = absolute pressure and temperature (in K) in state 1
p_2, T_2 = absolute pressure and temperature in state 2

EXAMPLE 4.15 A gas in a closed volume has a pressure of 120 psi at a temperature of 20°C. What will the pressure be at 100°C?

Solution

First we convert the temperature to the absolute scale of Kelvin using Equation (4.2). We get 293.15 K and 373.15 K. Then we use Equation (4.17) to find the pressure in state 2.

$$p_2 = \frac{T_2}{T_1} p_1$$

$$p_2 = \frac{373.15}{293.15}(120 \text{ psi})$$

$$p_2 = \mathbf{153\ psi}$$

Because the gas thermometer converts temperature information directly into pressure signals, it is particularly useful in pneumatic systems. Such transducers are also advantageous because there are no moving parts and no electrical stimulation is necessary. For electronic analog or digital process-control applications, however, it is necessary to devise systems for converting the pressure to electrical signals. This type of sensor is often used with Bourdon tubes (see Chapter 5) to produce direct-indicating temperature meters and recorders. The gas most commonly employed is nitrogen. Time response is slow in relation to electrical devices because of the greater mass that must be heated.

4.6.3 Vapor-Pressure Thermometers

A vapor-pressure thermometer converts temperature information into pressure, as does the gas thermometer, but it operates by a different process. If a closed vessel is partially filled with liquid, then the space above the liquid will consist of evaporated vapor of the liquid at a pressure that depends on the temperature. If the temperature is raised, more liquid will vaporize, and the pressure will increase. A decrease in temperature will result in condensation of some of the vapor, and the pressure will decrease. Thus, vapor pressure depends on temperature. Different materials have different curves of pressure versus temperature, and there is no simple equation like that for a gas thermometer. Figure 4.15 shows a curve of vapor pressure versus temperature for methyl chloride, which is often employed in these sensors. The pressure available is substantial as the temperature rises. As in the case of gas thermometers, the range is not great, and response time is slow (20 s and more) because the liquid and the vessel must be heated.

EXAMPLE 4.16 Two methyl chloride vapor-pressure temperature sensors will be used to measure the temperature difference between two reaction vessels. The nominal temperature is 85°C. Find the pressure difference per degree Celsius at 85°C from the graph in Figure 4.15.

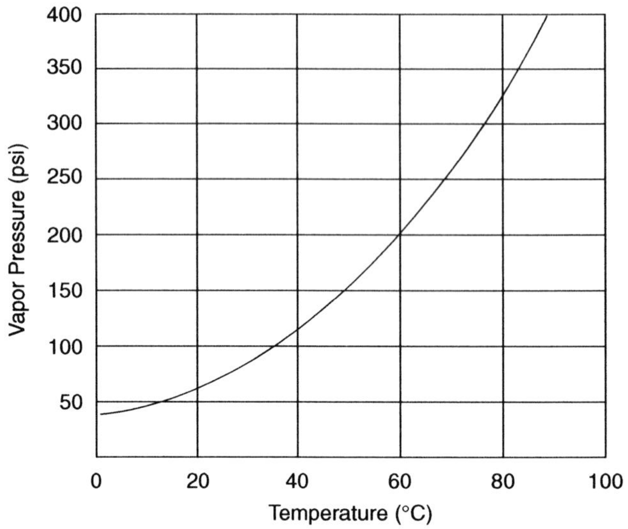

FIGURE 4.15
Vapor-pressure curve for methyl chloride.

Solution

We are just estimating the slope of the graph in the vicinity of 85°C. To do this, let us find the slope between 80°C and 90°C,

$$\frac{\Delta p}{\Delta T} = \frac{400 - 320 \text{ psi}}{90 - 80°\text{C}}$$

$$\frac{\Delta p}{\Delta T} = 8 \text{ psi/°C}$$

4.6.4 Liquid-Expansion Thermometers

Just as a solid experiences an expansion in dimension with temperature, a liquid also shows an expansion in volume with temperature. This effect forms the basis for the traditional liquid-in-glass thermometer that is so common in temperature measurement. The relationship that governs the operation of this device is

$$V(T) = V(T_0)[1 + \beta \Delta T] \tag{4.18}$$

where $V(T)$ = volume at temperature T
$V(T_0)$ = volume at temperature T_0
$\Delta T = T - T_0$
β = volume thermal expansion coefficient

In actual practice, the expansion effects of the glass container must be accounted for to obtain high accuracy in temperature indications. This type of temperature sensor is not commonly used in process-control work because further transduction is necessary to convert the indicated temperature into an electrical signal.

4.6.5 Solid-State Temperature Sensors

Many integrated circuit manufacturers now market solid-state temperature sensors for consumer and industrial applications. These devices offer voltages that vary linearly with temperature over a specified range. They function by exploiting the temperature sensitivity of doped semiconductor devices such as diodes and transistors. One common version is essentially a zener diode in which the zener voltage increases linearly with temperature.

The operating temperature of these sensors is typically in the range of $-50°$ to 150°C. The time constant in good thermal contact varies in the range of 1 to 5 seconds, whereas in poor thermal contact it may increase to 60 seconds or more. The dissipation constant is in the range of 2 to 20 mW/°C depending on the case, conditions, and heat sinking.

One of the simplest forms of solid-state temperature sensor operates electrically like a zener diode, but the zener voltage depends upon temperature. Generally, the voltage depends upon the absolute temperature in a linear way.

Figure 4.16 shows such a temperature sensor connected to provide an output voltage that depends upon temperature. Let's assume a typical transfer function of about 12 mV/K, for example. Therefore, at a nominal room temperature of 21°C, the absolute temperature is about 293 K, so the sensor output voltage would be 3.516 V.

Generally, these devices have an accuracy of better than 1°C and often can be calibrated to provide even better accuracy. Self-heating must be considered, because they do dissipate power as given by the zener voltage times the operating current. For example, in Figure 4.16 you can see that the zener current is

$$\frac{(5 - 3.516)\ \text{V}}{510\ \Omega} = 0.0029\ \text{A}$$

So with a current of 2.9 mA, the power dissipated by the zener will be

$$(0.0029\ \text{A})(3.516\ \text{V}) = 10.2\ \text{mW}$$

In still air, the dissipation constant is typically about 5 mW/°C, so there could be a 2°C self-heating error. Clearly we try to operate the sensor at a low current to avoid such errors. For example, if we change the resistor to 2000 Ω, we can show the self-heating is reduced to about 0.5°C.

These sensors are easy to interface to control systems and computers, and are becoming popular for measurements within the somewhat limited range they offer. An important application is to provide automatic reference temperature compensation for

FIGURE 4.16
Some solid-state temperature sensors operate like a zener diode whose voltage depends upon temperature.

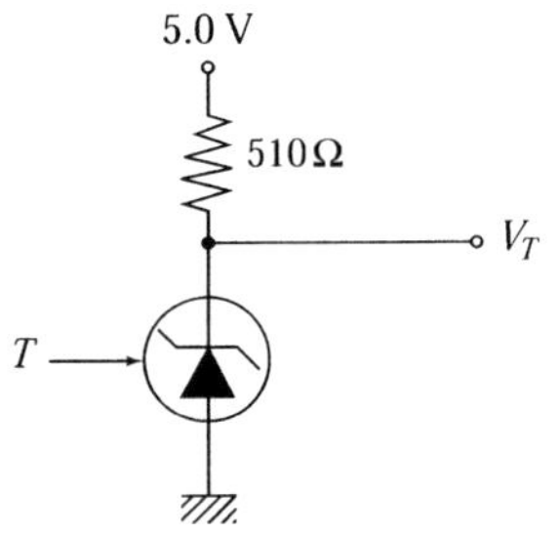

thermocouples. This is provided by connecting the sensor to the reference junction block of the TC and providing signal conditioning so that the reference corrections are automatically provided to the TC output. The following example illustrates a typical application.

EXAMPLE 4.17

A type J thermocouple is to be used in a measurement system that must provide an output of 2.00 V at 200°C. A solid-state temperature-sensor system will be used to provide reference temperature correction. The sensor has three terminals: supply voltage V_S, output voltage, V_T, and ground. The output voltage varies as 8 mV/°C.

Solution

A type J thermocouple with a 0°C reference will output 10.78 mV at 200°C. Therefore, the overall gain required will be

$$2.00 \text{ V}/0.01078 = 185.5$$

For compensation, the sensor will be physically connected to the reference block of the TC. The tables show that a type J thermocouple has a slope of approximately 50 μV/°C. Thus, the output of the sensor with temperature is $(8 \text{ mV/°C})/(50\ \mu\text{V/°C}) = 160$ times larger than the required correction. So we can provide the correction by amplifying the TC output by a gain of 160 and then adding the sensor reference correction. To make up the rest of the required gain of 185.5, we will need an amplifier with a gain of $185.5/160 = 1.159$. The output equation is given by

$$V_{\text{out}} = 1.159[160V_{TC} + V_C]$$

For the circuit we can use two differential amplifiers, one for the TC with a gain of 160 and the other to add the correction voltage and provide a gain of 1.159. We must be careful to get the polarities correct so that the correction factor is added. In Figure 4.17 the TC differential amplifier has an output of $-160V_{TC}$ since the iron is connected to the inverting side of the amplifier. This is fed to the inverting side of the second differential amplifier so the net effect provides a positive gain.

To see how well this works, we consider three cases: 50°C, 150°C, and 200°C. Suppose the actual reference temperature is 20°C. The following table shows (1) the expected output voltage if the reference was 0°C; (2) the compensation circuit output for a reference of 20°C as determined by the circuit of Figure 4.17 and the preceding equation; and (3) the percent difference.

T(°C)	V_{TC}(0°C)	V_{out}(0°C)	V_{out}(20°C)	Difference (%)
50	2.58 mV	0.479 V	0.475 V	−0.8
100	5.27 mV	0.976 V	0.974 V	−0.2
150	8.00 mV	1.484 V	1.480 V	−0.3
200	10.78 mV	1.999 V	1.995 V	−0.2

You can see that the compensated output differs from the output expected with an actual 0°C reference by less than 1%. This example illustrates the great success of using solid-state temperature sensors for reference correction when using thermocouples.

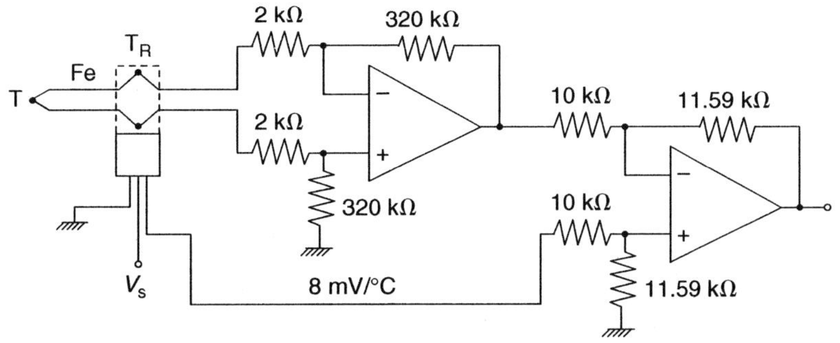

FIGURE 4.17
One possible solution to Example 4.17.

4.7 DESIGN CONSIDERATIONS

In the design of overall process-control systems, specific requirements are set up for each element of the system. The design of the elements themselves, which constitutes subsystems, involves careful matching of the elemental characteristics to the overall system design requirements. Even in the design of monitoring systems, where no integration of subsystems is required, it is necessary to match the transducer to the measurement environment and to the required output signal. In keeping with these considerations, we can treat temperature transducer design procedures by the following steps:

1. *Identify the nature of the measurement* This includes the nominal value and range of the temperature measurement, the physical conditions of the environment where the measurement is to be made, required speed of measurement, and any other features that must be considered.
2. *Identify the required output signal* In most applications, the output will be either a standard 4- to 20-mA current or a voltage that is scaled to represent the range of temperature in the measurement. There may be further requirements related to isolation, output impedance, or other factors. In some cases, a specific digital encoding of the output may be specified.
3. *Select an appropriate sensor* Based primarily on the results of the first step, a sensor that matches the specifications of range, environment, and so forth is selected. To some extent, factors such as cost and availability will be important in the selection of a sensor. The requirements of output signals also enter into this selection, but with lower significance because signal conditioning generally provides the required signal transformations.
4. *Design the required signal conditioning* Using the signal-conditioning techniques treated in Chapters 2 and 3, the direct transduction of temperature is converted into the required output signal. The specific type of signal conditioning depends, of course, on the type of sensor employed, as well as on the nature of the specified output signal characteristics.

In one form or another, these steps are required of any temperature sensor application, although they are not necessarily performed in the sequence indicated. The primary concern of this book is to enable the reader to do sensor selection and signal conditioning associated with a particular requirement. In a particular situation where a thermal sensor is required, there is no unique design to fit the application. There are, in fact, so many different designs possible that one must adjust his or her thinking from searching for *the* solution to searching for *a* solution to a design problem. A solution is any arrangement that satisfies *all* of the specified requirements of the problem. The following examples illustrate several problems in thermal design and typical solutions.

EXAMPLE 4.18

Develop a system that turns on an alarm LED when the temperature in a chamber reaches 10 ± 0.5°C. When the temperature drops below about 8°C, the LED should be turned off. For a temperature sensor use a thermistor which is 10 kΩ at 10°C and a slope at 10°C of −0.5%/°C. Its dissipation constant is 5 mW/°C.

Solution

This looks like a natural for a hysteresis comparator, as discussed in Chapter 3. What we need to do is develop a measurement to give a voltage that rises with temperature to trigger the comparator at 10°C and then allow the hysteresis to leave it on until the temperature falls to 8°C. We will work with three significant figures.

The thermistor nonlinearity will not matter, as we are interested in only two specific values of temperature/resistance. The thermistor has resistances of 10 kΩ at 10°C and 11 kΩ at 8°C. The ±0.5°C requirement means that the self-heating must be kept below 0.5°C; just to be sure, let us use 0.25°C. Because the dissipation constant for this thermistor was given as 5 mW/°C, we can determine the maximum allowable power dissipated by the transducer from Equation (4.13):

$$P = P_D \Delta T = (5 \text{ mW/°C})(0.25\text{°C})$$
$$P = 1.25 \text{ mW}$$

This represents a maximum current at 10°C of

$$I = [P/R]^{1/2} = [1.25 \text{ mW}/10 \text{ k}\Omega]^{1/2}$$
$$I = 0.354 \text{ mA}$$

Since this is a maximum, we will use $I = 0.35$ mA. We will convert the resistance variation to voltage variation using a divider, such as the one shown in Figure 2.4, and then feed the divider voltage to a hysteresis comparator. A common supply is +5 V, so using this for V_S, we will use the thermistor for R_1 (so V_D will increase with decreasing resistance, which is increasing temperature). R_2 will be determined by requiring the current to be 0.35 mA. Thus, the voltage dropped across the thermistor at 10°C is

$$V_{\text{TH}} = IR_1 = (0.35 \text{ mA})(10 \text{ k}\Omega) = 3.5 \text{ V}$$

Therefore, the value of R_2 is

$$R_2 = V/I = (5 - 3.5)/0.35 \text{ mA} = 4.28 \text{ k}\Omega$$

We will simply use 4.3 kΩ, since it is a common fixed-resistance value. We now have the two voltages of interest from the divider:

$$\text{For } 10\text{°C}, V_D = 1.50 \text{ V}$$
$$\text{For } 8\text{°C}, V_D = 1.41 \text{ V}$$

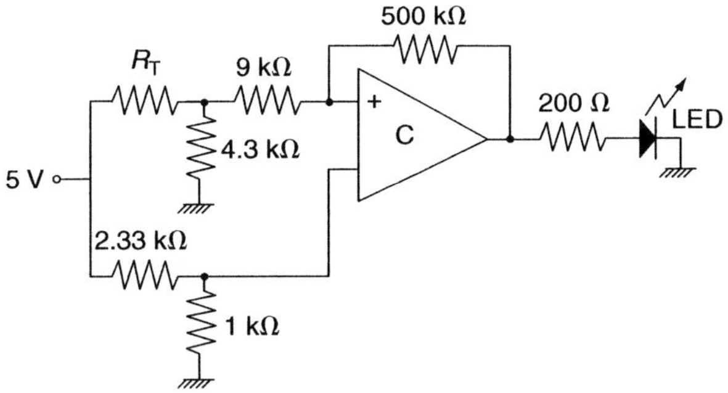

FIGURE 4.18
Circuit solution for Example 4.18. Trimmer resistors are used to obtain nonstandard resistance values.

Because the difference is 0.09 V, this must be the required hysteresis voltage. Assuming the comparator output is 5.0 V in the high state, the ratio of input to feedback resistance can be found from Equation (3.4), where $V_{\text{ref}} = 1.50$ V.

$$(R/R_f)(5.0\text{ V}) = 0.09\text{ V}$$
$$(R/R_f) = 0.018$$

We will pick $R_f = 500$ kΩ and then $R = 9$ kΩ. The final design is shown in Figure 4.18. The reference voltage has been achieved by a divider with a trimmer resistor. As usual, many other designs are possible.

EXAMPLE 4.19 Figure 4.19 shows an industrial process. Vapor flows through a chamber containing a liquid at 100°C. A control system will regulate the vapor temperature, so a measurement must be provided to convert 50° to 80°C into 0 to 2.0 V. The error should not exceed ± 1°C. If the liquid level rises to the tip of the transducer, its temperature will rise suddenly to 100°C. This event should cause an alarm comparator output to go high.

Solution

This is an example of a mid-range temperature measurement. Let us use an RTD, because the output over the 30° range will be substantially linear. Here are the specifications:

$$R \text{ at } 65°\text{C} = 150\ \Omega$$
$$\alpha \text{ at } 65°\text{C} = 0.004/°\text{C}$$
$$P_D = 30\text{ mW}/°\text{C}$$

The three resistances of interest are at 50°C, 80°C, and 100°C. From the linear RTD relation for resistance [Equation (4.9)], we find

$$\text{At } 50°\text{C, RTD} = 150[1 + 0.004(50 - 65)] = 141\ \Omega$$
$$\text{At } 80°\text{C, RTD} = 150[1 + 0.004(80 - 65)] = 159\ \Omega$$
$$\text{At } 100°\text{C, RTD} = 150[1 + 0.004(100 - 65)] = 171\ \Omega$$

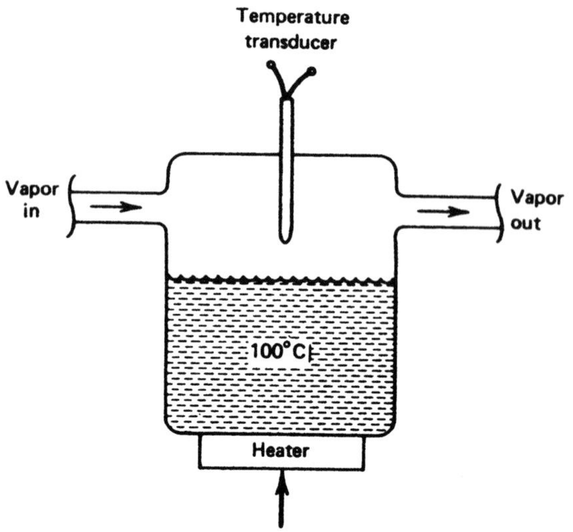

FIGURE 4.19
Vapor temperature-control process for Example 4.19.

For a 1°C error because of self-heating, we can now find the maximum current through the RTD. The maximum power is found from Equation (4.13):

$$P = P_D \Delta T = (30 \text{ mW/°C})(1\text{°C}) = 30 \text{ mW}$$

and then the maximum current from

$$I = [P/R]^{1/2} = [30 \text{ mW}/159\ \Omega]^{1/2}$$
$$I = 13.7 \text{ mA}$$

Although an op amp could be used, let us place the RTD in a bridge circuit and use the offset voltage for measurement. The small range of resistance will not cause any appreciable nonlinear effects, and the bridge can be nulled at 50°C, which will simplify the signal conditioning.

The bridge will be excited from a 5.0-V source, because this value is common. We will use the RTD as R_4 of Figure 2.4. The value of R_2 is determined by the requirement that the current be below 13.7 mA. The voltage across the RTD at 80°C will be

$$V = IR = (13.7 \text{ mA})(159\ \Omega) = 2.17 \text{ V}$$

Therefore, R_2 is found from

$$R_2 = (5 - 2.17)/13.7 \text{ mA} = 206.5\ \Omega$$

Let us use 220 Ω forR_2, because this is a standard value and will ensure that the current is low and the error condition is satisfied. To null the bridge at 50°C, we will make $R_1 = 220\ \Omega$ and use a trimmer to set R_3 to 141. The bridge is shown in Figure 4.20.

The bridge offset voltages will be found from Equation (2.8) for the endpoints and at 100°C.

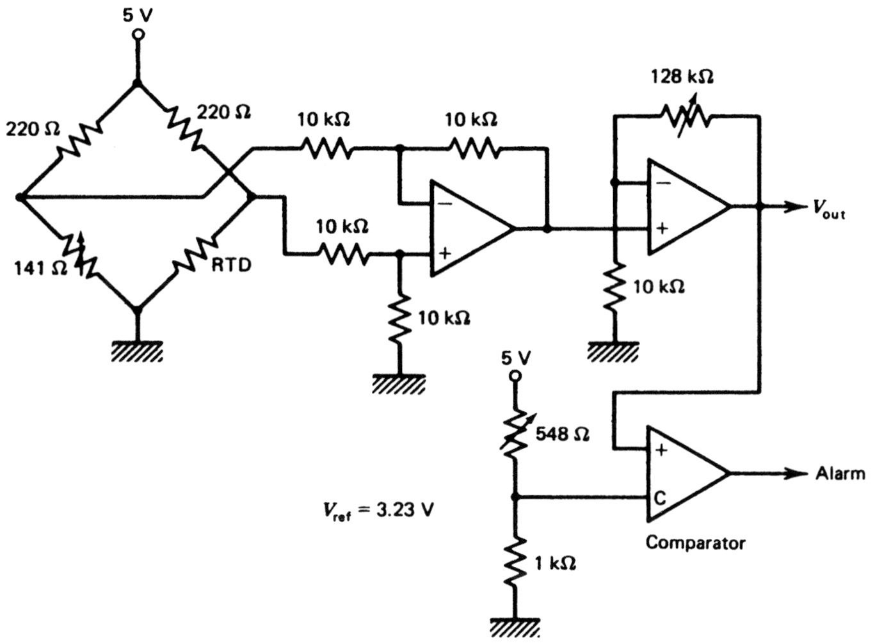

FIGURE 4.20
One possible solution for Example 4.19.

$$\text{At } 50°\text{C}, \ \Delta V = 5\frac{141}{220 + 141} - 5\frac{141}{220 + 141}$$
$$\Delta V = 0 \text{ (just as designed)}$$
$$\text{At } 80°\text{C}, \ \Delta V = 5\frac{159}{220 + 159} - 5\frac{141}{220 + 141}$$
$$\Delta V = 0.1447 \text{ V}$$
$$\text{At } 100°\text{C}, \ \Delta V = 5\frac{171}{220 + 171} - 5\frac{141}{220 + 141}$$
$$\Delta V = 0.2338 \text{ V}$$

All we need now is an amplifier to boost the 80°C voltage to 2.0 V. The required gain is (2/0.1447) = 13.8. Also, because the 5-V source used for the bridge is ground referenced, we must use a differential amplifier for the bridge offset voltage. Figure 4.20 shows the required amplifier with gain. The comparator reference voltage is V_{ref} = 13.8(0.2338) = 3.23 V.

EXAMPLE 4.20 Temperature for a plating operation must be measured for control within a range of 500° to 600°F. Develop a measuring system that scales this temperature into 0 to 5 V for input to an 8-bit ADC and a computer control system; measurement must be to within ±1°F.

Solution

The given temperature range corresponds to 260° to 315.6°C. For this high a temperature, we will use a type J thermocouple, although a platinum RTD could also be used.

The reference is assumed to be 25 ± 0.5°C to satisfy the required measurement accuracy. This can be supplied by a commercial reference, by a correction circuit, or by measuring the reference for input to the computer and software adjustment.

With a 25°C reference, the type J tables and interpolation can be used to find the voltages of the thermocouple as

$$\text{For } 260°\text{C}, V_{J25} = 12.84 \text{ mV}$$
$$\text{For } 315.6°\text{C}, V_{J25} = 15.90 \text{ mV}$$

Signal conditioning can be developed by first finding an equation for the final output from this input.

We have

$$V_{ADC} = mV_{J25} + V_0$$

using the two conditions

$$0 = m(0.01284) + V_0$$
$$5 = m(0.01590) + V_0$$

we find the values $m = 1634$ and $V_0 = -21$.

This gain is too high for a single amplifier; anyway, we need a differential amplifier on the front end for the thermocouple. Let us use a differential amplifier with a gain of 100 to produce an intermediate voltage, $V_1 = 100V_{J25}$. Then the equation becomes

$$V_{ADC} = 16.34V_1 - 21$$

or

$$V_{ADC} = 16.34(V_1 - 1.29)$$

This is simply a differential amplifier. We must be very careful to get the polarities correct. In this instance we chose to hook up the thermocouple to the first differential amplifier so that its output was negative (i.e., $-100\ V_{TC}$). When hooking it up to the second differential amplifier, we connect it to the inverting side and then −1.29 V from a divider to the noninverting side. The net result then is the required equation. The circuit is shown in Figure

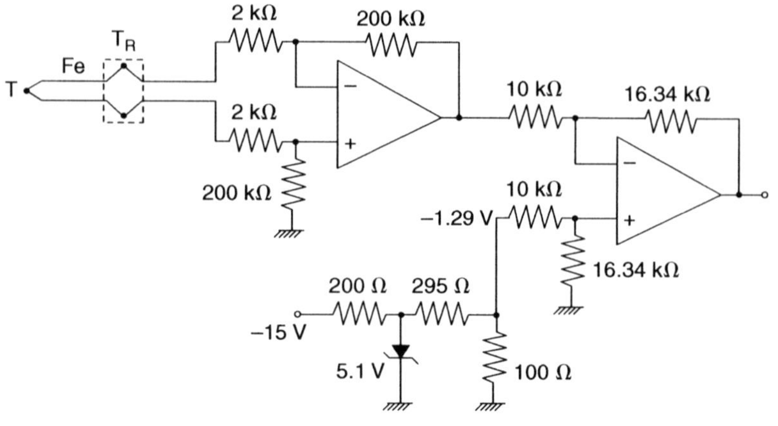

FIGURE 4.21
One possible solution for Example 4.20.

4.21. You should think about how this design could be improved in terms of reference correction. If the reference changed to 20°C, for example, the circuit output would no longer satisfy the specification.

SUMMARY

1. The measurement and control of temperature plays an important role in the process-control industry. The class of sensors that performs this measurement consists primarily of three types: (1) the resistance-temperature detector (RTD), (2) the thermistor, and (3) the thermocouple. In this chapter, the basic operating principles and application information have been provided for these sensors. Several other sensors have been briefly described.
2. The concept of temperature is contained in the representation of a body's thermal energy as the average thermal energy per molecule, expressed in units of degrees of temperature. Four units of temperature are in common use. Two of these scales are called *absolute,* because an indication of zero units corresponds to zero thermal energy. These scales, which are designated by kelvin (K) and degrees Rankine (°R), differ only in the amount of energy represented by each unit. The amount of energy represented by 1 K corresponds to 9/5°R. Thus, we can transform temperatures using

$$T(K) = \frac{5}{9}T(°\mathrm{R}) \tag{4.1}$$

3. The other two scales are called *relative,* because their zero does not occur at a zero of thermal energy. The Celsius scale (°C) corresponds in degree size to the kelvin, but has a shift of the zero so that

$$T(°\mathrm{C}) = T(K) - 273.15 \tag{4.2}$$

4. In a similar fashion, the Fahrenheit (°F) and Rankine scales are related by

$$T(°\mathrm{F}) = T(°\mathrm{R}) - 459.6 \tag{4.3}$$

5. The RTD is a sensor that depends on the increase in metallic resistance with temperature. This increase is very nearly linear, and analytical approximations are used to express the resistance versus temperature as either a linear equation,

$$R(T) = R(T_0)[1 + \alpha_0\Delta T] \tag{4.9}$$

or a quadratic relationship,

$$R(T) = R(T_0)[1 + \alpha_1\Delta T + \alpha_2(\Delta T)^2] \tag{4.12}$$

6. When a greater degree of accuracy is desired, tables or graphs of resistance versus temperature are used. Because of the small fractional change in resistance with temperature, the RTD is usually used in a bridge circuit with a high-gain null detector.
7. The thermistor is based on the decrease of semiconductor resistance with temperature. This device has a highly nonlinear resistance versus temperature curve and is not typically used with any analytical approximations. Such transducers can exhibit a very large change in resistance with temperature, and hence make very sensitive

temperature-change detectors. Many circuit configurations are used, including bridges and operational amplifiers.

8. A thermocouple is a junction of dissimilar metal wires, usually joined to a third metal wire through two reference junctions. A voltage is developed across the common metal wires that is proportional, almost linearly, to the difference in temperature between the measurement and reference junctions. Extensive tables of temperature versus voltage for numerous types of TCs using standard metals and alloys allow an accurate determination of temperature at the reference junctions. The voltage must be measured at high impedance to avoid loading effects on the measured voltage. Thus, potentiometric, operational amplifiers, or other high-impedance techniques are employed in signal conditioning.
9. A bimetal strip converts temperature into a physical motion of metal elements. This flexing can be used to close switches or cause dial indications.
10. Gas and vapor-pressure temperature sensors convert temperature into gas pressure, which then is converted to an electrical signal or is used directly in pneumatic systems.

PROBLEMS

Section 4.1

4.1 Convert 453.1°R into K, °F, and °C.

4.2 Convert −222°F into °C, °R, and K.

4.3 Convert 150°C into K and °F.

4.4 A process temperature is found to change by 33.4°F. Calculate the change in °C. *Hint:* A change in temperature does not involve a scale shift.

4.5 A sample of hydrogen gas has a temperature of 500°C. Calculate the average molecular speed in m/s. Express this also in ft/s. *Note:* Gaseous hydrogen exists as the molecule H_2 with a mass of 3.3×10^{-27} kg.

4.6 Temperature is to be controlled in the range 350° to 550°C. What is this expressed in °F?

Section 4.2

4.7 An RTD has $\alpha(20°C) = 0.004/°C$. If $R = 106\ \Omega$ at 20°C, find the resistance at 25°C.

4.8 The RTD of Problem 4.7 is used in the bridge circuit of Figure 4.4. If $R_1 = R_2 = R_3 = 100\ \Omega$ and the supply voltage is 10.0 V, calculate the voltage the detector must be able to resolve in order to resolve a 1.0°C change in temperature.

4.9 Use the values of RTD resistance versus temperature shown in the table to find the equations for the linear and quadratic approximations of resistance between 100°C and 130°C. Assume $T_0 = 115°C$. What is the error in percent between table resistance values and those determined from the two approximations?

T(°C)	R(Ω)
90.0	562.66
95.0	568.03
100.0	573.40
105.0	578.77
110.0	584.13
115.0	589.48
120.0	594.84
125.0	600.18
130.0	605.52

4.10 Suppose the RTD of Problem 4.7 has a dissipation constant of 25 mW/°C and is used in a circuit that puts 8 mA through the sensor. If the RTD is placed in a bath at 100°C, what resistance will the RTD have? What then is the indicated temperature?

Section 4.4

4.11 In Problem 4.8, the RTD is replaced by a thermistor with $R(20°\text{C}) = 100\ \Omega$ and an R versus T of −10%/°C near 20°C. Calculate the voltage resolution of the detector needed to resolve a 1.0°C change in temperature.

4.12 Modify the divider of Example 4.8 so that self-heating is reduced to 0.1°C. What is the divider voltage for 20°C? What is the divider voltage for 19°C and 21°C?

4.13 The thermistor of Figure 4.5 is used as shown in Figure 4.22 to convert temperature into voltage. Plot V_{out} versus temperature from 0° to 80°C. Is the result linear? What is the maximum self-heating if $P_D = 5$ mW/°C?

Section 4.5

4.14 A type J thermocouple measures 22.5 mV with a 0°C reference. What is the junction temperature?

4.15 A type S thermocouple with a 21°C reference measures 12.120 mV. What is the junction temperature?

4.16 If a type J thermocouple is to measure 500°C with a −10°C reference, what voltage will be produced?

4.17 A type K thermocouple with a 0°C reference will monitor an oven temperature at about 300°C.

a. What voltage would be expected?

b. Extension wires of 1000-ft length and 0.01 Ω/ft will be used to connect to the measurement site. Determine the minimum voltage measurement input impedance if the error is to be within 0.2%.

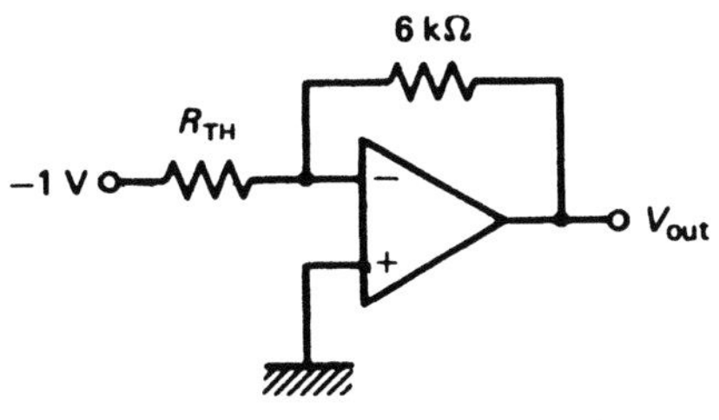

FIGURE 4.22
Circuit for Problem 4.13.

4.18 We need to get 1.5 V from a candle flame at about 700°C. How many type K thermocouples will be required in series if the reference is assumed to be nominal room temperature, 70°F?

Section 4.6

4.19 A thermoelectric switch composed of a copper rod 10 cm long at 20°C is to touch an electrical contact at a temperature of 150°C. What distance should there be between the rod end and the contact at 20°C?

4.20 A gas thermometer has a pressure of 125 kPa at 0°C and 215 kPa at some unknown temperature. Determine the temperature in °C.

4.21 A methyl chloride vapor-pressure thermometer will be used between 70°F and 200°F. What pressure range corresponds to this temperature range?

4.22 Find a linear approximation of methyl chloride pressure versus temperature from 70° to 90°C. Find the maximum error between the approximation and actual pressure in this range.

Section 4.7

4.23 Using an RTD with $\alpha = 0.0034/°C$ and $R = 100\ \Omega$ at 20°C, design a bridge and op amp system to provide a 0.0- to 10.0-V output for a 20° to 100°C temperature variation. The RTD dissipation constant is 28 mW/°C. Maximum self-heating should be 0.05°C.

4.24 A type K thermocouple with a 0°C reference will be used to measure temperature between 200°C and 350°C. Devise a system that will convert this temperature range into an 8-bit digital word with conversion from 00H to 01H at 200°C and the change from FEH to FFH occurring at 350°C. An ADC is available with a 2.500-V internal reference.

4.25 Solve Example 4.19 using a type K thermocouple with a reference compensated to 0°C.

4.26 Solve Example 4.20 using an RTD with $R = 500\ \Omega$ and $\alpha = 0.003/°C$ at 260°C. The dissipation constant is 25 mW/°C.

4.27 You have been commissioned to design a thermistor-based digital temperature measurement system. The ADC has a 5.00-V reference and is 8 bits. The thermistor specifications are $R = 5.00\ \text{k}\Omega$ at 90°F, $P_D = 5\ \text{mW}/°C$, and a slope between 90°F and 110°F of $-8\ \Omega/°C$. The design should be made so that 90°F gives an ADC output of 5AH (90_{10}) and 110°F gives 6EH(110_{10}).

4.28 Solve Example 4.19 using the RTD as the feedback element of an inverting amplifier instead of using a bridge circuit.

4.29 A type K thermocouple measurement system must provide an output of 0 to 2.5 V for a temperature variation of 500° to 700°C. A three-terminal solid-state sensor with 12 mV/°C will be used to provide reference compensation. Develop the complete circuit.

4.30 A calibrated RTD with $\alpha = 0.0041/°C$, $R = 306.5\ \Omega$ at 20°C, and $P_D = 30\ \text{mW}/°C$ will be used to measure a critical reaction temperature. Temperature must be measured between 50° and 100°C with a resolution of at least 0.1°C. Devise a signal-conditioning system that will provide an appropriate digital output to a computer. Specify the requirements on the ADC and appropriate analog signal conditioning to interface to your ADC.

SUPPLEMENTARY PROBLEMS

S4.1 Figure 4.23 shows a system that is proposed to power a radio from the heat of a portable lantern. There are eight "wings" on a shield around the lantern flame. Each wing has type J thermocouples facing the flame, with the cold junctions on the outer edge of the wings. A radio in the base requires a minimum of 5.0 V at 50 mA to operate. The following assumptions are made: (1) Each thermocouple segment has a resistance of 0.05 Ω, and (2) the flame heat at the thermocouple junction is at least 300°C and nominal room temperature is 25°C.

a. Determine how many thermocouples are required and how they are distributed on the wings. Be sure to consider the internal resistance of the thermocouples.

b. What is the maximum power that can be delivered into any load at 300°C and at 400°C?

c. On a cold winter night, the ambient temperature can drop to 10°C. If the hot junctions are still at 300°C, what effect does this have on the voltage applied to the radio?

S4.2 An RTD with $\alpha = 0.0035/°C$ and $R = 300\ \Omega$ at 25°C will be used to measure the temperature of hot gas flowing in a pipe. The dissipation constant is 25 mW/°C, and the time constant is 5.5 s. Normal gas temperature is in the range of 100° to 220°C.

a. Design a system by which the temperature variation is converted into a voltage of −2.0 to +2.0 V. Keep self-heating to 0.5°C.

b. Occasionally a turbulent shock wave will propagate down the pipe, causing a sudden reduction in temperature to less than 50°C. Devise a comparator alarm that will signal such an event within 3 s of dropping below 100°C.

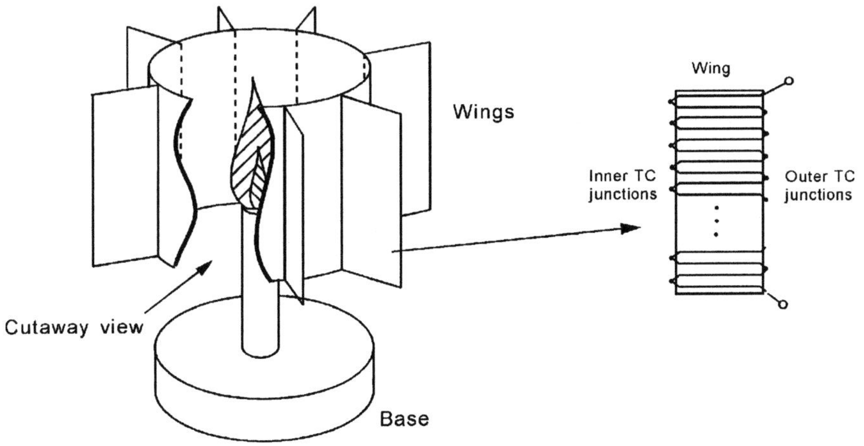

FIGURE 4.23
Electrical power from a lantern flame using thermocouples. See Problem S4.1.

FIGURE 4.24
Use of thermistors for measurement of wind speed in Problem S4.3.

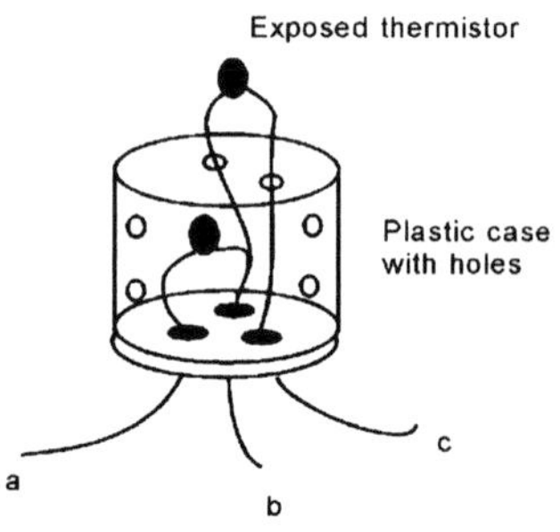

S4.3 Air-flow speed can be measured by comparing the temperatures of a shielded thermistor and one which is exposed to the moving air. A diagram of this sensor is shown in Figure 4.24. The thermistors are used in a bridge and both are operated at an elevated (no-wind) temperature of 45°C using self-heating. When wind blows, the exposed sensor will cool off and the other will not.

It has been experimentally found that the temperature difference between the thermistors are as shown in the following table for winds of 0 to 60 mph. Design a bridge and op amp system that will provide 0.0 V when the wind is 0 mph and 6.0 V when the wind is 60 mph so that voltage indicates speed directly. Conditions: (1) The thermistors have a resistance of 2.0 kΩ at 45°C with a slope of −24 Ω/°C within ±10°C, and the dissipation constant is 5.5 mW/°C, (2) use a 20-V supply voltage for the bridge, and (3) assume ambient temperature is 21°C.

Speed (mph)	ΔT(°C)	$\Delta R(\Omega)$	Bridge ΔV(V)	V_{out}(V)
0	0			0.0
10	−3.0			
20	−4.5			
30	−5.5			
40	−6.3			
50	−7.1			
60	−7.7			6.0

a. Complete the table from your design results.
b. Prepare a graph of output voltage versus wind speed.
c. What is the greatest error between voltage-indicated wind speed and actual speed? Why is there error?

CHAPTER 5

Mechanical Sensors

INSTRUCTIONAL OBJECTIVES

This chapter presents various types of sensors for the measurement of mechanical phenomena such as motion, force, pressure, and position. After you have read the chapter and worked through the examples and problems you will be able to:

- Describe three types of sensors for the measurement of displacement, location, or position.
- Explain the operating principle of an LVDT for measuring displacement.
- Calculate the strain experienced by a wire under the influence of a force.
- Design the application of a strain gauge for the measurement of stress.
- Explain the operation of a strain gauge-based load cell.
- Describe the operating principles of a spring-mass accelerometer.
- Explain the operating principle of a diaphragm pressure sensor.
- Describe the operating mechanism of an orifice plate flow sensor.

5.1 INTRODUCTION

The class of sensors used for the measurement of mechanical phenomena is of special significance because of the extensive use of these devices throughout the process-control industry. In many instances, an interrelation exists by which a sensor designed to measure some mechanical variable is used to measure another variable. To learn to use mechanical sensors, it is important to understand the mechanical phenomena themselves and the operating principles and application details of the sensor.

Our purposes here are to give an overview of the essential features associated with each variable and to make the reader conversant with the principal sensors used to measure mechanical variables, the characteristics of each, and appropriate application notes. As in previous chapters, an expert understanding of a phenomenon is not required to effectively employ sensors for its measurement.

5.2 DISPLACEMENT, LOCATION, OR POSITION SENSORS

The measurement of displacement, position, or location is an important topic in the process industries. Examples of industrial requirements to measure these variables are many and varied, and the required sensors are also of greatly varied designs. To give a few examples of measurement needs: (1) location and position of objects on a conveyor system, (2) orientation of steel plates in a rolling mill, (3) liquid/solid level measurements, (4) location and position of work piece in automatic milling operations, and (5) conversion of pressure to a physical displacement that is measured to indicate pressure. In the following sections, the basic principles of several common types of displacement, position, and location sensors are given.

5.2.1 Potentiometric Sensors

The simplest type of displacement sensor involves the action of displacement in moving the wiper of a potentiometer. This device then converts linear or angular motion into a changing resistance that may be converted directly to voltage and/or current signals. Such potentiometric devices often suffer from the obvious problems of mechanical wear, friction in the wiper action, limited resolution in wire-wound units, and high electronic noise.

Figure 5.1 shows a simple mechanical picture of the potentiometric displacement sensor. You will see that there is a wire wound around a form, making a wire-wound resistor with fixed resistance, R, between its endpoints, 1 and 2. A wiper assembly is connected in such a way that motion of an arm causes the wiper to slide across the wire-wound turns of the fixed resistor. An electrical connection is made to this wiper. Therefore, as the arm moves back and forth, the resistance between the wiper connection, 3, and either fixed resistor connection will change in proportion to the motion. This figure is highly pictorial. In actual sensors, the coil is very tightly wound of very fine wire.

In a schematic, the potentiometric sensor is simply a three-terminal variable resistor. The resolution of this sensor is limited to the distance, Δx, between individual turns of wire, with the resulting resistance change of the single turn, ΔR.

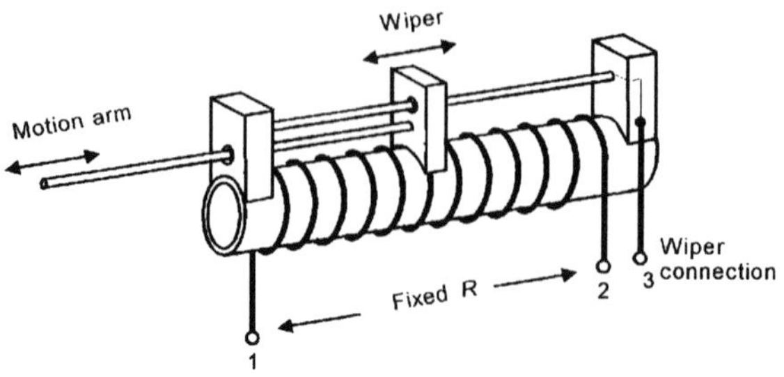

FIGURE 5.1
Potentiometric displacement sensor.

EXAMPLE 5.1 A potentiometric displacement sensor is to be used to measure work-piece motion from 0 to 10 cm. The resistance changes linearly over this range from 0 to 1 kΩ. Develop signal conditioning to provide a linear, 0- to 10-V output.

Solution

The key thing is to not lose the linearity of the resistance versus displacement. We cannot put the varying resistance in a divider to produce a varying voltage because the voltage varies nonlinearly with resistance. Remember though that the output voltage of an inverting amplifier varies linearly with the feedback resistance. Therefore, let's put the sensor in the feedback of a simple inverting amplifier. Then we would have something like

$$V_{out} = -\frac{R_2}{R_1}V_{In}$$

We can now get rid of that pesky negative by using V_{in} as a constant negative voltage, say -5.1 volts from a zener diode. Then we pick R_1 to give the desired output, 10 volts at 1 kΩ (10 cm),

$$10 = -\frac{1000}{R_1}(-5.1) \qquad \text{so} \qquad R_1 = 510\ \Omega$$

Figure 5.2 shows the circuit.

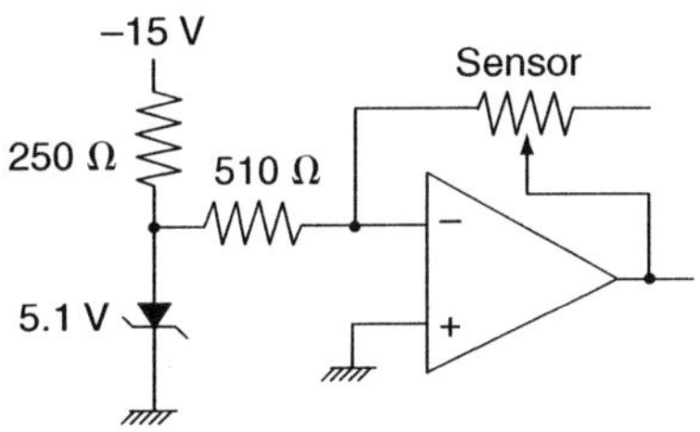

FIGURE 5.2
Circuit for Example 5.1.

5.2.2 Capacitive and Inductive Sensors

A second class of sensors for displacement measurement involves changes in capacity or inductance.

Capacitive The basic operation of a capacitive sensor can be seen from the familiar equation for a parallel-plate capacitor:

$$C = K\varepsilon_0\frac{A}{d} \tag{5.1}$$

where K = the dielectric constant
ε_0 = permittivity = 8.85 pF/m
A = plate common area
d = plate separation

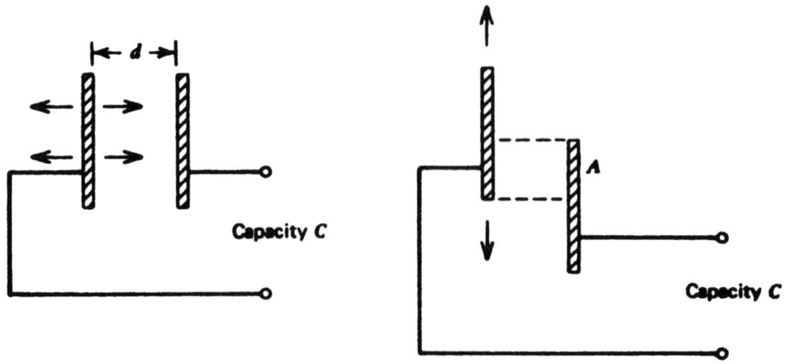

FIGURE 5.3
Capacity varies with the distance between the plates and the common area. Both effects are used in sensors.

There are three ways to change the capacity: variation of the distance between the plates (d), variation of the shared area of the plates (A), and variation of the dielectric constant (K). The former two methods are shown in Figure 5.3. The last method is illustrated in Example 5.4 later in this chapter. An ac bridge circuit or other active electronic circuit is employed to convert the capacity change to a current or voltage signal.

EXAMPLE 5.2 Figure 5.4 shows a capacitive-displacement sensor designed to monitor small changes in work-piece position. The two metal cylinders are separated by a plastic sheath/bearing of thickness 1 mm and dielectric constant at 1 kHz of 2.5. If the radius is 2.5 cm, find the sensitivity in pF/m as the upper cylinder slides in and out of the lower cylinder. What is the range of capacity if h varies from 1.0 to 2.0 cm?

Solution

The capacity is given by Equation (5.1). The net area is the area of the shared cylindrical area, which has a radius, r, and height, h. Thus, $A = 2\pi rh$, so the capacity can be expressed as

$$C = 2\pi K\varepsilon_0\frac{rh}{d}$$

FIGURE 5.4
Capacitive-displacement sensor for Example 5.2.

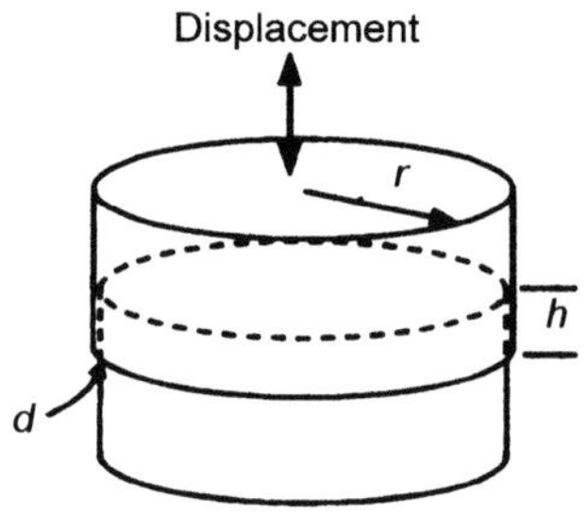

The sensitivity with respect to the height, *h,* is defined by how *C* changes with *h;* that is, it is given by the derivative

$$\frac{dC}{dh} = 2\pi K\varepsilon_0\frac{r}{d}$$

Substituting for the given values, we get

$$\frac{dC}{dh} = 2\pi(2.5)(8.85\,\text{pF/m})\,\frac{2.5 \times 10^{-2}\text{m}}{10^{-3}\text{m}} = 3475\,\text{pF/m}$$

Since the function is linear with respect to *h,* we find the capacity range as $C_{\text{min}} = (3475\,\text{pF/m})(10^{-2}\text{m}) = 34.75$ pF to $C_{\text{max}} = (3475\,\text{pF/m})(2 \times 10^{-2}\text{m}) = 69.50$ pF.

Inductive If a permeable core is inserted into an inductor as shown in Figure 5.5, the net inductance is increased. Every new position of the core produces a different inductance. In this fashion, the inductor and movable core assembly may be used as a displacement sensor. An ac bridge or other active electronic circuit sensitive to inductance then may be employed for signal conditioning.

5.2.3 Variable-Reluctance Sensors

The class of variable-reluctance displacement sensors differs from the inductive in that a moving core is used to vary the magnetic flux coupling between two or more coils, rather than changing an individual inductance. Such devices find application in many circumstances for the measure of both translational and angular displacements. Many configurations of this device exist, but the most common and extensively used is called a *linear variable differential transformer* (LVDT).

LVDT The LVDT is an important and common sensor for displacement measurement in the industrial environment. Figure 5.6 shows that an LVDT consists of three coils of wire wound on a hollow form. A core of permeable material can slide freely through the center of the form. The inner coil is the primary, which is excited by some ac source as

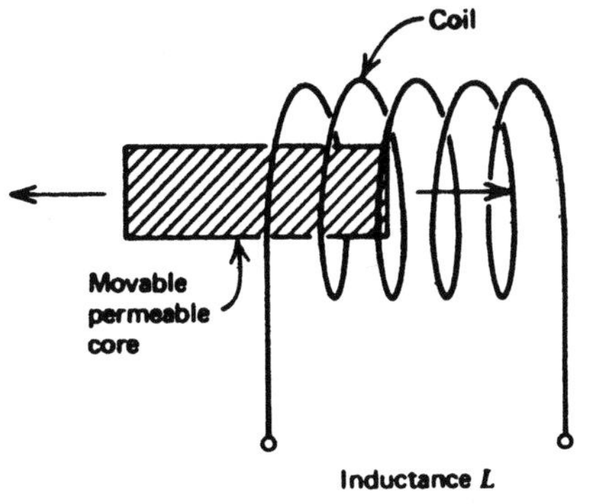

FIGURE 5.5
This variable-reluctance displacement sensor changes the inductance in a coil in response to core motion.

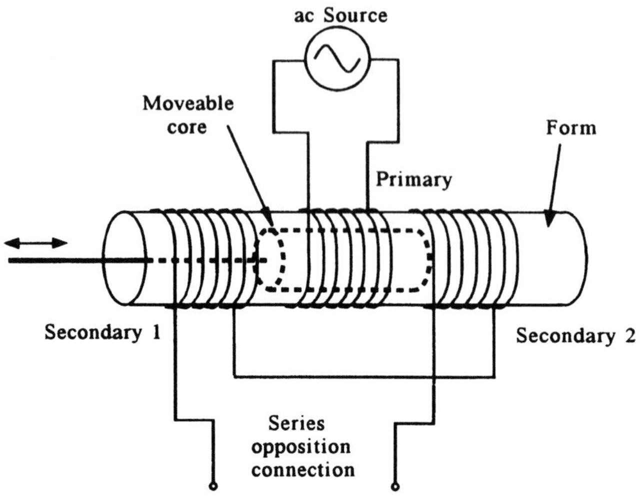

FIGURE 5.6
The LVDT has a movable core with the three coils as shown.

FIGURE 5.7
The LVDT secondary voltage amplitude for a series-opposition connection varies linearly with displacement.

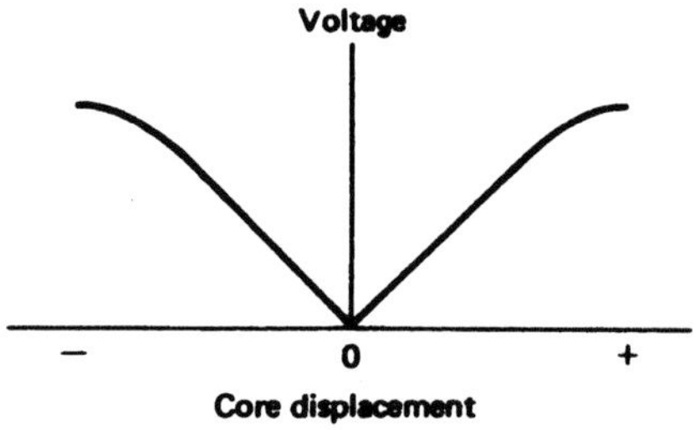

shown. Flux formed by the primary is linked to the two secondary coils, inducing an ac voltage in each coil.

When the core is centrally located in the assembly, the voltage induced in each primary is equal. If the core moves to one side or the other, a larger ac voltage will be induced in one coil and a smaller ac voltage in the other because of changes in the flux linkage associated with the core.

If the two secondary coils are wired in series opposition, as shown in Figure 5.6, then the two voltages will subtract; that is, the differential voltage is formed. When the core is centrally located, the net voltage is zero. When the core is moved to one side, the net voltage amplitude will increase. In addition, there is a change in phase with respect to the source when the core is moved to one side or the other.

A remarkable result, shown in Figure 5.7, is that the differential amplitude is found to increase linearly as the core is moved to one side or the other. In addition, as noted, there is a phase change as the core moves through the central location. Thus, by measurement of the voltage amplitude and phase, one can determine the direction and extent of the core motion—that is, the displacement.

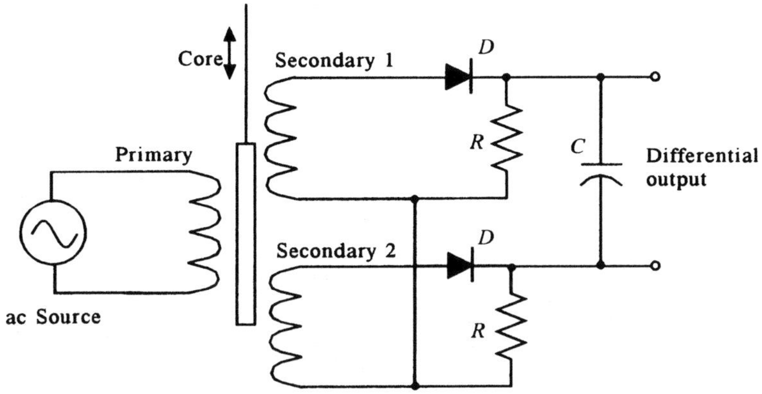

FIGURE 5.8
This simple circuit produces a bipolar dc voltage that varies with core displacement.

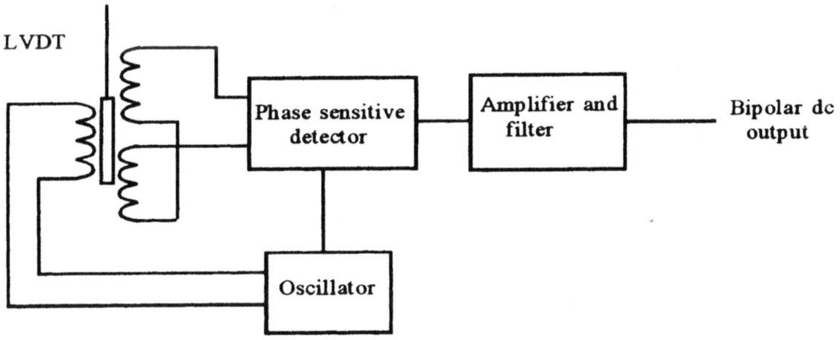

FIGURE 5.9
A more sophisticated LVDT signal-conditioning circuit uses phase-sensitive detection to produce a bipolar dc voltage output.

It turns out that a carefully manufactured LVDT can provide an output linear within $\pm 0.25\%$ over a range of core motion and with a very fine resolution, limited primarily by the ability to measure voltage changes.

The signal conditioning for LVDTs consists primarily of circuits that perform a phase-sensitive detection of the differential secondary voltage. The output is thus a dc voltage whose amplitude relates the extent of the displacement, and the polarity indicates the direction of the displacement. Figure 5.8 shows a simple circuit for providing such an output. An important limitation of this circuit is that the differential secondary voltage must be at least as large as the forward voltage drop of the diodes. The use of op amp detectors can alleviate this problem.

Figure 5.9 shows a more practical detection scheme, typically provided as a single integrated circuit (IC) manufactured specifically for LVDTs. The system contains a signal generator for the primary, a phase-sensitive detector (PSD), and amplifier/filter circuitry.

A variety of LVDTs are available with linear ranges at least from ±25 cm down to ±1 mm. The time response is dependent on the equipment to which the core is connected. The static transfer function is typically given in millivolts per millimeter (mV/mm) for a given primary amplitude. Also specified are the range of linearity and the extent of linearity.

EXAMPLE 5.3 An LVDT has a maximum core motion of ±1.5 cm with a linearity of ±0.3% over that range. The transfer function is 23.8 mV/mm. If used to track work-piece motion from −1.2 to +1.4 cm, what is the expected output voltage? What is the uncertainty in position determination due to nonlinearity?

Solution

Using the known transfer function, the output voltages can easily be found,

$$V(-1.2\,\text{cm}) = (23.8\,\text{mV/mm})(-12\,\text{mm}) = -285.6\,\text{mV}$$

and

$$V(1.4\,\text{cm}) = (23.8\,\text{mV/mm})(14\,\text{mm}) = 333\,\text{mV}$$

The linearity deviation shows up in deviations of the transfer function. Thus, the transfer function has an uncertainty of

$$(\pm 0.003)(23.8\,\text{mV/mm}) = \pm 0.0714\,\text{mV/mm}$$

This means that a measured voltage, V_m (in mV), could be interpreted as a displacement that ranges from $V_m/23.73$ to $V_m/23.87$ mm, which is approximately ±0.3%, as expected. Thus, if the sensor output was 333 mV, which is nominally 1.4 cm, the actual core position could range from 1.40329 to 1.39506 cm.

5.2.4 Level Sensors

The measurement of solid or liquid level calls for a special class of displacement sensors. The level measured is most commonly associated with material in a tank or hopper. A great variety of measurement techniques exist, as the following representative examples show.

Mechanical One of the most common techniques for level measurement, particularly for liquids, is a float that is allowed to ride up and down with level changes. This float, as shown in Figure 5.10a, is connected by linkages to a secondary displacement measuring system such as a potentiometric device or an LVDT core.

Electrical There are several purely electrical methods of measuring level. For example, one may use the inherent conductivity of a liquid or solid to vary the resistance seen by probes inserted into the material. Another common technique is illustrated in Figure 5.10b. In this case, two concentric cylinders are contained in a liquid tank. The level of the liquid partially occupies the space between the cylinders, with air in the remaining part. This device acts like two capacitors in parallel, one with the dielectric constant of air

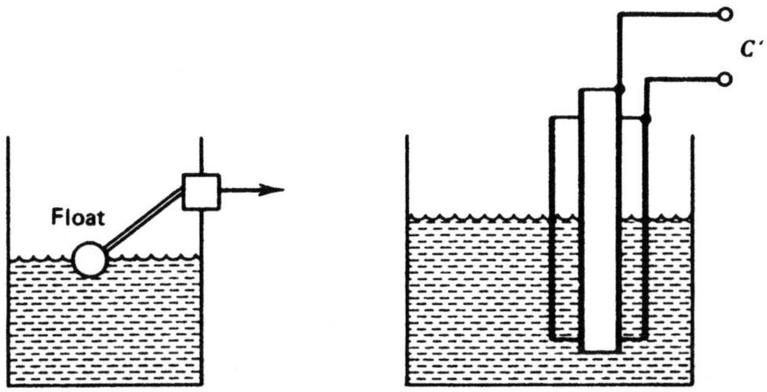

a) Level measurement by a float and linkage

b) Level measurement by a concentric cylindrical capacitor

FIGURE 5.10
There are many level-measurement techniques.

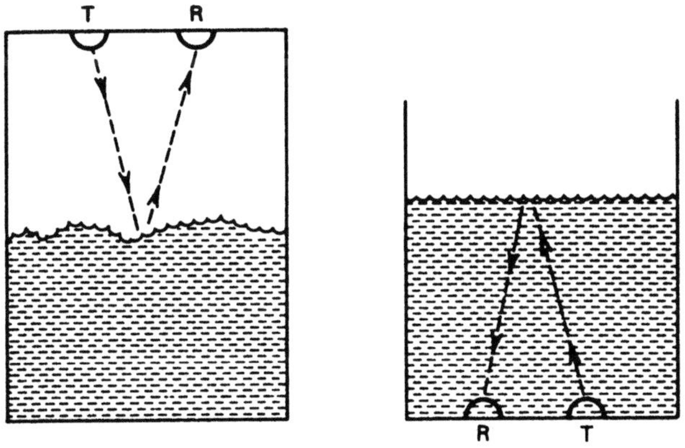

a) Solid or liquid, above surface measurement

b) Liquid material, below surface material

FIGURE 5.11
Ultrasonic level measurement needs no physical contact with the material, just a transmitter, *T*, and receiver, *R*.

($\approx$ 1) and the other with that of the liquid. Thus, variation of liquid level causes variation of the electrical capacity measured between the cylinders.

Ultrasonic The use of ultrasonic reflection to measure level is favored because it is a "noninvasive" technique; that is, it does not involve placing anything in the material. Figure 5.11 shows the external and internal techniques. Obviously, the external technique is better suited to solid-material level measurement. In both cases, the measurement depends on the length of time taken for reflections of an ultrasonic pulse from the surface of the material. Ultrasonic techniques based on reflection time also have become popular for ranging measurements.

Pressure For liquid measurement, it is also possible to make a noncontact measurement of level if the density of the liquid is known. This method is based on the well-known relationship between pressure at the bottom of a tank and the height and density of the liquid. This is addressed further in Section 5.5.1.

EXAMPLE 5.4

The level of ethyl alcohol is to be measured from 0 to 5 m using a capacitive system such as that shown in Figure 5.10b. The following specifications define the system:

$$\text{for ethyl alcohol: } K = 26 \text{ (for air, } K = 1)$$
$$\text{cylinder separation: } d = 0.5\,\text{cm}$$
$$\text{plate area: } A = 2\pi RL$$

where $R = 5.75\,\text{cm} = \text{average radius}$
$L = \text{distance along cylinder axis}$

Find the range of capacity variation as the alcohol level varies from 0 to 5 m.

Solution

We saw earlier that the capacity is given by $C = K\varepsilon_0(A/d)$. Therefore, all we need to do is find the capacity for the entire cylinder with *no* alcohol and then multiply that by 26.

$$A = 2\pi RL = 2\pi(0.0575\,\text{m})(5\,\text{m}) = 1.806\text{ m}^2$$

Thus, for air,

$$C = (1)(8.85\,\text{pF/M})(1.806\,\text{m}^2/0.005\,\text{m})$$
$$C = 3196\,\text{pF} \approx 0.0032\ \mu\text{F}$$

With the ethyl alcohol, the capacity becomes

$$C = 26(0.0032\ \mu\text{F})$$
$$C = \mathbf{0.0832\ \mu F}$$

The range is 0.0032 to 0.0832 μF.

5.3 STRAIN SENSORS

Although not obvious at first, the measurement of strain in solid objects is common in process control. The reason it is not obvious is that strain sensors are used as a secondary step in sensors to measure many other process variables, including flow, pressure, weight, and acceleration. Strain measurements have been used to measure pressures from over a million pounds per square inch to those within living biological systems. We will first review the concept of strain and how it is related to the forces that produce it, and then discuss the sensors used to measure strain.

5.3.1 Strain and Stress

Strain is the result of the application of forces to solid objects. The forces are defined in a special way described by the general term *stress.* For those readers needing a review of force principles, Appendix 4 discusses elementary mechanical principles, including force. In this section, we will define stress and the resulting strain.

Definition A special case exists for the relation between force applied to a solid object and the resulting deformation of that object. Solids are assemblages of atoms in which the atomic spacing has been adjusted to render the solid in equilibrium with all external forces acting on the object. This spacing determines the physical dimensions of the solid. If the applied forces are changed, the object atoms rearrange themselves again to come into equilibrium with the new set of forces. This rearrangement results in a change in physical dimensions that is referred to as a *deformation* of the solid.

The study of this phenomenon has evolved into an exact technology. The effect of applied force is referred to as a *stress,* and the resulting deformation as a *strain.* To facilitate a proper analytical treatment of the subject, stress and strain are carefully defined to emphasize the physical properties of the material being stressed and the specific type of stress applied. We delineate here the three most common types of stress-strain relationships.

Tensile Stress-Strain In Figure 5.12a, the nature of a tensile force is shown as a force applied to a sample of material so as to elongate or pull apart the sample. In this case, the stress is defined as

$$\text{tensile stress} = \frac{F}{A} \tag{5.2}$$

where F = applied force in N
A = cross-sectional area of the sample in m^2

We see that the units of stress are N/m^2 in SI units (or $lb/in.^2$ in English units), and they are like a pressure.

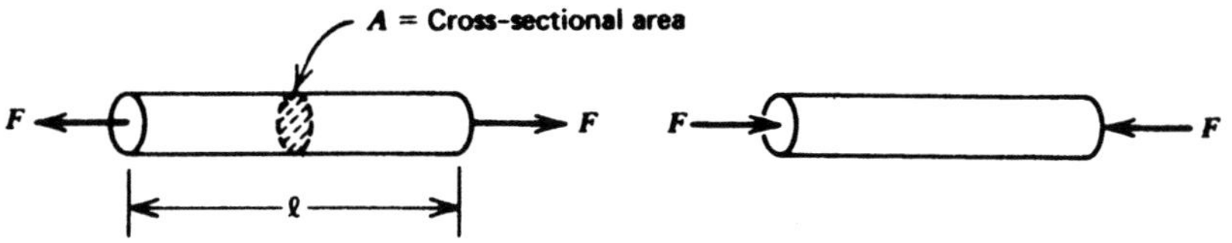

a) Tensile stress applied to a rod

b) Compressional stress applied to a rod

FIGURE 5.12
Tensile and compressional stress can be defined in terms of forces applied to a uniform rod.

The strain in this case is defined as the *fractional change in length* of the sample:

$$\text{tensile strain} = \frac{\Delta l}{l} \tag{5.3}$$

where Δl = change in length in m (in.)
l = original length in m (in.)

Strain is thus a unitless quantity.

Compressional Stress-Strain The only differences between *compressional* and *tensile* stress are the direction of the applied force and the polarity of the change in length. Thus, in a compressional stress, the force presses in on the sample, as shown in Figure 5.12b. The compressional stress is defined as in Equation (5.2):

$$\text{compressional stress} = \frac{F}{A} \tag{5.4}$$

The resulting strain is also defined as the fractional change in length as in Equation (5.3), but the sample will now decrease in length:

$$\text{compressional strain} = \frac{\Delta l}{l}$$

Shear Stress-Strain Figure 5.13a shows the nature of the shear stress. In this case, the force is applied as a *couple* (that is, *not* along the same line), tending to shear off the solid object that separates the force arms. In this case, the stress is again

$$\text{shear stress} = \frac{F}{A} \tag{5.5}$$

where F = force in N
A = cross-sectional area of sheared member in m^2

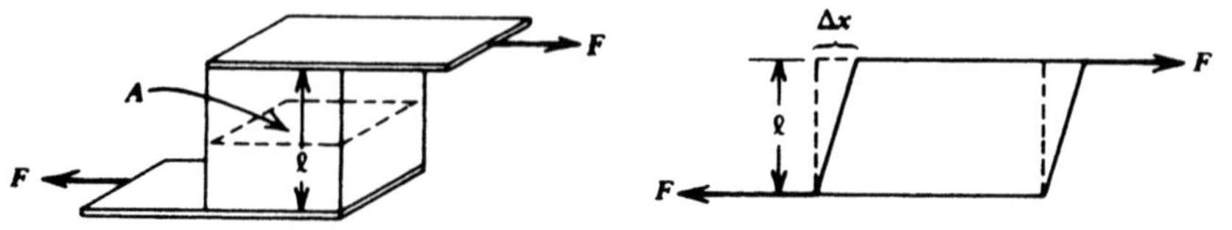

a) Shear stress results from a force couple b) Shear stress tends to deform an object as shown

FIGURE 5.13
Shear stress is defined in terms of forces not acting in a line (a couple), which deform a member linking the forces.

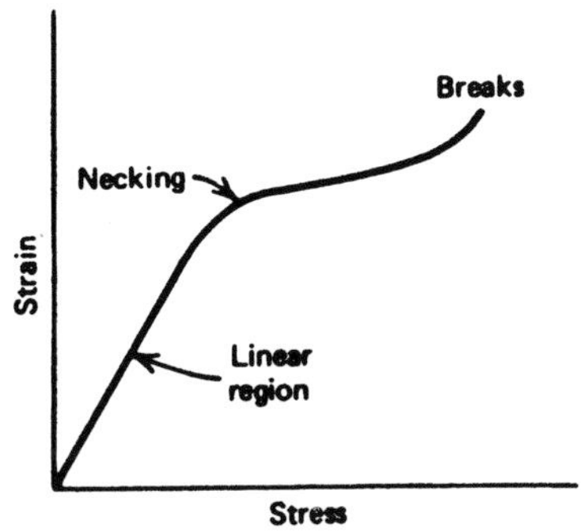

FIGURE 5.14
A typical stress-strain curve showing the linear region, necking, and eventual break.

The strain in this case is defined as the fractional change in dimension of the sheared member. This is shown in the cross-sectional view of Figure 5.13b.

$$\text{shear strain} = \frac{\Delta x}{l} \tag{5.6}$$

where Δx = deformation in m (as shown in Figure 5.13b)
l = width of a sample in m

Stress-Strain Curve If a specific sample is exposed to a range of applied stress and the resulting strain is measured, a graph similar to Figure 5.14 results. This graph shows that the relationship between stress and strain is linear over some range of stress. If the stress is kept within the linear region, the material is essentially *elastic* in that if the stress is removed, the deformation is also gone. But if the elastic limit is exceeded, permanent deformation results. The material may begin to "neck" at some location and finally break. Within the linear region, a specific type of material will always follow the same curves, despite different physical dimensions. Thus, we can say that the linearity and slope are a constant of the type of material only. In tensile and compressional stress, this constant is called the *modulus of elasticity,* or *Young's modulus,* as given by

$$E = \frac{\text{stress}}{\text{strain}} = \frac{F/A}{\Delta l/l} \tag{5.7}$$

where stress = F/A in N/m^2 (or lb/in.^{2})
strain = $\Delta l/l$ unitless
E = modulus of elasticity in N/m^2

The modulus of elasticity has units of stress—that is, N/m^2. Table 5.1 gives the modulus of elasticity for several materials. In an exactly similar fashion, the shear modulus is defined for shear stress-strain as

$$M = \frac{\text{stress}}{\text{strain}} = \frac{F/A}{\Delta x/l} \tag{5.8}$$

where Δx is defined in Figure 5.13b and all other units have been defined in Equation (5.7).

TABLE 5.1
Modulus of elasticity

Material	Modulus (N/m^2)
Aluminum	6.89×10^{10}
Copper	11.73×10^{10}
Steel	20.70×10^{10}
Polyethylene (plastic)	3.45×10^{8}

EXAMPLE 5.5 Find the strain that results from a tensile force of 1000 N applied to a 10-m aluminum beam having a 4×10^{-4}-m^2 cross-sectional area.

Solution
The modulus of elasticity of aluminum is found from Table 5.1 to be $E = 6.89 \times 10^{10}\,N/m^2$. Now we have, from Equation (5.7),

$$E = \frac{F/A}{\Delta l/l}$$

so that

$$\text{strain} = \frac{F}{EA}$$

$$= \frac{10^3 N}{(4 \times 10^{-4} m^2)(6.89 \times 10^{10} N/m^2)}$$

$$= \mathbf{3.63 \times 10^{-5}} \quad \text{or} \quad \mathbf{36.3\ \mu m/m} \qquad \text{(see next paragraph)}$$

Strain Units Although strain is a unitless quantity, it is common practice to express the strain as the ratio of two length units, for example, as m/m or in./in.; also, because the strain is usually a very small number, a micro (μ) prefix is often included. In this sense, a strain of 0.001 would be expressed as 1000 μin./in., or 1000 μm/m. In the previous example, the solution is stated as 36.3 μm/m. In general, the smallest value of strain encountered in most applications is 1 μm/m. Because strain is a unitless quantity, it is not necessary to do unit conversions. A strain of 153 μm/m could also be written in the form of 153 μin./in. or even 153 μfurlongs/furlong. Modern usage often just gives strain in "micros."

5.3.2 Strain Gauge Principles

In Section 4.2.1, we saw that the resistance of a metal sample is given by

$$R_0 = \rho \frac{l_0}{A_0} \tag{4.7}$$

where R_0 = sample resistance Ω
ρ = sample resistivity $\Omega \cdot$ m
l_0 = length in m
A_0 = cross-sectional area in m^2

Suppose this sample is now stressed by the application of a force, F, as shown in Figure 5.12a. Then we know that the material elongates by some amount, Δl, so that the new length is $l = l + \Delta l$. It is also true that in such a stress-strain condition, although the sample lengthens, its volume will remain nearly constant. Because the volume unstressed is $V = l_0 A_0$, it follows that if the volume remains constant and the length increases, then the area must decrease by some amount, ΔA:

$$V = l_0 A_0 = (l_0 + \Delta l)(A_0 - \Delta A) \tag{5.9}$$

Because both length and area have changed, we find that the resistance of the sample will have also changed:

$$R = \rho \frac{l_0 + \Delta l}{A_0 - \Delta A} \tag{5.10}$$

Using Equations (5.9) and (5.10), the reader can verify that the new resistance is approximately given by

$$R \simeq \rho \frac{l_0}{A_0}\left(1 + 2\frac{\Delta l}{l_0}\right) \tag{5.11}$$

from which we conclude that the change in resistance is

$$\Delta R \simeq 2R_0 \frac{\Delta l}{l_0} \tag{5.12}$$

Equation (5.12) is the basic equation that underlies the use of metal strain gauges because it shows that the strain $\Delta l / l$ converts directly into a *resistance change.*

EXAMPLE 5.6

Find the approximate change in a metal wire of resistance 120 Ω that results from a strain of 1000 μm/m.

Solution

We can find the change in gauge resistance from

$$\Delta R \simeq 2R_0 \frac{\Delta l}{l_0}$$

$$\Delta R \simeq (2)(120)(10^{-3})$$

$$\Delta R = \mathbf{0.24\ \Omega}$$

Example 5.6 shows a significant factor regarding strain gauges. The change in resistance is very small for typical strain values. For this reason, resistance change measurement methods used with strain gauges must be highly sophisticated.

Measurement Principles The basic technique of strain gauge (SG) measurement involves attaching (gluing) a metal wire or foil to the element whose strain is to be measured. As stress is applied and the element deforms, the SG material experiences the same deformation, if it is securely attached. Because strain is a fractional change in length, the change in SG resistance reflects the strain of both the gauge and the element to which it is secured.

Temperature Effects If not for temperature compensation effects, the aforementioned method of SG measurement would be useless. To see this, we need only note that the metals used in SG construction have linear temperature coefficients of $\alpha \cong 0.004/°\text{C}$, typical for most metals. Temperature changes of 1°C are not uncommon in measurement conditions in the industrial environment. If the temperature change in Example 5.6 had been 1°C, substantial change in resistance would have resulted. Thus, from Chapter 4,

$$R(T) = R(T_0)[1 + \alpha_0 \Delta T]$$

or

$$\Delta R_T = R_0 \alpha \Delta T$$

where
ΔR_T = resistance change because of temperature change
$\alpha_0 \simeq 0.004/°\text{C}$ in this case
$\Delta T \simeq 1°\text{C}$ in this case
$R(T_0) = 120\,\Omega$ nominal resistance

Then, we find $\Delta R_T = 0.48\,\Omega$, which is *twice* the change because of strain! Obviously, temperature effects can mask the strain effects we are trying to measure. Fortunately, we are able to compensate for temperature and other effects, as shown in the signal-conditioning methods in the next section.

5.3.3 Metal Strain Gauges

Metal SGs are devices that operate on the principles discussed earlier. The following items are important to understanding SG applications.

Gauge Factor The relation between strain and resistance change [Equation (5.12)] is only approximately true. Impurities in the metal, the type of metal, and other factors lead to slight corrections. An SG specification always indicates the correct relation through statement of a *gauge factor* (GF), which is defined as

$$\text{GF} = \frac{\Delta R/R}{\text{strain}} \tag{5.13}$$

where
$\Delta R/R$ = fractional change in gauge resistance because of strain
strain = $\Delta l/l$ = fractional change in length

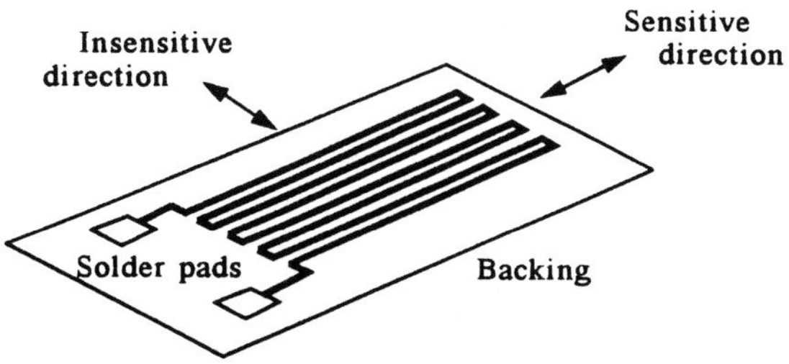

FIGURE 5.15
A metal strain gauge is composed of thin metal deposited in a pattern on a backing or carrier material.

For metal gauges, this number is always close to 2. For some special alloys and carbon gauges, the GF may be as large as 10. A high gauge factor is desirable because it indicates a larger change in resistance for a given strain and is easier to measure.

Construction Strain gauges are used in two forms, wire and foil. The basic characteristics of each type are the same in terms of resistance change for a given strain. The design of the SG itself is such as to make it very long in order to give a large enough nominal resistance (to be practical), and to make the gauge of sufficiently fine wire or foil so as not to resist strain effects. Finally, the gauge sensitivity is often made *unidirectional*; that is, it responds to strain *in only one direction.* In Figure 5.15, we see the common pattern of SGs that provides these characteristics. By folding the material back and forth as shown, we achieve a long length to provide high resistance. Further, if a strain is applied transversely to the SG length, the pattern will tend to unfold rather than stretch, with no change in resistance. These gauges are usually mounted on a paper backing that is bonded (using epoxy) to the element whose strain is to be measured. The nominal SG resistance (no strain) available are typically 60, 120, 240, 350, 500, and 1000 Ω. The most common value is 120 Ω.

Signal Conditioning Two effects are critical in the signal-conditioning techniques used for SGs. The first is the small, fractional changes in resistance that require carefully designed resistance measurement circuits. A good SG system might require a resolution of 2 μm/m strain. From Equation (5.12), this would result in a ΔR of only $4.8 \times 10^{-4}\ \Omega$ for a nominal gauge resistance of 120 Ω.

The second effect is the need to provide some compensation for temperature effects to eliminate masking changes in strain.

The bridge circuit provides the answer to both effects. The sensitivity of the bridge circuit for detecting small changes in resistance is well known. Furthermore, by using a dummy gauge as shown in Figure 5.16a, we can provide the required temperature compensation. In particular, the dummy is mounted in an insensitive orientation (Figure 5.16b), but in the same proximity as the active SG. Then, both gauges change in resistance from temperature effects, but the bridge does not respond to a change in both strain gauges. Only the active SG responds to strain effects. This is called a *one-arm bridge.*

The sensitivity of this bridge to strain can be found by considering the equation for bridge offset voltage. Suppose $R_1 = R_2 = R_D = R$, which is the nominal (unstrained) gauge resistance. Then the active strain gauge resistance will be given by

$$R_A = R\left(1 + \frac{\Delta R}{R}\right)$$

FIGURE 5.16
Strain gauges are used in pairs to provide temperature compensation. In some cases, such as this, only one gauge actually deforms during stress.

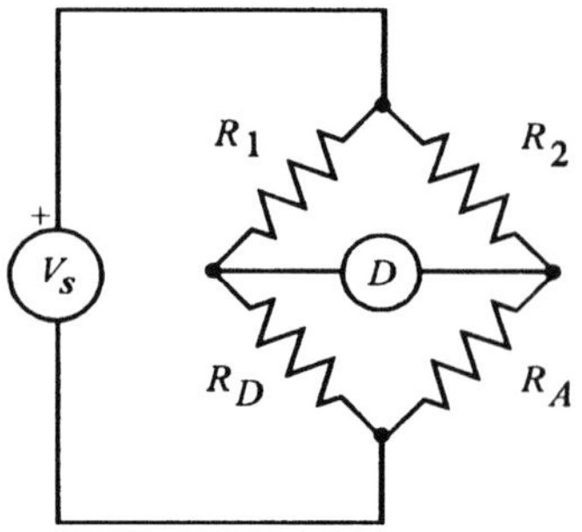

(a)

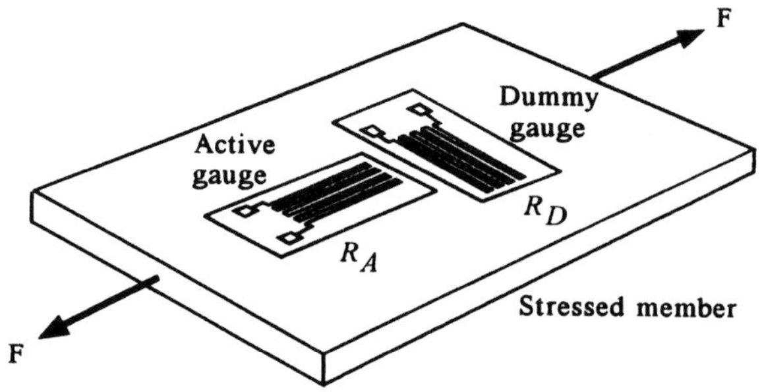

(b)

and the bridge off-null voltage will be given by

$$\Delta V = V_s\left[\frac{R_D}{R_D + R_1} - \frac{R_A}{R_A + R_2}\right]$$

If substitutions are made as defined previously, this voltage can be shown to be

$$\Delta V = -\frac{V_s}{4}\frac{\dfrac{\Delta R}{R}}{1 + \dfrac{\Delta R}{R}} \approx -\frac{V_s}{4}\frac{\Delta R}{R}$$

where the approximation is good for $(\Delta R/R) \ll 1$. Substituting from Equation (5.13) allows the expression for ΔV in terms of strain:

$$\Delta V = -\frac{V_s}{4} GF \frac{\Delta l}{l} \tag{5.14}$$

EXAMPLE 5.7

A strain gauge with GF = 2.03 and $R = 350\,\Omega$ is used in the bridge of Figure 5.16a. The bridge resistors are $R_1 = R_2 = 350\,\Omega$, and the dummy gauge has $R = 350\,\Omega$. If a tensile strain of $1450\,\mu\text{m/m}$ is applied, find the bridge offset voltage if $V_s = 10.0$ V. Find the relation between bridge off-null voltage and strain. How much voltage results from a strain of 1 micro?

Solution

With no strain, the bridge is balanced. When the strain is applied, the gauge resistance will change by a value given by

$$\text{GF} = \frac{\Delta R/R}{\text{strain}}$$

Thus,

$$\Delta R = (\text{GF})(\text{strain})(R)$$
$$\Delta R = (2.03)(1.45 \times 10^{-3})(350\,\Omega)$$
$$\Delta R = 1.03\,\Omega$$

Since it's tensile strain the resistance will increase to $R = 351\,\Omega$. The bridge offset voltage is

$$\Delta V = \frac{RV}{R_1 + R} - \frac{R_A V}{R_A + R_2}$$

Thus,

$$\Delta V = 5 - \frac{(351)(10)}{701}$$
$$\Delta V = \mathbf{-0.007\ V}$$

so that a 7-mV offset results.

The sensitivity is found from Equation (5.14):

$$\Delta V = -\frac{10}{2}(2.03)\frac{\Delta l}{l} = -10.15\frac{\Delta l}{l}$$

Thus, every micro of strain will supply only $10.15\,\mu\text{V}$.

Another configuration that is often employed uses active strain gauges in two arms of the bridge, and is thus called a *two-arm bridge.* All four arms are strain gauges, but two are for temperature compensation only. This has the added advantage of doubling the sensitivity. The bridge off-null voltage in terms of strain is given by

$$\Delta V = \frac{V_s}{2}\text{GF}\frac{\Delta l}{l} \tag{5.15}$$

Obviously, the placement of the active and dummy gauges in the environment and in the bridge circuit is important. Figure 5.17 shows a common application of strain gauges to

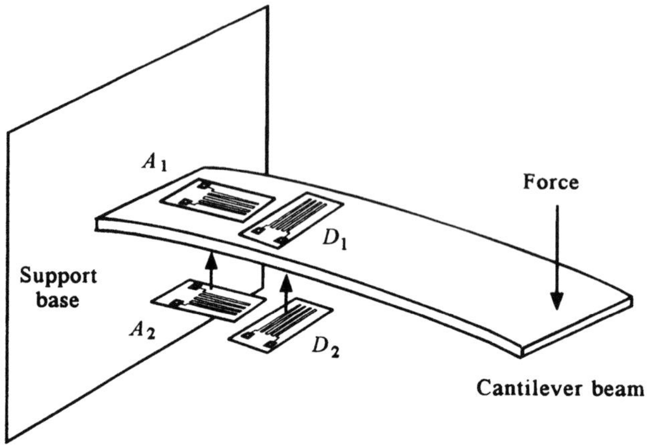

FIGURE 5.17
This structure shows how four gauges can be used to measure beam bending. Two respond to bending, and two are for temperature compensation.

measure deflections of a cantilever beam. This is a beam that is supported at only one end, and deflects as shown when a load is applied. In this application, it is common to use a two-arm bridge. One pair of active (A_1) and dummy (D_1) gauges is mounted on the top surface. The active gauge will experience tension with downward deflection of the beam, and its resistance will increase. The second pair, A_2 and D_2, are mounted on the bottom surface. The active gauge will experience compression with downward deflection, and its resistance will decrease.

EXAMPLE 5.8

Show how the four gauges in Figure 5.17 are connected into a bridge.

Solution

The gauges must be connected so that the off-null voltage increases with strain. Thus, one divider voltage should increase and the other should decrease so that the difference grows. This can be accomplished by using the active gauges in bridge resistor positions R_3 and R_4 of the standard bridge configuration shown in Figure 5.18.

It is also possible to wire strain gauges in a four-arm bridge, where all four gauges are active and temperature compensation is still supplied. In this case, the sensitivity is increased by another factor of two, from Equation (5.15).

FIGURE 5.18
Solution to Example 5.8.

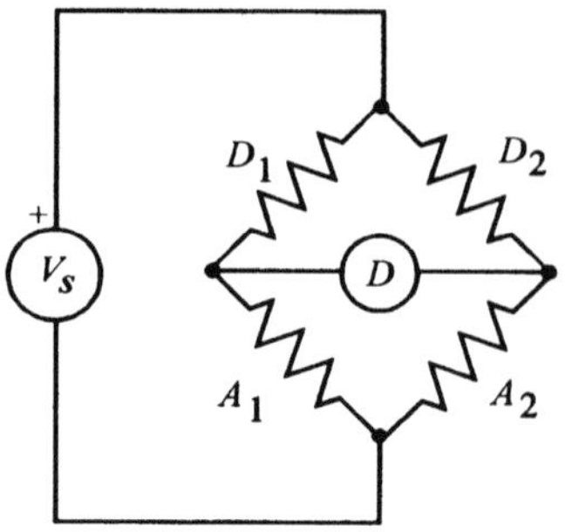

5.3.4 Semiconductor Strain Gauges (SGs)

The use of semiconductor material, notably silicon, for SG application has increased over the past few years. There are presently several disadvantages to these devices compared to the metal variety, but numerous advantages for their use.

Principles As in the case of the metal SGs, the basic effect is a change of resistance with strain. In the case of a semiconductor, the resistivity also changes with strain, along with the physical dimensions. This is due to changes in electron and hole mobility with changes in crystal structure as strain is applied. The net result is a much larger gauge factor than is possible with metal gauges.

Gauge Factor The semiconductor device gauge factor (GF) is still given by Equation (5.13):

$$\mathrm{GF} = \frac{\Delta R/R}{\text{strain}}$$

For semiconductor strain gauges, the GF is often negative, which means the resistance decreases when a tensile (stretching) stress is applied. Furthermore, the GF can be much larger than for metal strain gauges, in some cases as large as -200 with no strain. It must also be noted, however, that these devices are highly nonlinear in resistance versus strain. In other words, the gauge factor is not a constant as the strain takes place. Thus, the gauge factor may be -150 with no strain, but drop (nonlinearly) to -50 at $5000\,\mu\text{m/m}$. The resistance change will be nonlinear with respect to strain. To use the semiconductor strain gauge to measure strain, we must have a curve or table of values of gauge factor versus resistance.

Construction The semiconductor strain gauge physically appears as a band or strip of material with electrical connection, as shown in Figure 5.19. The gauge is either bonded

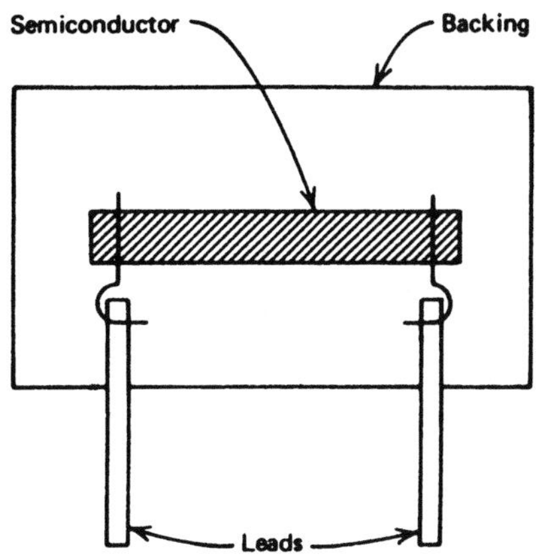

FIGURE 5.19
Typical semiconductor strain gauge structure.

directly onto the test element or, if encapsulated, is attached by the encapsulation material. These SGs also appear as IC assemblies in configurations used for other measurements.

Signal Conditioning The signal conditioning is still typically a bridge circuit with temperature compensation. An added problem is the need for linearization of the output because the basic resistance versus strain characteristic is nonlinear.

EXAMPLE 5.9

a. Contrast the resistance change produced by a 150-μm/m strain in a metal gauge with GF = 2.13 with

b. A semiconductor SG with GF = −151. Nominal resistances are both 120 Ω.

Solution

From the basic equation

$$\text{GF} = \frac{\Delta R/R}{\text{strain}}$$

a. We find for the metal gauge SG

$$\Delta R = (120\ \Omega)(2.13)(0.15 \times 10^{-3})$$
$$\Delta R = 0.038\ \Omega$$

b. For the semiconductor gauge, the change is

$$\Delta R = (120\ \Omega)(-151)(0.15 \times 10^{-3})$$
$$\Delta R = \mathbf{-2.72\ \Omega}$$

5.3.5 Load Cells

One important direct application of SGs is for the measurement of force or weight. These transducer devices, called *load cells* measure deformations produced by the force or weight. In general, a beam or yoke assembly is used that has several strain gauges mounted so that the application of a force causes a strain in the assembly that is measured by the gauges. A common application uses one of these devices in support of a hopper or feed of dry or liquid materials. A measure of the weight through a load cell yields a measure of the quantity of material in the hopper. Generally, these devices are calibrated so that the force (weight) is directly related to the resistance change. Forces as high as 5 MN (approximately 10^6 lb) can be measured with an appropriate load cell.

EXAMPLE 5.10

Figure 5.20 shows a simple load cell consisting of an aluminum post of 2.500-cm radius with a detector and compensation strain gauges. The 120 Ω strain gauges are used in the bridge of Figure 5.16, with $V = 2$ V, $R_1 = R_2 = R_D = 120.0\ \Omega$, and GF = 2.13. Find the variation of bridge offset voltage for a load of 0 to 5000 lb.

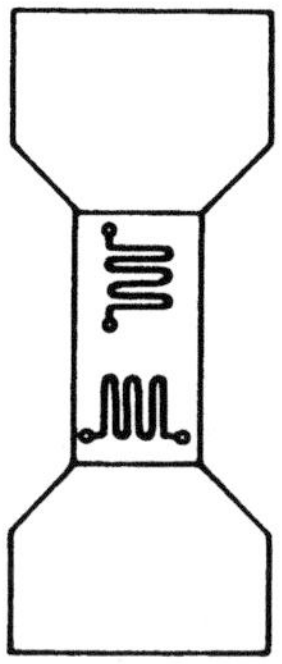

FIGURE 5.20
Load cell for Example 5.10.

Solution
We can find the strain for a 5000-lb load, then the resulting change in resistance, and from that, the bridge offset voltage. First, we change the force to newtons:

$$(5000\,\text{lb}/0.2248\,\text{lb/N}) = 22{,}240\,\text{N}$$

The cross-sectional area of the post is

$$A = \pi r^2 = \pi(0.025\,\text{m})^2 = 1.963 \times 10^{-3}\,\text{m}^2$$

From Table 5.1, the modulus of elasticity of aluminum is $E = 6.89 \times 10^{10}\,\text{N/m}^2$. From Equation (5.7), we find the strain

$$\Delta l/l = F/EA$$
$$\Delta l/l = (22{,}240\,\text{N})/[(6.89 \times 10^{10}\,\text{N/m}^2)(1.963 \times 10^{-3}\,\text{m}^2)]$$
$$\Delta l/l = 1.644 \times 10^{-4} = 164.4\,\mu\text{m/m}(\textit{or}\ \mu\text{in./in.})$$

The relationship between resistance and strain is given by Equation (5.13) ($\text{GF} = (\Delta R/R)/(\Delta l/l)$, so the resistance is given by

$$\Delta R/R = 2.13(1.644 \times 10^{-4})$$
$$= 3.502 \times 10^{-4}$$

Because $R = 120.0\ \Omega$, $\Delta R = 0.04203\ \Omega$. To get the bridge offset voltage, we note that the post is under compression and, therefore, the resistance will decrease. With no strain, the bridge is nulled. Under a 5000-lb load, the active gauge has $R = 119.958\ \Omega$. Thus, the offset voltage of the bridge is

$$\Delta V = 2\frac{120}{120 + 120} - 2\frac{119.958}{120 + 119.958}$$
$$\Delta V = 1.750 \times 10^{-4}\,\text{V} = 175.0\,\mu\text{V}$$

As the force varies from 0 to 5000 lb, the offset voltage varies from 0 to 175 μV.

The form of load cell considered in Figure 5.20 is fine for illustrating principles, but real load cells cannot be made in this simple way. The problem is that forces applied to the top of the load cell may cause it to lean or bend, instead of simply compressing. In such a case, one side surface of the beam may experience compression while the other side undergoes tension. Obviously, this will alter the correct interpretation of the result.

Practical load cells are made with yoke assemblies designed so that mounted strain gauges cannot be exposed to stresses other than those caused by the compressional force applied to the cell.

5.4 MOTION SENSORS

Motion sensors are designed to measure the rate of change of position, location, or displacement of an object that is occurring. If the position of an object as a function of time is $x(t)$, then the first derivative gives the speed of the object, $v(t)$, which is called the *velocity* if a direction is also specified. If the speed of the object is also changing, then the first derivative of the speed gives the acceleration. This is also the second derivative of the position.

$$v(t) = \frac{dx(t)}{dt} \tag{5.16}$$

$$a(t) = \frac{dv(t)}{dt} = \frac{d^2x(t)}{dt^2} \tag{5.17}$$

The primary form of motion sensor is the *accelerometer.* This device measures the acceleration, $a(t)$, of an object. By integrating Equations (5.16) and (5.17), it is easy to show that the accelerometer can be used to determine both the speed and position of the object as well:

$$v(t) = v(0) + \int_0^t a(t)dt \tag{5.18}$$

$$x(t) = x(0) + \int_0^t v(t)dt \tag{5.19}$$

Thus, in the accelerometer we have a sensor that can provide acceleration, speed (or velocity), and position information.

5.4.1 Types of Motion

The design of a sensor to measure motion is often tailored to the type of motion that is to be measured. It will help you understand these sensors if you have a clear understanding of the types of motion.

The proper unit of acceleration is meters per second squared (m/s^2). Then speed will be in meters per second (m/s) and position in meters (m). Often, acceleration is expressed by comparison with the acceleration due to gravity at the Earth's surface. This amount of acceleration, which is approximately 9.8 m/s^2, is called a *gee,* which is given as a bold **g** in this text.

EXAMPLE 5.11 An automobile is accelerating away from a stop sign at 26.4 ft/s^2. What is the acceleration in m/s^2 and in **g**s?

Solution

To find the acceleration in m/s^2, we simply convert the feet to meters according to 2.54 cm/in. and 12 in./ft:

$$a = (26.4\ \text{ft/s}^2)(12\ \text{in./ft})(2.54\ \text{cm/in.})(0.01\ \text{m/cm})$$
$$a = 8.05\ \text{m/s}^2$$

In terms of **g**s, we then have

$$a_{\mathbf{g}} = (8.05\ \text{m/s}^2)/(9.8\ \text{m/s}^2/\mathbf{g}) = 0.82\ \mathbf{g}\text{s}$$

where the subscript on a indicates the units are **g**s. Thus, this acceleration away from the stop sign provides an acceleration of about 80% of that caused by gravity at the Earth's surface.

Rectilinear This type of motion is characterized by velocity and acceleration which is composed of straight-line segments. Thus, objects may accelerate forward to a certain velocity, decelerate to a stop, reverse, and so on. There are many types of sensors designed to handle this type of motion. Typically, maximum accelerations are less than a few **g**s, and little angular motion (in a curved line) is allowed. If there is angular motion, then several rectilinear motion sensors must be used, each sensitive to only one line of motion. Thus, if vehicle motion is to be measured, two transducers may be used, one to measure motion in the forward direction of vehicle motion and the other perpendicular to the forward axis of the vehicle.

Angular Some sensors are designed to measure only rotations about some axis, such as the angular motion of the shaft of a motor. Such devices cannot be used to measure the physical displacement of the whole shaft, but only its rotation.

Vibration In the normal experiences of daily living, a person rarely experiences accelerations that vary from 1 **g** by more than a few percent. Even the severe environments of a rocket launching involve accelerations of only 1 **g** to 10 **g**. However, if an object is placed in periodic motion about some equilibrium position, as in Figure 5.21, very large *peak* accelerations may result that reach to 100 **g** or more. This motion is called *vibration*. Clearly, the measurement of acceleration of this magnitude is very important to industrial environments, where vibrations are often encountered from machinery operations. Often, vibrations are somewhat random in both the frequency of periodic motion and the magnitude of displacements from equilibrium. For analytical treatments, vibration is defined in terms of a regular periodic motion where the position of an object in time is given by

$$x(t) = x_0 \sin \omega t \tag{5.20}$$

FIGURE 5.21
An object in periodic motion about an equilibrium at $x = 0$. The peak displacement is x_0.

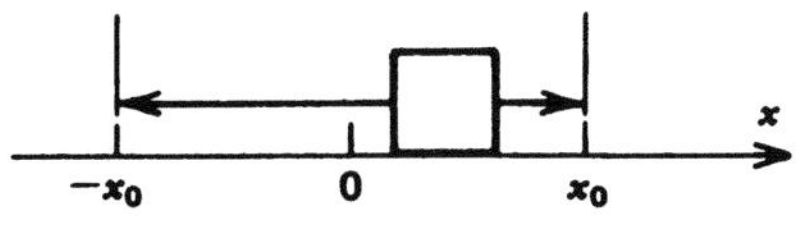

where $x(t)$ = object position in m
x_0 = peak displacement from equilibrium in m
ω = angular frequency in rad/s

The definition of ω as angular frequency is consistent with the reference to ω as angular speed. If an object rotates, we define the time to complete one rotation as a period T, that corresponds to a frequency $f = 1/T$. The frequency represents the number of revolutions per second and is measured in hertz (Hz), where 1Hz = 1 revolution per second. An angular rate of one revolution per second corresponds to an angular velocity of 2π rad/s, because one revolution sweeps out 2π radians. From this argument, we see that f and ω are related by

$$\omega = 2\pi f \tag{5.21}$$

Because f and ω are related by a constant, we refer to ω as both angular frequency and angular velocity.

Now we can find the vibration velocity as a derivative of Equation (5.20):

$$v(t) = \omega x_0 \cos \omega t \tag{5.22}$$

and we can get the vibration acceleration from a derivative of Equation (5.22):

$$a(t) = -\omega^2 x_0 \sin \omega t \tag{5.23}$$

Vibration position, velocity, and acceleration are all periodic functions having the same frequency. Of particular interest is the *peak* acceleration:

$$a_{peak} = \omega^2 x_0 \tag{5.24}$$

We see that the peak acceleration is dependent on ω^2, the angular frequency squared. This may result in very large acceleration values, even with modest peak displacements, as Example 5.12 shows.

EXAMPLE 5.12 A water pipe vibrates at a frequency of 10 Hz with a displacement of 0.5 cm. Find **(a)** the peak acceleration in m/s², and **(b)** **g** acceleration.

Solution

The peak acceleration will be given by

a. $a_{peak} = \omega^2 x_0$

where $\omega = 2\pi f = 20\pi$ rad/s and $x_0 = 0.5\,\text{cm} = 0.005\,\text{m}$

$a_{peak} = (20\pi)^2(0.005)$

$a_{peak} = \mathbf{19.7\ m/s^2}$

b. Noting that $1\mathbf{g} = 9.8\ \text{m/s}^2$, we get

$$a_{peak} = (19.7\,\text{m/s}^2)\left(\frac{1\,\mathbf{g}}{9.8\,\text{m/s}^2}\right)$$

$$a_{peak} = \mathbf{2.0\ g}$$

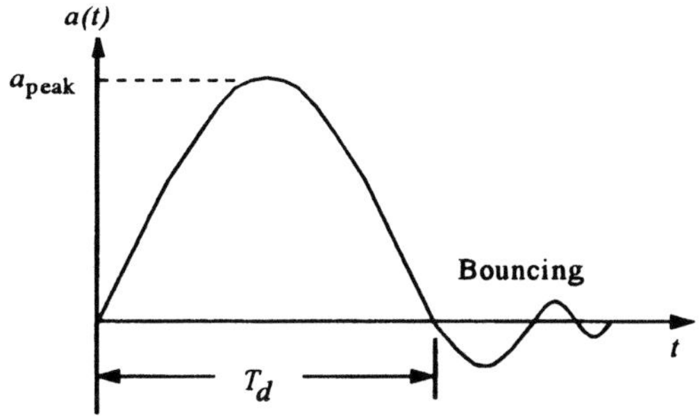

FIGURE 5.22
Typical shock acceleration profile.

A 2-**g** vibrating excitation of any mechanical element can be destructive, yet this is generated under the modest conditions of Example 5.12. A special class of sensors has been developed for measuring vibration acceleration.

Shock A special type of acceleration occurs when an object that may be in uniform motion or modestly accelerating is suddenly brought to rest, as in a collision. Such phenomena are the result of very large accelerations, or actually *decelerations,* as when an object is dropped from some height onto a hard surface. The name *shock* is given to decelerations that are characterized by very short times, typically on the order of milliseconds, with peak accelerations over 500 **g.** In Figure 5.22, we have a typical acceleration graph as a function of time for a shock experiment. This graph is characterized by a maximum or peak deceleration, a_{peak}, a shock duration, T_d, and bouncing. We can find an average shock by knowing the velocity of the object and the shock duration, as considered in Example 5.13.

EXAMPLE 5.13 A TV set is dropped from a 2-m height. If the shock duration is 5 ms, find the average shock in **g.**

Solution

The TV accelerates at 9.8 m/s^2 for 2 m. We find the velocity as

$$v^2 = 2\,gx$$
$$v^2 = (2)(9.8\,\text{m/s}^2)(2\,\text{m})$$
$$v = 6.3\,\text{m/s}$$

If the duration is 5 ms, we have

$$\bar{a} = \frac{6.3\,\text{m/s}}{5 \times 10^{-3}\text{s}}$$
$$\bar{a} = 1260\,\text{m/s}^2$$

or 128 **g.** No wonder that the TV breaks apart when it hits the ground!

5.4.2 Accelerometer Principles

There are several physical processes that can be used to develop a sensor to measure acceleration. In applications that involve flight, such as aircraft and satellites, accelerometers are based on properties of rotating masses. In the industrial world, however, the most common design is based on a combination of Newton's law of mass acceleration and Hooke's law of spring action.

Spring-Mass System Newton's law simply states that if a mass, m, is undergoing an acceleration, a, then there must be a force, F, acting on the mass and given by $F = ma$. Hooke's law states that if a spring of spring constant k is stretched (extended) from its equilibrium position for a distance Δx, then there must be a force acting on the spring given by $F = k\Delta x$.

In Figure 5.23a we have a mass that is free to slide on a base. The mass is connected to the base by a spring that is in its unextended state and exerts no force on the mass. In Figure 5.23b, the whole assembly is accelerated to the left, as shown. Now the spring extends in order to provide the force necessary to accelerate the mass. This condition is described by equating Newton's and Hooke's laws:

$$ma = k\Delta x \tag{5.25}$$

where k = spring constant in N/m
Δx = spring extension in m
m = mass in kg
a = acceleration in $\mathrm{m/s^2}$

Equation (5.25) allows the measurement of acceleration to be reduced to a measurement of spring extension (linear displacement) because

$$a = \frac{k}{m}\Delta x \tag{5.26}$$

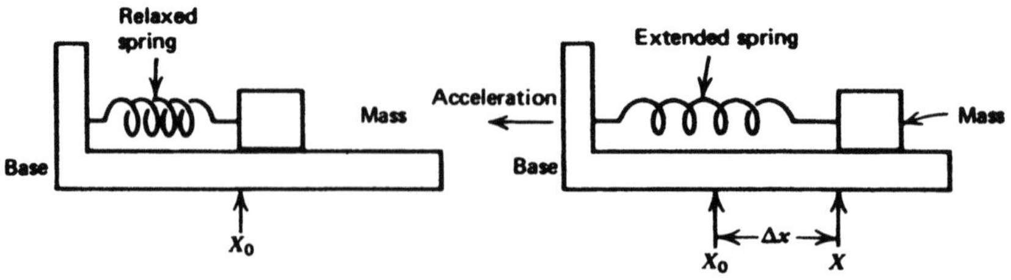

a) Spring-mass system with no acceleration b) Spring-mass system with acceleration

FIGURE 5.23
The basic spring-mass system accelerometer.

If the acceleration is reversed, the same physical argument would apply, except that the spring is compressed instead of extended. Equation (5.26) still describes the relationship between spring displacement and acceleration.

The *spring-mass principle* applies to many common accelerometer designs. The mass that converts the acceleration to spring displacement is referred to as the *test mass,* or *seismic mass.* We see, then, that acceleration measurement reduces to linear displacement measurement; most designs differ in how this displacement measurement is made.

Natural Frequency and Damping On closer examination of the simple principle just described, we find another characteristic of spring-mass systems that complicates the analysis. In particular, a system consisting of a spring and an attached mass always exhibits oscillations at some characteristic *natural frequency.* Experience tells us that if we pull a mass back and then release it (in the absence of acceleration), it will be pulled back by the spring, overshoot the equilibrium, and oscillate back and forth. Only friction associated with the mass and base eventually brings the mass to rest. Any displacement measuring system will respond to this oscillation as if an actual acceleration occurs. This natural frequency is given by

$$f_N = \frac{1}{2\pi}\sqrt{\frac{k}{m}} \qquad \textbf{(5.27)}$$

where f_N = natural frequency in Hz
k = spring constant in N/m
m = seismic mass in kg

The friction that eventually brings the mass to rest is defined by a *damping coefficient* α, which has the units of s^{-1}. In general, the effect of oscillation is called *transient response,* described by a periodic damped signal, as shown in Figure 5.24, whose equation is

$$X_T(t) = X_0 e^{-\alpha t}\sin(2\pi f_N t) \qquad \textbf{(5.28)}$$

where $X_T(t)$ = transient mass position
X_0 = peak position, initially
α = damping coefficient
f_N = natural frequency

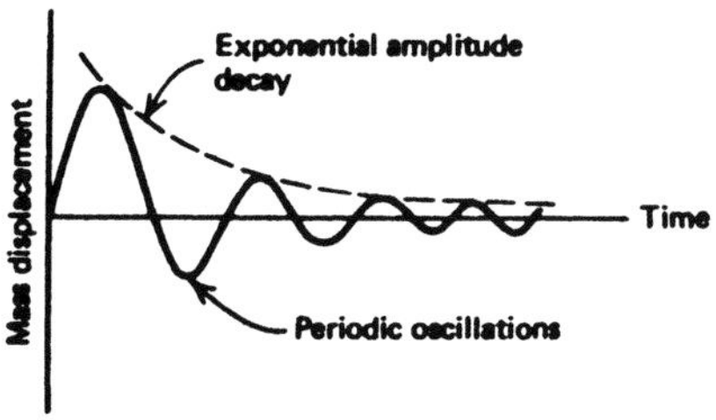

FIGURE 5.24
A spring-mass system exhibits a natural oscillation with damping as a response to an impulse input.

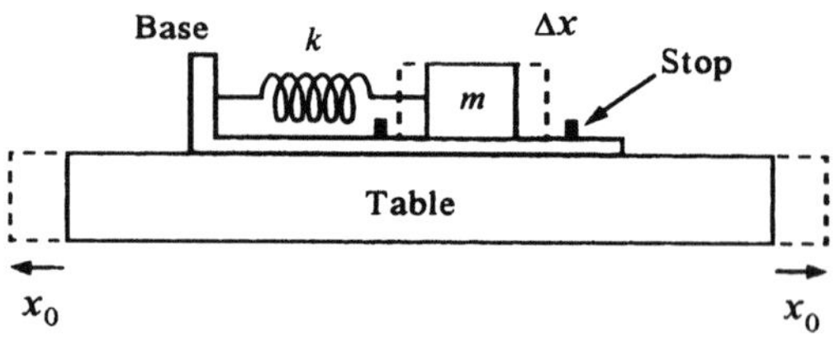

FIGURE 5.25
A spring-mass accelerometer has been attached to a table, which is vibrating. The table peak motion is x_0, and the mass motion is Δx.

The parameters, natural frequency, and damping coefficient in Equation (5.28) have a profound effect on the application of accelerometers.

Vibration Effects The effect of natural frequency and damping on the behavior of spring-mass accelerometers is best described in terms of an applied vibration. If the spring-mass system is exposed to a vibration, the resultant acceleration of the base is given by Equation (5.23):

$$a(t) = -\omega^2 x_0 \sin \omega t$$

If this is used in Equation (5.25), we can show that the mass motion is given by

$$\Delta x = -\frac{m x_0}{k} \omega^2 \sin \omega t \tag{5.29}$$

where all terms were previously defined, and $\omega = 2\pi f$, with f the applied frequency.

To make the predictions of Equation (5.29) clear, consider the situation presented in Figure 5.25. Our model spring-mass accelerometer has been fixed to a table that is vibrating. The x_0 in Equation (5.29) is the peak amplitude of the table vibration, and Δx is the vibration of the seismic mass within the accelerometer. Thus, Equation (5.29) predicts that the seismic-mass vibration peak amplitude varies as the vibration frequency squared, but linearly with the table-vibration amplitude. However, this result was obtained without considering the spring-mass system natural vibration. When this is taken into account, something quite different occurs.

Figure 5.26a shows the actual seismic-mass vibration peak amplitude versus table-vibration frequency compared with the simple frequency-squared prediction.

You can see that there is a resonance effect when the table frequency equals the natural frequency of the accelerometer—that is, the value of Δx goes through a peak. The amplitude of the resonant peak is determined by the amount of damping. The seismic-mass vibration is described by Equation (5.29) only up to about $f_N/2.5$.

Figure 5.26b shows two effects. The first is that the actual seismic-mass motion is limited by the physical size of the accelerometer. It will hit "stops" built into the assembly that limit its motion during resonance. The figure also shows that for frequencies well above the natural frequency, the motion of the mass is proportional to the table peak motion, x_0, but not to the frequency. Thus, it has become a displacement sensor. To summarize:

1. $f < f_N$—For an applied frequency less than the natural frequency, the natural frequency has little effect on the basic spring-mass response given by Equations (5.25) and (5.29). A rule of thumb states that a safe maximum applied frequency is $f < 1/2.5 f_N$.

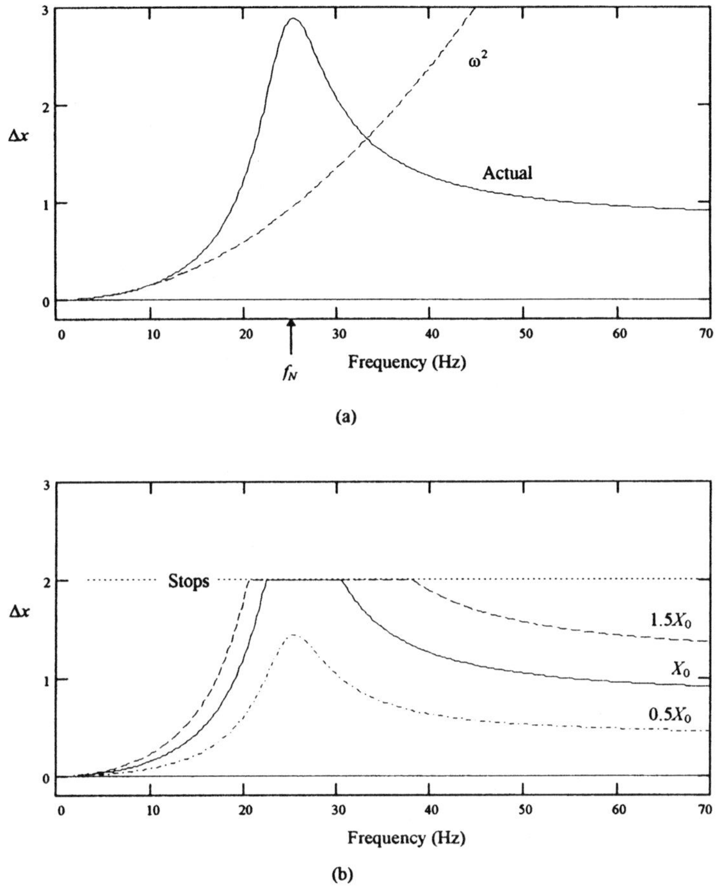

FIGURE 5.26
(a) The actual response of a spring-mass system to vibration is compared to the simple ω^2 prediction. (b) The effect of actual response with stops and various table peak motions is shown.

2. $f > f_N$—For an applied frequency much larger than the natural frequency, the accelerometer output is independent of the applied frequency. As shown in Figure 5.26b, the accelerometer becomes a measure of *vibration displacement*, x_0, of Equation (5.20) under these circumstances. It is interesting to note that the seismic mass is stationary in space in this case, and the housing, which is driven by the vibration, moves about the mass. A general rule sets $f > 2.5f_N$ for this case.

Generally, accelerometers are not used near the resonance at their natural frequency because of high nonlinearities in output.

EXAMPLE 5.14 An accelerometer has a seismic mass of 0.05 kg and a spring constant of 3.0×10^3 N/m. Maximum mass displacement is ± 0.02 m (before the mass hits the stops). Calculate **(a)** the maximum measurable acceleration in **g**, and **(b)** the natural frequency.

Solution

We find the maximum acceleration when the maximum displacement occurs, from Equation (5.26):

a.

$$a = \frac{k}{m}\Delta x$$

$$a = \left(\frac{3.0 \times 10^3 \text{ N/m}}{0.05 \text{ kg}}\right)(0.02 \text{ m})$$

$$a = 1200 \text{ m/s}^2$$

or because

$$1 \mathbf{g} = 9.8 \text{ m/s}^2$$

$$a = (1200 \text{ m/s}^2)\left(\frac{1 \mathbf{g}}{9.8 \text{ m/s}^2}\right)$$

$$a = 122 \mathbf{g}$$

b. The natural frequency is given by Equation (5.27):

$$f_N = \frac{1}{2\pi}\sqrt{\frac{k}{m}}$$

$$f_N = \frac{1}{2\pi}\sqrt{\frac{3.0 \times 10^3 \text{ N/m}}{0.05 \text{ kg}}}$$

$$f_N = \mathbf{39 \text{ Hz}}$$

5.4.3 Types of Accelerometers

The variety of accelerometers used results from different applications with requirements of range, natural frequency, and damping. In this section, various accelerometers with their special characteristics are reviewed. The basic difference is in the method of mass displacement measurement. In general, the specification sheets for an accelerometer will give the natural frequency, damping coefficient, and a scale factor that relates the output to an acceleration input. The values of test mass and spring constant are seldom known or required.

Potentiometric This simplest accelerometer type measures mass motion by attaching the spring mass to the wiper arm of a potentiometer. In this manner, the mass posi-

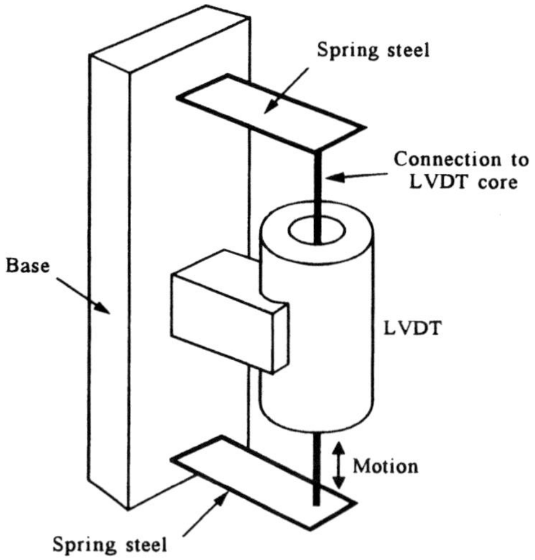

FIGURE 5.27
An LVDT is often used as an accelerometer, with the core serving as the mass.

tion is conveyed as a changing resistance. The natural frequency of these devices is generally less than 30 Hz, limiting their application to *steady-state* acceleration or *low-frequency* vibration measurement. Numerous signal-conditioning schemes are employed to convert the resistance variation into a voltage or current signal.

LVDT A second type of accelerometer takes advantage of the natural linear displacement measurement of the LVDT (see Section 5.2.3) to measure mass displacement. In these instruments, the LVDT core itself is the seismic mass. Displacements of the core are converted directly into a linearly proportional ac voltage. These accelerometers generally have a natural frequency of less than 80 Hz and are commonly used for steady-state and low-frequency vibration. Figure 5.27 shows the basic structure of such an accelerometer.

Variable Reluctance This accelerometer type falls in the same general category as the LVDT in that an inductive principle is employed. Here, the test mass is usually a permanent magnet. The measurement is made from the voltage induced in a surrounding coil as the magnetic mass moves under the influence of an acceleration. This accelerometer is used in vibration and shock studies only, because it has an output *only* when the mass is in motion. Its natural frequency is typically less than 100 Hz. This type of accelerometer often is used in oil exploration to pick up vibrations reflected from underground rock strata. In this form, it is commonly referred to as a *geophone*.

Piezoelectric The piezoelectric accelerometer is based on a property exhibited by certain crystals where a voltage is generated across the crystal when stressed. This property is also the basis for such familiar sensors as crystal phonograph cartridges and crystal microphones. For accelerometers, the principle is shown in Figure 5.28. Here, a piezoelectric

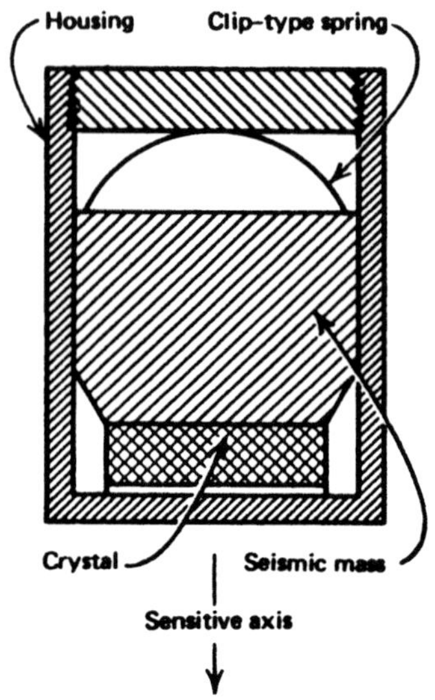

FIGURE 5.28
A piezoelectric accelerometer has a very high natural frequency.

crystal is spring-loaded with a test mass in contact with the crystal. When exposed to an acceleration, the test mass stresses the crystal by a force ($F = ma$), resulting in a voltage generated across the crystal. A measure of this voltage is then a measure of the acceleration. The crystal per se is a very high-impedance source, and thus requires a high-input-impedance, low-noise detector. Output levels are typically in the millivolt range. The natural frequency of these devices may exceed 5 kHz, so that they can be used for vibration and shock measurements.

5.4.4 Applications

A few notes about the application of accelerometers will help in understanding how the selection of a sensor is made in a particular case.

Steady-State Acceleration In steady-state accelerations, we are interested in a measure of acceleration that may vary in time but that is nonperiodic. Thus, the stop-go motion of an automobile is an example of a steady-state acceleration. For these steady-state accelerations, we select a sensor having (1) adequate range to cover expected acceleration magnitudes and (2) a natural frequency sufficiently high that its period is shorter than the characteristic time span over which the measured acceleration changes. By using electronic integrators, the basic accelerometer can provide both velocity (first integration) and position (second integration) information.

EXAMPLE 5.15 An accelerometer outputs 14 mV per **g**. Design a signal-conditioning system that provides a velocity signal scaled at 0.25 V for every m/s, and determine the gain of the system and the feedback resistance ratio.

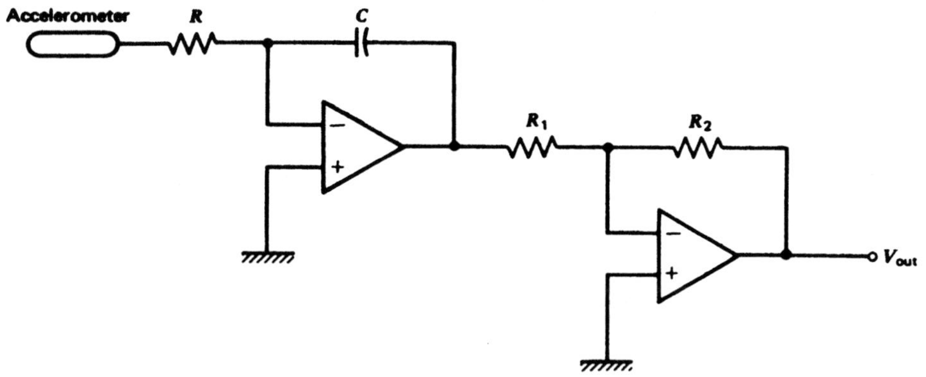

FIGURE 5.29
An integrator can be used to obtain velocity information from an accelerometer.

Solution
First we note that 14 mV/**g** becomes

$$\left(14\,\frac{\text{mV}}{\mathbf{g}}\right)\left(\frac{1\ \mathbf{g}}{9.8\,\text{m/s}^2}\right) = 1.43\,\frac{\text{mV}}{\text{m/s}^2}$$

Thus we can write the output voltage of the sensor as $V_a = Ka$, where K is the 1.43 mV(m/s^2) and a is the acceleration in m/s^2. If this is used as input to an op amp integrator, the result is:

$$V_v = -\frac{1}{RC}\int V_a dt = -\frac{K}{RC}\int a dt = -\frac{K}{RC}v$$

where V_v is the integrator output voltage and v is the velocity in m/s. We must get rid of the negative sign and provide the correct scale factor of 0.25 V/(m/s). Therefore, we can use an inverting amplifier on the output of the integrator. The circuit is shown in Figure 5.29. The output is $V_{out} = \left(\frac{R_2}{R_1}\right)\frac{K}{RC}v$. Notice the output is now positive as required. Any combination of quantities can be used that give the desired result. For example, we can take $R = 1$ MΩ and $C = 1$ μF, which makes $RC = 1$. Then we use a gain (R_2/R_1) to provide the correct scale factor,

$$0.25 = \left(\frac{R_2}{R_1}\right)\frac{1.43x10^{-3}}{1} \text{ so, } \frac{R_2}{R_1} = 175$$

Therefore, so we could use $R_1 = 1$ kΩ and then $R_2 = 175$ kΩ.

Vibration The application of accelerometers for vibration first requires that the applied frequency is less than the natural frequency of the accelerometer. Second, one must be sure the stated range of acceleration measured will never exceed that of the specification for the device. This assurance must come from a consideration of Equation (5.29) under circumstances of maximum frequency and vibration displacement.

Shock The primary elements of importance in shock measurements are that the device has a natural frequency that is greater than 1 kHz and a range typically greater than 500 **g.** The primary accelerometer that can satisfy these requirements is the piezoelectric type (see Section 5.4.3).

5.5 PRESSURE SENSORS

The measurement and control of fluid (liquid and gas) pressure has to be one of the most common in all the process industries. Because of the great variety of conditions, ranges, and materials for which pressure must be measured, there are many different types of pressure sensor designs. In the following paragraphs, the basic concepts of pressure are presented, and a brief description is given of the most common types of pressure sensors. You will see that pressure measurement is often accomplished by conversion of the pressure information to some intermediate form, such as displacement, which is then measured by a sensor to determine the pressure.

5.5.1 Pressure Principles

Pressure is simply the force per unit area that a fluid exerts on its surroundings. If it is a gas, then the pressure of the gas is the force per unit area that the gas exerts on the walls of the container that holds it. If the fluid is a liquid, then the pressure is the force per unit area that the liquid exerts on the container in which it is contained. Obviously, the pressure of a gas will be uniform on all the walls that must enclose the gas completely. In a liquid, the pressure will vary, being greatest on the bottom of the vessel and zero on the top surface, which need not be enclosed.

Static Pressure The statements made in the previous paragraph are explicitly true for a fluid that is not moving in space, that is not being pumped through pipes or flowing through a channel. The pressure in cases where no motion is occurring is referred to as *static* pressure.

Dynamic Pressure If a fluid is in motion, the pressure that it exerts on its surroundings *depends* on the motion. Thus, if we measure the pressure of water in a hose with the nozzle closed, we may find a pressure of, say, 40 lb per square inch (*Note:* force per unit area). If the nozzle is opened, the pressure in the hose will drop to a different value, say, 30 lb per square inch. For this reason, a thorough description of pressure must note the circumstances under which it is measured. Pressure can depend on flow, compressibility of the fluid, external forces, and numerous other factors.

Units Since pressure is force per unit area, we describe it in the SI system of units by newtons per square meter. This unit has been named the *pascal* (Pa), so that $1\ \text{Pa} = 1\ \text{N/m}^2$. As will be seen later, this is not a very convenient unit, and it is often used in conjunction with the SI standard prefixes as kPa or MPa. You will see the combination N/cm^2 used, but use of this combination should be avoided in favor of Pa with the appropriate prefix. In the English system of units, the most common designation is the pound per square inch, lb/in^2, usually written *psi*. The conversion is that 1 psi is approximately 6.895 kPa. For very low pressures, such as may be found in vacuum systems, the unit *torr* is often used. One torr is approximately 133.3 Pa. Again, use of the pascal with an appropriate prefix is preferred. Other units that you may encounter in the pressure description are the *atmosphere* (atm), which is 101.325 kPa or $\simeq$ 14.7 psi, and the *bar,* which is 100 kPa. The use of inches or feet of water and millimeters of Mercury will be discussed later.

Gauge Pressure In many cases, the absolute pressure is not the quantity of major interest in describing the pressure. The atmosphere of gas that surrounds the earth exerts a pressure, because of its weight, at the surface of the earth of approximately 14.7 psi, which defines the "atmosphere" unit. If a closed vessel at the earth's surface contained a gas at an absolute pressure of 14.7 psi, then there would be no *net* pressure on the walls of the container because the atmospheric gas exerts the same pressure from the outside. In cases like this, it is more appropriate to describe pressure in a relative sense—that is, compared to atmospheric pressure. This is called *gauge pressure* and is given by

$$p_g = p_{\text{abs}} - p_{\text{at}} \tag{5.30}$$

where p_g = gauge pressure
p_{abs} = absolute pressure
p_{at} = atmospheric pressure

In the English system of units, the abbreviation *psig* is used to represent the gauge pressure.

Head Pressure For liquids, the expression *head pressure,* or *pressure head,* is often used to describe the pressure of the liquid in a tank or pipe. This refers to the static pressure produced by the weight of the liquid above the point at which the pressure is being described. This pressure depends *only* on the height of the liquid above that point and the liquid density (mass per unit volume). In terms of an equation, if a liquid is contained in a tank, the pressure at the bottom of the tank is given by

$$p = \rho g h \tag{5.31}$$

where p = pressure in Pa
ρ = density in kg/m^3
g = acceleration due to gravity ($9.8\,\text{m/s}^2$)
h = depth in liquid in m

This same equation could be used to find the pressure in the English system, but it is common to express the density in this system as the weight density, ρ_w, in lb/ft^3, which includes the gravity term of Equation (5.31). In this case, the relationship between pressure and depth becomes

$$p = \rho_w h \tag{5.32}$$

where p = pressure in lb/ft^2
ρ_w = weight density in lb/ft^3
h = depth in ft

If the pressure is desired in psi, then the ft^2 would be expressed as $144\,\text{in}^2$. Because of the common occurrence of liquid tanks and the necessity to express the pressure of such systems, it has become common practice to describe the pressure directly in terms of the *equivalent* depth of a particular liquid. Thus, the term *mm of Mercury* means that the pressure is equivalent to that produced by so many millimeters of Mercury depth, which could be calculated from Equation (5.31) using the density of Mercury. In the same sense, the expression "inches of water" or "feet of water" means the pressure that is equivalent to some particular depth of water using its weight density.

Now you can see the basis for level measurement on pressure mentioned in Section 5.2.4. Equation (5.31) shows that the level of liquid of density p is directly related to the pressure. From level measurement we pass to pressure measurement, which is usually done by some type of displacement measurement.

EXAMPLE 5.16 A tank holds water with a depth of 7.0 ft. What is the pressure at the tank bottom in psi and Pa (density $= 10^3\ \text{kg/m}^3$)?

Solution

We can find the pressure in Pa directly by converting the 7.0 ft into meters; thus, (7.0 ft) (0.3048 m/ft) = 2.1 m. From Equation (5.31),

$$p = (10^3\ \text{kg/m}^3)(9.8\ \text{m/s}^2)(2.1\ \text{m})$$

$$p = 21\ \text{kpa} \qquad \text{(note significant figures)}$$

To find the pressure in psi, we can convert the pressure in Pa to psi or use Equation (5.32). Let's use the latter. The weight density is found from

$$\rho_w = (10^3\ \text{kg/m}^3)(9.8\ \text{m/s}^2) = 9.8 \times 10^3\ \text{N/m}^3$$

or

$$\rho_w = (9.8 \times 10^3\ \text{N/m}^3)(0.3048\ \text{m/ft}^3)(0.2248\ \text{lb/N})$$

$$\rho_w = 62.4\ \text{lb/ft}^3$$

The pressure is

$$p = (62.4\ \text{lb/ft}^3)(7.0\ \text{ft}) = 440\ \text{lb/ft}^2$$

$$p = \mathbf{3\ psi}$$

5.5.2 Pressure Sensors (p > 1 atmosphere)

In general, the design of pressure sensors employed for measurement of pressure higher than one atmosphere differs from those employed for pressure less than 1 atmosphere (atm). In this section the basic operating principles of many types of pressure sensors used for the higher pressures are considered. You should be aware that this is not a rigid separation, because you will find many of these same principles employed in the lower (vacuum) pressure measurements.

Most pressure sensors used in process control result in the transduction of pressure information into a physical displacement. Measurement of pressure requires techniques for producing the displacement and means for converting such displacement into a proportional electrical signal. This is *not* true, however, in the very low pressure region ($p < 10^{-3}$ atm), where many purely electronic means of pressure measurement may be used.

Diaphragm One common element used to convert pressure information into a physical displacement is the diaphragm (thin, flexible piece of metal) shown in Figure 5.30.

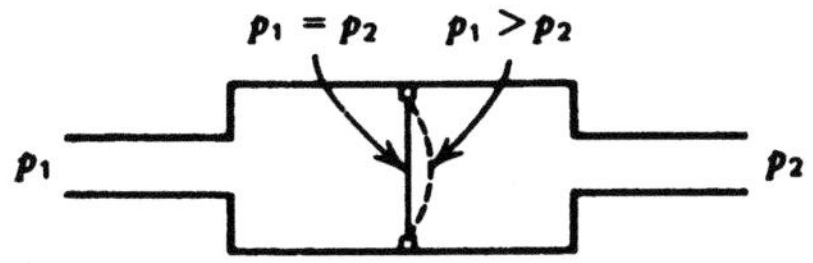

FIGURE 5.30
A diaphragm is used in many pressure sensors. Displacement varies with pressure difference.

If a pressure p_1 exists on one side of the diaphragm and p_2 on the other, then a net force is exerted given by

$$F = (p_2 - p_1)A \quad \textbf{(5.33)}$$

where A = diaphragm area in m^2
p_1, p_2 = pressure in N/m^2

A diaphragm is like a spring and therefore extends or contracts until a Hooke's law force is developed that balances the pressure difference force. This is shown in Figure 5.30 for p_1 greater than p_2. Notice that since the force is greater on the p_1 side of the diaphragm, it has deflected toward the p_2 side. The extent of this deflection (i.e., the diaphragm displacement) is a measure of the pressure difference.

A *bellows,* shown in Figure 5.31, is another device much like the diaphragm that converts a pressure differential into a physical displacement, except that here the displacement is much more a straight-line expansion. The accordion-shaped sides of the bellows are made from thin metal. When there is a pressure difference, a net force will exist on the flat, front surface of the bellows. The bellows assembly will then collapse like an accordion if p_2 is greater than p_1 or expand if p_2 is less than p_1. Again, we have a displacement which is proportional to pressure difference. This conversion of pressure to displacement is very nearly linear. Therefore, we have suggested the use of an LVDT to measure the displacement. This sensor will output an LVDT voltage amplitude that is linearly related to pressure.

FIGURE 5.31
A bellows is another common method of converting pressure to displacement. Here an LVDT is used to convert the displacement to voltage amplitude.

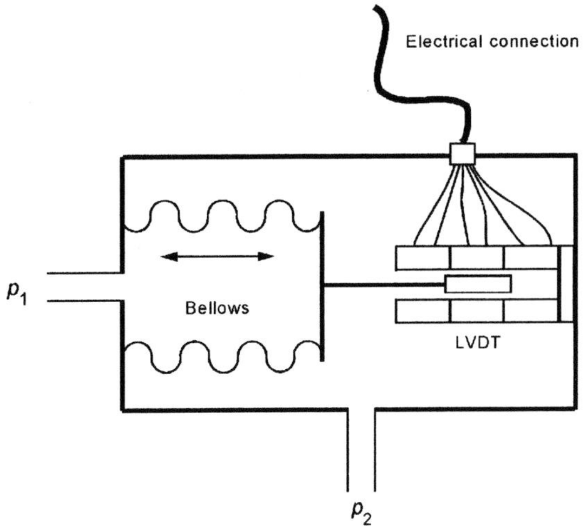

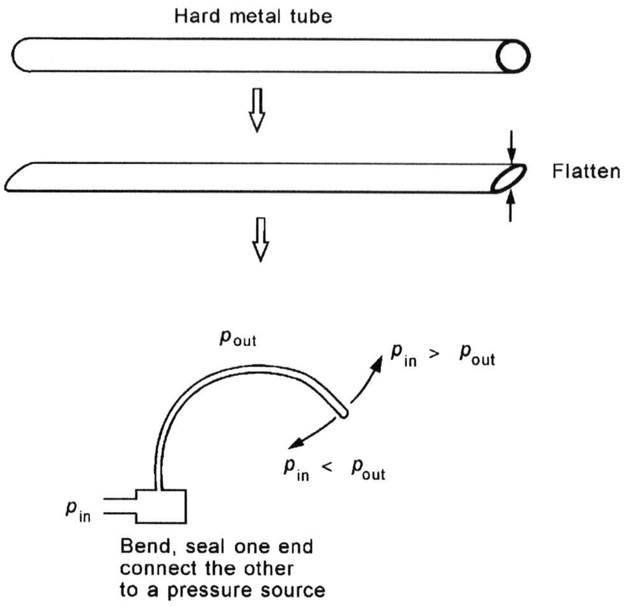

FIGURE 5.32
The Bourdon tube is probably the most common pressure-to-displacement element.

Figure 5.31 also shows how an LVDT can be connected to the bellows so that pressure measurement is converted directly from displacement to a voltage. In addition, the displacement and pressure are nearly linearly related, and because the LVDT voltage is linear with displacement, the voltage and pressure are also linearly related.

Bourdon Tube Probably the most common pressure sensor in universal use is based upon the Bourdon tube concept. Figure 5.32 shows the process for making a Bourdon tube and how it measures pressure. A hard metal tube, usually a type of bronze or brass, is flattened, and one end is closed off. The tube is then bent into a curve or arc, sometimes even a spiral. The open end is attached to a header by which a pressure can be introduced to the inside of the tube. When this is done, the tube will deflect when the inside applied pressure is different from the outside pressure. The tube will tend to straighten out if the inside pressure is higher than the outside pressure and to curve more if the pressure inside is less than that outside.

Most of the common, round pressure gauges with a meter pointer that rotates in proportion to pressure are based on this sensor. In this case, the deflection is transformed into a pointer rotation by a system of gears. Of course, for control applications, we are interested in converting the deflection into an electrical signal. This is accomplished by various types of displacement sensors to measure the deflection of the Bourdon tube.

Electronic Conversions Many techniques are used to convert the displacements generated in the previous examples into electronic signals. The simplest technique is to use a mechanical linkage connected to a potentiometer. In this fashion, pressure is related to a resistance change. Other methods of conversion employ strain gauges directly on a di-

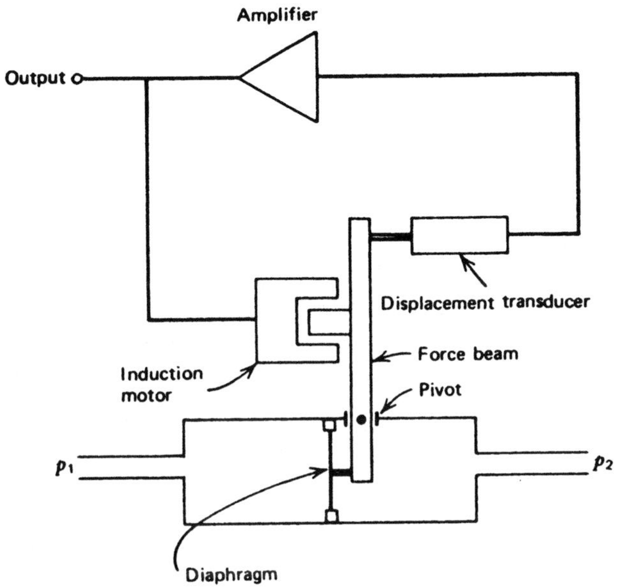

FIGURE 5.33
A differential pressure (DP) cell measures pressure difference with a diaphragm. A feedback system minimizes actual diaphragm deflection.

aphragm. LVDTs and other inductive devices are used to convert bellows or Bourdon tube motions into proportional electrical signals.

Often, pressure measurement is accomplished using a diaphragm in a special feedback configuration, shown in Figure 5.33. The feedback system keeps the diaphragm from moving, using an induction motor. The error signal in the feedback system provides an electrical measurement of the pressure.

Solid-State Pressure Sensors Integrated circuit technology has led to the development of solid-state (SS) pressure sensors that find extensive application in the pressure ranges of 0 to 100 kPa (0 to 14.7 psi). These small units often require no more than three connections—dc power, ground, and the sensor output. Pressure connection is via a metal tube, as shown in Figure 5.34a. Generally, manufacturers provide a line of such sensors with various ranges of pressure and configurations.

The basic sensing element is a small wafer of silicon acting as a diaphragm that, as usual, deflects in response to a pressure difference. However, as suggested in Figure 5.34b, in this case the deflection is sensed by semiconductor strain gauges grown directly on the silicon wafer; furthermore, signal-conditioning circuitry is grown directly on the wafer as well. This signal conditioning includes temperature compensation and circuitry that provides an output voltage that varies linearly with pressure over the specified operating range.

The configuration shown in Figure 5.34b is for measuring *gauge* pressure, since one side of the diaphragm is open to the atmosphere. Figure 5.35 shows that simple modifications are used to convert the basic sensor to an absolute or differential-type gauge. For

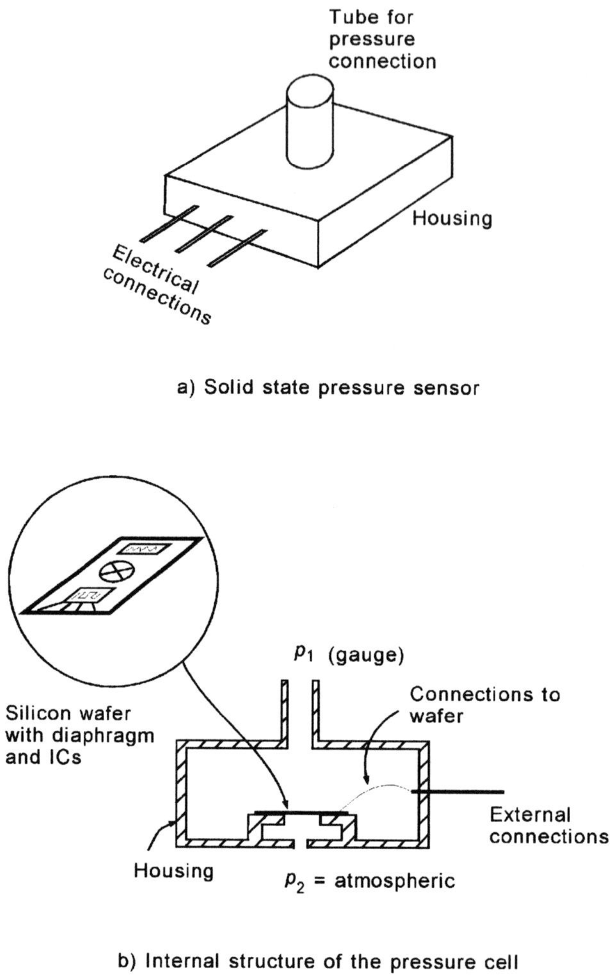

FIGURE 5.34
Solid-state pressure sensors employ integrated circuit technology and silicon diaphragms. This example measures gauge pressure.

absolute pressure measurement, one side of the wafer is sealed off and evacuated. For differential measurement, facilities are provided to allow application of independent pressures p_1 and p_2 to the two sides of the diaphragm.

SS pressure sensors are characterized by

1. Sensitivities in the range of 10 to 100 mV/kPa.
2. Response times on the order of 10 ms. These are not first-order time-response devices, so response time is defined as the time for a change from 10% to 90% of the final value following a step change in input pressure.

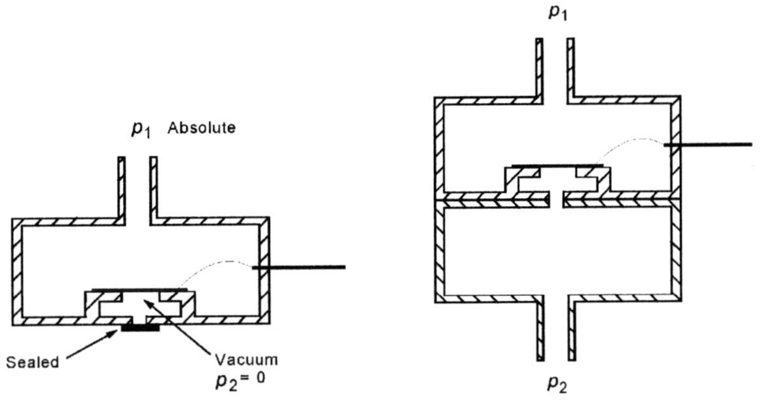

a) Measurement of absolute pressure

b) Measurement of differential pressure, $p_1 - p_2$

FIGURE 5.35
Simple modifications allow SS pressure sensors to measure absolute or differential pressure.

3. Linear voltage versus pressure within the specified operating range.
4. Ease of use, with often only three connections: dc power (typically 5 V), ground, and the sensor output voltage.

SS pressure sensors find application in a broad sector of industry and control, wherever low pressures are to be measured. Another important application is in the commercial field, where such sensors are employed, for example, in home appliances such as dishwashers and washing machines.

EXAMPLE 5.17

A SS pressure sensor that outputs 25 mV/kPa for a pressure variation of 0.0 to 25 kPa will be used to measure the level of a liquid with a density of $1.3 \times 10^3\,\text{kg/m}^3$. What voltage output will be expected for level variations from 0 to 2.0 m? What is the sensitivity for level measurement expressed in mV/cm?

Solution

The pressure sensor will be attached to the bottom of the tank holding the liquid. Therefore, the pressure measured will be given by Equation (5.31). Clearly, when empty the pressure will be zero and output voltage will be zero as well. At 2.0 m, the pressure will be

$$p = \rho g h = (1.3 \times 10^3\,\text{kg/m}^3)(9.8\,\text{m/s}^2)(2.0\,\text{m}) = 25.48\,\text{kPa}.$$

Therefore, the sensor output will be

$$V = (25\,\text{mV/kPa})(25.48\,\text{kPa}) = 0.637\,\text{V}$$

So the sensitivity will be

$$S = 637\,\text{mV}/200\,\text{cm} = 3.185\,\text{mV/cm}$$

5.5.3 Pressure Sensors ($p < 1$ atmosphere)

Measurements of pressure less than 1 atm are most conveniently made using purely electronic methods. There are three common methods of electronic pressure measurements.

The first two devices are useful for pressure less than 1 atm, down to about 10^{-3} atm. They are both based on the rate at which heat is conducted and radiated away from a heated filament placed in the low-pressure environment. The heat loss is proportional to the number of gas molecules per unit volume, and thus, under constant filament current, the filament temperature is proportional to gas pressure. We have thus transduced a pressure measurement to a temperature measurement.

Pirani Gauge This gauge determines the filament temperature through a measure of filament *resistance* in accordance with the principles established in Section 4.3. Filament excitation and resistance measurement are both performed with a bridge circuit. The response of resistance versus pressure is highly nonlinear.

Thermocouple A second pressure transducer or gauge measures filament temperature using a thermocouple directly attached to the heated filament. In this case, ambient room temperature serves as a reference for the thermocouple, and the voltage output, which is proportional to pressure, is highly nonlinear. Calibration of both Pirani and thermocouple gauges also depends on the type of gas for which the pressure is being measured.

Ionization Gauge This device is useful for the measurement of very low pressures from about 10^{-3} atm to 10^{-13} atm. This gauge employs electrons, usually from a heated filament, to ionize the gas whose pressure is to be measured, and then measures the current flowing between two electrodes in the ionized environment, as shown in Figure 5.36. The number of ions per unit volume depends on the gas pressure, and hence the current also depends on gas pressure. This current is then monitored as an approximately linear indication of pressure.

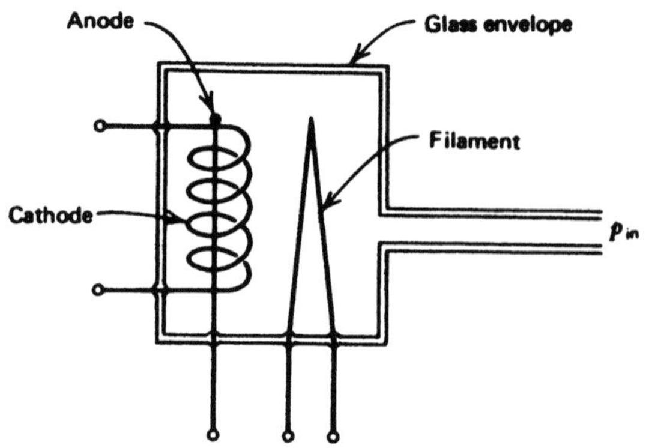

FIGURE 5.36
The ionization gauge is used to measure very low pressures, down to about 10^{-13} atm.

5.6 FLOW SENSORS

The measurement and control of flow can be said to be the very heart of process industries. Continuously operating manufacturing processes involve the movement of raw materials, products, and waste throughout the process. All such functions can be considered flow, whether automobiles through an assembly line or methyl chloride through a pipe. The methods of measurement of flow are at least as varied as the industry. It would be unreasonable to try to present every type of flow sensor, and in this section we will consider flow on three broad fronts—solid, liquid, and gas. As with pressure, we will find that flow information is often translated into an intermediate form, that is then measured using techniques developed for that form.

5.6.1 Solid-Flow Measurement

A common solid-flow measurement occurs when material in the form of small particles, such as crushed material or powder, is carried by a conveyor belt system or by some other host material. For example, if solid material is suspended in a liquid host, the combination is called a *slurry,* which is then pumped through pipes like a liquid. We will consider the conveyor system and leave slurry to be treated as liquid flow.

Conveyor Flow Concepts For solid objects, the flow usually is described by a specification of the mass or weight per unit time that is being transported by the conveyor system. The units will be in many forms—for example, kg/min or lb/min. To make a measurement of flow, it is only necessary to weigh the quantity of material on some fixed length of the conveyor system. Knowing the speed of the conveyor allows calculation of the material flow rate.

Figure 5.37 shows a typical conveyor system where material is drawn from a hopper and transported by the conveyor system. A mechanical valve controls the rate at which material can flow from the hopper onto the conveyor belt. The belt is driven by a motor system. Flow rate is measured by weighing the amount of material on a platform of length L at any instant. The conveyor belt slides over the platform, which deflects slightly due to the weight of material. A load cell measures this deflection as an indication of weight. In this case, flow rate can be calculated from

$$Q = \frac{WR}{L} \tag{5.34}$$

where
Q = flow (kg/min or lb/min)
W = weight of material on section of length L (kg or lb)
R = conveyor speed (m/min or ft/min)
L = length of weighing platform (m or ft)

Flow Sensor In the example with which we are working in Figure 5.37, it is evident that the flow sensor is actually the assembly of the conveyor, hopper opening, and weighing platform. It is the actual weighing platform that performs the measurement from

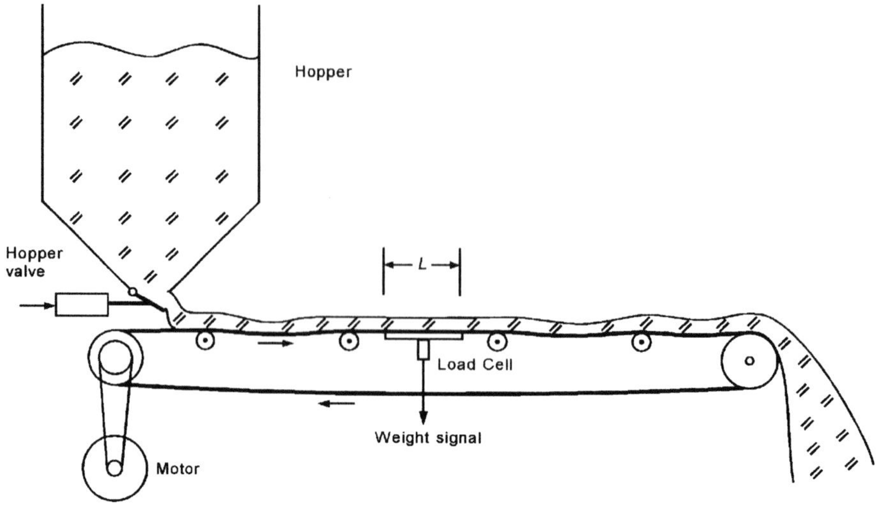

FIGURE 5.37
Conveyor system for illustrating solid-flow measurement.

which flow rate is determined, however. We see that flow measurement becomes weight measurement. In this case, we have suggested that this weight is measured by means of a load cell, which is then a strain gauge measurement. Another popular device for weight measurement of moving systems like this is an LVDT that measures the droop of the conveyor at the point of measurement because of the material that it carries.

EXAMPLE 5.18 A coal conveyor system moves at 100 ft/min. A weighing platform is 5.0 ft in length, and a particular weighing shows that 75 lb of coal are on the platform. Find the coal delivery in lb/h.

Solution

We can use Equation (5.34) directly to find the flow:

$$Q = \frac{(75\,\text{lb})(100\,\text{ft/min})}{5\,\text{ft}}$$

$$Q = 1500\,\text{lb/min}$$

Then, converting to lb/h by multiplying by 60 min/h,

$$Q = \mathbf{90{,}000\ lb/h}$$

5.6.2 Liquid Flow

The measurement of liquid flow is involved in nearly every facet of the process industry. The conditions under which the flow occurs and the vastly different types of material that flow result in a great many types of flow measurement methods. Indeed, entire books are written devoted to the problems of measuring liquid flow and how to interpret the results of flow measurements. It is impractical and not within the scope of this book to present a comprehensive study of liquid flow; only the basic ideas of liquid flow measurement will be presented.

Flow Units The units used to describe the flow measured can be of several types, depending on how the specific process needs the information. The most common descriptions are the following:

1. *Volume flow rate* Expressed as a volume delivered per unit time. Typical units are gals/min, m^3/h, or ft^3/h (1 gal = 231 $in.^3$).
2. *Flow velocity* Expressed as the distance the liquid travels in the carrier per unit time. Typical units are m/min or ft/min. This is related to the volume flow rate by

$$V = \frac{Q}{A} \tag{5.35}$$

where V = flow velocity
Q = volume flow rate
A = cross-sectional area of flow carrier (pipe, and so on)

3. *Mass or weight flow rate* Expressed as mass or weight flowing per unit time. Typical units are kg/h or lb/h. This is related to the volume flow rate by

$$F = \rho Q \tag{5.36}$$

where F = mass or weight flow rate
ρ = mass density or weight density
Q = volume flow rate

EXAMPLE 5.19 Water is pumped through a 1.5-in. diameter pipe with a flow velocity of 2.5 ft/s. Find the volume flow rate (ft^3/min) and weight flow rate (lb/min). The weight density is 62.4 lb/ft^3.

Solution

The flow velocity is given as 2.5 ft/s, so the volume flow rate can be found from Equation (5.35), $Q = VA$. The area is given by

$$A = \pi d^2/4$$

where the diameter $d = (1.5 \text{ in.})(1/12 \text{ ft/in.}) = 0.125 \text{ ft}$
so that

$$A = (3.14)(0.125)^2/4 = 0.0122 \text{ ft}^2$$

Then, the volume flow rate is

$$Q = (2.5 \text{ ft/s})(0.0122 \text{ ft}^2)(60 \text{ s/min})$$
$$Q = 1.8 \text{ ft}^3/\text{min} \quad \text{or} \quad 13.5 \text{ gal/min}$$

The weight flow rate is found from Equation (5.36):

$$F = (62.4\ \text{lb/ft}^3)(1.8\ \text{ft}^3/\text{min})$$
$$F = \mathbf{112\ lb/min}$$

Pipe Flow Principles The flow rate of liquids in pipes is determined primarily by the pressure that is forcing the liquid through the pipe. The concept of pressure head, or simply *head,* introduced in the previous sections is often used to describe this pressure, because it is easy to relate the forcing pressure to that produced by a depth of liquid in a tank from which the pipe exits. In Figure 5.38, flow through pipe P is driven by the pressure in the pipe, but this pressure is caused by the weight of liquid in the tank of height h (head). The pressure is found by Equation (5.31) or (5.32). Many other factors affect the actual flow rate produced by this pressure, including liquid viscosity, pipe size, pipe roughness (friction), turbulence of flowing liquid, and others. It is beyond the scope of this book to detail exactly how these factors determine the flow. Instead, it is our objective to discuss how such flow is measured, regardless of those features that may determine exactly what the flow is relative to the conditions.

Restriction Flow Sensors One of the most common methods of measuring the flow of liquids in pipes is by introducing a restriction in the pipe and measuring the pressure drop that results across the restriction. When such a restriction is placed in the pipe, the velocity of the fluid through the restriction *increases,* and the pressure in the restriction *decreases.* We find that there is a relationship between the pressure drop and the rate of flow such that, as the flow increases, the pressure drops. In particular, one can find an equation of the form

$$Q = K\sqrt{\Delta p} \qquad \textbf{(5.37)}$$

where
Q = volume flow rate
K = a constant for the pipe and liquid type
Δp = drop in pressure across the restriction

The constant, K, depends on many factors, including the type of liquid, size of pipe, velocity of flow, temperature, and so on. The type of restriction employed also will change the value of the constant used in this equation. The flow rate is linearly dependent not on the

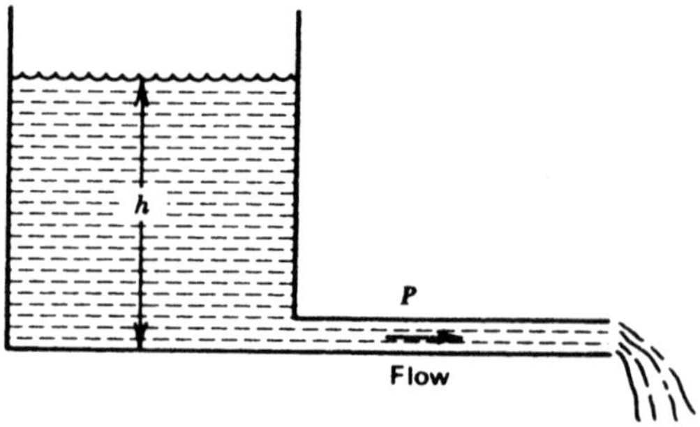

FIGURE 5.38
Flow through the pipe, P, is determined in part by the pressure due to the head, h.

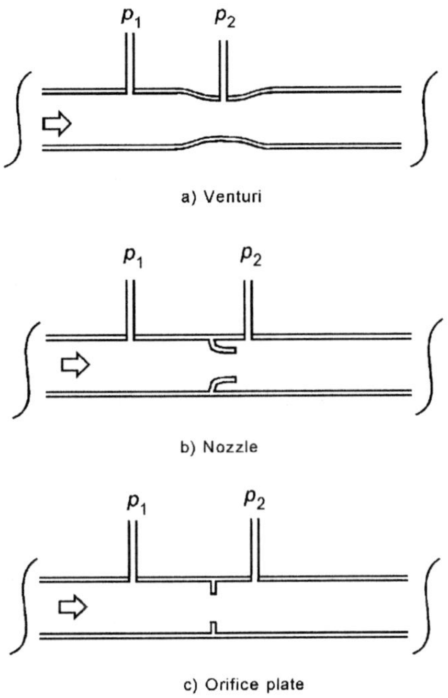

FIGURE 5.39
Three different types of restrictions are commonly used to convert pipe flow to a pressure difference, $p_1 - p_2$.

pressure drop, but on the square root. Thus, if the pressure drop in a pipe increased by a factor of 2 when the flow rate was increased, the flow rate will have increased only by a factor of 1.4 (the square root of 2). Certain standard types of restrictions are employed in exploiting the pressure-drop method of measuring flow.

Figure 5.39 shows the three most common methods. It is interesting to note that having converted flow information to pressure, we now employ one of the methods of measuring pressure, often by conversion to displacement, which is measured by a displacement sensor before finally getting a signal that will be used in the process-control loop. The most common method of measuring the pressure drop is to use a differential pressure sensor similar to that shown in Figure 5.35b. These are often described by the name *DP cell.*

EXAMPLE 5.20

Flow is to be controlled from 20 to 150 gal/min. The flow is measured using an orifice plate system such as that shown in Figure 5.39c. The orifice plate is described by Equation (5.37), with $K = 119.5(\text{gal/min})/\text{psi}^{1/2}$. A bellows measures the pressure with an LVDT so that the output is 1.8 V/psi. Find the range of voltages that result from the given flow range.

Solution

From Equation (5.37), we find the pressures that result from the given flow:

$$\Delta p = (Q/K)^2$$

For 20 gal/min,

$$\Delta p = (20/119.5)^2 = 0.0280\,\text{psi}$$

and for 150 gal/min,

$$\Delta p = (150/119.5)^2 = 1.5756 \text{ psi}$$

Because there are 1.8 V/psi, the voltage range is easily found.

$$\text{For 20 gal/min, } V = 0.0280(1.8) = 0.0504 \text{ V}$$
$$\text{For 150 gal/min, } V = 1.5756(1.8) = 2.836 \text{ V}$$

Pitot Tube The pitot tube is a common way to measure flow rate at a particular point in a flowing fluid (liquid or gas). Figure 5.40 shows a tube placed in the flowing fluid with its opening directed into the direction of flow. The principle is that the fluid will be brought to rest in the tube, and therefore its pressure will be the sum of the static fluid pressure plus the effective pressure of the flow. The pressure in the pitot tube is measured in differential to the static pressure of the flowing fluid in the same vicinity as the tube. This differential pressure will be proportional to the square root of the flow rate. The flow rate in a pipe varies across the pipe, so the pitot tube determines the flow rate only at the point of insertion.

Obstruction Flow Sensor Another type of flow sensor operates by the effect of flow on an obstruction placed in the flow stream. In a *rotameter,* the obstruction is a float that rises in a vertical tapered column. The lifting force, and thus the distance to which the float rises in the column, is proportional to the flow rate. The lifting force is produced by the differential pressure that exists across the float, because it is a restriction in the flow. This type of sensor is used for both liquids and gases. A *moving vane* flow meter has a vane target immersed in the flow region, which is rotated out of the flow as the flow velocity increases. The angle of the vane is a measure of the flow rate. If the rotating vane shaft is attached to an angle-measuring sensor, the flow rate can be measured for use in a process-control application. A *turbine* type of flow meter is composed of a freely spinning turbine blade assembly in the flow path. The rate of rotation of the turbine is proportional to the flow rate. If the turbine is attached to a tachometer, a convenient electrical signal can

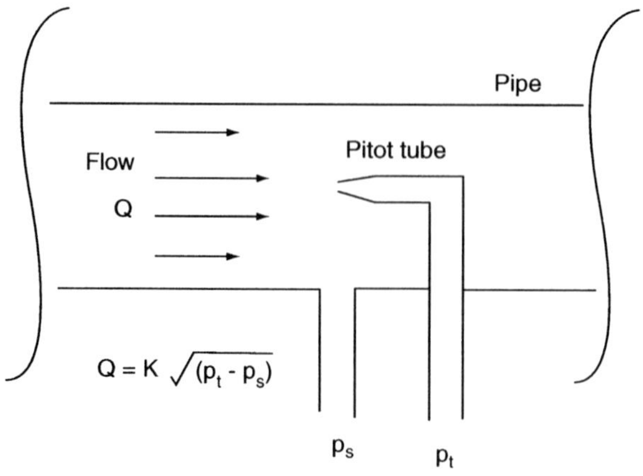

FIGURE 5.40
The pitot tube measures flow at a point in the gas or liquid.

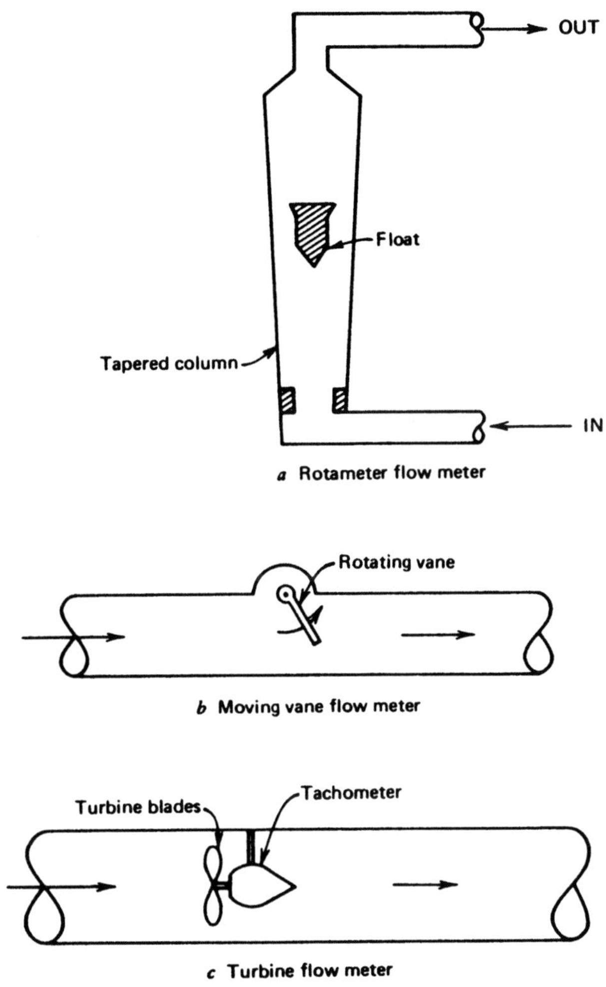

FIGURE 5.41
Three different types of obstruction flow meters.

be produced. In all of these methods of flow measurement, it is necessary to present a substantial obstruction into the flow path to measure the flow. For this reason, these devices are used only when an obstruction does not cause any unwanted reaction on the flow system. These devices are illustrated in Figure 5.41.

Magnetic Flow Meter It can be shown that if charged particles move across a magnetic field, a potential is established across the flow, perpendicular to the magnetic field. Thus, if the flowing liquid is also a conductor (even if not necessarily a good conductor) of electricity, the flow can be measured by allowing the liquid to flow through a magnetic field and measuring the transverse potential produced. The pipe section in which this measurement is made must be insulated, and a nonconductor itself, or the potential produced, will be

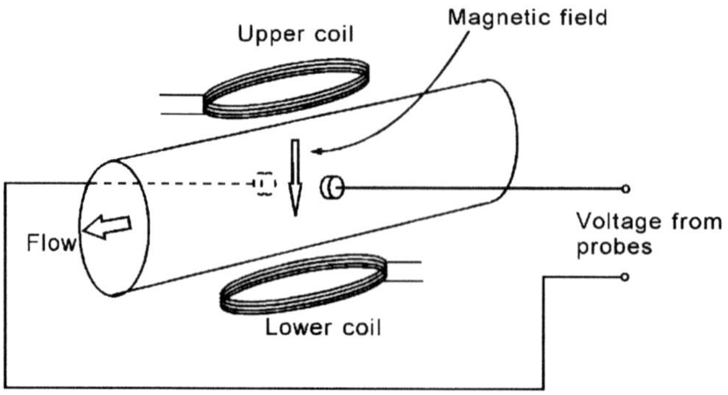

FIGURE 5.42
A magnetic flow meter will work only with conducting fluids such as blood.

cancelled by currents in the pipe. A diagram of this type of flow meter is presented in Figure 5.42. This type of sensor produces an electrical signal directly and is convenient for process-control applications involving conducting fluid flow.

SUMMARY

In this chapter, an assortment of measurement systems that fall under the general description of mechanical sensors has been studied. The objective was to gain familiarity with the essential features of the variables themselves and the typical measurement methods.

Topics covered were the following:

1. Position, location, and displacement sensors, including the potentiometric, capacitive, and LVDT. The LVDT converts displacement linearly into a voltage.
2. The strain gauge measures deformation of solid objects resulting from applied forces called stress. The strain gauge converts strain into a change of resistance.
3. Accelerometers are used to measure the acceleration of objects because of rectilinear motion, vibration, and shock. Most of them operate by the spring-mass principle, which converts acceleration information into a displacement.
4. Pressure is the force per unit area that a fluid exerts on the walls of a container. Pressure sensors often convert pressure information into a displacement. Examples include diaphragms, bellows, and the Bourdon tube. Electronic measures are often used for low pressures.
5. For gas pressures less than 1 atm, purely electrical techniques are used. In some cases, the temperature of a heated wire is used to indicate pressure.
6. Flow sensors are very important in the manufacturing world. Typically, solid flow is mass or weight per unit time.
7. Fluid flow through pipes or channels typically is measured by converting the flow information into pressure by a restriction in the flow system.

PROBLEMS

Section 5.2

5.1 A 50-kΩ, wire-wound pot is used to measure the displacement of a work piece. A linkage is employed so that as the work piece moves over a distance of 12 cm, the pot varies by the full 50-kΩ. The pot is wound on a 3-cm-diameter form, 5 cm long, and the distance between wires in the pot is 0.25 mm. What is the resolution in work-piece motion? What resistance change corresponds to this resolution?

5.2 Develop signal conditioning for Problem 5.1 so the output is −6 to +6 V as the work piece moves over the 0- to 12-cm motion limit.

5.3 A capacitive displacement sensor is used to measure rotating shaft wobble, as shown in Figure 5.43. The capacity is 880 pF with no wobble. Find the change in capacity for a +0.02- to −0.02-mm shaft wobble.

5.4 Develop an ac bridge for Problem 5.3. Use 880 pF for all the bridge capacitors, and assume a 10-V rms, 10-kHz excitation. What is the maximum bridge offset voltage amplitude?

FIGURE 5.43
Figure for Problem 5.3.

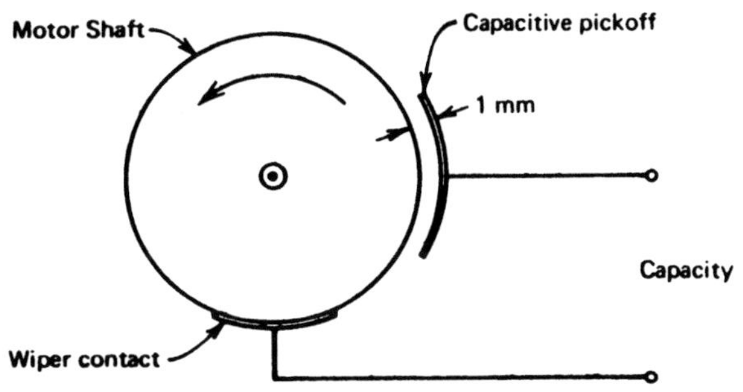

FIGURE 5.44
Figure for Problem 5.5.

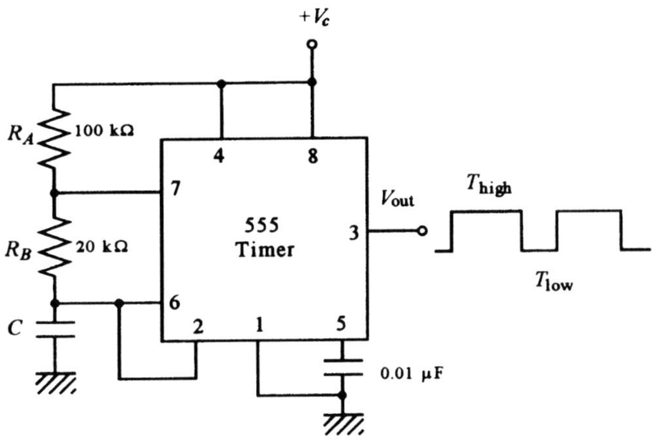

5.5 Figure 5.44 shows how a 555 IC timer can be connected to make a frequency generator. The output low time is given by $T_{low} = 0.693R_BC$, whereas the high time is given by $T_{high} = 0.693(R_A + R_B)C$. Suppose the capacitor in Example 5.2 is used in parallel with a 1000-pF fixed capacitor for C in this circuit. Then displacement will be converted to a variation of output frequency.

a. What range of frequency corresponds to the 1- to 2-cm displacement of the sensor?

b. Plot the frequency change versus displacement. Is it linear?

5.6 An LVDT with associated signal conditioning will be used to measure work-piece motion from −20 to +20 cm. The static transfer function is 2.5 mV/mm. The output will be interfaced to a computer via an ADC.

a. What is the range of output voltage?

b. If the desired resolution is 0.5 mm, how many bits must the ADC have?

c. Design analog signal conditioning to provide interface to a bipolar ADC with a 5-V reference.

5.7 Design a linkage system such that as a float for liquid level measurement moves from 0 to 1 m, an LVDT core moves over its linear range of 3 cm. Suppose the LVDT output is interfaced to a 10-bit ADC. What is the resolution in level measurement?

5.8 For Example 5.4, what capacity change would need to be measured to have a resolution of 2 cm? If the measurement will ultimately go to a computer, how many bits must the ADC have to support this resolution?

5.9 Design an ac bridge like that in Figure 2.11 to convert the capacity change of Example 5.4 into an ac offset voltage. The bridge should null at 0 m of level. Use $R_1 = R_3 = 1\ \text{k}\Omega$, $C_3 = 0.02\ \mu\text{F}$, and an excitation of 5 V rms at 1 kHz. Plot the voltage versus level.

Section 5.3

5.10 An aluminum beam supports a 550-kg mass. If the beam diameter is 6.2 cm, calculate **(a)** the stress and **(b)** the strain of the beam.

5.11 Using Equations (5.9) and (5.10), prove that Equation (5.11) is valid to first order in $\Delta l/l$. Note that $1/(1-x) \sim 1 + x$ for $x \ll 1$.

5.12 A strain gauge has GF = 2.06 and $R = 120\ \Omega$, and is made from wire with $\alpha_0 = 0.0034/^\circ\text{C}$ at 25°C. The dissipation factor is given as $P_D = 25\ \text{mW}/^\circ\text{C}$. What is the maximum current that can be placed through the SG to keep self-heating errors below 1 μ of strain?

5.13 A strain gauge has GF = 2.14 and a nominal resistance of 120 Ω. Calculate the resistance change resulting from a strain of 144 μin./in.

5.14 A strain gauge with GF = 2.03 and 120 Ω nominal resistance is to be used to measure strain with a resolution of 5 μs. Design a bridge and detector that provides this over five switched ranges of 1000-μ spans (i.e., 0 to 1000 μ, 1000 to 2000 μ, etc.). The idea is that a bridge null is found by a combination of switching to the appropriate range and then making a smooth null adjustment within that range. If the strain were 3390 μ, it would be necessary to switch to the 3000- to 4000-μ range and then adjust the smooth pot until a null occurred, which would be at 390 μ.

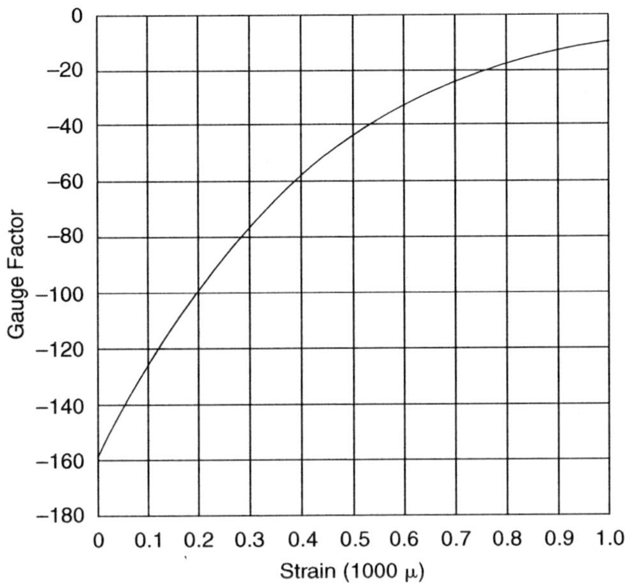

FIGURE 5.45
Semiconductor GF versus strain for Problem 5.16.

5.15 Derive the Equation (5.14), giving the approximate bridge off-null voltage versus strain. How much error does this equation have from the exact off-null voltage if the GF $= 2.05, R_A = 120\ \Omega, V_s = 10$ V, and the strain is 500 μ? Assume all other bridge resistors are 120 Ω also.

5.16 A semiconductor strain gauge has $R = 300\ \Omega$ and a GF versus strain given in Figure 5.45. This gauge is used in a bathroom scale for which the strain varies from 0 to 1000 μ as weight varies from 0 to 300 lb. Plot the gauge resistance change versus weight.

5.17 For Example 5.10, develop signal conditioning to provide input to a 10-bit unipolar ADC with a 5.000-V reference. How many pounds does each LSB represent? Plot the ADC output in hex versus force. Evaluate the linearity. Do not use a gain of more than 500 for any single op amp circuit stage in the design.

5.18 We will weigh objects by a strain gauge of $R = 120\ \Omega$, GF $= 2.02$ mounted on a copper column of 6-in. diameter. Find the change in resistance per pound placed on the column. Is this change an increase or decrease in resistance? Draw a diagram of the system showing how the active and dummy gauges should be mounted.

5.19 Figure 5.46 shows a micro-miniature cantilever beam with four strain gauges mounted. All are active gauges. Show how the gauges are wired into a bridge circuit to provide temperature compensation. If each gauge has the same gauge factor, GF, derive an equation for the bridge off-null voltage as a function of strain.

5.20 Show how to add another active gauge and dummy gauge to the system of Problem 5.18 to provide increased sensitivity. Assuming a bridge excitation of 5.0 volts, determine the bridge off-null voltage per 1000 pounds placed on the beam.

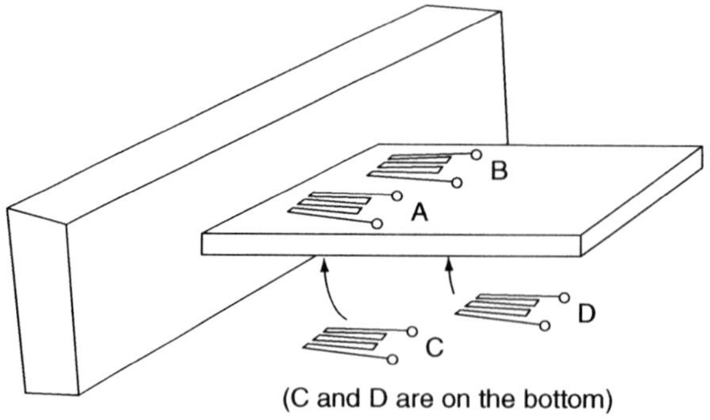

FIGURE 5.46
Four-gauge system for Problem 5.19.

Section 5.4

5.21 Calculate the rotation rate of a 10,000-rpm motor in rad/s.

5.22 An object falls from rest near the Earth's surface, accelerating downward at 1 **g.** After 5 s, what is the speed and the distance moved?

5.23 A force of 2.7 lb is applied to a 5.5-kg mass. Find the resulting acceleration in m/s^2.

5.24 Calculate the average shock in **g**s experienced by a transistor that falls 1.5 m from a tabletop, if it takes 2.7 ms to decelerate to zero when reaching the floor.

5.25 An automobile fender vibrates at 16 Hz with a peak-to-peak amplitude of 5 mm. Calculate the peak acceleration in **g**s.

5.26 A spring-mass system has a mass of 0.02 kg and a spring constant of 140 N/m. Calculate the natural frequency.

5.27 An LVDT is used in an accelerometer to measure seismic-mass displacement. The LVDT and signal-conditioning output is 0.31 mV/mm with a ± 2 cm maximum core displacement. The spring constant is 240 N/m, and the core mass is 0.05 kg. Find **(a)** the relation between acceleration in m/s^2 and output voltage, **(b)** the maximum acceleration that can be measured, and **(c)** the natural frequency.

5.28 For the accelerometer in Problem 5.27, design a signal-conditioning system that provides velocity information at 2 mV/(m/s) and position information at 0.5 V/m.

5.29 A piezoelectric accelerometer has a transfer function of 61 mV/g and a natural frequency of 4.5 kHz. In a vibration test at 110 Hz, a reading of 3.6 V-peak results. Find the vibration peak displacement.

5.30 An accelerometer for shock is designed as shown in Figure 5.47. Find a relation between strain gauge resistance change and shock in **g** (i.e., the resistance change per **g**). The force rod cross-sectional area is $2.0 \times 10^{-4}\,m^2$.

5.31 Design a signal-conditioning scheme for the accelerometer in Problem 5.30 using a bridge circuit. Plot the bridge offset voltage versus shock in **g** from 0 to 5000 **g**s.

Section 5.5

5.32 Calculate **(a)** the pressure in atmospheres that a water column 3.3 m high exerts on its base and **(b)** the pressure if the liquid is Mercury. Convert these results to pascals. Mercury density is $13.546\ g/cm^3$.

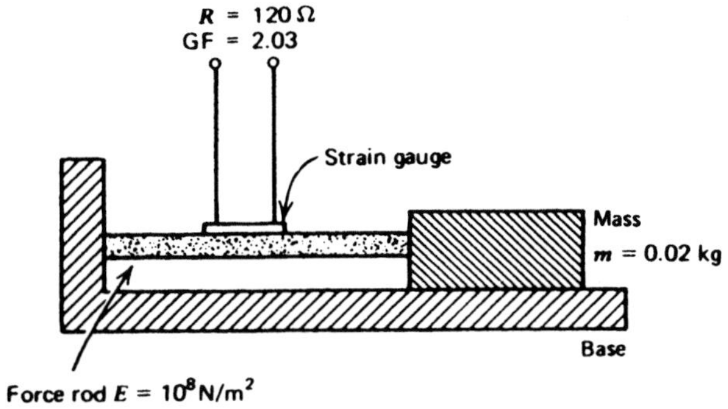

FIGURE 5.47
Figure for Problem 5.28.

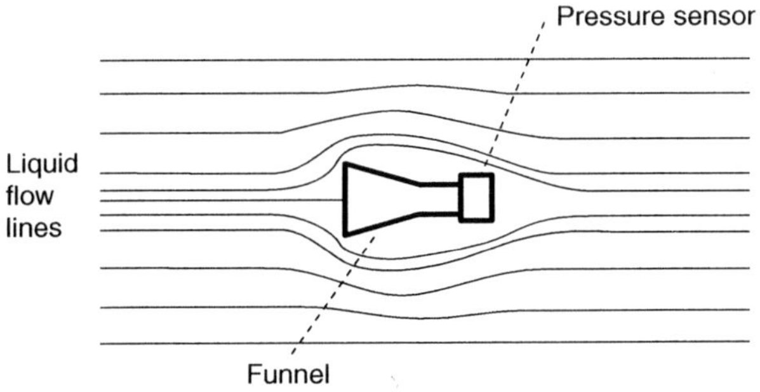

FIGURE 5.48
Figure for Problem 5.35.

5.33 A welding tank holds oxygen at 1500 psi. What is the tank pressure expressed in Pa? What is the pressure in atmospheres?

5.34 A diaphragm has an effective area of 25 cm^2. If the pressure difference across the diaphragm is 5 psi, what force is exerted on the diaphragm?

5.35 Figure 5.48 shows a proposed sensor for measuring the speed of liquid flowing in an open channel. An SS pressure sensor is connected to a funnel as shown. A pressure is formed when the funnel has its open end pointed upstream so that the liquid is brought to rest against the funnel opening. The pressure is given by $p = \frac{1}{2}\rho v^2$, where ρ is the liquid density in kg/m^3 and v is the liquid speed in m/s. The SS pressure sensor has a range of 0 to 5 kPa with a transfer function of 40 mV/kPa. Suppose the liquid is water with a density of $1 g/cm^3$. What is the maximum speed which can be measured? Plot a graph of sensor output voltage versus liquid speed. Comment on linearity.

5.36 The bellows, diaphragm, and Bourdon tube pressure sensors all exhibit second-order time response. This means that a sudden change in pressure will cause an oscillation in the displacement and, therefore, in sensor output. Because they are like springs, they have an effective spring constant and mass, so the frequency can be estimated by Equation (5.27). Consider a bellows with an effective spring constant of 3500 N/m and mass of 50 g. The effective area against which the pressure acts is 0.5 in^2. Calculate **(a)** the bellows deflection for a pressure of 20 psi and **(b)** the natural frequency of oscillation.

Section 5.6

5.37 A grain conveyor system finds the weight on a 1.0-m platform to be 258 N. What conveyor speed is needed to get a flow of 5200 kg/h?

5.38 Convert water flow of 52.2 gal/h into kg/h and velocity in m/s through a 2-in.-diameter pipe.

5.39 For an orifice plate in a system pumping alcohol, we find $K = 0.4\ \text{m}^3/\text{min}\,(\text{kPa})^{1/2}$. Plot the pressure versus flow rate from 0 to 100 m^3/h.

SUPPLEMENTARY PROBLEMS

S5.1 An ultrasonic system will be used to measure the level of grain from 1 to 9 m in a bin, as shown in Figure 5.49. This will be done by measuring the time delay between transmitting a short ultrasonic pulse and receiving the pulse echo from the grain surface. A 50-kHz transmitter can be triggered on and off by a logic high/low input, as shown. The transmitted pulse is to have a duration of 6 ms, and the propagation speed is 300 m/s. The receiver is connected to a comparator that goes high when a signal is received. The receiver must be disabled while the transmitter is sending a pulse using a logic high input, as shown. An 8-bit counter will start counting when the pulse

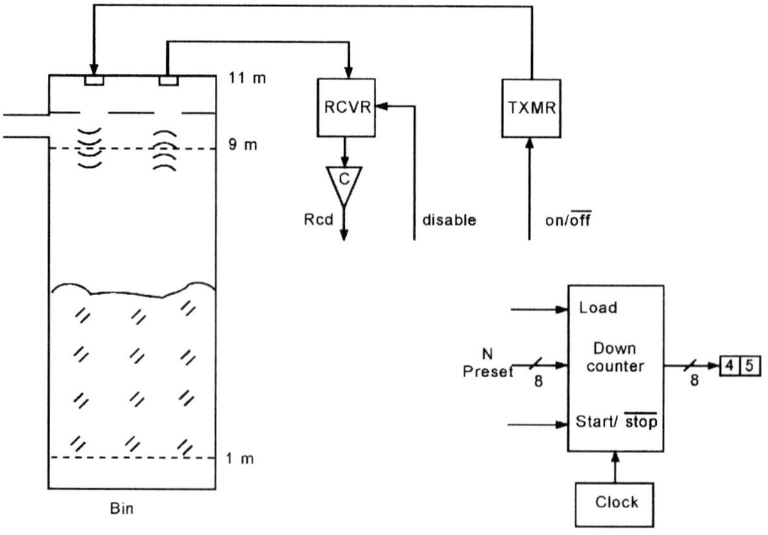

FIGURE 5.49
Ultrasonic system of level measurement for Problem S5.1.

is sent and stop when the echo is received. Its count is thus a measure of the time delay and the grain level. A reading is to be taken every second.

a. Calculate the expected time delay for grain levels between 1 m and 9 m.

b. Since the ultrasonic system measures from the bin top, the time delay will be reversed (i.e., long time is low level and short time is high level). To account for this, the counter will count down from some preset value. Determine the preset value and the correct counter clock speed so that the digital display is equal to the grain level (i.e., a level of 4.5 m would produce a binary count of $\mathbf{00101101}_2$, so the display would show 45).

c. Use one-shot pulse generators, flip/flops, and any other digital logic devices to complete the design shown in Figure 5.49.

S5.2 Show how the gauge mounting in Figure 5.17 can be changed so that all four gauges are active but temperature compensation is still provided. Show how the gauges are connected in a bridge.

S5.3 Figure 5.45 shows the variation of GF with strain for a semiconductor strain gauge with $R = 300\ \Omega$ under no strain. This gauge is used to measure solid material flow on a conveyor using a system like that in Figure 5.37. The load cell structure is shown in Figure 5.50. The conveyor speed is $R = 0.3$ m/s, and the platform length is $L = 1.5$ m.

a. Design a bridge and signal conditioning for the strain gauges that is temperature compensated and provides an output of 0.0 to 5.0 V for a strain of 0 to 500 μ.

b. Prepare a plot of output voltage versus flow rate from 0.0 to 200 kg/s.

S5.4 An SS pressure sensor will be used to measure the specific gravity of a flowing production liquid. Specific gravity is simply the ratio of liquid density to water density. Figure 5.51 shows how this can be measured by the difference in head pressure for equal heights of the liquid and water. A level control system maintains the flowing production liquid at the same level as the water. The differential SS sensor has a sensitivity of 45 mV/kPa. The specific gravity will be measured in the range of 1.0 to 2.0. The liquid/water level is 1.0 m, and the density of water is 1 g/cm^3.

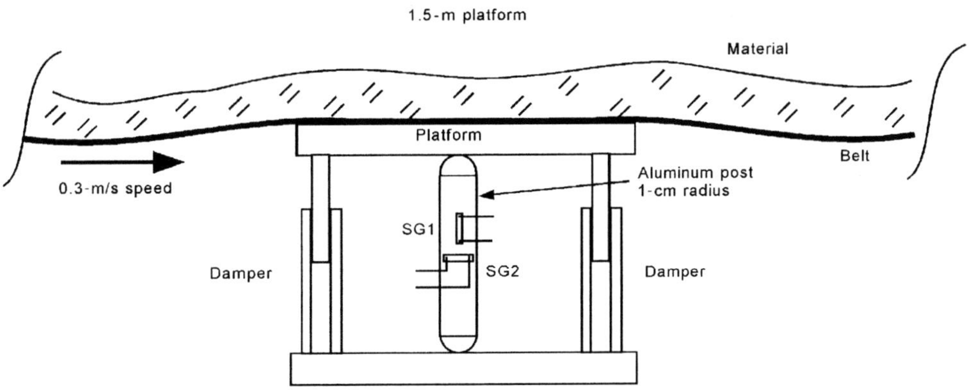

FIGURE 5.50
Load cell for Problem S5.3.

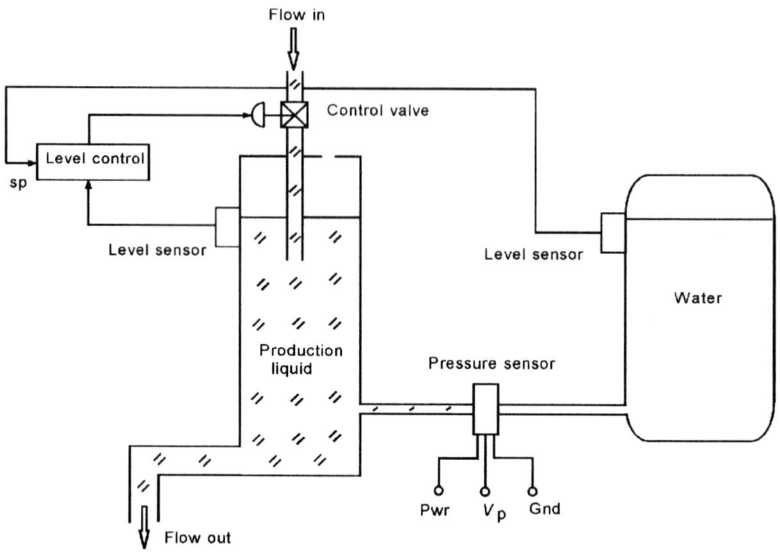

FIGURE 5.51
System for Problem S5.4.

a. What is the pressure of water at the sensor? What is the range of pressure of the liquid at the sensor?
b. What sensor voltage results for the range of 1.0 to 2.0 specific gravity?
c. Develop signal conditioning to interface the sensor output to an 8-bit ADC with a 5.00-V reference.
d. What is the digital resolution in specific gravity measurement?
e. If the level control system has a ±2-cm error about the setpoint, how much error will there be in specific gravity measurement?

FIGURE 5.52
Pirani gauge for Problem S5.5.

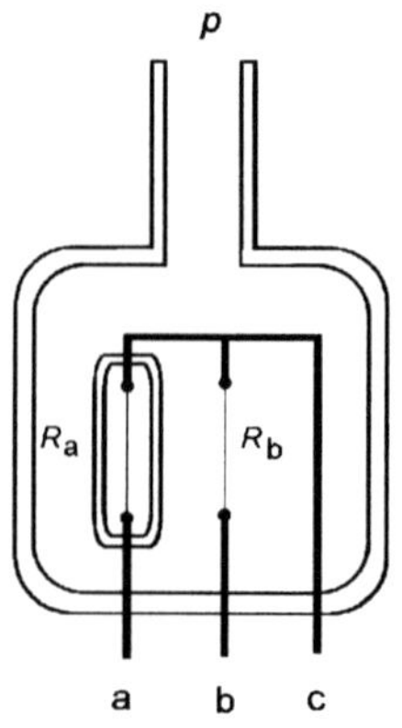

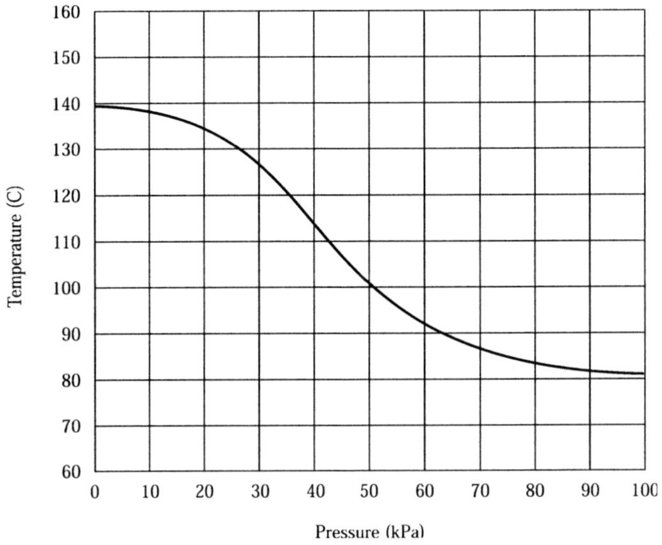

FIGURE 5.53
Resistance versus pressure for Problem S5.5.

S5.5 Figure 5.52 shows a Pirani gauge for measuring vacuum pressures from about 0 to 100 kPa (atmospheric pressure). The two resistive filaments are operated at an elevated temperature of 80°C at 1 atm pressure. The exposed filament temperature is a function of pressure, but the encapsulated filament temperature is not. Figure 5.53 shows how the exposed filament temperature varies with pressure. Variation of resistance with temperature is given by the RTD relation with $\alpha = 0.035/^\circ\text{C}$ and $R = 20\ \Omega$ at 20°C and a self-heating dissipation constant of 30 mW/°C. Design a bridge circuit that provides the necessary self-heating current and an off-null voltage that depends on pressure. Plot the off-null voltage versus pressure from 0.0 to 100 kPa.

CHAPTER 6

Optical Sensors

INSTRUCTIONAL OBJECTIVES

After you have read this chapter, you should be able to

- Describe electromagnetic (EM) radiation in terms of frequency, wavelength, speed of propagation, and spectrum.
- Define the energy of EM radiation in terms of power, intensity, and the effects of divergence.
- Compare photoconductive, photovoltaic, and photoemissive-type photodetectors.
- Describe the principles and structure of both total radiation and optical pyrometers.
- Distinguish incandescent, atomic, and laser light sources by the characteristics of their light.
- Design the application of optical techniques to process-control measurement applications.

6.1 INTRODUCTION

A desirable characteristic of sensors is that they have a negligible effect on the measured environment—that is, the process. Thus, if a resistance-temperature detector (RTD) heats up its own temperature environment, there is less confidence that the RTD resistance truly represents the environmental temperature. Much effort is made in sensor and transducer design to reduce backlash from the measuring instrument on its environment.

When electromagnetic (EM) radiation is used to perform process-variable measurements, transducers that do not affect the system measured emerge. Such systems of measurement are called *nonlocal* or *noncontact* because no physical contact is made with the environment of the variable. Noncontact characteristic measurements often can be made from a distance.

In process control, EM radiation in either the visible or infrared light band is frequently used in measurement applications. The techniques of such applications are called *optical* because such radiation is close to visible light.

A common example of optical transduction is measurement of an object's temperature by its emitted EM radiation. Another example involves radiation reflected off the surface to yield a level or displacement measurement.

Optical technology is a vast subject covering a span from geometrical optics, including lenses, prisms, gratings, and the like, to physical optics, with lasers, parametric frequency conversion, and nonlinear phenomena. These subjects are all interesting, but all that is required for our purposes is a familiarity with optical principles and a knowledge of specific transduction and measurement methods.

6.2 FUNDAMENTALS OF EM RADIATION

We are all familiar with EM radiation as *visible light.* Visible light is all around us. EM radiation is also familiar in other forms, such as radio or TV signals and ultraviolet or infrared light. Most of us falter, however, if asked to give a general technical description of such radiation including criteria for measurement and units.

This section covers a general method of characterizing EM radiation. Although much of what follows is valid for the complete range of radiation, particular attention is given to the infrared, visible, and ultraviolet, because most sensor applications are concerned with these ranges.

6.2.1 Nature of EM Radiation

EM radiation is a form of energy that is always in motion—that is, it propagates through space. An object that releases, or *emits,* such radiation *loses* energy. One that *absorbs* radiation *gains* energy. Thus, we must describe how this energy appears as EM radiation.

Frequency and Wavelength Because we use the term *electromagnetic* radiation to name this form of energy, it is no surprise that it is intimately tied to electricity and magnetism. Indeed, careful study shows that electrical and magnetic phenomena produce EM radiation. The radiation propagates through space in a manner similar to waves in water propagating from some disturbance. As such, we can define both a frequency and a wavelength of the radiation. The *frequency* represents the oscillation per second as the radiation passes some fixed point in space. The *wavelength* is the spatial distance between two successive maxima or minima of the wave in the direction of propagation.

Speed of Propagation EM radiation propagates through a vacuum at a constant speed independent of both the wavelength and frequency. In this case, the velocity is

$$c = \lambda f \tag{6.1}$$

where $c = 2.998 \times 10^8\ \text{m/s} \approx 3 \times 10^8\ \text{m/s}$ = speed of EM radiation in a vacuum
λ = wavelength in meters
f = frequency in hertz (Hz) or cycles per second (s^{-1})

EXAMPLE 6.1 Given an EM radiation frequency of 10^6 Hz, find the wavelength.

Solution
We have

$$c = \lambda f$$

and

$$\lambda = \frac{c}{f} = \frac{3 \times 10^8 \text{ m/s}}{10^6 \text{ s}^{-1}}$$

$$\lambda = \mathbf{300\ m}$$

Note: This EM radiation is used to carry AM radio signals.

When such radiation moves through a nonvacuum environment, the propagation velocity is reduced to a value less than *c*. In general, the new velocity is indicated by the *index of refraction* of the medium. The index of refraction is a ratio defined by

$$n = \frac{c}{v} \tag{6.2}$$

where n = index of refraction
v = velocity of EM radiation in the material (m/s)

The index of refraction often varies with the radiation wavelength for some sample of material.

EXAMPLE 6.2 Find the velocity of EM radiation in glass that has an index of refraction of $n = 1.57$.

Solution
We know that

$$n = \frac{c}{v}$$

so that

$$v = \frac{c}{n} = \frac{3 \times 10^8 \text{ m/s}}{1.57}$$

$$v = \mathbf{1.91 \times 10^8\ m/s}$$

Wavelength Units For many applications, specification of EM radiation is made through the frequency of the radiation, as in a 1-MHz radio signal or a 1-GHz microwave signal. In other cases, however, it is common to describe EM radiation by the wavelength. This is particularly true near the visible light band. The proper unit of measurement is the

FIGURE 6.1
The electromagnetic radiation spectrum covers everything from very low frequency (VLF) radio to X-rays and beyond.

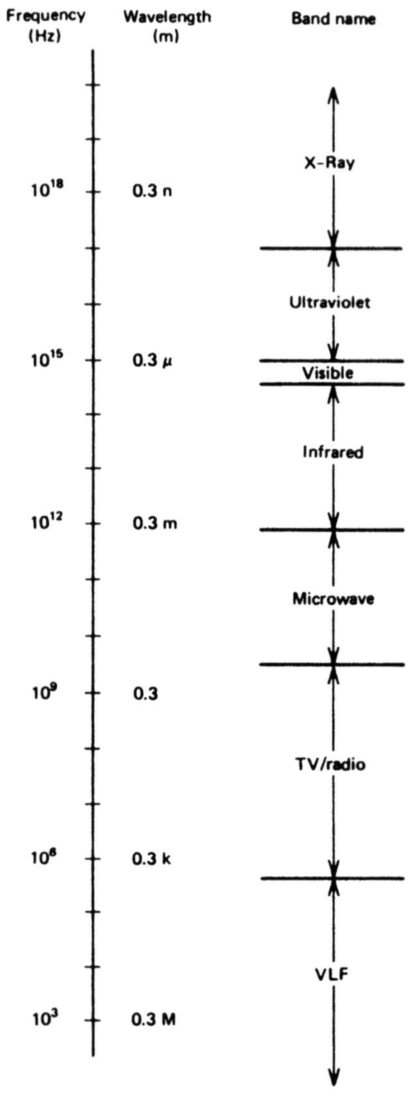

length in meters with associated prefixes. Thus, a 10-GHz signal is described by a 30-mm wavelength. Red light is emitted at about 0.7 μm wavelength.

Another unit often employed is the *Angstrom* (Å), defined as 10^{-10} m or 10^{-10} m/Å. Thus, the red light previously described has a wavelength of 7000 Å. The conversion is left as an exercise for the reader. (See also Problem 6.2.)

EM Radiation Spectrum We have seen that EM radiation is a type of energy that propagates through space at a constant speed or velocity if we specify the direction. The oscillating nature of this radiation gives rise to a different interpretation of this radiation in relation to our environment, however. In categorizing radiation by wavelength or frequency, we are describing its position in the *spectrum* of radiation. Figure 6.1 shows the range of EM radiation from very low frequency to very high frequency, together with the associated wavelength in meters from Equation (6.1) and how the bands of frequency relate to our world.

This one type of energy ranges from radio signals and visible light to X-rays and penetrating cosmic rays and all through the smooth variation of frequency. In process-control instrumentation, we are particularly interested in two of the bands, infrared and visible light.

Even though Figure 6.1 presents distinct boundaries between EM radiation descriptions, in reality the boundaries are quite indistinct. Thus, the transition between microwave and infrared, for example, is gradual, so that over a considerable band, the radiation could be described by either term.

Visible Light The small band of radiation between approximately 400 nm and 760 nm represents *visible light* (Figure 6.1). This radiation band covers those wavelengths to which our eyes (or radiation detectors in our heads) are sensitive.

Infrared Light The longer-wave radiation band that extends from the limit of eye sensitivity at 0.76 μm to approximately 100 μm is called *infrared* (IR) radiation. In some cases, the band is further subdivided so that radiation of wavelength 3 to 100 μm is called *far-infrared.*

Again, the limits of these bands are not distinct and serve only to roughly separate the described radiation into broad categories. Our treatment for the rest of this chapter refers simply to *light,* meaning either IR or visible, because our concern is with these bands.

EXAMPLE 6.3 Describe (**a**) the wavelength (in m and Å), and (**b**) the nature of EM radiation of 5.4×10^{13} Hz frequency.

Solution

a. The wavelength is given by

$$\lambda = \frac{c}{f}$$

$$\lambda = (3 \times 10^{8}\ \text{m/s})/(5.4 \times 10^{13}\ \text{Hz})$$

$$\lambda = \mathbf{5.56\ \mu m}$$

$$\lambda = 5.56 \times 10^{-6}\ \text{m} \times \frac{1\ \text{Å}}{10^{-10}\ \text{m}} = \mathbf{55{,}600\ Å}$$

b. From Figure 6.1, we see that this radiation lies in the infrared band, generally, and is designated as far-infrared, specifically.

6.2.2 Characteristics of Light

Because light has been described as a source of energy, it is natural to inquire about the energy content and its relation to the spectrum.

Photon No description of EM radiation is complete without a discussion of the photon. EM radiation at a particular frequency can propagate only in *discrete* quantities of energy. Thus, if some source is emitting radiation of one frequency, then in fact it is emitting this energy as a large number of discrete units or *quanta*. These quanta are called *photons*. The actual energy of one photon is related to the frequency by

$$W_p = hf = \frac{hc}{\lambda} \tag{6.3}$$

where W_p = photon energy (J)
$h = 6.63 \times 10^{-34}$ J·s (known as Planck's constant)
f = frequency (s^{-1})
λ = wavelength (m)

The energy of one photon is very small compared to electric energy, as shown by Example 6.4.

EXAMPLE 6.4 A microwave source emits a pulse of radiation at 1 GHz with a total energy of 1 J. Find (**a**) the energy per photon and (**b**) the number of photons in the pulse.

Solution

a. We find the energy per photon where 1 GHz $= 10^9\ s^{-1}$

$$W_p = hf$$
$$W_p = (6.63 \times 10^{-34}\ \text{J}\cdot\text{s})(10^9\ \text{s}^{-1})$$
$$W_p = \mathbf{6.63 \times 10^{-25}\ J}$$

b. The number of photons is

$$N = \frac{W}{W_p} = \frac{1\ \text{J}}{6.63 \times 10^{-25}\ \text{J/photon}}$$
$$N = \mathbf{1.5 \times 10^{24}\ photons}$$

Figure 6.2 shows the energy carried by a single photon at various wavelengths. This energy is expressed in electron volts where ($1\ \text{eV} = 1.602 \times 10^{-19}$ J). This unit is conventionally employed and provides convenient magnitude for photon energy discussion.

Energy When dealing with typical sources and detectors of light, it is impractical and unnecessary to consider the discrete nature of the radiation. Instead, we deal with *macroscopic* properties that result from the collective behavior of a vast number of photons moving together. The general energy principles involve a description of the net energy of the radiation as it propagates through a region of space. This description is then given in *joules* of energy in the propagating light. A simple statement of energy is insufficient, however, because of the motion of the energy and the spatial distribution of the energy.

Power Because EM radiation is energy in motion, a more complete description is the joules per second, or *watts*, of power carried. Thus, one might describe a situation where

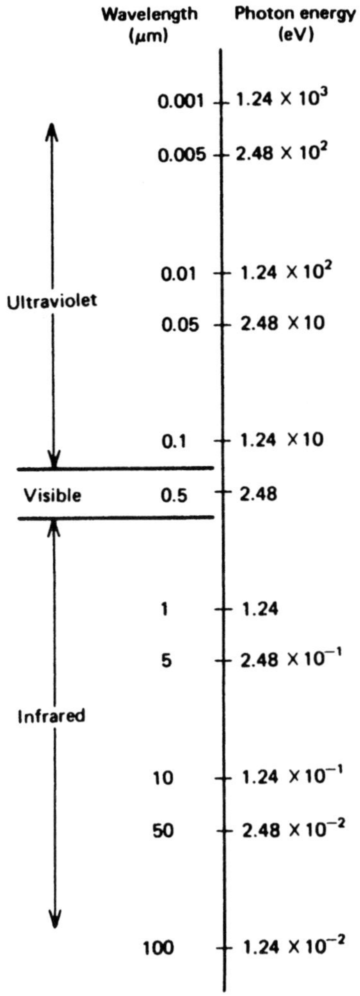

FIGURE 6.2
The energy carried by one photon varies inversely with the wavelength of the EM radiation.

a source emits 10 W of light; that is, 10 J of energy in the form of light radiation are emitted every second. Even this description is incomplete without specifying how the power is spatially distributed.

Intensity A more complete picture of the radiation emerges if we also specify the spatial distribution of the power *transverse* to the *direction* of propagation. Thus, if the 10-W source just discussed was concentrated in a beam with a cross-sectional area of 0.2 m^2, then we can specify the *intensity* as the watts per unit area, in this case as (10 W)/(0.2 m^2), or 50 W/m^2. In general, the intensity is

$$I = \frac{P}{A} \tag{6.4}$$

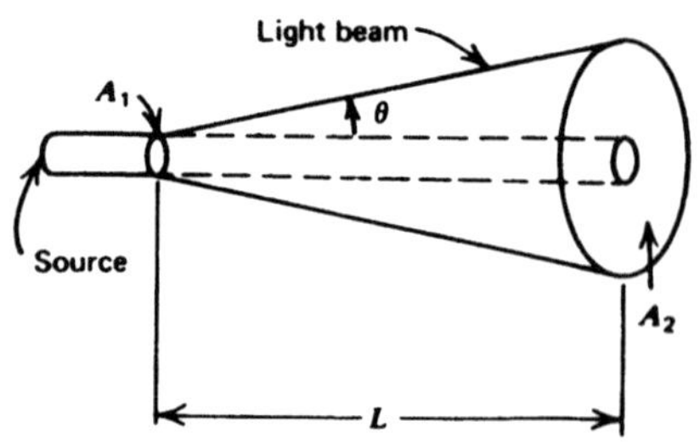

FIGURE 6.3
Sources of EM radiation exhibit divergence through the spreading of the beam with distance from the source.

where I = intensity in W/m^2
P = power in W
A = beam cross-sectional area in m^2

One problem with expressing the intensity as W/m^2 is that it is sometimes difficult to imagine how the distribution of power over a square meter relates to a small-scale problem. Thus, if a sensor has an area of only 2 cm^2, some people have difficulty relating this size to an intensity of, say, 10 W/m^2. For this reason, the unit mW/cm^2 is in common use. This quantity is equal to 10 W/m^2, so they are not so much different in size but in visualization. In the former case of a 2-cm^2 sensor, the intensity would be 1 mW/cm^2. Another motivation for using mW/cm^2 is that light-source power in measurement situations is often in the milliwatt range.

Divergence We still have not quite exhausted the necessary descriptors of the energy because of the tendency of light to travel in straight lines. Because the radiation travels in straight lines, it is possible for the intensity of the light to change even though the power remains constant. This is best seen in Figure 6.3. We have a 10-W source with an area, A_1, at the source. Because of the nature of the source and the straight-line propagation, however, we see that some distance away the same 10 W is distributed over a larger area, A_2, and hence the *intensity* is *diminished.* Such spreading of radiation is called *divergence* and specified as the angle θ, which is made by the outermost edge of the beam to the central direction of propagation, as shown in Figure 6.3.

Thus far, a description of the energy state of light propagating through a region of space demands knowledge of the power carried, the cross-sectional area over which the power is distributed, and the divergence.

EXAMPLE 6.5 Find the intensity of a 10-W source (**a**) at the source and (**b**) 1 m away for the case shown in Figure 6.3 if the radius at A_1 is 0.05 m and the divergence is 2°.

Solution

The intensity at the source is simply the power over the area. Thus, for a circular cross-section,

$$A_1 = \pi R_1^2 = (3.14)(0.05\,\text{m})^2$$
$$A_1 = 7.85 \times 10^{-3}\,\text{m}^2$$

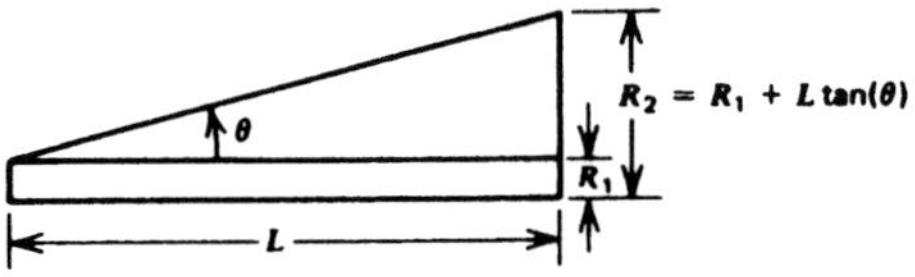

FIGURE 6.4
Diagram to aid in solving divergence problems, as in Example 6.5.

a. The intensity at the source is

$$I_1 = \frac{P}{A_1}$$

$$I_1 = \frac{10\,\text{W}}{7.85 \times 10^{-3}\,\text{m}^2} = \mathbf{1273\ W/m^2}$$

b. We first must find the beam area at 1 m to get A_2 and then the intensity at 1 m. We can find the radius using Figure 6.4 as a guide. From elementary trigonometry, we see that, for the right triangle formed,

$$R_2 = R_1 + L\tan(\theta) \qquad \textbf{(6.5)}$$

Then

$$R_2 = 0.05\,\text{m} + (1\,\text{m})\tan(2^\circ)$$
$$R_2 = 0.085\,\text{m}$$
$$A_2 = \pi R_2^2 = (3.14)(0.085\,\text{m})^2$$
$$A_2 = 0.0227\,\text{m}^2$$
$$I_2 = \frac{10\,\text{W}}{0.0227\,\text{m}^2}$$
$$I_2 = 440.53\ \text{W/m}^2,\ \text{or}\ \mathbf{441\ W/m^2}$$

When dealing with sources originating from a very small point but propagating in all directions, we have a condition of *maximum divergence.* Such *point sources* have an intensity that decreases as the *inverse square* of the distance from the point. This can be seen from a consideration of the divergence. Suppose a source delivers a power, *P,* of light as shown in Figure 6.5. The intensity at a distance *R* is found by dividing the total power by the surface area of a sphere of radius *R* from the source:

$$I = \frac{P}{A} \qquad \textbf{(6.4)}$$

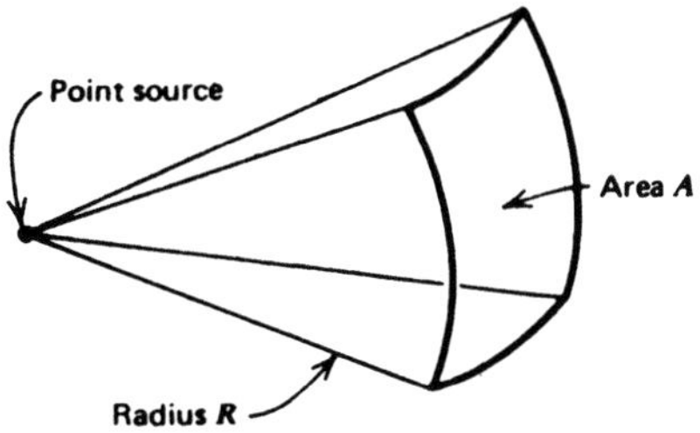

FIGURE 6.5
The intensity of light from a point source depends on the distance from the source, *R,* and the area considered, *A.*

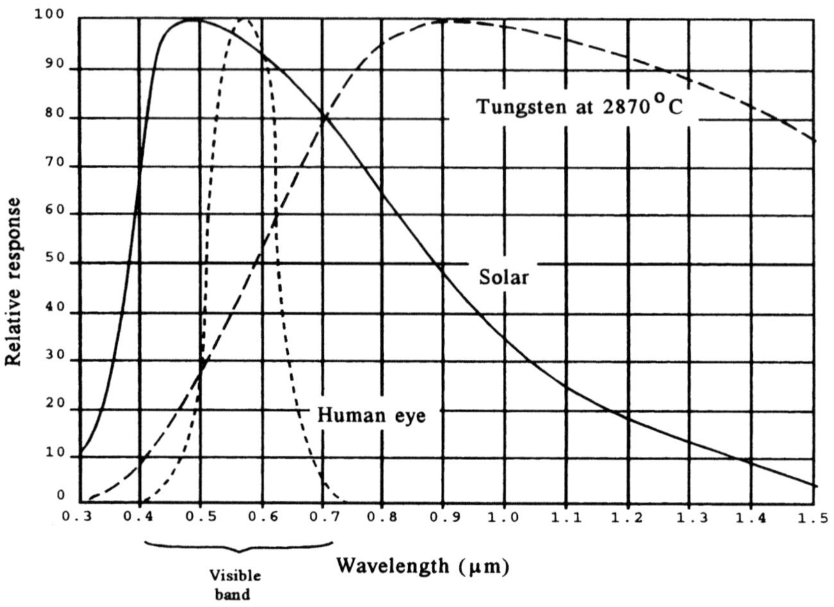

FIGURE 6.6
Comparison of EM radiation emitted by the sun and heated tungsten filament, as well as the spectral sensitivity of the human eye.

The surface area of a sphere of radius R is $A = 4\pi R^2$; thus, over the entire surface surrounding a point source,

$$I = \frac{P}{4\pi R^2} \tag{6.6}$$

This equation shows that the intensity of a point source decreases as the inverse square of the distance from the point.

Spectrum Another factor of significance in the description of light includes the *spectral content* of the radiation. A source such as a laser beam, which delivers light of a single wavelength (or very nearly), is called a *monochromatic* source. A source such as an incandescent bulb may deliver a broad spectrum of radiation and is referred to as a *polychromatic source.*

In general, the spectrum of a source is described by a curve showing how the power is distributed as a function of wavelength (or frequency) of the radiation. Figure 6.6 compares the spectrum of the EM radiation from the sun to that from a standard tungsten source at a temperature of 2870°C (white hot). This curve gives the spectrum in terms of the percent of the emitted power relative to the maximum as a function of radiation wavelength.

Coherency A less familiar characteristic of the radiation is its coherency. We have seen that light is described through electric and magnetic effects that oscillate in time and space. Whenever we consider oscillating phenomena, it is of interest to determine the

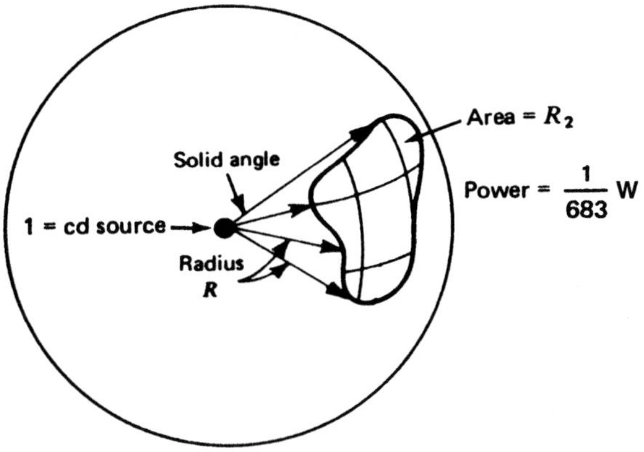

FIGURE 6.7
The candela is defined in terms of uniform monochromatic radiation from a point source.

phase relation of oscillations of different parts of the beam. When all points along some cross-section of a radiation beam are in phase, the beam has *spatial coherence.* When the radiation in a line along the beam has a fixed-phase relation, the beam has *temporal coherence.* In general, conventional sources of light, such as incandescent and fluorescent light bulbs, produce beams with no coherence. A laser is the only convenient source of coherent radiation available.

6.2.3 Photometry

The conventional units described previously would seem to be satisfactory for a complete description of optical processes. For designs in human engineering, however, these traditional units are insufficient. This is because the human eye responds not only to the intensity of light but also to its spectral content. Figure 6.6 also shows the nominal spectral sensitivity of the human eye. You can see that it is peaked in the visible band, as you would expect.

Suppose it is known that a human needs an intensity of 1 W/m^2 to read. If an infrared source at that intensity were used, the human would still not be able to read, because the eye does not respond to infrared.

Because of these problems and others, special sets of units for EM radiation have been developed to be used in such human-related design problems. The most basic unit is the SI *candela* (cd). A 1-cd source is defined as one that emits monochromatic radiation at 340×10^{12} Hz ($\approx$ 555 nm wavelength) at such an intensity that there is 1/683 W passing through one steradian of solid angle. This radiation is roughly in the middle of the visible spectrum.

Figure 6.7 shows how a 1-cd source is pictorially defined. The 1-cd source at the center of a sphere of radius R will emit radiation of the aforementioned frequency with an energy of 1/683 W through any part of the surface with an area of R^2.

There are many other units employed to describe visible or luminous energy. These units do not find significant application in instrumentation and measurement applications using optical technology in control, so they will not be covered.

6.3 PHOTODETECTORS

An important part of any application of light to an instrumentation problem is how to measure or detect radiation. In most process-control-related applications, the radiation lies in the range from IR through visible and sometimes UV bands. The measurement sensors generally used are called *photodetectors* to distinguish them from other spectral ranges of radiation, such as RF detectors in radio frequency (RF) applications.

In this section, we will study the principal types of photodetectors through a description of their operation and specifications.

6.3.1 Photodetector Characteristics

Several characteristics of photodetectors are particularly important in typical applications of these devices in instrumentation. In the following discussions, the various types of detectors are described in terms of these characteristics.

The particular characteristic related to EM radiation detection is the *spectral sensitivity.* This is given as a graph of sensor response relative to the maximum as a function of radiation wavelength. Obviously, it is important to match the spectral response of the sensor to the environment in which it is to be used.

As usual, the standard characteristics of sensors, such as transfer function, linearity, time constant, and signal conditioning, will be given.

6.3.2 Photoconductive Detectors

One of the most common photodetectors is based on the change in *conductivity* of a semiconductor material with radiation *intensity.* The change in conductivity appears as a change in *resistance,* so that these devices also are called *photoresistive* cells. Because resistance is the parameter used as the transduced variable, we describe the device from the point of view of *resistance changes* versus *light intensity.*

Principle In Section 4.4.1, we noted that a semiconductor is a material in which an energy gap exists between conduction electrons and valence electrons. In a semiconductor photodetector, a photon is absorbed and thereby excites an electron from the valence to the conduction band. As many electrons are excited into the conduction band, the semiconductor resistance decreases, making the resistance an inverse function of radiation intensity. For the photon to provide such an excitation, it must carry at least as much energy as the gap. From Equation (6.3), this indicates a maximum wavelength:

$$E_p = \frac{hc}{\lambda_{max}} = \Delta W_g$$
$$\lambda_{max} = \frac{hc}{\Delta W_g} \tag{6.7}$$

where h = Planck's constant, 6.63×10^{-34} J·s
ΔW_g = semiconductor energy gap (J)
λ_{max} = maximum detectable radiation wavelength (m)

Any radiation with a wavelength greater than that predicted by Equation (6.7) cannot cause any resistance change in the semiconductor.

EXAMPLE 6.6 Germanium has a band gap of 0.67 eV. Find the maximum wavelength for resistance change by photon absorption. Note that $1.6 \times 10^{-19}\text{ J} = 1\text{ eV}$.

Solution

We find the maximum wavelength from

$$\lambda_{\max} = \frac{hc}{\Delta W_g}$$

$$\lambda_{\max} = \frac{(6.63 \times 10^{-34}\text{ J}\cdot\text{s})(3 \times 10^{8}\text{ m/s})}{(0.67\text{ eV})(1.6 \times 10^{-19}\text{ J/eV})}$$

$$\lambda_{\max} = 1.86\ \mu\text{m}$$

which lies in the IR.

It is important to note that the operation of a thermistor involves *thermal*-energy-exciting electrons in the conduction band. To prevent the photoconductor from showing similar thermal effects, it is necessary either to operate the devices at a controlled temperature or to make the gap too large for thermal effects to produce conduction electrons. Both approaches are employed in practice. The upper limit of the cell spectral response is determined by many other factors, such as reflectivity and transparency to certain wavelengths.

Cell Structure Two common photoconductive semiconductor materials are cadmium sulfide (CdS), with a band gap of 2.42 eV, and cadmium selenide (CdSe), with a 1.74-eV gap. Because of these large gap energies, both materials have a very high resistivity at room temperature. This gives bulk samples a resistance much too large for practical applications. To overcome this, a special structure is used, as shown in Figure 6.8, that minimizes resistance geometrically and provides maximum surface area for the detector. This result is based on Equation (4.7):

$$R = \rho l/A$$

where R = resistance (Ω)
ρ = resistivity ($\Omega\cdot\text{m}$)
l = length (m)
A = cross-sectional area (m^2)

Figure 6.8a shows the basic idea behind the structure of a photoconductive cell. Essentially, we make the photoconductor in a wide, thin layer, as shown. The area presented to light, $A = WL$, is large for maximum exposure, whereas the electrical length, $l = L$, is small, which reduces the nominal resistance. Of course, such a structure is unwieldy, so the strip is wound back and forth on an insulating base, as shown in Figure 6.8b.

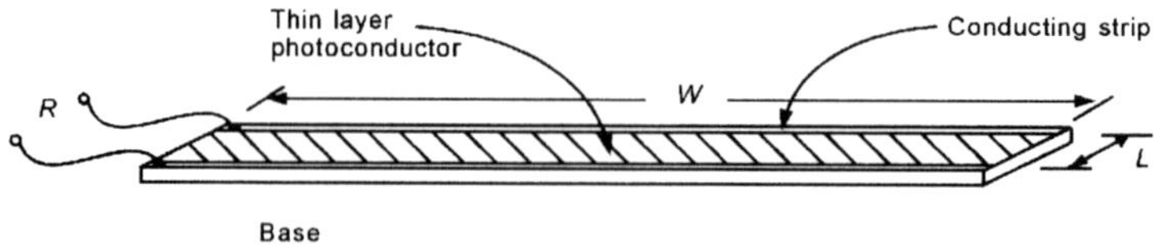

a) Cell configuration to have large area and low resistance

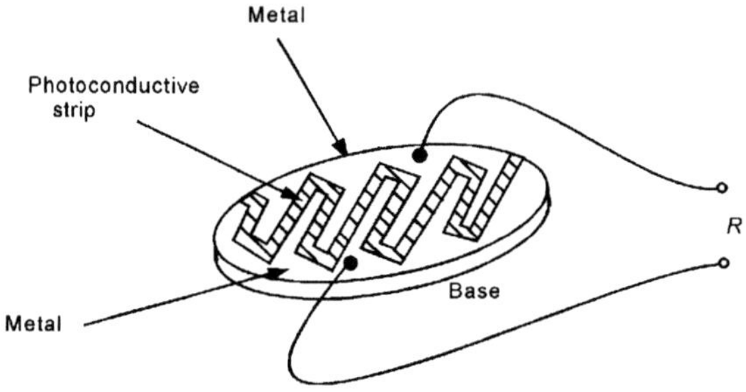

b) Folding the configuration back and forth
reduces the physical size

FIGURE 6.8
The photoconductive cell has a structure to maximize exposure and minimize resistance.

Cell Characteristics The characteristics of photoconductive detectors vary considerably when different semiconductor materials are used as the active element. These characteristics are summarized for typical values in Table 6.1.

The nominal dark resistance and variation of resistance with intensity are usually provided in terms of a graph or table of resistance versus intensity at a particular wavelength within the spectral band. Typical values at dark resistance vary from hundreds of ohms to several MΩ for various types of photoconductors. The variation with radiation intensity is usually nonlinear, with resistance decreasing as the radiation intensity increases. Figure 6.9

TABLE 6.1
Photoconductor characteristics

Photoconductor	Time Constant	Spectral Band
CdS	~100 ms	0.47 to 0.71 μm
CdSe	~10 ms	0.6 to 0.77 μm
PbS	~400 μs	1 to 3 μm
PbSe	~10 μs	1.5 to 4 μm

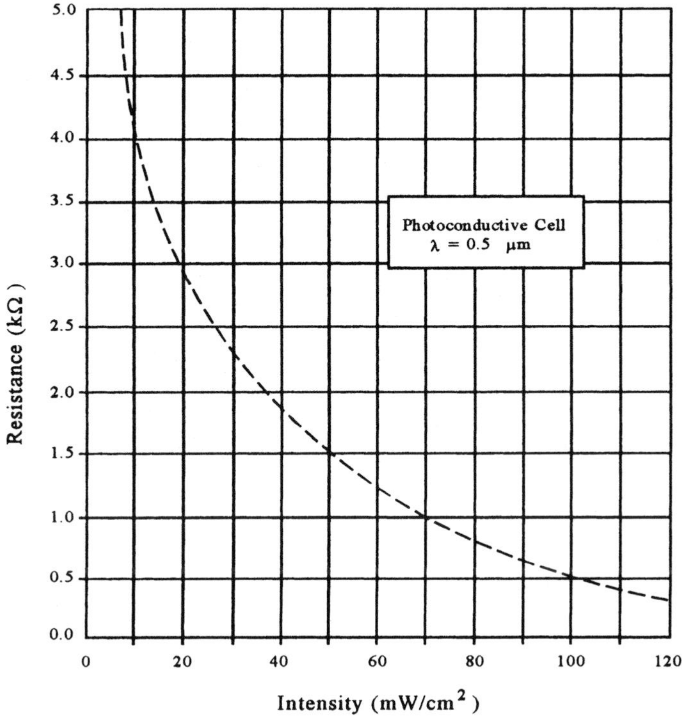

FIGURE 6.9
A photoconductive cell resistance changes nonlinearly with radiation intensity.

shows a typical graph of resistance change versus EM radiation intensity. This graph is for the wavelength of maximum response. The response is less as the wavelength goes above or below this value.

Both the PbS and PbSe cells are able to detect infrared radiation, but they are therefore also susceptible to thermal resistance changes. They are usually used in temperature-controlled enclosures to prevent thermal resistance changes from masking radiation effects. The CdS cell is the most common photoconductive cell in commercial use. Its spectral response is similar to that of the human eye, as shown in Figure 6.6.

Signal Conditioning Like the thermistor, a photoconductive cell exhibits a resistance that decreases nonlinearly with the dynamic variable, in this case, radiation intensity. Generally, the change in resistance is pronounced where a resistance can change by several hundred orders of magnitude from dark to normal daylight. If an absolute-intensity measurement is desired, calibration data are used in conjunction with any accurate resistance-measurement method.

Sensitive control about some ambient-radiation intensity is obtained using the cell in a bridge circuit adjusted for a null at the ambient level.

Various op amp circuits using the photoconductor as a circuit element are used to convert the resistance change to a current or voltage change.

It is important to note that the cell is a variable resistor, and therefore has some maximum power dissipation that cannot be exceeded. Most cells have a rating from 50 to 500 mW, depending on size and construction.

EXAMPLE 6.7 A CdS cell has a dark resistance of 100 kΩ and a resistance in a light beam of 30 kΩ. The cell time constant is 72 ms. Devise a system to trigger a 3-V comparator within 10 ms of the beam interruption.

Solution

There are many possible solutions to this problem. Let us first find the cell resistance at 10 ms using

$$R(t) = R_1 + (R_f - R_1)[1 - e^{-t/\tau}]$$
$$R(10\text{ ms}) = 30\text{k} + 70\text{ k}\,(0.1296) = 39.077\text{ k}\Omega$$

so that we must have a +3-V signal to the comparator when the cell resistance is 39.077 kΩ. The circuit in Figure 6.10 will accomplish this. The cell is R_2 in the feedback of an inverting amplifier with a −1.0-V constant input. The output is

$$V_{out} = -\left(\frac{R_2}{R_1}\right)(-1\text{ V}) = \frac{R_2}{R_1}$$

when $R_2 = 39.077\text{ k}\Omega$; we make $V_{out} = 3\text{ V}$ so that

$$R_2 = 39.077\text{ k}\Omega$$
$$R_1 \cong \mathbf{13\text{ k}\Omega}$$

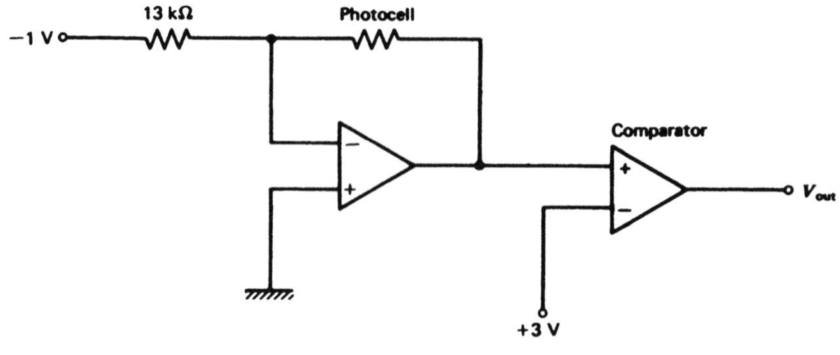

FIGURE 6.10
This circuit is the solution for Example 6.7.

which ensures that the comparator will trigger at 10 ms from beam interruption. To see that the comparator will *not* trigger with the beam present, set $R_2 = 30\ \mathrm{k\Omega}$ and the amplifier output is

$$V_{\text{out}} = \frac{-30\ \mathrm{k\Omega}}{13\ \mathrm{k\Omega}}(-1\ \mathrm{V})$$

$$V_{\text{out}} = 2.3\ \mathrm{V}$$

which is insufficient to trigger the comparator.

6.3.3 Photovoltaic Detectors

Another important class of photodetectors generates a voltage that is proportional to incident EM radiation intensity. These devices are called *photovoltaic cells* because of their *voltage-generating* characteristics. They actually convert the EM energy into electrical energy. Applications are found as both EM radiation detectors and power sources converting solar radiation into electrical power. The emphasis of our consideration is on instrumentation-type applications.

Principle Operating principles of the photovoltaic cell are best described by Figure 6.11. We see that the cell is actually a giant diode that is constructed using a *pn* junction between appropriately doped semiconductors. Photons striking the cell pass through the thin *p*-doped upper layer and are absorbed by electrons in the *n* layer, which causes formation of conduction electrons and holes. The depletion-zone potential of the *pn* junction

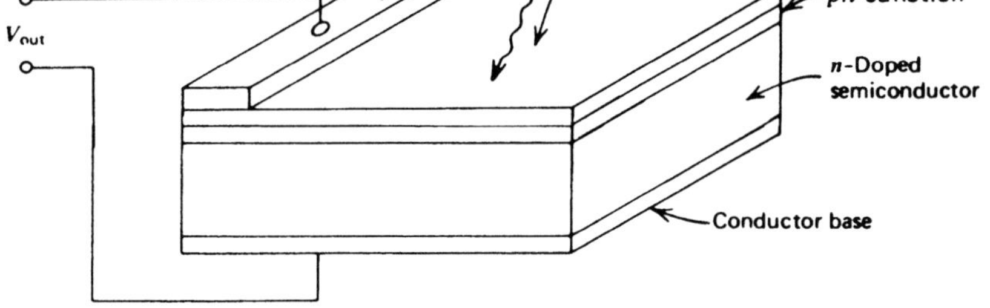

FIGURE 6.11
A photovoltaic "solar" cell is a giant *pn* junction diode.

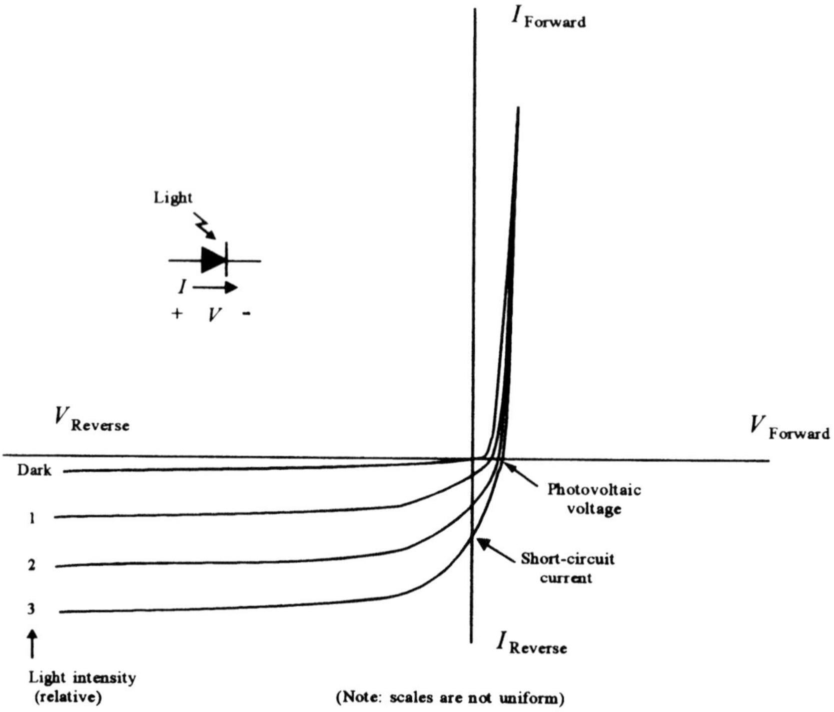

FIGURE 6.12
The IV curves of a *pn* junction diode vary with exposure to EM radiation.

then separates these conduction electrons and holes, which causes a difference of potential to develop across the junction. The upper terminal is positive, and the lower negative. It is also possible to build a cell with a thin, *n*-doped layer on top so that all polarities are opposite.

Electrical characteristics of the photovoltaic cell can be understood by reference to the *pn* junction diode IV characteristics. In Figure 6.12, a family of IV curves is shown for a photosensitive diode as a function of light intensity on the junction. You will note that when the junction is illuminated, a voltage is generated across the diode, as shown by the IV curve crossing the zero current axis with a nonzero voltage. This is the photovoltaic voltage.

Figure 6.12 also shows that the reverse current through the diode, when it is reverse biased, also increases with radiation intensity. This is the basis of the photodiode, to be discussed next.

Photovoltaic cells also have a range of spectral response within which a voltage will be produced. Clearly, if the frequency is too small, the individual photons will have insufficient energy to create an electron-hole pair, and no voltage will be produced. There is also an upper limit to the frequency because of optical effects such as radiation penetration through the cell.

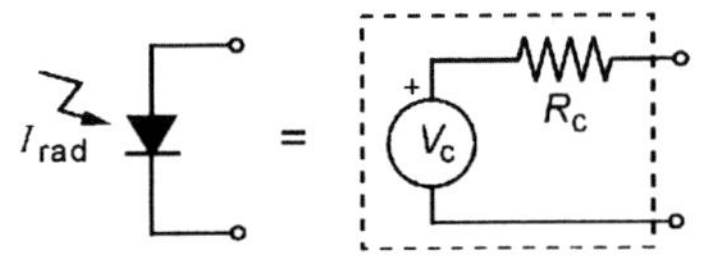

FIGURE 6.13
The Thévenin equivalent circuit for a photovoltaic cell. The resistance also varies with radiation.

Since the photovoltaic cell is a battery, it can be modeled as an ideal voltage source, V_c, in series with an internal resistance, R_c, as shown in Figure 6.13. It turns out that the voltage source varies with light intensity in an approximately logarithmic fashion:

$$V_c \approx V_0 \log_e(1 + I_R) \tag{6.8}$$

where V_c = open-circuit cell voltage
V_0 = constant, dependent on cell material
I_R = light intensity

The internal resistance of the cell also varies with light intensity. At low intensity, the resistance may be thousands of ohms, whereas at higher intensities it may drop to less than 50 ohms. This complicates the design of systems to derive maximum power from the cell, since the optimum load is equal to the internal resistance. Fortunately, at higher intensities, the internal resistance is nearly constant.

The short-circuit current, I_{SC}, which is simply the cell voltage divided by its internal resistance, V_c/R_c, varies linearly with light intensity. This is easy to understand, since the current is proportional to the number of charge carriers, which is proportional to the number of photons striking the cell and thus the radiation intensity.

Figure 6.12 shows the short-circuit current as the values of current when the IV curve crosses the zero-voltage line.

Signal Conditioning Since the short-circuit current, I_{SC}, is linearly related to the radiation intensity, it is preferable to measure this current when using the cell in measurement and instrumentation. Figure 6.14 shows how I_{SC} can be obtained by connecting the cell directly to an op amp. Since the current to the op amp input must be zero, the feedback current through R must equal I_{SC}. Therefore, the output voltage is given by

$$V_{out} = -RI_{SC} \tag{6.9}$$

Since the current is linearly proportional to light intensity, so is the output voltage.

FIGURE 6.14
This circuit converts the cell short-circuit current into a proportional voltage.

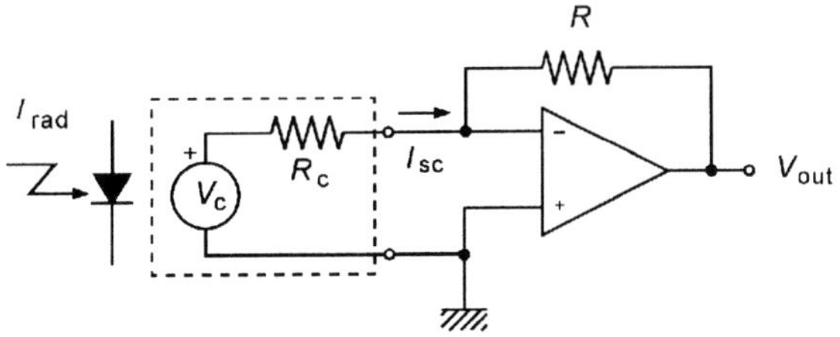

EXAMPLE 6.8 A photovoltaic cell is to be used with radiation of intensity from 5 to 12 mW/cm^2. Measurements show that its unloaded output voltage ranges from 0.22 to 0.41 V over this intensity while it delivers current from 0.5 to 1.7 mA into a 100-Ω load.

a. Find the range of short-circuit current.
b. Develop signal conditioning to provide a linear voltage from 0.5 to 1.2 V as the intensity varies from 5 to 12 mW/cm^2.

Solution

a. To find the short-circuit current, we need to find the cell resistance at the given intensities. This can be done by noting that the load current delivered into a 100-Ω load is given by

$$I_L = \frac{V_c}{100 + R_c}$$

where V_c is the cell open-circuit voltage and R_c is the cell internal resistance. Solving for the resistance, we find

$$R_c = (V_c - 100I_L)/I_L$$

So, at 5 mW/m^2, we find

$$R_c = (0.22 - 0.05)/0.0005 = 340\ \Omega$$

and at 12 mW/cm^2,

$$R_c = (0.42 - 0.17)/0.0017 = 147\ \Omega$$

Now the short-circuit current is given by $I_{SC} = V_c/R_c$, so

$$I_{SC}(5\,\text{mW/cm}^2) = 0.22/340 = 0.65\,\text{mA}$$

and

$$I_{SC}(12\,\text{mW/cm}^2) = 0.42/147 = 2.86\,\text{mA}$$

b. The signal conditioning will consist first of an op amp current-to-voltage converter, as in Figure 6.14. Further signal conditioning will provide the required output. Let's just find an equation for the whole signal conditioning:

$$V_{\text{out}} = mI_{SC} + V_0$$

Then, to find m and V_0, we form two equations from the given facts:

$$1.2 = 0.00286m + V_0$$
$$0.5 = 0.00065m + V_0$$

Subtracting,

$$0.7 = 0.0021m \quad \text{or} \quad m = 316.7$$

and

$$V_0 = 1.2 - 0.00286(316.7) = 0.294$$

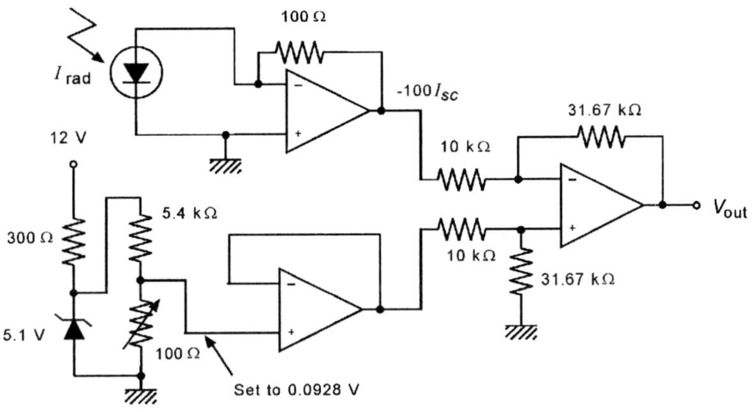

FIGURE 6.15
One solution to Example 6.8.

So the whole signal conditioning is

$$V_{out} = 316.7I_{SC} + 0.294$$

For the first stage, let's use a 100-Ω feedback resistor:

$$V_{out} = 3.167(100I_{SC}) + 0.294$$

What remains can be provided by a differential amplifier, since

$$V_{out} = 3.167(100I_{SC} + 0.0928)$$

Figure 6.15 shows the final circuit.

Cell Characteristics The properties of photovoltaic cells depend on the materials employed for the cell and the nature of the doping used to provide the *n* and *p* layers. Some cells are used only at low temperatures to prevent thermal effects from obscuring radiation detection. The silicon photovoltaic cell is probably the most common. Table 6.2 lists several types of cells and their typical specifications.

Thermal effects can be pronounced, producing changes of the order of mV/°C in output voltage of fixed intensity.

TABLE 6.2
Typical photovoltaic cell characteristics

Cell Material	Time Constant	Spectral Band
Silicon (Si)	~20 μs	0.44 μm to 1 μm
Selenium (Se)	~2 ms	0.3 μm to 0.62 μm
Germanium (Ge)	~50 μs	0.79 μm to 1.8 μm
Indium arsenide (InAs)	~1 μs	1.5 μm to 3.6 μm (cooled)
Indium antimonide (InSb)	~10 μs	2.3 μm to 7 μm (cooled)

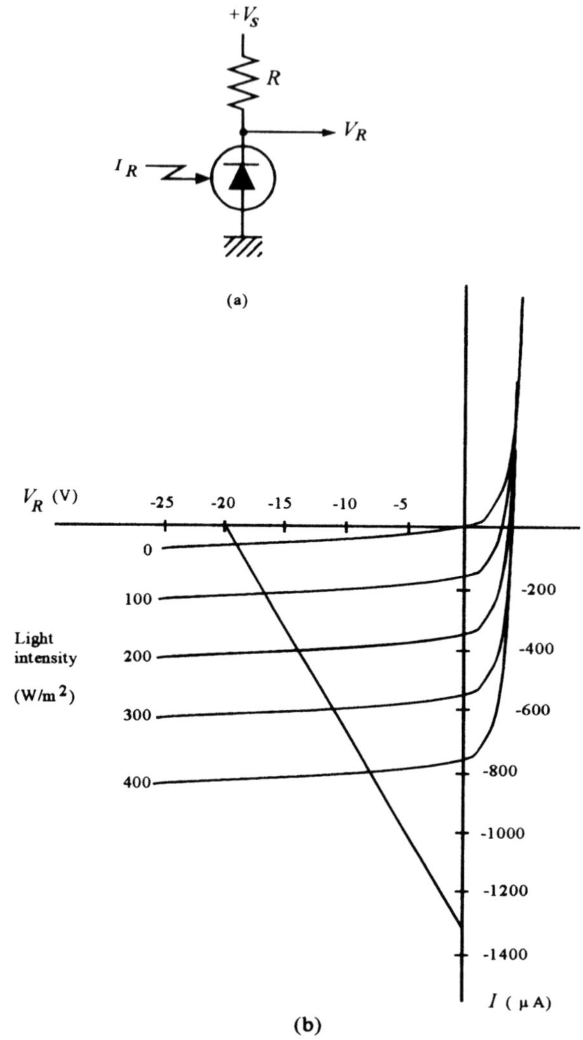

FIGURE 6.16
The photodiode uses the *pn* junction reverse current to measure radiation.

6.3.4 Photodiode Detectors

The previous section showed one way that the *pn* junction of a diode is sensitive to EM radiation, the *photovoltaic effect.* A *pn* diode is sensitive to EM radiation in another way as well, which gives rise to photodiodes as sensors.

The *photodiode effect* refers to the fact that photons impinging on the *pn* junction also alter the reverse current-versus-voltage characteristic of the diode. In particular, the reverse current will be increased almost linearly with light intensity. Thus, the photodiode is operated in the reverse-bias mode. Figure 6.16a shows a basic reverse-bias connection of such a diode.

FIGURE 6.17
Photodiodes are very small and often use an internal lens to focus light on the junction.

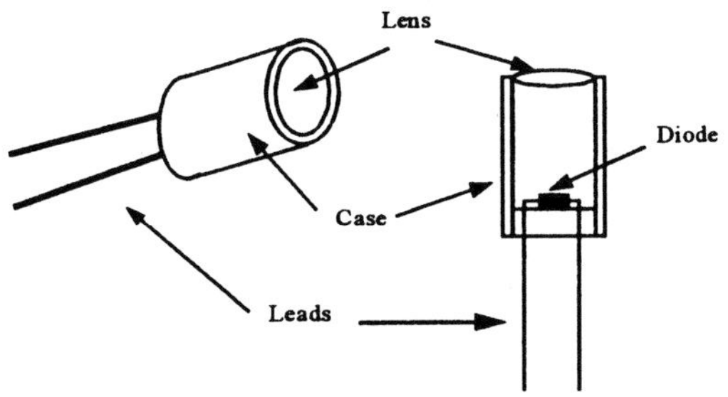

Figure 6.12 already showed how the reverse current increases with light intensity. In Figure 6.16b, a load line has been drawn on the reverse-current IV curves of the diode. The load line is used to determine the voltage changes across the diode as a function of light intensity. The dark-current response is the basic diode reverse current, with no illumination of the *pn* junction. Curves like those in Figure 6.12 and Figure 6.16b are given for radiation intensity from a polychromatic standard source such as tungsten heated to 2870 K (white hot).

One of the primary advantages of the photodiode is its fast time response, which can be in the nanosecond range. Generally, photodiodes are very small, like regular diodes, so that a lens must be used to focus light on the *pn* junction. Often, the lens is built into the photodiode casing, as shown in Figure 6.17.

Spectral response of photodiodes is typically peaked in the infrared, but with usable response in the visible band. The standard sources, like the tungsten given previously, provide radiation in a broad range of visible to infrared and far-infrared radiation.

EXAMPLE 6.9

The photodiode in Figure 6.16 is used in the circuit shown in Figure 6.18. What range of output voltage will result from light intensities from 100 to 400 W/m^2?

Solution

A load line is drawn on the curves of Figure 6.16b, extending from the supply voltage of 20 V to the current with no voltage drop across the diode, $20\text{ V}/15\text{ k}\Omega = 1.33$ mA. This load line shows that the reverse current will range from -200 to -800 μA as the intensity ranges from 100 to 400 W/m^2. The op amp circuit is a simple current-to-voltage converter with an output of $V_{out} = 1000I$. Therefore, the range of output voltage will be from $1000(200\ \mu\text{A}) = 0.2$ V to $1000(800\ \mu\text{A}) = 0.8$ V. Note that there is a dark current of about 75 μA, so there will be an output of about 0.075 V with no light present.

FIGURE 6.18
Circuit for Example 6.9.

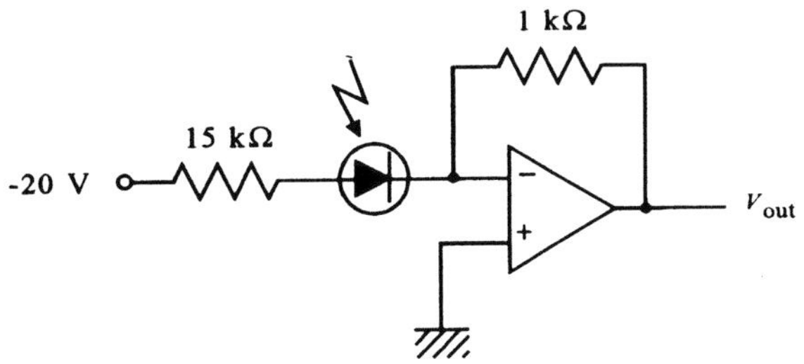

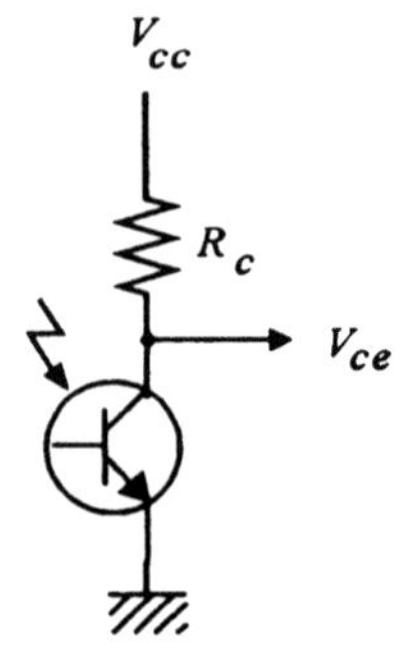

FIGURE 6.19
A phototransistor does not need base current because it is effectively supplied by incoming light intensity.

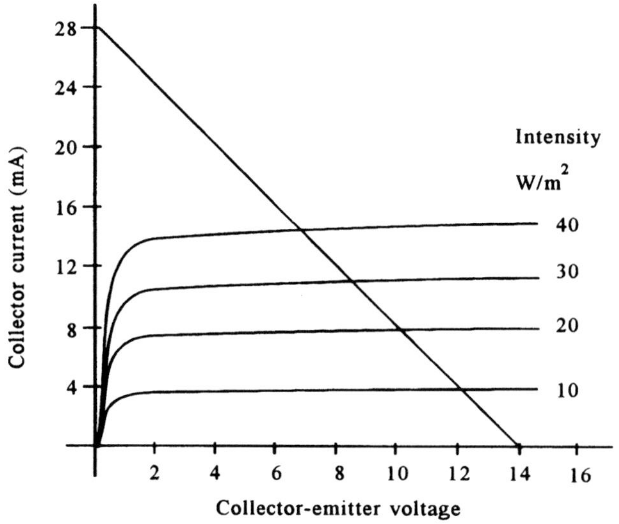

FIGURE 6.20
IV curves of phototransistor collector current and collector-emitter voltage form a family of curves with light intensity as the parameter.

In some cases, a photodiode is operated in the photovoltaic mode, since the output will then be zero when the voltage is zero. In this mode, the speed is slower, however, and the short-circuit currents are very small.

Phototransistor An extension of the photodiode concept is the *phototransistor.* In this sensor, the intensity of EM radiation impinging on the collector-base junction of the transistor acts much like a base current in producing an amplified collector-emitter current. Figure 6.19 shows the schematic symbol and a typical grounded emitter connection of a phototransistor. Figure 6.20 illustrates how EM radiation intensity can be represented as a family of collector-emitter current versus voltage curves, in much the same way that base current can for regular bipolar transistor action.

The phototransistor is not as fast as the photodiode, but still offers response times in microseconds. As usual, there is a limit to the spectral response, with maximum response in the infrared but usable range in the visible band. This device can be used much like a transistor, except that no base current is required. A load line using the collector resistor and supply voltage will show response as a function of light intensity.

Often, the phototransistor has a built-in lens, like that in Figure 6.17, to concentrate radiation on the transistor junction.

EXAMPLE 6.10

The phototransistor with characteristics shown in Figure 6.20 is used in a circuit like that in Figure 6.19, with a 14-V supply and a 500-Ω collector resistance. What range of V_{ce} results from light intensity ranging from 10 to 40 W/m^2?

Solution

A load line runs from the 14-V value on the V_{ce} axis to a current of 14 V/500 Ω = 28 mA on the I_c axis. You can see that this crosses the 10 W/m^2 line at about 12 V across the transistor. When the intensity is 40 mW/cm^2, the load line crosses for a voltage of about 7 V. Thus, the transistor voltage range will be 12 to 7 V.

6.3.5 Photoemissive Detectors

This type of photodetector was developed many years ago, but it is still one of the most sensitive types. A wide variety of spectral ranges and sensitivities can be selected from the many types of photoemissive detectors available.

Principles To understand the basic operational mechanism of photoemissive devices, let us consider the two-element vacuum phototube shown in Figure 6.21. Such photodetectors have been largely replaced by other detectors in modern measurements. In

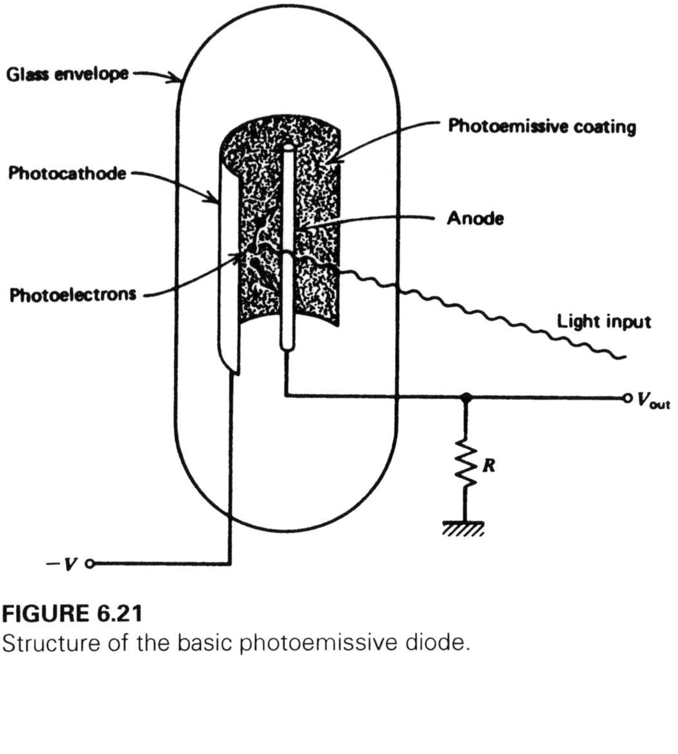

FIGURE 6.21
Structure of the basic photoemissive diode.

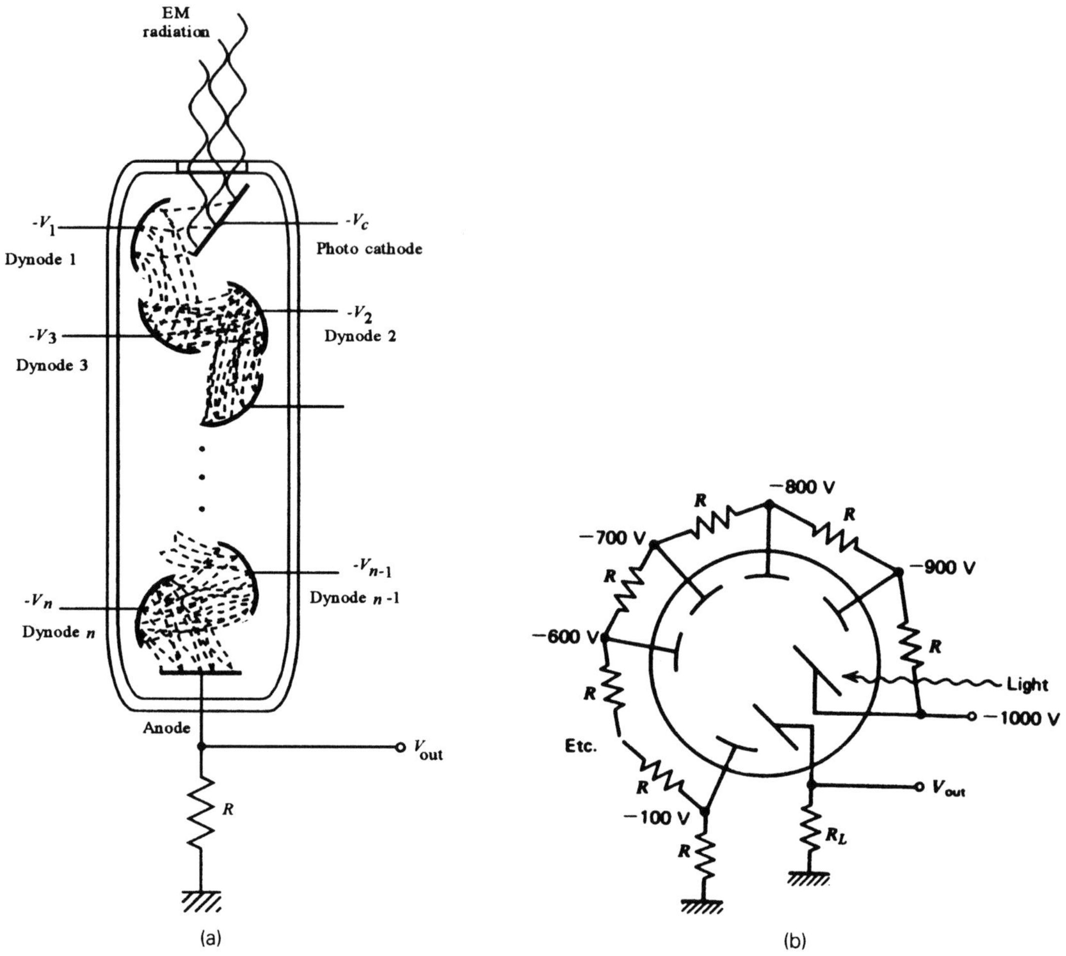

FIGURE 6.22
A photomultiplier depends on multiplication of photoelectrons to achieve a high gain. Each dynode is maintained at successively more positive voltages to accelerate the electrons.

Figure 6.21, we note that the cathode is maintained at some negative voltage with respect to the wire anode that is grounded through resistor R. The inner surface of the cathode has been coated with a *photoemissive* agent. This material is a metal for which electrons are easily detached from the metal surface. ("Easily" means it does not take much energy to cause an electron to leave the material.) In particular, then, a photon can strike the surface and impart sufficient energy to an electron to eject it from this coating. The electron will then be driven from the cathode to the anode by the potential, and then through resistor R. Thus, we have a current that depends on the intensity of light striking the cathode.

Photomultiplier Tube The simple diode described previously is the basis for one of the most sensitive photodetectors available, as shown in Figure 6.22a. As previously noted, a cathode is maintained at a large negative voltage and coated with a photoemissive material. In this case, however, we have many following electrodes, called *dynodes,* maintained at successively more positive voltages. The final electrode is the *anode,* which is grounded through a resistor, R. A photoelectron from the cathode strikes the first dynode with sufficient energy to eject several electrons. All of these electrons are accelerated to the second dynode, where each strikes the surface with sufficient energy to eject several more electrons. This process is repeated for each dynode until the electrons that reach the anode are greatly multiplied in number, where they constitute a current through R. Thus, the photomultiplier (unlike some other transducers) has a *gain* associated with its detection. One single photon striking the cathode may result in a million electrons at the anode! It is this effect that gives the photomultiplier its excellent sensitivity. Many other electrode design arrangements are used with the same principle of operation shown in Figure 6.22a.

Specifications The specifications of photomultiplier tubes depend on several features:

1. The number of dynodes and the material from which they are constructed determine the amplification or current gain. Gains of 10^5 to 10^7 (relating direct photoelectrons from the cathode to electrons at the anode) are typical.
2. The spectral response is determined by two factors. The first is the spectral response of the photoemissive material coated on the cathode. The second is the transparency of the glass envelope or window through which the EM radiation must pass. Using various materials, it is possible to build different types of photomultipliers that, taken together, span wavelengths from 0.12 to 0.95 μm.

The combination of cathode coating and window material is described by a standard system to indicate the spectral response. Thus, a designation of *S* and a number identifies a particular band. For example, S-3 designates flat response from about 0.35 to 0.7 μm.

The time constant for photomultiplier tubes ranges from 20 μs down to 0.1 μs, typically.

Signal Conditioning The signal conditioning is usually a high-voltage negative supply directly connected to the photocathode. A resistive voltage divider (Figure 6.22b) provides the respective dynode voltages. Usually, the anode is grounded through a resistor, and a voltage drop across this resistor is measured. The cathode typically requires -1000 to -2000 V, and each dynode is then divided evenly from that. A 10-dynode tube with a -1000-V cathode thus has -900 V on the first dynode, -800 V on the second, and so on, as shown in Figure 6.22b.

6.4 PYROMETRY

One of the most significant applications of optoelectronic transducers is in the noncontact measurement of temperature. The early term *pyrometry* has been extended to include any of several methods of temperature measurement that rely on EM radiation. These methods depend on a direct relation between an object's temperature and the EM radiation emitted. In this section, we consider the mechanism by which such radiation and temperature are related and how it is used for temperature measurement.

6.4.1 Thermal Radiation

All objects having a finite absolute temperature emit EM radiation. The nature and extent of such radiation depend on the temperature of the object. The understanding and description of this phenomenon occupied the interest and attention of physicists for many years, but we can briefly note the results of this research through the following argument. It is well known that EM radiation is generated by the acceleration of electrical charges. We also have seen that the addition of thermal energy to an object results in vibratory motion of the molecules of the object. A simple marriage of these concepts, coupled with the fact that molecules consist of electrical charges, led to the conclusion that an object with finite thermal energy emits EM radiation because of *charge* motion.

Because an object is *emitting* EM radiation and such radiation is a form of energy, the object must be *losing* energy, but if this were true its temperature would decrease as the energy radiates away. (In fact, for an isolated small sample, this does occur.) In general, however, a state of equilibrium is reached where the object gains as much energy as it radiates, and so remains at a fixed temperature. This energy gain may be because of thermal contact with another object or the absorption of EM radiation from surrounding objects. It is most important to note this interplay between EM radiation emission and heat absorption by an object in achieving *thermal equilibrium.*

Blackbody Radiation To develop a quantitative description of thermal radiation, consider first an *idealized* object. An idealized object absorbs *all* radiation impinging on it, regardless of wavelength, and therefore becomes an *ideal absorber.* This object also emits radiation without regard to any special peculiarities of particular wavelength and therefore becomes an *ideal emitter.*

Assume this ideal object is now placed in thermal equilibrium so that its temperature is controlled. In Figure 6.23, the EM radiation emitted from such an ideal object is plotted to show the intensity and spectral content of the radiation for several temperatures. The abscissa is the radiation wavelength, and the ordinate is the energy emitted per second (power dissipated) per unit area at a particular wavelength. The area under the curve indicates the total energy per second (power dissipated) per unit area emitted by the object. Several curves are shown on the plot for different temperatures. We see that, at low temperatures, the radiation emitted is predominantly in the long-wavelength (far-infrared to microwave) region. As the temperature is increased, the maximum emitted radiation is in the shorter wavelengths, and finally, at very high temperatures, the maximum emitted radiation is near the visible band. Because of this shift in emission peak with temperature, an object begins to glow as its temperature increases. For the blackbody, the temperature and emitted radiation are in one-to-one correspondence in the following respects:

1. *Total radiation* A study of blackbody radiation shows that the total emitted radiation energy per second for all wavelengths increases with the fourth power of the temperature, or

$$E \alpha T^4 \tag{6.10}$$

where E = radiation emission in J/s per unit area or W/m^2
T = temperature (K) of object

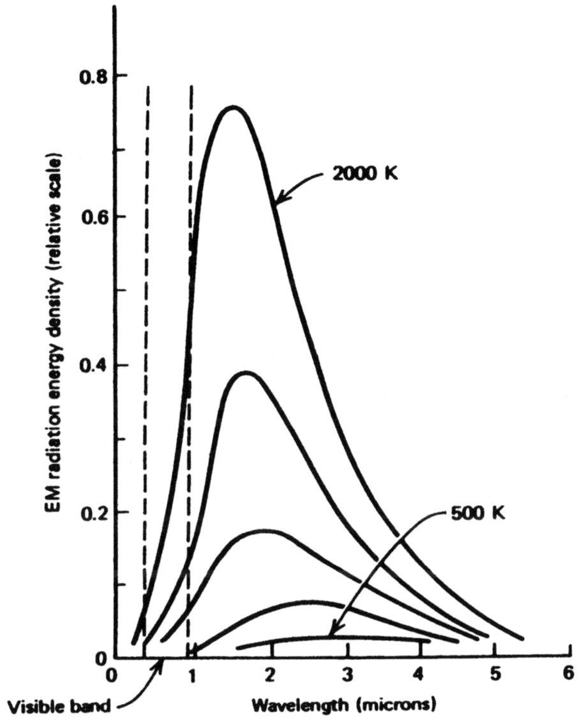

FIGURE 6.23
Ideal curves of EM radiation as a function of temperature.

2. *Monochromatic radiation* It is also clear from Figure 6.23 that the radiation energy emitted at any particular wavelength increases as a function of temperature. Thus, the J/s per area at some given wavelength increases with temperature. This is manifested by the object getting brighter at the (same) wavelength as its temperature increases.

Blackbody Approximation Most materials emit and absorb radiation at *preferred* wavelengths, giving rise to color, for example. Thus, these objects cannot display a radiation energy versus wavelength curve like that of an ideal blackbody. Correction factors are applied to relate the radiation curves of *real* objects to that of an ideal blackbody. For calibration purposes, a blackbody is constructed as shown in Figure 6.24. The radiation emitted from a small hole in a metal enclosure is close to an ideal blackbody.

6.4.2 Broadband Pyrometers

One type of temperature measurement system based on emitted EM radiation uses the relation between total emitted radiation energy and given temperature. This shows that the total EM energy emitted for all wavelengths, expressed as joules per second per unit area,

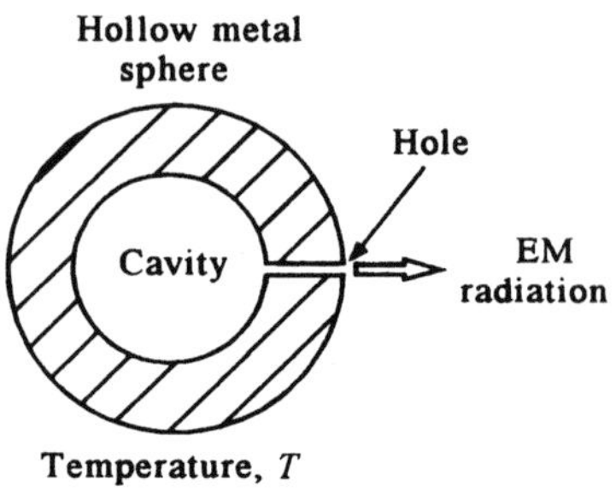

FIGURE 6.24
A blackbody can be simulated by a hole in a metal sphere at a temperature, T.

varies as the fourth power of the temperature. A system that responded to this energy could thus measure the temperature of the emitting object. In practice, it is virtually impossible to build a detection system to respond to radiation of all wavelengths. A study of the curves in Figure 6.23, however, shows that most of the energy is carried in the IR and visible bands of radiation. Collection of radiation energy in these bands provides a good approximation of the total radiated energy.

Total Radiation Pyrometer One type of broadband pyrometer is designed to collect radiation extending from the visible through the infrared wavelengths, and is referred to as a *total radiation pyrometer.* One form of this device is shown in Figure 6.25. The radiation from an object is collected by the spherical mirror S and focused on a broadband detector, D. The signal from this detector, then, is a representation of the incoming radiation intensity, and thus the object's temperature. In these devices, the detector is often a series of microthermocouples attached to a blackened platinum disc. The radiation is absorbed by the disc, which heats up, and results in an emf developed by the thermocouples. The advantage of such a detector is that it responds to visible and IR radiation with little regard to wavelength.

IR Pyrometer Another popular version of the broadband pyrometer is one that is mostly sensitive to IR wavelengths. This device often uses a lens formed from silicon or

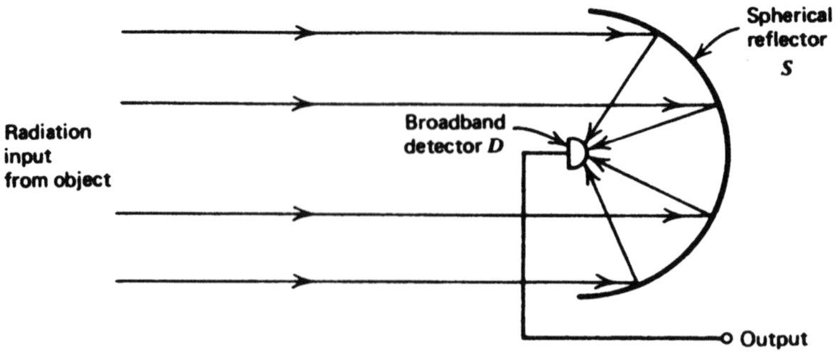

FIGURE 6.25
A total radiation pyrometer determines an object's temperature by input of radiation of a broad band of wavelengths.

germanium to focus the IR radiation on a suitable detector. IR pyrometers are often handheld, pistol-shaped devices that read the temperature of an object toward which they are pointed.

Characteristics Broadband pyrometers often have a readout directly in temperature, either analog or digital. Generally, the switchable range is 0° to 1000°C with an accuracy of ±5°C to ±0.5°C, depending on cost. Accurate measurements require the input of emissivity information by variation in a reading scale factor. Such a correction factor accounts for the fact that an object is not an ideal blackbody and does not conform exactly to the radiation curves.

Applications As the technology of IR pyrometers has advanced, these devices have experienced a vast growth in industrial applications. Some applications are as follows:

1. *Metal production facilities* In the numerous industries associated with the production and working of metals, temperatures in excess of 500°C are common. Contact temperature measurement elements usually have a very limited lifetime. With broadband pyrometers, however, a noncontact measurement can be made and the result converted into a process-control loop signal.
2. *Glass industries* Another area where high temperatures must be controlled is in the production, working, and annealing of glasses. The broadband pyrometers find ready application in process-control loop situations. In carefully designed control systems, glass furnace temperatures have been regulated to within ±0.1 K.
3. *Semiconductor processes* The extensive use of semiconductor materials in electronics has resulted in a need for carefully regulated, high-temperature processes producing pure crystals. In these applications, the pyrometer measurements are used to regulate induction heating equipment, crystal pull rates, and other related parameters.

As the accuracy of IR pyrometers is improved in the temperature ranges below 500 K, many applications that have historically used contact measurements will change to IR pyrometers.

6.4.3 Narrowband Pyrometers

Another class of pyrometer depends on the variation in *monochromatic* radiation energy emission with temperature. These devices often are called *optical pyrometers* because they generally involve wavelengths only in the visible part of the spectrum. We know that the intensity at any particular wavelength is proportional to temperature. If the intensity of one object is matched to another, the temperatures are the same. In the optical pyrometer, the intensity of a heated platinum filament is varied until it matches an object whose temperature is to be determined. Because the temperatures are now the same and filament temperature is calibrated versus a heat setting, the temperature of the object is determined.

Figure 6.26 shows a typical system for implementation of an optical pyrometer. The system is focused on the object whose temperature is to be determined, and the filter picks

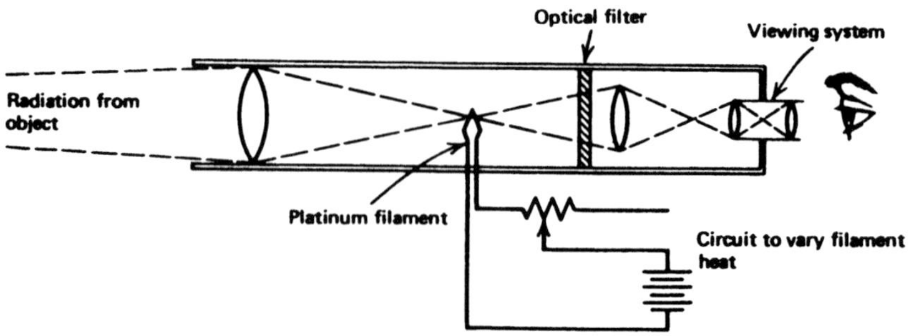

FIGURE 6.26
An optical pyrometer matches the intensity of the object to a heated, calibrated filament. Comparison is made in the red, using red filters.

out only the desired wavelength, which is usually in the red. The viewer also sees the platinum filament superimposed on an image of the object. At low heating, the filament appears dark against the background object, as in Figure 6.27a. As the filament is heated, it eventually appears as a bright filament against the background object, as in Figure 6.27c. Somewhere in between is the point when the brightness of the filament and the measured object match. At this setting, the filament disappears with respect to the background object, and the object temperature is read from the filament heating dial.

The range of optical pyrometer devices is determined in the low end at the point where an object becomes visible in the red (~500 K) and is virtually limited by the melting point of platinum at the upper end (~3000 K). Accuracy is typically ±5 K to ±10 K and is a function of operator error in matching intensities and emissivity corrections for the object. These devices are not easily adapted to control processes because they require acute optical comparisons, usually by a human operator. Applications are predominantly in spot measurements where constant monitoring or control of temperature is not required.

6.5 OPTICAL SOURCES

One limitation in the application of EM radiation devices to process control has been the lack of convenient characteristics of available optical sources. Often, complicated collimating lens systems are required, heat dissipation may be excessive, wavelength characteristics may be undesirable, or a host of other problems may arise. The development of sources relying on *l*ight *a*mplification by *s*timulated *e*mission of *r*adiation (LASER) has provided EM radiation sources having good characteristics for application to process-control measurement. In this section, we will consider the general characteristics of both conventional and laser light sources and their applications to measurement problems. Our discussion is confined to sources in the visible or IR wavelength bands, although it should be noted that many applications exist in other regions of the EM radiation spectrum.

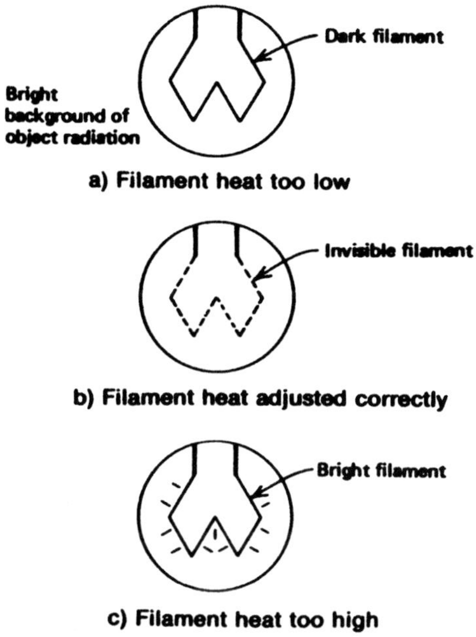

FIGURE 6.27
Examples of the appearance of the filament during use of an optical pyrometer.

6.5.1 Conventional Light Sources

Before the development of the laser, two primary types of light sources were employed. Both of these are fundamentally *distributed,* because radiation emerges from a physically distributed source. They also are both *divergent,* incoherent, and often not particularly monochromatic.

Incandescent Sources A common light source is based on the principle of thermal radiation discussed in Section 6.4. Thus, if a fine current-carrying wire is heated to a very high temperature by I^2R losses, it emits considerable EM radiation in the visible band. A standard lamp is an example of this type of source, as are flashlight lamps, automobile headlights, and so on. Because the light is distributed in a very broad wavelength spectrum (see Figure 6.23), it clearly is not monochromatic. Such light actually results from molecular vibrations induced by heat, and light from one section of the wire is not associated with the light from another section. From this argument we see that the light is incoherent. The divergent nature of the light is inherent in the observation that no direction of emission is preferential. In fact, the employment of lenses or mirrors to collimate light is familiar to anyone who uses a flashlight. A large fraction of the emitted radiation lies in the IR spectrum, which shows up as radiant heat loss rather than effective lighting. In fact, to a great extent, the elevated temperature of the glass bulb of an incandescent lamp is caused through absorption by the glass of the IR radiation emitted by the filament.

FIGURE 6.28
A representation of electron transitions in an atom with the emission of EM radiation.

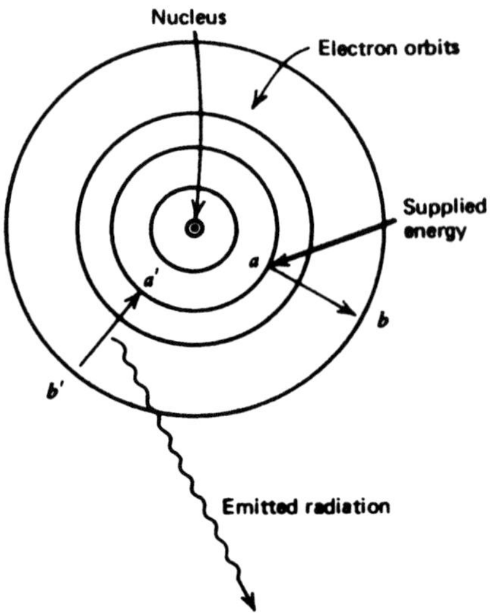

From this we see that an incandescent source is *polychromatic, divergent, incoherent,* and *inefficient* for visible-light production. Yet this source has been a workhorse for lighting for many years. It is most deserving, but for use as a measurement transducer, its limitations are severely restrictive.

Atomic Sources The light sources that provide the red neon signs used in advertising and the familiar fluorescent lighting are examples of another type of light source. Such light sources are atomic in that they depend on rearrangements of electrons within atoms of the material from which the light originates. Figure 6.28 shows a schematic representation of an atom with the nucleus and associated electrons. If one of these electrons is excited from its normal energy level to a different position (as indicated by an $a \rightarrow b$ transition), energy must be provided to the atom. This electron returns to its normal level in a very short time ($\sim 10^{-8}$ s for most atoms). In doing so, it gives up energy in the form of emitted EM radiation, as indicated by the $b' \rightarrow a'$ transition in Figure 6.28. This process is often represented by an *energy diagram,* as in Figure 6.29. Here the lines represent all possible positions or energies of electron-excited states. The excitation $a \rightarrow b$ of the previous example is indicated by the upward arrow. The de-excitation and resulting radiation is shown as $b' \rightarrow a'$. The other arrows show that a multitude of possible excitations and de-excitations may occur. The wavelength of the emitted radiation is inversely proportional to the energy of the transition (see Equation 6.3). Thus, $b' \rightarrow a'$ will have a shorter wavelength than $c' \rightarrow d'$.

In atomic sources, a mechanism is provided to cause excitation of electron states, so that the EM radiation emitted by the resulting de-excitation appears as visible light.

In a neon light, the excitation is provided by collisions between electrons and ions in the gas, where the electrons are provided by an electric current through the gas. In the case of neon gas, the de-excitation results in light emission whose wavelength is predominantly in the red-orange part of the visible spectrum. Thus, the light is nearly

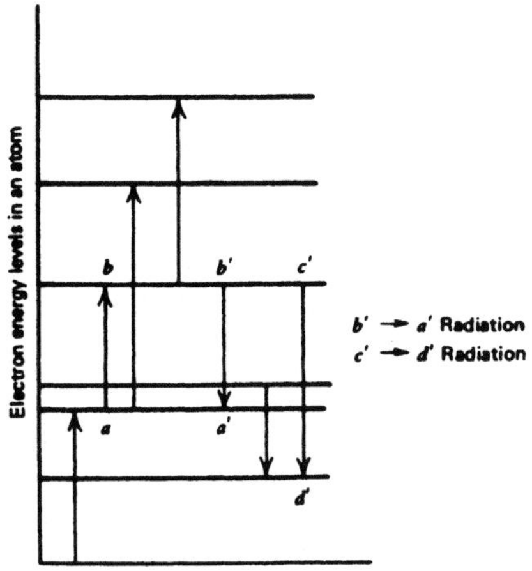

FIGURE 6.29
An energy-level diagram schematically shows the electron orbit energies and possible transitions.

monochromatic, although light of other wavelengths representing different de-excitation modes is still present.

In a fluorescent light, a two-step process occurs. The initial de-excitation produces light predominantly in the ultraviolet (UV) part of the spectrum that is absorbed on an inner coating of the bulb. Electrons of atoms in the coating material are excited by UV radiation and then de-excited by many levels of transitions, producing radiation of a broad band of wavelengths in the visible region. Thus, the radiation emitted is *polychromatic.* The radiation in these sources is also divergent and incoherent.

Fluorescence Certain materials exhibit a peculiar characteristic with regard to the de-excitation transition time of electrons; that is, a transition may take much longer on the average than the normal 10^{-8} s. In some cases, the average transition time may be even hours or days. Such levels are called *long-lived states* and show up in materials that fluoresce or "glow in the dark" following exposure to an intense light source. What actually happens is that the material is excited by exposure to the light source and fills electrons into some long-lived excited states. Because the transition time may be several minutes, the object continues to emit light when taken into a dark room until the excited levels are finally depleted. Such long-lived-state materials actually form a basis for development of a laser.

6.5.2 Laser Principles

Stimulation Emission The basic operation of the laser depends on a principle formulated by Albert Einstein regarding the emission of radiation by excited atoms. He found that if several atoms in a material are excited to the same level and one of the atoms emits its radiation before the others, then the passage of this radiation by such excited atoms can also stimulate them to de-excite. It is significant that when stimulated to de-excite, the emitted radiation will be *inphase* and *in the same direction* as the stimulating radiation. This effect is shown in Figure 6.30, where atom *a* emits radiation spontaneously. When this radiation passes

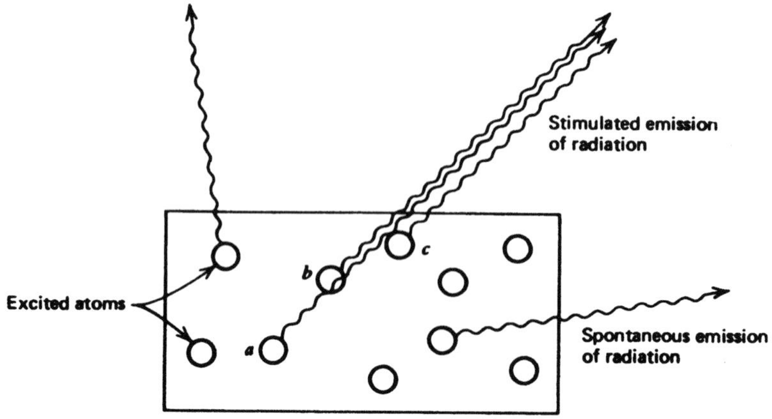

FIGURE 6.30
Stimulated emission of radiation gives rise to monochromatic, coherent radiation pulses moving in random directions.

by atoms indicated by *b, c,* and so on, they also are stimulated to emit in the same direction and inphase (coherently). Such radiation is also monochromatic because only a single transition energy is involved. Such stimulated emission is the first requirement in the realization of a laser.

Laser Structure To see how the concept of stimulated emission is employed in a laser, consider Figure 6.31. We have a host material that also contains atoms having the long-lived states described earlier. If some of these atoms spontaneously de-excite, their radiation stimulates other atoms in the radiation path to de-excite, giving rise to pulses of radiation indicated by P_1, P_2, and so on. Now consider one of these pulses directed perpendicularly to mirrors M_1 and M_2. This pulse reflects between the mirrors at the speed of light, stimulating atoms in its paths to emit. The majority of excited atoms are quickly de-excited in this fashion. If mirror M_2 is only 60% reflecting, some of this pulse in each reflection will be passed. The overall result is that, following excitation, a pulse of light emerges from M_2 that is monochromatic, coherent, and has very little divergence. This system can also be made to operate continuously by providing a continuous excitation of the atoms to replenish those de-excited by stimulated emission.

Properties of Laser Light The light that comes from the laser is characterized by the following properties:

1. *Monochromatic* Laser light comes predominantly from a particular energy-level transition and is therefore almost monochromatic. (Thermal vibration of the atoms and the presence of impurities cause some other wavelengths to be present.)
2. *Coherent* Laser light is coherent as it emerges from the laser output mirror and remains so for a certain distance from the laser; this is called the *coherence length.* (Slight variations in coherency induced by thermal vibrations and other effects eventually cause the beam to lose coherency.)

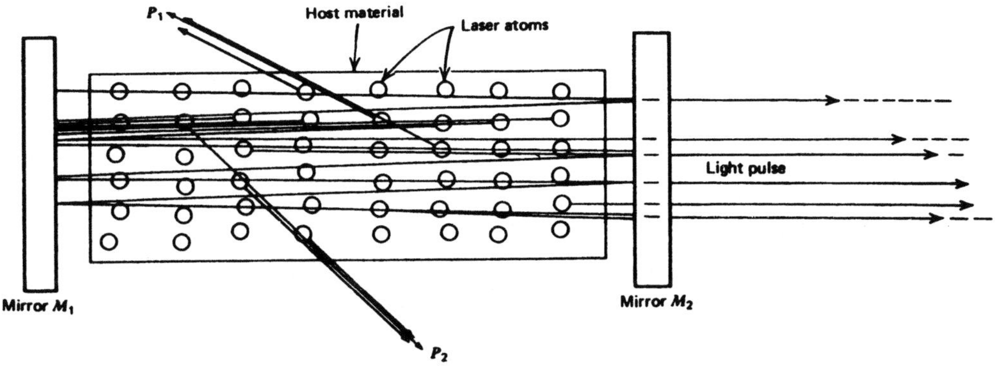

FIGURE 6.31
A laser gives preference to radiation pulses emitted perpendicular to reflecting surfaces.

3. *Divergence* Because the laser light emerges perpendicular to the output mirror, the beam has very little divergence. Typical divergency may be 0.001 rad.
4. *Power* Continuous-operation lasers may have power outputs of 0.5 mW to 100 W or more. Pulse-type lasers have power levels up to terrawatts, but for only short time pulses—microseconds or even nanoseconds in duration.

Table 6.3 summarizes some characteristics of typical industrial lasers. The workhorses of measurement applications are the He-Ne continuous-operation lasers. These lasers are relatively cheap and operate in the visible (red) wavelength regions.

TABLE 6.3
Laser characteristics

Material	Wavelength (μm)	Power	Applications
Helium-neon (gas)	0.6328 (red)	0.5 to 100 mW (cw)	General purpose, ranging, alignment, communication, etc.
Argon (gas)	0.4880 (green)	0.1 to 5 W (cw) 10 to 100 W (pulsed)	Heating, small-part welding, communication
Carbon dioxide (gas)	10.6 (IR)	0 to 1 kW (cw) 0 to 100 kW (pulse)	Cutting, welding, communication, vaporization, drilling
Ruby (solid)	0.6943 (red)	0 to 1 GW (pulse)	Cutting, welding, vaporization, drilling, ranging
Neodymium (solid)	1.06 (IR)	0 to 1 GW (pulse)	Cutting, welding, vaporization, drilling, communication

EXAMPLE 6.11

A He-Ne laser with an exit diameter of 0.2 cm, power of 7.5 mW, and divergence of 1.7×10^{-3} rad is to be used with a detector 150 m away. If the detector has an area of 5 cm^{-2}, find the power of the laser light to which the detector must respond.

Solution

To solve this problem, we must ultimately find the intensity of the laser beam 150 m away. We first find the beam area at 150 m using the known divergence. From Figure 6.4, we find (see Example 6.5) the radius at 150 m by

$$R_2 = R_1 + L\tan(\theta)$$
$$R_2 = 10^{-3} + (150 \text{ m})\tan(1.7 \times 10^{-3} \text{ rad})$$
$$R_2 = 0.256 \text{ m}$$

Thus, the area is

$$A_2 = \pi R_2^2 = (3.14)(0.256 \text{ m})^2$$
$$A_2 = 0.206 \text{ m}^2$$

The intensity at 150 m is

$$I = P/A$$
$$I = \frac{7.5 \times 10^{-3} \text{ W}}{0.206 \text{ m}^2}$$
$$I = 0.036 \text{ W/m}^2$$

Then we find the power intercepted by the detector as

$$P_{det} = IA_{det} = (0.036 \text{ W/m}^2)(5 \times 10^{-4} \text{ m}^2)$$
$$P_{det} = \mathbf{1.82 \times 10^{-5}\ W}$$

Thus, the detector must respond to a power of 18.2 μW of power. Although small, such low power is measured by many detectors. The reader can show (by similar calculation) that a source such as a flashlight with a divergence of 2° or 34.9×10^{-3} rad would have resulted in a much lower power at the detector of $\mathbf{4.35 \times 10^{-8}\ W}$.

6.6 APPLICATIONS

Several applications of optical transduction techniques in process control will be discussed. The intention is to demonstrate only the typical nature of such application, and not to design details. Pyrometry for temperature measurement has already been discussed, and is not considered in these examples.

6.6.1 Label Inspection

In many manufacturing processes, a large number of items are produced in batch runs where an automatic process attaches labels to the items. Inevitably, some items are either missing labels or have the labels incorrectly attached. The system in Figure 6.32 examines

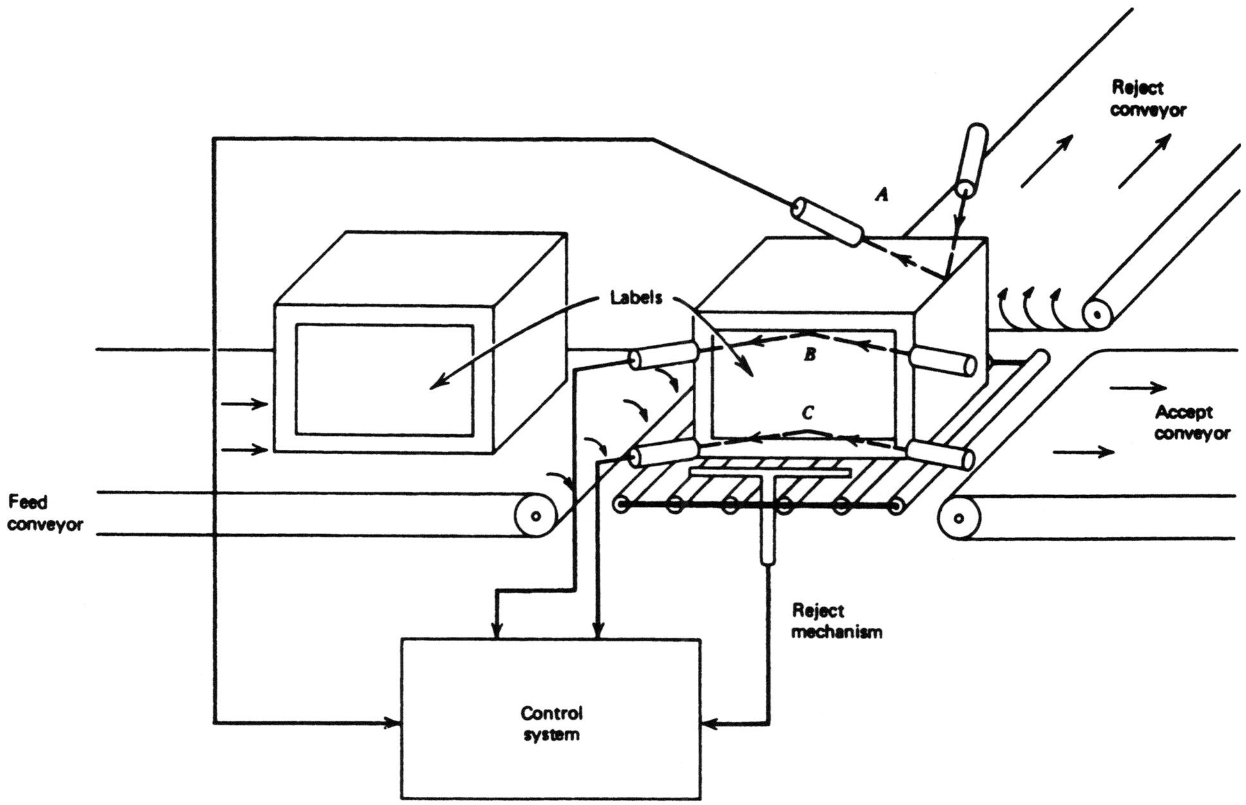

FIGURE 6.32
Label-inspection system as an example of optical technique.

the presence and alignment of labels on boxes moving on a conveyor belt system. If the label is missing or improperly aligned, the photodetector signals are incorrect in terms of light reflected from the sources, and a solenoid pushout rejects the item from the conveyor. The detectors in this case could be CdS cells, and the sources either focused incandescent lamps or a small He-Ne laser. Source/detector system *A* detects the presence of a box and initiates measurements by source/detector systems *B* and *C*. If the signals received by detectors *B* and *C* are identical and at a preset level, the label is correct, and the box moves on to the accept conveyor. In any other instance, a misalignment or missing label is indicated, and the box is ejected onto the reject conveyor.

EXAMPLE 6.12 Devise signal-conditioning circuitry for the application of Figure 6.32 using CdS cells as the detectors. If both cells have resistance of 1000 $\pm$ 100 Ω or less, the label is considered correct. No label produces 2000 Ω or more.

Solution

One of the many possible methods for implementing this solution is shown in Figure 6.33. If the label is misaligned or missing, one or both comparator outputs is high, thus driving

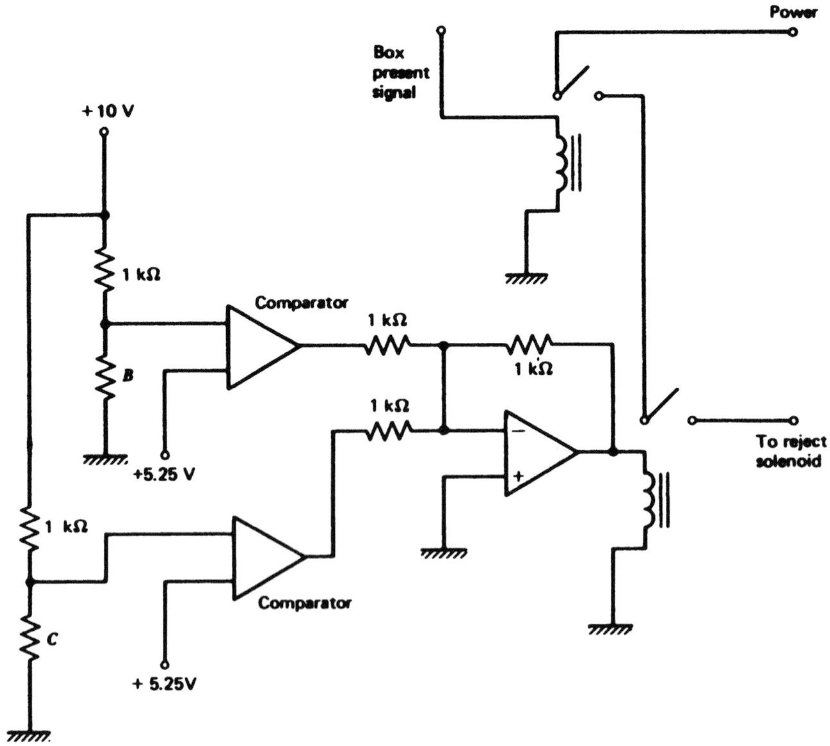

FIGURE 6.33
One possible circuit to implement Example 6.12.

the summing amplifier to close the reject relay. If the box is present, then power is applied to the reject solenoid and the box is ejected. The resistors are chosen so that if the cell resistance exceeds 1100 Ω, the comparator outputs go high. The relay is chosen to close if either comparator signal (or both) is present.

6.6.2 Turbidity

One of the many characteristics of liquids involved in process industries is called *turbidity.* Turbidity refers to the lack of clarity of a liquid, which can be caused by suspended particulate material. Turbidity can be an indication of a problematic condition because of impurities or improperly dissolved products. It also can be intentional as, for example, when some material is suspended in a liquid for ease of transport through pipes. Turbidity can be measured optically because it affects the propagation of light through the liquid.

It is also possible to measure the turbidity of liquids in a process in-line—that is, without taking periodic samples, by a method similar to that in Figure 6.34. In this case, a laser beam is split and passed through two samples to matched photodetectors. One sample is a carefully selected *standard* of allowed (acceptable) turbidity. The other is an in-line sample of the process liquid itself. If the in-line sample attenuates the light more than the standard, the signal-conditioning system triggers an alarm or takes other appropriate action to reduce turbidity.

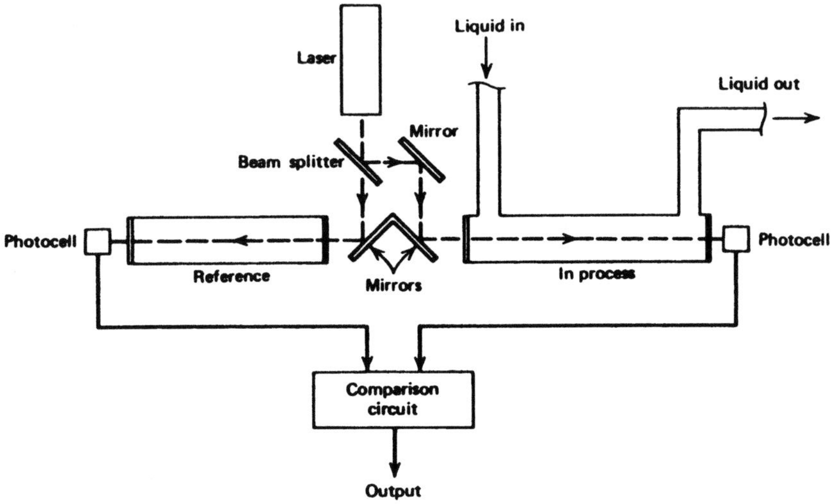

FIGURE 6.34
Turbidity measurement can be made in-line with this optical system.

6.6.3 Ranging

The development of the laser and fast (small τ) photodetectors has introduced a number of methods for measuring distances and the rate of travel of objects by noncontact means. Distances can be measured by measuring the time of flight of light pulses scattered off a distant object. Because the speed of light is constant, we use a simple equation to find distance, providing the time of flight, *T*, is known. Thus, if a pulse of light is directed at a distant object and the reflection is detected a time *T* later, then the distance is

$$D = cT/2 \tag{6.11}$$

where D = distance to the object (m)
c = speed of light (m/s)
T = time for light round trip (s)

EXAMPLE 6.13 An object is approximately 300 m away. Find the approximate time difference to calculate the distance, using a light pulse reflected from the object.

Solution
From

$$D = \frac{cT}{2}$$

we have

$$T = \frac{2D}{c} = \frac{(2)(3 \times 10^2 \text{ M})}{3 \times 10^8 \text{ m/s}}$$

$$T = \mathbf{2\ \mu s}$$

This ranging method can be employed for measuring shorter distances limited by time measurement capability and to detecting the reflected signal for longer distances. Surveying instruments for measuring distance have been developed by this method. Velocity or rate of motion can be measured by an electronic computing system that records the changing reflected-pulse travel time and computes velocity. These and the interferometric methods, such as those used in Doppler radar, are beyond the scope of this book text.

SUMMARY

EM radiation measurement allows noncontact measurement techniques for many variables, such as temperature, level, and others. This chapter presented the essential elements of EM radiation and its application to process-control measurement. This included the following:

1. EM radiation is defined as a form of energy characterized by constant propagation speed. The wavelength and frequency are related by

$$c = \lambda f \tag{6.1}$$

2. As a form of energy, EM radiation is best described by the intensity in watts per unit area, divergence (spreading) in radians, and spectral content.
3. When associated with human eye detection, light is described by a set of luminous units that are all relative to a standard, visible source.
4. There are four basic photodetectors: photoconductive, photovoltaic, photoemissive, and photodiode. Each has its special characteristics relative to spectral sensitivity, detectable power, and response time.
5. Pyrometry relates to measurement of temperature by measuring the intensity of EM radiation from an object as a function of its temperature.
6. Total radiation and IR pyrometers may be used in process-control applications for measurement of temperature from about 300 K to almost no limit. Accuracy varies from ± 10 K to ± 0.5 K or better in some cases.
7. Conventional light sources are usually divergent and incoherent. Incandescent types are polychromatic, but atomic sources may be almost monochromatic.
8. A laser is based on a concept of stimulated emission where one photon can stimulate excited atoms to emit many more photons.
9. The light from a laser is nearly nondivergent, coherent, and monochromatic.
10. Applications of optical techniques are particularly useful where contact measurement is difficult.

PROBLEMS

Section 6.2

6.1 Find the frequency of 3-cm wavelength EM radiation. What band of phenomena does this radiation represent?

6.2 A source of green light has a frequency of 6.5×10^{14} Hz. What is its wavelength in nanometers (nm) and in Å?

6.3 A light beam is passed through 100 m of liquid with an index of refraction $n = 1.7$. How long does it take the light to traverse the 100 m with and without the liquid?

6.4 A flashlight beam has an exit power of 100 mW, an exit diameter of 4 cm, and a divergence of 1.2°. Calculate the intensity in mW/cm^2 at 60 m and the size of the beam at that distance.

6.5 A detector can just resolve a light intensity of 25 mW/m^2. How far away can it be placed from a 10-W point source?

6.6 How many watts of power are emitted by a 1-cd source?

6.7 A laser with a 0.03-mrad divergence, 1000 W of power, and 2-cm exit diameter is aimed at the moon (238,000 mi). What is the beam size on the moon? What is the radiation intensity?

6.8 Suppose a 700-nm beam with a power of 0.2 mW and a 5-cm diameter strikes a detector with a 0.2-cm diameter. How many photons strike the detector per second?

Section 6.3

6.9 A CdS cell has a time constant of 73 ms and a dark resistance of 150 kΩ. A light pulse that is only 20 ms in length strikes the cell. If the intensity of the light pulse is such that the final resistance would be 45 kΩ, plot the resistance versus time for 100 ms. What is the resistance at 20 ms?

6.10 Suppose there is another 20-ms light pulse that may strike the cell in Problem 6.9, but it is of lower intensity, so that the final resistance would be 85 kΩ. Call this one P_1 and that of Problem 6.9 P_2. Devise a system that will latch on a red LED if P_1 strikes the cell, or a green LED if P_2 strikes the cell. Each time a pulse strikes the cell, the latches are cleared and the appropriate LED turned on.

6.11 The photoconductive cell in Figure 6.9 is used to measure distance from a stable source of 0.5-μm wavelength radiation. The source has 75 mW of power, 2 mr divergence, and a 0.5-cm exit radius. Construct a graph of sensor resistance versus distance from 0 to 2.5 m.

6.12 Design a system using the photoconductive cell in Figure 6.9 to measure and display light intensity. Make the design such that 20 to 100 mW/cm^2 produces an output of 0.2 to 1.0 V. What is the readout error when the intensity is 60 mW/cm^2?

6.13 A silicon photovoltaic cell is employed in the circuit shown in Figure 6.35. The cell has an internal resistance of 65 Ω. A light pulse of 20-ms duration and 2 W/m^2 intensity strikes the detector. Sketch the output voltage versus time, and find the maximum voltage produced. The cell calibration voltage is $V_0 = 0.60$ V.

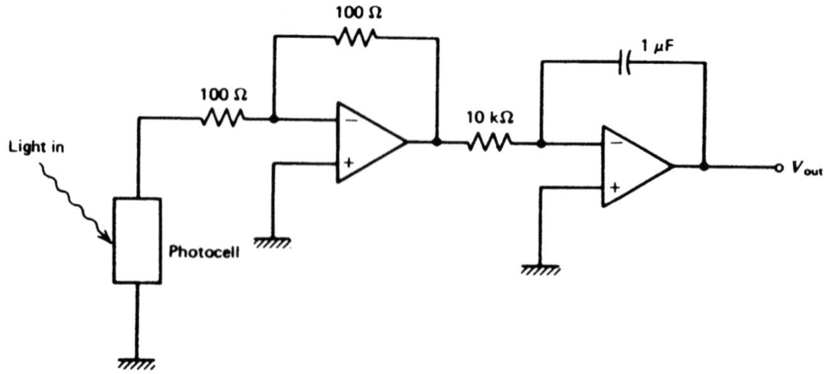

FIGURE 6.35
Circuit for Problem 6.13.

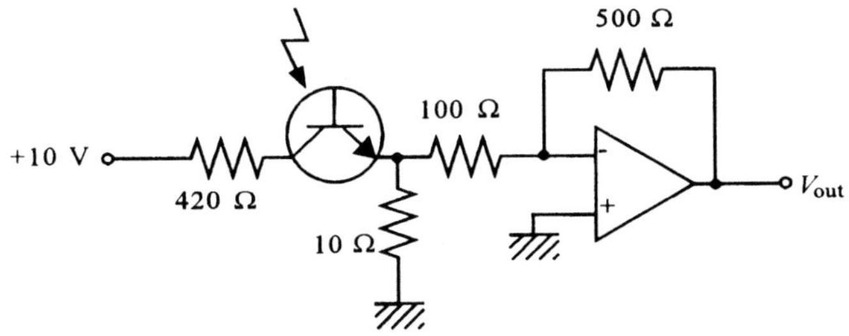

FIGURE 6.36
Circuit for Problem 6.16.

6.14 A single silicon photovoltaic cell is found to have an open-circuit voltage of 0.6 V and a short-circuit current of 15 mA in full sunlight. Show how a collection of these cells can be arranged to deliver 500 mW at 9.0 V into a load.

6.15 Design a system using the photodiode in Figure 6.16 to interface a 100- to 400-W/m^2 intensity to a computer. The design must use a 10-bit, bipolar ADC with a 5.0-V reference. The design must be such that 100 W/m^2 produces an ADC output of 100_{10}, and 400 W/m^2 produces 400_{10}.

6.16 The phototransistor in Figure 6.20 is used in the circuit of Figure 6.36. Calculate the output voltage versus light intensity from 10 to 40 W/m^2.

6.17 Devise a system by which the phototransistor in Figure 6.20 can trigger a comparator when the light intensity rises above 20 W/m^2.

6.18 A photomultiplier has a current gain of 3×10^6. A weak light beam produces 50 electrons/s at the photocathode. What anode-to-ground resistance must be used to get a 3-μV voltage from this light pulse? The charge of an electron is 1.6×10^{-19} coulomb.

Section 6.5

6.19 A 10-W argon laser has a beam diameter of 1.5 cm. If focused to a 1-mm diameter, find the intensity of the radiation beam at the focus.

6.20 In a turbidity system such as that of Figure 6.34, the tanks are 2 m in length. A laser with a 2.2-mrad divergence, 2.1-mW power, and 1-mm exit radius is employed. If the sample nominally detracts from the beam power by 12% per meter, find the intensity of the beam at the detector. The laser is 1.5 m from the beam splitter. Note that the splitter halves the power and does not affect the divergence.

6.21 A special timing circuit can resolve 2.4-ns time difference. Find the closest distance that could be measured using laser ranging.

6.22 For the turbidity system in Figure 6.34, two matched photoconductive cells are used with R vs. I_L as given in Figure 6.37. Design a signal-conditioning system that outputs the deviation of the flowing system turbidity in volts and triggers an alarm if the intensity is reduced by 10% from the nominal of 15 mW/cm^2.

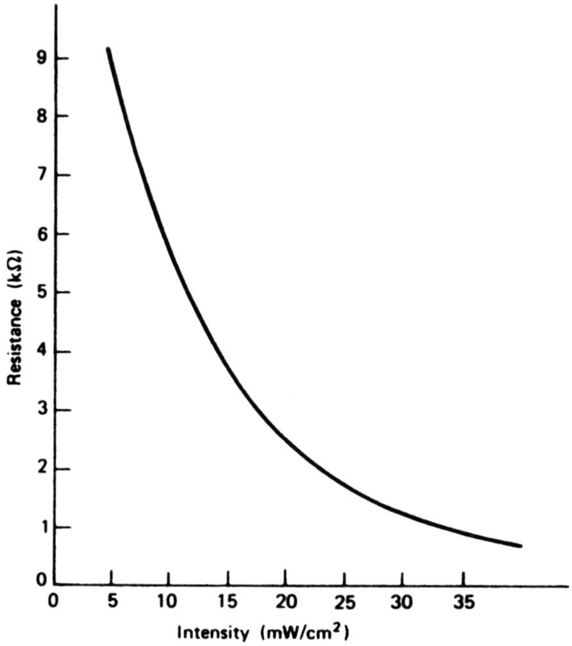

FIGURE 6.37
Figure for Problem 6.22.

SUPPLEMENTARY PROBLEMS

S6.1 Figure 6.38 shows a laser with a 60-mW output power, 1.0-cm output beam diameter, and fixed divergence of 0.2°. A photodiode will be used to measure distance from this source from 0.0 to 5.0 m using the known decrease in intensity with distance. The photodiode is used in the circuit in Figure 6.16a with $V_s = 20$ V and $R = 15$ kΩ. For that circuit, the relation between diode reverse current and light intensity is $I(\mu\text{A}) = 1.429 I_L(\text{W/m}^2) + 57$.

a. Design signal conditioning so that the distance from 0.0 to 5.0 m is converted to an 8-bit digital signal. Use an 8-bit unipolar ADC with a 10.0-V reference.

b. Plot the ADC output, expressed as a base 10 number versus distance. Comment on the linearity.

S6.2 An automatic security camera must operate as follows when an intrusion occurs:

1. If the ambient light level is greater than 20 mW/cm^2, the shutter is activated.
2. If ambient light is less than 20 mW/cm^2, a light source is triggered.
3. When the source lighting reaches 20 mW/cm^2, the shutter is triggered.

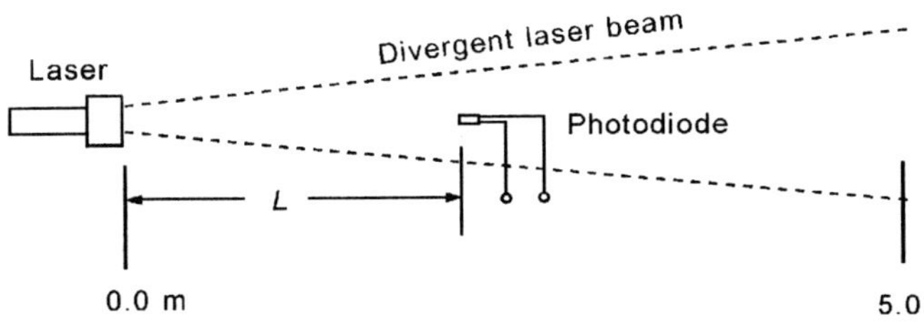

FIGURE 6.38
Setup for Problem S6.1.

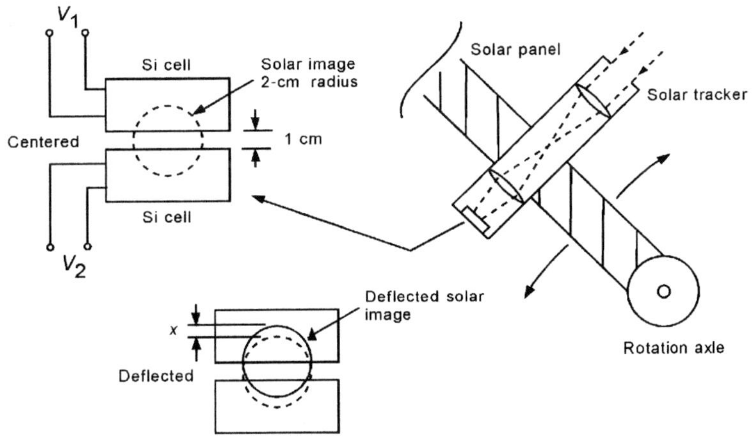

FIGURE 6.39
Solar tracking system.

Design the system and signal conditioning using the following assumptions:

a. The shutter and light source are triggered by TTL logic high signals.

b. A CdS cell with characteristics given in Figure 6.9 is used to monitor ambient light level.

c. A photodiode with the characteristics given in Figure 6.16b will be used to measure light intensity after the light source is triggered. Explain why the CdS cell could not be used here also.

S6.3 A one-dimensional solar panel sun-tracking system uses two silicon photocells, as shown in Figure 6.39. When the solar image is centered between the cells, their exposed areas will be the same, and so their short-circuit currents (SSCs) will be the same. As the sun moves, the image will displace by an amount x, as shown. Now the

SSC of one will increase because of increased exposure area, whereas the other will decrease. The SSC varies with x by

$$I_{SC}(x) = \left[2\pi - \frac{1 - 2x}{8}\sqrt{15 + 4x - 4x^2} - \sin^{-1}\left(\frac{1 - 2x}{4}\right)\right] \quad \text{in mA}$$

The difference in SSCs is an error signal that can be used to drive the panel until the SSCs are the same again. Use op amp circuits to convert the SSC of each cell to a proportional voltage. Then devise a differential amplifier system that will output an error voltage of ± 2.5 V as the solar image deflects by ± 1 cm. Plot a graph of output voltage versus deflection. Comment on the linearity.

CHAPTER 7

Final Control

INSTRUCTIONAL OBJECTIVES

This chapter presents the general techniques and features of final control elements used in process-control systems. After reading the chapter and working through the examples and problems at the end of the chapter, you will be able to:

- Define the three parts of final control element operation.
- Explain the basic principles of the pneumatic nozzle/flapper system.
- Define the basic operation of the power electronics devices: SCR, GTO, and TRIAC.
- Explain why BJT switching results in less power dissipation than simple amplification.
- Provide descriptions of the power MOSFET and IGBT.
- Explain how a dc motor works.
- Describe the two types of ac motors.
- Explain the operation of a pneumatic valve stem positioning actuator.
- Describe the difference between three control valve types.
- Explain the process of control valve size determination.

7.1 INTRODUCTION

In a typical process-control application, the measurement and evaluation of some process variable is carried out using a low-energy analog or digital signal to represent the variable. The control signal that carries feedback information back to the process for necessary corrective action is expressed by the same low level of representation. In general, the controlled process itself may involve a high-energy condition, such as the flow of thousands of cubic meters of liquid or several hundred thousand newton hydraulic forces, as in a steel rolling mill. The function of the final control element is to translate low-energy control signals into a level of action commensurate with the process under control. This can be considered an amplification of the control signal, although in many cases the signal is also converted into an entirely different form.

In this chapter, the general concepts used to implement the final control element function are presented, together with specific examples in several areas of process control.

A sensor used to measure some variable in a process-control application should have negligible effect on the process itself. It follows that sensor selection is based mainly on required measurement specifications and necessary protections (of the sensor) from harmful effects of the process environment. In sensor selection, the process-control technologist need not have intimate knowledge of the mechanisms of the process itself.

These arguments do not apply, however, when considering the final control element. The final control element necessarily has a profound effect on the process, and therefore must be selected after detailed considerations of the possessive operational mechanisms. Such a selection, therefore, cannot be the responsibility of the process-control technologist alone. In this view, the process-control technologist should have sufficient background on the final control element and its associated signal conditioning to know how such devices interface with preceding process controllers and transducers. The technologist should be able to communicate and work closely with process engineers on these subjects. The objectives of this chapter were selected to enable the future technologist to fulfill this responsibility.

7.2 FINAL CONTROL OPERATION

Final control element operations involve the steps necessary to convert the control signal (generated by a process controller) into proportional action on the process itself. Thus, to use a typical 4- to 20-mA control signal to vary a large flow rate from, say, 10.0 m^3/min to 50.0 m^3/min certainly requires some intermediate operations. The specific intermediate operations vary considerably, depending on the process-control design, but certain generalizations can be made regarding the steps leading from the control signal to the final control element. For a typical process-control application, the conversion of a process-controller signal to a control function can be represented by the steps shown in Figure 7.1. The input control signal may take many forms, including an electric current, a digital signal, or pneumatic pressure.

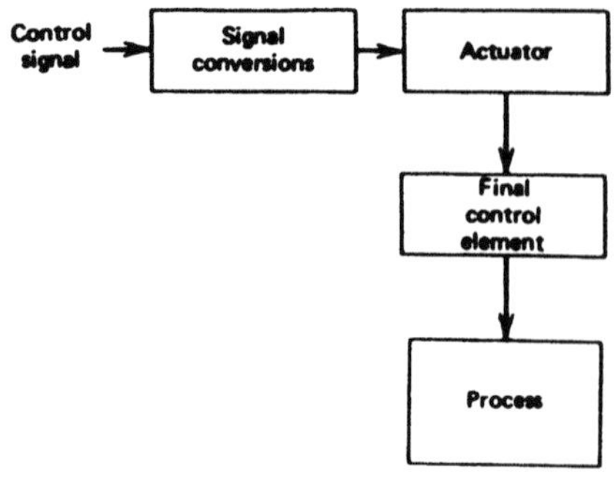

FIGURE 7.1
Elements of the final control operation.

7.2.1 Signal Conversions

This step refers to the modifications that must be made to the control signal to properly interface with the next stage of control—that is, the actuator. Thus, if a valve control element is to be operated by an electric motor actuator, then a 4- to 20-mA dc control signal must be modified to operate the motor. If a dc motor is used, modification might be current-to-voltage conversion and amplification.

Standard types of signal modification are discussed in Section 7.3. The devices that perform such signal conversions are often called *transducers* because they convert control signals from one form to another, such as current to pressure, current to voltage, and the like.

A corollary to the signal-conversion process is the development of special electronic devices that provide a high-energy output under the control of a low-energy input. This is generally described by the term *industrial electronics.* Section 7.4 presents an overview of this important topic.

7.2.2 Actuators

The results of signal conversions provide an amplified and/or converted signal designed to operate (actuate) a mechanism that changes a controlling variable in the process. The direct effect is usually implemented by something in the process, such as a valve or heater that must be operated by some device. The *actuator* is a translation of the (converted) control signal into action on the control element. Thus, if a valve is to be operated, then the actuator is a device that converts the control signal into the physical action of opening or closing the valve. Several examples of actuators in common process-control use are discussed in Section 7.5.

7.2.3 Control Element

At last we get to the final control element itself. This device has direct influence on the process dynamic variable and is designed as an integral part of the process. Thus, if flow is to be controlled, then the control element, a valve, must be built directly into the flow system. Similarly, if temperature is to be controlled, then some mechanism or control element that has a direct influence on temperature must be involved in the process. This could be a heater/cooler combination that is electrically actuated by relays or a pneumatic valve to control influx of reactants.

In Figure 7.2, a control system is shown to control the degree of baking of, say, crackers, as determined by the cracker *color.* The optical measurement system produces a 4- to 20-mA conditioned signal that is an analog representation of cracker color (and, therefore, proper baking). The controller compares the measurement to a setpoint and outputs a 4- to 20-mA signal that regulates the conveyer belt feed-motor speed to adjust baking time. The final control operation is then represented by a *signal conversion* that transforms the 4- to 20-mA signal into a 50- to 100-V signal as required for motor speed control. The motor itself is the *actuator,* and the conveyer belt assembly is the *control element.*

Because applications of process-control techniques in industry are as varied as the industry itself, it is impractical to consider more than a few final control techniques. By studying some examples, the reader should be prepared to analyze and understand many other techniques that arise in industry.

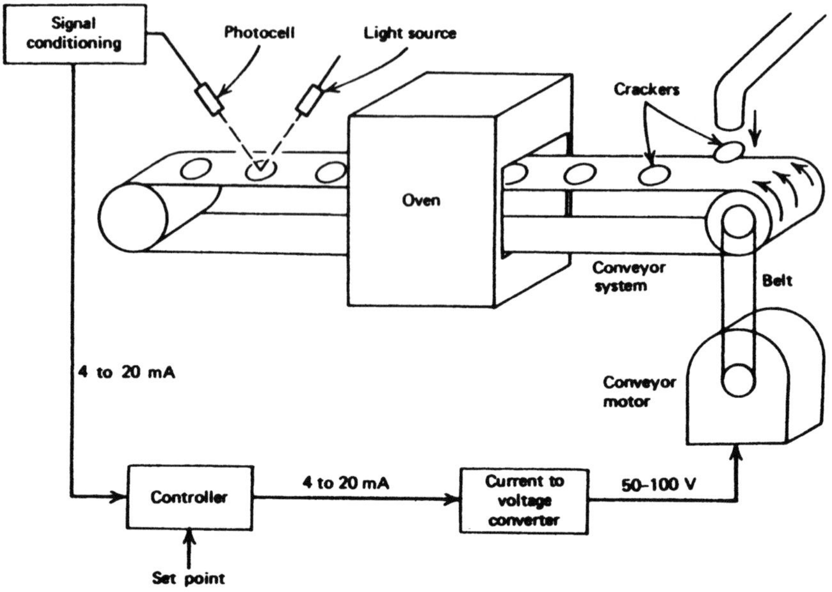

FIGURE 7.2
A process-control system showing the final control operations.

7.3 SIGNAL CONVERSIONS

The principal objective of signal conversion is to convert the low-energy control signal to a high-energy signal to drive the actuator. Controller output signals are typically in one of three forms: (1) electrical current, usually 4- to 20-mA; (2) pneumatic pressure, usually 3 to 15 psi or 20 to 100 kPa; and (3) digital signals, usually TTL-level voltages in serial or parallel format. There are many different schemes for conversion of these signals to other forms, depending on the desired final form and evolving technology. In the following sections, a number of the more common conversion schemes are presented. You should always be receptive to the advances of technology and the subsequent new methods of signal conditioning and conversion.

7.3.1 Analog Electrical Signals

Many of the methods of analog signal conditioning discussed in Chapter 2 are used in conversions necessary for final control. The following paragraphs summarize some of the more common approaches.

Relays A common conversion is to use the controller signal to activate a relay when a simple ON/OFF or two-position control is sufficient. In some cases, the low-current signal is insufficient to drive a heavy industrial relay, and an amplifier is used to boost the control signal to a level sufficient to do the job.

In many instances, the electromechanical relay has been replaced by high-power industrial electronic devices called *solid-state relays*. There are no moving parts, yet they perform the same function as an electromechanical relay. Generally, low-power ac or dc signals will turn the solid-state relay on and off. In some cases, these devices can switch hundreds of amps using only a milliamp control signal. The solid-state relay is usually implemented using SCRs and TRIACs. No special knowledge of these devices is necessary to use such a relay, however. SCRs and TRIACs are discussed in Section 7.4.

Amplifiers High-power ac or dc amplifiers often can provide the necessary conversion of the low-energy control signal to a high-energy form. Such amplifiers may serve for motor control, heat control, light-level control, and a host of other industrial needs.

EXAMPLE 7.1 A magnetic amplifier requires a 5- to 10-V input signal from a 4- to 20-mA control signal. Design a signal-conversion system to provide this relationship.

Solution

We first must convert the current to a voltage, and then provide the required gain and bias. We can get a voltage using a resistor in the current line of, say, 100 Ω. Then the 4 to 20 mA becomes 0.4 to 2.0 V. The amplifier system must provide an output given by

$$V_{\text{out}} = KV_{\text{in}} + V_B$$

where K is the gain and V_B is an appropriate bias voltage. We know that 0.4-V input must provide 5-V output, and 2-V input must provide 10-V output. This allows us to find K and V_B, using simultaneous equations, as

$$5 = 0.4K + V_B$$
$$10 = 2K + V_B$$

Subtracting, we get

$$5 = 1.6K$$
$$K = \mathbf{3.125}$$

that we use in either equation to find

$$V_B = \mathbf{3.75}$$

Thus, the result is

$$V_{\text{out}} = 3.125V_{\text{in}} + 3.75$$
$$V_{\text{out}} = 3.125(V_{\text{in}} + 1.2)$$

The circuit of Figure 7.3 shows how this can be implemented using an op amp configuration. The rest of this design would involve determining if the op amp can deliver enough current to drive the magnetic amplifier. If not, a booster would be required.

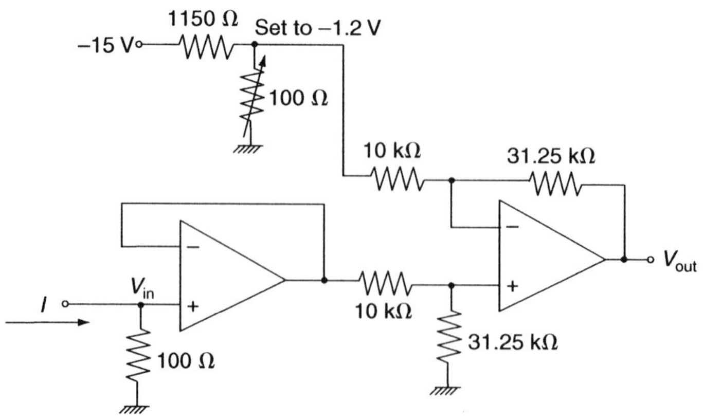

FIGURE 7.3
An op amp circuit for Example 7.1.

Motor Control Many motor control circuits are designed as packaged units that accept a low-level dc signal directly to control motor speed. If such a system is not available, it is possible to build circuits using amplifiers along with SCRs or TRIACs to perform this control. The nature of SCRs and TRIACs are presented in the next section. The basic elements of electrical motors and some words about their control are discussed later in this chapter.

7.3.2 Digital Electrical Signals

Conversions of digital signals to forms required by final control operations usually are carried out using systems already discussed in Chapter 3. We mention again, however, the basic elements of the output interface between computer and final control.

ON/OFF Control There are many cases in process control where the control algorithm is accomplished by simple commands to outside equipment to change speed, turn on (or off), move up, and so on. In such cases, the computer can simply load a latched output line with a **1** or **0** as appropriate. Then it is a simple matter to use this signal to close a relay or activate some other outside circuit.

DAC When the digital output must provide a smooth control, as it does in valve positioning, the computer must provide an input to a DAC that then determines an appropriate analog output. When a computer must provide outputs to many final control elements, a data output module or system such as that described in Chapter 3 can be employed. These integrated modules contain channel addressing, DAC, and other required elements of a self-contained output interface system.

EXAMPLE 7.2 A 4-bit digital word is intended to control the setting of a 2-Ω dc resistive heater. Heat output varies as a 0 to 24 V input to the heater. Using a 10-V DAC followed by an amplifier with a high-current output, calculate **(a)** the settings from minimum to maximum heat dissipation, and **(b)** how the power varies with LSB changes.

Solution

a. The 4-bit word has a total of 16 states; thus, the DAC outputs voltage from 0 V for a $\mathbf{0000}_2$ input to 9.375 V for a $\mathbf{1111}_2$ input. If we use a gain of 2.56, then the heater input will be 0 to 24 V in 1.6-V steps. The heat dissipation is found from

$$P = \frac{V^2}{R} \tag{7.1}$$

Then, for the minimum, $P = 0$ because $V = 0$, and for the maximum

$$P_{\text{max}} = \frac{(24\ \text{V})^2}{2\ \Omega}$$

$$P_{\text{max}} = 288\ \text{W}$$

b. The variation in heating with voltage is not linear, because the power varies as the square of the voltage. The increase in power for a step change in voltage can be found by taking the difference in power between the two cases:

$$\Delta P = \frac{V^2}{R} - \frac{(V - \Delta V)^2}{R} = \frac{\Delta V(2V - \Delta V)}{R}$$

Because $\Delta V = 1.6$ V, we have

$$\Delta P = \frac{1.6(2V - 1.6)}{R} = 0.8(2V - 1.6)$$

The first bit change produces a power change of 1.28 W, and the last bit, a power change of 37.12 W.

Direct Action As the use of digital and computer techniques in process control has become more widespread, new methods of final control have been developed that can be actuated directly by the computer. Thus, a stepping motor, to be discussed later, interfaces easily to the digital signals that a computer outputs. In another development, special integrated circuits are made that reside within the final control element and allow the digital signal to be connected directly.

7.3.3 Pneumatic Signals

The general field of pneumatics covers a broad spectrum of applications of gas pressure to industrial needs. One of the most common applications is to provide a force by the gas pressures acting on a piston or diaphragm. Later in this chapter, we will deal with this application in process control. In this section, however, we are interested in pneumatics as a means of propagating information—that is, as a signal carrier—and how that signal can be converted to other forms.

Principles In a pneumatic system, information is carried by the pressure of gas in a pipe. If we have a pipe of any length and raise the pressure of gas in one end, this increase in pressure will propagate down the pipe until the pressure throughout is raised to the new value. The pressure signal travels down the pipe at a speed in the range of the speed of sound in the gas (say, air), which is about 330 m/s (1083 ft/s). Thus, if a transducer varies gas pressure at one end of a 330-m pipe (about 360 yd) in response to some controlled variable, then that same pressure occurs at the other end of the pipe after a delay of approximately 1 s. For many process-control installations, this delay time is of no consequence, although it is slow compared to an electrical signal. This type of signal propagation was used for many years in process control before electrical/electronic technology advanced to a level of reliability and safety to enable its use with confidence. Pneumatics is still employed in many installations either because of danger from electrical equipment or as a carryover from previous years, where conversion to electrical methods would not be cost effective. In general, pneumatic signals are carried with dry air as the gas and signal information are adjusted to lie within the range of 3 to 15 psi. In SI unit systems, the range of 20 to 100 kPa is used. There are three types of signal conversion of primary interest.

Amplification A pneumatic amplifier, also called a *booster* or *relay,* raises the pressure and/or air flow volume by some linearly proportional amount from the input signal. Thus, if the booster has a pressure gain of 10, the output would be 30 to 150 psi for an input of 3 to 15 psi. This is accomplished via a regulator that is activated by the control signal. A schematic diagram of one type of pressure booster is shown in Figure 7.4. As the sig-

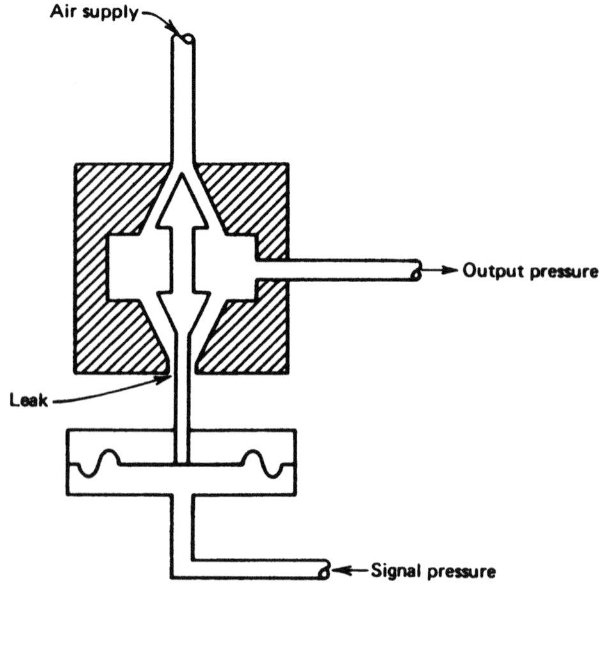

FIGURE 7.4
A pneumatic amplifier or booster converts the signal pressure to a higher pressure or the same pressure but with greater gas volume.

nal pressure varies, the diaphragm motion will move the plug in the body block of the booster. If motion is down, the gas leak is reduced and pressure in the output line is increased. The device shown is reverse acting, because a high-signal pressure will cause output pressure to decrease. Many other designs are also used.

Nozzle/Flapper System An important signal conversion is from pressure to mechanical motion and vice versa. This conversion can be provided by a nozzle/flapper system (sometimes called a nozzle/baffle system). A diagram of this device is shown in Figure 7.5a. A regulated supply of pressure, usually over 20 psig, provides a source of air through the restriction. The nozzle is open at the end where the gap exists between the nozzle and flapper,

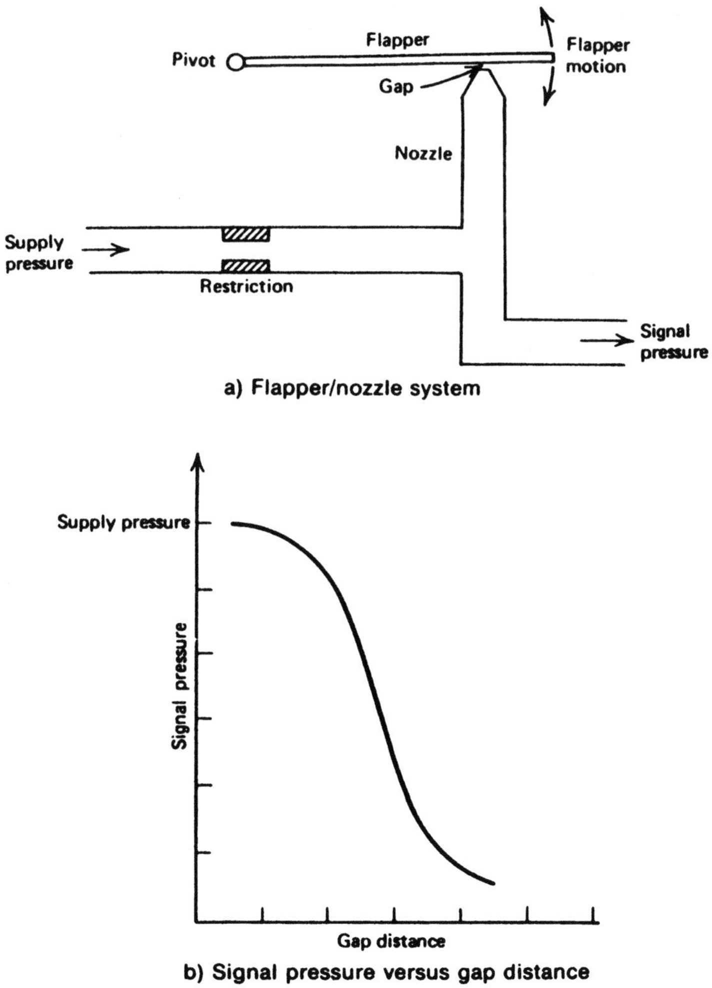

FIGURE 7.5
Principles of the nozzle/flapper system.

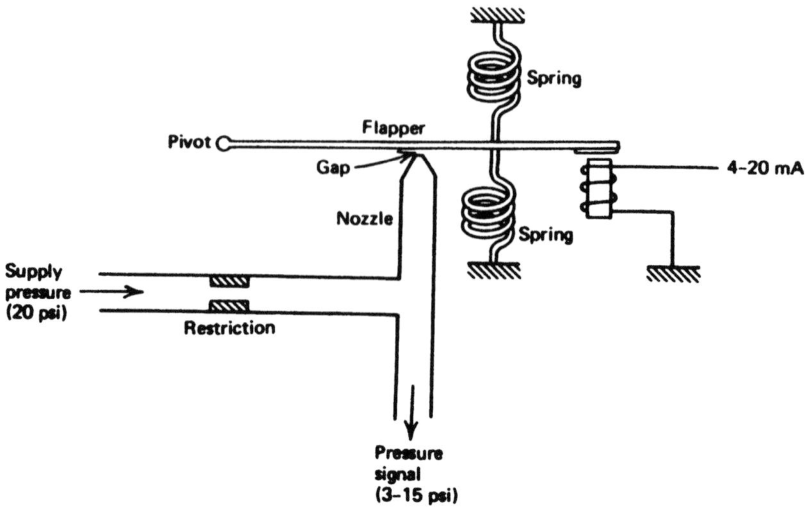

FIGURE 7.6
Using a nozzle/flapper for a current-to-pressure converter.

and air escapes in this region. If the flapper moves down and closes off the nozzle opening so that no air leaks, the signal pressure will rise to the supply pressure. As the flapper moves away, the signal pressure will drop because of the leaking gas. Finally, when the flapper is far away, the pressure will stabilize at some value determined by the maximum leak through the nozzle. Figure 7.5b shows the relationship between signal pressure and gap distance. Note the great sensitivity in the central region. A nozzle/flapper is designed to operate in the central region, where the slope of the line is greatest. In this region, the response will be such that a very small motion of the flapper can change the pressure by an order of magnitude. This system is covered more extensively in Chapter 10 under the discussion of pneumatic controllers.

Current-to-Pressure Converters The current-to-pressure converter, or simply *I/P converter,* is an important element in process control. Often, when we want to use the low-level electric current signal to do work, it is easier to let the work be done by a pneumatic signal. The I/P converter gives us a linear way of translating the 4- to 20-mA current into a 3- to 15-psig signal. There are many designs for these converters, but the basic principle almost always involves the use of a nozzle/flapper system. Figure 7.6 illustrates a simple way to construct such a converter. Notice that the current through a coil produces a force that will tend to pull the flapper down and close off the gap. A high current produces a high pressure so that the device is direct acting. Adjustment of the springs and perhaps the position relative to the pivot to which they are attached allows the unit to be calibrated so that 4 mA corresponds to 3 psig and 20 mA corresponds to 15 psig.

7.4 POWER ELECTRONICS

An important part of final control operations is electrical signal energy conversion. This means taking the low-energy control signal (e.g., 4-20 mA) and using this to impart a large energy impact on the plant (e.g., controlling a 100-Hp (74.6 kW) motor). This kind of signal conversion is carried out using high-power electronics devices. Generally power electronics refers to an assortment of very special semiconductor devices that have been designed and developed to allow control of hundreds to thousands of amperes at hundreds to thousands of volts. In this section an assortment of the most common devices will be described.

These devices fall within two categories: switching devices and control devices. Switching in this context means devices that have only two output states, *on* and *off.* Control devices, by contrast, can have an output that varies continuously over some range. It is important to realize, however, that even these devices are usually used in a switching mode. Both types of devices are used for a variety of control applications including: control of large electric heaters, high-power motor speed control, high-energy power supplies, large solenoid actuation, and a host of other applications.

7.4.1 Switching Devices

Some solid-state semiconductor devices have the property that they can either be in an *off* state or an *on* state, much like a mechanical switch. Of course the common diode acts like this in the sense that it can conduct current in one direction (forward biased) but not in the opposite direction (reverse biased). The difference is that the devices to be considered in this section can be turned on and off, even when forward biased.

Silicon-Controlled Rectifier (SCR) The first device we will consider, an SCR, is also called a *thyristor.* Figure 7.7 shows the schematic symbol for the SCR and a typical curve of current versus voltage. The device is much like a diode but with an extra terminal called the *gate.* Like any diode the SCR will not conduct if it is reverse-biased (i.e., if the anode voltage is more negative than the cathode). However, unlike the diode, if the SCR is forward-biased it will still not conduct unless and until a positive trigger voltage is applied to the gate, as indicated in Figure 7.7. Then it will turn on and conduct like a diode. It is important to note that it will continue to conduct even when the gate voltage is taken away. In fact, the only way to get

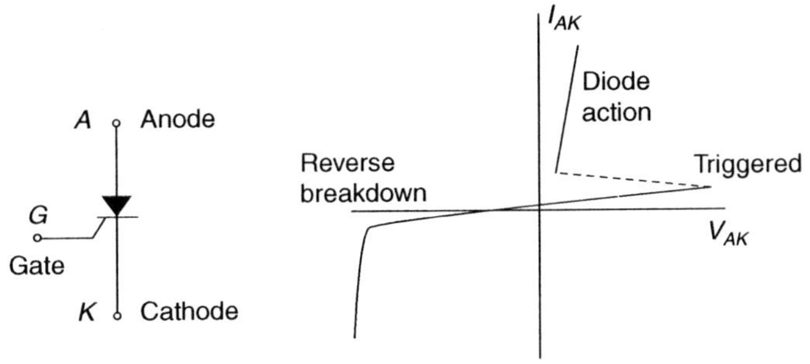

FIGURE 7.7
The silicon-controlled-rectifier (SCR) symbol and characteristic curve.

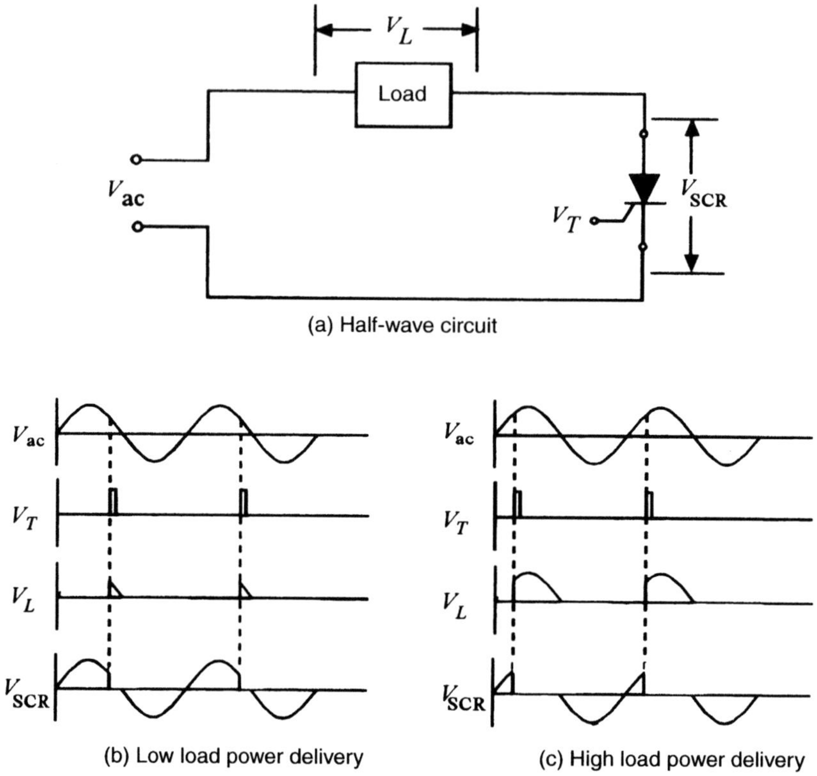

FIGURE 7.8
An SCR for variation of load voltage using the half-wave circuit. Only the positive half-source cycle delivers power to the load.

it to turn back off is to remove the forward-biased condition. For this reason its application in dc problems is very limited unless some external method is provided to interrupt the current.

Notice in Figure 7.7 that even when switched into the conducting, *on* state, the SCR exhibits some forward voltage drop that increases with current. This means it acts like a resistance. The SCR, and all the power electronic devices we will study, are specified with some *on* resistance, R_{on}. Since there is some effective resistance and since a current (a large current) is flowing, there will be power dissipated in the SCR, I^2R_{on}.

When used in an ac application the reverse-bias condition for turn-off is provided automatically when the ac voltage cycles. Figure 7.8 shows a simple half-wave application of the SCR to vary the power delivered to a load. The trigger voltage, V_T, is derived from a circuit (not shown) that is able to change the time during the forward-biased half cycle during which the trigger voltage is applied to the gate. Figures 7.8b and 7.8c show how the effective rms voltage across the load varies with the time at which the pulse is applied to the gate. The time is often referred to as the *firing angle,* where angle refers to the 180° of the half cycle.

Figure 7.9 shows an illustration of how the SCR can be used in a full-wave fashion to vary power delivered to a load. You can see that a full-wave diode bridge converts the ac into pulsating dc, going to zero every 180° of the source cycle. In this case the SCR is au-

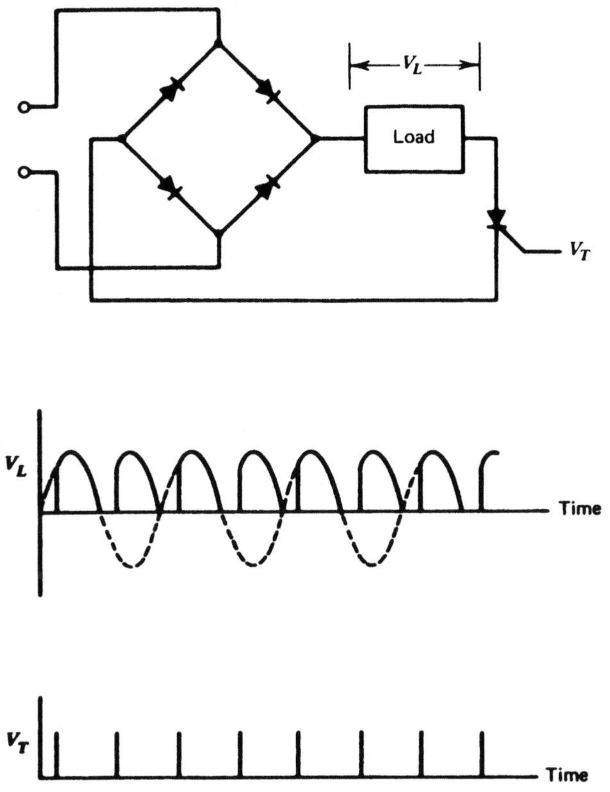

FIGURE 7.9
A full-wave SCR circuit. The effective rms dc voltage applied to the load is greater than that of the half-wave because the entire cycle is used.

tomatically turned off when the voltage drops to zero each half cycle. Again, a special timing circuit adjusts when the trigger voltage is applied in the half cycle.

Of course, in a control application some means must be found by which the low energy control signal changes the firing angle of the trigger voltage. Figure 7.10 shows an elementary way by which this can be done. The opto-coupler allows the control voltage to change the current through the phototransistor. This in turn changes the charging rate of the capacitor. When the voltage on the capacitor reaches the required gate trigger voltage the capacitor discharges through the resistor R and turns on the SCR. When the control voltage is small the transistor current is small, so the charging rate is small and the trigger voltage is not reached until late in the cycle. A large control voltage means a high charging rate, and, thus, early triggering in the cycle.

SCRs are characterized like diodes by specification of the *maximum reverse voltage* and *maximum forward current* as well as the *turn-off time* and *gate voltage.* Modern high power SCRs are available with up to 6 kV reverse voltage and 3.5 kA forward current rating. Turn-off time, which is the major limitation on switching frequency, is typically 100 to 200 μs but no faster than 10 μs. The higher power units tend to have the slower turn-off time. Another

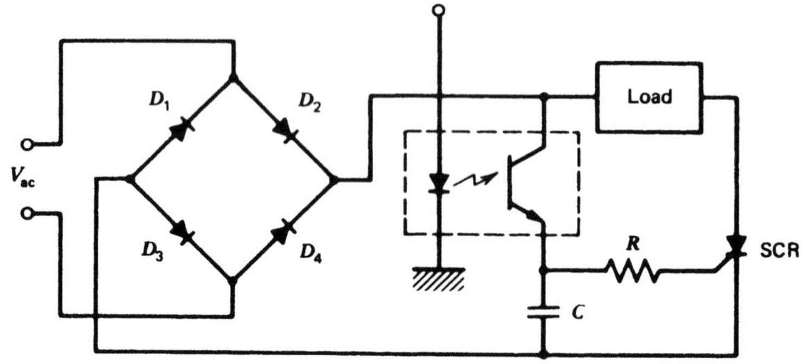

FIGURE 7.10
An optical coupler can be used to control triggering of the SCR from an external circuit.

SCR parameter is the *holding current,* which is the minimum forward current necessary to maintain the SCR in the on state. In general SCRs can be operated in switching modes up to at most a few kilohertz. Typical values of R_{on} for high-power SCRs are in the range of 0.25 mΩ. This means that even at high current the forward voltage drop is small, so the power dissipation by the SCR is not too great. For example, a current of 100 A would mean a voltage of V_{SCR} = (100 A)(0.00025 Ω) = 0.025 V and a power dissipation of P_{SCR} = (100)(0.025) = 2.5 W.

SCRs suffer from a problem associated with the rate of change of voltage across its anode and cathode terminals. If this rate of change exceeds a certain threshold value the SCR will spontaneously go into conduction, even with no gate trigger voltage. The effect is referred to as the derivative effect (*dV/dt*) since the rate of change is just the derivative.

EXAMPLE 7.3 An SCR with a 4.0-V trigger is used in the circuit of Figure 7.11 as a light-dimmer control. What resistance, *R,* should be used to provide approximately 10% to 90% on time?

FIGURE 7.11
Circuit for Example 7.3.

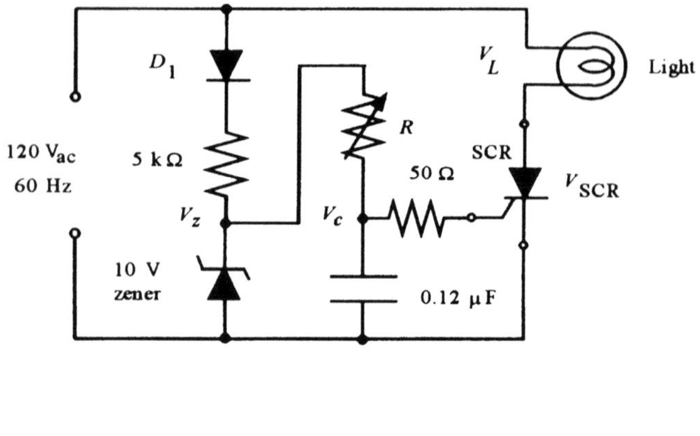

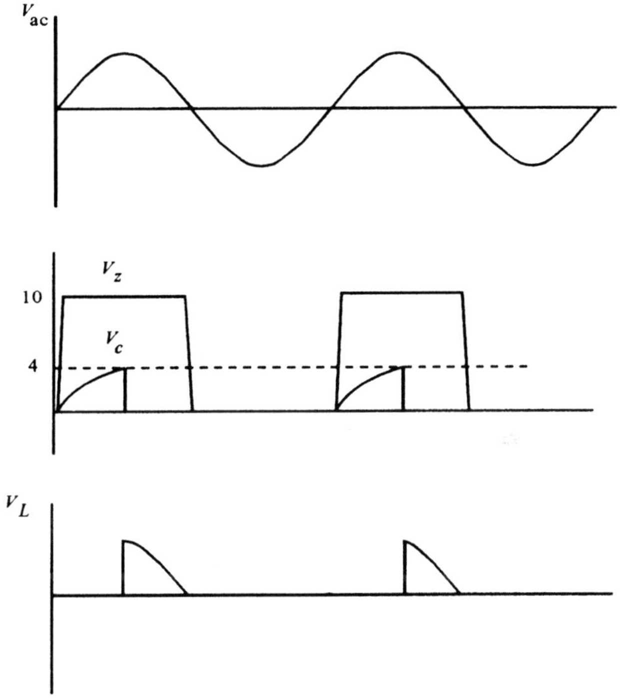

FIGURE 7.12
Voltages versus time for Example 7.3.

Solution

Let's start by understanding each part of the circuit. Diode D_1, the 5-kΩ resistor, and the 10-V zener diode provide the source of about 10 V, but this goes to zero every cycle, as shown in Figure 7.12. This serves to charge the 0.12-μF capacitor through resistor R. When the ac voltage on the SCR is on the positive half of the cycle and the capacitor charge reaches the trigger value of 4 V, the SCR will go on, discharging the capacitor and turning on the light. When the SCR voltage goes to the negative half cycle, the SCR will turn off. Figure 7.12 shows the voltage waveforms in Figure 7.11 for a 50% on-time setting of R.

When will the capacitor charge reach the trigger of 4.0 V? This can be found by the simple capacitor-charging equation, $V_c(t) = V_0(1 - e^{-t/RC})$, where V_0 is the zener voltage. In this case, we are looking for a charge of 4.0 V, so

$$4.0 = 10(1 - e^{-t/RC})$$

Solving for t, we find $t = 0.511RC$. Thus, whatever R and C are, the SCR will trigger at the time given by this expression.

The period of a half cycle is $1/120 = 8.33$ ms. Therefore, for a 10% on time we use $t = 0.833$ ms, and for a 90% on time we use $t = 7.5$ ms. Since the capacitor is given as 0.12 μF, we can find the required resistance for the two cases:

$$R_{\text{min}} = \frac{0.833 \text{ ms}}{(0.511)(0.12\ \mu\text{F})} = 13.6 \text{ k}\Omega$$

and

$$R_{max} = \frac{7.5 \text{ ms}}{(0.511)(0.12\ \mu\text{F})} = 122\ \text{k}\Omega$$

This gives the basic range of resistance. Let's use a 15-kΩ resistor in series with a 100-kΩ variable. The range is then 15 to 115 kΩ, which corresponds to a range of about 11% to 95% on time.

Gate Turn-off Thyristor (GTO) A significant improvement in SCR technology has been a modification that allows the SCR to also be turned *off* by application of a gate signal. Two symbols in common use for this device are shown in Figure 7.13. Like the SCR this device can be turned *on* when forward biased by application of a positive trigger voltage to the gate terminal. The SCR cannot be turned off except by taking away the forward-bias condition. With the GTO, if sufficient negative current is applied to the gate, the GTO can be made to go *off*. Thus, we have the ability to turn the device *on* and *off*. The amount of negative current required to effect turn-off is defined by the current gain parameter, β_{OFF}, like the beta of a transistor. One disadvantage of this device is that the turn-off current gain is not very large, perhaps only about 5. This means if the forward current was 100 A then the gate current required for turn-off would be 100/5 = 20 A. Of course this can be just a pulse but it is still a significant current. Generally the turn-off time of the GTO is shorter than an SCR so that switching at higher speeds is possible.

Since the GTO can be turned *on* and *off* it can be used in dc circuits. Figure 7.14 shows an elementary illustration suggesting how a GTO can be used to vary the average power dissipated by a load. The controlling square wave has fixed frequency (period = T) but the duty cycle can be varied; hence, the time, t_x, is positive. The square wave amplitude is $\pm V_s$. During the positive part of the cycle transistor T1 is turned on and a positive voltage and current is applied to the GTO gate, turning it *on*. During the negative part of the cycle transistor T2 is turned on and current is drawn from the gate, turning the GTO back *off* again. Thus, the battery voltage will be applied to the load only while the square wave is high. The voltage across the load when the GTO is *on* will be $V_L = V_B - V_{GTO}$, where $V_{GTO} = IR_{on}$ is the voltage across the GTO and R_{on} is the on resistance. We can find the average power dissipated by the load as follows,

$$P_{AVE} = \frac{1}{T}\int_0^T i(t)v(t)dt = \frac{1}{T}\int_0^{t_x} I(IR)dt = \frac{t_x}{T}I^2R \tag{7.2}$$

where the ratio t_x/T is the duty cycle.

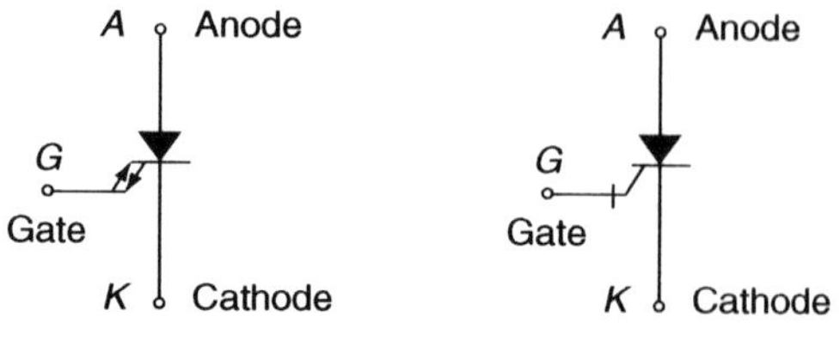

FIGURE 7.13
Two common symbols for the Gate Turn-off (GTO).

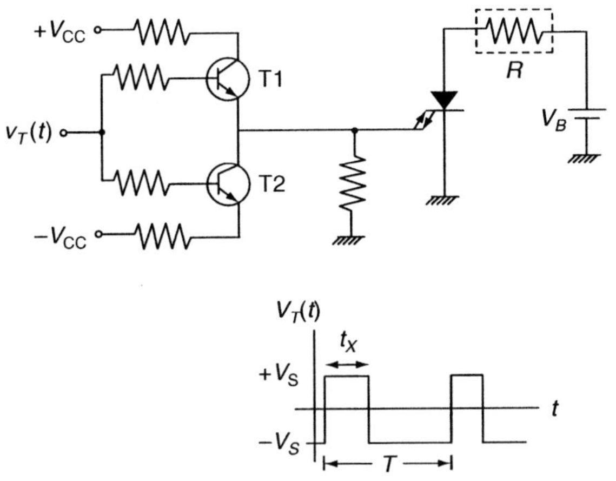

FIGURE 7.14
GTO circuit for Example 7.4.

EXAMPLE 7.4 How much current must be drawn through T2 to turn the GTO of Figure 7.14 *off* if the following specifications apply to the GTO: $\beta_{OFF} = 7$, $R_{on} = 2.5\ m\Omega$, $V_B = 75$ V, and $R_L = 2\ \Omega$? What average power is dissipated by the load and by the GTO for a 50% duty cycle?

Solution

First we find the current through the load when the GTO is on,

$$I_L = \frac{V_B - I_L R_{on}}{R_L} \quad \text{or} \quad I_L = \frac{V_B}{R_L + R_{on}} = \frac{75}{2 + 0.0025} = 37.5\ A$$

Since the gain is 7, the current through T2 must be: 37.5 A/7 = 5.4 A.

When the GTO is *on* the load dissipates an average power found from Equation (7.2),

$$P_{AVE} = (0.5)(37.5\text{ A})^2(2\ \Omega) = 1{,}406\text{ W}$$

Power dissipated by the GTO can be found by using the given value of R_{on},

$$P_{GTO} = (0.5)(37.5\text{ A})^2(0.0025\ \Omega) = 1.75\text{ W}$$

Like the SCR, the GTO is susceptible to unwanted turn-on if the rate-of-change of the voltage across its terminals is too large. Special "snubber" circuits are used to limit the rate of change. Also, the GTO requires a small amount of gate current even when *on* to assure proper operation.

FIGURE 7.15
A TRIAC can be triggered to conduct in either direction so a true ac voltage can be applied to a load.

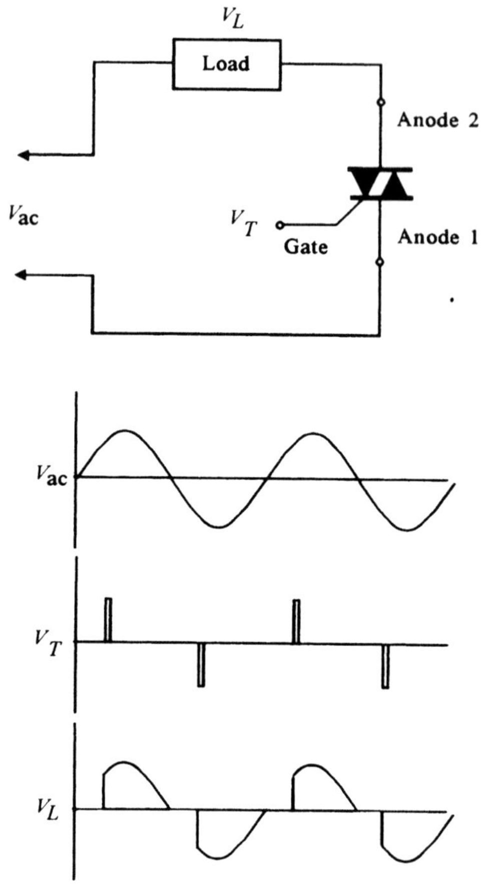

GTOs are available with forward voltage hold-off up to 4,500 volts and currents up to 3,000 A. They generally can be used in switching applications up to about 10 kHz. The on-state resistance of the GTO is in the range of 2 to 3 mΩ.

TRIAC Another type of switching power electronic device is the *TRIAC,* which has the schematic symbol shown in Figure 7.15. Note that this symbol looks like two diodes in parallel but reversed, along with a third gate terminal. Indeed the properties of the TRIAC are such that it conducts current in both directions. Like the SCR and GTO, conduction does not begin until a trigger voltage is applied to the gate terminal. In the case of the TRIAC, application of a positive voltage to the gate allows conduction of current in one direction through the device while a voltage of negative polarity applied to the gate allows conduction in the opposite direction. Figure 7.15 shows waveforms across the source and the load in response to pulses applied to the TRIAC gate.

The TRIAC has somewhat limited application in power electronics because its switching speed is low in comparison to other devices. In addition, like the SCR, it may be

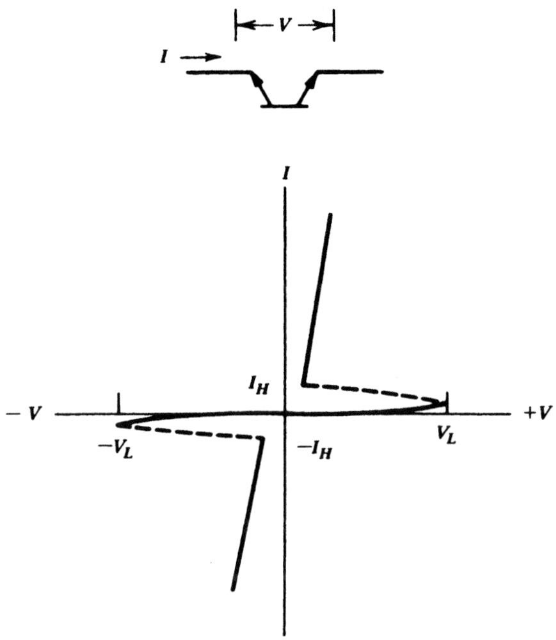

FIGURE 7.16
The DIAC has a bipolar breakover voltage, $\pm V_L$. After the voltage reaches $\pm V_L$, the DIAC conducts like a diode.

driven into conduction if the rate of change of voltage across its terminals is too large (*dV/dt* effect). Also, like the SCR once turned *on* in either direction, it can only be turned *off* by taking the voltage across its terminals to zero.

In order to provide the necessary bipolar gate voltage, the TRIAC is often used in conjunction with a *DIAC* that has one commonly employed schematic symbol, shown in Figure 7.16. This two-terminal device can conduct in both directions but only after the voltage across its terminals exceeds a *breakover* value in either direction. Typical breakover voltage for families of DIACs range from 25 to 75 volts. The following example illustrates the application of a TRIAC and DIAC to a light dimmer.

EXAMPLE 7.5

A DIAC is available that has a 28-V breakover voltage, V_L. Show how this can be used to make the light dimmer of Example 7.3 using a TRIAC. Again, we want approximately a 10% to 90% range of adjustment using R, with $C = 0.12\ \mu\text{F}$.

Solution

The light dimmer can be implemented so that power is applied every cycle instead of every other cycle. We can use the circuit of Figure 7.17 with the load replaced by a light and the source given by the 120-V_{ac} rms, 60-Hz line. This is given by

$$V_{ac} = 169.7 \sin(120\pi t)$$

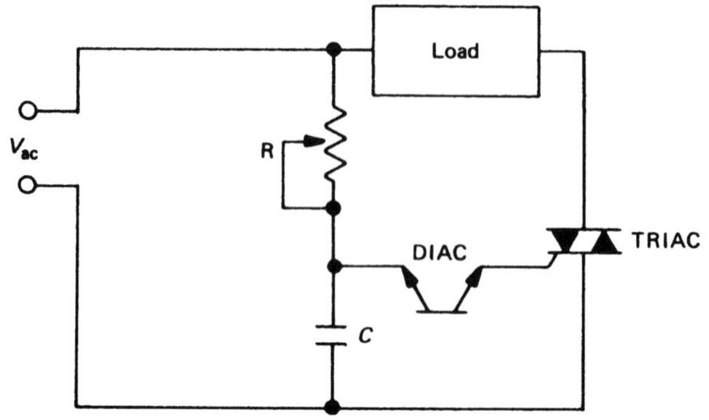

FIGURE 7.17
The TRIAC can be triggered from the ac line using a DIAC.

The first question is, When does the line voltage reach 28 so that it would even be possible to trigger the DIAC and thus the TRIAC? This can be found from

$$28 = 169.7\sin(120\pi t)$$

Solving for *t*,

$$t = [\sin^{-1}(28/169.7)]/(120\pi) = 0.44 \text{ ms}$$

Since the half period is 8.33 ms, the light cannot be turned on before about 5% of the period has passed.

The next problem is to find the values of R that will provide a 10% and 90% on time, respectively. As before, this means the voltage must reach 28 V at 0.833 ms and 7.5 ms, respectively. Unfortunately, there is not such a simple relation between capacitor voltage and time as in Example 7.3, because the charging voltage is changing in time—that is, $169.7\sin(120\pi t)$. An exact equation can be found, but let's try a simpler approach.

We can modify the circuit as shown in Figure 7.18 to provide a clamped ac voltage of about 40 V for charging the capacitor. The question now is, When does the ac voltage first reach 40 V? This is again found from

$$40 = 169.7\sin(120\pi t)$$

so,

$$t = [\sin^{-1}(40/169.7)/(120\pi)] = 0.63 \text{ ms}$$

or about 7.5% into the half cycle. So we can now estimate the R to give about 10% and 90% as in Example 7.3.

$$28 = 40[1 - e^{-t/RC}]$$

or $t = 1.2RC$. This is an approximation, because the voltage is not 40 V from the start of the period. Using $C = 0.12\ \mu\text{F}$ and times of 0.833 and 7.5 ms gives

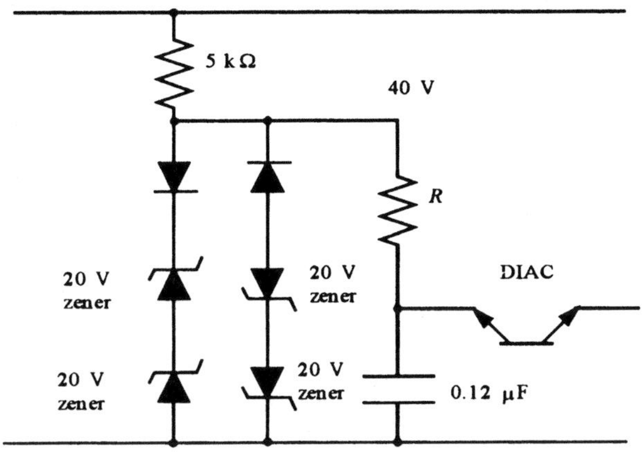

FIGURE 7.18
This modification of the triggering circuit eases calculation of a solution for Example 7.4.

$$R_{\text{min}} = 0.833 \text{ ms}/[(1.2)(0.12\ \mu\text{F})] = 5.8\ \text{k}\Omega$$

and

$$R_{\text{max}} = 7.5 \text{ ms}/[(1.2)(0.12\ \mu\text{F})] = 52\ \text{k}\Omega$$

Thus, we could use a 6-kΩ fixed resistor in series with a 50-kΩ pot. This would give on times of 10.4% to 97%. Figure 7.19 shows the waveforms of source, load, and capacitor.

High power TRIACs can be generally used up to 1,200 V and a current of 300 A. Maximum switching frequency is only about 400 Hz due in part to the slow turn-off time of 200 to 400 μs. The effective on resistance runs 3 to 4 mΩ.

7.4.2 Controlling Devices

The previous section presented three common power electronics semiconductor devices based upon diode action. The characteristics of these devices are that they are either conducting (*on*) or not conducting (*off*). Another class of power electronic devices is based upon semiconductors for which a high current or voltage is controlled by a small current or voltage. These devices can be used to execute smooth control of current but also can be used as switches (*on* or *off*). For high-power applications it is much more efficient to operate the devices in the switched mode. We will use an example of the bipolar junction transistor to see why this is true.

Bipolar Junction Transistor (BJT) The common bipolar junction transistor can be used for modest power electronic applications. The schematic symbol of the bipolar transistor for an NPN type is shown in Figure 7.20. Figure 7.20 also shows the basic operating curves of the BJT. The collector current is plotted versus the collector-emitter voltage as a family of curves for different base currents. In the linear regime the BJT is a current amplifier wherein the collector-to-emitter current is given by some current gain factor, β, times the base-to-emitter current, $I_c = \beta I_b$.

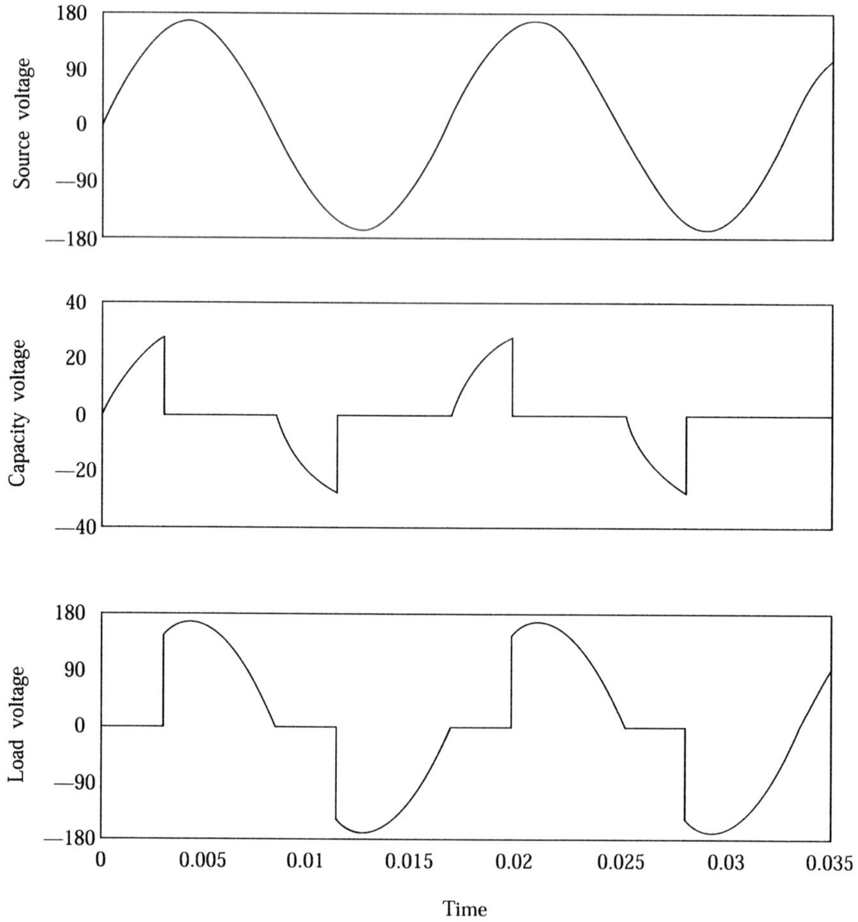

FIGURE 7.19
These are the voltage waveforms for Example 7.4.

FIGURE 7.20
Bipolar Junction Transistor (BJT) symbol and characteristic curves.

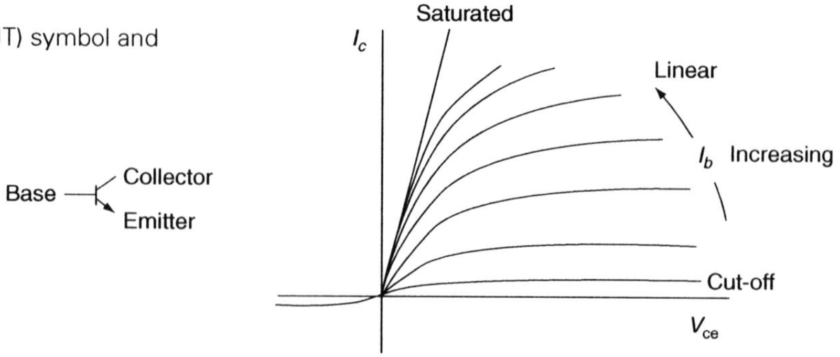

In the switched mode the BJT is either *on* or *off.* When *off* the base current is zero and the transistor is cut-off, meaning no collector current is flowing. Thus, it is not dissipating any power, sort of like an open circuit (i.e., an open switch). In the *on* state the BJT acts more like a resistance, R_{on}, as in the case of SCRs and GTOs. In this case, as the curves in Figure 7.20 show, there is some voltage drop across the BJT called the *saturation voltage,* $V_{sat} = I_c R_{on}$, so the BJT is dissipating a power, $I_c V_{sat}$. Often the specifications of the BJT simply give the saturation voltage for some collector current instead of the effective saturation *on* resistance.

Let's contrast using the BJT as a linear amplifier versus a switching amplifier. Suppose a resistive, 10-Ω heater is to have a power output which varies from 25.6 W to 640 W. The heater output is to be adjusted by the common 4- to 20-mA control signal. A power BJT is available with a linear current gain of 400, and a saturation voltage of 1.5 V at about 10 A. A 100-V dc power supply is available.

Figure 7.21 shows how a linear circuit could be used to provide the required variation. To dissipate 640 W would require a maximum current given from $P = IR^2$,

$$I = \sqrt{\frac{P}{R}} = \sqrt{\frac{640}{10}} = 8A$$

You can quickly calculate that the minimum current would be 1.6 A. As the BJT base current varies from 4 to 20 mA, the collector current will vary from 1.6 to 8 A as required (remember $\beta = 400$). Let's look at the power dissipation. At full power the voltage across the heater would have to be $V_{Hmax} = P/I = 640/8 = 80$ V. So there must be 20 V across the transistor (80 + 20 = 100). This means that although the heater is dissipating 640 W the transistor itself is dissipating (20 V)(8 A) = 160 W! This is wasted dissipation. At the low end when the heater is dissipating 25.6 W the voltage across the heater is 16 V so there must be 84 V across the transistor. Then the BJT is dissipating (84 V)(1.6 A) = 134.4 W. Thus, the BJT is dissipating more than the heater! This is terribly inefficient.

Now let's see what happens when we operate the BJT in a switching mode by switching the base current between zero (cut-off, no collector current) and some value that leads to saturation, wherein the collector current is at a maximum determined essentially by external components rather than the β of the device. In this case the voltage from collector-to-emitter is fixed at some saturation forward voltage drop. Typically this value is in the range of fractions of a volt to several volts. The only power dissipated in the BJT is at a maximum current but small voltage across the transistor.

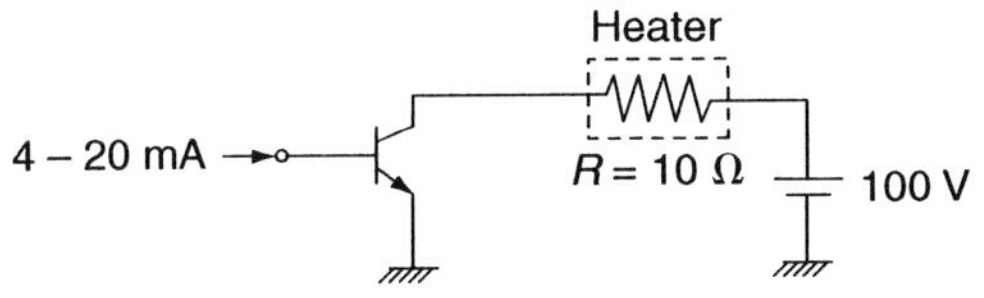

FIGURE 7.21
Operating the BJT in the linear mode.

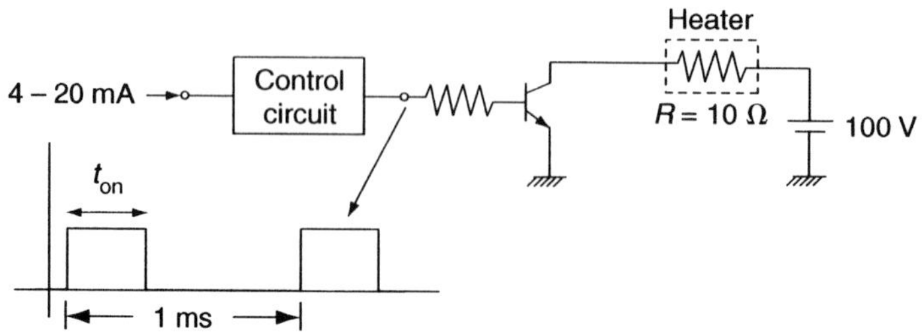

FIGURE 7.22
Operating the BJT in the switched mode.

Figure 7.22 shows the slightly modified circuit. Notice that the 4- to 20-mA signal is fed into a control circuit that varies the duty cycle of a fixed 1 kHz square wave. The BJT will be *on* when the square wave is high (for t_{on} seconds) and *off* for the rest of the period. The *average* power delivered to the heater is thus given by

$$P_{AVE} = \frac{1}{T}\int_0^T p(t)dt = \frac{1}{T}\int_0^{t_{on}} v(t)i(t)dt = \frac{t_{on}}{T}VI \tag{7.3}$$

where V and I are the constant voltage across and current through the heater during the *on* part of the cycle. Of course since $I = V/R$ we can write

$$P_{AVE} = \frac{t_{on}}{T}\frac{V^2}{R} \tag{7.4}$$

For this particular problem we need a duty cycle so that the average power is 640 W. When the BJT is *on* it drops 1.5 volts so the heater voltage is $V = 100 - 1.5 = 98.5$ V. Now we can solve for the duty cycle (i.e., t_{on}) from Equation (7.4),

$$\left.\frac{t_{on}}{T}\right|_{640W} = \frac{(640)(10)}{98.5^2} = 0.66$$

or a 66% duty cycle so that $t_{on} = 0.66$ ms. You can quickly show that for the case of 25.6 W the voltage across the load is still about 98.5 V but the duty cycle (on time) is now only 0.03 or 3% (or 0.03 ms). Thus, the control circuit will be designed so that 4 mA produces a 3% duty cycle and 20 mA produces a 66% duty cycle. Thus, the average power will be from 25.6 W to 640 W as required. What about dissipation by the BJT? The current through the BJT when it is *on* is $I_c = V/R = 98.5/10 = 9.85$ A at the 640 W setting and 2.56 A at the 25.6 W setting. The BJT saturation is about 1.5 V; thus, the average power dissipated by the BJT from Equation (7.3) is,

$$\text{At 640 W, } P_{BJT} = \frac{0.66}{1}(1.5)(9.85) = 9.8 \text{ W!!}$$

$$\text{At 25.6 W, } P_{BJT} = 0.44 \text{ W!!}$$

Thus, in the switching mode much less power is lost through heating of the BJT.

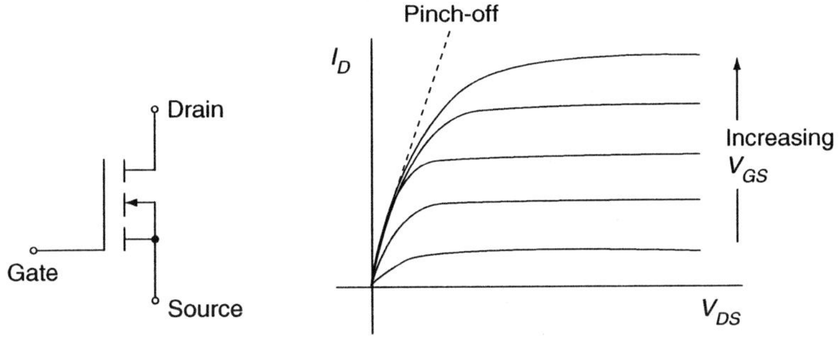

FIGURE 7.23
Power MOSFET symbol and characteristic curves.

Generally power BJT devices are limited to less than 250 A and 600 V applications because of thermal problems and with switching speeds not exceeding about 25 kHz. The switching turn-off time generally runs from 2 to 9 μs and the effective *on* resistance about 20 mΩ BJT devices suffer from a condition called *thermal runaway* that limits their application in power electronics. Thermal runaway works like this. Suppose a BJT is operating at some collector current, I_0. If the forward voltage is V_{ce} then the device is dissipating power $P = I_0 V_{ce}$. This dissipation heats up the BJT. It is a property of the BJT that when its temperature rises its collector current will also rise, $I_0 + \Delta I$. This means the dissipated power increases, $P + \Delta P = P + \Delta I V_{ce}$. But then its temperature would rise more, so its current would rise more, and so on until the unit fails.

Power MOSFET The schematic symbol of an n-channel, enhancement mode MOSFET is shown in Figure 7.23 along with typical characteristic curves. You can see that the MOSFET is a device for which the value of gate-source voltage controls the drain-source current. The transfer function is called the transconductance,

$$g_m = \left.\frac{\Delta I_D}{\Delta V_{GS}}\right|_{\text{constant } V_{DS}} \tag{7.5}$$

Therefore, the MOSFET, like the BJT, can be used in linear circuits as an amplifier. But again there will be inefficient power losses by the MOSFET itself. For high-power applications, then, we use switching circuits that avoid excessive power dissipation by the MOSFET.

One important advantage of the MOSFET is that it essentially draws no current from the driving signal source since the gate is essentially insulated from the drain and source. Contrast this with the BJT where the effective base current can be significant if the β is not high.

The power MOSFET has very fast switching speeds so it can be used in high-frequency power applications. The general specifications for MOSFETs show units are available with hold-off voltages as high as 1,000 volts and maximum current capacity of 50 A. Switching speeds can be as high as 100 kHz and turn-off times range from 0.5 to 1.0 μs. The effective on resistance of the MOSFET runs high, from 0.02 to 2 Ω for power types. This high on resistance limits their effectiveness in high current applications because the heating would become excessive.

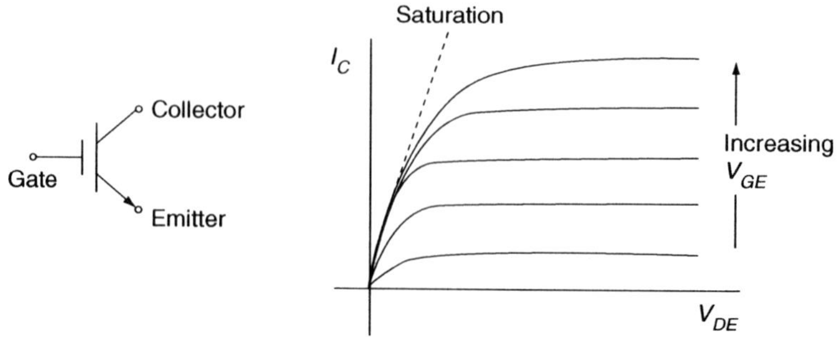

FIGURE 7.24
IGBT symbol and characteristic curves.

Insulated Gate Bipolar Transistor (IGBT) The final semiconductor device that we will consider is a kind of a combined BJT and MOSFET. In fact, the schematic symbol shown in Figure 7.24 indicates a collector and emitter, but now a gate that is insulated from the collector and emitter replaces the base. The characteristic curves shown in Figure 7.24 indicate that this device uses an input voltage to control current through the collector-emitter. Furthermore, the gate presents very high impedance so very little gate current flows. Like the BJT and the MOSFET, the IGBT is used in the switching mode to avoid power losses in the device. The IGBT and MOSFET are the predominant players in the power electronics applications.

The IGBT can be used in applications up to 1,200 volts and 400 A. It has a relatively high switching capacity, up to about 20 kHz, and the turn-off time runs around 2.5 μs. The on resistance is modest at around 50 mΩ so that some heating at high currents occurs.

7.5 ACTUATORS

If a valve is used to control fluid flow, some mechanism must physically open or close the valve. If a heater is to warm a system, some device must turn the heater on or off or vary its excitation. These are examples of the requirements for an *actuator* in the process-control loop. Notice the distinction of this device from both the input control signal and the control element itself (valve, heater, and so on, as shown in Figure 7.1). Actuators take on many diverse forms to suit the particular requirements of process-control loops. We will consider several types of electrical and pneumatic actuators.

7.5.1 Electrical Actuators

The following paragraphs give a short description of several common types of electrical actuators. The intention is to present only the essential features of the devices and not an in-depth study of operational principles and characteristics. In any specific application, one would be expected to consult detailed product specifications and books associated with each type of actuator.

Solenoid A *solenoid* is an elementary device that converts an electrical signal into mechanical motion, usually rectilinear (in a straight line). As shown in Figure 7.25, the solenoid consists of a coil and plunger. The plunger may be freestanding or spring loaded. The coil will have some voltage or current rating and may be dc or ac. Solenoid specifications include the electrical rating and the plunger pull or push force when excited by the specified voltage. This force may be expressed in newtons or kilograms in the SI system and in pounds or ounces in the English system. Some solenoids are rated only for intermittent duty because of thermal constraints. In this case, the maximum duty cycle (percentage on time to total time) will be specified. Solenoids are used when a large, sudden force must be applied to perform some job. In Figure 7.26, a solenoid is used to change the gears of a two-position transmission. An SCR is used to activate the solenoid coil.

Electrical Motors Electrical motors are devices that accept electrical input and produce a continuous rotation as a result. Motor styles and sizes vary as demands for rotational speed (revolutions per minute, or rpm), starting torque, rotational torque, and other specifications vary. There are numerous cases where electrical motors are employed

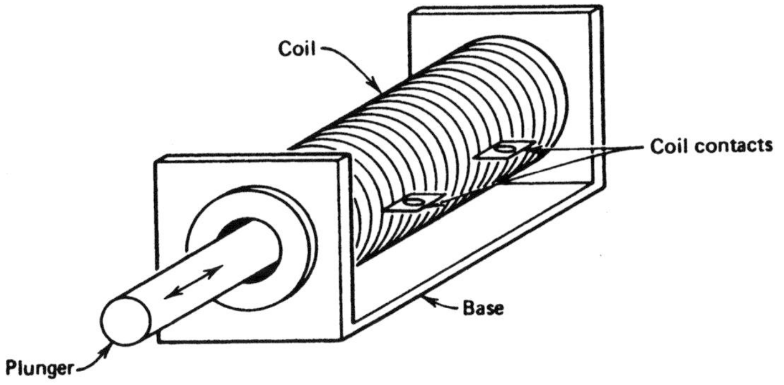

FIGURE 7.25
A solenoid converts an electrical signal to a physical displacement.

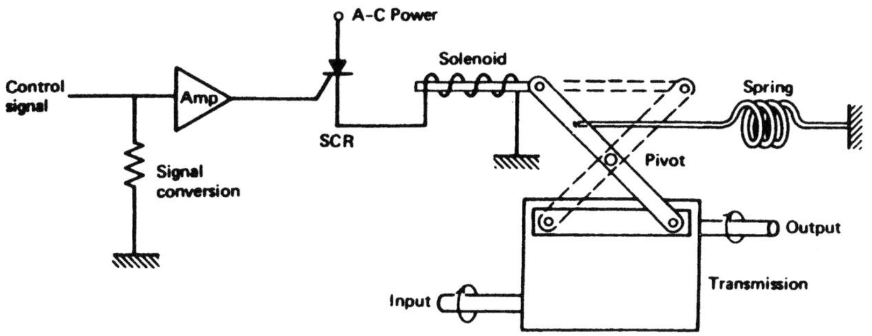

FIGURE 7.26
A solenoid used to change gears.

as actuators in process control. Probably the most common control situation is where motor speed drives some part of a process, and must be controlled to control some variable in the process—the drive of a conveyor system, for example. There are many types of electrical motors, each with its special set of characteristics. We will simply discuss the three most common varieties: the dc motor, ac motor, and stepping motor.

dc Motor The rotation of a dc motor is produced by the interaction of two constant magnetic fields. Figure 7.27a shows one type of dc motor that employs a permanent magnet (PM) to form one of the magnetic fields. The second magnetic field is formed by passing a current through a coil of wire contained within the PM field. This coil of wire, called the *armature,* is free to rotate. Notice that the coil is connected to the current source through slip rings and brushes (called a *commutator*) but the slip rings are split so that the current reverses direction as the armature rotates. To see how rotation occurs, look at Figure 7.27b. Notice that the north (N) and south (S) poles of the PM and the armature are not aligned. Thus, there will be a torque driving the N from the N and the S from the S. Therefore, the armature will rotate counterclockwise as shown. If it were not for the split slip rings the armature would rotate until N and N and of course S and S were aligned. But the

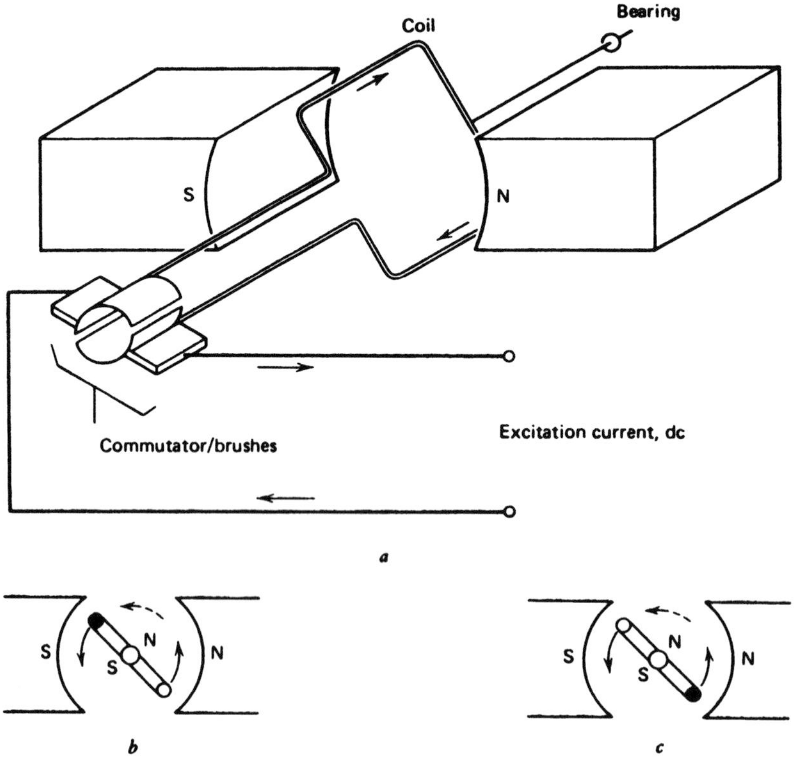

FIGURE 7.27
Permanent magnet dc motor.

split slip rings are arranged so that the current direction and therefore N-S orientation of the armature reverses as the alignment occurs. This is shown in Figure 7.27c, where again there is a torque on the armature. So rotation continues. The speed will depend on the current. Actually, the armature current is not determined by the coil resistance, because of a counter-emf produced by the rotating wire in a magnetic field. Thus, the effective voltage, which determines the current from the wire resistance and Ohm's law, is the difference between the applied voltage and the counter-emf produced by the rotation.

Many dc motors use an electromagnet instead of a PM to provide the static field. The coil used to produce this field is called the *field coil.* This kind of dc motor is called a *wound field* motor. The current for this field coil can be provided by placing the coil in series with the armature or in parallel (shunt). In some cases, the field is composed of two windings, one of each type. This is a *compound dc motor.* The schematic symbols of each type of motor are shown in Figure 7.28. Characteristics of dc motors with a field coil are as follows.

1. *Series field* This motor has large starting torque but is difficult to speed control. Good in applications for starting heavy, nonmobile loads and where speed control is not important, such as for quick-opening valves.

FIGURE 7.28
Three dc motor configurations.

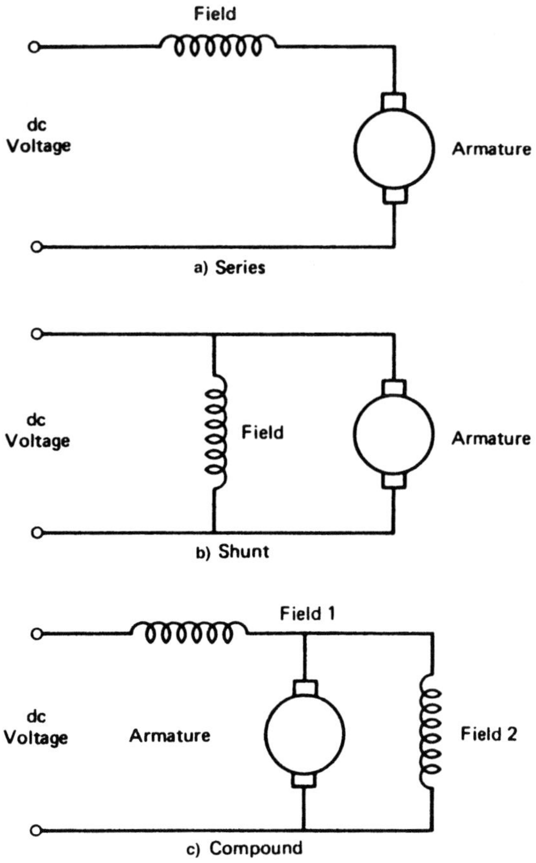

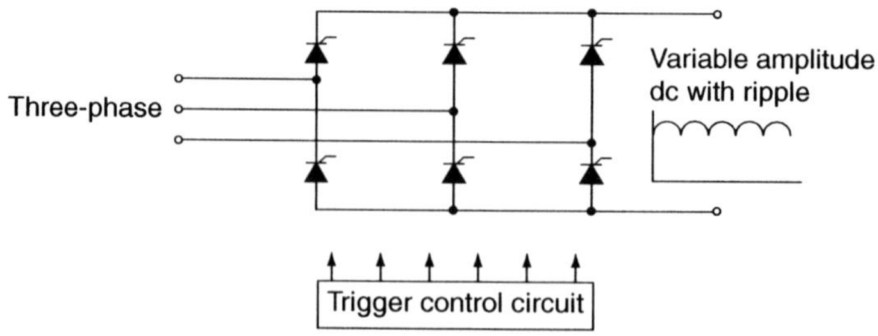

FIGURE 7.29
A three-phase rectifier for high power dc motors implemented with SCRs.

2. *Shunt field* This motor has a smaller starting torque, but good speed-control characteristics produced by varying armature excitation current. Good in applications where speed is to be controlled, such as in conveyor systems.
3. *Compound field* This motor attempts to obtain the best features of both of the two previous types. Generally, starting torque and speed-control capability fall predictably between the two pure cases.

The use of dc motors in control systems ranges from very low energy, delicate control applications, to heavy-duty control operations in elevators and vehicles. In general, PM types are used for motors of less than 10 hp (~7.5 kW) and wound field types for units up to about 100 to 200 hp (~75 to 150 kW). Control of the speed and torque of these large machines requires very high power dc electricity. Such power is derived from the power electronics devices described earlier in this chapter. In general three-phase ac power is rectified using switching technology to produce the required high-voltage, high-current dc electricity. Control is often made possible by variation of the voltage amplitude. Figure 7.29 shows a diagram of how such a three-phase rectifier is implemented using SCRs. By variation of the SCRs' firing angle the amplitude of the dc voltage and current of the armature, and hence speed of the motor, can be varied. Excitation of the field coils can be provided by a fixed amplitude rectified ac or another variable supply to provide further control.

ac Motors The basic operating principle of ac motors still involves the interaction between two magnetic fields. In this case, however, both fields are varying in time in consonance with the ac excitation voltage. Therefore, the force between the fields is a function of the angle of the rotor but also the phase of the current passing through the coils. There are two basic types of ac motors, synchronous and induction. The primary motor for application to the control industry is the induction motor.

In a synchronous motor the ac voltage is applied to the field coils, called the *stator* in an ac motor. This means the magnetic field is changing in time in phase with the impressed ac voltage. The armature, called the *rotor* for ac motors, is either a permanent magnet or a dc electromagnet, and possesses a fixed magnetic field. In the synchronous motor the rotor essentially follows the ac magnetic field of the stator. Figure 7.30 shows a simple synchronous motor with two poles and a permanent magnet rotor. A four-pole motor would have a rotor with two N-S bars for the rotor and four poles to the stator. The speed of rotation, n_s, of a synchronous motor is related to the frequency of the ac excitation and the number of poles by,

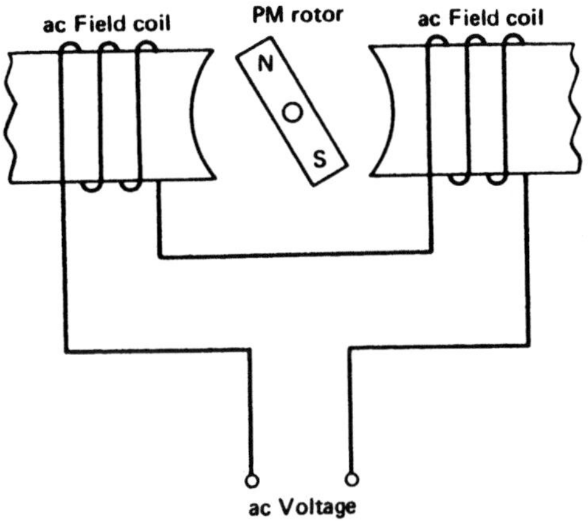

FIGURE 7.30
Simple ac motor with a permanent magnet rotor.

$$n_s = \frac{120f}{p} \quad \text{(in rpm)} \tag{7.6}$$

where f is the excitation frequency in Hertz and p is the number of poles.

Synchronous motors can be operated using single-phase ac but such units are used for only very low power (~0.1 hp) and suffer from very low starting torque. When operated from three-phase, ac synchronous motors can be operated at very high power, up to 50,000 hp.

Induction ac motors are characterized by a rotor which is neither a PM nor a dc excited electromagnet. Instead current induced in a coil wound on the rotor generates the interacting magnetic field of the rotor. This current is induced from the stator coils. Figure 7.31 illustrates the basic concept. You can see that the ac field of the stator produces a changing magnetic field passing through the closed loop of the rotor. Faraday's law teaches us that this changing flux will induce current in the loop. This in turn creates a magnetic field in the rotor coil, which interacts with the field of the stator. The bottom line is that a torque exists on the rotor caused by these two fields.

Single-phase induction motors are used for applications of relatively low power, say less than 5 hp (< 3.7 kW). Such motors are typical of those found in household appliances, for example. For higher power we use three-phase ac excitation. Such motors are available up to 10,000 hp.

In a control environment we want to have control over the speed of these motors. The induction motor speed is dependent upon the ac excitation frequency, much like Equation (7.6) for the synchronous motor, but with some small difference referred to as the slip (in

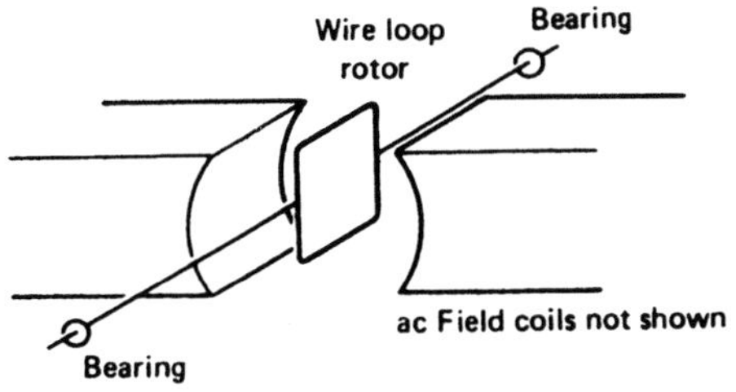

FIGURE 7.31
The induction motor depends on a rotor field from current induced by ac field coils, as shown in Figure 7.30.

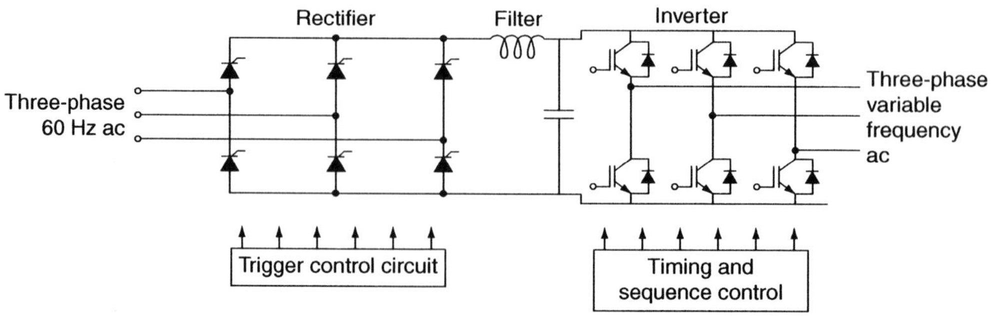

FIGURE 7.32
ac motor control with a variable amplitude, variable frequency circuit using power electronics.

%). Thus, speed control of both types can be affected by a variable frequency ac excitation. In order to provide this we turn to power electronics. Figure 7.32 provides a typical scheme where the three-phase ac (at 60 Hz) is rectified to produce a variable amplitude dc voltage by using SCRs with variable firing angle from the control circuit. This voltage is filtered and then applied to a IGBT-based *inverter* whose purpose is to produce a variable frequency ac voltage. The diodes across each IGBT provide for current flow in reverse of the IGBT. Of course the sequence and speed of switching provided by the control circuit determine the frequency. The circuit could equally well use GTOs, MOSFETs, or even BJT devices.

Stepping Motor The stepping motor has increased in importance in recent years because of the ease with which it can be interfaced with digital circuits. A *stepping motor* is a rotating machine that actually completes a full rotation by sequencing through a series of discrete rotational steps. Each step position is an equilibrium position in that, without further excitation, the rotor position will stay at the latest step. Thus, continuous rotation is achieved by the input of a train of pulses, each of which causes an advance of one step. It is not really continuous rotation, but discrete, stepwise rotation. The rotational rate is determined by the number of steps per revolution and the rate at which the pulses are applied. A driver circuit is necessary to convert the pulse train into proper driving signals for the motor.

EXAMPLE 7.6 A stepper motor has 10° per step and must rotate at 250 rpm. What input pulse rate, in pulses per second, is required?

Solution

A full revolution has 360°, so with 10° per step it will take 36 steps to complete one revolution. Thus,

$$\left(250\,\frac{\text{rev}}{\text{min}}\right)\left(36\,\frac{\text{pulses}}{\text{rev}}\right) = 9000\ \text{pulses/min}$$

Therefore,

$$(9000\ \text{pulses/min})(1\ \text{min}/60\ \text{s}) = \mathbf{150\ pulses/s}$$

The operation of a stepping motor can be understood from the simple model shown in Figure 7.33, which has 90° per step. In this motor, the rotor is a PM that is driven by a particular set of electromagnets. In the position shown, the system is in equilibrium and no motion occurs. The switches are typically solid-state devices, such as transistors, SCRs, or TRIACs. The switch sequencer will direct the switches through a sequence of positions as the pulses are received. The next pulse in Figure 7.33 will change S2 from C to D, resulting in the poles of that electromagnet reversing fields. Now, because the pole north/south orientation is different, the rotor is repelled and attracted so that it moves to the new position of equilibrium shown in Figure 7.34b. With the next pulse, S1 is changed to B, causing the same kind of pole reversal and rotation of the PM to a new position, as shown in Figure 7.34c. Finally, the next pulse causes S2 to switch to C again, and the PM rotor again steps to a new equilibrium position, as in Figure 7.34d. The next pulse will send the system back to the original state and the rotor to the original position. This sequence is then repeated as the pulse train comes in, resulting in a stepwise continuous rotation of the rotor PM. Although this example illustrates the principle of operation, the most common stepper motor does not use a PM,

FIGURE 7.33
An elementary stepping motor.

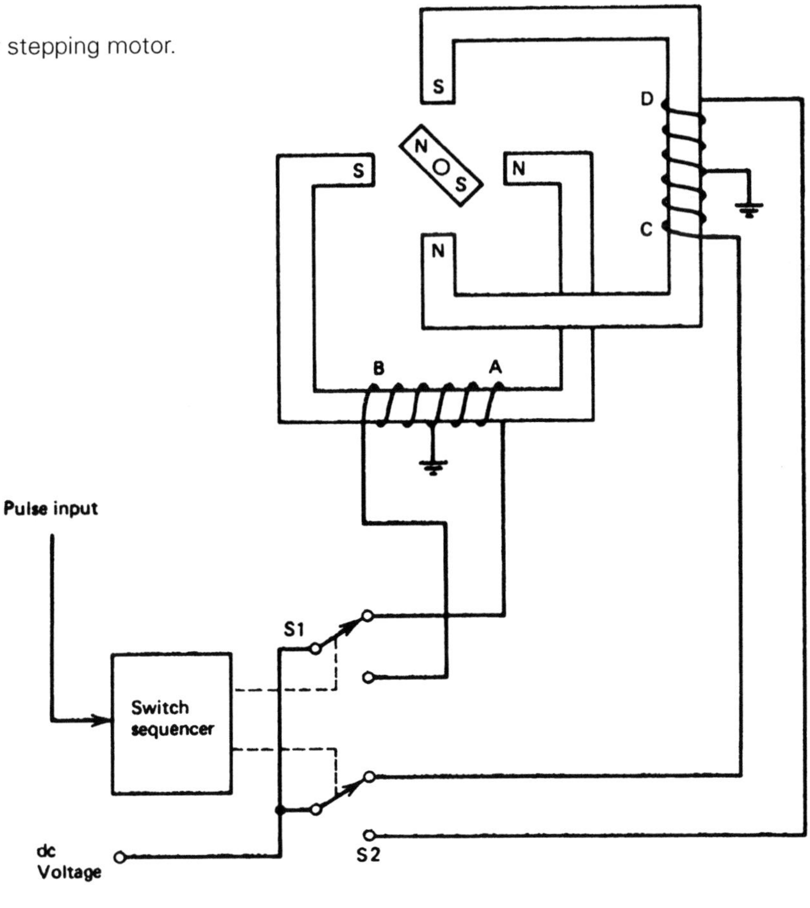

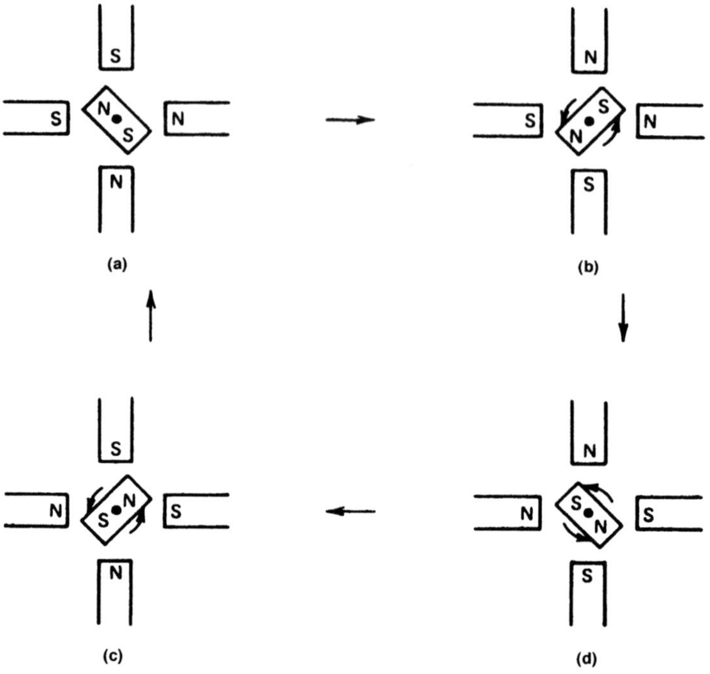

FIGURE 7.34
The four positions of the elementary stepper rotor.

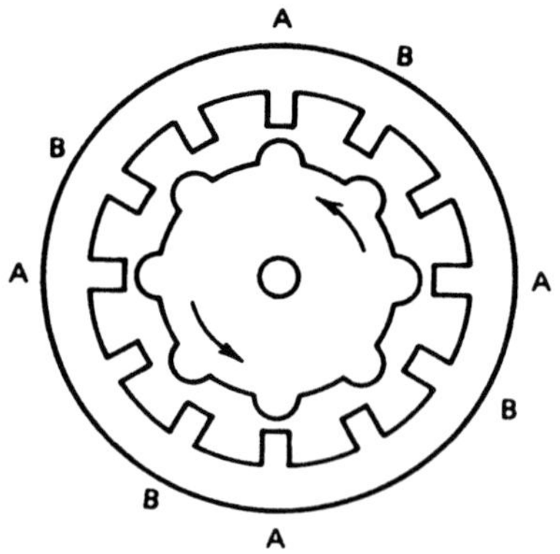

FIGURE 7.35
Stepper with 8 rotor teeth and 12 stator poles. Note that the rotor teeth line up with the A poles. With the next step, the rotor teeth will line up with the B poles.

but rather a rotor of magnetic material (not a magnet) with a certain number of teeth. This rotor is driven by a phased arrangement of coils with a different number of poles so that the rotor can never be in perfect alignment with the stator. Figure 7.35 illustrates this for a rotor with 8 "teeth" and a stator with 12 "poles." One set of four teeth is aligned, but the other four are not. If excitation is placed on the next set of poles (B) and taken off the first set (A), then the rotor will step once to come into alignment with the

B set of poles. The direction of rotation of stepper motors can be changed just by changing the order in which different poles are activated and deactivated.

7.5.2 Pneumatic Actuators

The actuator often translates a control signal into a large force or torque as required to manipulate some control element. The *pneumatic actuator* is most useful for such translation. The principle is based on the concept of pressure as force per unit area. If we imagine that a net pressure difference is applied across a diaphragm of surface area *A*, then a net force acts on the diaphragm given by

$$F = A(p_1 - p_2) \tag{7.7}$$

where $p_1 - p_2$ = pressure difference (Pa)
A = diaphragm area (m^2)
F = force (N)

If we need to double the available force for a given pressure, it is merely necessary to double the diaphragm area. Very large forces can be developed by standard signal-pressure ranges of 3 to 15 psi (20 to 100 kPa). Many types of pneumatic actuators are available, but perhaps the most common are those associated with *control valves.* We will consider these in some detail to convey the general principles.

The action of a direct pneumatic actuator is shown in Figure 7.36. Figure 7.36 on the left shows the condition in the low signal-pressure state, where the spring, *S,* maintains the diaphragm and the connected control shaft in a position as shown. The pressure on the opposite (spring) side of the diaphragm is maintained at atmospheric pressure by the open hole, *H.* Increasing the control pressure (gauge pressure) applies a force on the diaphragm, forcing the diaphragm and connected shaft down against the spring force. Figure 7.36 on the right shows this in the case of maximum control pressure and maximum travel of the shaft. The pressure and force are linearly related, as shown in Equation (7.7), and the compression of a spring is linearly related to forces, as

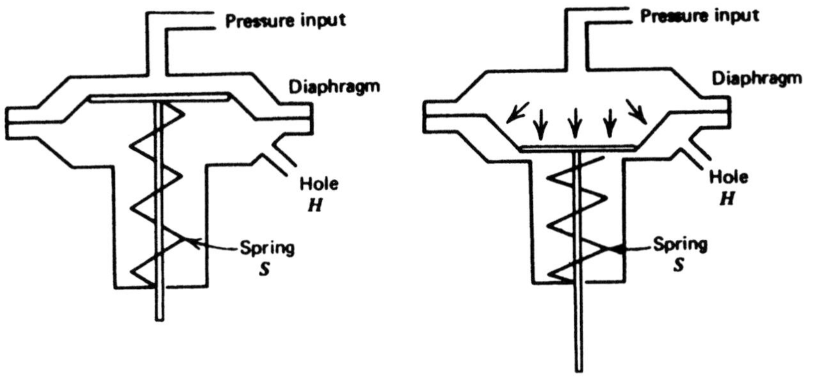

FIGURE 7.36
A direct pneumatic actuator for converting pressure signals into mechanical shaft motion.

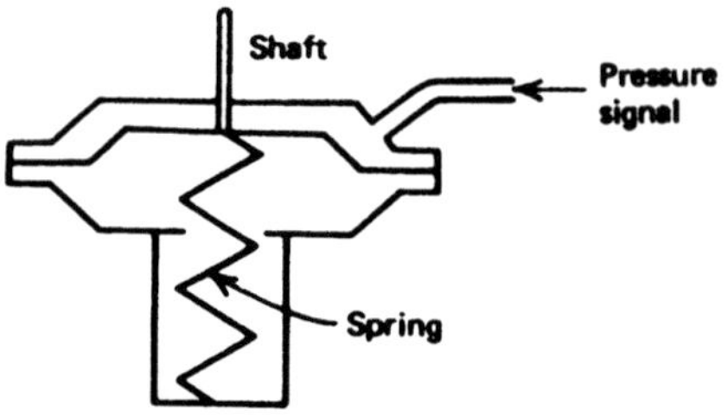

FIGURE 7.37
A reverse-acting pneumatic actuator.

discussed in Chapter 5. Then we see that the shaft position is linearly related to the applied control pressure

$$\Delta x = \frac{A}{k}\Delta p \tag{7.8}$$

where
Δx = shaft travel (m)
Δp = applied gauge pressure (Pa)
A = diaphragm area (m^2)
k = spring constant (N/m)

A *reverse* actuator, shown in Figure 7.37, moves the shaft in the opposite sense from the direct actuator but obeys the same operating principle. Thus, the shaft is pulled in by the application of a control pressure.

EXAMPLE 7.7

Suppose a force of 400 N must be applied to open a valve. Find the diaphragm area if a control gauge pressure of 70 kPa (~ 10 psi) must provide this force.

Solution

We must calculate the area from

$$F = A(p_1 - p_2)$$

where our applied pressure is $p_1 - p_2$ because a gauge pressure is specified. Then

$$A = \frac{F}{p} = \frac{400 \text{ N}}{7 \times 10^4 \text{ Pa}}$$

$$A = 5.71 \times 10^{-3} \text{ m}^2$$

or about 8.5 cm in diameter.

The inherent compressibility of gases causes an upper limit to the usefulness of gas for propagating force. Consider the pneumatic actuator of Figure 7.36. When we want motion to occur under low shaft load, we simply increase pressure in the actuator via a regulator. The regulator allows more gas from a high-pressure reservoir to enter the actuator. This increases the diaphragm force until it is able to move the shaft.

Suppose the shaft is connected to a very high load—that is, something requiring a very large force for movement. In principle, it is simply a matter of increasing the pressure of the input gas until the pressure × diaphragm area equals the required force. What we find, however, is that as we try to raise the input gas pressure, large volumes of gas must be passed into the actuator to bring about any pressure rise because the gas is compressing; that is, its density is increasing.

7.5.3 Hydraulic Actuators

We have seen that there is an upper limit to the forces that can be applied using gas as the working fluid. Yet there are many cases when large forces are required. In such cases, a hydraulic actuator may be employed. The basic principle is shown in Figure 7.38. The basic idea is the same as for pneumatic actuators, except that an incompressible fluid is used to provide the pressure, which can be made very large by adjusting the area of the forcing piston, A_1. The hydraulic pressure is given by

$$p_H = F_1/A_1 \tag{7.9}$$

where
p_H = hydraulic pressure (Pa)
F_1 = applied piston force (N)
A_1 = forcing piston area (m^2)

This pressure is transferred equally throughout the liquid, so the resulting force on the working piston is

$$F_w = p_H A_2 \tag{7.10}$$

where
F_w = force of working piston (N)
A_2 = working piston area (m^2)

Thus, the working force is given in terms of the applied force by

$$F_w = \frac{A_2}{A_1} F_1 \tag{7.11}$$

EXAMPLE 7.8

a. Find the working force resulting from 200 N applied to a 1-cm-radius forcing piston if the working piston has a radius of 6 cm.
b. Find the hydraulic pressure.

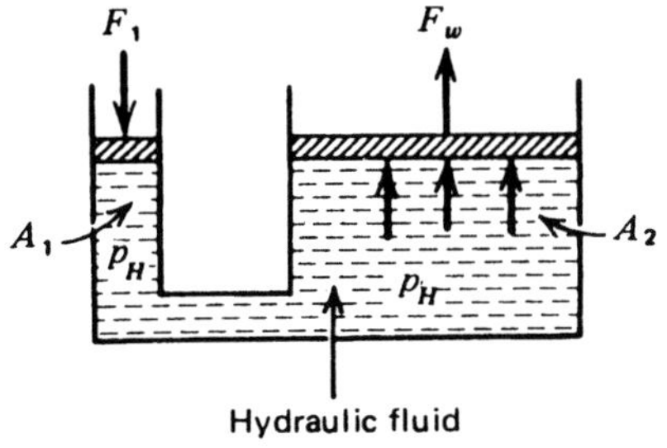

FIGURE 7.38
A hydraulic actuator converts a small force, F_1, into an amplified force, F_w.

Solution

a. We can find the working force from

$$F_w = \frac{A_2}{A_1} F_1$$

or

$$F_w = \left(\frac{R_2}{R_1}\right)^2 F_1 \quad F = \left(\frac{6\text{ cm}}{1\text{ cm}}\right)^2 (200\text{ N})$$

$$F_w = \mathbf{7200\ N}$$

b. Thus, the 200-N force provides 7200 N of force. The hydraulic pressure is

$$p_H = \frac{F_w}{A_2} = \frac{7200\text{ N}}{(\pi)(6 \times 10^{-2}\text{ m})^2}$$

$$p_H = \mathbf{6.4 \times 10^5\ Pa}$$

This pressure is approximately 93 lb/in.^2

Hydraulic Servos In some cases, it is desired to control the position of large loads as part of the control system. This often can be done by using the low-energy controller output as the setpoint input to a hydraulic control system. This concept is illustrated in Figure 7.39.

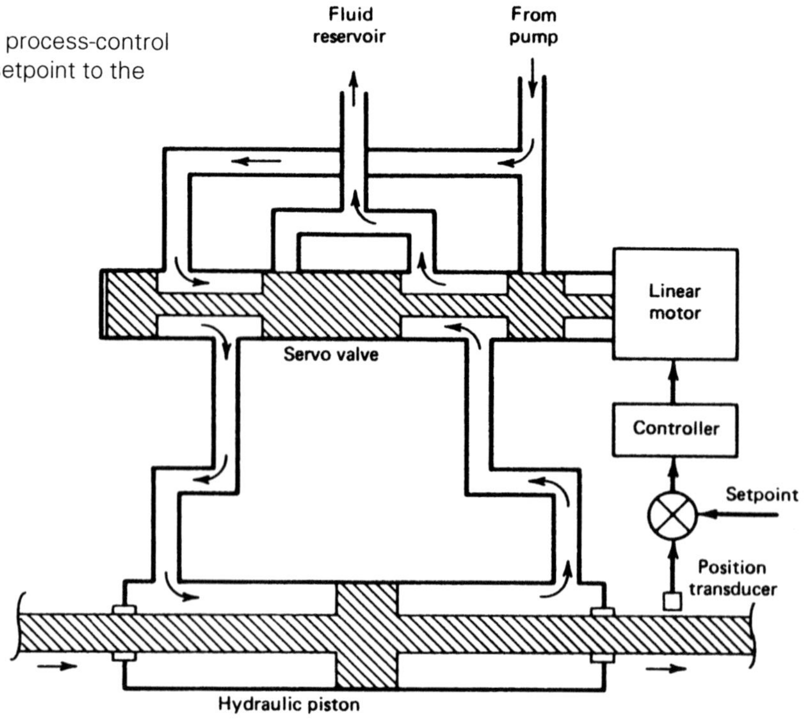

FIGURE 7.39
A hydraulic servo system. The process-control system provides input of the setpoint to the hydraulic servo.

In this system, high-pressure hydraulic fluid can be directed to either side of a force piston, which causes motion in either direction. The direction is determined by the position of a control valve piston in the *hydraulic servo valve.* The position of this valve piston is controlled by a linear motor driven by the output of an amplifier and error detector. The inputs to the error detector are the process-controller output, which forms the setpoint of the hydraulic servo, and a feedback from the force piston shaft. Thus, the amplifier will drive the hydraulic servo until the feedback matches the setpoint input.

7.6 CONTROL ELEMENTS

The actual control element (which is a part of the process itself) can be many different devices. It is not the intention of this book to present many of these devices, but a general survey of standard devices is valuable for a complete picture of process control. Several examples of control elements are described later in terms of different control problems.

7.6.1 Mechanical

Control elements that perform some mechanical operation in a process (by virtue of operations) are called *mechanical control elements.* Examples of these types follow.

Solid-Material Hopper Valves Consider the grain supply bin of Figure 7.40. The control system is to maintain the flow of grain from the storage bin to provide a constant flow

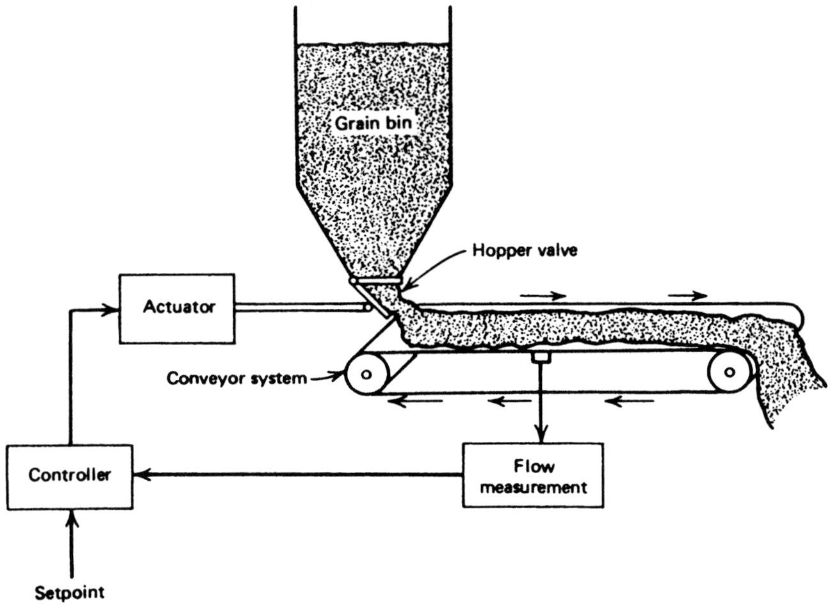

FIGURE 7.40
An example of a mechanical control element in the form of a hopper valve and conveyor.

rate on the conveyor. This flow depends on the height of grain in the bin, and hence the *hopper valve* must open or close to compensate for the variation. In this case, an actuator operates a vane-type valve to control the grain flow rate. The actuator could be a motor to adjust shaft position, a hydraulic cylinder, or some other mechanism.

Paper Thickness In Figure 7.41, the essential features of a system for controlling paper thickness are shown. The paper is in a wet fiber suspension and is passed between rollers. By varying the roller separation, paper thickness is regulated. The mechanical control element shown is the movable roller. The actuator, which could be electrical, pneumatic, or hydraulic, adjusts roller separation based on a thickness measurement.

7.6.2 Electrical

There are numerous cases where a direct electrical effect is impressed in some process-control situation. The following examples illustrate some typical cases of electrical control elements.

Motor-Speed Control The speed of large electrical motors depends on many factors, including supply voltage level, load, and others. A process-control loop regulates this speed through direct change of operating voltage or current, as shown in Figure 7.42 for a dc motor. Voltage measurements of engine speed from a tachometer are used in a process-control loop to determine the power applied to the motor brushes. In some cases, motor speed control is an intermediate operation in a process-control application. Thus, in the operation of a kiln for solid chemical reaction, the rotation (feed) rate may be varied by motor speed control based on, for example, reaction temperature, as shown in Figure 7.43.

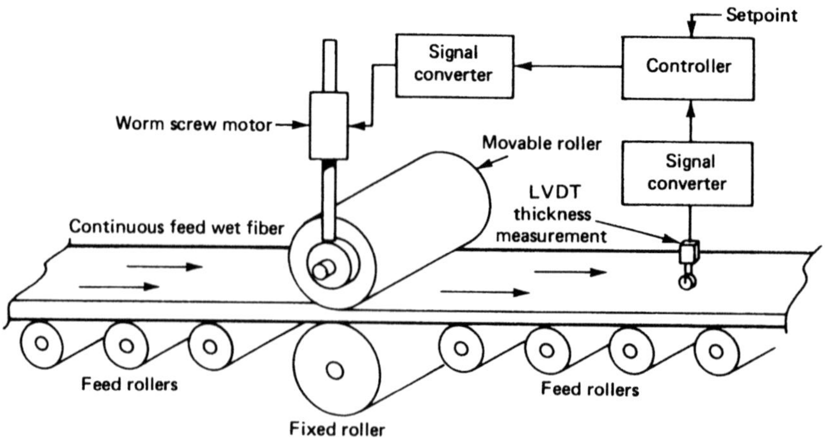

FIGURE 7.41
A continuous-operation paper-thickness-controlling system using mechanical final control elements.

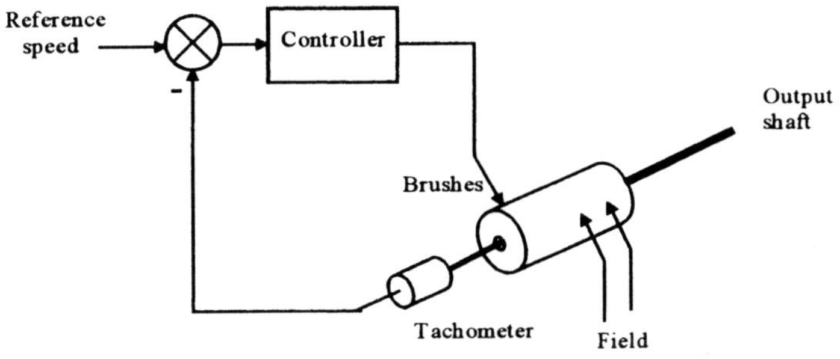

FIGURE 7.42
Basic control system for motor speed using a tachometer to sense the motor speed.

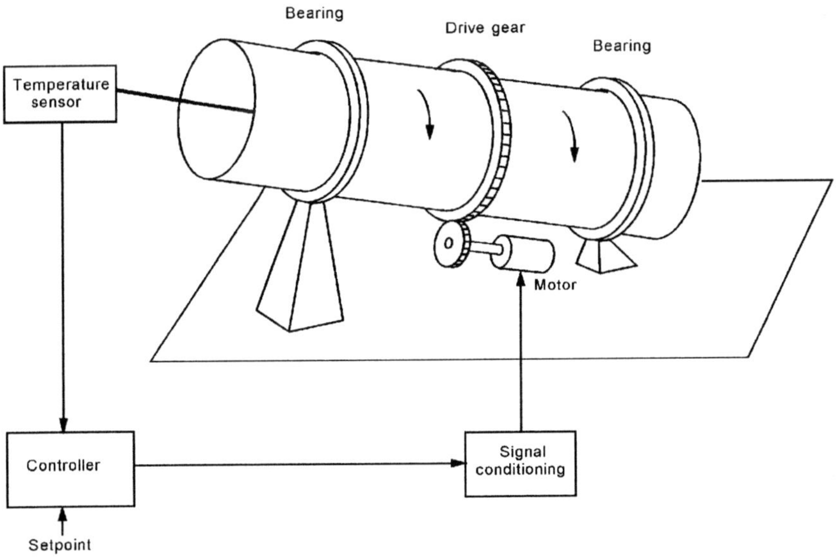

FIGURE 7.43
A control system varies the rotation rate of a reaction kiln based upon temperature.

Temperature Control Temperature often is controlled by using electrical heaters in some application of industrial control. Thus, if heat can be supplied through heaters electrically in an endothermic reaction, then the process-control signal can be used to ON/OFF cycle a heater or set the heater within a continuous span of operating voltages, as in Figure 7.44. In this example, a reaction vessel is maintained at some constant temperature using an electrical heater. The process-control loop provides this by smoothly varying excitation to the heater.

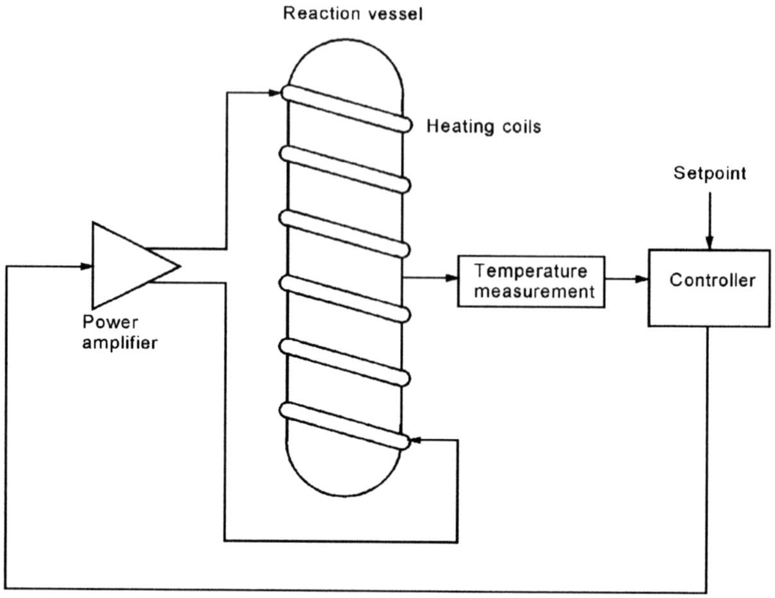

FIGURE 7.44
Control of heat into a reaction vessel through an electrical power amplifier.

7.6.3 Fluid Valves

The chemical and petroleum industries have many applications that require control of fluid processes. Many other industries also depend in part on operations that involve fluids and the regulation of fluid parameters. The word *fluid* here represents either gases, liquids, or vapors. Many principles of control can be equally applied to any of these states of matter, with only slight corrections. Many fluid operations require regulation of such quantities as density and composition, but by far the most important control parameter is flow rate. A regulation of flow rate emerges as the regulatory parameter for reaction rate, temperature, composition, or a host of other fluid properties. We will consider in some detail that process-control element specifically associated with flow—the control valve.

Control-Valve Principles Flow rate in process control is usually expressed as volume per unit time. If a mass flow rate is desired, it can be calculated from the particular fluid density. If a given fluid is delivered through a pipe, then the volume flow rate is

$$Q = Av \tag{7.12}$$

where Q = flow rate (m^3/s)
A = pipe area (m^2)
v = flow velocity (m/s)

EXAMPLE 7.9 Alcohol is pumped through a pipe of 10-cm diameter at 2 m/s flow velocity. Find the volume flow rate.

Solution

A pipe of 10-cm diameter has a cross-sectional area of

$$A = \frac{\pi D^2}{4} = \frac{(\pi)(10^{-1}\text{ m})^2}{4}$$

$$A = 7.85 \times 10^{-3}\text{ m}^2$$

Thus, the flow rate is

$$Q = Av = (7.85 \times 10^{-3}\text{ m}^2)(2\text{ m/s})$$

$$Q = \mathbf{0.0157\ m^3/s}$$

The purpose of the control valve is to regulate the flow rate of fluids through pipes in the system. This is accomplished by placing a variable-size restriction in the flow path, as shown in Figure 7.45. You can see that as the stem and plug move up and down, the size of the opening between the plug and the seat changes, thus changing the flow rate. Note the direction of flow with respect to the seat and plug. If the flow were reversed, force from the flow would tend to close the valve further at small openings.

There will be a drop in pressure across such a restriction, and the flow rate varies with the square root of this pressure drop, with an appropriate constant of proportionality, shown by

$$Q = K\sqrt{\Delta p} \tag{7.13}$$

where K = proportionality constant $(\text{m}^3/\text{s}/\text{Pa}^{1/2})$
$\Delta p = p_2 - p_1$ = pressure difference (Pa)

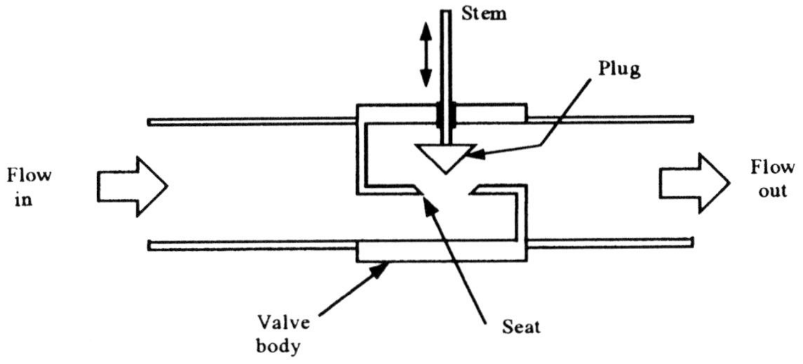

FIGURE 7.45
A basic control-valve cross-section. The direction of flow is important for proper valve action.

EXAMPLE 7.10 A pressure difference of 1.1 psi occurs across a constriction in a 5-cm-diameter pipe. The constriction constant is 0.009 m^3/s per $kPa^{1/2}$. Find **(a)** the flow rate in m^3/s and **(b)** the flow velocity in m/s.

Solution

First we note that 1.1 psi gives

$$\Delta p = (1.1 \text{ psi})(6.895 \text{ kPa/psi})$$
$$\Delta p = 7.5845 \text{ kPa}$$

a. The flow rate is

$$Q = K\sqrt{\Delta p} = (0.009)(7.5845)^{1/2}$$
$$Q = \mathbf{0.025\ m^3/s}$$

b. The flow velocity is found from

$$Q = Av$$
$$v = \frac{Q}{A} = 4\left[\frac{0.025 \text{ m}^3/\text{s}}{\pi(5 \times 10^{-2})^2}\right]$$
$$v = \mathbf{12.7\ m/s}$$

The constant, *K*, depends on the size of the valve, the geometrical structure of the delivery system, and, to some extent, on the material flowing through the valve. Now the actual pressure of the entire fluid delivery (and sink) system in which the valve is used (and, hence, the flow rate) is not a predictable function of the valve opening only. But because the valve opening does change flow rate, it provides a mechanism of flow control.

Control-Valve Types The different types of control valves are classified by a relationship between the valve stem position and the flow rate through the valve. This *control-valve characteristic* is assigned with the assumptions that the stem position indicates the extent of the valve opening and that the pressure difference is determined by the valve alone. Correction factors allow one to account for pressure differences introduced by the whole system. Figure 7.46 shows a typical control valve using a pneumatic actuator attached to drive the stem and hence open and close the valve. There are three basic types of control valves, whose relationship between stem position (as a percentage of full range) and flow rate (as a percentage of maximum) is shown in Figure 7.47.

The types are determined by the shape of the plug and seat, as shown in Figure 7.45. As the stem and plug move with respect to the seat, the shape of the plug determines the amount of actual opening of the valve.

1. *Quick Opening* This type of valve is used predominantly for full ON/full OFF control applications. The valve characteristic of Figure 7.47 shows that a relatively small motion of the valve stem results in maximum possible flow rate through the valve. Such a valve, for example, may allow 90% of maximum flow rate with only a 30% travel of the stem.

FIGURE 7.46
A pneumatic actuator connected to a control valve. The actuator is driven by a current through an I/P converter.

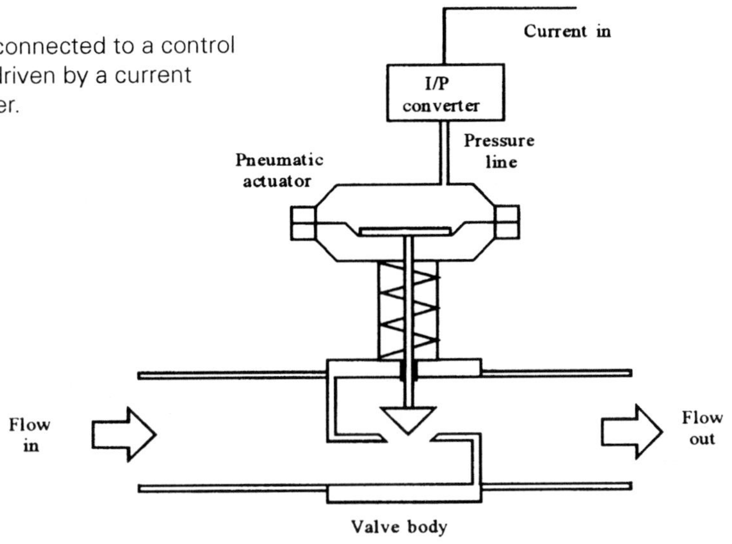

FIGURE 7.47
Three types of control valves open differently as a function of valve stem position.

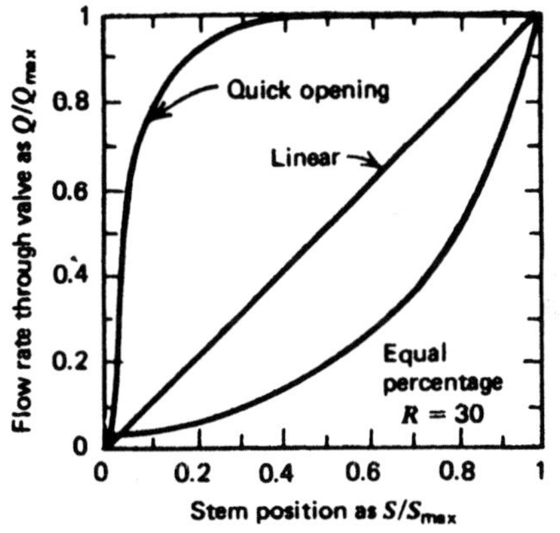

2. *Linear* This type of valve, as shown in Figure 7.47, has a flow rate that varies linearly with the stem position. It represents the ideal situation where the valve alone determines the pressure drop. The relationship is expressed as

$$\frac{Q}{Q_{max}} = \frac{S}{S_{max}} \tag{7.14}$$

where Q = flow rate (m^3/s)
Q_{max} = maximum flow rate (m^3/s)
S = stem position (m)
S_{max} = maximum stem position (m)

3. *Equal Percentage* A very important type of valve employed in flow control has a characteristic such that a given percentage change in stem position produces an equivalent change in flow—that is, an equal percentage. Generally, this type of valve does not shut off the flow completely in its limit of stem travel. Thus, Q_{min} represents the minimum flow when the stem is at one limit of its travel. At the other extreme, the valve allows a flow Q_{max} as its maximum, open-valve flow rate. For this type, we define *rangeability, R,* as the ratio

$$R = \frac{Q_{\text{max}}}{Q_{min}} \tag{7.15}$$

The curve in Figure 7.47 shows a typical equal percentage curve that depends on the rangeability for its exact form. The curve shows that increase in flow rate for a given change in valve opening depends on the extent to which the valve is already open. This curve is typically exponential in form and is represented by

$$Q = Q_{\text{min}} R^{S/S_{\text{max}}} \tag{7.16}$$

where all terms have been defined previously.

EXAMPLE 7.11 An equal percentage valve has a maximum flow of 50 cm^3/s and a minimum of 2 cm^3/s. If the full travel is 3 cm, find the flow at a 1-cm opening.

Solution

The rangeability is

$$R = Q_{\text{max}}/Q_{\text{min}}$$
$$R = (50\ \text{cm}^3/\text{s})/(2\ \text{cm}^3/\text{s}) = 25$$

Then the flow at a 1-cm opening is

$$Q = Q_{\text{min}} R^{S/S_{\text{max}}}$$
$$Q = (2\ \text{cm}^3/\text{s})(25)^{1\ \text{cm}/3\ \text{cm}}$$
$$Q = \mathbf{5.85\ cm^3/s}$$

Control-Valve Sizing Another important factor associated with all control valves involves corrections to Equation (7.13) because of the nonideal characteristics of the materials that flow. A standard nomenclature is used to account for these corrections, depending on the liquid, gas, or steam nature of the fluid. These correction factors allow selection of the proper size of valve to accommodate the rate of flow that the system must support. The correction factor most commonly used at present is measured as the number of U.S. gallons of water per minute that flow through a fully open valve with a pressure differential of 1 lb per square inch. The correction factor is called the *valve flow coefficient* and is designated as C_v. Using this factor, a liquid flow rate in U.S. gallons per minute is

TABLE 7.1
Control-valve flow coefficients

Valve Size (inches)	C_v
$\frac{1}{4}$	0.3
$\frac{1}{2}$	3
1	14
$1\frac{1}{2}$	35
2	55
3	108
4	174
6	400
8	725

$$Q = C_v\sqrt{\frac{\Delta p}{S_G}} \tag{7.17}$$

where Δp = pressure across the valve (psi)
S_G = specific gravity of liquid

Typical values of C_v for different-size valves are shown in Table 7.1. Similar equations are used for gases and vapors to determine the proper valve size in specific applications.

EXAMPLE 7.12 Find (**a**) the proper C_v for a valve that must allow 150 gal of ethyl alcohol per minute with a specific gravity of 0.8 at maximum pressure of 50 psi, and (**b**) the required valve size.

Solution

a. We find the correct sizing factor from

$$Q = C_v\sqrt{\frac{\Delta p}{S_G}}$$

Then

$$C_v = Q\sqrt{\frac{S_G}{\Delta p}}$$

$$C_v = \left(150\,\frac{\text{gal}}{\text{min}}\right)\sqrt{\frac{0.8}{50\ \text{lb/in.}^2}}$$

$$C_v = \mathbf{18.97}$$

b. A $1\frac{1}{2}$-in.-diameter valve (3.8 cm) is selected from Table 7.1.

Fluid-Control Example The chemical and process-control industry uses fluid-control systems extensively. Examples of such applications are many and varied. Consider, for example, control of distillation column composition by regulation of a fixed-point column temperature. Such regulation is achieved by controlling the feed rate as shown in Figure 7.48. A thermocouple measures temperature that is transmitted to the controller as a

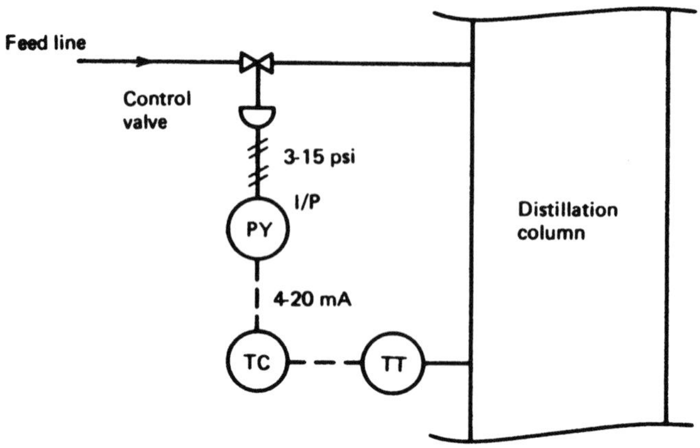

FIGURE 7.48
Feed control to a distillation column based on temperature.

4- to 20-mA control signal. The controller outputs a 4- to 20-mA signal proportional to proper control-valve position. This is converted to a 3- to 15-psi (20- to 100 kPa) pneumatic signal by an I/P converter that, in turn, operates a pneumatic actuator connected to the control valve. The valve size is determined by the characteristics of the gas or vapor that is flowing. The size of the required actuator is determined from the valve size.

SUMMARY

The operation of the final control element has three separate functions; the ultimate goal is to translate a low-level control signal into a large-scale process. The following specific details were considered:

1. The final control function can be implemented by signal conditioning, an actuator, and a final control element.
2. Signal conditioning involves changing a control signal into that form and power necessary to energize the actuator. Simple electronic amplification, digital-to-analog conversion, electrical-to-pneumatic conversion, and pneumatic-to-hydraulic conversion are all typical signal-conditioning operations.
3. The current-to-pressure converter is frequently employed in process-control systems. This device is based on a flapper/nozzle (nozzle/baffle) system that converts linear displacement into a pressure change.
4. Special power electronics switches such as SCRs, TRIACs, and GTOs are used to control high power from low-energy control signals.
5. Power electronics devices such as BJTs, MOSFETs, and IGBTs allow control of high power motor speed and many other control needs.
6. Actuators are an intermediate step between the converted control signal and the final control element. Common electrical actuators are solenoids, digital stepping motors, and ac and dc motors.

7. Pneumatic and hydraulic actuators are often used in process control because they allow very large forces to be produced from modest pressure systems. A pneumatic actuator converts a pressure signal to a shaft extension according to

$$\Delta x = \frac{A}{k}\Delta p \tag{7.8}$$

where the force that causes this extension is given by

$$F = A(p_1 - p_2) \tag{7.7}$$

8. Actual final control elements are as varied as the applications of process control in industry. Examples include motor-driven conveyor belts, paper-thickness roller assemblies, and heating systems.
9. The most general type of final control element is a control valve. This device is designed for use in process-control applications involving liquid, gas, or vapor-flow rate control. Three types are commonly used: quick opening, linear, and equal percentage.

PROBLEMS

Section 7.3

7.1 A 4- to 20-mA control signal is loaded by a 100-Ω resistor and must produce a 20- to 40-V motor drive signal. Find an equation relating the input current to the output voltage.

7.2 Implement the equation of Problem 7.1 if a power amplifier is available that can output 0 to 100 V and has a gain of 10.

7.3 A motor to be driven by a digital signal has a speed variation of 200 rev/min per volt with a minimum rpm at 5 V and a maximum at 10 V. Find the minimum-speed word, maximum-speed word, and the speed change per LSB change. Use a 5-bit, 15-V reference DAC.

Section 7.4

7.4 Figure 7.49 shows an SCR used in a circuit to provide variable electrical power to a resistive load from a high-power square-wave generator. The SCR trigger voltage is 2.5 V.

a. What is the range of R to provide 5% to 95% on time of the SCR?

b. Plot the voltages across the capacitor, load, and SCR for a 50% on time.

c. What is the average power delivery to the load at 50%?

FIGURE 7.49
Figure for Problem 7.4.

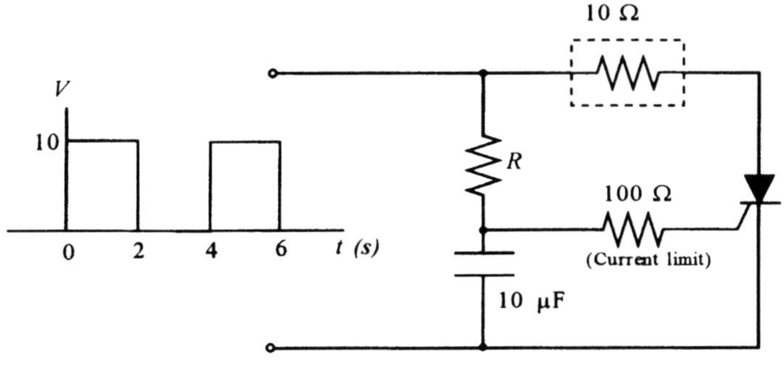

7.5 Modify Problem 7.4 by using a DIAC with a ±2-V breakover voltage, in place of the 100-Ω resistor, to trigger a TRIAC. The source is now a bipolar ±5-V square wave of the same period as Figure 7.49.

a. Find the range of R to provide a 5% to 95% on time.

b. Plot the voltages across the capacitor, load, and TRIAC.

c. What is the average power delivered to the load?

7.6 Determine and plot the average power delivered to a 100-Ω load in place of the light bulb in Example 7.3 as a function of resistor R. (*Hint:* When the SCR is on you can write an equation for the instantaneous power across the load. The average is found by integrating the instantaneous power over one period.)

7.7 Figure 7.50 shows a system to regulate dc motor speed by temperature. A thermistor is used to vary the capacitor charging rate, and thus when in the cycle the SCR is turned on. The thermistor resistance versus temperature is given in Figure 4.5.

a. Determine the values for the resistor and capacitor to provide 10% on time at 0°C and 90% on time at 60°C.

b. Determine the required power ratings of the zener diode and 2.5-kΩ resistor.

c. Construct a plot of the average motor voltage versus temperature.

7.8 In Example 7.4, the actual capacitor voltage with no clamp is given by the expression

$$V_c(t) = \frac{V_0}{1 + (\omega RC)^2}\left[\sin(\omega t) + \omega RCe^{-t/RC} - \omega RC\cos(\omega t)\right]$$

where $\omega = 2\pi f$ and, of course, $f = 60$ Hz. The DIAC will trigger when this voltage reaches ±28 V. Following the example, find the exact values of R to provide 10% and 90% on time. (*Hint:* It is necessary to find the solution using numerical or graphical methods and/or computer simulations.)

7.9 A solenoid plunger has a 5-cm stroke and a 6-N pull-in force. Devise a system using a pivot/lever to provide a stroke of 8 cm. What is the force for the 8-cm stroke? (*Hint:* Consider the torque about the pivot.)

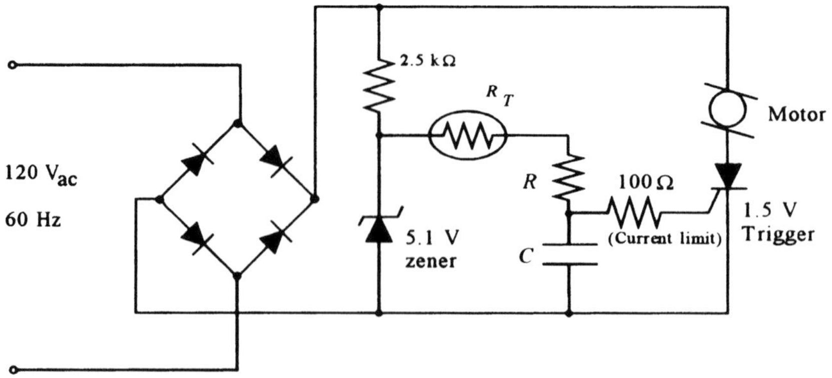

FIGURE 7.50
Figure for Problem 7.7.

Section 7.5

7.10 A stepping motor has 130 steps per revolution. Find the digital input rate that produces 10.5 rev/s.

7.11 A stepping motor has 7.5° per step. Find the rpm produced by a pulse rate of 2000 pps on the input.

7.12 What force is generated by 90 kPa acting on a 30-cm^2-area diaphragm?

7.13 A hydraulic system uses pistons of diameter 2 cm and 40 cm. What force on the small piston will raise a 500-kg mass?

7.14 What pneumatic pressure is required on the small piston of Problem 7.13 to produce the necessary force?

7.15 The SCR in Figure 7.26 requires a 4-V trigger. Design a system by which the gears are shifted when a CdS photocell resistance drops below 2.5 kΩ.

7.16 Design a system by which a control signal of 4 to 20 mA is converted into a force of 200 to 1000 N. Use a pneumatic actuator and specify the required diaphragm area if the pressure output is to be in the range of 20 to 100 kPa. An I/P converter is available that converts 0 to 5 V into 20 to 100 kPa.

7.17 A feed hopper requires 30 lb of force to open. Find the pneumatic actuator area to provide this force from a 9-psi input signal.

Section 7.6

7.18 Find the proper valve size in inches and centimeters for pumping a liquid flow rate of 600 gal/min with a maximum pressure difference of 55 psi. The liquid specific gravity is 1.3.

7.19 An equal percentage control valve has a rangeability of 32. If the maximum flow rate is 100 m^3/h, find the flow at 2/3 and 4/5 open settings.

7.20 The level of water in a tank is to be controlled at 20 m, and the output flow rate is nominally 65 m^3/h through a control valve, as shown in Figure 7.51. Under nominal conditions, determine the required valve size in inches and centimeters.

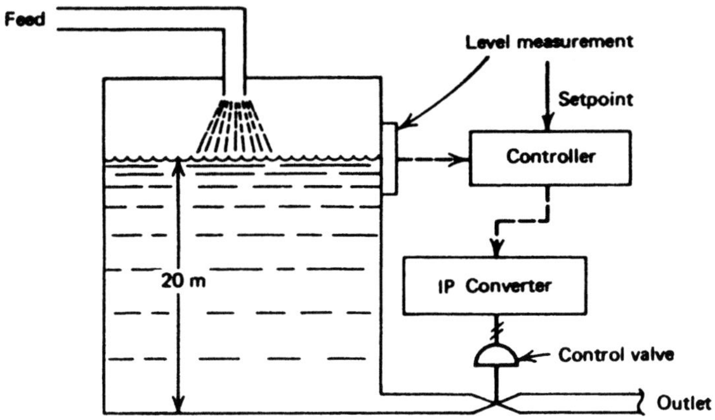

FIGURE 7.51
Figure for Problem 7.20.

7.21 If the valve actuator of Problem 7.20 has a rangeability of 30 and a maximum stem travel of 5 cm and is to be half-open under the nominal conditions, find the minimum flow, maximum flow, and stem opening for 100 m^3/h flow.

7.22 A quick-opening valve moves from closed to maximum open with five turns of a shaft. The shaft is driven through a 10:1 reducer from a stepping motor of 3.6° per step. If the maximum input pulse rate to the motor is 250 steps per second, find the fastest time for the valve to move from closed to open.

7.23 A control valve operates from a 3- to 15-psi control signal. To have a 40-gal/min flow rate, express the signal input in both psi and percent of range if **(a)** it is a linear valve from 0 to 90 gal/min, and **(b)** it is an equal percentage valve with $R = 6$ and $Q_{min} = 15$ gal/min.

SUPPLEMENTARY PROBLEMS

S7.1 A stepper motor must control conveyor speed in the range 100 to 200 rpm. The stepper control unit accepts a TTL square wave of frequency *f* and translates this into appropriately sequenced stepper pulses. The stepper has 2.5° per step. Design a system that accepts a 4- to 20-mA input to produce the 100- to 200-rpm speed. Use a V/F converter as presented in Chapter 3.

S7.2 Figure 7.52 shows a valve that is operated by a worm gear and motor. Each turn of the worm gear moves the stem by 0.18 cm, and the total range of stem travel is 4.0 cm. The worm gear motor has three states: A TTL high to the *up* terminal causes rotation at 15 rpm that moves the valve stem up, a TTL high to the *down* terminal causes rotation to reverse, moving the valve stem down (closing the valve), and with TTL low to both terminals the motor is off. The valve body is equal percentage with a rangeability of 20 and a maximum flow of 0.6 m^3/min.

a. How many rotations of the worm gear are required for the full stem travel?

b. How long does it take to move the stem over the full range of travel? What is the minimum flow rate?

c. What is the stem position for flow at 90% of the maximum?

FIGURE 7.52
Figure for Problems S7.2 and S7.3.

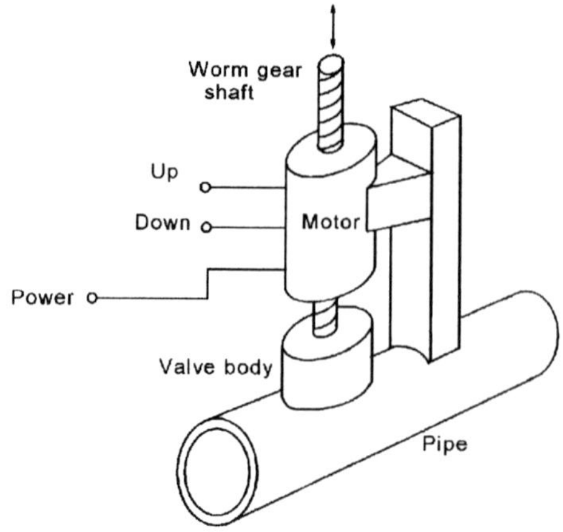

S7.3 Suppose the worm gear of Figure 7.52 has a potentiometer attached as shown in Figure 7.53, so that the stem position is converted to resistance. A controller outputs a 4- to 20-mA signal, where 4 mA should correspond to minimum flow (stem = 0 cm) and 20 mA should correspond to maximum flow (stem = 4 cm). Design a system so that comparators are used to activate the *up* or *down* terminals of the motor and drive the stem to the correct position required by the current. That is, suppose the controller outputs a current of 10 mA. This should cause the motor to move the stem to 1.5 cm. (*Hint:* Convert the current to a voltage that is provided to two comparators.) The reference terminals of the comparators have a voltage from a divider using the potentiometer connected to the worm gear. Be sure to include some hysteresis on the comparators.

FIGURE 7.53
Figure for Problem S7.3.

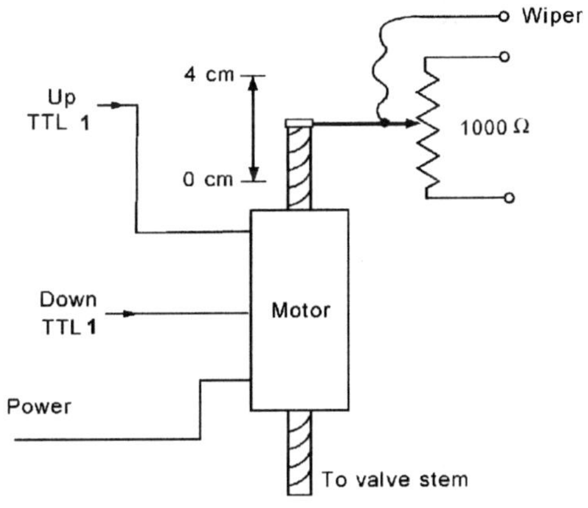

CHAPTER 8

Discrete-State Process Control

INSTRUCTIONAL OBJECTIVES

The objectives of this chapter are to provide an understanding of process-control operations for which the process variables can take on only discrete values. After you have read this chapter and developed solutions to the problems, you should be able to

- Define the nature of discrete-state process-control systems.
- Give three examples of applications of discrete-state process control in industry.
- Explain how a discrete-state process can be described in terms of the objectives and hardware of the process.
- Construct a table of ladder diagram symbols with an explanation of the function of each symbol.
- Develop a ladder diagram from the narrative event-sequence description of a discrete-state control system.
- Describe the nature of a programmable logic controller and how it is used in discrete-state process control.
- Develop a programmable logic controller program from the ladder diagram of a discrete-state process-control application.

8.1 INTRODUCTION

The majority of industrial process-control installations involve more than simply regulating a controlled variable. The requirement of regulation means that some variable tends to vary in a continuous fashion because of external influences. But there are a great many processes in industry in which it is not a variable that has to be controlled but a sequence of events.

This sequence of events typically leads to the production of some product from a set of raw materials. For example, a process to manufacture toasters inputs various metals and plastics and outputs toasters. The term *discrete state* expresses that each event in the sequence can be described by specifying the condition of all operating units of the process. Such condition descriptions are presented by expressions such as: valve A is open, valve B is closed, conveyor C is on, limit switch S_1 is closed, and so on. A particular set of conditions is described as a *discrete state* of the whole system.

In this chapter, we will examine the nature of discrete-state process control. In addition to the nature of such control, a special technique for designing and describing the sequence of process events, called a *ladder diagram,* will be presented. The ladder diagram evolved from the early use of electromechanical relays to control the sequence of events in such processes. Relay control systems have mostly given way to computer-based methods of control, the most common of which is called a *programmable logic controller* (PLC). The characteristics and programming of PLCs will be studied in this chapter, along with numerous applications.

8.2 DEFINITION OF DISCRETE-STATE PROCESS CONTROL

To better understand the material of this chapter, it is helpful to have a general definition of a discrete-state control system. Then you can see how the detailed considerations of the characteristics of such control systems fit within the overall scheme.

Discrete-State Process Control Figure 8.1 is a symbolic representation of a manufacturing process and the controller for the process. Let us suppose that all measurement input variables (S_1, S_2, S_3) and all control output variables (C_1, C_2, C_3) of the process can take on or be assigned only two values. For example, valves are open/closed, motors are on/off, temperature is high/low, limit switches are closed/open, and so on.

Now we define a *discrete state* of the process at any moment to be the set of all input and output values. Each state is discrete in the sense that there is only a discrete number of possible states. If there are three input variables and three output variables, then a state consists of specification of all six values. Because each variable can take on two values, there is a total of 64 possible states.

An *event* in the system is defined by a particular state of the system—that is, particular assignment of all output values and a particular set of the input variables. The event lasts for as long as the input variables remain in the same state and the output variables are

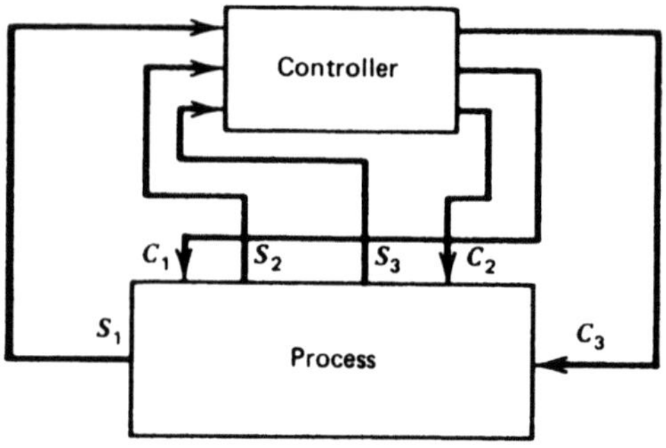

FIGURE 8.1
Process and controller.

left in the assigned state. For a simple oven, we can have the temperature low and the heater on. This state is an event that will last until the temperature rises.

With these definitions in mind, *discrete-state process control* is a particular *sequence of events* through which the process accomplishes some objective. For a simple heater, such a sequence might be

1. Temperature low, heater off
2. Temperature low, heater on
3. Temperature high, heater on
4. Temperature high, heater off

The objective of the controller of Figure 8.1 is to direct the discrete-state system through a specified event sequence. In the following sections, we will consider how the event sequence is specified, how it is described, and how a controller can be developed to direct the sequence of events.

8.3 CHARACTERISTICS OF THE SYSTEM

The objective of an industrial process-control system is to manufacture some product from the input raw materials. Such a process will typically involve many operations or steps. Some of these steps must occur in series and some can occur in parallel. Some of the events may involve the discrete setting of states in the plant—that is, valves open or closed, motors on or off, and so on. Other events may involve regulation of some continuous variable over time or the duration of an event. For example, it may be necessary to maintain the temperature in some vat at a setpoint for a given length of time. In the sense of the previous statements, the discrete-state process-control system is the *master control system* for the entire plant operation.

EXAMPLE 8.1 Use the definitions in this section to construct a description of the frost-free refrigerator/freezer shown in Figure 8.2 as a process with a discrete-state control system. Define the input variables, output variables, and sequence of serial/parallel events.

Solution

The discrete-state input variables are

1. Door open/closed
2. Cooler temperature high/low
3. Freezer temperature high/low
4. Frost eliminator timer time-out/not time-out
5. Power switch on/off
6. Frost detector on/off

The discrete-state output variables are

1. Light on/off
2. Compressor on/off
3. Frost eliminator timer started/not started
4. Frost eliminator heater and fan on/off
5. Cooler baffle open/closed

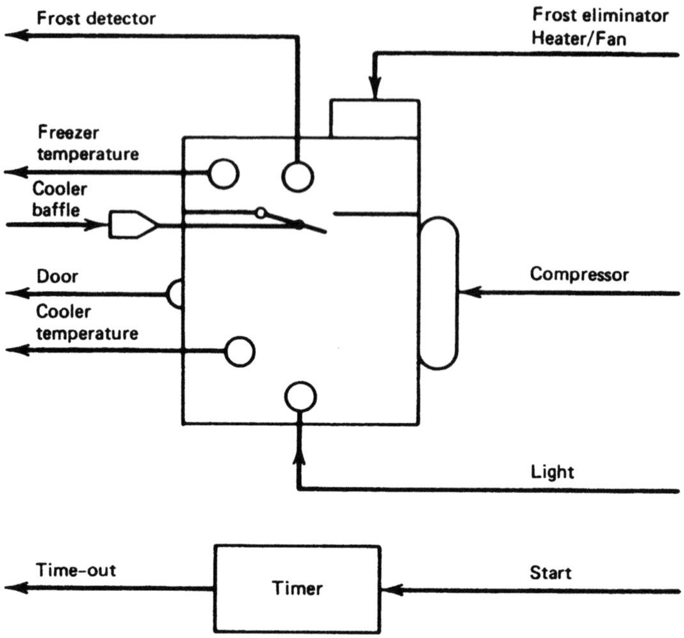

FIGURE 8.2
Refrigerator/freezer system inputs and outputs.

This is a total of 11 two-state variables. In principle, there are $2^{11} = 2048$ possible states or events. Of course, only a few of these are necessary. The event sequences are

a. If the door is opened, the light is turned on.
b. If the cooler temperature is high and the frost eliminator is off, the compressor is turned on and the baffle is opened until the cooler temperature is low.
c. If the freezer temperature is high and the frost eliminator is off, the compressor is turned on until the temperature is low.
d. If the frost detector is on, the timer is started, the compressor is turned off, and the frost eliminator heater/fan are turned on until the timer times out.

The events of (a) can occur in parallel with any of the others. The events of (b) and (c) can occur in parallel. Event (d) can only be serial with (b) or (c).

8.3.1 Discrete-State Variables

It is important to be able to distinguish between the nature of variables in a discrete-state system and those in continuous control systems. To define the difference carefully, we will consider an example contrasting a continuous variable situation with a discrete-state vari-

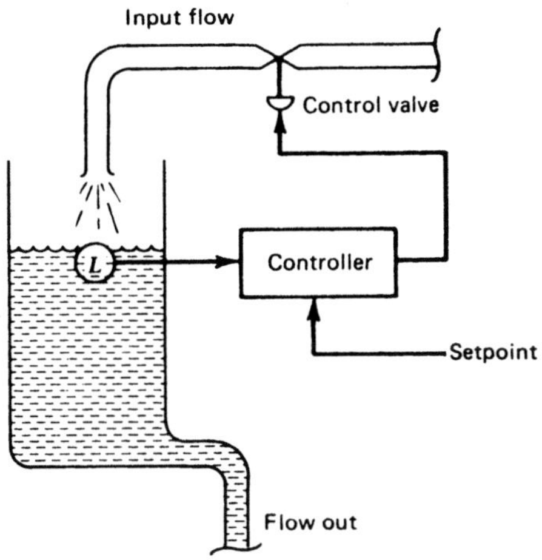

FIGURE 8.3
Continuous control of level.

able situation for the same application. Later it will be shown that continuous variable regulation can be itself a part of a discrete-state system.

Continuous Control Consider for a moment the problem of liquid level in a tank. Figure 8.3 shows a tank with a valve that controls flow of liquid into the tank and some unspecified flow out of the tank. A transducer is available to measure the level of liquid in the tank. Also shown is the block diagram of a control system whose objective is to maintain the level of liquid in the tank at some preset or setpoint value.

The controller will operate according to some mode of control to maintain the level against variations induced from external influences. Thus, if the outflow increases, the control system will increase the opening of the input valve to compensate by increasing the input flow rate. The level is thus *regulated.* This is a continuous variable control system because both the level and the valve setting can vary over a range. Even if the controller is operating in an ON/OFF mode, there is still variable regulation, although the level will now oscillate as the input valve is opened and closed to compensate for output flow variation.

Discrete-State Control Now consider the revised problem shown in Figure 8.4. We have the same situation as in Figure 8.3, but the objectives are different and the variables, level and valve settings, are discrete because they can take on only two values. This means that the valves can only be open or closed, and the level is either above or below the specified value.

Now the objective is to fill the tank to a certain level with no outflow. To do this, we specify an event sequence:

1. Close the output valve.
2. Open the input valve and let the tank fill to the desired level, as indicated by a simple switch.
3. Close the input valve.

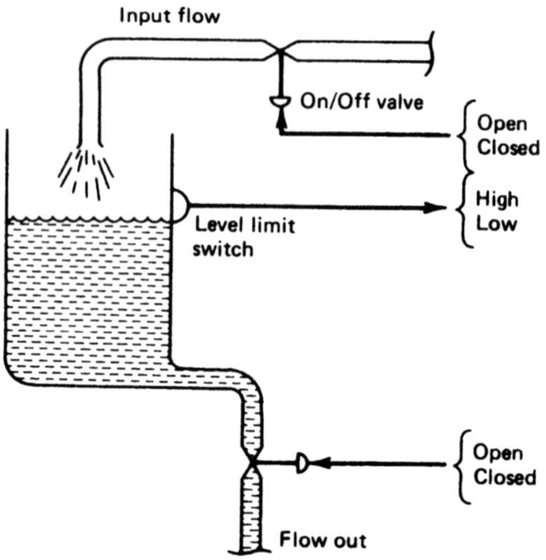

FIGURE 8.4
Discrete control of level.

The level is certainly not going to change until, at some later time, the output valve is opened to let the liquid flow out. Notice that the variables—level measurement, input valve setting, and output valve setting—are two-state quantities. There is no continuous measurement or output over a range.

Composite Discrete/Continuous Control It is possible for a continuous control system to be part of a discrete-state process-control system. As an example, consider the problem of the tank system described in Figure 8.3. In this case, we specify that the outlet valve is to be closed and the tank filled to the required level as in Figure 8.4. We now specify, however, that periodically a bottle comes into position under the outlet valve, as shown in Figure 8.5. The level must be maintained at the setpoint *while the outlet valve is opened and the bottle filled.* This requirement may be necessary to ensure a constant pressure head during bottle filling.

This process will require that a continuous-level control system be used to adjust the input flow rate during bottle-fill through the output valve. The continuous control system will be turned on or off just as would a valve or motor or other discrete device. You can see that the continuous control process is but a part of the overall discrete-state process.

8.3.2 Process Specifications

Specification of the sequence of events in some discrete-state process is directly tied to the process itself. The process is specified in two parts. The first part consists of the objectives of the process, and the second is the nature of the hardware assembled to achieve the objectives. To participate in the design and development of a control system for the process, it is essential that you understand both parts.

Process Objectives The objectives of the process are simply statements of what the process is supposed to accomplish. Objectives are usually associated with knowledge

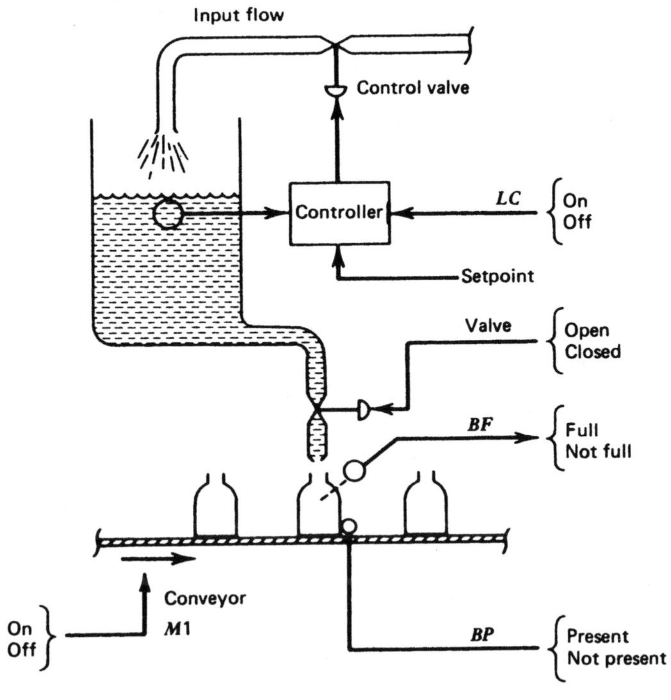

FIGURE 8.5
Composite continuous and discrete control.

of the industry. Often a *global objective* is defined as the end result of the plant. This is then broken down into individual, mostly independent *secondary objectives* to which the actual control is applied.

For example, in a food industry plant, a particular global objective might be to produce crackers. Clearly, this means that the plant takes in raw materials, processes them in specified ways, and outputs packaged and labeled crackers, ready for sale.

The overall objective can be broken down into many secondary objectives. Figure 8.6 suggests some of the secondary objectives that might be involved. There can be further subdivisions into simpler operations. The objectives of the process are formed by the objectives of each independent part of the whole operation. A discrete-state control system then will be applied to each independent part. Thus, in Figure 8.6 the operations within cracker batter preparation can probably be viewed as a stand-alone process.

A process-control specialist typically will not be responsible for the development of the objectives. That is the job of the industry experts. Thus, for crackers, we need experts in food chemistry; for petrochemical industries, we need chemical engineers; for steel production, we need metal specialists; and so on. Nevertheless, it is important for the control system specialist to study the industry and come to an understanding of the products, the process, and the objectives of the process.

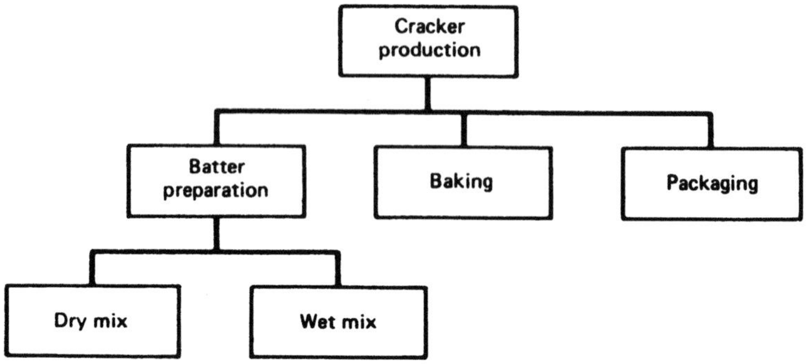

FIGURE 8.6
Objectives and subobjectives of a process.

Process Hardware With determination of the objectives of the process comes the design of hardware to implement these objectives. This hardware is closely tied to the nature of the industry, and its design must come from the joint efforts of process, production, and control personnel. For the control system specialist, the essential thing is to develop a good understanding of the nature of the hardware and its characteristics.

Figure 8.7 shows a pictorial representation of process hardware for a conveyor system. The objective is to fill boxes moving on two conveyors from a common feed hopper and material-conveyor system. A process-control system specialist may not have been involved in the development of this system. To develop the control system, he or she must study the hardware carefully and understand the characteristics of each element.

In general, the specialist analyzes the hardware by considering how each part is related to the control system. There are really only two basic categories.

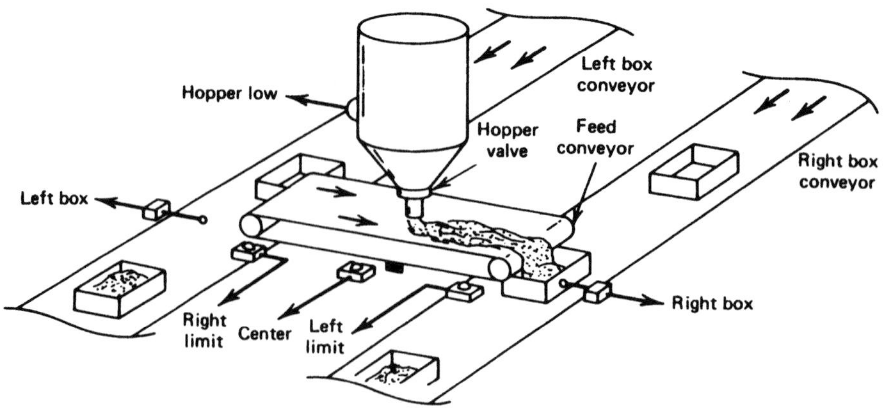

FIGURE 8.7
A discrete-control process.

1. *Input devices* provide inputs *to* the control system. The operation of these devices is similar to the measurement function of continuous control systems. In the case of discrete-state process control, the inputs are two-state specifications, such as

   Limit switches: open or closed

   Comparators: high or low

   Push buttons: depressed or not depressed

2. *Output devices* accept output commands *from* the control system. The final control element of continuous control systems does the same thing. In discrete-state process control, these output devices accept only two-state commands, such as

   Light: on or off

   Motor: rotating or not rotating

   Solenoid: engaged or not engaged

EXAMPLE 8.2

Study the pictorial process of Figure 8.7. Identify the input and output devices and the characteristics of each device.

Solution

A study of the system described in Figure 8.7 shows the following distribution of elements:

Input Devices (All Switches)

Right box present

Left box present

Feed conveyor right travel limit

Feed conveyor left travel limit

Hopper low

Feed conveyor center

Output Devices

Hopper valve solenoid

Feed stock conveyor motor off

Feed stock conveyor motor right

Feed stock conveyor motor left

Right box conveyor motor

Left box conveyor motor

It is not enough simply to identify the input and output devices. In addition, it is important to note how the two states of the devices relate to the process. Thus, if a level-limit switch is open, does that mean the level is low or at the required value? If a command is to be used to turn on a cooler, does this require a high- or low-output command?

Finally, a full study of the hardware also should include the nature of the electrical, pneumatic, or hydraulic signals required for the element. Thus, a motor may be started by application of a 110-V ac, low-current signal to a motor start relay, or it may be started by a 5-V dc TTL-type signal to an electronic starter. Obviously, this type of information will be essential to the development of the control system interface to the process hardware.

8.3.3 Event Sequence Description

Now that the subobjectives of a process and the necessary hardware have been defined, the job remains to describe how this hardware will be manipulated to accomplish the objective. A sequence of events must be described that will direct the system through the operations to provide the desired end result.

Narrative Statements Specification of the sequence of events starts with narrative descriptions of what events must occur to achieve the objective. In many cases, this first attempt at specification reveals modifications that must be made in the hardware, such as extra limit switches. This specification describes in narrative form what must happen during the process operation. In systems that run continuously, there are typically a *start-up,* or *initialization, phase* and a *running phase.*

As an example, consider the system described by Figure 8.7. The start-up phase is used to place the feed conveyor in a known condition. This initialization might be accomplished by the following specification:

- **I.** Initialization Phase
  - **A.** All motors off, feed valve solenoid off
  - **B.** Test for right limit switch
    - **1.** If engaged, go to C
    - **2.** If not, set feed motor for right motion
    - **3.** Start feed-conveyor motor
    - **4.** Test for right limit switch
      - **a.** If engaged, go to C
      - **b.** If not, go to 4
  - **C.** Set feed motor for left motion and start
  - **D.** Test for center switch
    - **1.** If engaged, go to E
    - **2.** If not, go to D
  - **E.** Open hopper-feed valve
  - **F.** Test for left limit switch
    - **1.** If engaged, go to G
    - **2.** If not, go to F
  - **G.** All motors off, hopper-feed valve closed
  - **H.** Go to running phase

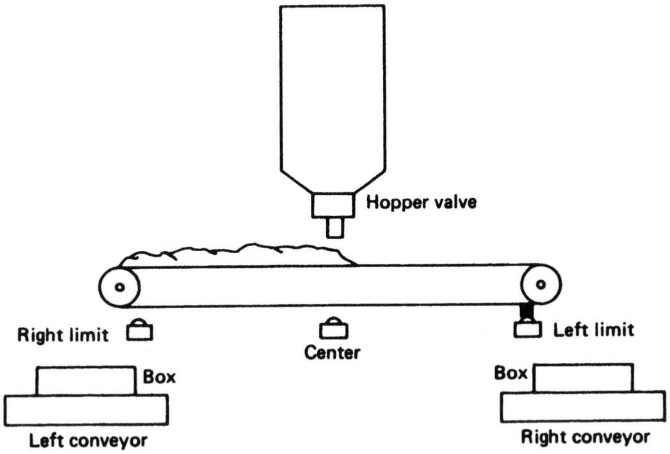

FIGURE 8.8
Completion of the initialization phase of the process shown in Figure 8.7.

Completion of this phase means that the feed conveyor is positioned at the left limit position and the right half of the conveyor has been filled from the feed hopper. The system is in a known configuration, as shown in Figure 8.8.

The running phase is described by a similar set of statements of the sequence of events. For the example of Figure 8.7, this phase might be described as follows:

I. Running
- **A.** Start right box conveyor
- **B.** Test right box present switch
  - **1.** If set, go to C
  - **2.** If not, go to B
- **C.** Start feed-conveyor motor, right motion
- **D.** Test center switch
  - **1.** If engaged, go to E
  - **2.** If not, go to D
- **E.** Open hopper-feed valve
- **F.** Test right limit switch
  - **1.** If engaged, go to G
  - **2.** If not, go to F
- **G.** Close hopper-feed valve, stop feed conveyor
- **H.** Start left box conveyor
- **I.** Test left box present switch
  - **1.** If set, go to J
  - **2.** If not, go to I
- **J.** Start feed conveyor, left motion

K. Test center switch
  1. If engaged, go to L
  2. If not, go to K

L. Open hopper-feed valve

M. Test left limit switch
  1. If engaged, go to II.A
  2. If not, go to M

Note that the system cycles from step M to step A. The description is constructed by simple analysis of what events must occur and what the input and outputs must be to support these events.

EXAMPLE 8.3 Construct a narrative statement outline of the event sequence for the system shown in Figure 8.5. The objective is to fill bottles moving on a conveyor.

Solution

We assume that when a command is given to stop the continuous level control system, the input valve is driven to the closed position. Then the sequence would be

I. Initialization (prefill of tank)
  A. Conveyor stopped, output valve closed
  B. Start the level-control system
    1. Operate for a sufficient time to reach the setpoint, or
    2. Add another sensor so that the system knows when the setpoint has been reached
  C. Go to the running phase

II. Running
  A. Start the bottle conveyor
  B. When a bottle is in position (BP true):
    1. Stop the conveyor (M1 off)
    2. Open the output valve
  C. When the bottle is full (BF true):
    1. Close the output valve
  D. Go to step II.A and repeat

Notice that hardware was added to the system when the event sequence was constructed. Hardware and software are often developed in conjunction, because the development of one demonstrates extra needs in the other.

Flowcharts of the Event Sequence It is often easier to visualize and construct the sequence of events if a flowchart is used to pictorially present the flow of events. Although there are many sophisticated types of flowcharts, the concept can be presented easily by using the three symbols shown in Figure 8.9.

FIGURE 8.9
Basic flowchart symbols.

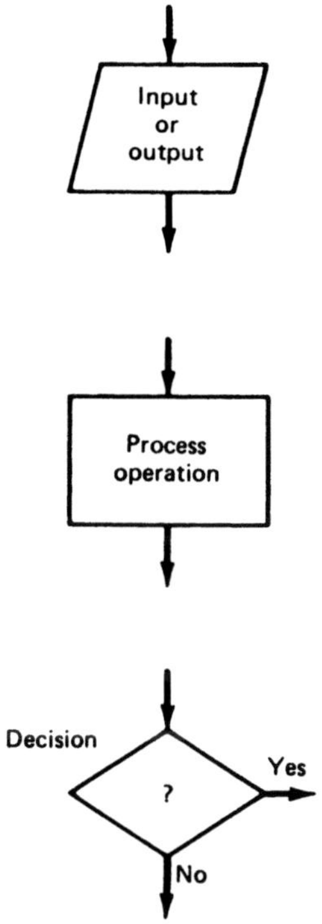

The narrative statements are then simply reformatted into flowchart symbols. Often it is easier to express the sequence of events directly in terms of the flowchart symbols. Figure 8.10 shows part of the initialization phase of the conveyor system of Figure 8.7 expressed in the flowchart format.

Binary-State Variable Descriptions Each event that makes up the sequence of events described by the narrative scheme corresponds to a *discrete state* of the system. Thus, it is also possible to describe the sequence of events in terms of the sequence of discrete states of the system. To do this simply requires that for each event the state, including both input and output variables, be specified.

The input variables cause the state of the system to change because operations within the system cause a change of one of the state variables; for example, a limit switch becomes engaged. The output variables, in contrast, are changes in the system state that are caused by the control system itself.

The control system works like a look-up table. The input state variables with the output become like a memory address, and the new output state variables are the contents of that memory.

FIGURE 8.10
Part of the initialization flowchart for Figure 8.7.

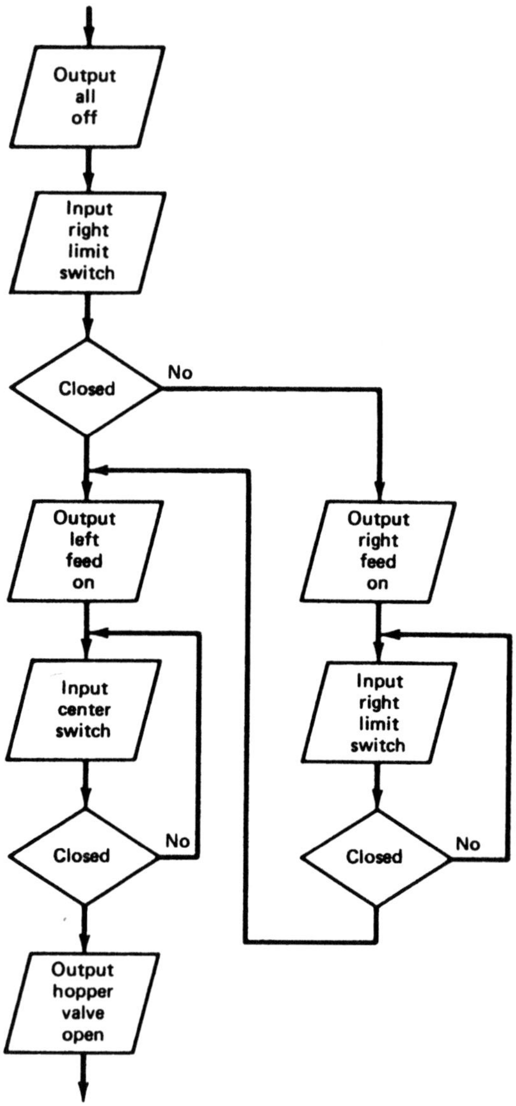

EXAMPLE 8.4

Construct a state variable description of the process shown in Figure 8.11. The timer output (*TU*) is initially low when its input (*TM*) is low. When *TM* is taken high the output stays low for five minutes and then goes high. It resets to low when *TM* is taken low. All level sensors become true when the level is reached. The process sequence is:

1. Fill the tank to level *A* (*LA*) from valve *A* (*VA*)
2. Fill the tank to level *B* (*LB*) from valve *B* (*VB*)
3. Start the timer (*TM*), stir (*S*), and heater (*H*)
4. When five minutes are up take stir (*S*) and heater (*H*) off

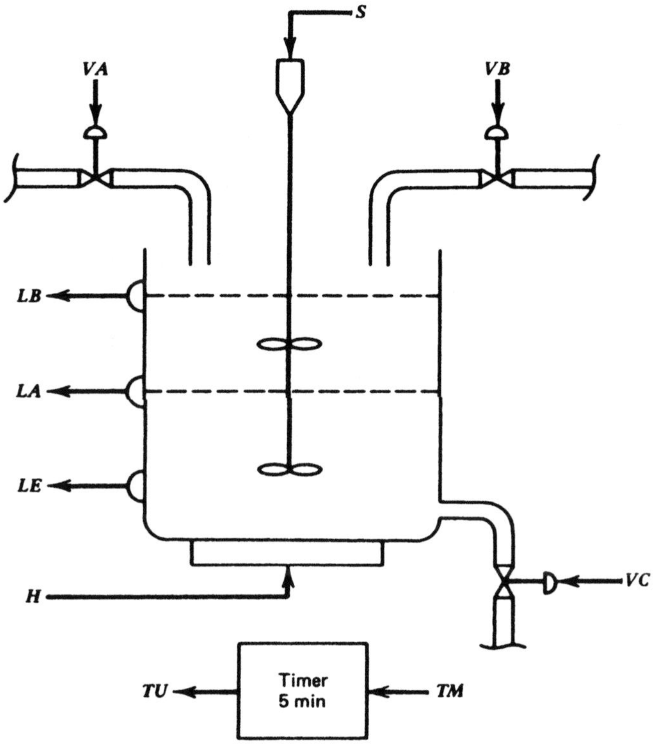

FIGURE 8.11
Tank process for Example 8.4.

5. Open output valve (*VC*) until the tank is empty (*LE*)
6. Take the timer low (*TM*) and go to step 1.

Solution

To provide the solution, we first form the state variable representation of the system by assignment of binary states. There are four input variables (*LA, LB, LE, TU*) and six output variables (*VA, VB, VC, TM, S, H*).

A discrete state of the system is defined by specifying these variables. Because each variable is a two-state variable, we use a binary representation: true = **1** and false = **0**. Thus, for input, if level *A* has not been reached, then $LA = 0$, and if it has been reached, then $LA = 1$. Also, for output, if valve *C* is to be closed, then we take $VC = 0$, and if it is commanded to be open, then $VC = 1$. Let us take the binary "word" describing the state of the system to be defined by bits in the order

$$(LA)(LB)(LE)(TU)(VA)(VB)(VC)(TM)(S)(H)$$

The sequence of events is now translated into an expression of the discrete state as a binary word per state.

The system is processed through the proper sequence by having the input state uniquely select an output state. The following table shows the states of the machine along with brief descriptions. The bit order is the same as previously given.

Input		Output	Description
0000	→	100000	Starting state, open valve *A*
0010	→	100000	Reaches level *LE*, continue with *A* fill
1010	→	010000	Reaches level *A*, close valve *A*, open valve *B*
1110	→	000111	Reaches level *B*, close valve *B*, start timer, heater, stir
1111	→	001100	Time up, stop stir and heater, open valve *C* to empty
1011	→	001100	Reaches level *B*, continue with empty
0011	→	001100	Reaches level *A*, continue with empty
0001	→	000000	Tank empty, turn off timer, go to first state

Typically, this approach to specification of the event sequence is used when a computer will be used to implement the control functions.

Boolean Equations Because the discrete state of the system is described by variables that can take on only two values, it is natural to think of using binary numbers to represent these variables, as in the previous example. It is also natural to consider use of Boolean algebra techniques to deduce the output states from the input states. Although this technique is used, there are generally easier ways to view and solve the problems than with traditional Boolean techniques.

When this technique is used, it is necessary to write a Boolean equation for each output variable in the system. This equation will then determine when that variable is taken to its true state. The equation may depend not only on the set of input variables, but on some of the other output variables. Problems of this type are often considered in digital electronics courses—for example, the common stoplight sequence problem. The following simple example applies this description technique to a more traditional control problem.

EXAMPLE 8.5

Figure 8.12 shows a pictorial view of an oven, along with the associated input and output signals. All of the inputs and outputs are two-state variables, and the relation of the states and the variables is indicated. Construct Boolean equations that implement the following events:

1. The heater will be on when the power-switch is activated, the door is closed, and the temperature is below the limit.
2. The fans will be turned on when the heater is on or when the temperature is above the limit and the door is closed.
3. The light will be turned on if the light switch is on or whenever the door is opened.

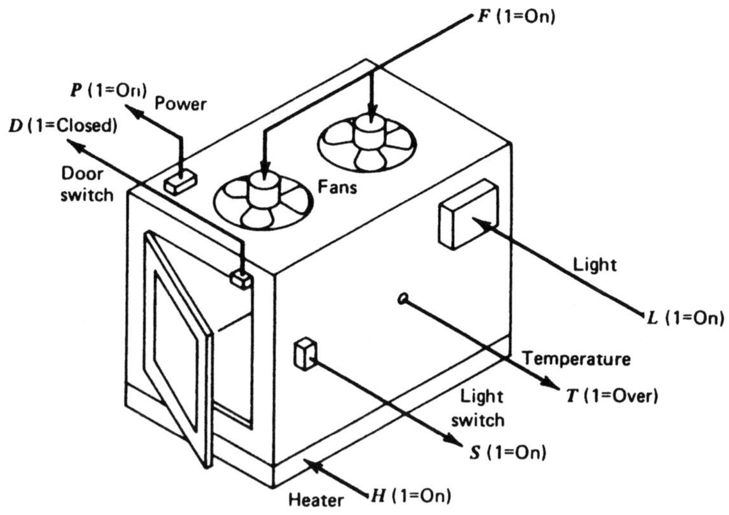

FIGURE 8.12
Oven system for discrete control.

Solution
The solution for problems of this type is developed by simply translating the narrative statements of the events into Boolean equations. In this case, referring to the variables defined in Figure 8.12, you can see that the solution is

Heater: $H = D \cdot \overline{T} \cdot P$
Fans: $F = H + D \cdot T$
Light: $L = \overline{D} + S$

8.4 RELAY CONTROLLERS AND LADDER DIAGRAMS

The previous section showed how a discrete-state control system is described in terms of the hardware of the system and the sequence of events through which that hardware is taken. These two elements are now combined to show how the hardware should be driven so that the proper sequence of events can be accomplished. In essence, this amounts to a "program" for the system written with symbols for the hardware.

A special schematic representation of the hardware and its connection has been developed that makes combination of the hardware and event sequence description clear. This schematic is called a *ladder diagram*. It is an outgrowth of early controllers that operated from ac lines and used relays as the primary switching elements.

8.4.1 Background

An industrial control system typically involves electric motors, solenoids, heaters or coolers, and other equipment that is operated from the ac power line. Thus, when a control

FIGURE 8.13
Use of a relay and switch to start a motor.

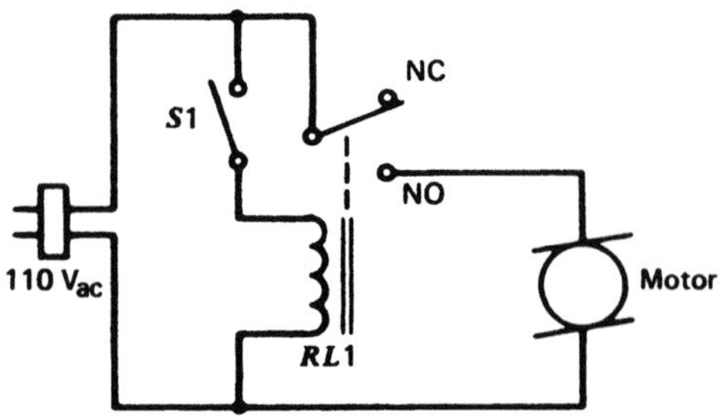

system specifies that a "conveyor motor be turned on," it may mean starting a 50-HP motor. This is not done by a simple toggle switch. Instead, one would logically assume that a small switch may be used to energize a relay with contact ratings that can handle the heavy load, such as that shown in Figure 8.13. In this way, the relay has become the primary control element of discrete-state control systems.

Control Relays Relays can be used for much more than just an energy-level translator. For example, Figure 8.14 shows a relay used as a latch where a green light is on when the relay is not latched and a red light is on when the relay is latched. In this case, when the normally open (NO) push-button switch *PB*1 is depressed, control relay *RL*1 is energized. But then its normally open (NO) contact closes, bypassing *PB*1, so that the relay stays closed. Thus, it is latched. To de-energize or unlatch the relay, the normally closed (NC) push button *PB*2 is depressed. *PB*2 opens the circuit, and the relay is released.

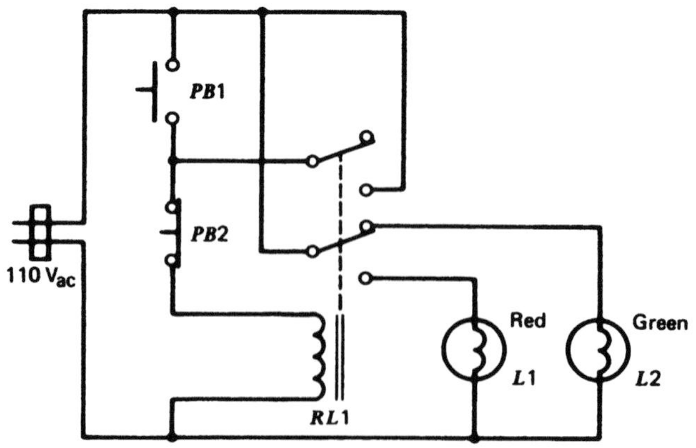

FIGURE 8.14
Use of momentary push-button switches and a relay to implement a latch.

When an entire control system is implemented using relays, the system is called a *relay sequencer*. A relay sequencer consists of a combination of many relays, including special time-delay types, wired up to implement the specified sequence of events. Inputs are switches and push buttons that energize relays, and outputs are closed contacts that can turn lights on or off, start motors, energize solenoids, and so on.

Schematic Diagrams The wiring of a relay control system can be described by traditional schematic diagrams, such as those shown in Figures 8.13 and 8.14. Such diagrams are cumbersome, however, when many relays, each with many contacts, are used in a system. Simplified diagrams have been gradually adopted by the industry over the years. As an example of such simplification, a relay's contacts need not be placed directly over the coil symbol, but can go anywhere in the circuit diagram with a number to associate them with a particular coil. These simplifications resulted in the ladder diagram in use today.

8.4.2 Ladder Diagram Elements

The ladder diagram is a symbolic and schematic way of representing both the system hardware and the process controller. It is called a ladder diagram because the various circuit devices connected in parallel across the ac line form something that looks like a ladder, with each parallel connection a "rung" on the ladder.

In the construction of a ladder diagram, it is understood that each rung of the ladder is composed of a number of conditions or input states and a single command output. The nature of the input states determines whether the output is to be energized or not energized.

Special symbols are used to represent the various circuit elements in a ladder diagram. The following sections present these symbols.

Relays A relay coil is represented by a circle identified as *CR* (for control relay) and an associated identifying number. The contacts for that relay will be either normally open (NO) or normally closed (NC) and can be identified by the same number. A NO contact is one that is open when the relay coil is not energized and becomes closed when the relay is energized. Conversely, the NC contact is closed when the relay coil is not energized and opens when the relay is energized. The symbols for the coil and the NO and NC contacts are shown in Figure 8.15a.

It is also possible to designate a *time-delay relay* as one for which the contacts do not activate until a specified time delay has occurred. The coil is still indicated by a circle, but with the designation of *TR* to indicate timer relay. The contacts, as shown in Figure 8.15b, have an arrow to indicate NO-to-close after delay or NC-to-open after delay. This is called an *on-delay* timer relay. When the coil is energized, the contacts are not energized until the time delay has lapsed.

There is also an *off-delay* timer relay. In this case, the contacts engage when the coil is energized. When the coil is de-energized, however, there is a time delay before the contacts go to the de-energized state.

FIGURE 8.15
Symbols of input devices used in ladder diagrams.

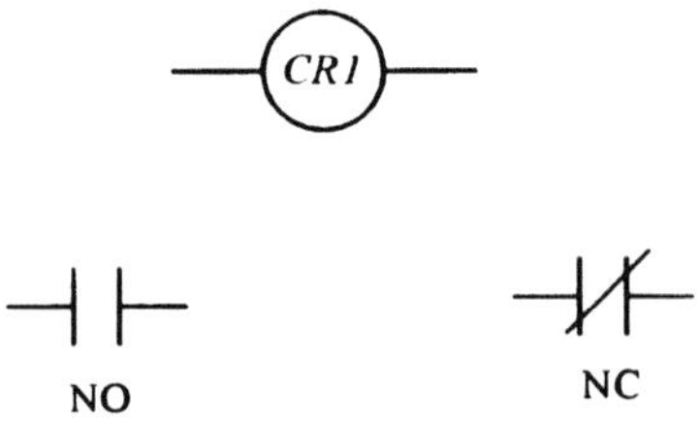

(a) Control relay

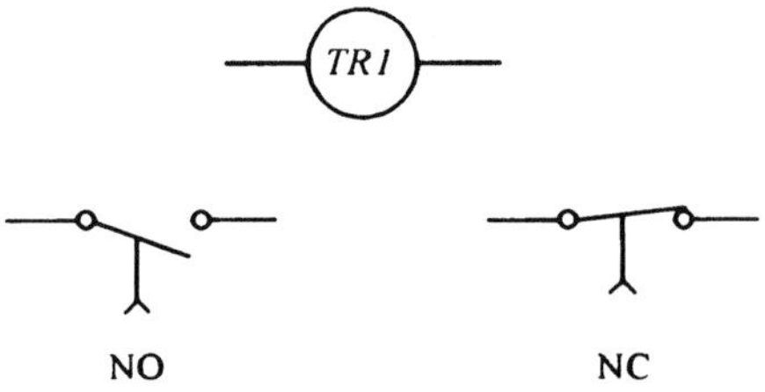

(b) Timer relay

Motors and Solenoids The symbol for a motor is a circle with a designation of *M* followed by a number, as shown in Figure 8.16a. The control system treats this circle as the actual motor, although in fact, this may be a motor start system. The control system uses this symbol to represent the *fact* of the motor, even though other operations may be necessary in the actual hardware to start the motor.

The solenoid symbol is shown in Figure 8.16b. Of course, the symbol itself tells nothing of what function the solenoid plays in the process. For example, it may be a solenoid to

FIGURE 8.16
Symbols of output devices used in ladder diagrams.

(*a*) Motor

(*b*) Solenoid

(*c*) Light (red)

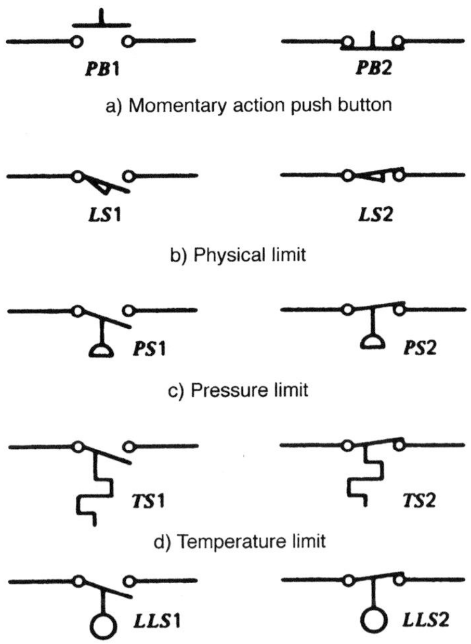

FIGURE 8.17
Symbols for switches used in ladder diagrams.

open a flow valve, or move material off a conveyor, or a host of other possibilities. The solenoid is designated by *SOL* and a number.

Lights A light symbol, such as that shown in Figure 8.16c, is used to give operators information about the state of the system. The color of the light is indicated by a capital letter in the circle; for example, *R* stands for red, *G* for green, *A* for amber, and *B* for blue.

Switches One of the primary input elements in a discrete-state control system is a switch. The switch may be normally open (NO) or normally closed (NC) and may be activated from many sources. In the ladder diagram, different symbols are used to distinguish among different types of switches.

Figure 8.17a shows the symbols for *push-button switches.* Both the NO and NC types are employed. These switches are typically used for operator input, such as to stop and/or start a system.

Figure 8.17b shows the symbols for the NO and NC *limit switches.* These devices are used to detect physical motion limits within the process.

Figure 8.17c shows the symbol for *pressure switches,* both NO and NC. *Thermally activated switches,* such as for ovens or overheating protection of a motor, are indicated by the symbol shown in Figure 8.17d. Figure 8.17e shows the symbols for *level switches.* In all these switches, the NO is closed by rising pressure, temperature, or level.

FIGURE 8.18
Ladder diagram to control two lights.

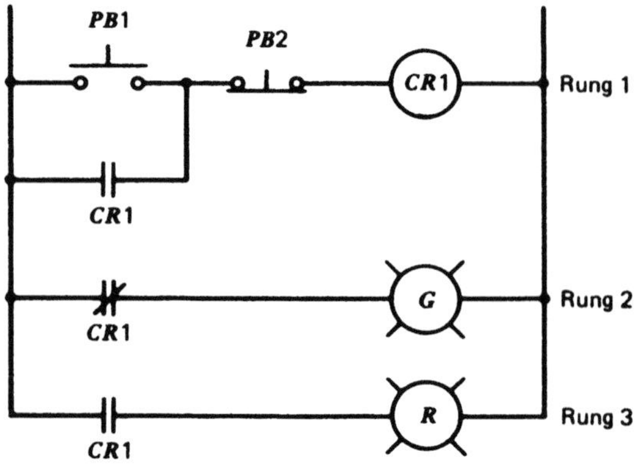

8.4.3 Ladder Diagram Examples

In many cases, it is possible to prepare a ladder diagram directly from the narrative description of a control event sequence. An elementary and common example is the relay latch illustrated by the electrical schematic of Figure 8.14. In terms of a ladder diagram, the same situation is described in Figure 8.18. This diagram has three rungs. The first is a latch involving control relay *CR*1; the second rung is for the green OFF light; the third rung is for the red ON light.

The following example illustrates many features of ladder diagram construction and its application to control problems.

EXAMPLE 8.6

The elevator shown in Figure 8.19 employs a platform to move objects up and down. The global objective is that when the UP button is pushed, the platform carries something to the up position, and when the DOWN button is pushed, the platform carries something to the down position.

The following hardware specifications define the equipment used in the elevator:

Output Elements

$M1$ = Motor to drive the platform up

$M2$ = Motor to drive the platform down

Input Elements

$LS1$ = NC limit switch to indicate UP position

$LS2$ = NC limit switch to indicate DOWN position

START = NO push button for START

STOP = NO push button for STOP

UP = NO push button for UP command

DOWN = NO push button for DOWN command

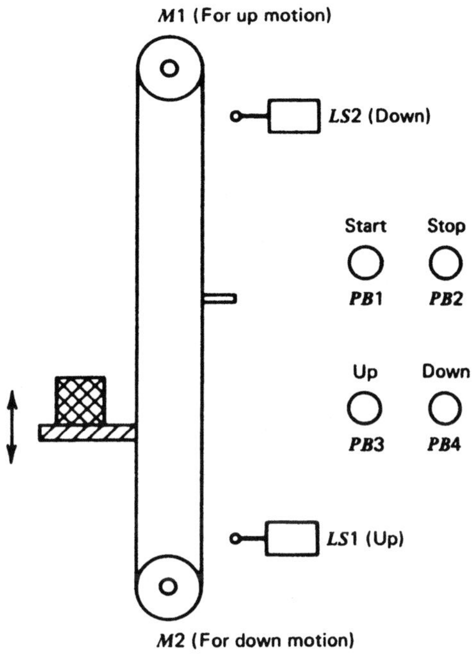

FIGURE 8.19
Elevator system for Example 8.6.

The following narrative description indicates the required sequence of events for the elevator system.

1. When the START button is pushed, the platform is driven to the down position.
2. When the STOP button is pushed, the platform is halted at whatever position it occupies at that time.
3. When the UP button is pushed, the platform, if it is not in downward motion, is driven to the up position.
4. When the DOWN button is pushed, the platform, if it is not in upward motion, is driven to the down position.

Prepare a ladder diagram to implement this control function.

Solution

Let us prepare a solution by breaking the requirements into individual tasks. For example, the first task is to move the platform to the down position when the START button is pushed.

This task can be done by using the START button to latch a relay, whose contacts also energize *M*2 (the down motor). The relay is released, stopping *M*2, when the *LS*2 limit switch opens. Figure 8.20 shows ladder rungs 1 and 2 that provide these functions. Pushing START energizes *CR*1 if *LS*2 is not open (platform not down). *CR*1 is latched by the contacts across the START button. Another set of *CR*1 contacts starts *M*2 to drive the platform down. When *LS*2 opens, indicating the full down position has been reached, *CR*1 is released and unlatched, and *M*2 stops. These two rungs will operate only when the START button is pushed.

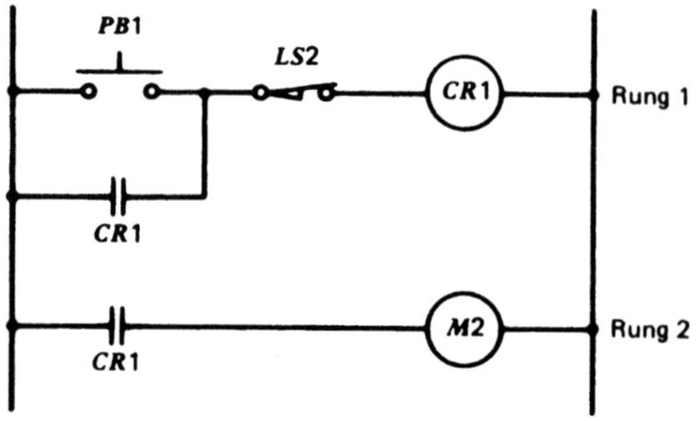

FIGURE 8.20
Initialization to move platform down when the START button is pushed for Example 8.6.

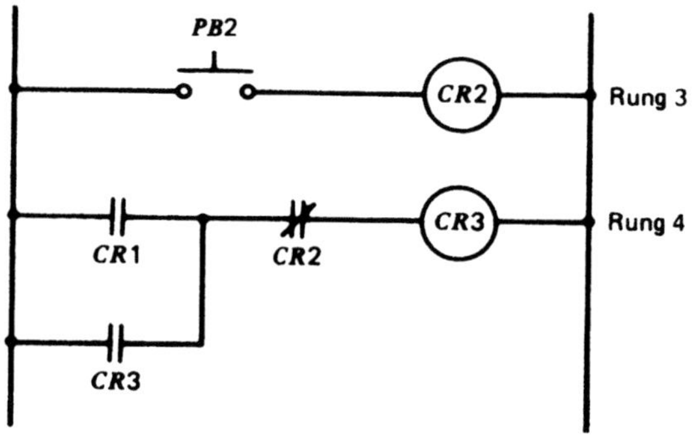

FIGURE 8.21
Ladder diagram for the STOP sequence of Example 8.6.

For the STOP sequence, let us assume a relay $CR3$ is the master control for the rest of the system. Because STOP is a NO switch, we cannot use it to release $CR3$ in the same sense used in previous examples. Instead, we use STOP to energize another relay, $CR2$, and use the NC contacts of that relay to release $CR3$. This is shown in Figure 8.21. You can see that when START is pushed, $CR3$ in rung 4 is energized by the latching of the $CR1$ contact and the NC contact of $CR2$. When STOP is pushed, $CR2$ in rung 3 is energized, which causes the NC $CR2$ contact in rung 4 to open and release $CR3$.

Finally, we come to the sequences for up and down motion of the platform. In each case, a relay is latched to energize a motor if $CR3$ is energized, the appropriate button has been pushed, the limit has not been reached, and the other direction is not energized. The entire ladder diagram is shown in Figure 8.22. A NC relay connection is used to ensure that the up motor is not turned on if the down motor is on, and vice versa. Also, it was necessary to add a contact to rung 2 to be sure $M2$ could not start if there was up motion and some joker pushed the START button.

The solution to Example 8.6 can be simplified by considering the fact that $M1$ and $M2$ are actually relays used to turn on the motors via contacts. If we assume that these relays can have added contacts to drive other ladder diagram operations, then some of the control

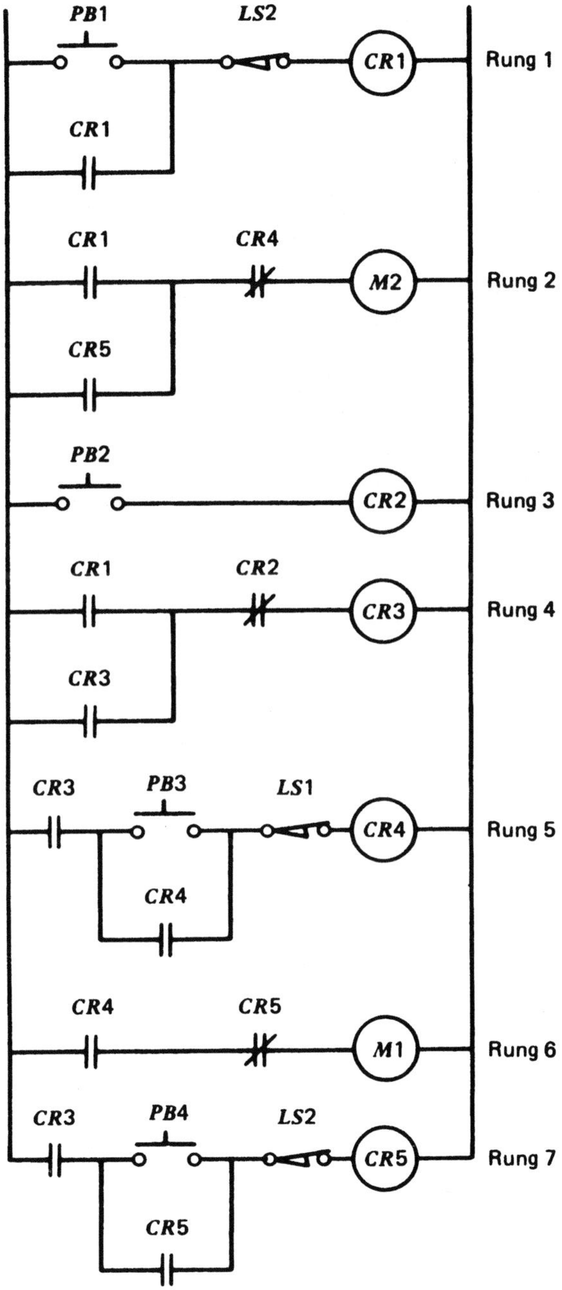

FIGURE 8.22
The complete ladder diagram for Example 8.6.

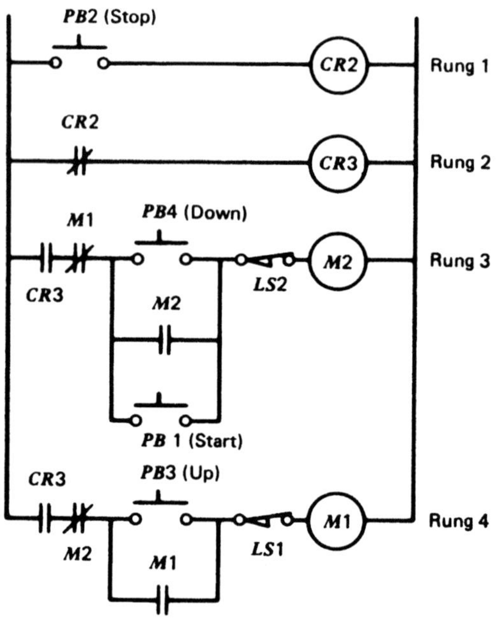

FIGURE 8.23
A simplified ladder diagram for Example 8.6.

relays can be eliminated. Figure 8.23 shows a simplified solution for Example 8.6. We use the *M*1 and *M*2 designations for contacts in other parts of the diagram, just as with control relays. You should work through this diagram to see how the sequence of events is satisfied.

EXAMPLE 8.7 Construct the ladder diagram that will provide a solution to the discrete-state control problem defined by Figure 8.5 and Example 8.3. Assume that when the level-control system is commanded off, the input valve is closed and a 1-min prefill is required for initialization.

Solution

A START/STOP latch is provided to define the initial start-up of the system. The ladder diagram is shown in Figure 8.24.

Initialization is accomplished by a 60-s timer in rung 2 that turns on the level-control system for 1 min following the start button. It is never energized again during running.

Rung 3 drives the conveyor motor until a bottle is in position, as indicated by the bottle position switch opening. Rung 4 is used to detect the bottle-full condition by energizing *CR*2. The contacts of *CR*2 turn on both the valve solenoid (rung 5) and the level-control system (rung 6). Note the timer in rung 6 for initialization. Rung 7 is necessary to detect that the bottle is full and to restart the conveyor until the bottle is moved out of position and the bottle-present switch is opened. Continuous running now occurs between rung 3 and rung 7.

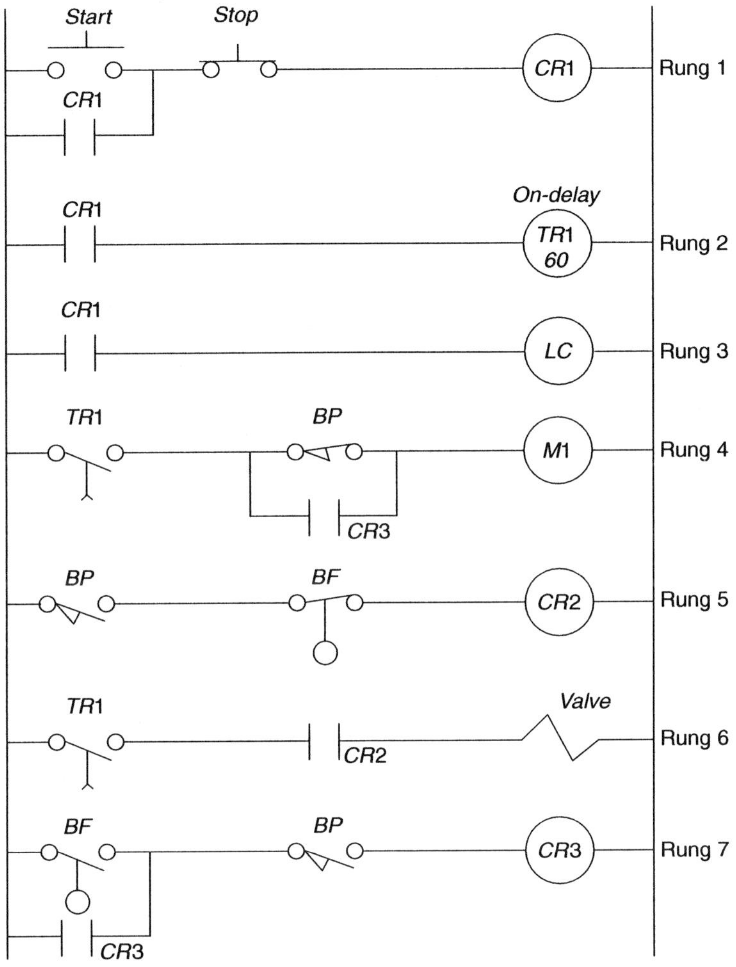

FIGURE 8.24
Solution for Example 8.7.

8.5 PROGRAMMABLE LOGIC CONTROLLERS (PLCS)

The previous sections of this chapter have explained what a discrete-state control system is and why such a system is needed in industrial processes. The last section shows how ladder diagrams are used to describe the event sequence that makes up such a control system. Finally, in this section, you will learn how to actually provide the control system—for example, using the controller shown in Figure 8.1 that inputs the state of the system and generates the required output states to make the process follow the proper event sequence.

8.5.1 Relay Sequencers

One way to provide a discrete-state controller is to use physical relays to put together a circuit that satisfies the requirements of the ladder diagram. Such a control system is called a *relay sequencer* or *relay logic panel.* In the early days of industrial control processes, this was the only way to provide control. It is still used in many applications today, although modern computer-based controllers have replaced many relay-based systems.

The ladder diagram technique of describing discrete-state control systems originated from relay logic systems, which is why the diagram contains so many relay-related terms and symbols. The ladder diagram continues to be used today because it has evolved into an efficient method of defining the event sequence required in a discrete-state control system.

It is important to realize that with relay control each rung of the ladder is evaluated simultaneously and continuously, because the switches and relays are all hardwired to ac power. If any switch anywhere in the ladder diagram changes state, the consequences are immediate. This is not true for computer-based programmable controllers, to be discussed in the next section.

Special Functions To build a relay-based control system, it is necessary to provide certain kinds of special functions not normally associated with relays. These functions are often provided using analog and digital electronic techniques. Included in the special functions are such features as time-delay relays, up/down counters, and real-time clocks.

Hard-wired Programming When a relay panel has been wired to implement a ladder diagram, we say that it has been *programmed* to satisfy the ladder diagram; that is, the event sequence required and described by the ladder diagram will be provided by the relay system when power is applied. Thus, the program has been wired into the relays that make up the relay logic panel.

If the event sequence is to be changed, it is necessary to rewire all or part of the panel. It may even be necessary to add more relays to the system or to use more relays than in the previous program.

Obviously, such a task is quite troublesome and time consuming. A number of ingenious methods are used to ease some of the problems of changing the relay program. One is the use of *patch panels* for the programming. In these systems, all relay contacts and coils are brought to an array of sockets. Cords with plugs in each end are then used to patch the required coils, contacts, inputs, and outputs together in the manner required by the ladder diagram.

The patch panel acts like a memory in which the program is placed. With the development of reliable computers, it was an easy decision to replace relay logic-based systems with computer-controlled systems.

8.5.2 Programmable Logic Controller Design

The modern solution for the problem of how to provide discrete-state control is to use a computer-based device called a *programmable controller* (PC) or *programmable logic controller* (PLC).

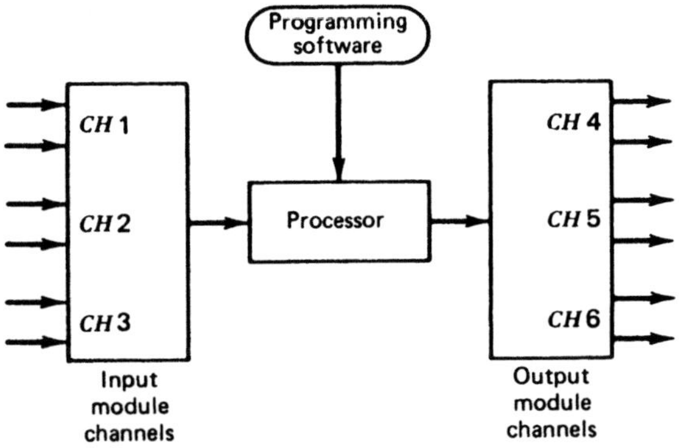

FIGURE 8.25
Basic structure of the programmable logic controller (PLC).

The move from relay logic controllers to computer-based controllers was an obvious one because

1. The input and output variables of discrete-state control systems are binary in nature, just as with a computer.
2. Many of the "control relays" of the ladder diagram can be replaced by software, which means less hardware failure.
3. It is easy to make changes in a programmed sequence of events when it is only a change in software.
4. Special functions, such as time-delay actions and counters, are easy to produce in software.
5. The semiconductor industry developed solid-state devices that can control high-power ac/dc in response to low-level commands from a computer, including SCRs and TRIACs.

A programmable controller can be studied by considering the basic elements shown in Figure 8.25: the processor, the input/output modules, and the software.

Processor The processor is a computer that executes a program to perform the operations specified in a ladder diagram or a set of Boolean equations. The processor performs arithmetic and logic operations on input variable data and determines the proper state of the output variables. The processor functions under a permanent supervisory operating system that directs the overall operations from data input and output to execution of user programs.

Of course, the processor, being a computer, can only perform one operation at a time. That is, like most computers, it is a serial machine. Thus, it must sequentially sample each of the inputs, evaluate the ladder diagram program, provide each output, and then repeat the whole process. The speed of the processor is important. This is discussed further under Scan and Execution Modes.

The heart of a PLC is a microprocessor, much like the ones used in modern personal computers. Because much of the data in PLC operation is processed bit by bit, special microprocessors optimized for such operation, such as the AMD 2901 and 2903, are often

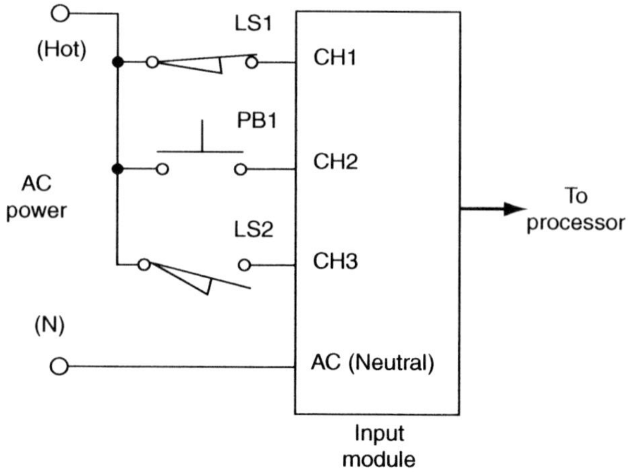

FIGURE 8.26
Typical wiring to a PLC input module.

employed. With the great increases in processor speed, it is now possible to employ a desktop personal computer with data I/O boards running PLC software to emulate PLC operation.

Input Modules The input modules examine the state of physical switches and other input devices and put their state into a form suitable for the processor. The PLC is able to accommodate a number of inputs, called *channels.*

As noted at the beginning of this section, the origin of the PLC was relay logic systems which operated directly off of the common 120 vac line. A switch would deliver 120 vac to the relay logic panel if closed or an open circuit if not closed. Early versions of solid-state PLCs continued this familiar association by having an input module which was connected to input devices through the ac line, as shown in Figure 8.26. If a switch such as PB1 was closed, the ac voltage was impressed on the input module that would internally provide a conversion to a digital signal (**1** or **0**) as required by the computer-based PLC processing unit.

Modern PLCs and the input sensors to which they are connected are much more versatile. The input module can usually be configured so that the input device is simply connected with no reference power. Therefore, the limit switch LS1 in Figure 8.26 would simply be connected between CH1 and the common connection at the bottom, which would now be labeled *Common,* for example. The same process would apply for PB1 and LS2. This also has a great advantage with respect to electrical safety since the ac power is not connected to the switches. Many input modules can also be used in a networked environment wherein the input signal arrives as serial digital data encoded with the input device address and state. Regardless of how the process input devices are connected to the input module, the result is that the state of each device is presented to the PLC processor for evaluation by the ladder program.

Output Modules The objective of the output module is ultimately to supply power to an external device such as a motor, light, solenoid, and so on, as required by the ladder diagram. Early relay-logic sequencers were able to provide 120 vac directly to devices as long as the power requirement was not too great. When solid-state PLCs were introduced this was no longer possible. The output of the PLC was now a high (**1**) or low (**0**) signal of low power dc. Thus, the modern output module is designed to input the processor

FIGURE 8.27
Typical wiring to a PLC output module.

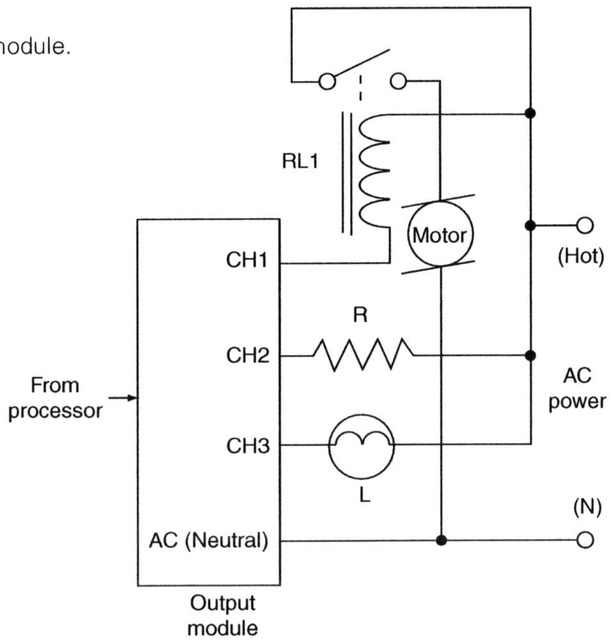

digital output and use this to activate relays, SCRs, TRIACs, BJTs, and other solid-state switches that can handle the high-power ac requirements. In this case the PLC output module can be wired directly to the output device and the ac power source, as shown in Figure 8.27. This shows how the channels of the output module are connected between the ac power and common. If the device, such as the motor in Figure 8.27, requires a current that exceeds the module output capability, then external relays or solid-state switches can be used.

Modern output modules can now provide many other kinds of outputs such as dc voltages, pulse outputs, and even serial digital outputs that can be transmitted to the output device over a network.

8.5.3 PLC Operation

Let us consider the typical operation of a PLC. First of all, the operation is *not* simultaneous for the entire ladder diagram and is *not* continuous as it is for relay sequencers. This is very important and can have significant impact if not taken into consideration in a design. Operation of the programmable controller can be considered in two modes, the I/O scan mode and the execution mode.

I/O Scan Mode During the I/O scan mode, the processor updates all outputs and inputs the state of all inputs one channel at a time. The time required for this depends on the speed of the processor.

Execution Mode During this mode, the processor evaluates each rung of the ladder diagram program that is being executed sequentially, starting from the first rung and proceeding to the last rung. As a rung is evaluated, the last known state of each switch and relay contact in the rung is considered, and if any TRUE path to the output device is detected, then that output is indicated to be energized—that is, set to ON.

At the end of the ladder diagram, the I/O mode is entered again, and all output devices are provided with the ON or OFF state determined from execution of the ladder diagram program. All inputs are sampled, and the execution mode starts again.

Scan Time An important characteristic of the programmable controller is how much time is required for one complete cycle of I/O scan and execution. Of course, this depends on how many input and output channels are involved and on the length of the ladder diagram program. A typical maximum scan/execution time is 5 to 20 ms.

The speed of the controller depends on the clock frequency of the processor. The higher the clock frequency, the greater the speed, and the faster the scan/execution time.

The length of time for one scan consists of three parts: (1) input time, (2) execution time, and (3) output time. Most of the scan time comes from the execution phase.

The scan time may have an impact on the ability of the PLC to detect events that occur on the inputs. For example, if some limit switch goes to the ON state for less than a scan time, it may be missed by the PLC. The high speed of modern PLCs makes this less of a problem than in the old days, but the following example illustrates how it can become an issue.

EXAMPLE 8.8

A complex manufacturing operation results in a 30-ms PLC scan time. The PLC must detect individual 2-cm objects moving on a high-speed conveyor. Figure 8.28a shows that the object breaks a light beam, and this provides a high input to the PLC. What is the highest speed of the conveyor to be sure the object is detected?

Solution

The photodetector output will be a high pulse that will last for a time determined by the conveyor speed:

$$t_{\text{pulse}} = (2\text{ cm})/(S\text{ cm/s}) = (2/S)\text{ s}$$

where S is the conveyor speed. Figure 8.28b shows the relationship between the PLC scan time and the object pulse. Here we suppose that an object breaks the beam just after the input phase of the scan. To assure detection, we must be sure the signal lasts at least until the end of the next scan input phase. Thus, the pulse time must be:

$$t_{\text{pulse}} = 1\text{ scan time} = 30\text{ ms}$$

Thus, the maximum conveyor speed can be determined as,

$$(2/S) = 0.03\text{ s or } S = 66.7\text{ cm/s}$$

Programming Unit The programming unit is an external electronic package that is connected to the programmable controller when programming occurs. The unit usually allows input of a program in ladder diagram symbols. The unit then transmits that program into the memory of the programmable controller.

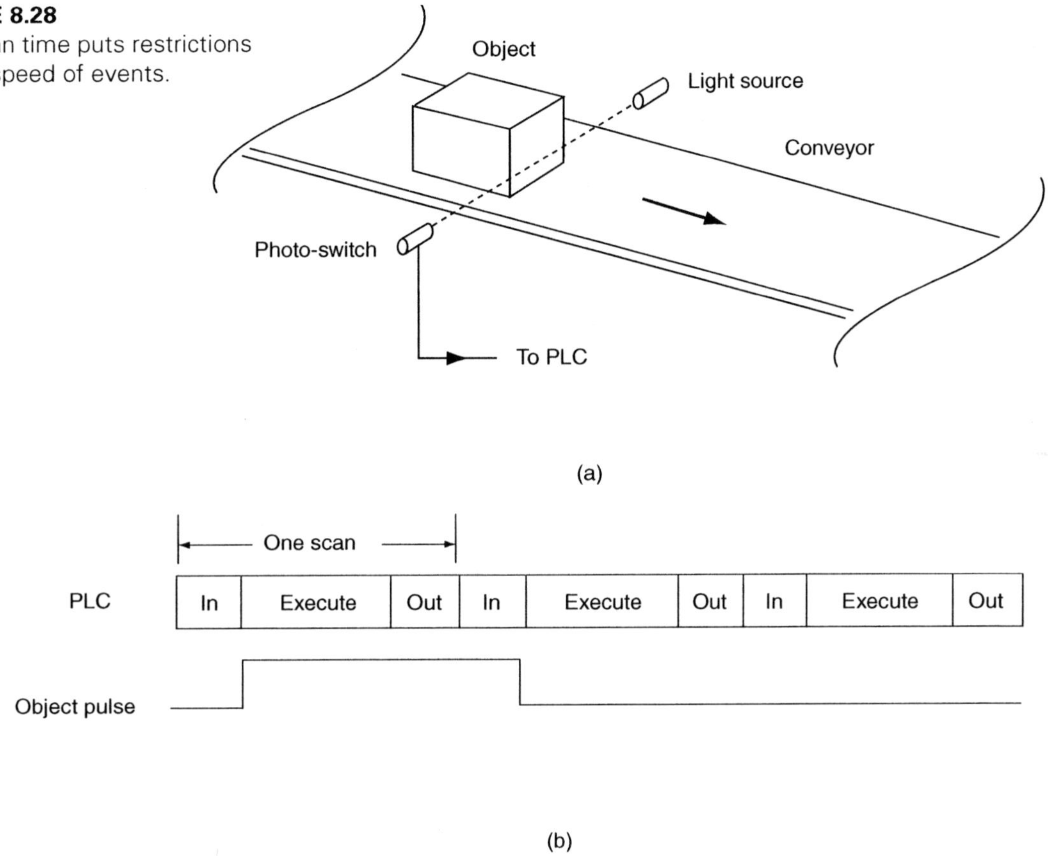

FIGURE 8.28
PLC scan time puts restrictions on the speed of events.

Programming units may be small, self-contained units, such as that shown in Figure 8.29. In this unit, the ladder diagram is displayed one rung at a time in a special liquid crystal display (LCD). The user can enter a program, perform diagnostic tests, run the program through the programmable controller, and perform editing of the installed program. The installed program is stored in a temporary memory that will be lost without ac power or battery backup. The program can be permanently "burned" into a ROM for final installation.

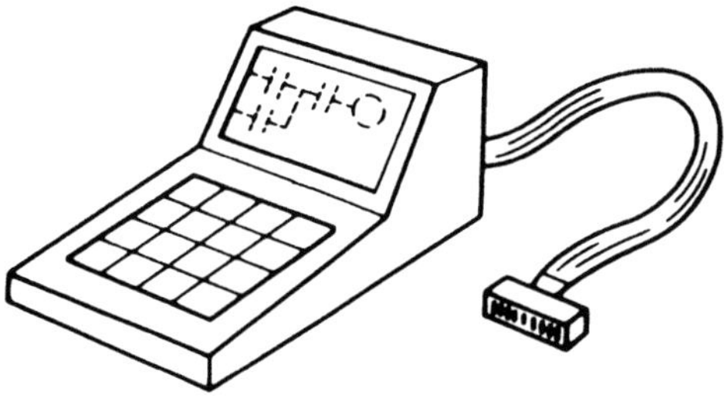

FIGURE 8.29
Hand-held PLC programming unit.

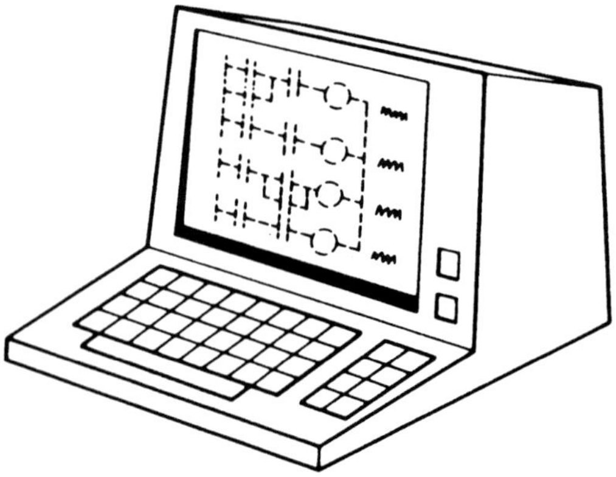

FIGURE 8.30
Desktop terminal PLC programming unit.

Another type of programming unit is able to display many rungs of the ladder diagram. Such a unit is shown in Figure 8.30. The same capabilities are available for entering, testing, and changing a program. The program is still in temporary memory.

Once the program has been debugged, the programming unit can be disconnected, and the programmable controller can now operate the process according to the ladder diagram program. There is the danger of loss of the program because of power failure, but this can be prevented by placing the program into the permanent memory.

The dedicated PLC programming unit can be replaced by software and I/O boards on common desktop or laptop personal computers. Special software allows the pc to program the PLC and then download the program into the PLC unit.

PLC Networks Modern plants employ computer network technology to link PLCs and other control hardware via network communication systems. These systems may use custom installations of Ethernet-type local area networks (LANs) or dedicated control networks such as Foundation Fieldbus or Profibus (see Chapter 11). In any case, it is often possible to download the PLC programming from a development computer to the PLC in which it will be used. The network can also be used to edit an existing PLC program and to transfer process data to a network computer to analyze plant operation.

RAM/ROM The temporary memory used during ladder diagram program testing and evaluation is called RAM, for read and write memory. Once the program is stored in RAM, it can be easily modified. As in any computer program, it is necessary to perform testing and evaluation of programmable controller programs, and corrections are usually necessary.

When a program has been debugged and is considered finished, it may be "burned" into a ROM. This is a read-only memory that cannot be changed and is not affected by power failure. The ROM often can be programmed directly by the controller programming unit. When the ROM is plugged into the programmable controller, the device is ready to be placed into service in the industrial setting.

In the case of network programmable PLCs, the final PLC program is downloaded into special re-programmable ROM in the PLC that can be erased and rewritten but that still retains its contents after power-down.

8.5.4 Programming

Although the programmable controller can be programmed directly in ladder diagram symbols through the programming unit, there are some special considerations. These considerations include the availability of special functions and the relation between external I/O devices and their programmed representations.

The programmable controller has no "real" relays or relay contacts. The only real devices are those that are actually part of the process being controlled—that is, limit switches, motors, solenoids, and so on. We continue to use symbols for relays and relay contacts, even though they are software symbols.

Addressing When the ladder diagram for some event sequence was developed in a previous section, each switch device, output device, and relay was referred to by a label. For example, $CR1$ referred to control relay 1, and the contacts for that relay were referred to by the same label. Other designations included $LS1$ for a limit switch, $M1$ for a motor relay, and so on.

The programmable controller uses a similar method of identifying devices, but it is referred to as the device *address* or *channel.* The addresses are used to identify both the physical and software devices according to the following categories:

1. Physical input devices—ON or OFF
2. Physical output devices—energized (ON) or de-energized (OFF)
3. Programmed control relay coils and contacts
4. Programmed time-delay relay coils and contacts
5. Programmed counters and contacts
6. Special functions

The address designation depends on the type of programmable controller. Some controllers may reserve certain addresses for physical I/O devices, other addresses for software control relays, and yet others for special functions.

For the examples and problems to be considered in this chapter, we will use the definitions of addresses shown in Table 8.1.

Programmed Diagram Interpretation There is an important difference between the interpretation of a physical ladder diagram and a programmed ladder diagram. This difference arises from the fact that the programmed diagram bases the state of a rung on a logical interpretation of the symbol rather than its physical state.

In a programmed diagram rung, the ON or OFF state of the output of the rung is determined by testing the elements of the rung for a TRUE or FALSE condition. If a

TABLE 8.1
PLC addressing

Function	Address
Input channels	00 to 07
Output channels	08 to 15
Internal relays	16 to 31
Timers and counters	32 to 39

complete TRUE element path to the output exists in the rung, then the output will be made TRUE or ON.

In a *physical diagram,* the symbol for a NO contact indicates a normally open contact through which current cannot flow unless the contact has been closed. If it is a push-button switch, then someone must close the contacts by pushing. If it is the contact of a relay, then the relay coil must be energized.

For the NC contact, the idea is that current will flow until the contact has been opened. If it is a push-button switch, then someone must open the contact and stop the current flow by pushing. If it is the contact of a relay, then the relay must be energized to open the contact and stop current flow through the contacts.

In a *programmed diagram,* the symbol for a NO contact indicates that the device should be interpreted as FALSE if the contact is tested and found to be open, and TRUE if it is found to be closed. We often say it is to be "examined ON," and if ON, it is TRUE.

Consider the programmed NC symbol. This means if it is tested and found to be closed, then it is FALSE, and if tested and found open, it is ON. This is not like the physical. We often say this is an "examine OFF," and if it is OFF, it is TRUE.

The diagrams of Figure 8.31 illustrate this concept. Suppose we have a physical NC push-button switch and we want to turn on a red light when the switch is pushed. First let us look at the physical interpretation. Figure 8.31a shows that we cannot simply wire the light to the switch. In this case, the light will normally be ON and go out (OFF) when the switch is pushed.

Figure 8.31b shows how to provide the answer in a physical system with a control relay, *CR*1. Now, *CR*1 is normally energized, so its NC contacts are open and the light will

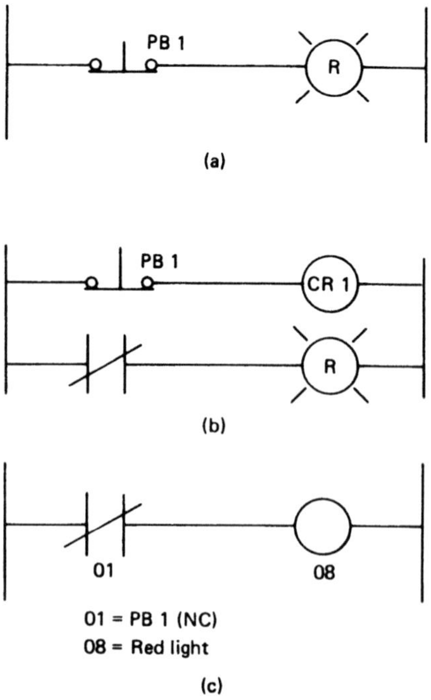

FIGURE 8.31
Using an NC switch to turn on a light with relay logic and with PLC programming: (a) Wrong, the button turns the light OFF; (b) with relay logic, the button turns the light ON; and (c) with a PLC program, the button turns the light ON.

be OFF. When the button is pushed, *CR*1 is de-energized and its NC contact, which was open because *CR*1 was energized, closes and the red light comes ON. So this works.

Now, in the programmed system we do not need either physical or programmed control relay to do this. Figure 8.31c shows that we simply refer to the push button with an "examine OFF" symbol connected directly to the light. So, if the switch has not been pushed, a test of the symbol shows it to be closed; therefore, it interprets logically as FALSE, and the light is not energized by the program. When pushed, it is tested to be open, and therefore there is a logic TRUE and the light is energized.

Suppose we want to implement a latch to turn on and off a motor using two NO push-button switches. Figure 8.32a shows how the switches and motor are physically connected to the input and output channels of a PLC.

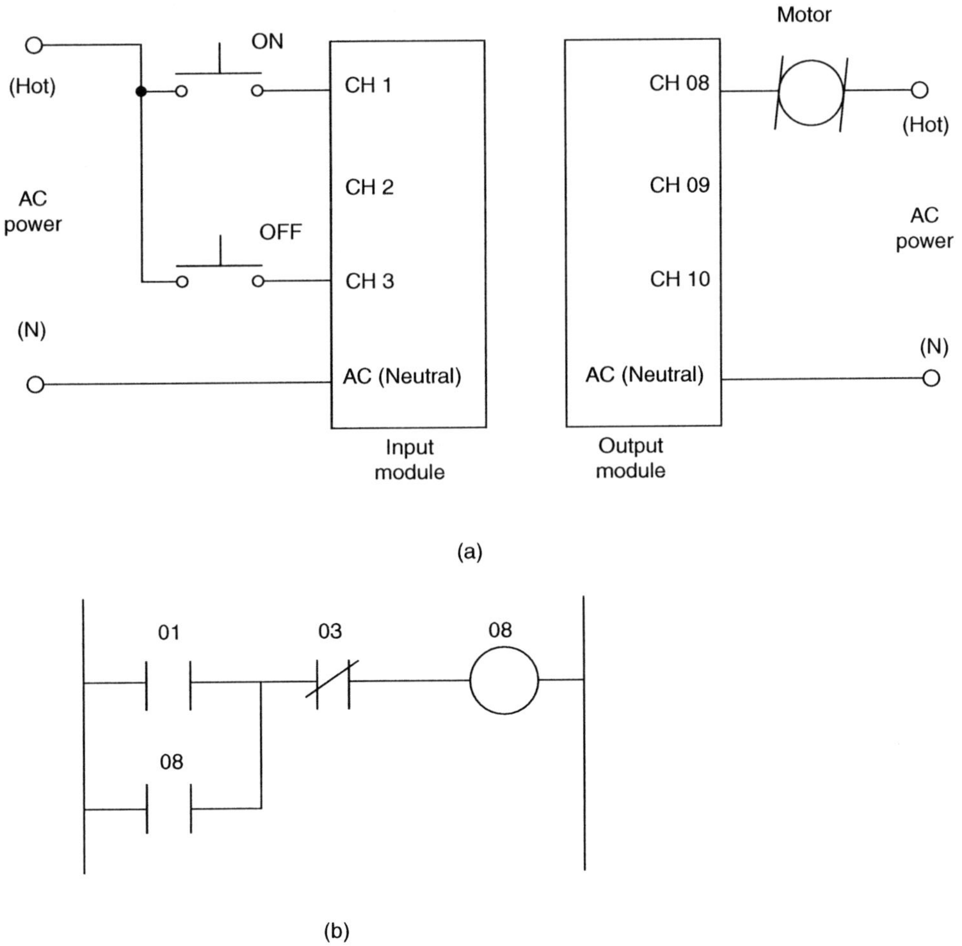

FIGURE 8.32
Physical connections to the PLC (a) and the programmed ladder diagram (b) for a motor start/stop system.

Now, for the programmed ladder diagram, we want the motor latched on if *PB*1 is depressed. This would make input channel 1 high. Thus, we want an EXAMINE ON condition for this element of the rung. We want the motor to be turned on, and the rung latched, even though no power is applied to input channel 3 (that is, *PB*2 had not been depressed). Thus, for this channel we want an EXAMINE OFF. (Is it OFF? If yes, then the channel is TRUE.) So the ladder diagram of Figure 8.32b will provide the required solution.

The following example further illustrates the concepts of programmed interpretation of switch states.

EXAMPLE 8.9

The objective of the system shown in Figure 8.33 is as follows:

1. If *PB*1 alone is pushed, the red light turns on.
2. If *PB*2 alone is pushed, the green light turns on.
3. If both buttons are pushed at once, neither light turns on.

Show the wiring connection to the PC and the ladder program that will accomplish this task. Channels 01 and 02 are inputs, and 08 and 09 are outputs.

Solution

Following the example of Figures 8.27 and 8.28, the lights and switches are wired as shown in Figure 8.34. The ac hot lead is connected to the objects, which are then connected to the PLC.

To accomplish the first objective, we need the red light on when *PB*1 is pushed and *PB*2 is not. Since *PB*1 is NO, true will be ON (*PB*1 has been pushed), and so we need an EXAMINE ON. Since *PB*2 is NO, for it *not* to be pushed, it will be OFF, so we need an EXAMINE ON again (i.e., if it is ON, then it must not have been pushed).

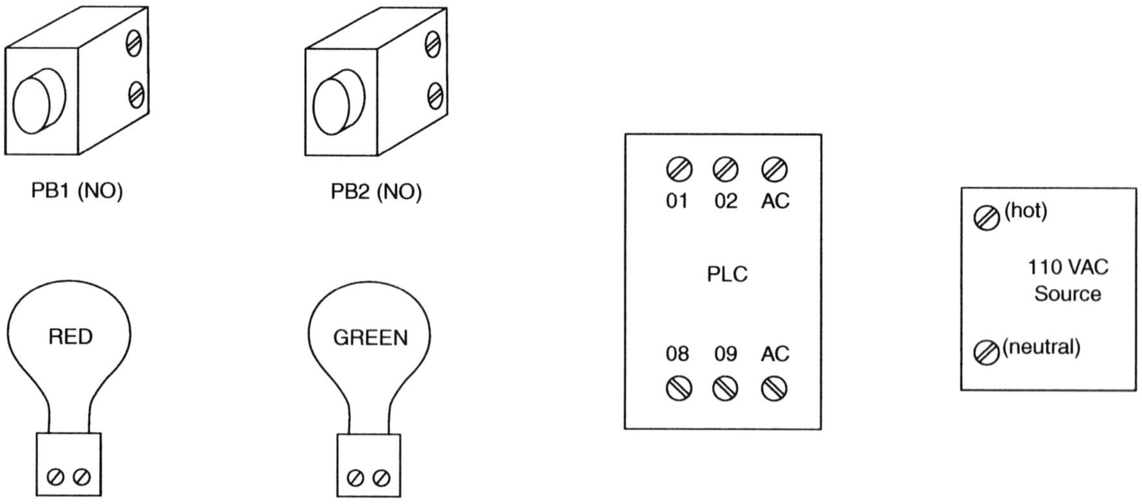

FIGURE 8.33
Physical elements for Example 8.9.

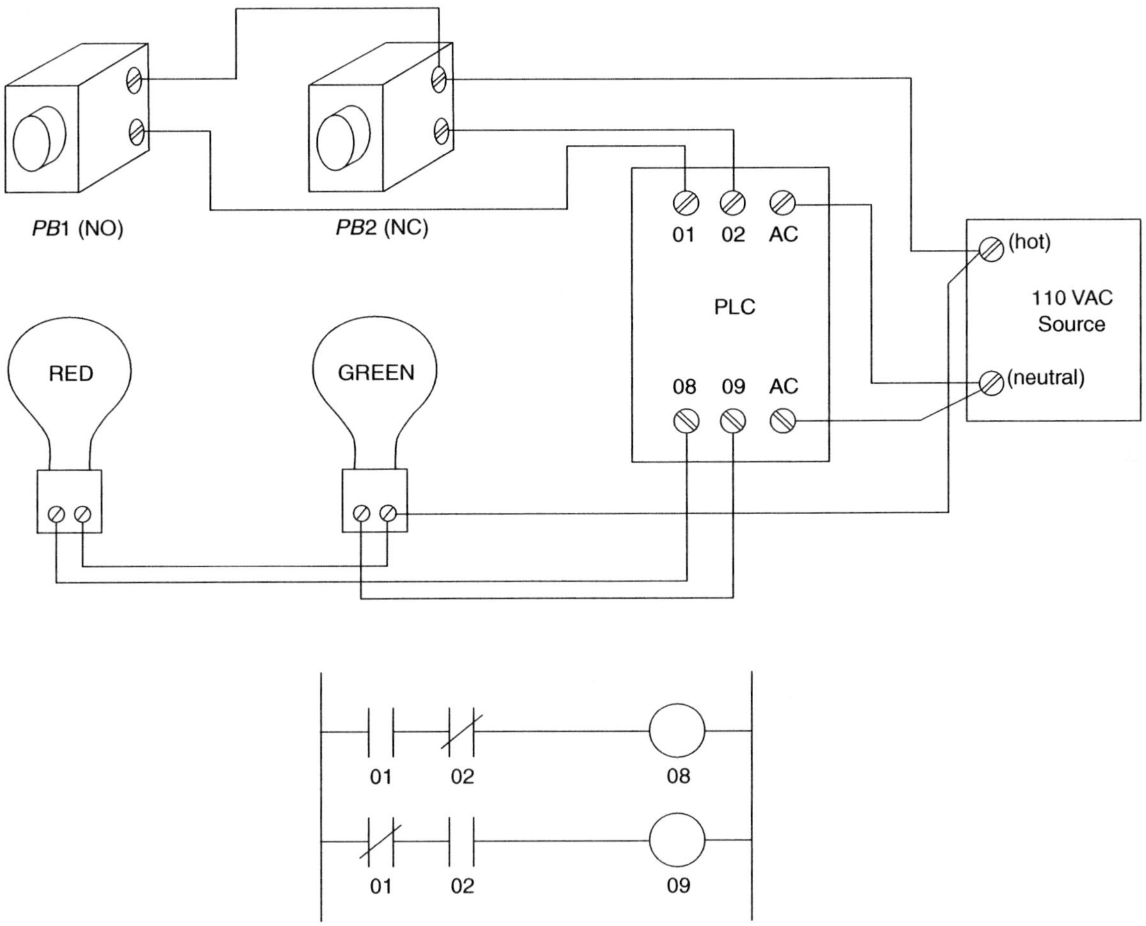

FIGURE 8.34
Wiring and program solution for Example 8.9.

To accomplish the second objective, we need the green light to be on when *PB*1 is not pushed but *PB*2 is pushed. This is accomplished by using EXAMINE OFF for *PB*1 (*PB*1 has not been pushed) and EXAMINE OFF for *PB*2 also (*PB*2 has been pushed and so is open). Figure 8.34 shows the ladder diagram program.

8.5.5 PLC Software Functions

A modern PLC has many functions implemented in the software associated with the processor. In this section, some of the more common functions of PLCs are presented, with examples of their application. Rather generic descriptions are given so you will understand

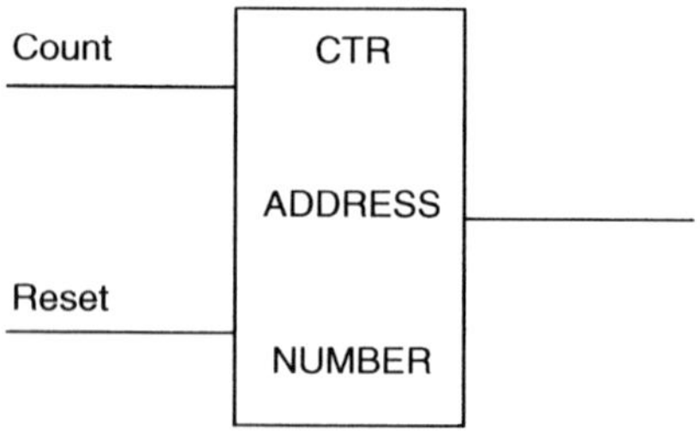

FIGURE 8.35
The basic PLC counter.

the basic concept of the functions. Individual PLC manufacturers will have their own details on actual implementation. The addressing scheme shown in Table 8.1 will be used in the following descriptions.

Control Relay The software control relay is denoted by a circle with an identification number. It can have any number of NO or NC contacts, which are identified by the same number as the relay. Previous examples show software control relays in use.

Counter A counter is a programmed function that counts (increments) every time the input changes from False to True. This means that, if in one scan the input is False and in the next scan the input is True, the counter increments. No further counts will occur until the input goes False again and then True.

Counters are often drawn as a rectangle in programmed ladder diagrams, like that shown in Figure 8.35. You can see that there are two input lines, one for the count input and another to reset the counter. The counter has an address and a preset number of counts. When the preset number of counts have been accumulated, the counter becomes True and can activate some other part of the ladder diagram.

There are also counters that start with some preset value and count down to zero, at which time the counter becomes True. Different PLC manufacturers have their individual ways of addressing and describing the counter function. Some treat the counter as a device with a logic output, in which case the counter output will be a True or False and fed to some output device.

The following example illustrates a counter application.

EXAMPLE 8.10 Suppose a counter is to be used to count objects in the conveyor delivery system of Figure 8.28. Show how a counter can be set up to count 200 objects and then turn off the conveyor motor. What is the maximum conveyor speed to assure that no objects will be missed in the count if the objects are at least 1 cm apart? The scan time is 30 ms.

Solution

Figure 8.36 shows the conveyor with multiple objects and how a counter can be connected to the object detector so that the conveyor is turned off after counting 200 objects. The object detector is connected to input channel 01, and the conveyor motor to output channel 08. The counter is assigned address 32, and the count is preset to 200. This means that, when a count of 200 occurs, the counter will become true. When it becomes true, the EXAMINE OFF contact 32 in the third rung will become false, and the conveyor motor will turn off. Input channel 02 is unspecified but would be used to reset the counter to start another count of 200.

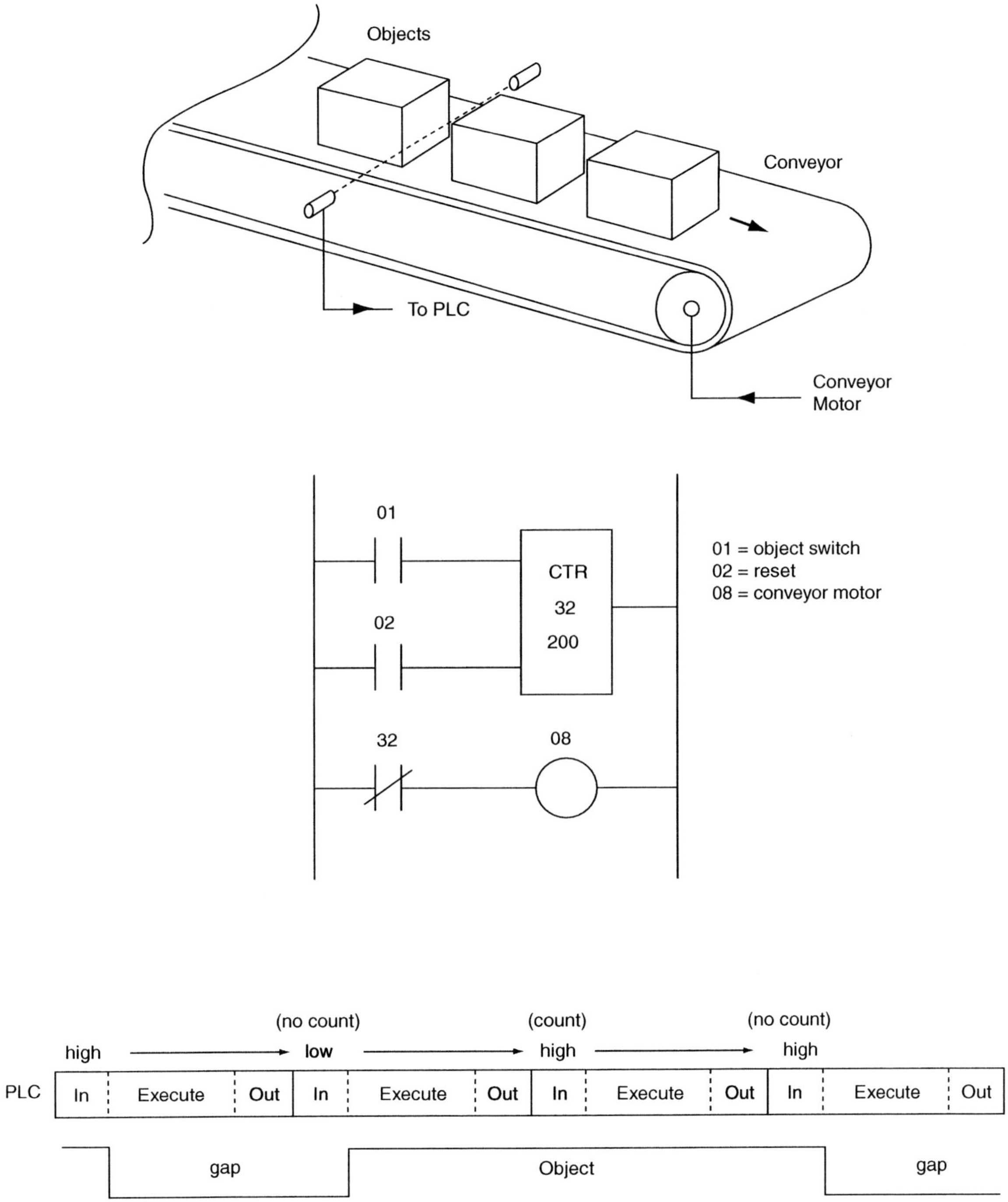

FIGURE 8.36
Program and timing solution to Example 8.10.

Figure 8.36 also shows a timing diagram of the PLC scan cycles and the pulses coming in from detected objects. Remember that there must be a return to zero (false) before the next object can be detected. This means that even the shortest time, which is the 1 cm between objects, must last one full scan to assure a return to zero.

$$t_{\text{count}} = (1\ \text{cm})/(S\ \text{cm/s}) = 30\ \text{ms}$$
$$S = 33.3\ \text{cm/s}$$

Some PLCs also provide a counter that can count either up or down depending upon the state of an input.

Timers The programmed timer function plays an important role in PLC applications to provide for needed delays in some manufacturing sequence and to specify the period of time that some operation is to last. While activated by a true path, the timer begins to accumulate time in the form of "ticks." Each tick is worth a certain amount of time. The timer is pre-loaded with a specified number of these ticks. When the accumulated time ticks equal the pre-load value, the timer itself becomes true.

Figure 8.37a shows that a common representation of the simplest timer function in the ladder diagram is a rectangular box. The timer will have an address and the preset number of ticks to count. The amount of time per tick is a function of the type of PLC being used. Typical values are 10 ms and 100 ms. The timer only counts while it has a true input. If the input becomes False and then True again, the timer will reset to zero and start to count again.

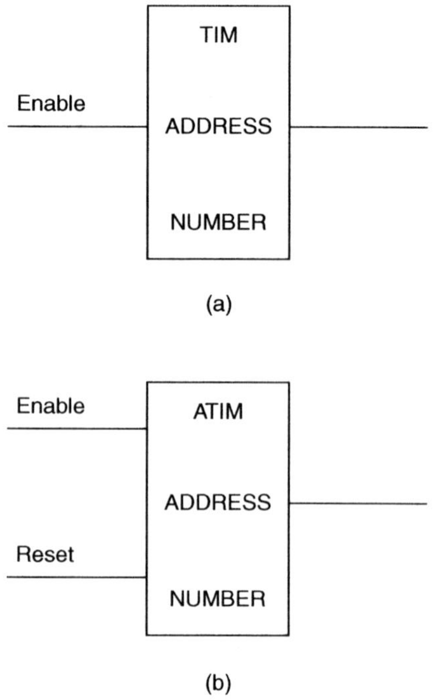

FIGURE 8.37
PLC timer functions: (a) simple timer, (b) accumulating timer.

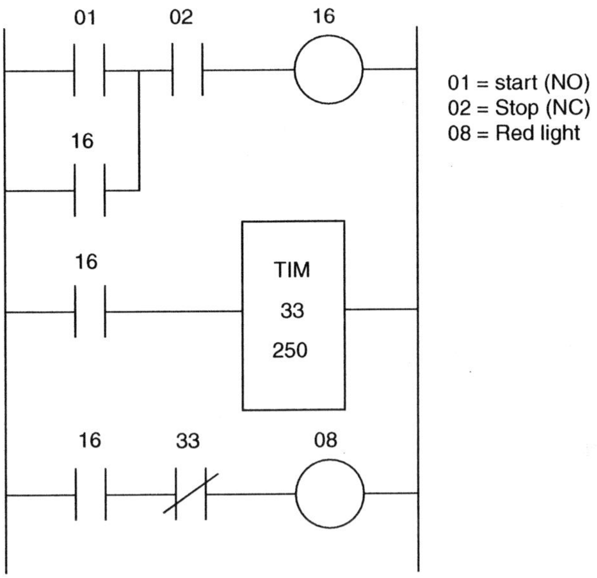

FIGURE 8.38
Solution to Example 8.11.

EXAMPLE 8.11

Show how a timer can be used to turn a red light on for 2500 ms when a NO start push button is pushed. The PLC timer tick is 10 ms. An NC stop button resets the system.

Solution

We need a preset tick count of 2500 ms/10 ms = 250. Let's use address 33 for the timer, 01 for the push button, and 08 for the red light. We need to latch the effect of the push button, since it will likely not be held for the full 2500 ms. Figure 8.38 shows the programmed ladder diagram. The stop button is used to reset the process. Without the latch, the timer would only count while the push button is held down.

Figure 8.37b shows an *accumulating timer,* which will retain a tick count when its input goes false. When the input goes true again, the tick count will pick up where the previous one left off. It is necessary to have a reset input to this timer so that, when desired, the timer can be reset back to zero.

EXAMPLE 8.12

Figure 8.39 shows a P&ID of a chemical vat in which a mixture must be cooked at a temperature greater than 100°C for 10 minutes. Due to external influences, the temperature might fall below 100°C periodically, and this should not be counted in the cooking time. After the vat is filled, a Start push button (NC) starts the cooking. It is terminated by an NC Stop push button. A thermal switch goes high when the temperature is above 100°C. When the mixture has been cooked for 10 minutes at 100°C, the heater should be turned off and the drain valve opened. The PLC tick time is 10 ms.

Solution

Since the tick marks are 10 ms, the required count is (60 s/min)(10 min)/(10 ms) = 6000 counts. A start/stop latch is used with a programmed control relay to start the process with

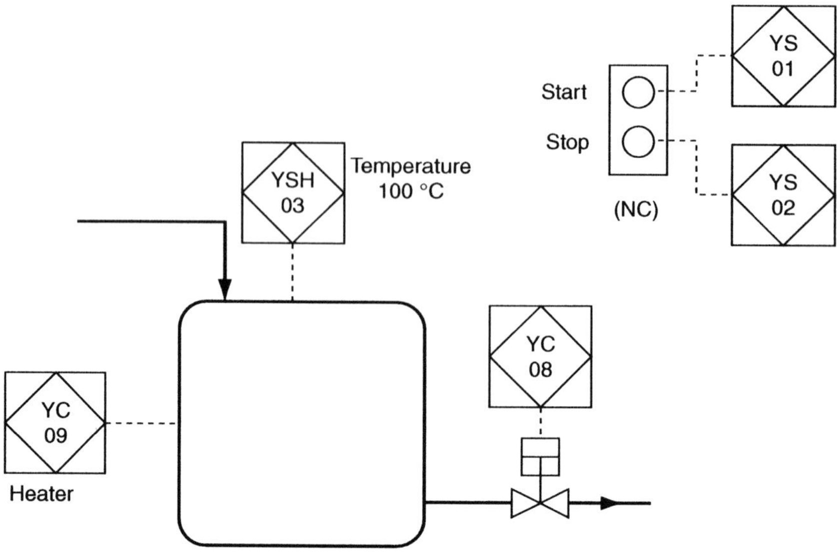

FIGURE 8.39
Chemical system for Example 8.12.

the NC switches addressed at 01 and 02. The accumulating timer (address 32) will count so long as the start is latched and the temperature is above 100°C as indicated by the thermal switch, address 03. As long as the timer has not timed out, the heater is on and the valve is closed. When the timer times out, the heater is turned off and the valve is opened. The timer is reset by pressing the stop button. Figure 8.40 shows the programmed ladder diagram for this system.

Scan-time/Timer Relationship Depending upon the size of the program and the type of PLC, a scan time can range from 5 to 20 ms or more. Care must be taken not to design timer functions with reactions less than a scan time. Suppose a PLC has a 1 ms tick time and we want a light to be triggered 15 ms after a limit switch goes true. In principle, we could use a timer with a preset count of 15 since that would make 15 ms. But if the PLC scan time were 20 ms, there would be an inconsistency. A hardware timer can be used in these instances.

Special Functions The following paragraphs describe briefly some other functions that are available in most PLCs.

The *one-shot* is an output that will go true for only one scan when its input path is True. This device is used for initialization and other instances when an event is to occur only once, even though the conditions that make it True may occur again.

A *shift-register* is used to remember a certain number of previous states of a rung. A four-event shift register could be used to remember the state of a rung for the latest four scans.

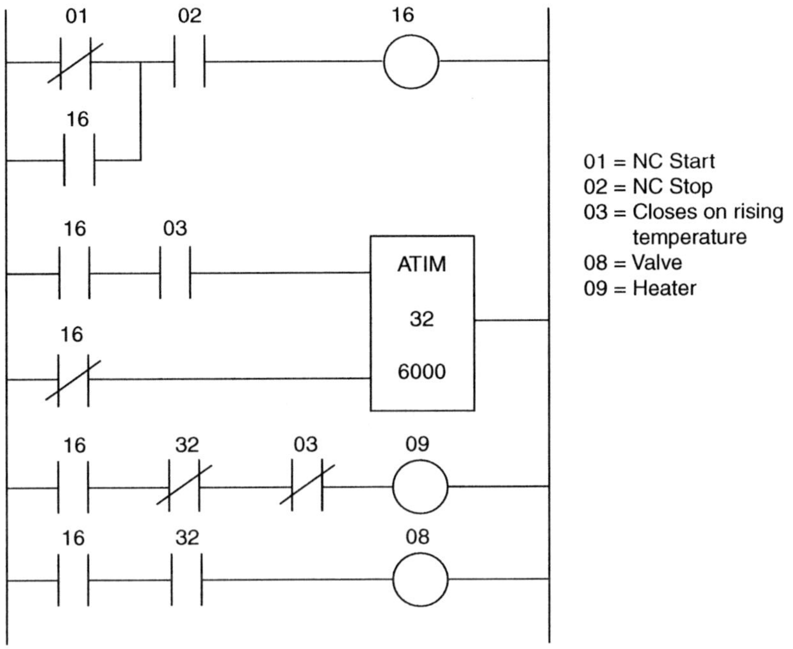

FIGURE 8.40
PLC program solution for Example 8.12.

Modern PLCs are able to handle multi-digit numbers as well as single-bit, binary numbers. Thus, it is possible for a PLC to accept input from an 8-bit ADC and store the result in a known memory location or register. These PLCs have the ability to perform a variety of math operations on this data, from simple addition, subtraction, multiplication, and division, to exponential and trigonometric operations. In this way, the PLC can be made to process continuous plant data such as PID control.

EXAMPLE 8.13 Prepare the physical and programmed ladder diagram for the control problem shown in Figure 8.41. The global objective is to heat a liquid to a specified temperature and keep it there with stirring for 30 min.

The hardware has the following characteristics:

1. START push button is NO, STOP is NC.
2. NO and NC are available for the limit switches.

The event sequence is

1. Fill the tank.
2. Heat and stir the liquid for 30 min.
3. Empty the tank.
4. Repeat from step 1.

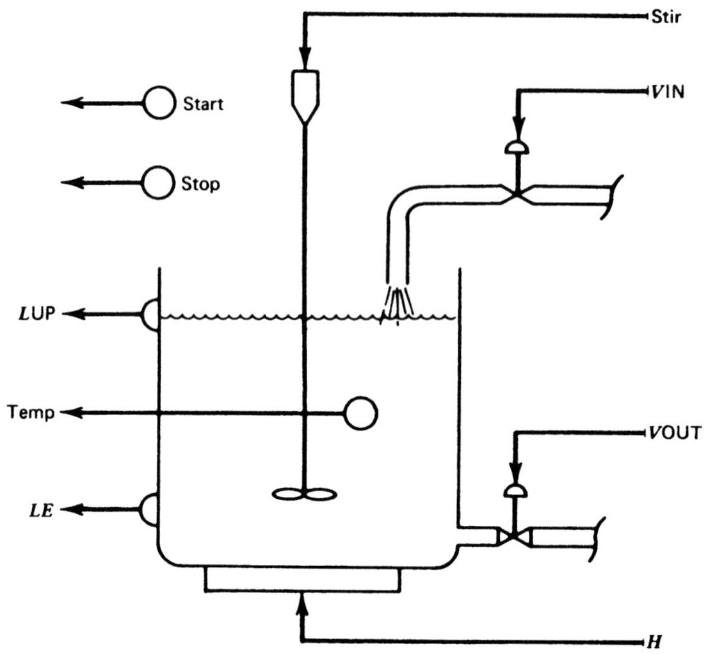

FIGURE 8.41
Tank system for Example 8.13.

Solution

The solution is provided by first constructing the physical ladder diagram. Once this is done, addresses are assigned to all the elements, and the ladder diagram program is prepared.

Figure 8.42 shows the six-rung physical ladder diagram. Rung 2 opens the input valve, provided the output valve is not open, until the full level is reached. When the full level is reached, rung 3 turns on the stir, provided the output valve is not open. Rung 4 starts a 30-min timer. The heater is controlled by rung 5. The rung is energized and de-energized as the temperature goes below and above the limit. When the timer times out, the rung is de-energized, and rung 6 is energized to open the output valve. The output valve remains open until the empty limit switch opens. The output valve cannot be opened as long as the input valve is open.

The programmed ladder diagram is shown in Figure 8.43. Addresses for the input, output, and internal devices have been assigned. Notes indicate how the physical contacts change in response to the control variable. The timer becomes True after 1800 ticks or 30 minutes, so we use an EXAMINE OFF for the timer in rung 5 to assure that the heater can be on only during the 30 minutes the timer is not True. Since the empty valve closes on a rising level, it will be closed until the tank empties. Therefore, we use an EXAMINE ON to assure that the output valve stays open until the tank is empty.

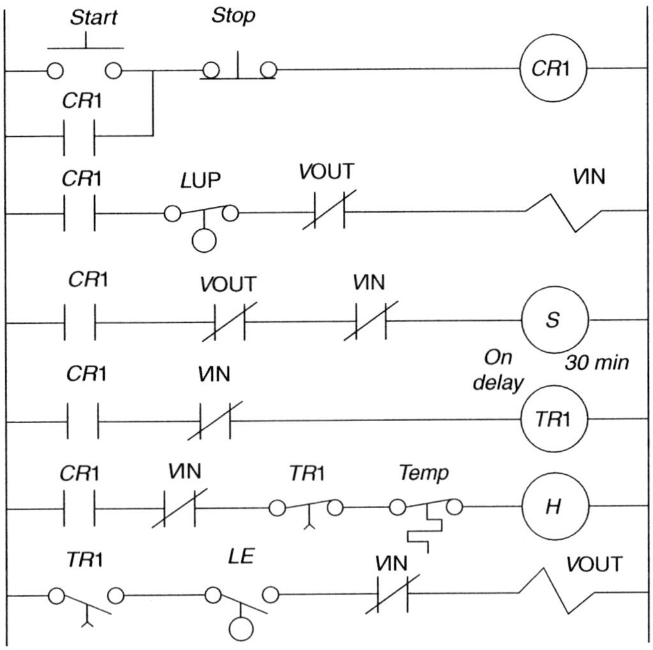

FIGURE 8.42
Physical ladder diagram for Example 8.13.

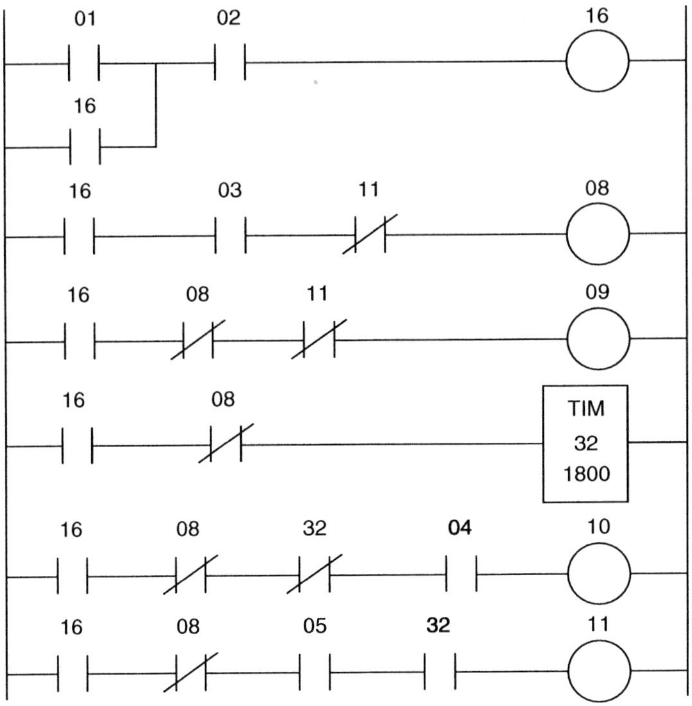

Inputs:
01 = NO Start switch
02 = NC Stop switch
03 = Full level switch (opens on rising level)
04 = Temperature switch (opens on rising temperature)
05 = Empty switch (closes on rising level)

Outputs:
08 = Input valve
09 = Stir motor
10 = Heater
11 = Output valve

Timer tick = 1 second

FIGURE 8.43
Programmed ladder diagram for Example 8.13.

SUMMARY

This chapter has presented the concepts of discrete-state control systems. The topics covered include the following:

1. A discrete-state process is one for which the process variables can take on only two states.
2. A discrete-state process-control system is one that causes the process to pass through a sequence of events. Each event is described by a unique specification of the process variables.
3. The hardware of the process must be carefully defined in terms of the nature of its two states and its relation to the process.
4. The sequence of events can be described in narrative fashion, as a flowchart, or in terms of Boolean equations.
5. A ladder diagram is a schematic way of describing the sequence of events of a discrete-state control system.
6. A programmable controller is a computer-based device that implements the required sequence of events of a discrete-state process.

PROBLEMS

Section 8.3

8.1 Describe a typical microwave oven in the framework of a discrete-state process. Define input variables, output variables, and the sequence of serial/parallel events.

8.2 Develop the control system of an automatic coffee-vending machine. Insertion of a coin and pushing of buttons provides a paper cup with coffee that can be black, with sugar, with cream, or with both. Describe the features of the machine as a discrete-state system.

8.3 Develop a flowchart to describe the event sequence for the running phase of the bottle-filling system of Example 8.3 and Figure 8.5.

8.4 Prepare a flowchart of the operations required to support the coffee-vending machine of Problem 8.2.

8.5 Design a state variable solution to the system of Figure 8.5 such as in Example 8.4. Assume another binary input, R, that is **1** when the system is running.

8.6 Develop Boolean equations to satisfy the requirement of the process of Example 8.4 and Figure 8.11. *Note:* You may need to specify additional hardware.

Section 8.4

8.7 Develop the physical ladder diagram for a motor with the following: NO start button, NC stop button, thermal overload limit switch opens on high temperature, green light when running, red light for thermal overload.

8.8 When turned ON, the tank system of Figure 8.44 alternately fills to level L and then empties to level E. The level switches are activated on a rising level. Both NO and NC connections are available for the level switches and the ON/OFF push buttons. Prepare a physical ladder diagram for this system.

8.9 Develop a ladder diagram that provides for the running phase of the process described in Example 8.3 and Figure 8.5. Assume the switches have both NO and NC contacts.

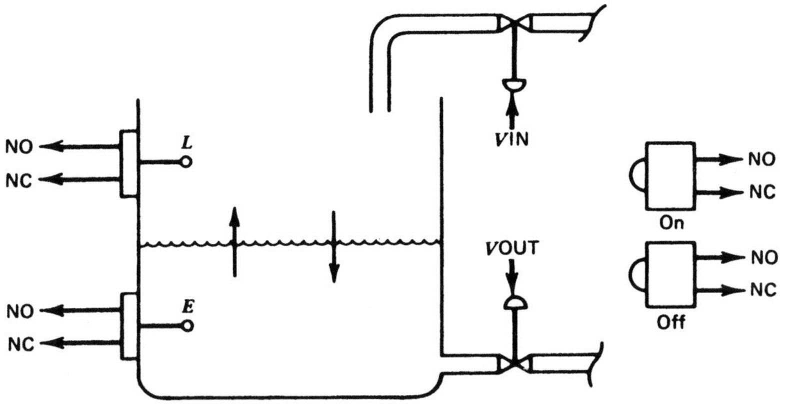

FIGURE 8.44
System for use in Problems 8.8 and 8.14.

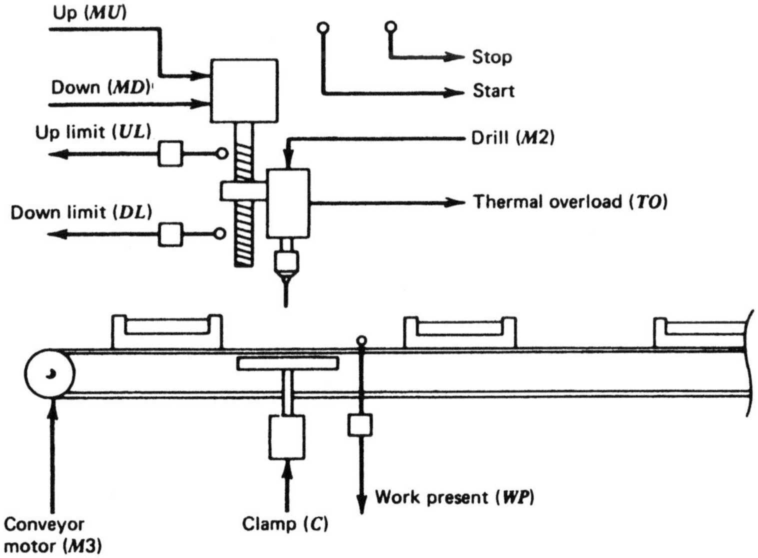

FIGURE 8.45
Process for Problems 8.11 and 8.16.

8.10 Prepare a ladder diagram for the process described in Example 8.4 and Figure 8.11. Assume a logic off-delay 5-minute timer (i.e., goes true when enabled and stays true for 5 min). Assume a NO start push button and an NC stop push button.

8.11 The system of Figure 8.45 has the following functions: move a work piece into position, clamp it, start the drill, move it down to drill a hole in the work piece (as long as a thermal overload does not occur), back the drill out, turn the drill off, and then repeat for a new work piece. Assume master NO start and NC stop buttons and that limit switches have both NO and NC contacts. The system stops if a thermal overload occurs and turns on a red light (not shown). Develop the physical ladder diagram.

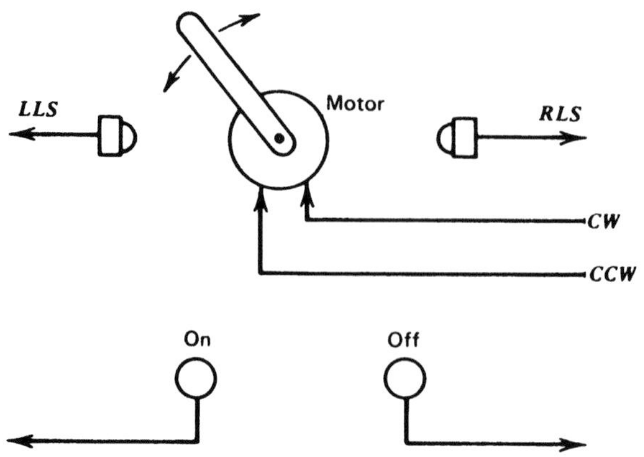

FIGURE 8.46
Motor and switches for Problem 8.15.

Section 8.5

Note: For the following problems, use the programmable controller addresses given in Section 8.5.4.

8.12 Design a programmable controller program to solve the control problem for the process described in Example 8.4 and Figure 8.11.

8.13 Prepare a programmable controller ladder diagram for Problem 8.7.

8.14 Prepare a PLC ladder program for the system of Problem 8.8 and Figure 8.44 with the following added requirements: (1) The system should cycle 100 times and then quit, and (2) during each cycle, there should be a 1.5-minute delay after filling before the empty phase starts. The PLC tick time is 10 ms.

8.15 When the system of Figure 8.46 is turned ON, the motor is to alternate rotation *CW* and then *CCW*, cycling, as the shaft extension contacts the two limit switches, *RLS* and *LLS*. All four switches have only normally closed positions. Prepare a PLC ladder program with the following requirements:

1. When the On button is pushed, the system motor moves the arm to the right-limit switch position and waits 30 seconds.
2. The system then cycles 75 times between the right- and left-limit switches, and then stops.
3. The Off button stops the system at any time or, after the 75 cycles have been made, resets the system.

8.16 Prepare a programmed ladder diagram for Problem 8.11.

SUPPLEMENTARY PROBLEMS

S8.1 Figure 8.47 shows a system for batch processing. The system operations can be described as follows: (1) A weighed quantity of dry material is added to liquid that has

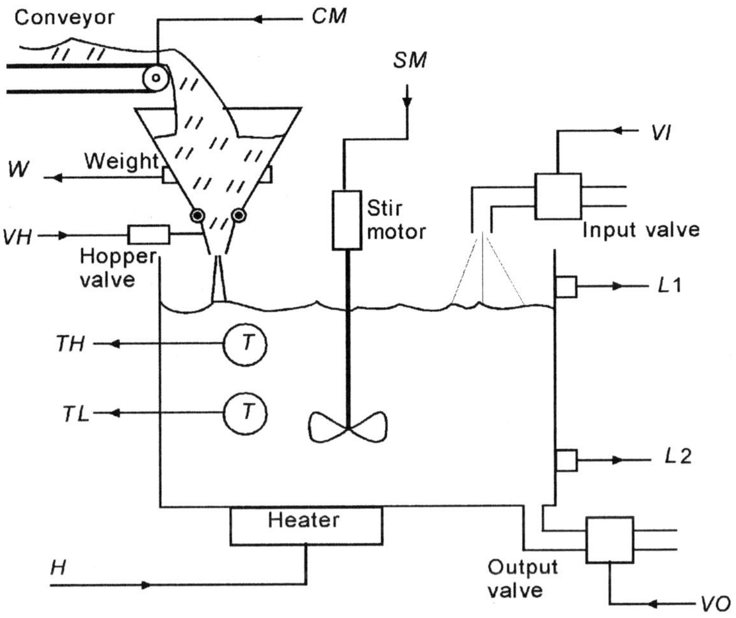

FIGURE 8.47
System for Problems S8.1, S8.2, and S8.3.

filled the tank to level *L*1. (2) The mixture is stirred, and heat is applied to maintain the temperature between *TL* and *TH* for a period of 10 minutes. (3) With the stir and heater off, the batch is emptied down to level *L*2. The whole process then starts again. Prepare a flowchart that describes the sequence of events. Be sure to consider any events that could occur in parallel.

S8.2 All the sensors of Figure 8.47 are NO switches that close when the variable reaches the trip value. Thus, *TH* represents a NO connection that closes when the temperature exceeds *TH*, *L*1 closes when the level exceeds *L*1, and so forth. The conveyor and stir are motors; the valves and heater are all solenoids. Prepare a physical ladder diagram for the system of Figure 8.47 as described in Problem S8.1. Be sure to consider any operations that can be executed simultaneously.

S8.3 A PLC will be used with channels assigned as given in Section 8.5.4. Prepare a programmed ladder diagram for the system of Figure 8.47 as described in Problem S8.1. All switches are NO. Thus, *TH* closes when the temperature reaches *TH*, *L*1 closes when the level exceeds *L*1, and so forth.

S8.4 Figure 8.48 shows an automated hydroponics system. A PLC will be used to provide the following sequence of light and nutrient. The PLC has a 10-ms timer tick time, and the maximum count of timer and counters is 9999.

1. The growing vat must have light of minimum intensity for at least 8 hours. A photo-switch, which closes on rising light level, monitors natural light during

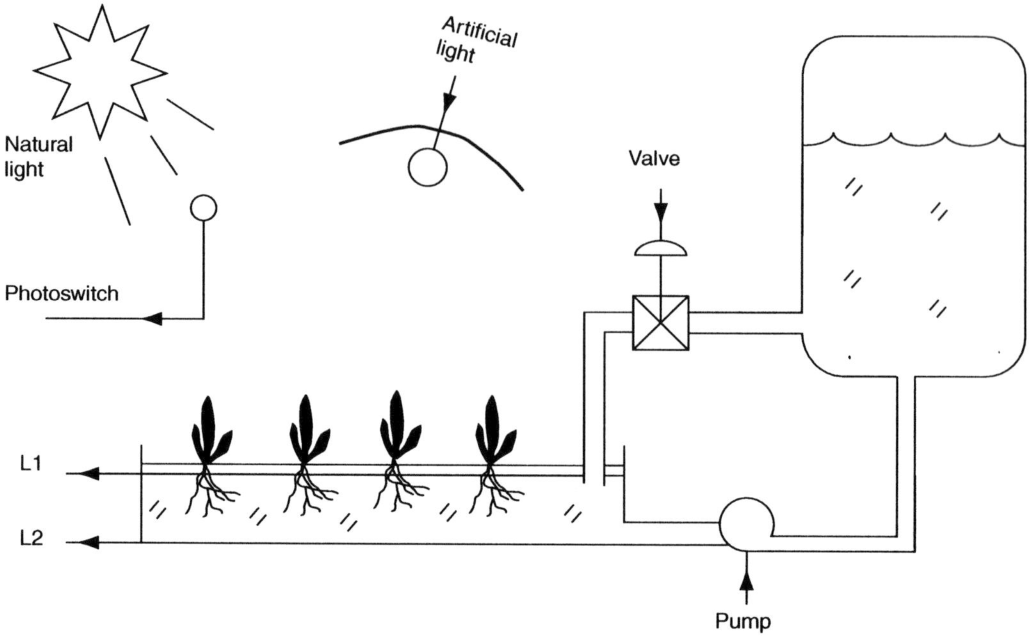

FIGURE 8.48
Hydroponic system for Problem S8.4.

this period. If the light intensity falls below the switch level for more than 10 minutes, an artificial light is turned on until the natural light brightens again.

2. During the 8-hour light period, nutrient must be cycled into the vat every 2 hours and left for 2 hours. Nutrient is filled to $L1$ (closes on rising level) and emptied to $L2$ (opens with dropping level).
3. During the remaining 16 hours, nutrient is pumped in, left for 3 hours, pumped out and left out for 1 hour, cycling for the 16 hours.

CHAPTER 9

Controller Principles

INSTRUCTIONAL OBJECTIVES

This chapter presents the operational modes of a process-control loop. A mode is determined by the nature of controller responses to a controlled variable measurement and setpoint comparison. After you have read this chapter, you should be able to

- Define process load, process lag, and self-regulation.
- Describe two-position and floating-control mode.
- Define the proportional controller mode.
- Give an example and description of an integral-control mode.
- Describe the derivative-control mode.
- Contrast proportional-integral and proportional-derivative control modes.
- Describe three-mode controllers.
- Provide a description of the controller output for a fixed error input of any of the controller modes.

9.1 INTRODUCTION

In Chapter 8, we studied the nature and implementation of controllers for discrete-state control processes. In those processes, the operations and variables were in one of only two states: ON or OFF. In the present chapter, we consider the nature of controller action for systems with operations and variables that range over continuous values. The controller inputs the result of a measurement of the controlled variable and determines an appropriate output to the final control element. Essentially, the controller is some form of computer—either analog or digital, pneumatic or electronic—that, using input measurements, solves certain equations to calculate the proper output. The equations necessary to obtain control exist in only a few forms, independent of both the process itself and whether the controller function is provided by an analog or digital computer. These equations describe the modes or action of controller operation. The nature of the process itself and the particular variable controlled determine which mode or modes of control are to be used and the value of certain constants in the mode equations. In this chapter, we will study the various modes of controller operation. Later chapters will examine how the modes are implemented by analog or digital means and by pneumatic and electronic means.

9.2 PROCESS CHARACTERISTICS

The selection of what controller modes to use in a process is a function of the characteristics of the process. It is not our intention to discuss how the modes are selected but to define the meaning of each mode. At the same time, it is helpful in understanding the modes if certain pertinent characteristics of the process are considered. In this section, we will define a few properties of processes that are important for selecting the proper modes.

9.2.1 Process Equation

A process-control loop regulates some *dynamic variable* in a process. This controlled variable, a process parameter, may depend on many other parameters (in the process) and thus suffer changes from many different sources. We have selected one of these other parameters to be our controlling parameter. If a measurement of the controlled variable shows a deviation from the setpoint, then the controlling parameter is changed, which in turn changes the controlled variable.

As an example, consider the control of liquid temperature in a tank, as shown in Figure 9.1. The *controlled variable* is the liquid temperature, T_L. This temperature depends on many parameters in the process—for example, the input flow rate via pipe *A*, the output flow rate via pipe *B*, the ambient temperature, T_A, the steam temperature, T_S, inlet temperature, T_0, and the steam flow rate, Q_S. In this case, the steam flow rate is the *controlling parameter* chosen to provide control over the variable (liquid temperature). If one of the other parameters changes, a change in temperature results. To bring the temperature back to the setpoint value, we change only the steam flow rate—that is, heat input to the process. This process could be described by a *process equation* where liquid temperature T_L is a function as

$$T_L = F(Q_A, Q_B, Q_S, T_A, T_S, T_0) \tag{9.1}$$

where
Q_A, Q_B = flow rates in pipes *A* and *B*
Q_S = steam flow rate
T_A = ambient temperature
T_0 = inlet fluid temperature
T_S = steam temperature

To provide control via Q_S, we do not need to know the functional relationship exactly, nor do we require linearity of the function. The control loop adjusts Q_S and thereby regulates T_L, regardless of how the other parameters in Equation (9.1) vary with each other. In many cases, the relationship of Equation (9.1) is not even analytically known.

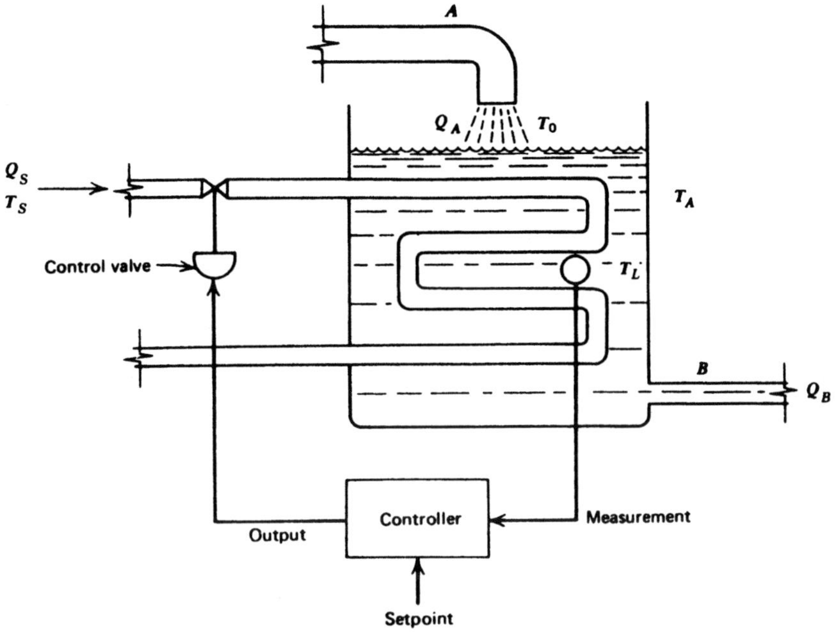

FIGURE 9.1
Control of temperature by process control.

9.2.2 Process Load

From the process equation, or knowledge of and experience with the process, it is possible to identify a set of values for the process parameters that results in the controlled variable having the setpoint value. This set of parameters is called the *nominal set.* The term *process load* refers to this set of all parameters, excluding the controlled variable. When all parameters have their nominal values, we speak of the *nominal load* on the system. The required controlling variable value under these conditions is the *nominal value* of that parameter. If the setpoint is changed, the control parameter is altered to cause the variable to adopt this new operating point. The load is still nominal, however, because the other parameters are assumed to be unchanged. Suppose one of the parameters changes from nominal, causing a corresponding shift in the controlled variable. We then say that a *process load change* has occurred. The controlling variable is adjusted to compensate for this load change and its effect on the dynamic variable to bring it back to the setpoint. In the example of Figure 9.1, a process load change is caused by a change in any of the five parameters affecting liquid temperature. The extent of the load change on the controlled variable is formally determined by process equations such as Equation (9.1). In practice, we are concerned only that variation in the controlling parameter brings the controlled variable back to the setpoint. We are not necessarily concerned with the cause, nature, or extent of the load change.

Transient Another type of change involves a temporary variation of one of the load parameters. After the excursion, the parameter returns to its nominal value. This variation is called a *transient.* A transient causes variations of the controlled variable, and the control system must make equally transient changes of the controlling variable to keep error to a minimum. A transient is not a load change because it is not permanent.

9.2.3 Process Lag

As previously noted, process-control operations are essentially a time-variation problem. At some point in time, a process-load change or transient causes a change in the controlled variable. The process-control loop responds to ensure that, some finite time later, the variable returns to the setpoint value. Part of this time is consumed by the process itself and is called the *process lag.* Thus, referring to Figure 9.1, assume the inlet flow is suddenly doubled. Such a large process-load change radically changes (reduces) the liquid temperature. The control loop responds by opening the steam inlet valve to allow more steam and heat input to bring the liquid temperature back to the setpoint. The loop itself reacts faster than the process. In fact, the physical opening of the control valve is the slowest part of the loop. Once steam is flowing at the new rate, however, the body of liquid must be heated by the steam before the setpoint value is reached again. This time delay or process lag in heating is a function of the process, *not* the control system. Clearly, there is no advantage in designing control systems many times faster than the process lag.

9.2.4 Self-Regulation

A significant characteristic of some processes is the tendency to adopt a specific value of the controlled variable for nominal load with no control operations. The control operations may be significantly affected by such *self-regulation.* The process of Figure 9.1 has self-regulation, as shown by the following argument.

(1) Suppose we fix the steam valve at 50% and open the control loop so that no changes in valve position are possible. (2) The liquid heats up until the energy carried away by the liquid equals that input energy from the steam flow. (3) If the load changes, a new temperature is adopted (because the system temperature is not controlled). (4) The process is self-regulating, however, because the temperature will not "run away," but stabilizes at some value under given conditions.

An example of a process without self-regulation is a tank from which liquid is pumped at a fixed rate. Assume that the influx just matches the outlet rate. Then the liquid in the tank is fixed at some nominal level. If the influx increases slightly, however, the level rises until the tank overflows. No self-regulation of the level is provided.

9.3 CONTROL SYSTEM PARAMETERS

We have just described the basic characteristics of the process that are related to control. Let us now examine the general properties of the controller shown in Figure 9.2.

To review: (1) Inputs to the controller are a measured indication of both the controlled variable and a setpoint representing the desired value of the variable, expressed in the same

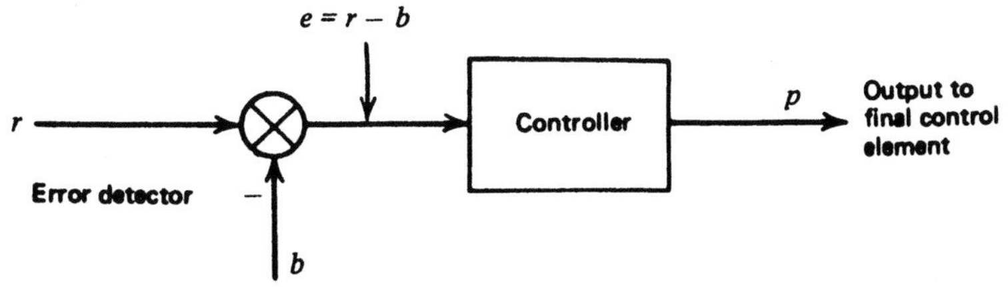

FIGURE 9.2
The error detector and controller.

fashion as the measurement; (2) the controller output is a signal representing action to be taken when the measured value of the controlled variable deviates from the setpoint.

The measured indication of a variable is denoted by *b*, and the actual variable is denoted by *c*. Thus, if a sensor measures temperature by conversion to resistance, the actual variable is temperature in degrees Celsius, but the measured indication is resistance in ohms.

Further conversion may be performed by transducers or transmitters to provide a current in mA, for example. In such a case, the current becomes the measured indication of the variable.

9.3.1 Error

The deviation or error of the controlled variable from the setpoint is given by

$$e = r - b \tag{9.2}$$

where e = error
b = measured indication of variable
r = setpoint of variable (reference)

Equation (9.2) expresses error in an absolute sense, or in units of the measured analog of the control signal. Thus, if the setpoint in a 4- to 20-mA range corresponds to 9.9 mA and the measured value is 10.7 mA, we have an error of −0.8 mA. Obviously, this current error has little direct meaning unless related to the controlled variable. We could work back through the loop and prove that it corresponds to a flow rate of 1.1 m^3/h, for example. This would show the significance of the error relative to the actual process-control loop.

To describe controller operation in a general way, it is better to express the error as percent of the measured variable range (i.e., the span). The measured value of a variable can be expressed as percent of span over a range of measurement by the equation

$$c_p = \frac{c - c_{min}}{c_{max} - c_{min}} \times 100$$

where c_p = measured value as percent of measurement range
c = actual measured value
c_{max} = maximum of measured value
c_{min} = minimum of measured value

The previous equation is in terms of the actual measured variable, c, but the same equation can be expressed in terms of the measured indication, b. It is necessary only to translate the measured minimum and maximum to b_{min} and b_{max}.

To express error as percent of span, it is necessary only to write both the setpoint and measurement in terms of percent of span and take the difference according to Equation (9.2). The result is

$$e_p = \frac{r - b}{b_{max} - b_{min}} \times 100 \tag{9.3}$$

where e_p = error expressed as percent of span

You can see the convenience of using a standard measured indication range like 4 to 20 mA, because the span is always 16 mA. Suppose we have a setpoint of 10.5 mA and a measurement of 13.7 mA. Then, without even knowing what is being measured, we know the error is

$$e_p = \frac{10.5\text{ mA} - 13.7\text{ mA}}{20\text{ mA} - 4\text{ mA}} \times 100$$

$$e_p = -20\%$$

A positive error indicates a measurement below the setpoint, and a negative error indicates a measurement above the setpoint.

EXAMPLE 9.1 The temperature in Figure 9.1 has a range of 300 to 440 K and a setpoint of 384 K. Find the percent of span error when the temperature is 379 K.

Solution

The percent error is

$$e_p = \frac{r - b}{b_{max} - b_{min}} \times 100$$

$$e_p = \frac{384 - 379}{440 - 300} \times 100$$

$$e_p = \mathbf{3.6\%}$$

9.3.2 Variable Range

Generally, the variable under control has a range of values within which control is to be maintained. This range can be expressed as the minimum and maximum value of the variable or the nominal value plus and minus the spread about this nominal value. If a standard 4- to 20-mA signal transmission is employed, then 4 mA represents the minimum value of the variable and 20 mA the maximum.

When a computer-based control system is used, the dynamic variable is converted to an n-bit digital signal. Often, the transformation is made so that all **0**'s are the minimum value of the variable and all **1**'s are the maximum value.

9.3.3 Control Parameter Range

Another range is associated with the controller output. Here we assume the final control element has some minimum and maximum effect on the process. The controller output range is the translation of output to the range of possible values of the final control element. This range is also expressed as the 4- to 20-mA standard signal, again with the minimum and maximum effects in terms of the minimum and maximum current.

Similarly, in computer-based control, the output will range over all states of the *n*-bit output. Generally, all **0**'s are the minimum output and all **1**'s the maximum. These numbers do not necessarily represent the minimum and maximum of the final control element, however. We may wish a valve never to be fully closed, for example; therefore, all **0**'s might represent some percentage of full open.

Often, the output is expressed as a percentage where 0% is the minimum controller output and 100% the maximum (obviously). Thus, in the example of Figure 9.1, the valve in the fully open position corresponds to a 100% controller signal output. Often, however, the minimum does *not* correspond to zero effect. For example, it may be that the steam flow should never be less than that flow which results with the valve half open. In this case, a 0% minimum controller corresponds to the flow rate with a half-open valve.

The controller output as a percent of full scale when the output varies between specified limits is given by

$$p = \frac{u - u_{min}}{u_{max} - u_{min}} \times 100 \quad \textbf{(9.4)}$$

where p = controller output as percent of full scale
u = value of the output
u_{max} = maximum value of controlling parameter
u_{min} = minimum value of controlling parameter

EXAMPLE 9.2

A controller outputs a 4- to 20-mA signal to control motor speed from 140 to 600 rpm with a linear dependence. Calculate **(a)** current corresponding to 310 rpm, and **(b)** the value of (a) expressed as the percent of control output.

Solution

a. We find the slope m and intersect S_0 of the linear relation between current I and speed S, where

$$S_p = mI + S_0$$

Knowing S_p and I at the two given positions, we write two equations:

$$140 = 4m + S_0$$
$$600 = 20m + S_0$$

Solving these simultaneous equations, we get $m = 28.75$ rpm/mA and $S_0 = 25$ rpm. Thus, at 310 rpm we have $310 = 28.75I + 25$, which gives $I = 9.91$ mA.

b. Expressed as a percentage of the 4- to 20-mA range, this controller output is

$$p = \frac{u - u_{\min}}{u_{\max} - u_{\min}} \times 100$$

$$p = \left[\frac{9.91 - 4}{20 - 4}\right] \times 100$$

$$p = \mathbf{36.9\%}$$

9.3.4 Control Lag

The control system also has a lag associated with its operation that must be compared to the process lag (see Section 9.2.3). When a controlled variable experiences a sudden change, the process-control loop reacts by outputting a command to the final control element to adopt a new value to compensate for the detected change. *Control lag* refers to the time for the process-control loop to make necessary adjustments to the final control element. Thus, in Figure 9.1, if a sudden change in liquid temperature occurs, it requires some finite time for the control system to physically actuate the steam control valve.

9.3.5 Dead Time

Another time variable associated with process control is a function of both the process-control system and the process. This is the elapsed time between the instant a deviation (error) occurs and when the corrective action first occurs. An example of *dead time* occurs in the control of a chemical reaction by varying reactant flow rate through a long pipe. When a deviation is detected, a control system quickly changes a valve setting to adjust flow rate. But if the pipe is particularly long, there is a period of time during which no effect is felt in the reaction vessel. This is the time required for the new flow rate to move down the length of the pipe. Such dead times can have a profound effect on the performance of control operations on a process.

9.3.6 Cycling

We frequently refer to the behavior of the dynamic variable error under various modes of control. One of the most important modes is an oscillation of the error about zero. This means the variable is *cycling* above and below the setpoint value. Such cycling may continue indefinitely, in which case we have *steady-state cycling.* Here we are interested in both the peak amplitude of the error and the period of the oscillation.

If the cycling amplitude decays to zero, however, we have a *cyclic transient error.* Here we are interested in the initial error, the period of the cyclic oscillation, and decay time for the error to reach zero.

9.3.7 Controller Modes

The controller was defined in Section 9.3 by the statement that a controller generates a control signal to the final element, based on a measured deviation of the controlled variable from the setpoint. It is natural to ask how the controller responds to the deviation. In a thermostatically controlled temperature system used in the home, the controller response is simple. If the temperature drops below the thermostat setpoint, a bimetallic relay turns on a heater.

But consider the case of the system shown in Figure 9.1. No simple ON/OFF decision can be made because the setting of the steam valve can be smoothly varied from one extreme to another. Thus, if a deviation from liquid temperature setpoint occurs, what should the controller do? Should it open the valve a little or a lot? Should it open the valve fast or slowly?

These questions are answered by specifying the *mode* of the controller operation. One distinction is clear from the earlier examples. The domestic thermostat involves a mode that is *discontinuous,* where the controller command initiates a discontinuous change in the control parameter. The process of Figure 9.1 is *continuous,* because smooth variation of the control parameter is possible. Section 9.4 covers various controller modes in detail.

The choice of operating mode for any given process-control system is a complicated decision. It involves not only process characteristics but cost analysis, product rate, and other industrial factors. At the outset, the process-control technologist should have a good understanding of the operational mechanism of each mode and its advantages and disadvantages. The operation of each mode is defined later, and examples are given with some general statements of application details. In each case, the output of the controller is described by a factor p. This is the *percent of controller output* relative to its total range, as defined in Equation (9.4).

For example, if a controller outputs a 4- to 20-mA current signal to the final control element and has a $p = 25\%$, then the corresponding current is

$$I = I_{\min} + p(I_{\max} - I_{\min})$$
$$I = 4\text{ mA} + (0.25)(20 - 4)\text{mA}$$
$$I = 8\text{ mA}$$

If this current is used to drive a value actuator for which 4 mA is closed and 20 mA is full open, then the valve is 25% open. If the valve is an equal-percentage type with a rangeability of 30, then the flow rate (from Section 7.6.3) is

$$Q = Q_{\min} R^{S/S_{\max}}$$
$$Q = Q_{\min}(30)^{0.25} \qquad \textbf{(7.16)}$$
$$Q = 2.34\, Q_{\min}$$

The example shows that if the percentage output of the controller is known, then the actual value of the controlled variable can be determined.

The input of the controller is described by the error, e_p, defined in Equation (9.3) as the percentage error of the measured variable from the setpoint relative to range. In general, the controller operation is expressed as a relation:

$$p = F(e_p) \qquad \textbf{(9.5)}$$

where $F(e_p)$ represents the relation by which the appropriate controller output is determined. In some cases, a graph of p versus e_p is also employed to aid in a definition of the control mode.

Reverse and Direct Action The error that results from measurement of the controlled variable may be positive or negative, because the value may be greater or less than the setpoint. How this polarity of the error changes the controller output can be selected according to the nature of the process.

A controller operates with *direct action* when an increasing value of the controlled variable causes an increasing value of the controller output. An example would be a level-control system that outputs a signal to an output valve. Clearly, if the level rises (increases), the valve should be opened (i.e., its drive signal should be increased).

Reverse action is the opposite case, where an increase in a controlled variable causes a decrease in controller output. An example of this would be a simple temperature control from a heater. If the temperature increases, the drive to the heater should be decreased.

9.4 DISCONTINUOUS CONTROLLER MODES

This section discusses the various controller modes that show discontinuous changes in controller output as controlled variable error occurs. It is important that you understand these modes, both because of their frequent use in process control, and because they form the basis of the continuous modes to be discussed in Section 9.5.

9.4.1 Two-Position Mode

The most elementary controller mode is the ON/OFF, or two-position, mode. This is an example of a discontinuous mode. It is the simplest and the cheapest, and often suffices when its disadvantages are tolerable. Although an analytic equation cannot be written, we can, in general, write

$$p = \begin{cases} 0\% & e_p < 0 \\ 100\% & e_p > 0 \end{cases} \tag{9.6}$$

This relation shows that when the measured value is less than the setpoint, full controller output results. When it is more than the setpoint, the controller output is zero. A space heater is a common example. If the temperature drops below a setpoint, the heater is turned ON. If the temperature rises above the setpoint, it turns OFF.

Neutral Zone In virtually any practical implementation of the two-position controller, there is an overlap as e_p increases through zero or decreases through zero. In this span, no change in controller output occurs. This is best shown in Figure 9.3, which plots p versus e_p for a two-position controller. We see that until an increasing error changes by Δe_p above zero, the controller output will not change state. In decreasing, it must fall Δe_p below zero before the controller changes to the 0% rating. The range 2 Δe_p, which is referred to as the *neutral zone* or *differential gap,* is often purposely designed above a certain minimum quantity to prevent excessive cycling. The existence of such a neutral zone is an example of desirable hysteresis in a system.

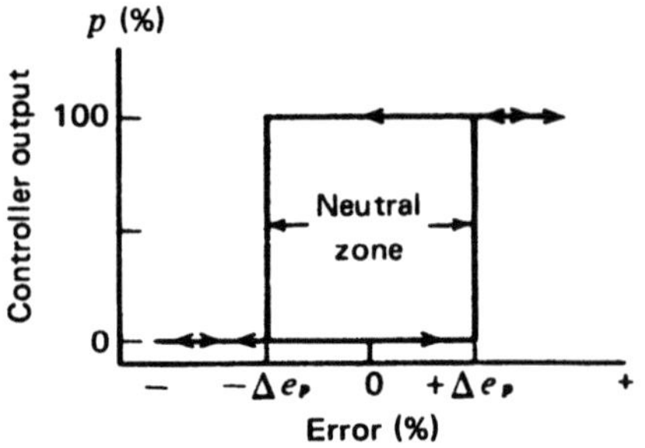

FIGURE 9.3
Two-position controller action with neutral zone.

EXAMPLE 9.3 A liquid-level control system linearly converts a displacement of 2 to 3 m into a 4- to 20-mA control signal. A relay serves as the two-position controller to open or close an inlet valve. The relay closes at 12 mA and opens at 10 mA. Find **(a)** the relation between displacement level and current, and **(b)** the neutral zone or displacement gap in meters.

Solution

a. The relation between level and a current is a linear equation such as

$$H = KI + H_0$$

We find K and H_0 by writing two equations:

$$2\text{ m} = K(4\text{ mA}) + H_0$$
$$3\text{ m} = K(20\text{ mA}) + H_0$$

Solving these simultaneous equations yields $K = 0.0625\text{ m/mA}$ and $H_0 = 1.75\text{ m}$, at the intersection of the linear relations.

b. The relay closes at 12 mA, which is a high level, H_H, of

$$H_H = (0.0625\text{ m/mA})(12\text{ mA}) + 1.75\text{ m}$$
$$H_H = 2.5\text{ m}$$

The low level, H_L, occurs at 10 mA, which is

$$H_L = (0.0625\text{ m/mA})(10\text{ mA}) + 1.75\text{ m}$$
$$H_L = 2.375\text{ m}$$

Thus, the neutral zone is $H_H - H_L = (2.5 - 2.375)$ m, or **0.125 m.**

Applications Generally, the two-position control mode is best adapted to large-scale systems with relatively slow process rates. Thus, in the example of either a room heating or air-conditioning system, the capacity of the system is very large in terms of air volume, and the overall effect of the heater or cooler is relatively slow. Sudden, large-scale changes are not common to such systems. Other examples of two-position control applications are liquid bath-temperature control and level control in large-volume tanks. The process under two-position control must allow continued oscillation in the controlled variable because, by its very nature, this mode of control always produces such oscillation. For large systems, these oscillations are of long duration, which is partly a function of the neutral-zone size. To illustrate this, consider the following example.

EXAMPLE 9.4 The temperature of water in a tank is controlled by a two-position controller. When the heater is *off* the temperature drops at 2 K per minute. When the heater is *on* the temperature rises at 4 K per minute. The setpoint is 323 K and the neutral zone is $\pm 4\%$ of the setpoint. There is a 0.5-min lag at both the *on* and *off* switch points. Find the period of oscillation and plot the water temperature versus time.

Solution

The neutral zone in temperature is $\Delta T = \pm 4\% \ (323 \text{ K}) = \pm 12.92$ K. Therefore, the switching temperatures are about 336 K and 310 K. When the system is cooling the temperature can be written,

$$T_c(t) = T(t_0) - 2(t - t_0) \tag{9.7}$$

where t_0 is the starting time and $T(t_0)$ is the temperature at that time. Similarly, when heating the temperature rises according to the relation,

$$T_h(t) = T(t_1) + 4(t - t_1) \tag{9.8}$$

The effect of the overshoot is found as follows. Suppose the heater is *on* and the upper trip point, 336 K, was reached at some time t. The system will continue to heat for another 0.5 min at 4 K/min (i.e., another 2 K). Therefore, when the system really starts cooling the temperature is 338 K. In the same sense, when cooling to 310 K the system will continue to cool for another 0.5 min at -2 K/min or -1 K, so the actual heating starts from 309 K.

Suppose at $t = 0$ the temperature has reached the "real" upper trip point of 338 K and is now cooling. It will then drop according to Equation (9.7), $T_c(t) = 338 - 2t$. We can find the time when it reaches 309 K by solving, $309 = 338 - 2t_1$, so $t_1 = 14.5$ min. Therefore, at 14.5 minutes it starts heating. We can find the time it reaches 338 K again from Equation (9.8), for heating, $338 = 309 + 4(t_2 - 14.5)$, so $t_2 = 21.75$ min. Thus, the total time for one cycle, which is the period, is 21.75 minutes. Figure 9.4 shows a plot of the temperature variation. (The plot was constructed with a starting point of the setpoint, 323 K and cooling, instead of the upper trip point used in the previous calculations.)

FIGURE 9.4
Figure for Example 9.4.

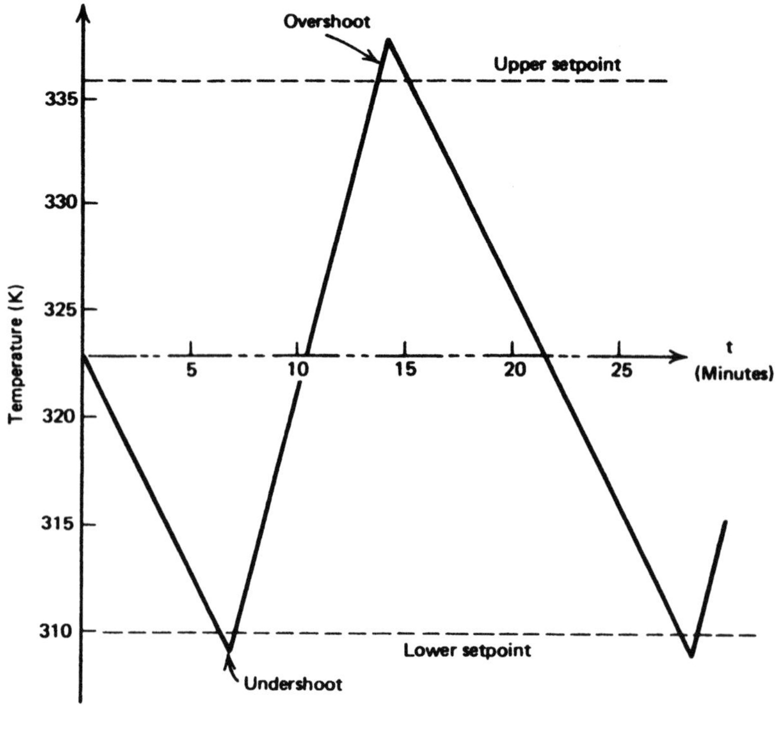

In general, some overshoot and undershoot of the controlled variable will occur, as in Example 9.4. This is due to the finite time required for the control element to impress its full effect on the process. Thus, the finite warm-up time and cool-off time of the heater (included in Example 9.4) caused some overshoot and undershoot of temperature. In some cases, if the final control element lag is large, substantial errors can result, and the neutral zone must be reduced to lower these errors. In general, the cycling, as noted in the previous example, is a function of the neutral zone. If the neutral zone of this example is reduced to $\pm 2\%$, the reader can verify that, although tighter control is now maintained, the cycling period is reduced to 11.91 min. The solution for these values is left for the reader to calculate.

9.4.2 Multiposition Mode

A logical extension of the previous two-position control mode is to provide several intermediate, rather than only two, settings of the controller output. This discontinuous control mode is used in an attempt to reduce the cycling behavior and overshoot and undershoot inherent in the two-position mode. In fact, however, it is usually more expedient to use some other mode when the two-position is not satisfactory. This mode is represented by

$$p = p_i \qquad e_p > |e_i| \; i = 1, 2, \ldots, n \tag{9.9}$$

As the error exceeds certain set limits $\pm e_i$, the controller output is adjusted to preset values, p_i. The most common example is the three-position controller where

$$p = \begin{cases} 100 & e_p > e_2 \\ 50 & -e_1 < e_p < e_2 \\ 0 & e_p < -e_1 \end{cases} \tag{9.10}$$

As long as the error is between e_2 and e_1 of the setpoint, the controller stays at some nominal setting indicated by a controller output of 50%. If the error exceeds the setpoint by e_1 or more, then the output is increased by 100%. If it is less than the setpoint by $-e_1$ or more, the controller output is reduced to zero.

Figure 9.5 illustrates this mode graphically. Some small neutral zone usually exists about the change points, but not by design; thus, it is not shown. This type of control mode

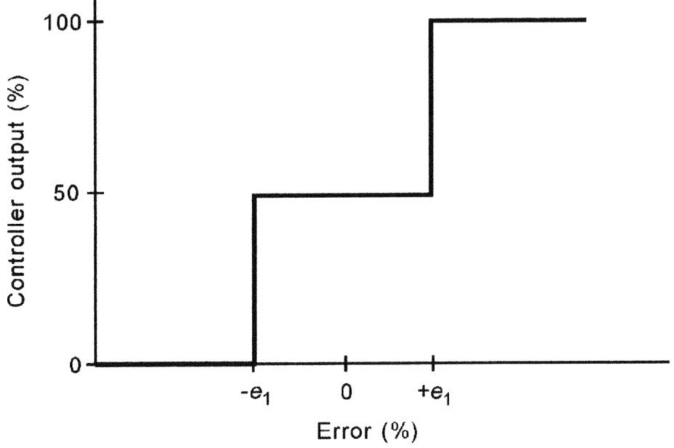

FIGURE 9.5
Three-position controller action.

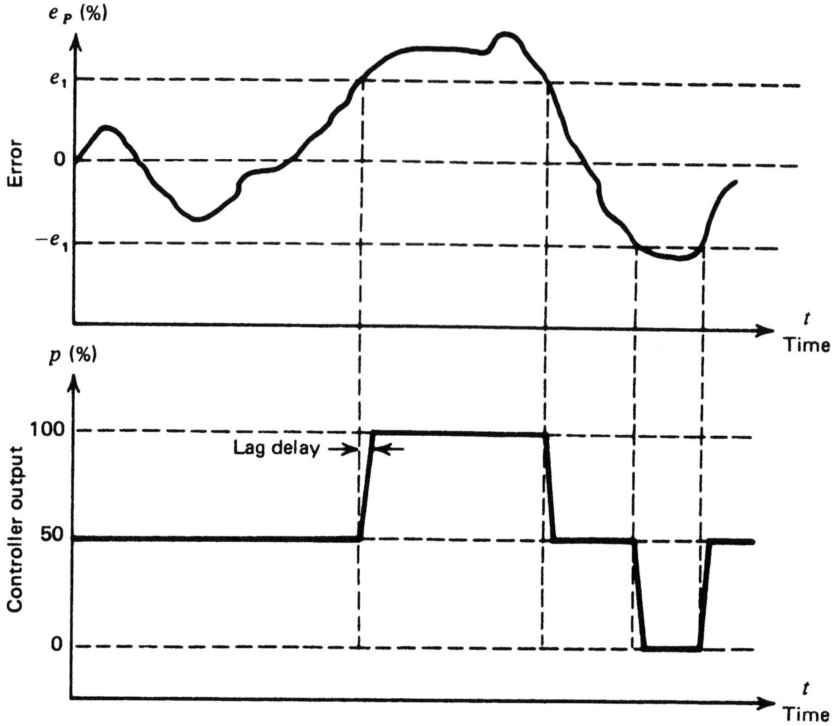

FIGURE 9.6
Relationship between error and three-position controller action, including the effects of lag.

usually requires a more complicated final control element, because it must have more than two settings. Figure 9.6 shows a graph of dynamic variable and final control element setting versus time for a hypothetical case of three-position control. Note the change in control element setting as the variable changes about the two trip points. On this graph, the finite time required for the final control element to change from one position to another also is shown. Notice the overshoot and undershoot of the error around the upper and lower setpoints. This is due to both the process lag time and the controller lag time, indicated by the finite time required for the control element to reach a new setting.

9.4.3 Floating-Control Mode

In the two previous modes of controller action, the output was uniquely determined by the magnitude of the error input. If the error exceeded some preset limit, the output was changed to a new setting as quickly as possible. In floating control, the specific output of the controller is not uniquely determined by the error. If the error is zero, the output does not change but remains (floats) at whatever setting it was when the error went to zero. When the error moves off zero, the controller output again begins to change. Actually, as with the two-position mode, there is typically a neutral zone around zero error where no change in controller position occurs.

Single Speed In the single-speed floating-control mode, the output of the control element changes at a fixed rate when the error exceeds the neutral zone. An equation for this action is

$$\frac{dp}{dt} = \pm K_F \qquad |e_p| > \Delta e_p \tag{9.11}$$

where $\frac{dp}{dt}$ = rate of change of controller output with time
K_F = rate constant (%/s)
Δe_p = half the neutral zone

If Equation (9.11) is integrated for the actual controller output, we get

$$p = \pm K_F t + p(0) \qquad |e_p| > \Delta e_p \tag{9.12}$$

where $p(0)$ = controller output at $t = 0$

which shows that the present output depends on the time history of errors that have previously occurred. Because such a history usually is not known, the actual value of p floats at an *undetermined* value. If the deviation persists, then Equation (9.12) shows that the controller saturates at either 100% or 0% and remains there until an error drives it toward the opposite extreme. A graph of single-speed floating control is shown in Figure 9.7a.

EXAMPLE 9.5 Suppose a process error lies within the neutral zone with $p = 25\%$. At $t = 0$, the error falls below the neutral zone. If $K = +2\%$ per second, find the time when the output saturates.

Solution

The relation between controller output and time is

$$p = K_F t + p(0)$$

When $p = 100$

$$100\% = (2\%/\text{s})(t) + 25\%$$

that, when solved for t, yields

$$t = \mathbf{37.5\ s}$$

In Figure 9.7b, a graph shows controller output versus time and error versus time for a hypothetical case illustrating typical operation. In this example, we assume the controller is reverse acting, which means the controller output decreases when the error exceeds the neutral

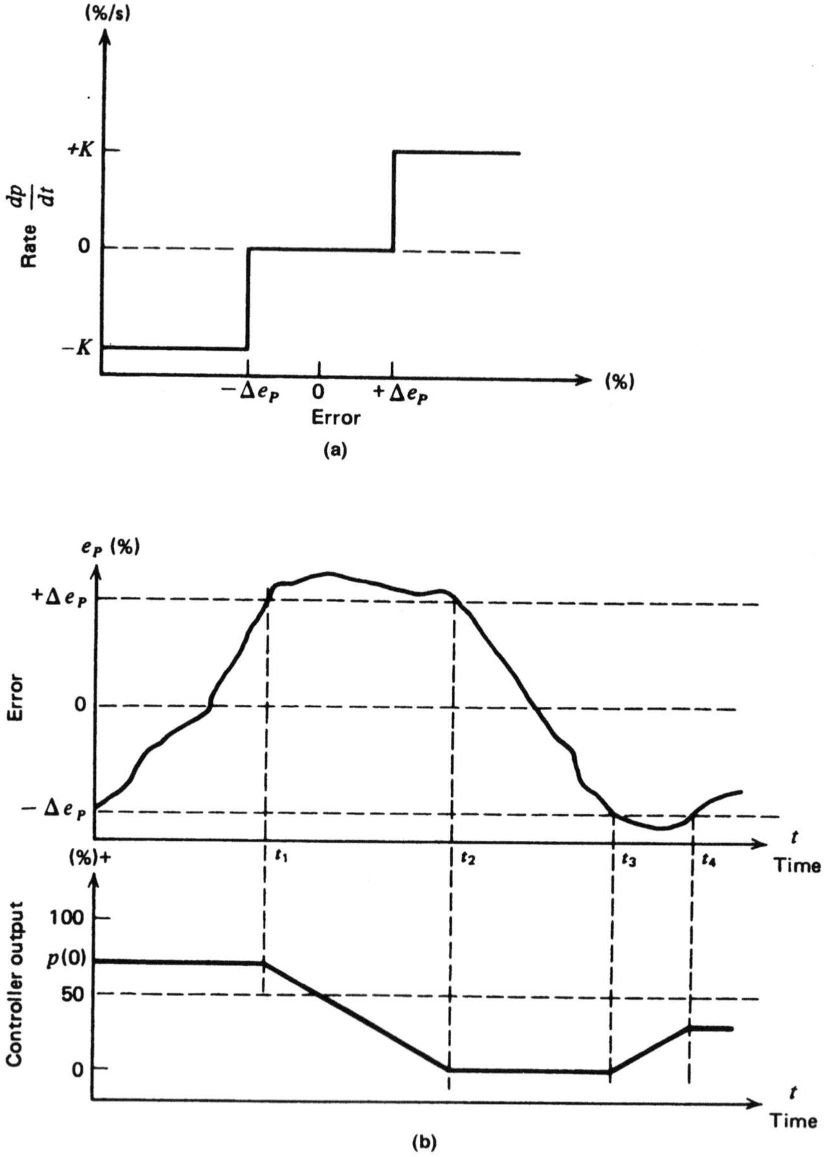

FIGURE 9.7
Single-speed floating controller as shown in: (a) single-speed controller action as the output rate of change to input error, and (b) an example of error and controller response.

zone. This corresponds to a negative K_F in Equation (9.11). Most controllers can be adjusted to act in either the reverse or direct mode. Here the controller starts at some output $p(0)$. At time t_1, the error exceeds the neutral zone. The controller output decreases at a constant rate until t_2, when the error again falls below the neutral zone limit. At t_3, the error falls below the lower limit of the neutral zone, causing controller output to change until the error again moves within the allowable band.

Multiple Speed In the floating multiple-speed control mode, not one but several possible speeds (rates) are changed by controller output. Usually, the rate increases as the deviation exceeds certain limits. Thus, if we have certain speed change points, e_{pi}, depending on the error, then each has its corresponding output rate change, K_i. We can then say

$$\frac{dp}{dt} = \pm K_{Fi} \qquad |e_p| > e_{pi} \tag{9.13}$$

If the error exceeds e_{pi}, then the speed is K_{Fi}. If the error rises to exceed e_{p2}, the speed is increased to K_{F2}, and so on. Actually, this mode is a discontinuous attempt to realize an integral mode (see Section 9.5.4). A graph of this mode is shown in Figure 9.8.

Applications Primary applications of the floating-control mode are for the single-speed controllers with a neutral zone. This mode has an inherent cycle nature much like the two-position, although this cycling can be minimized, depending on the application. Generally, the method is well suited to self-regulation processes with very small lag

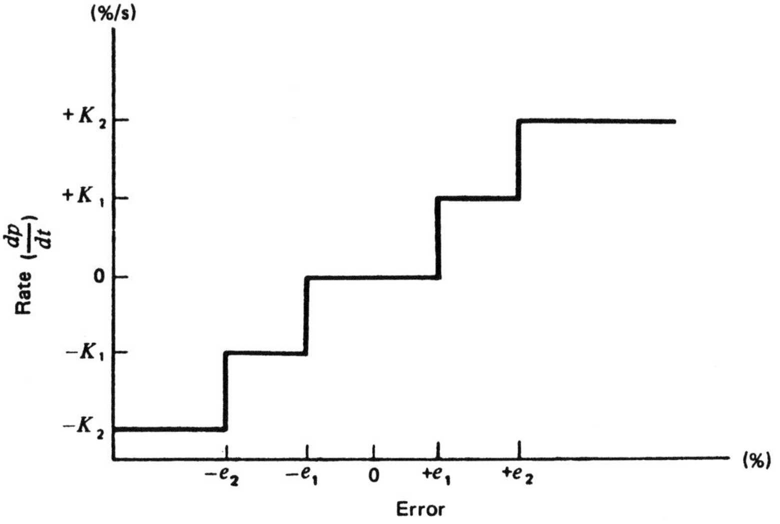

FIGURE 9.8
Multiple-speed floating mode control action.

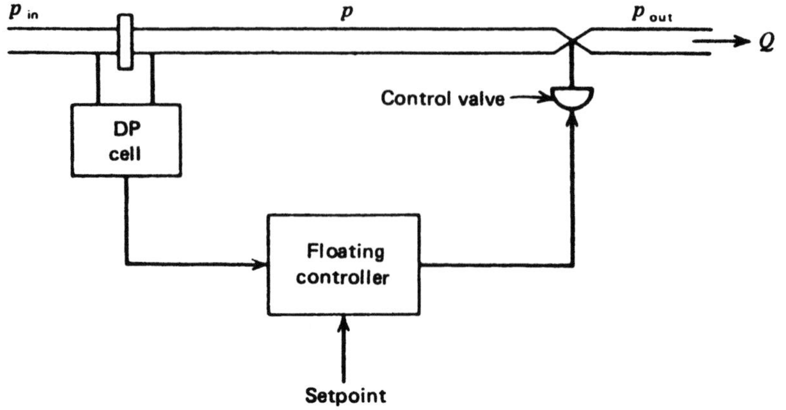

FIGURE 9.9
Single-speed floating-control action applied to a flow-control system.

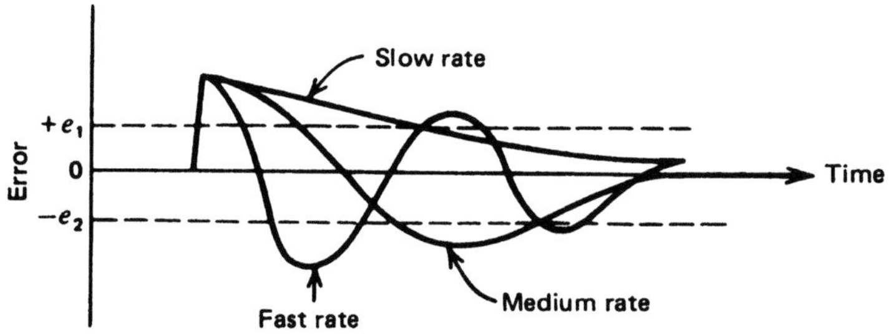

FIGURE 9.10
The rate of controller output change has a strong effect on error recovery in a floating controller.

or dead time, which implies small-capacity processes. When used with large-capacity systems, the inevitable cycling must be considered.

An example of single-speed floating control is a liquid flow rate through a control valve. Such a system is shown in Figure 9.9. The load is determined by the inlet and outlet pressures p_{in} and p_{out}, and the flow is determined in part by the pressure, p, within the *DP* cell and control valve. This is an example of a system with self-regulation. We assume some valve opening has been found commensurate with the desired flow rate. If the load changes (either p_{in} or p_{out}), then an error occurs. If larger than the neutral zone, the valve begins to open or close at a constant rate until an opening is found that supports the proper flow rate at the new load conditions. Clearly, the rate is very important, because especially fast process lags cause the valve to continue opening (or closing) beyond that optimum self-regulated position. This is shown in Figure 9.10, where the response to a sudden deviation is shown for various floating rates.

9.5 CONTINUOUS CONTROLLER MODES

The most common controller action used in process control is one or a combination of continuous controller modes. In these modes, the output of the controller changes smoothly in response to the error or rate of change of error. These modes are an extension of the discontinuous types discussed in the previous section.

9.5.1 Proportional Control Mode

The two-position mode had the controller output of either 100% or 0%, depending on the error being greater or less than the neutral zone. In multiple-step modes, more divisions of controller outputs versus error are developed. The natural extension of this concept is the *proportional mode,* where a smooth, linear relationship exists between the controller output and the error. Thus, over some range of errors about the setpoint, each value of error has a unique value of controller output in one-to-one correspondence. The range of error to cover the 0% to 100% controller output is called the *proportional band,* because the one-to-one correspondence exists only for errors in this range. This mode can be expressed by

$$p = K_P e_p + p_0 \tag{9.14}$$

where K_P = proportional gain between error and controller output (% per %)
p_0 = controller output with no error (%)

Direct and Reverse Action Recall that the error in Equation (9.14) is expressed using the difference between setpoint and the measurement, $r - b$. This means that as the measured value increases above the setpoint, the error will be negative and the output will decrease. That is, the term $K_P e_p$ will subtract from p_0. Thus, Equation (9.14) represents reverse action. Direct action would be provided by putting a negative sign in front of the correction term.

A plot of the proportional mode output versus error for Equation (9.14) is shown in Figure 9.11. In this case, p_0 has been set to 50% and two different gains have been used. Note that the proportional band is dependent on the gain. A high gain means large response to an error, but also a narrow error band within which the output is not saturated.

In general, the proportional band is defined by the equation

$$PB = \frac{100}{K_P} \tag{9.15}$$

Let us summarize the characteristics of the proportional mode and Equation (9.14).

1. If the error is zero, the output is a constant equal to p_0.
2. If there is error, for every 1% of error, a correction of K_P percent is added to or subtracted from p_0, depending on the sign of the error.
3. There is a band of error about zero of magnitude PB within which the output is not saturated at 0% or 100%.

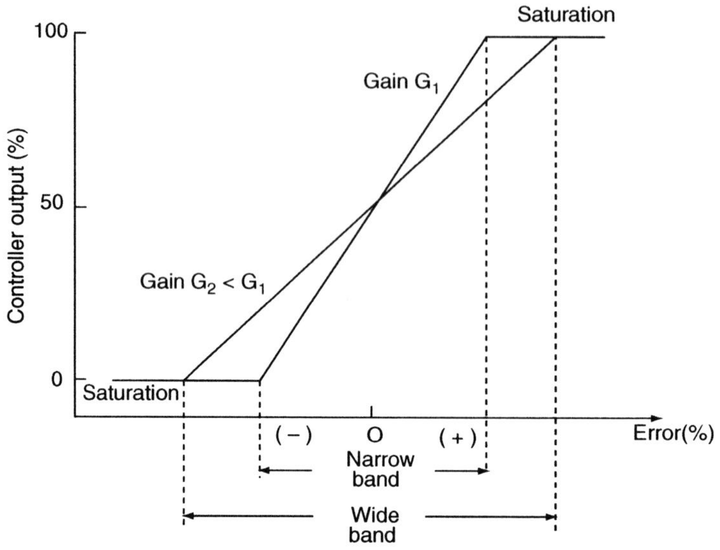

FIGURE 9.11
The proportional band of a proportional controller depends on the inverse of the gain.

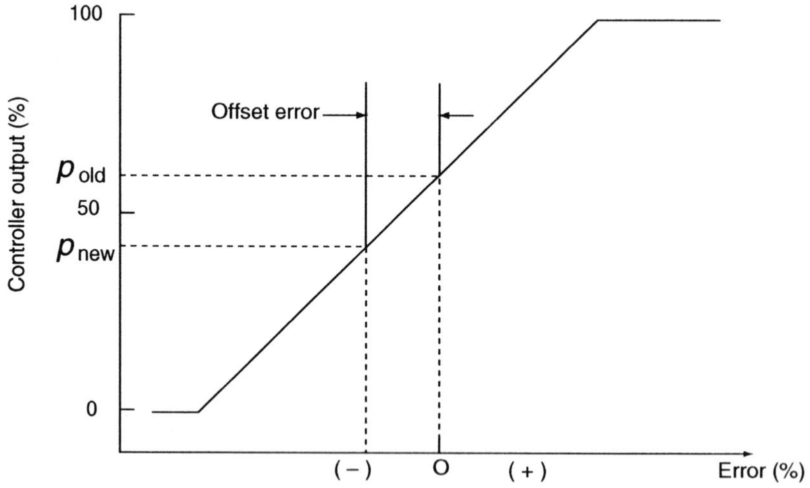

FIGURE 9.12
An offset error must occur if a proportional controller requires a new zero-error output following a load change.

Offset An important characteristic of the proportional control mode is that it produces a permanent *residual error* in the operating point of the controlled variable when a change in load occurs. This error is referred to as *offset.* It can be minimized by a larger constant, K_P, which also reduces the proportional band. To see how offset occurs, consider a system under nominal load with the controller at 50% and the error zero, as shown in Figure 9.12.

If a transient error occurs, the system responds by changing controller output in correspondence with the transient to effect a return-to-zero error. Suppose, however, a load change occurs that requires a permanent change in controller output to produce the zero-error state. Because a one-to-one correspondence exists between controller output and error, it is clear that a new, zero-error controller output can *never* be achieved. Instead, the system produces a small permanent offset in reaching a compromise position of controller output under new loads.

EXAMPLE 9.6 Consider the proportional-mode level-control system of Figure 9.13. Value A is linear, with a flow scale factor of 10 m^3/h per percent controller output. The controller output is nominally 50% with a constant of $K_P = 10\%$ per %. A load change occurs when flow through valve B changes from 500 m^3/h to 600 m^3/h. Calculate the new controller output and offset error.

Solution

Certainly, valve A must move to a new position of 600 m^3/h flow or the tank will empty. This can be accomplished by a 60% new controller output because

$$Q_A = \left(\frac{10\ \text{m}^3/\text{h}}{\%}\right)(60\%) = 600\ \text{m}^3/\text{h}$$

as required. Because this is a proportional controller, we have

$$p = K_P e_p + p_0$$

with the nominal condition $p_0 = 50\%$. Thus

$$e_p = \frac{p - p_0}{K_P} = \frac{60 - 50}{10}\%$$

$$e_p = 1\%$$

so a 1% offset error occurred because of the load change.

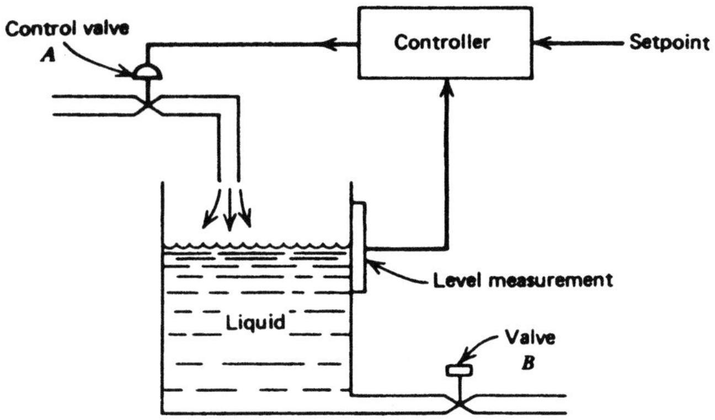

FIGURE 9.13
Level-control system for Example 9.6.

Application The offset error limits use of the proportional mode to only a few cases, particularly those where a manual reset of the operating point is possible to eliminate offset. Proportional control generally is used in processes where large load changes are unlikely or with moderate to small process lag times. Thus, if the process lag time is small, the proportional band can be made very small (large K_P), which reduces offset error. Figure 9.11 shows that if K_P is made very large, the PB becomes very small, and the proportional mode acts just like an ON/OFF mode. Remember that the ON/OFF mode exhibited oscillations about the setpoint. From these statements it is clear that, for high gain, the proportional mode causes oscillations of the error.

9.5.2 Integral-Control Mode

The offset error of the proportional mode occurs because the controller cannot adapt to changing external conditions—that is, changing loads. In other words, the zero-error output is a fixed value. The integral mode eliminates this problem by allowing the controller to adapt to changing external conditions by changing the zero-error output.

The need for integral action shows up when it is noted that even with proportional action correction, the error does not go to zero in time. Suppose a system has some error, e_p, and the proportional mode provides a change in controller output, $K_P e_p$. As we watch the error in time, we note that the error may reduce, but *it does not go to zero;* in fact, it may become constant. Integral action is needed.

Integral action is provided by summing the error over time, multiplying that sum by a gain, and adding the result to the present controller output. You can see that if the error makes random excursions above and below zero, the net sum will be zero, so the integral action will not contribute. But if the error becomes positive or negative for an extended period of time, the integral action will begin to accumulate and make changes to the controller output.

In the mathematics of continuous functions, such as error, summation is represented by integration. Therefore, this mode is represented by an integral equation

$$p(t) = K_I \int_0^t e_P \, dt + p(0) \tag{9.16}$$

where $p(0)$ is the controller output when the integral action starts. The gain K_I expresses how much controller output in percent is needed for every percent-time accumulation of error.

Another way of thinking of integral action is found by taking the derivative of Equation (9.16). In that case, we find a relation for the rate at which the controller output changes,

$$\frac{dp}{dt} = K_I e_p \tag{9.17}$$

This equation shows that when an error occurs, the controller begins to increase (or decrease) its output at a rate that depends upon the size of the error and the gain. If the error is zero, the controller output is not changed. If there is positive error, the controller output begins to ramp up at a rate determined by Equation (9.17). Figure 9.14 illustrates this for two different values of gain.

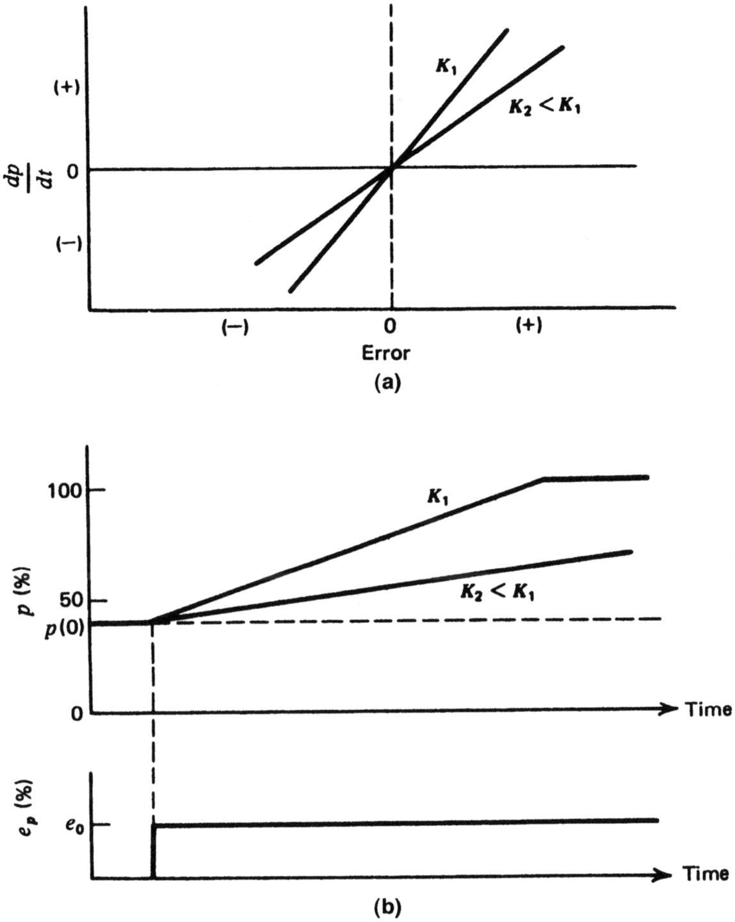

FIGURE 9.14
Integral mode controller action: (a) The rate of output change depends on error, and (b) an illustration of integral mode output and error.

Figure 9.14a shows how the *rate of change* of controller output depends upon the value of error and the size of the gain. Figure 9.14b shows how the actual controller output would look if a constant error occurred. You can see how the controller output begins to ramp up at a rate determined by the gain. In the case of gain K_1, the output finally saturates at 100%, and no further action can occur (perhaps a control valve is fully open, for example).

Let us summarize the characteristics of the integral mode and Equation (9.16).

1. If the error is zero, the output stays fixed at a value equal to what it was when the error went to zero.
2. If the error is not zero, the output will begin to increase or decrease at a rate of K_I percent/second for every 1% of error.

Area Accumulation From calculus we learn that an integral determines the area of the function being integrated. Thus, Equation (9.16) can be interpreted as providing a controller output equal to the net area under the *error-time curve* multiplied by K_I. We often say that the integral term *accumulates* error as a function of time. Thus, for every $1\% - s$ of accumulated error-time area, the output will be K_I percent.

EXAMPLE 9.7 An integral controller is used for speed control with a setpoint of 12 rpm within a range of 10 to 15 rpm. The controller output is 22% initially. The constant $K_I = -0.15\%$ controller output per second per percentage error. If the speed jumps to 13.5 rpm, calculate the controller output after 2 s for a constant e_p.

Solution

We find e_p from Equation (9.3):

$$e_p = \frac{r - b}{b_{\max} - b_{\min}} \times 100$$

$$e_p = \frac{12 - 13.5}{15 - 10} \times 100$$

$$e_p = -30\%$$

The rate of controller output change is then given by Equation (9.17),

$$\frac{dp}{dt} = K_I e_p = (-0.15\ \mathrm{s}^{-1})(-30\%)$$

$$\frac{dp}{dt} = 4.5\%/\mathrm{s}$$

The controller output for constant error will be found from Equation (9.16)

$$p = K_I \int_0^t e_p\, dt + p(0)$$

but because e_p is constant,

$$p = K_I e_p t + p(0)$$

After 2 s, we have

$$p = (0.15)(30\%)(2) + 22$$

$$p = \mathbf{31\%}$$

The integral gain, K_I, is often represented by the inverse, which is called the *integral time*, or the *reset action*, $T_I = 1/K_I$. This is often expressed in minutes instead of seconds because this unit is more typical of many industrial process speeds.

The integral controller constant K_I may be expressed in percentage change per *minute* per percentage error, whenever a typical process-control loop has characteristic response time in minutes rather than seconds. Thus, an integral mode controller with reset action at 5.7 minutes means that K_I for our equations would be

$$K_I = \frac{1}{(5.7\ \text{min})(60\ \text{s/min})}$$

$$K_I = 2.92 \times 10^{-3}\ \text{s}^{-1}$$

Applications Use of the integral mode is shown by the flow control system in Figure 9.9, except that we now assume a reverse-acting integral controller mode. Operation can be understood using Figure 9.15. A load change-induced error occurs at $t = 0$.

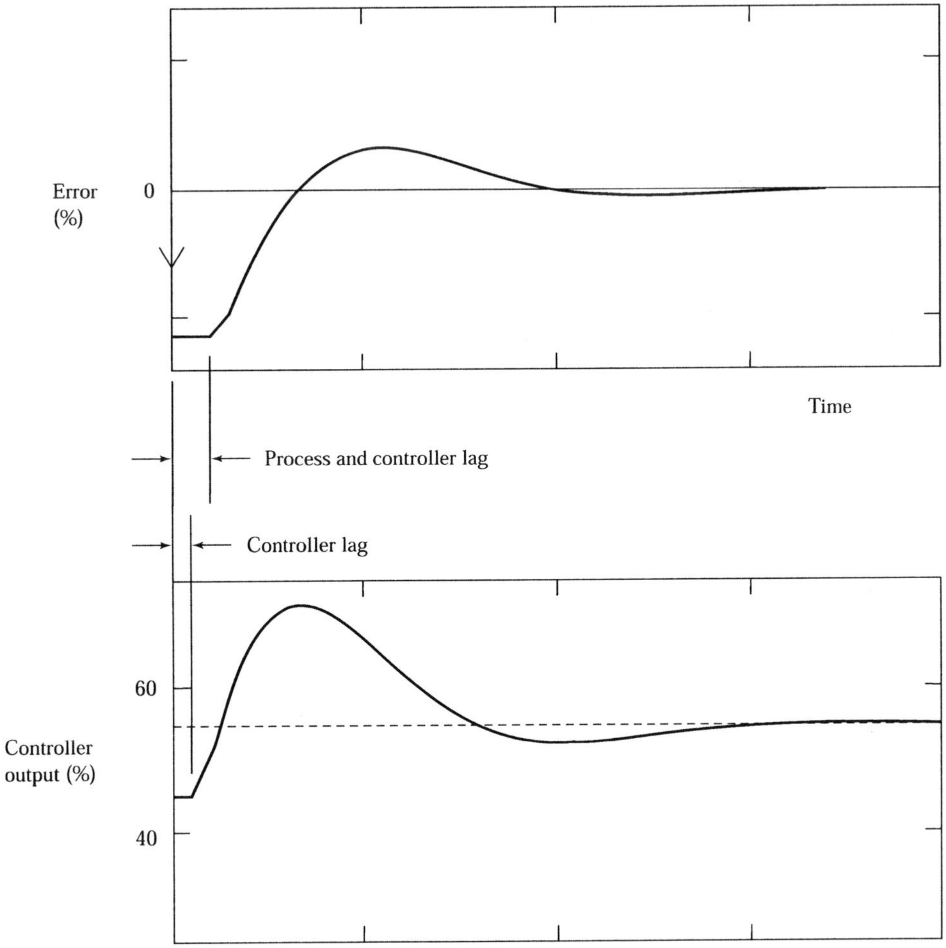

FIGURE 9.15
Illustration of integral mode output and error, showing the effect of process and control lag.

The proper valve position under the new load to maintain the constant flow rate is shown as a dashed line in the controller output graph of Figure 9.15. In the integral mode, the value initially begins to change rapidly, as predicted by Equation (9.17). As the valve opens, the error decreases and slows the valve opening rate as shown. The ultimate effect is that the system drives the error to zero at a slowing controller rate. The effect of process and control system lag is shown as simple delays in the controller output change and in the error reduction when the controller action occurs. If the process lags are too large, the error can oscillate about zero or even be cyclic. Typically, the integral mode is not used alone, but can be used for systems with small process lags and correspondingly small capacities.

9.5.3 Derivative-Control Mode

Suppose you were in charge of controlling some variable, and at some time, t_0, your helper yelled out, "The error is zero. What action do you want to take?" Well, it would seem perfectly rational to answer "None" because, after all, the error was zero. But suppose you have a screen that shows the variation of error in time and that it looks like Figure 9.16.

You can clearly see that even though the error at t_0 is zero, it is changing in time and will certainly not be zero in the following time. Therefore, some action should be taken even though the error is zero! This scenario describes the nature and need for derivative action.

Derivation controller action responds to the rate at which the error is changing—that is, the derivative of the error. Appropriately, the equation for this mode is given by the expression

$$p(t) = K_D \frac{de_p}{dt} \tag{9.18}$$

where the gain, K_D, tells us by how much percent to change the controller output for every percent-per-second rate of change of error. Derivative action is not used alone because it provides no output when the error is constant.

Derivative controller action is also called *rate action* and *anticipatory control.*

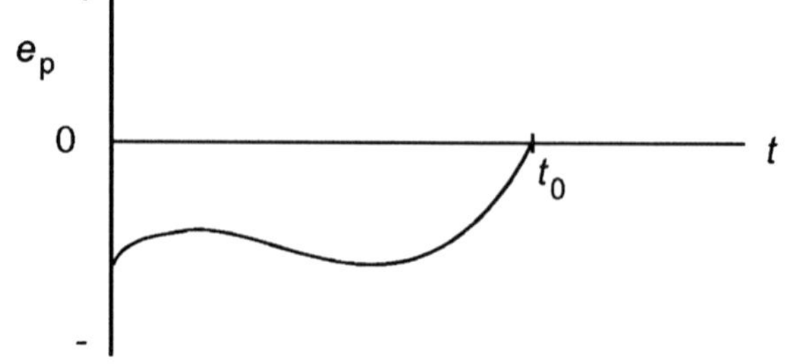

FIGURE 9.16
The error can be zero but the rate of change very large.

Figure 9.17 illustrates how derivative action changes the controller output for various rates of change of error. For this example, it is assumed that the controller output with no error or rate of change of error is 50%. When the error changes very rapidly with a positive slope, the output jumps to a large value, and when the error is not changing, the output returns to 50%. Finally, when the error is decreasing—that is, has a negative slope—the output discontinuously changes to a lower value.

The derivative mode must be used with great care and usually with a small gain, because a rapid rate of change of error can cause very large, sudden changes of controller output. Such an event can lead to instability.

Let us summarize the characteristics of the derivative mode and Equation (9.18).

1. If the error is zero, the mode provides no output.
2. If the error is constant in time, the mode provides no output.
3. If the error is changing in time, the mode contributes an output of K_D percent for every 1%-per-second rate of change of error.
4. For direct action, a positive rate of change of error produces a positive derivative mode output.

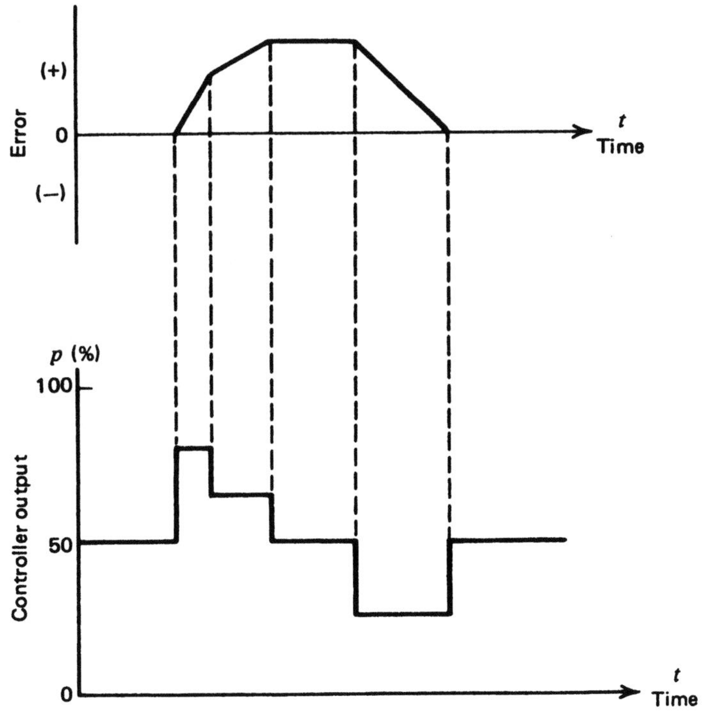

FIGURE 9.17
Derivative mode controller action changes depending on the rate of error.

9.6 COMPOSITE CONTROL MODES

It is common in the complex of industrial processes to find control requirements that do not fit the application norms of any of the previously considered controller modes. It is both possible and expedient to combine several basic modes, thereby gaining the advantages of each mode. In some cases, an added advantage is that the modes tend to eliminate some limitations they individually possess. We will consider only those combinations that are commonly used and discuss the merits of each mode.

9.6.1 Proportional-Integral Control (PI)

This is a control mode that results from a combination of the proportional mode and the integral mode. The analytic expression for this control process is found from a series combination of Equations (9.14) and (9.16):

$$p = K_P e_p + K_P K_I \int_0^t e_p \, dt + p_I(0) \tag{9.19}$$

where $p_I(0)$ = integral term value at $t = 0$ (initial value)

The main advantage of this composite control mode is that the one-to-one correspondence of the proportional mode is available and the integral mode eliminates the inherent offset. Notice that the proportional gain, by design, also changes the net integration mode gain, but that the integration gain, through K_I, can be independently adjusted. Recall that the proportional mode offset occurred when a load change required a new nominal controller output that could not be provided except by a fixed error from the setpoint. In the present mode, the integral function provides the required new controller output, thereby allowing the error to be zero after a load change. The integral feature effectively provides a reset of the zero error output after a load change occurs. This can be seen by the graphs of Figure 9.18. At time t_1, a load change occurs that produces the error shown. Accommodation of the new load condition requires a new controller output. We see that the controller output is provided through a sum of proportional plus integral action that finally leaves the error at zero. The proportional part is obviously just an image of the error.

Let us summarize the characteristics of the PI mode and Equation (9.19).

1. When the error is zero, the controller output is fixed at the value that the integral term had when the error went to zero. This output is given by $p_I(0)$ in Equation (9.19) simply because we chose to define the time at which observation starts as $t = 0$.
2. If the error is not zero, the proportional term contributes a correction, and the integral term begins to increase or decrease the accumulated value [initially, $p_I(0)$], depending on the sign of the error and the direct or reverse action.

The integral term cannot become negative. Thus, it will saturate at zero if the error and action try to drive the area to a net negative value.

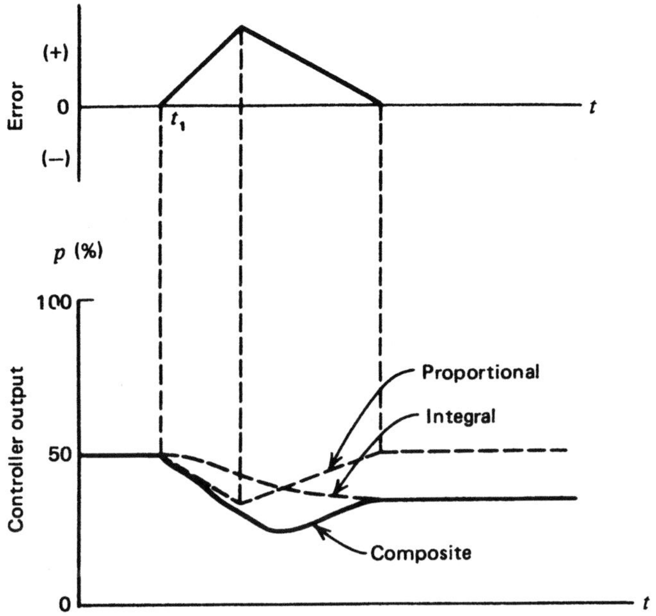

FIGURE 9.18
Proportional-integral (PI) action showing the reset action of the integral contribution. This example is for reverse action.

Application As noted, this composite proportional-integral mode eliminates the offset problem of proportional controllers. It follows that the mode can be used in systems with frequent or large load changes. Because of the integration time, however, the process must have relatively slow changes in load to prevent oscillations induced by the integral overshoot. Another disadvantage of this system is that during start-up of a batch process, the integral action causes a considerable overshoot of the error and output before settling to the operation point. This is shown in Figure 9.19, where we see the proportional band as a dashed band. The effect of the integral action can be viewed as a shifting of the whole proportional band. The proportional band is defined as that positive and negative error for which the output will be driven to 0% and 100%. Therefore, the presence of an integral accumulation changes the amount of error that will bring about such saturation by the proportional term. In Figure 9.19, the output saturates whenever the error exceeds the *PB* limits. The *PB* is constant, but its location is shifted as the integral term changes.

EXAMPLE 9.8 Given the error of Figure 9.20 (top), plot a graph of a proportional-integral controller output as a function of time.

$$K_P = 5,\ K_I = 1.0\ \text{s}^{-1},\ \text{and}\ p_I(0) = 20\%$$

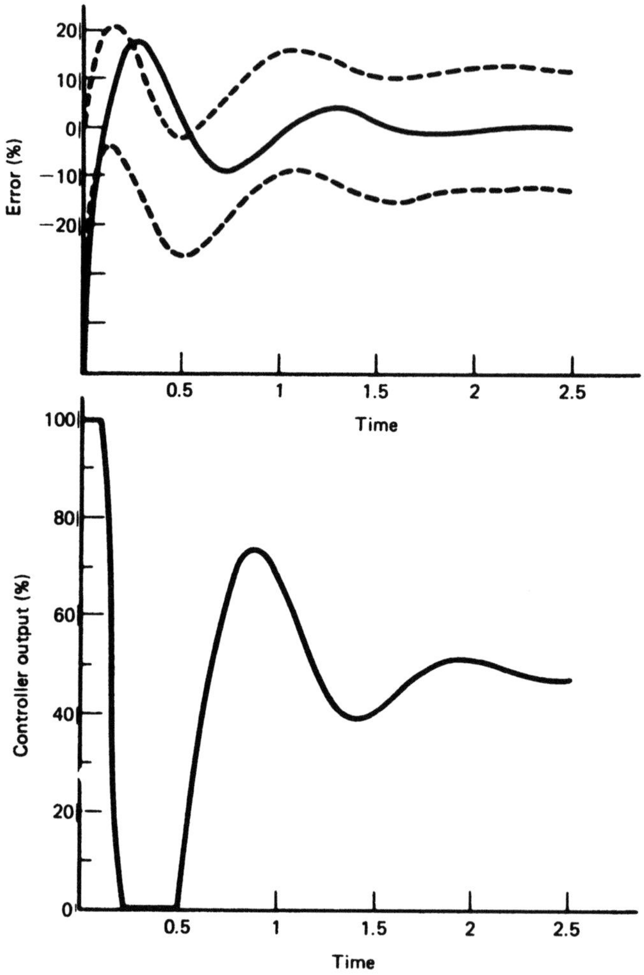

FIGURE 9.19
Overshoot and cycling often result when PI mode control is used in start-up of batch processes. The dashed lines show the proportional band.

Solution

We find the solution by an application of

$$p = K_P e_p + K_P K_I \int_0^t e_p \, dt + p_I(0)$$

To find the controller output, we solve Equation (9.19) in time. The error can be expressed in three time regions.

$$0 \le t \le 1 \qquad (t \text{ between 0 and 1 s})$$

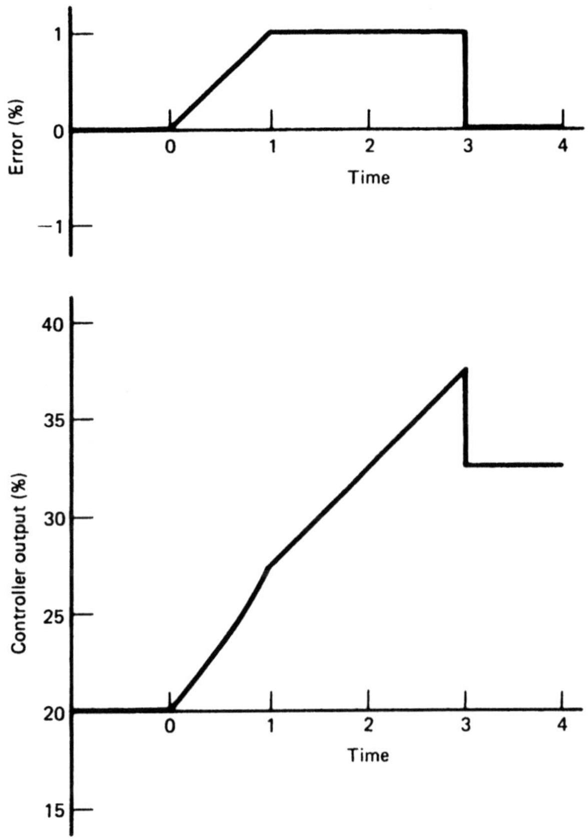

FIGURE 9.20
Solution for Example 9.8.

The error rises from 0% to 1% in 1 s. Thus, it is given by $e_p = t$.

$$1 \le t \le 3$$

For this time span, the error is constant and equal to 1%; therefore, it is given by $e_p = 1$.

$$t \ge 3$$

For this time, the error is zero, $e_p = 0$.

We now write out and solve Equation (9.19) for each of these time spans.

$$0 \le t \le 1 \qquad e_p = t$$

$$p_1 = 5t + 5\int_0^t t\,dt + 20$$

$$p_1 = 5t + 5\left[\frac{t^2}{2}\right]\Big|_0^t + 20$$

$$p_1 = 5t + 2.5t^2 + 20$$

This is plotted in Figure 9.20 (bottom) from 0 to 1 s. Notice the curvature because of the squared term. Remember that only the integral term accumulates values, so in finding the output at 1 s, the contribution of the proportional term, $5t$, is not included. Therefore, the starting value for the next time span is given by $p_1(1) = 2.5t^2 + 20 = 22.5\%$.

$$1 \le t \le 3 \qquad e_p = 1$$

$$p_2 = 5 + 5\int_1^t 1\ dt + 22.5$$

The integral term accumulation from 0 to 1 s forms the initial condition for this new equation.

$$p_2 = 5 + 5[t]_1^t + 22.5$$
$$p_2 = 5 + 5(t - 1) + 22.5$$

This function is plotted in Figure 9.20 from 1 to 3 s. At the end of this period, the integral term has accumulated a value of $p_2(3) = 32.5\%$.

$$t \ge 3 \qquad e_p = 0$$

$$p_3 = 5[0] + 5\int_3^t 0\ dt + 32.5$$

$$p_3 = 32.5$$

Figure 9.20 (bottom) shows that the output will stay constant at 32.5% from 3 s. The sudden drop of 5% is due to the sudden change of error from 1% to 0% at $t = 3$ s.

9.6.2 Proportional-Derivative Control Mode (PD)

A second combination of control modes has many industrial applications. It involves the serial (cascaded) use of the proportional and derivative modes. The analytic expression for this mode is found from a combination of Equations (9.14) and (9.18):

$$p = K_P e_p + K_P K_D \frac{de_p}{dt} + p_0 \qquad \textbf{(9.20)}$$

where the terms are all defined in terms given by previous equations.

It is clear that this system cannot eliminate the offset of proportional controllers. It can, however, handle fast process load changes as long as the load change offset error is acceptable. An example of the operation of this mode for a hypothetical load change is shown in Figure 9.21. Note the effect of derivative action in moving the controller output in relation to the error rate change.

EXAMPLE 9.9 Suppose the error, Figure 9.22a, is applied to a proportional-derivative controller with $K_P = 5$, $K_D = 0.5$ s, and $p_0 = 20\%$. Draw a graph of the resulting controller output.

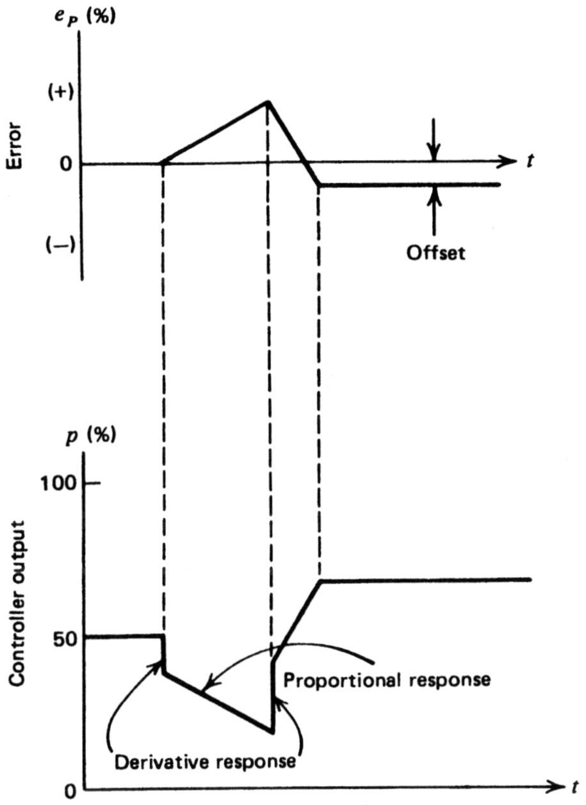

FIGURE 9.21
Proportional-derivative (PD) action showing the offset error from the proportional mode. This example is for reverse action.

Solution

In this case, we evaluate

$$p = K_P e_p + K_D K_P \frac{de_P}{dt} + p_0$$

over the two spans of the error. In the time of 0 to 1 s where $e_p = at$, we have

$$p_1 = K_P at + K_D K_P a + p_0$$

or, because $a = 1\%/\text{s}$,

$$p_1 = 5t + 2.5 + 20$$

Note the instantaneous change of 2.5% produced by this error. In the span from 1 to 3 s, we have

$$p_2 = 5 + 20 = 25$$

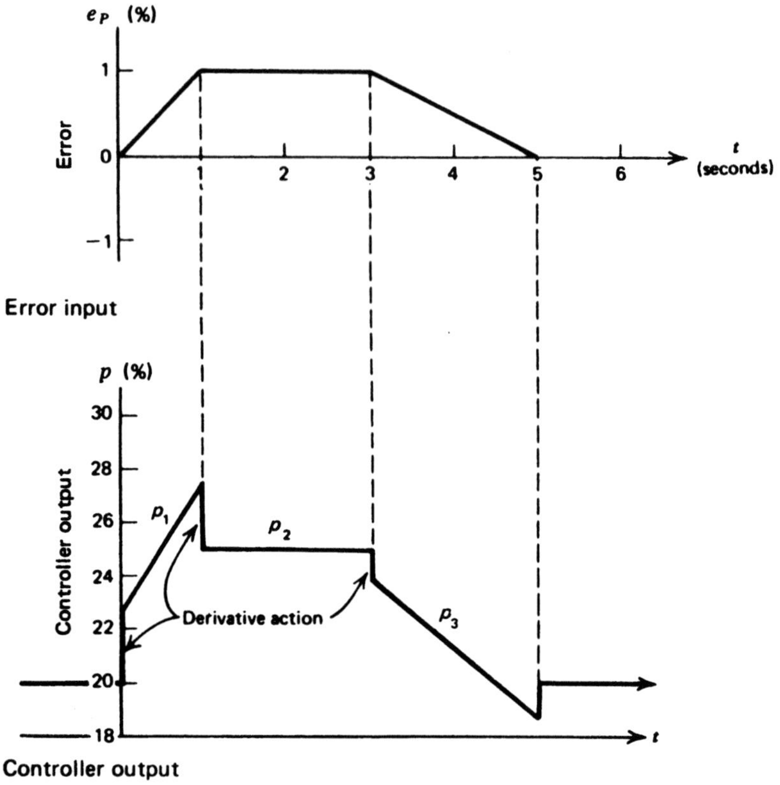

FIGURE 9.22
Solution for Example 9.9.

The span from 3 to 5 s has an error of $e_p = -0.5t + 2.5$, so that we get for 3 to 5 s

$$p_3 = -2.5t + 12.5 - 12.5 + 20$$

or

$$p_3 = -2.5t + 31.25$$

This controlled output is plotted in Figure 9.22b.

9.6.3 Three-Mode Controller (PID)

One of the most powerful but complex controller mode operations combines the proportional, integral, and derivative modes. This system can be used for virtually any process condition. The analytic expression is

$$p = K_P e_p + K_P K_I \int_0^t e_p \, dt + K_P K_D \frac{de_p}{dt} + p_I(0) \qquad \textbf{(9.21)}$$

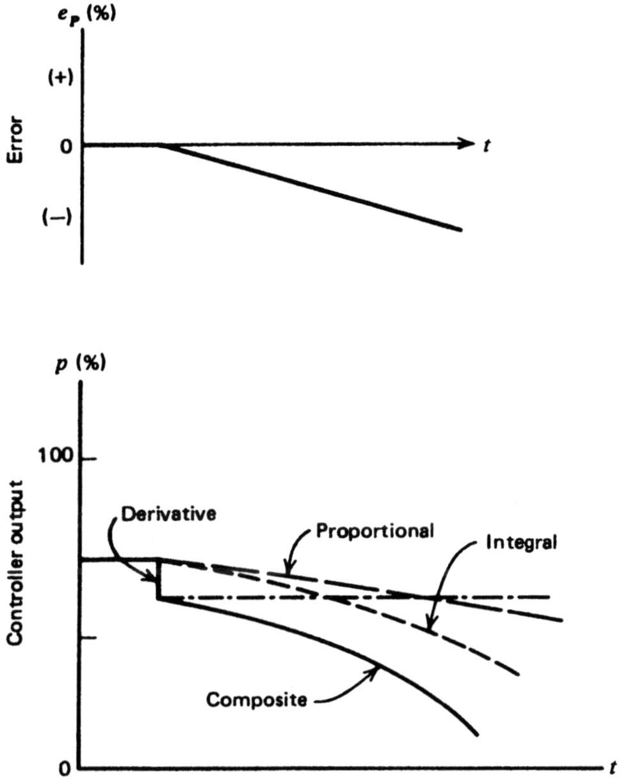

FIGURE 9.23
The three-mode controller action exhibits proportional, integral, and derivative action.

where all terms have been defined earlier.

This mode eliminates the offset of the proportional mode and still provides fast response. In Figure 9.23, the response of the three-mode system to an error is shown.

EXAMPLE 9.10 Let us combine everything and see how the error of Figure 9.22a produces an output in the three-mode controller with $K_P = 5$, $K_I = 0.7\ \text{s}^{-1}$, $K_D = 0.5$ s, and $p_I(0) = 20\%$. Draw a plot of the controller output.

Solution

From Figure 9.22a, the error can be expressed as follows:

$$0\text{–}1\ \text{s} \qquad e_p = t\%$$

$$1\text{–}3\ \text{s} \qquad e_p = 1\%$$

$$3\text{–}5\ \text{s} \qquad e_p = -\frac{1}{2}t + 2.5\%$$

We must apply each of these spans to the three-mode equation for controller output:

$$p = K_P e_p + K_P K_I \int_0^t e_p\, dt + K_P K_D \frac{de_p}{dt} + p_I(0)$$

or

$$p = 5e_p + 3.5 \int_0^t e_p\, dt + 2.5 \frac{de_p}{dt} + 20$$

From 0 to 1 s, we have

$$p_1 = 5t + 3.5 \int_0^t t\ dt + 2.5 + 20$$

or

$$p_1 = 5t + 1.75t^2 + 22.5$$

This is plotted in Figure 9.24 in the span of 0 to 1 s. At the end of 1 s, the integral term has accumulated to $p_I(1) = 21.75\%$. Now, from 1 to 3 s, we have

$$p_2 = 5 + 3.5 \int_1^t (1)\, dt + 21.75$$

or

$$p_2 = 3.5(t - 1) + 26.75$$

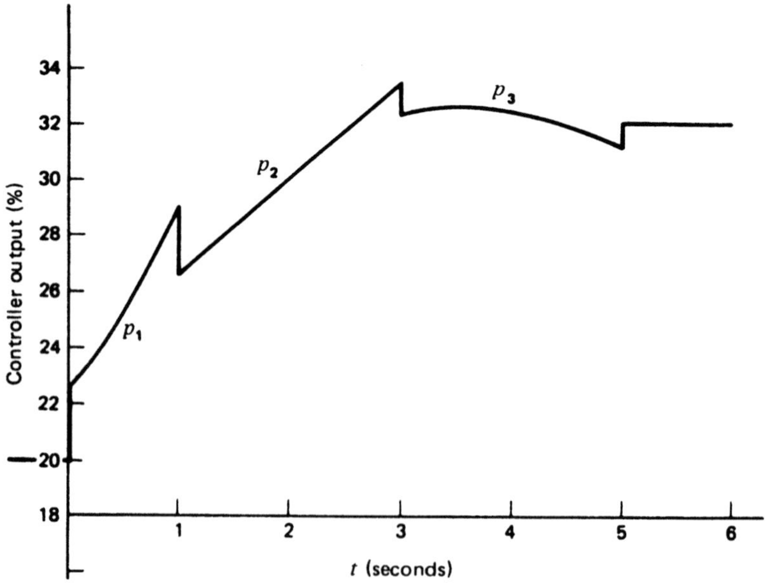

FIGURE 9.24
Solution for Example 9.10.

This controller variation is shown in Figure 9.24 from 1 to 3 s. At the end of 3 s, the integral term has accumulated to a value of $p_I(3) = 28.75\%$. Finally, from 3 to 5 s, we have

$$p_3 = 5\left(-\frac{1}{2}t + 2.5\right) + 3.5\int_3^t\left(-\frac{1}{2}t + 2.5\right)dt - \frac{2.5}{2} + 28.75$$

or

$$p_3 = -0.875t^2 + 6.25t + 21.625$$

This is plotted in Figure 9.24 from 3 to 5 s. After 5 s, the error is zero. Therefore, the output will simply be the accumulated integral response providing a constant output of $p_I = 32.25\%$.

The examples used in this chapter are idealized in terms of the sudden way that errors change. In the real world, changes are not instantaneous, and therefore the sharp breaks in output, such as those shown in Figure 9.24, do not occur.

9.6.4 Special Terminology

A number of special terms are used in process control for discussing the controller modes. The following summary defines some of these terms and shows how they relate to the equations presented in this chapter.

1. *Proportional band (PB)* Although this term was defined earlier, let us note again that this is the percentage error that results in a 100% change in controller output.
2. *Repeats per minute* This term is another expression of the integral gain for PI and PID controller modes. The term derives from the observation that the integral gain, K_I, has the effect of causing the controller output to change every unit time by the proportional mode amount. You can also see this by taking the derivative of the integral term in the controller equation. This gives a change in controller output Δp of

   $$\Delta p = K_I K_P e_p \Delta t$$

   Because $K_P e_p$ is just the proportional contribution, in a unit time interval $\Delta t = 1$, K_I just repeats the proportional term. For example, if $e_p = 0.5\%$ and $K_P = 10\%$, then $K_P e_p = 5\%$. If $K_I = 10\%/(\%\text{-min})$, then every minute, the output would increase by 5% times 10%/(%-min) or 50% or "10 repeats per minute." It repeats the proportional amount 10 times per minute.
3. *Rate gain* This is just another way of saying the derivative gain, K_D. Because K_D has the units of %-s/% (or %-min/%), one often expresses the gain as time directly. Thus, a rate gain of 0.05 min or a *derivative time* of 0.05 min both mean $K_D = 0.05\%\text{-min}/\%$.
4. *Direct/reverse action* This specifies whether the controller output should increase (direct) or decrease (reverse) for an increasing controlled variable. The action is specified by the sign of the proportional gain; $K_P < 0$ is direct, and $K_P > 0$ is reverse.

SUMMARY

This chapter covers the general characteristics of controller operating modes without considering implementation of these functions. Numerous terms that are important to an understanding of controller operations are defined. The highlighted items are as follows:

1. In considering controller operating modes, it is important to know the *process load,* which is the nominal value of all process parameters, and the *process lag,* which represents a delay in reaction of the controller variable to a change of load variable.
2. Some processes exhibit *self-regulation*—that is, the characteristic that a dynamic variable adopts some nominal value commensurate with the load with no control action.
3. The controller operation is defined through a relationship between percentage *error* or *deviation* relative to full scale

$$e_p = \frac{r - b}{b_{\max} - b_{\min}} \times 100 \qquad \textbf{(9.3)}$$

   and the controller output as a percentage of the controlling parameter

$$p = \frac{u - u_{\min}}{u_{\max} - u_{\min}} \times 100 \qquad \textbf{(9.4)}$$

4. Control lag and dead time, respectively, refer to a delay in controller response when a deviation occurs and a period of no response of the process to a change in the controlling variable.
5. Discontinuous controller modes refer to instances where the controller output does not change smoothly for input error. Examples are two-position, multiposition, and floating.
6. Continuous controller modes are modes where the controller output is a smooth function of the error input or rate of change. Examples are proportional, integral, and derivative modes.
7. Composite controller modes combine the continuous modes. Examples are the proportional-integral (PI), the proportional-derivative (PD), and the proportional-integral-derivative (PID) (or three-mode).

PROBLEMS

Section 9.2

9.1 Define the variables in the system of Figure 9.1 that constitute the process load.

9.2 Analyze each of the following control systems and determine whether they have self-regulation, what the process load would be, what would constitute a transient, and whether a process lag would be expected:

a. A home air-conditioning system

b. The cracker-baking system of Figure 7.2

c. The level-control system of Figure 8.3

Section 9.3

9.3 A velocity control system has a range of 220 to 460 mm/s. If the setpoint is 327 mm/s and the measured value is 294 mm/s, calculate the error as percentage of span.

9.4 A controlling variable is a motor speed that varies from 800 to 1750 rpm. If the speed is controlled by a 25- to 50-V dc signal, calculate **(a)** the speed produced by an input of 38 V, and **(b)** the speed calculated as a percent of span.

Section 9.4

9.5 A 5-m-diameter cylindrical tank is emptied by a constant outflow of 1.0 m^3/min. A two-position controller is used to open and close a fill valve with an open flow of 2.0 m^3/min. For level control, the neutral zone is 1 m and the setpoint is 12 m.

a. Calculate the cycling period.

b. Plot the level versus time.

9.6 For Example 9.4, verify that a ±2% neutral zone produces the results of limits of oscillation and period given in the text.

9.7 A floating controller with a rate gain of 6%/min and $p(0) = 50\%$ has a ±5-gal/min deadband. Plot the controller output for an input given by Figure 9.25. The setpoint is 60 gal/min.

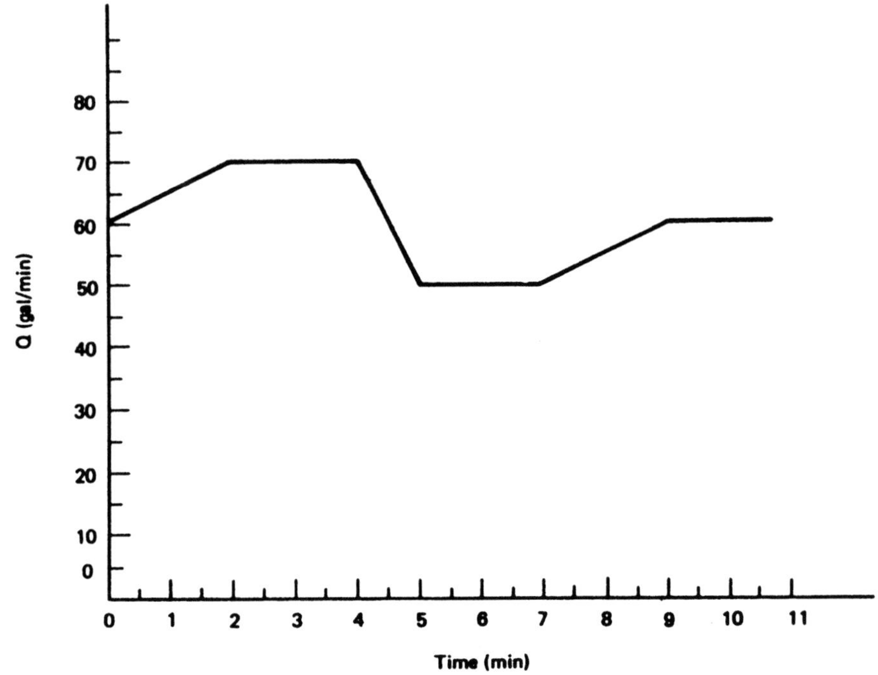

FIGURE 9.25
Figure for Problems 9.7 and 9.18.

Section 9.5

9.8 For a proportional controller, the controlled variable is a process temperature with a range of 50 to 130°C and a setpoint of 73.5°C. Under nominal conditions, the setpoint is maintained with an output of 50%. Find the proportional offset resulting from a load change that requires a 55% output if the proportional gain is **(a)** 0.1, **(b)** 0.7, **(c)** 2.0, and **(d)** 5.0.

9.9 For the applications of Problem 9.8, find the percentage controller output with the 73.5°C setpoint and a proportional gain of 2.0 if the temperature is **(a)** 61°C, **(b)** 122°C, and **(c)** a ramping temperature of $(82 + 5t)$°C.

9.10 An integral controller has a reset action of 2.2 minutes. Express the integral controller constant in s^{-1}. Find the output of this controller to a constant error of 2.2%.

9.11 How would a derivative controller with $K_D = 4$ s respond to an error that varies as $e_p = 2.2 \sin(0.04t)$?

9.12 A proportional controller has a gain of $K_P = 2.0$. Plot the controller output for the error given by Figure 9.26 if $p_0 = 50\%$.

Section 9.6

9.13 A PI controller has $K_P = 2.0$, $K_I = 2.2\ s^{-1}$, and $p_I(0) = 40\%$. Plot the output for an error given by Figure 9.26.

9.14 A PD controller has $K_P = 2.0$, $K_D = 2$ s, and $p_0 = 40\%$. Plot the controller output for the error input of Figure 9.26.

9.15 A PID controller has $K_P = 2.0$, $K_I = 2.2\ s^{-1}$, $K_D = 2$ s, and $p_I(0) = 40\%$. Plot the controller output for the error of Figure 9.26.

9.16 A PI controller is reverse acting, PB = 20, 12 repeats per minute. Find **(a)** the proportional gain, **(b)** the integral gain, and **(c)** the time that the controller output will

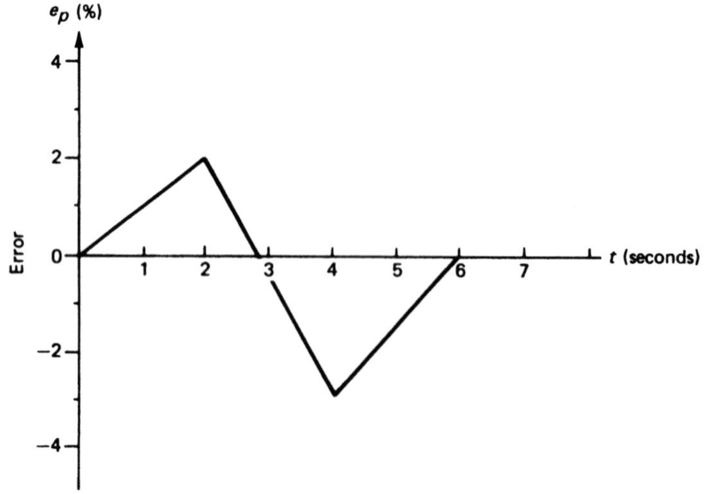

FIGURE 9.26
Figure for Problems 9.12 to 9.15.

reach 0% after a constant error of −1.5% starts. The controller output when the error occurred was 72%.

9.17 Suppose rate action was added to the controller of Problem 9.16 with a rate gain of 0.2 min. Specify the derivative gain and determine the time at which the controller output reaches 0% with this added mode if the input error is $e_p = 0.9t^2$.

9.18 A PI controller is used to control flow within a range of 20 to 100 gal/min. The setpoint is 60 gal/min, and the controller output drives a valve with a 3- to 15-psi signal. The controller settings are direct action, $K_P = -0.9$, $K_I = 0.4\ \text{min}^{-1}$. Plot the pneumatic pressure for the flow of Figure 9.25. Assume an initial pressure output of 10.8 psi.

9.19 A PI controller has $K_P = 4.5$ and $K_I = 7\ \text{s}^{-1}$. Find the controller output for an error given by $e_p = 3\sin(\pi t)$. What is the phase shift between error and controller output?

SUPPLEMENTARY PROBLEMS

The response of a controller can be affected greatly by nonlinear measurement data. Consider a PID controller with gains of $K_P = 2\%/\%$, $K_I = 0.5\%/(\%\text{-}s)$, and $K_D = 5\%/(\%/s)$ used to control temperature in the range of 20 to 100°C. Signal conditioning has provided a current that varies from 4 to 20 mA for that temperature range, but it varies nonlinearly by the relation

$$I = 1.91\left(1 + \frac{T}{44.72}\right)^2$$

The controller reacts to the current, not the actual temperature. The next two problems study the reaction of the controller to a uniform temperature increase of 5°C over 10 s but for two different setpoints: (1) with a setpoint of 30°C, and (2) with a setpoint of 80°C.

S9.1 First let us study the error.

a. Plot the current versus temperature and note the nonlinearity. In both cases, the temperature changes by 5°C, which is 6.25% of the range. So in both cases, the error expressed as temperature percent of range is described by the same equation, $e_p(t) = 0.625t$.

b. How much does the current change in the two cases? What is this in percent of current range?

c. For the two cases, express the error in current as percent of range using equations of the form $e_p(t) = Kt$; assume the current changes linearly over the 10 s.

S9.2 Set up the PID controller equations for the two cases of Problem S9.1 and solve to obtain an equation for the controller output as a function of time. Use the 2% error-time equations found in Problem S9.1. Assume the initial controller output is 30% in the first case and 50% in the second case. Plot these functions to compare response. What is the change in controller output over 10 s in the two cases?

S9.3 Suppose a PID controller has $K_P = K_I = K_D = 1$. Set up the equation for controller output to some error, $e_p(t)$. Assume $p(0) = 0$. Now assume the error has the form $e_p(t) = \sin(2\pi f t)$, where f is some frequency. Find an equation for the controller output $p(t)$ for this case. Using the identity

$$A \sin(\theta) + B \cos(\theta) = K \sin(\theta + \phi)$$

where $K = \sqrt{A^2 + B^2}$ and $\phi = \tan^{-1}\left(\frac{B}{A}\right)$

express $p(t)$ in the phase shift form on the right. Plot amplitude and phase shift versus frequency on a semilog plot over the range of 0.001 to 100 Hz. What happens to the amplitude and phase at very low and very high frequencies? This shows that special care must be taken if high or low frequencies occur with integral and derivative modes.

CHAPTER 10

Analog Controllers

INSTRUCTIONAL OBJECTIVES

The objectives of this chapter are to provide an understanding of how controller modes are implemented using electronic circuits. After a comprehensive study of this chapter, you should be able to

- Diagram the physical appearance of an analog controller.
- Diagram and describe how a two-position control can be implemented using op amps.
- Draw schematics and describe how op amps can be used to implement the proportional, integral, and derivative modes of analog control.
- Describe an op amp circuit that will implement the proportional-integral, proportional-derivative, and proportional-integral-derivative control modes.
- Describe how the nozzle/flapper system can be used to implement proportional control using pneumatics.

10.1 INTRODUCTION

Chapter 9 presented the controller principles in a process-control system. We saw that the controller must perform a variety of math operations such as multiplication, integration, and differentiation of measurement data. In this chapter and the next, the methods of actually implementing controller action are presented. Modern implementation of controller action is provided by computers using software to perform the required math operations. Prior to the widespread use of computers, analog electronic circuits and pneumatic systems provided controller action. Some control system implementations still use analog electronics for special purposes and there remains a vast array of equipment in continued use in the process industry.

This chapter presents the basic principles by which controller action can be provided by analog electronics and pneumatics. This is useful for a more in-depth understanding of controller action but also of value for understanding and working with the continued use of these technologies. The operational amplifier (op amp) is used as the basic functional unit in describing controller action by analog electronics. For pneumatics the nozzle/flapper system forms the basis of controller mode implementation.

Existing standards for signals in process control allow controllers to be interfaced with a variety of sensors and final control units. Thus, electronic controllers are designed to input and output the standard 4- to 20-mA signal. For pneumatic controllers the signal standard is 3 to 15 psi for U.S. installations and 20 to 100 kPa in many other locations throughout the world.

10.2 GENERAL FEATURES

An analog controller is a device that implements the controller modes described in Chapter 9, using analog signals to represent the loop parameters. The analog signal may be in the form of an electric current or a pneumatic air pressure. The controller accepts a measurement expressed in terms of one of these signals, calculates an output for the mode being used, and outputs an analog signal of the same type. Because the controller does solve equations, we think of it as an analog computer. The controller must be able to add, subtract, multiply, integrate, and find derivatives. It does this by working with analog voltages or pressures. In this section, we will examine the general physical layout of typical analog controllers.

10.2.1 Typical Physical Layout

Analog controllers are usually designed to fit into a panel assembly as a slide in/out module, as shown in Figure 10.1. The front displays all necessary information and provides adjustment capability for the operator. When the unit is pulled out partway but still connected, other, less frequently required adjustments are available. When the controller is pulled still further out, an extension cable can be disconnected and the entire unit removed from the panel for replacement, if necessary.

10.2.2 Front Panel

The front panel of an analog controller displays information for operators and allows adjustment of the setpoint. Figure 10.1 shows a typical front panel. The setpoint knob moves a sliding scale under the fixed setpoint indicator. Thus, a fixed span of measurement above and below the setpoint is visible, as indicated by the measurement-value indicator. The error is the difference between the setpoint indicator and the measurement meter. The display is typically expressed in percentage of span (4 to 20 mA or 3 to 15 psi). The lower meter shows the controller output, again expressed in percentage of span. Of course, the output is actually 4 to 20 mA or 3 to 15 psi, so that 0% would mean 4 mA, for example. There is often a switch on the front panel by which the controller can be placed in a manual control, which means that the output can be adjusted independently of the input using

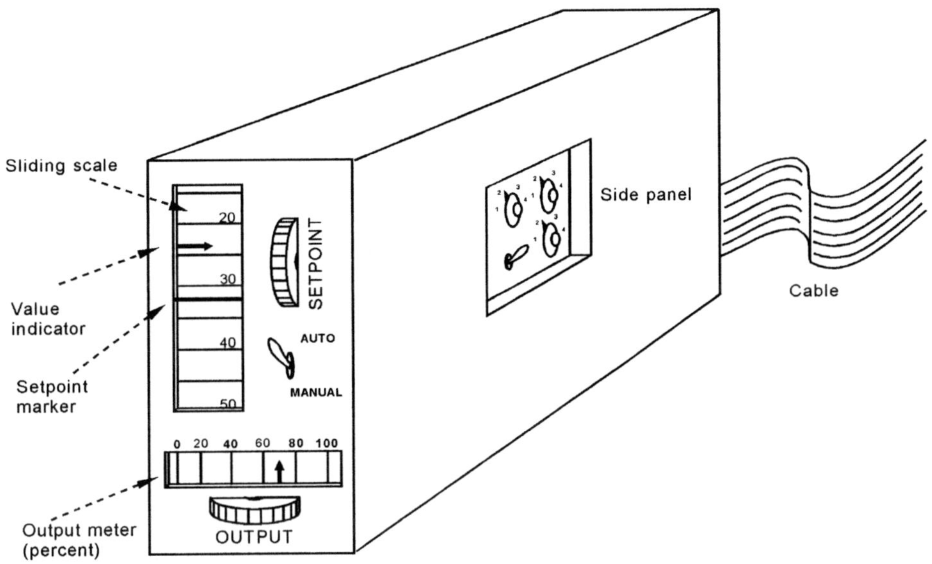

FIGURE 10.1
Typical physical appearance of a controller.

the output-adjust knob. In automatic mode, this knob has no effect on the output. Connections to the controller are made through electrical or pneumatic cables connected to the rear of the unit.

10.2.3 Side Panel

On the side of the controller, when partially pulled out, knobs are available to adjust operation of the controller modes. On this panel, as shown in Figure 10.1, the proportional, integral (reset), and derivative (rate) gains can be adjusted. In addition, filtering action and reverse/direct operation can often be selected.

10.3 ELECTRONIC CONTROLLERS

In the following treatment of electronic methods of realizing controller modes, emphasis is on the use of op amps as the primary circuit element. Discrete electronic components also are used to implement this function, but the basic principles are best illustrated using op amp circuits. Op amp circuits other than the ones described also can be developed.

10.3.1 Error Detector

The detection of an error signal is accomplished in electronic controllers by taking the difference between voltages. One voltage is generated by the process signal current passed through a resistor. The second voltage represents the setpoint. This is usually generated by

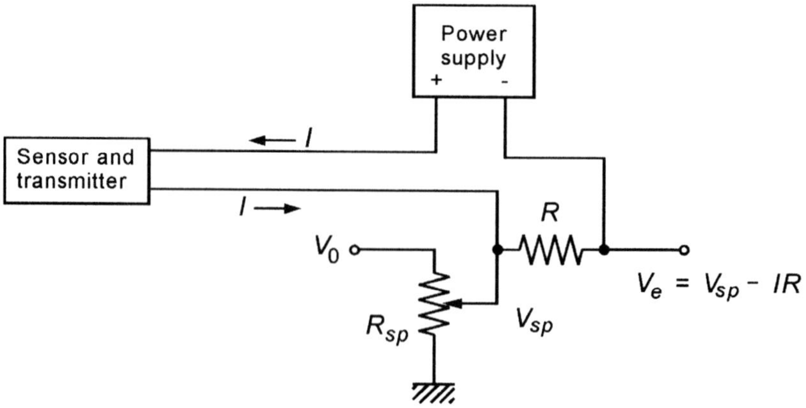

FIGURE 10.2
Error detection for systems using a floating power supply.

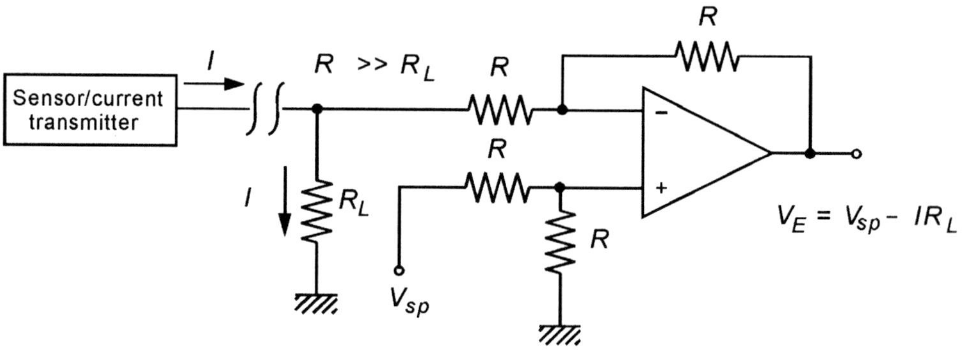

FIGURE 10.3
Error detector using a ground-based current and a differential amplifier.

a voltage divider using a constant voltage as a source. An example is shown in Figure 10.2. We assume a two-wire system is in use so that the current drawn from the floating power supply is the 4- to 20-mA signal current. The signal current is used to produce a voltage, IR, across the resistor, R. This is placed in series opposition to a voltage, V_{sp}, tapped from a variable resistor, R_{sp}, connected to a constant positive source, V_0. The result is an error voltage, $V_e = V_{sp} - IR$. This is then used in the process controller to calculate controller output.

An error detector also can be made from a differential amplifier. Such a system can be used only if the current from the transducer is referenced to ground. Figure 10.3 shows one typical configuration. The sensor signal current passes to ground through R_L, providing a signal voltage, $V_m = IR_L$. The differential amplifier then subtracts this from the setpoint voltage.

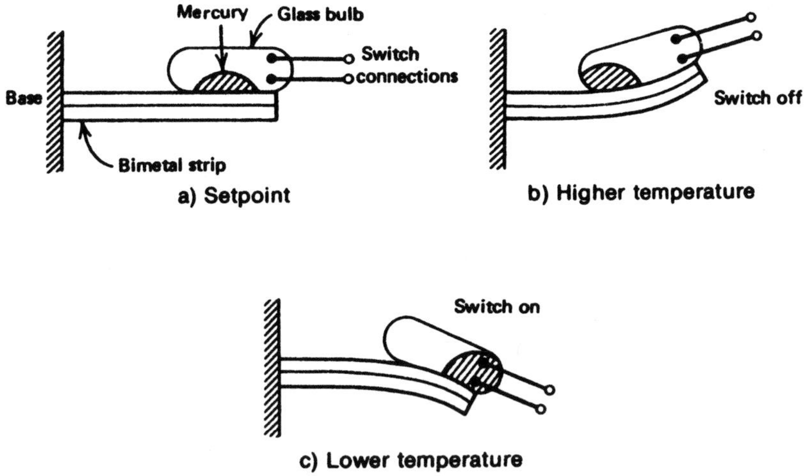

FIGURE 10.4
A mercury switch on a bimetal strip is often used as a two-position temperature controller.

10.3.2 Single Mode

The following systems and op amp circuits illustrate methods of implementing the pure modes of controller action.

Two-Position A two-position controller can be implemented by a great variety of electronic and electromechanical designs. In the past, many household air-conditioning and heating systems employed a two-position controller constructed from a bimetal strip and mercury switch, as shown in Figure 10.4. We see that as the bimetal strip bends because of a temperature decrease, it reaches a point where the mercury slides down to close an electrical contact. The inertia of the mercury tends to keep the system in that position until the temperature increases to a value above the setpoint temperature. This provides the required neutral zone to prevent excessive cycling of the system.

A method using op amp implementation of ON/OFF control with adjustable neutral zone is given in Figure 10.5. For this circuit, we assume that if the controller input voltage, V, reaches a value V_H, then the comparator output should go to the ON state, which is defined as some voltage, V_0. When the input voltage falls below a value V_L, the comparator output should switch to the OFF state, which is defined as 0 V. The comparator output switches states when the voltage on its input, V_1, is equal to the setpoint value, V_{sp}. Analysis of this circuit shows that the high (ON) switch voltage is

$$V_H = V_{sp} \tag{10.1}$$

and the low (OFF) switching voltage is

$$V_L = V_{sp} - \frac{R_1}{R_2} V_0 \tag{10.2}$$

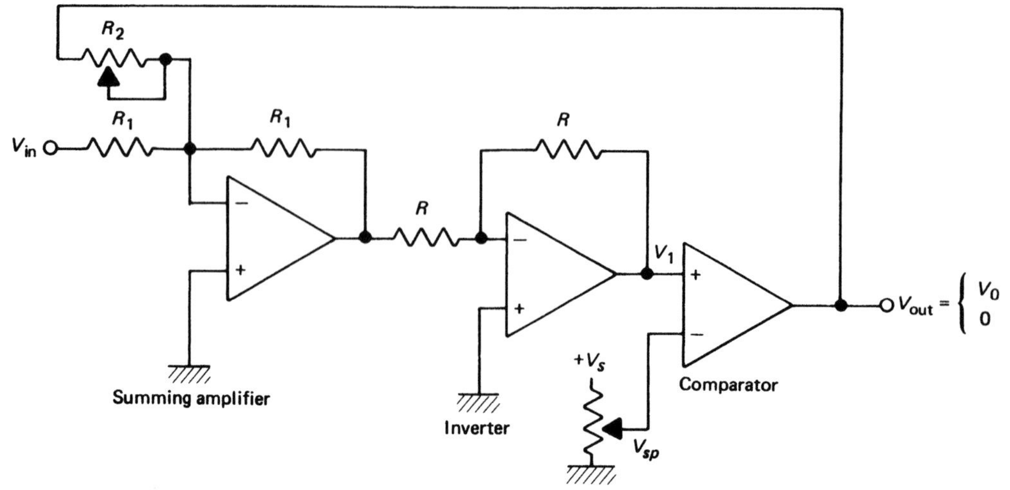

FIGURE 10.5
A two-position controller with neutral zone made from op amps and a comparator.

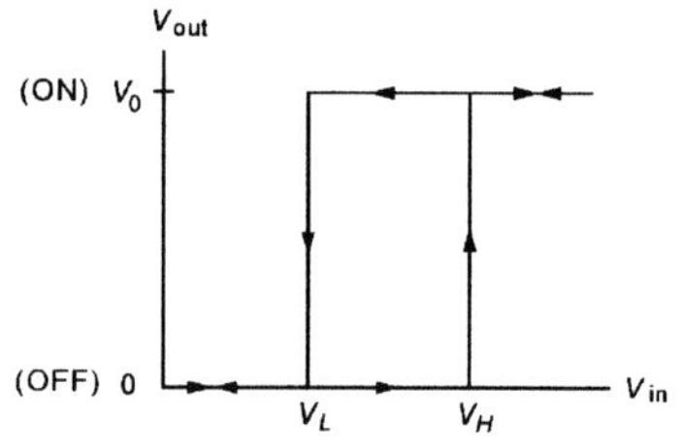

FIGURE 10.6
The circuit of Figure 10.5 shows the characteristic two-position response in terms of voltage.

Figure 10.6 shows the typical two-position relationship between input and output voltage for this circuit. The width of the neutral zone between V_L and V_H can be adjusted by variation of R_2. The relative location of the neutral zone is calculated from the difference between Equations (10.1) and (10.2).

The inverter resistance in Figure 10.5 can be chosen as any convenient value. Typically, it is in the 1- to 100-kΩ range.

EXAMPLE 10.1 Level measurement in a sump tank is provided by a transducer scaled as 0.2 V/m. A pump is to be turned on by application of +5 V when the sump level exceeds 2.0 m. The pump is to be turned back off when the sump level drops to 1.5 m. Develop a two-position controller.

Solution

Let us use the circuit of Figure 10.5. The high and low trip voltages will be determined by the conditions of the problem. From these, the values of the resistances can be determined.

$$V_H = (0.2 \text{ V/m})(2.0 \text{ m}) = 0.4 \text{ V}$$

and

$$V_L = (0.2 \text{ V/m})(1.5 \text{ m}) = 0.3 \text{ V}$$

This gives the following relations for the resistances and V_{sp}:

$$0.4 \text{ V} = V_{sp}$$

$$0.3 \text{ V} = V_{sp} - \frac{R_1}{R_2} V_0$$

Therefore, $V_{sp} = 0.4$ V and, from the second equation,

$$0.3 = 0.4 - (R_1/R_2)(5)$$

$$(R_1/R_2) = 0.02$$

Since there are two unknowns and only one condition, one unknown can be selected. Picking $R_1 = 5 \text{ k}\Omega$, for example, means that $R_2 = 250 \text{ k}\Omega$.

Proportional Mode Implementation of this mode requires a circuit that has a response given by

$$p = K_P e_p + p_0 \tag{9.14}$$

where
p = controller output 0–100%
K_P = proportional gain
e_p = error in percent of variable range
p_0 = controller output with no error

If we consider both the controller output and error to be expressed in terms of voltage, we see that Equation (9.14) is simply a *summing amplifier.* The op amp circuit in Figure 10.7 shows such an electronic proportional controller. In this case, the analog electronic equation for the output voltage is

$$V_{\text{out}} = G_P V_e + V_0 \tag{10.3}$$

where
V_{out} = output voltage
$G_P = R_2/R_1$ = gain
V_e = error voltage
V_0 = output with zero error

The design of a proportional controller calls for specification of the proportional gain described by K_P in Equation (9.14) that expresses the percent of output for an error of 1% of the measurement range. Alternatively it could be described as the proportional band, $PB = 100/K_P$. This must now be expressed in terms of the voltage gain, G_P, in Equation (10.3). The relationship between G_P and K is given by,

$$G_P = K_P \frac{\Delta V_{\text{out}}}{\Delta V_{\text{m}}} \tag{10.4}$$

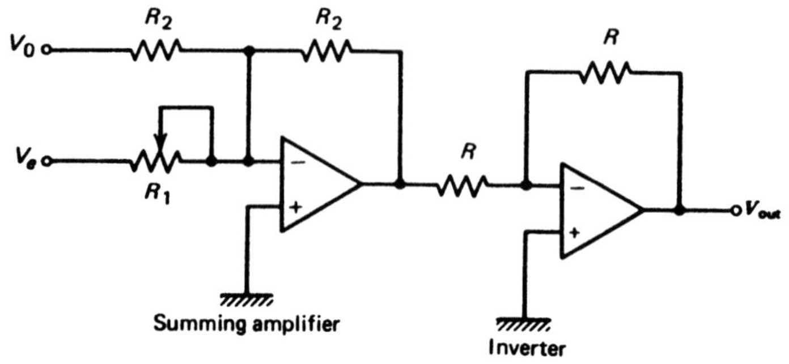

FIGURE 10.7
An op amp proportional-mode controller.

where ΔV_{out} = the range of output voltage
ΔV_m = the range of measurement voltage

The following example illustrates how Equation (10.4) is used to find the proportional gain in terms of volts/volts instead of %/%.

EXAMPLE 10.2 A temperature controller controls temperature from 100° to 200°C. A sensor provides an output of 2 to 8 V for this temperature range. The controller output drives a heater with an output of 0 to 5 volts. What circuit gain is needed if the controller of Figure 10.7 is to be used with a proportional gain of 4%/%?

Solution

The range of measurement voltage will be $\Delta V_m = 8\text{ V} - 2\text{ V} = 6\text{ V}$. The output range is 5 volts. Thus, the circuit gain will be,

$$G_P = (4\%/\%)\frac{5\text{ V}}{6\text{ V}} = 3.33\text{ V/V}$$

Notice that if the input and output ranges are the same, then $G_P = K_P$.

Another way of looking at this is using the concept of proportional band, $PB = 100/K_P$. The PB is that percent of error that will cause a 100% change of output. For the example of the last paragraph, $PB = 100/4 = 25\%$. Therefore, a 25% change of error must produce a 100% change of output. Twenty-five percent of the error is simply $(0.25)(6\text{ V}) = 1.5$ V, and 100% of the output is 5 V. Thus, $G_P = (5\text{ V}/1.5\text{ V}) = 3.33$, as before.

EXAMPLE 10.3 A controller is shown in Figure 10.7 with scaling so that 0–10 V corresponds to a 0–100% output. If $R_2 = 10\text{ k}\Omega$ and full-scale error range is 10 V, find the values of V_0 and R_1 to support a 20% proportional band about a 50% zero-error controller output.

Solution

The value of V_0 is simply 50% of 10 V, or 5 V, to provide the zero-error controller output. To design for a 20% proportional band means that a change of error of 20% must cause the controller output to vary 100%. Thus, from

$$V_{out} = G_P V_e + V_0$$

we note that when the error has changed 20% of 10 V, or 2 V, we must have full controller output change. Thus,

$$G_P = \frac{\Delta V_{out}}{\Delta V_e} = \frac{10}{2}$$
$$G_P = 5$$

so that if $R_2 = 10\ \text{k}\Omega$, then

$$R_1 = R_2/G_P = \mathbf{2\ k\Omega}$$

EXAMPLE 10.4 If the load in the previous example changes such that a new controller output of 40% is required, find the corresponding offset error.

Solution

In this case, we need a negative error so that the output is 40% of 10 V = 4 V.

$$V_{out} = G_P V_e + V_0$$
$$4 = 5V_e + 5$$

$\therefore V_e = 1/5$ V and, because the full-scale error signal is 10 V, we have an error of

$$\frac{-0.2}{10} \times 100 = \mathbf{-2\%}$$

Often, a voltage-to-current converter is used on the output to convert the output voltages to a 4- to 20-mA range of current signals to drive the final control element.

Integral Mode In Chapter 9, we saw that the integral mode was characterized by an equation of the form

$$p(t) = K_I \int_0^t e_p\, dt + p(0) \qquad \textbf{(9.16)}$$

where $p(t)$ = controller output in percent of full scale
K_I = integration gain (s^{-1})
e_p = deviations in percent of full-scale variable value
$p(0)$ = controller output at $t = 0$

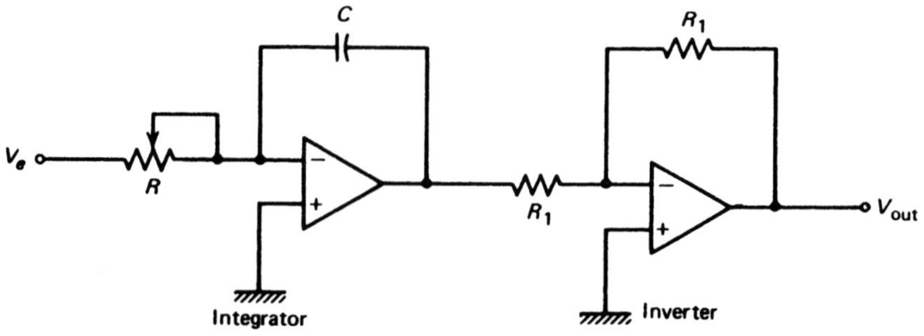

FIGURE 10.8
An op amp integral-mode controller.

This function is easy to implement when op amps are used as the building blocks. A diagram of an integral controller is shown in Figure 10.8. The corresponding equation relating input to output is

$$V_{\text{out}} = G_I \int_0^t V_e \, dt + V_{\text{out}}(0) \tag{10.5}$$

where
V_{out} = output voltage
$G_I = 1/RC$ = integration gain
V_e = error voltage
$V_{\text{out}}(0)$ = initial output voltage

The values of R and C can be adjusted to obtain the desired integration time. The initial controller output is the integrator output at $t = 0$. As we noted earlier, the integration time constant determines the *rate* at which controller output increases when the error is constant. If K_I is made too large, the output rises so fast that overshoots of the optimum setting occur and cycling is produced.

The actual value of G_I, and therefore of R and C, is determined from K_I and the input and output voltage ranges. One way to do this is to recognize that the integral gain says that an input error of 1% must produce an output that changes as K_I percent per second. Another way is to know that if an error of 1% lasts for 1 s, the output must change by K_I percent.

Suppose we have an input range of 6 V, an output range of 5 V, and $K_I = 3.0\%/(\% - \text{min})$. Integral gain is often given in minutes because industrial processes are slow, compared to a time of seconds. This gain is often expressed as integration time, T_I, which is just the inverse of the gain, so $T_I = 3.33\,\text{min}$.

We must first convert the time units to seconds. Therefore

$$[3\%/(\%\text{–min})][1\text{ min}/60\text{ s}] = 0.05\%/(\%\text{–s})$$

An error of 1% for 1 s is found from

$$(0.01)(6\text{ V})(1\text{ s}) = 0.06\text{ V} - \text{s}$$

Furthermore, K_I percent of the output (using the seconds expression for gain) is

$$(0.0005)(5\ \text{V}) = 0.0025\ \text{V}$$

Therefore, the integral gain in terms of voltage must be

$$G_I = (0.0025\ \text{V})/(0.06\ \text{V}\cdot\text{s}) = 0.0417\ \text{s}^{-1}$$

The values of R and C can be selected from this.

EXAMPLE 10.5 An integral control system will have a measurement range of 0.4 to 2.0 V and an output range of 0 to 6.8 V. Design an op amp integral controller to implement a gain of $K_I = 4\%/(\%\text{-min})$. Specify the values of G_I, R, and C.

Solution

The input range is $2.0 - 0.4 = 1.6$ V, and the output range is 6.8 V. We must convert K_I to units of seconds

$$[4\%/(\%\text{-min})][1\ \text{min}/60\ \text{s}] = 0.0667\%/(\%\text{-s})$$
$$1\%\ \text{of the input for 1 s} = (0.01)(1.6\ \text{V})(1\ \text{s}) = 0.016\ \text{V}\cdot\text{s}$$
$$0.0667\%\ \text{of the output} = (0.000667)(6.8\ \text{V}) = 0.00454\ \text{V}$$

Thus, the gain is

$$G_I = (0.00454\ \text{V}/0.016\ \text{V} - \text{s}) = 0.283\ \text{s}^{-1}$$

Because $G_I = 1/(R/C)$, we have

$$RC = 3.53\ \text{s}$$

If we pick $C = 100\ \mu\text{F}$, then $R = \mathbf{35.3\ k\Omega}$.

Derivative Mode The derivative mode is never used alone because it cannot provide a controller output when the error is zero (see Section 9.5.3). Nevertheless, we show here how it is implemented with op amps so it can be combined with other modes in the next section. The control mode equation was given earlier as

$$p(t) = K_D \frac{de_p}{dt} \tag{9.18}$$

where
p = controller output in percent of full output
K_D = derivative time constant (s)
e_p = error in percent of full-scale range

In principle, this mode could be implemented by the op amp circuit presented in Chapter 2 and shown in Figure 2.42. Indeed, the theoretical transfer function for this circuit given in Equation (10.6) looks just like Equation (9.18) with appropriate identifications in terms of circuit elements:

$$V_{\text{out}} = -RC \frac{dV_e}{dt} \tag{10.6}$$

where the input voltage has been set equal to the controller error voltage.

From a practical perspective, this circuit cannot be used because it tends to be unstable; that is, it may begin to exhibit spontaneous oscillations in the output voltage. The reason for this instability is the very large gain at high frequencies where the derivative is very large.

To study this effect, let us assume that the input voltage is given by a sinusoidal voltage oscillating with some frequency, f; then $V_e = V_0 \sin(2\pi f t)$. So, from Equation (10.6) we can write the amplitude of the output as,

$$|V_{out}| = 2\pi f RC\,|V_e| \tag{10.7}$$

This equation shows that the magnitude of the output voltage increases linearly with frequency. So, in principle, as the frequency goes to infinity, so does the output! Clearly this is unacceptable in our control system. A little high-frequency noise will cause large excursions in output voltage.

In order to make a practical circuit, a modification is provided that essentially "clamps" the gain above some frequency to a constant value. We make sure that the clamped frequencies are well above anything that could occur in the actual control system. This way, the circuit provides a derivative output in the frequencies of practical interest but simply acts like a fixed-gain amplifier at higher frequencies. Figure 10.9 shows that the simple modification is to place a resistor in series with the capacitor.

The actual transfer function for this circuit can be shown (see Appendix 6) to be given by

$$V_{out} + R_1 C\frac{dV_{out}}{dt} = -R_2 C\frac{dV_e}{dt} \tag{10.8}$$

You can see that the output depends upon the derivative of the input voltage, but there is now an extra term involving the derivative of the output voltage. Essentially, we have a first-order differential equation relating input and output voltage.

For very high frequencies the impedance of the capacitor becomes very small and can be neglected. Then the circuit becomes just an inverting amplifier with a gain $-(R_2/R_1)$. At

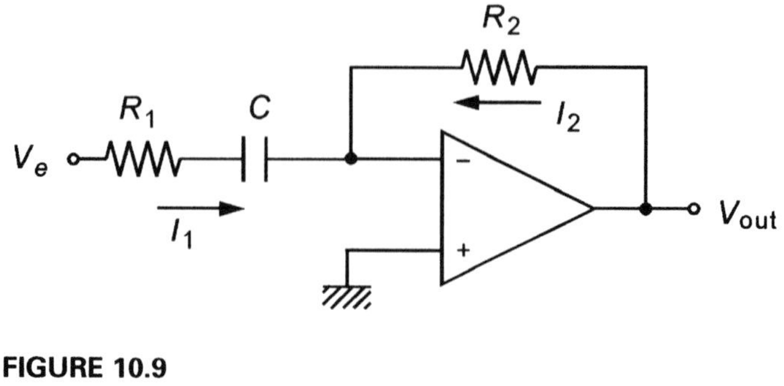

FIGURE 10.9
A practical derivative-mode op amp controller.

low frequency the impedance of the capacitor will be large so R_1 can be neglected. Then, the response of Equation (10.6) prevails. The circuit exhibits a derivative response provided the following inequality is satisfied,

$$2\pi f R_1 C << 1 \tag{10.9}$$

Therefore, when using a derivative action circuit, we must estimate the maximum physical frequency at which the system can respond, f_{max}, and pick R_1 so that for frequencies much higher than this, the inequality of Equation (10.9) is satisfied.

The following derivative mode design guidelines can be followed:

1. Estimate the maximum frequency at which the physical system can respond, f_{max}.
2. Set $2\pi f_{max} R_1 C = 0.1$ and solve for R_1. (C is found from the mode derivative gain requirement.)

Assuming this criterion has been met, we can ignore R_1 for the controller design and define the circuit derivative gain or derivative time in seconds as $G_D = R_2C$. G_D will be determined from the design controller derivative gain, K_D.

EXAMPLE 10.6

Derivative control action with a gain of $K_D = 0.04\%/(\%/\text{min})$ is needed to control flow through a pipe. The flow surges with a minimum period of 2.2 s. The input signal has a range of 0.4 to 2.0 V, and the output varies from 0.0 to 5.0 V. Develop the op amp derivative action circuit.

Solution

First we find the appropriate circuit gain, G_D. The derivative gain should first be converted to the units of seconds:

$$[0.04\%/(\%/\text{min})](60 \text{ s/min}) = 2.4\%/(\%/\text{s})$$

This result says that for every 1%/s rate of change of input, the output should change by 2.4%. So, 1%/s of the input is $(0.01)(2.0 - 0.4)$ V/s $= 0.016$ V/s. Then 2.4% of the output is simply $(0.024)(5) = 0.12$ V, so

$$G_D = (0.12 \text{ V}/0.016 \text{ V/s}) = 7.5 \text{ s}$$

This result allows us to pick R_2 and C from Figure 10.6:

$$R_2C = 7.5 \text{ s}$$

If we pick $C = 20\ \mu\text{F}$, then $R_2 = 375\ \text{k}\Omega$.

To find R_1, we need the maximum frequency. If the minimum period is 2.2 s, then the maximum frequency is $f_{max} = 1/2.2 \text{ s} = 0.45$ Hz. From the design guidelines, we set

$$2\pi f_{max} R_1 C = 2\pi(0.45)R_1(20\ \mu\text{F}) = 0.1$$

We can now solve for $R_1 \approx 1800\ \Omega$.

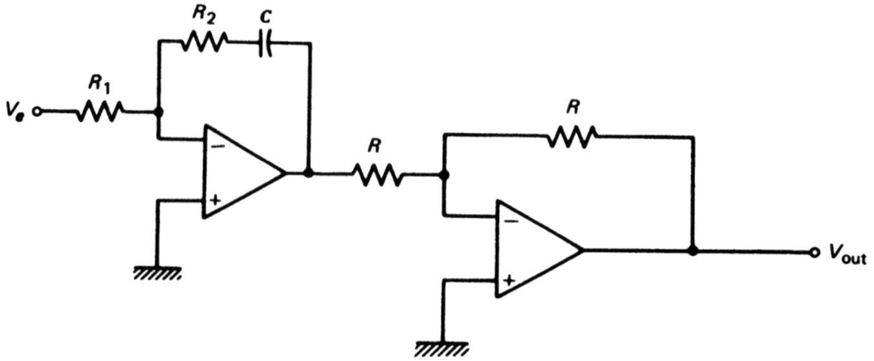

FIGURE 10.10
An op amp proportional-integral (PI) mode controller.

The value of $G_D = R_2C$ is determined from K_D and knowledge of the measurement and output voltage ranges. For this mode, the interpretation of K_D is that, for an error change of 1% in 1 s, the output should change by K_D percent. Thus, G_D is found from the quotient of K_D percent of the output voltage and 1% of the input voltage. Of course, K_D must be expressed in seconds.

The circuits of this section show that the pure modes of controller operation are easily constructed from op amps. As stated in Chapter 9, a pure mode is seldom used in process control because of the advantages of composite modes in providing good control. In the next section, implementation of composite modes using op amps is considered.

10.3.3 Composite Controller Modes

In Chapter 9, the combination of several controller modes was found to combine the advantages of each mode and, in some cases, eliminate disadvantages. Composite modes are implemented easily using op amp techniques. Basically, this consists of simply combining the mode circuits introduced in the previous section.

Proportional-Integral A simple combination of the proportional and integral circuits provides the proportional-integral mode of controller action. The resulting circuit is shown in Figure 10.10. For this case, the relation between input and output is most easily found by applying op amp circuit analysis. We get (including the inverter)

$$V_{out} = \left(\frac{R_2}{R_1}\right) V_e + \frac{1}{R_1C}\int_0^t V_e\, dt$$

The definition of the proportional-integral controller mode includes the proportional gain in the integral term, so we write

$$V_{out} = \left(\frac{R_2}{R_1}\right) V_e + \left(\frac{R_2}{R_1}\right) \frac{1}{R_2C}\int_0^t V_e\, dt + V_{out}(0) \qquad \textbf{(10.10)}$$

Equation (10.10) has the same form as Equation (9.19) for this mode. The adjustments of this controller are the *proportional band* through $G_p = R_2/R_1$, and the *integration gain* through $G_I = 1/R_2C$.

EXAMPLE 10.7 Design a proportional-integral controller with a proportional band of 30% and an integration gain of 0.1%/(%–s). The 4- to 20-mA input converts to a 0.4- to 2-V signal, and the output is to be 0–10 V. Calculate values of G_p, G_I, R_2, R_1, and C, respectively.

Solution

A proportional band of 30% means that when the input changes by 30% of range, or 0.48 V, the output must change by 100%, or 10 V. This gives a gain of

$$G_p = \frac{R_2}{R_1} = \frac{10\ \text{V}}{0.48} = \mathbf{20.83}$$

A K_I of 0.1%/(% – s) says that a 1% error for 1 s should produce an output change of 0.1%. One percent of 1.6 V is 0.016 V, and 0.1% of 10 V is 0.01 V, so

$$G_I = \frac{1}{R_2C} = \frac{0.01}{0.016} = 0.625\ \text{s}^{-1}$$

or

$$R_2C = 1.6\ \text{s}$$

As an example of values to do this, we could pick

$$C = \mathbf{10\ \mu F}, \qquad \text{which requires}$$

$$R_2 = \frac{1.6\ \text{s}}{10^{-5}\ \text{F}} = \mathbf{160\ k\Omega}$$

Then, to get the proportional gain, we use

$$R_1 = \frac{160\ \text{k}\Omega}{20.83} = \mathbf{7.68\ k\Omega}$$

Proportional-Derivative A powerful combination of controller modes is the proportional and derivative modes (see Section 9.6.2). This combination is implemented using a circuit similar to that shown in Figure 10.11. Analysis shows that this circuit responds according to the equation

$$V_{\text{out}} + \left(\frac{R_1}{R_1 + R_3}\right) R_3C\frac{dV_{\text{out}}}{dt} = \left(\frac{R_2}{R_1 + R_3}\right) V_e + \left(\frac{R_2}{R_1 + R_3}\right) R_3C\frac{dV_e}{dt} + V_0$$

where the quantities are defined in the figure and the output inverter has been included.

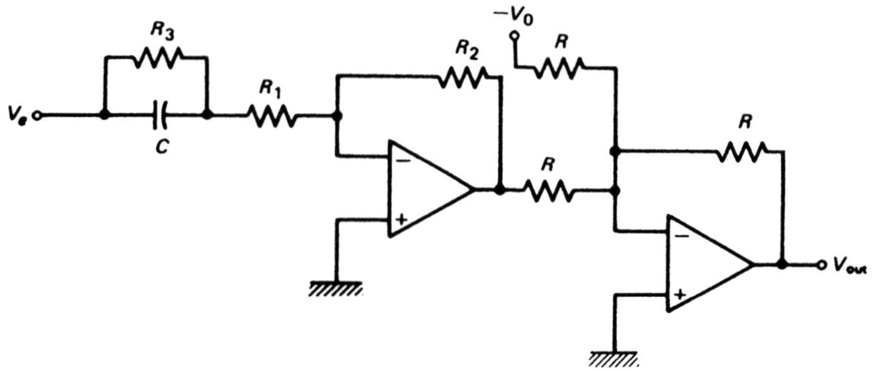

FIGURE 10.11
An op amp proportional-derivative (PD) mode controller.

This circuit includes the clamp to protect against high gain at high frequency in the derivative term. In this case, the condition represented by Equation (10.9) is modified slightly to use an effective resistance given by

$$R = \frac{R_1 R_3}{R_1 + R_3}$$

Then the condition becomes as usual, $2\pi f_{max} RC = 0.1$. Assuming this criterion has been met, the equation for the proportional-derivative response becomes

$$V_{out} = \left(\frac{R_2}{R_1 + R_3}\right) V_e + \left(\frac{R_2}{R_1 + R_3}\right) R_3 C \frac{dV_e}{dt} + V_0 \qquad \textbf{(10.11)}$$

where the proportional gain is $G_p = R_2/(R_1 + R_3)$, and the derivative gain is $G_D = R_3C$. This equation now corresponds to the form given by Equation (9.20) for the proportional-derivative controller. Of course, this mode still has the offset error of a proportional controller because the derivative term cannot provide reset action.

EXAMPLE 10.8 A proportional-derivative controller has a 0.4- to 2.0-V input measurement range, a 0- to 5-V output, $K_p = 5\%/\%$, and $K_D = 0.08\%$ per (%/min). The period of the fastest expected signal change is 1.5 s. Implement this controller with an op amp circuit.

Solution

To use the circuit of Figure 10.11, we first find the appropriate circuit gains, G_p and G_D.

A $K_p = 5\%/\%$ means a 20% *PB*. So we can write

$$G_p = \frac{(100\%)(5\text{ V})}{(20\%)(1.6\text{ V})} = \frac{5\text{ V}}{0.32\text{ V}} = 15.625$$

To find G_D, we must first change K_D to seconds:

$$K_D = 0.08\%/(\%/\text{min}) \times 60 \text{ s/min} = 4.8\%/(\%/\text{s})$$

Now we can write

$$G_D = \frac{(4.8\%)(5 \text{ V})}{(1\%)(1.6 \text{ V})} = 15 \text{ s}$$

The period limitation allows us to write

$$\frac{R_1}{R_1 + R_3} R_3 C = \frac{0.1}{2\pi}(1.5 \text{ s}) = 0.024 \text{ s}$$

Now we have three relations:

$$\frac{R_2}{R_1 + R_3} = 15.625, \qquad R_3 C = 15, \qquad \frac{R_1}{R_1 + R_3} = 0.0016$$

The last relation comes from combining the limitation equation with $R_3C = 15$. With these three relations, we have four unknowns. One can be selected. Let us try $C = 100\ \mu$F. Then the equations give

$$R_1 = 240\ \Omega, \qquad R_2 = 2.35\ \text{M}\Omega, \qquad R_3 = 150\ \text{k}\Omega$$

PID (Three-Mode) The ultimate process controller is the one that exhibits proportional, integral, and derivative response to the process-error input. In Chapter 9, we saw that this mode was characterized by the equation

$$p = K_p e_p + K_p K_I \int_0^t e_p \, dt + K_p K_D \frac{de_p}{dt} + p_I(0) \qquad \textbf{(9.21)}$$

where
- p = controller output in percent of full scale
- e_p = process error in percent of the maximum
- K_p = proportional gain
- K_I = integral gain
- K_D = derivative gain
- $p_I(0)$ = initial controller integral output

The zero-error term of the proportional mode is not necessary because the integral automatically accommodates for offset and nominal setting. This mode can be provided by a straight application of op amp circuits, resulting in the circuit of Figure 10.12. It must be noted, however, that it is possible to reduce the complexity of the circuitry of Figure 10.12 and still realize the three-mode action, but in these cases an interaction results between derivative and integral gains. We will use the circuit of Figure 10.12 because it is easy to follow in illustrating the principles of implementing this mode. Analysis of the circuit shows that the output is

$$-V_{\text{out}} = \left(\frac{R_2}{R_1}\right) V_e + \left(\frac{R_2}{R_1}\right) \frac{1}{R_I C_I} \int V_e \, dt + \left(\frac{R_2}{R_1}\right) R_D C_D \frac{dV_e}{dt} + V_{\text{out}}(0) \qquad \textbf{(10.12)}$$

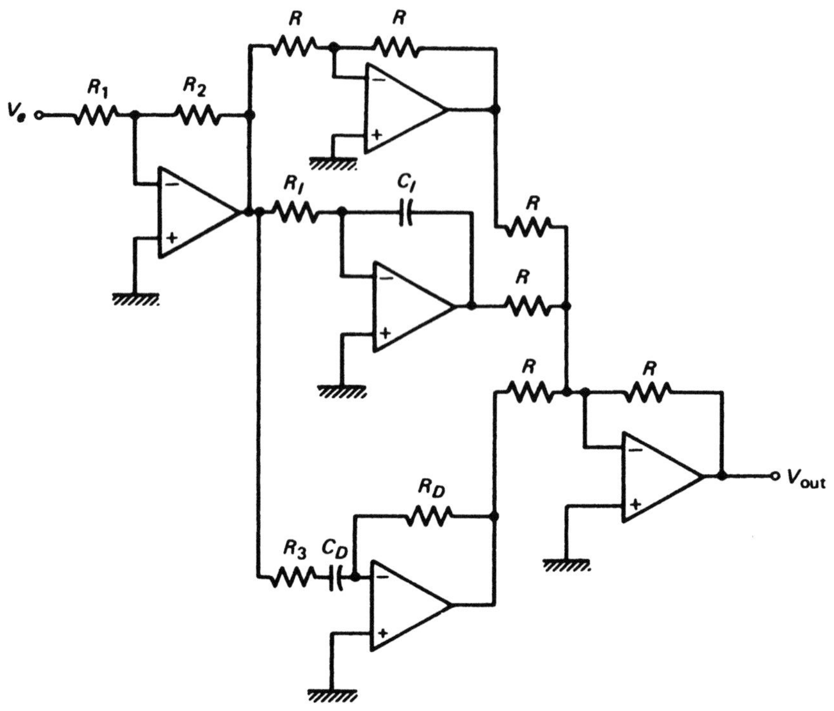

FIGURE 10.12
Direct implementation of a three-mode (PID) controller with op amps. Circuits with fewer op amps are often used.

where R_3 has been chosen from $2\pi f_{max} R_3 C_D = 0.1$ for stability. Comparison with Equation (9.21) shows that this implements the three-mode controller if

$$G_p = \frac{R_2}{R_1}, \qquad G_D = R_D C_D, \qquad G_I = \frac{1}{R_I C_I}$$

EXAMPLE 10.9 A temperature-control system inputs the controlled variable as a range from 0 to 4 V. The output is a heater requiring 0 to 8 V. A PID is to be used with $K_p = 2.4\%/\%$, $K_I = 9\%/(\%/\text{min})$, $K_D = 0.7\%/(\%/\text{min})$. The period of the fastest expected change is estimated to be 8 s. Develop the PID circuit.

Solution

The input range is 4 V, and the output range is 8 V. Let us figure the circuit gains.

For the proportional mode, a 1% error means a voltage change of $(0.01)\cdot(4\text{V}) = 0.04$ V. This should cause an output change of 2.4% or $(0.024)(8\text{ V}) = 0.192$ V. Thus,

$$G_p = (0.192\text{ V}/0.04\text{ V}) = 4.8$$

For the integral term, an error of 1% should cause the output to change by 9%/min, which is $(9/60) = 0.15\%/\text{s}$. Thus,

$$G_I = (0.0015\ \text{s}^{-1})(8\ \text{V})/(0.04\ \text{V})$$
$$G_I = 0.3\ \text{s}^{-1}$$

For the derivative term, an error change of 1% per min or $(0.04\ \text{V}/60) = 6.67 \times 10^{-4}\ \text{V/s}$ should cause an output change of 0.7% or $(0.007)(8\ \text{V}) = 0.056\ \text{V}$. Thus,

$$G_D = (0.056\ \text{V}/6.67 \times 10^{-4}\ \text{V/s})$$
$$G_D = 84\ \text{s}$$

These results provide the following relations:

$$(R_2/R_1) = 4.8$$
$$1/(R_I C_I) = 0.3\ \text{s}^{-1}$$
$$R_D C_D = 84\ \text{s}$$

From the fastest period specification, we form the relationship

$$2\pi R_3 C_D = (0.1)(8\ \text{s}) = 0.8$$

which gives seven unknowns and four equations. We can pick three quantities. Let us try $R_1 = 10\ \text{k}\Omega, C_I = C_D = 10\ \mu\text{F}$. This gives

$$R_2 = 4.8 R_1 = 48\ \text{k}\Omega$$
$$R_I = 1/(0.3\, C_I) = 333\ \text{k}\Omega$$
$$R_D = 84/C_D = 8.4\ \text{M}\Omega$$
$$R_3 = 0.8/(2\pi C_D) = 12.7\ \text{k}\Omega$$

8.4 MΩ seem too large for practical consideration. Let us change C_D to 100 μF. Now we get

$$R_D = 840\ \text{k}\Omega$$
$$R_3 = 1270\ \Omega$$

which seems more reasonable.

These circuits have shown that the direct implementation of controller modes can be provided by standard op amp circuits. It is necessary, of course, to scale the measurement as a voltage within the range of operation selected by the circuit. Furthermore, the outputs of the circuits shown have been voltages that may be converted to currents for use in an actual process-control loop.

These circuits are only examples of basic circuits that implement the controller modes. Many modifications are employed to provide the controller action with different sets of components.

10.4 PNEUMATIC CONTROLLERS

Historically, the reason for using pneumatics in process control was probably that electronic methods were not yet competitive in cost or reliability. Safety was and still is a factor where the danger of explosion from electrical malfunctions exists. It is also true that the final control element is often pneumatically or hydraulically operated, which suggests that an all-pneumatic process-control loop might be advantageous. It appears that analog or digital electronic methods will eventually replace most pneumatic installations. But we will still have pneumatic equipment for many years until these are depreciated in industry. A good understanding of process-control principles can be applied to either electronic or pneumatic techniques, but it is necessary to consider some special features of pneumatic technology. This section provides a brief description of operations by which controller modes are pneumatically implemented.

10.4.1 General Features

The outward appearance of a pneumatic controller is typically the same as that for the electronic controller shown in Figure 10.1. The same readout of setpoint, error, and controller output appears, and adjustments of gain, rate, and reset are available. The working signal is most typically the 3- to 15-psi standard pneumatic process-control signal, usually derived from a regulated air supply of 20 to 30 psi. As usual, we use the English system unit of pressure because its use is so widespread in the process-control industry. Eventual conversion to the SI unit of N/m^2 or Pa will require some alteration in scale (of measurement) to a range of 20 to 100 kPa.

The pneumatic controller is based on the nozzle/flapper described in Section 7.3.3 as the basic mechanism of operation, much as the op amp is used in electronics. The schematic drawings of controller mode implementation are intended to convey the operating principles. Specific designs may vary considerably from the systems shown, however.

10.4.2 Mode Implementation

The following discussions present the essential features of controller-mode implementation using pneumatic techniques. The equations are stated in general form with units in SI, but the reader should be prepared to work with English units when necessary.

Proportional A proportional mode of operation can be achieved with the system shown in Figure 10.13. Operation is understood by noting that if the input pressure increases, then the input bellows forces the flapper to rotate to close off the nozzle. When this happens, the output pressure increases so that the feedback bellows exerts a force to balance that of the input bellows. A balance condition then occurs when torques exerted by each about the pivot are equal, or

$$(p_{\text{out}} - p_0)A_2x_2 = (p_{\text{in}} - p_{sp})A_1x_1$$

This equation is solved to find the output pressure

$$p_{\text{out}} = \frac{x_1}{x_2}\frac{A_1}{A_2}(p_{\text{in}} - p_{sp}) + p_0 \tag{10.13}$$

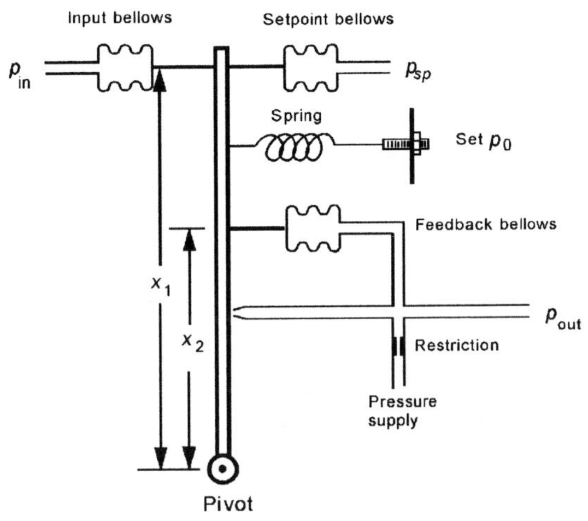

FIGURE 10.13
The pneumatic proportional-mode controller.

where
p_0 = pressure with no error
p_{in} = input pressure (Pa)
A_1 = input and setpoint bellows effective area (m^2)
x_1 = level arm of input (m)
p_{out} = output pressure (Pa)
A_2 = feedback bellows effective area (m^2)
x_2 = feedback lever arm (m)
p_{sp} = setpoint pressure

This relation is based on the notion of torque equaling force times lever arm, and the fact that a pressure in a bellows produces a force that is effectively the pressure times bellows area, much like a diaphragm. Equation (10.13) displays the standard response of a proportional mode in that output is directly proportional to input. The gain in this case is given by

$$K_p = \left(\frac{x_1}{x_2}\right)\left(\frac{A_1}{A_2}\right) \quad \textbf{(10.14)}$$

Because the bellows are usually of fixed geometry, the gain is varied by changing the lever arm length. In this simple representation, the gain is established by the distance between the bellows. If this separation is changed, the forces are no longer balanced, and for the same pressure a new controller output will be formed, corresponding to the new gain.

EXAMPLE 10.10 Suppose a proportional pneumatic controller has $A_1 = A_2 = 5\ \text{cm}^2$, $x_1 = 8$ cm, and $x_2 = 5$ cm. The input and output pressure ranges are 3 to 15 psi. Find the input pressures that will drive the output from 3 to 15 psi. The setpoint pressure is 8 psi, and $p_0 = 10$ psi. Find the proportional band.

Solution
First we find the gain from

$$K_p = \left(\frac{x_1}{x_2}\right)\left(\frac{A_1}{A_2}\right) = \left(\frac{8\text{ cm}}{5\text{ cm}}\right)\left(\frac{5\text{ cm}^2}{5\text{ cm}^2}\right)$$
$$K_p = 1.6$$

Now we have

$$p_{\text{out}} = K_p(p_{\text{in}} - p_{sp}) + p_0$$
$$p_{\text{out}} = 1.6(p_{\text{in}} - 8) + 10$$

The low input occurs when $p_{\text{out}} = 3$ psi, so

$$3 = 1.6(p_L - 8) + 10$$

which gives

$$p_L = \mathbf{3.625\ psi}$$

The high is found from

$$15 = 1.6(p_H - 8) + 10$$

which gives

$$p_H = \mathbf{11.125\ psi}$$

The proportional band (PB) is

$$PB = \left(\frac{11.125 - 3.625}{15 - 3}\right)100$$
$$PB = \mathbf{62.5\%}$$

Note that this checks with

$$PB = \frac{100}{K_p} = \frac{100}{1.6} = 62.5\%$$

which could be used because the input and output ranges are the same.

Proportional-Integral This control mode is also implemented using pneumatics by the system shown in Figure 10.14. In this case, an extra bellows with a variable restriction is added to the proportional system. Suppose the input pressure shows a sudden increase. This drives the flapper toward the nozzle, increasing output pressure until the proportional bellows balances the input as in the previous case. The integral bellows is still at the original output pressure, because the restriction prevents pressure changes from being transmitted immediately. As the increased pressure on the output bleeds through the restriction, the integral bellows slowly moves the flapper closer to the nozzle, thereby causing a steady increase in output pressure (as dictated by the integral mode). The variable restriction allows for variation of the leakage rate, and hence the integration time.

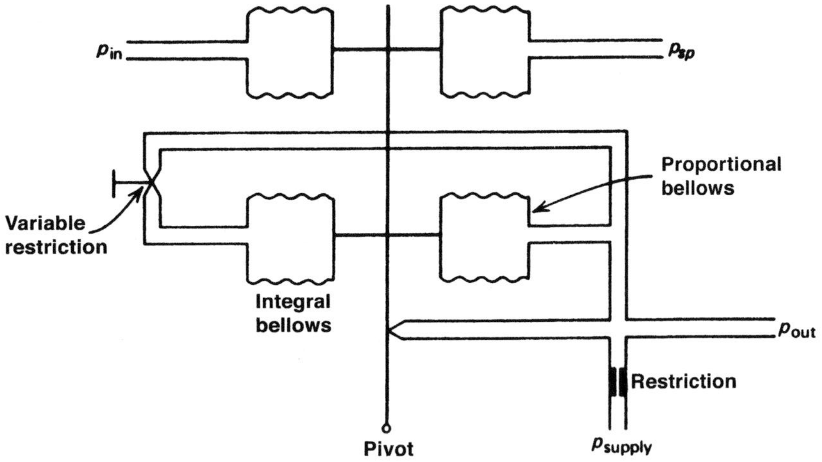

FIGURE 10.14
Pneumatic proportional-integral controller.

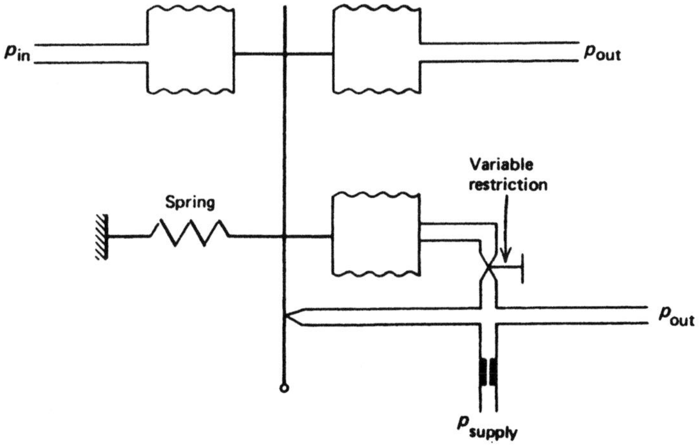

FIGURE 10.15
Pneumatic proportional-derivative controller.

Proportional-Derivative This controller action can be accomplished pneumatically by the method shown in Figure 10.15. A variable restriction is placed on the line leading to the balance bellows. Thus, as the input pressure increases, the flapper is moved toward the nozzle with no impedance, because the restrictions prevent an immediate response of the balance bellows. Thus, the output pressure rises very fast and then, as the increased pressure leaks into the balance bellows, decreases as the balance bellows moves the flapper back away from the nozzle. Adjustment of the variable restriction allows for changing the derivative time constraint.

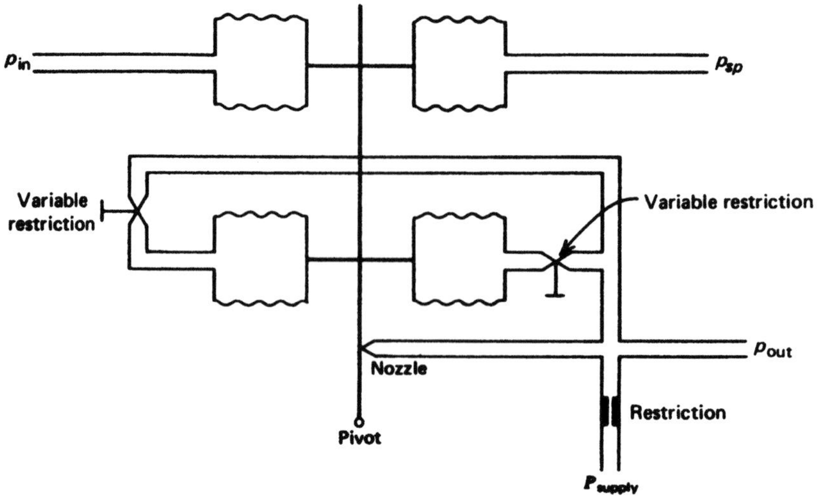

FIGURE 10.16
Pneumatic three-mode (PID) controller.

Three-Mode The three-mode controller is actually the most common type produced, because it can be used to accomplish any of the previous modes by setting of restrictions. This device is shown in Figure 10.16, and, as can be seen, it is simply a combination of the three systems presented.

By opening or closing restrictions, the three-mode controller can be used to implement the other composite modes. Proportional gain, reset time, and rate are set by adjustment of bellows separation and restriction size.

10.5 DESIGN CONSIDERATIONS

To illustrate some of the facets involved in setting up a process-control loop, it would be valuable to follow through some hypothetical examples. The following examples assume that a process-control loop is required, and that the controller operation must be provided by electronic analog circuits.

EXAMPLE 10.11 Design a process-control system that regulates light level by outputting a 0–10-V signal to a lighting system that provides 30–180 lux. The sensor has a transfer function of $-120\ \Omega/\text{lux}$ with a 10-kΩ resistance at 100 lux. The setpoint is to be 75 lux, and proportional control with a 75% proportional band has been selected.

Solution

We solve such problems by first establishing the characteristics of each part of the system.

1. The illumination varies from 30 to 180 lux. We find the resistance changes according to

$$R = 10\ \text{k}\Omega - 0.12\ \text{k}\Omega(I - 100)$$

where I is the illumination in lux.

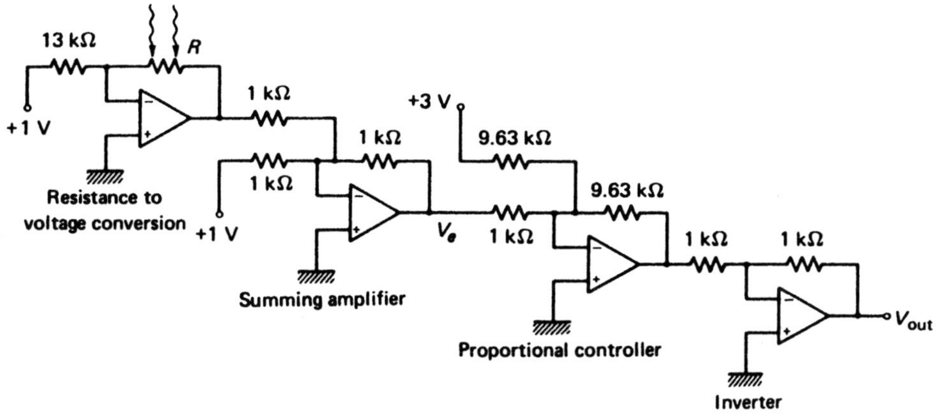

FIGURE 10.17
Circuit for Example 10.11.

2. This allows us to find the resistance at 30 lux as

$$R = 10 \text{ k}\Omega - 0.12 \text{ k}\Omega(30 - 100)$$
$$R = \mathbf{18.4 \text{ k}\Omega}$$

and at 180 lux we get **0.4 kΩ**. The setpoint (75 lux) has a resistance of **13 kΩ**.

3. We can convert this resistance variation to voltage using the photocell in an op amp circuit. In Figure 10.17, we use an inverting amplifier with a gain of 1 at the setpoint and a constant −1-V input. The resistance to voltage conversion gives

$$V = -\frac{R}{13 \text{ k}\Omega}(-1 \text{ V}) = \frac{R}{13 \text{ k}\Omega}$$

Using this equation, we find the output voltage at 18.4 kΩ to be

$$V = -\frac{18.4 \text{ k}\Omega}{13 \text{ k}\Omega}(-1 \text{ V}) = \mathbf{+1.42 \text{ V}}$$

and at 0.4 kΩ, we get

$$V = -\frac{0.4 \text{ k}\Omega}{13 \text{ k}\Omega}(-1 \text{ V}) = \mathbf{0.031 \text{ V}}$$

so the input voltage range is 1.42 − 0.031 = 1.389 V.

4. Now we use a summing amplifier to find the error in Figure 10.17.

$$V_e = \frac{R}{13 \text{ k}\Omega} - 1$$

A 75% proportional band controller with a 75-lux setpoint requires a zero-error output of

$$V_0 = \frac{75 - 30}{180 - 30} 10 \text{ V} = \mathbf{3 \text{ V}}$$

5. The 75% band means that when the illumination changes by 75% of $(180 - 30) = 112.5$ lux, the output should swing by 10 V. Thus, in terms of resistance, this corresponds to 13.5 kΩ, and in terms of error voltage, it is 13.5 kΩ/13 kΩ, or 1.038 V.
6. Finally, the gain must be

$$G_p = \frac{10 \text{ V}}{1.038} = \mathbf{9.63}$$

The overall response is

$$V_{\text{out}} = 9.63V_e + 3$$

or

$$V_{\text{out}} = 9.63\left(\frac{R}{13 \text{ k}\Omega} - 1\right) + 3$$

The rest of the circuit in Figure 10.17 accomplishes this function. When $V_{\text{out}} = 0, R = 8.9 \text{ k}\Omega$ or 90.83 lux, and for $V_{\text{out}} = 10 \text{ V}, R = 22.4 \text{ k}\Omega$ or 203.33 lux, so that the output swings 100% as the input swings

$$\frac{203.33 - 90.83}{180 - 30} = 0.75, \quad \text{or } 75\%$$

as required.

The proportional band could not be used to find the gain directly in Example 10.11 because the input and output were not expressed to the same scale—that is, as 0–100% or 0–10 V, and so on.

EXAMPLE 10.12 A type-J thermocouple (TC) with a 0°C reference is used to control temperature between 100°C and 200°C. Design a proportional-integral controller with a 40% band and a 0.08-min reset (integral) time. The final control element requires a 0–10-V range.

Solution

a. In this problem, we must perform the following steps:

1. Amplify the low TC voltage to a more convenient value than the TC mV output.
2. Use this amplifier output as input to the proportional-integral controller and pick a proportional gain that swings the output 0–10 V as the input swings 40% of full scale.
3. Select values to provide a 0.08-min (4.8-s) integral time.

b. The solution is shown in Figure 10.18.

1. We note that a type-J TC produces a voltage of 5.27 mV at 100°C and 10.78 mV at 200°C. An amplifier with a gain of 100 will convert these to 0.527 and 1.078 V, respectively.

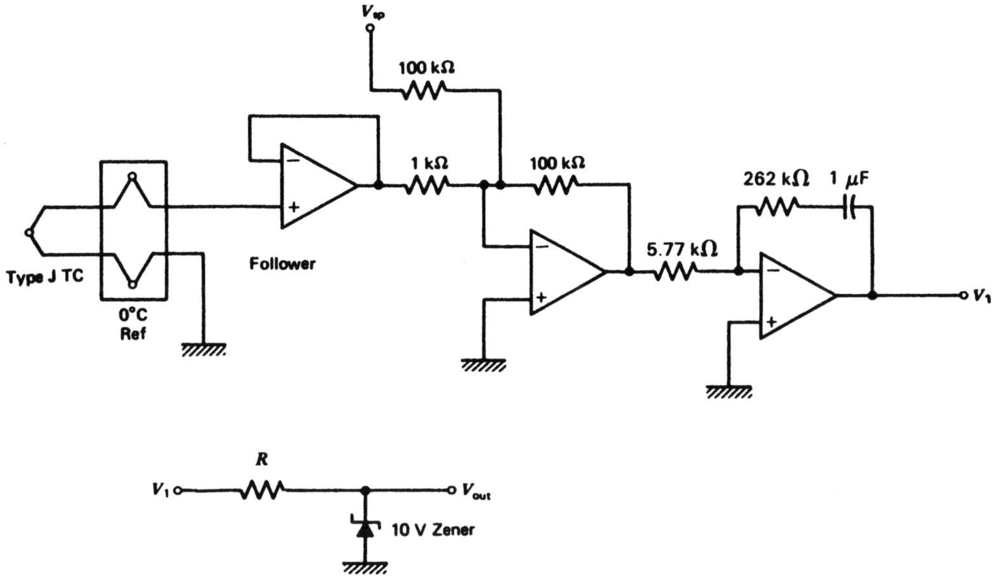

FIGURE 10.18
Circuit for Example 10.12.

2. Now we sum this output to a properly scaled setpoint voltage to get an error signal. The setpoint value is obtained from a voltage divider. To get the proper controller values, we note that 40% of the input swing is

$$0.4(1.078 - 0.527) = 0.2204 \text{ V}$$

Thus, the proportional gain is

$$G_p = \frac{10 \text{ V}}{0.2204 \text{ V}} = 45.37$$

So values are to be chosen to provide this gain.

3. For the integral term, an 0.08-min reset means $K_I = 12.5\%/(\%\text{-min})$ or $(12.5/60) = 0.21\%/(\%\text{-s})$. Thus, an error of 1% for 1 s must produce a change in output of 0.21%.

$$G_I = \frac{(0.0021)(10 \text{ V})}{(0.01)(0.551 \text{ V})} = 3.81 \text{ s}^{-1}$$

so

$$\frac{R_2}{R_1} = 45.37, R_2C = 0.262 \text{ s}$$

Let us try $C = 1\ \mu\text{F}$. Then

$$R_2 = 262 \text{ k}\Omega \quad \text{and} \quad R_1 = 5.77 \text{ k}\Omega$$

The overall transfer function for the final circuit shown in Figure 10.18 is found to be

$$V_{\text{out}} = 45.37V_e + 173.2\int V_e\,dt$$

where

$$V_e = 100V_{TC} - V_{sp}$$

The output diode and zener limit the swing from 0 to 10 V.

EXAMPLE 10.13 A differential pressure gauge is used to measure flow that varies as the square root of the pressure difference (Equation 7.8). The pressure signal is a 0–2-V range for minimum to maximum flow. A *square root extractor* circuit is available that accepts from 0 to 10 V and outputs the square root of the input. Design a proportional controller with a 15% proportional band having a 0–10-V output and a nominal (zero-error) output of 5 V.

Solution

The circuit of Figure 10.19 implements this function. The controller input is a 0–3.162-V signal. A 15% proportional band means that if the input changes by (0.15) (3.162 V) = 0.474 V, the output must change by 10 V. Thus, the gain is

$$G_p = \frac{10}{0.474} = \mathbf{21.1}$$

This is provided by the 1- kΩ and 21.1- kΩ resistors.

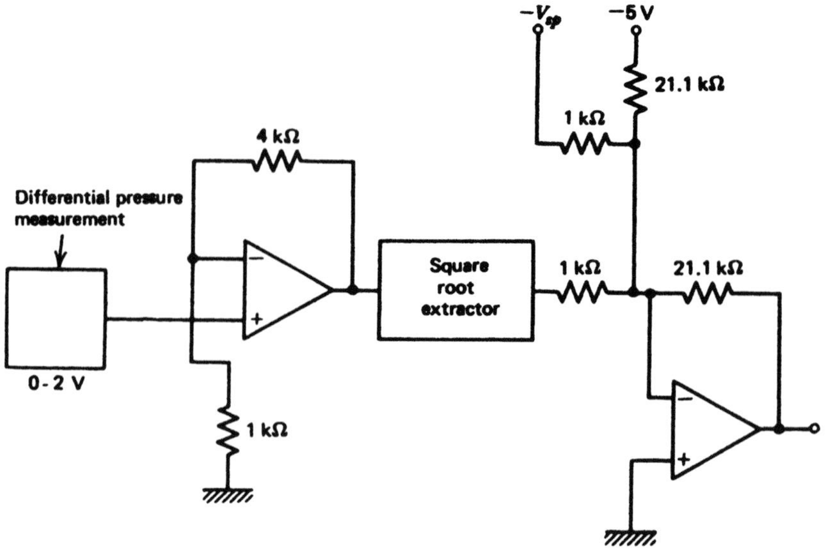

FIGURE 10.19
Circuit for Example 10.13.

SUMMARY

This chapter presented numerous methods of implementing the controller function of a process-control loop. It shows typical methods of obtaining controller modes from analog, electronic, and pneumatic approaches. If the reader understands these typical methods, then other specific methods of implementation can be analyzed and understood by analogy.

The topics covered are summarized by the following:

1. Realization of controller modes with op amps is obtained by a straight application of amplifier, integrator, and differentiator circuits using standard op amp techniques. The gains are found by the external resistors and capacitors used with the op amps.
2. Pneumatic controller mode implementation is made possible by a combination of a flapper/nozzle system, appropriate bellows, and variable-flow restrictions. In general, given a three-mode controller, any of the other composite modes is obtained by opening the restrictions.

PROBLEMS

Section 10.3

10.1 A sensor converts position from 0 to 2.0 m into a 4- to 20-m A current. An error detector such as that shown in Figure 10.2 is used with $R = 100\ \Omega, V_0 = 5.0$ V, and $R_{sp} = 1\ \text{k}\Omega$ pot.

a. If the setpoint is 0.85 m, what is V_{sp}?

b. If $V_{sp} = 1.5$ V, what is the range of error voltage as position varies from 0 to 2.0 m?

10.2 Show how the circuit of Figure 10.3 can be applied to find the error voltage for Problem 10.1.

10.3 Using the system of Figure 10.5, design a two-position controller with a 0- to 10-V input and a 0- or 10-V output. The setpoint is 4.3 V and the neutral zone is to be ± 1.1 V about this setpoint.

10.4 Design a two-position controller that turns a 5-V light relay ON when a silicon photocell output drops to 0.22 V and OFF when the cell voltage reaches 0.78 V.

10.5 Design a two-position controller that provides an output of 5 V when a type-J TC junction reaches 250°C and drops to a low of 0 V when the temperature has fallen to 240°C. Assume a 0°C reference.

10.6 Figure 10.20 shows a drying oven for which the oven is either *off* with 0 V input or *on* with 8 volts input. The thermistor properties are: resistance of 4.7 kΩ at 35°C and 1.4 kΩ at 60°C. It has a 5-mW/°C-dissipation constant. Design a two-position controller with trip points of 35°C and 60°C. Keep self-heating below 0.5°C.

10.7 **a.** Design a 45% PB controller for motor-speed control. The motor speed varies from 100 to 150 rpm for an input control voltage of 0 to 5 V. A speed sensor linearly changes from 2.0 to 5.0 kΩ over the speed range. A setpoint of 125 rpm is desired, for which the motor control circuit input is 2.5 V.

b. Suppose the setpoint is changed to 120 rpm with no other adjustments. What offset error will occur?

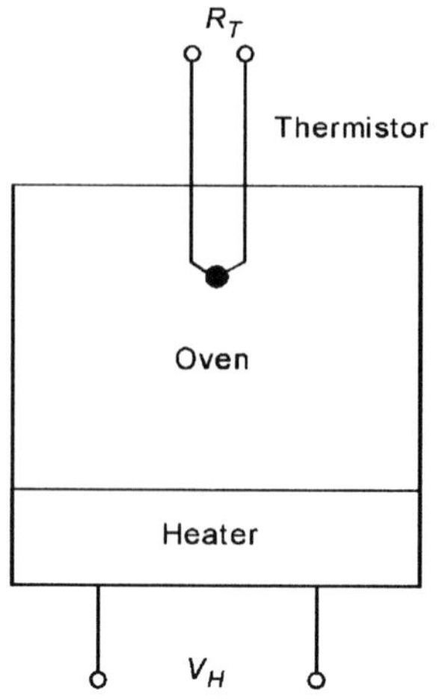

FIGURE 10.20
System for Problem 10.6.

10.8 A type-J TC with a 0°C reference is used in a proportional-mode temperature-control system with a 140°C setpoint and a range of 100–180°C. The zero-error output should be 45%, and the PB = 35%. The output is 0–10 V, and the full-scale input range is 0 to 1 V. Design a controller according to the circuit of Figure 10.7.

10.9 An integral controller has an input range of 1 to 8 V and an output range of 0 to 12 V. If K_I = 12%/(%/min), find G_I, R, and C.

10.10 Rate (derivative) action is needed for steering a boat. The rate gain should be K_D = 0.02%/(%/min). The error voltage range is −4.0 to +4.0 V, and the output signal varies from 0 to 2.5 V. The fastest physical turning period is 0.4 min. Find the component values of a derivative-mode op amp circuit such as Figure 10.9.

10.11 Design a proportional-integral controller with an 80% PB and a 0.03-min reset time. Use a 0–5-V input and a 0–12-V output. See Figure 10.10 for the circuit.

10.12 Design a PD controller with a 140% PB and a 0.2-min derivative time. The fastest signal speed is 1 min. Measurement range is 0.4 to 2 V, and the output is 0 to 10 V.

10.13 A liquid-level system converts a 4–10-m level into a 4- to 20-mA current. Design a three-mode controller that outputs 0–5 V with a 50% PB, 0.03-min reset time, and 0.05-min derivative time. Fastest expected change time is 0.8 min.

Section 10.4

10.14 A proportional pneumatic controller has equal area bellows. If 3- to 15-psi signals are used on input and output, find the ratio of pivot distances that provides a 23% PB.

10.15 If the setpoint in Problem 10.14 is 7 psi and the zero-error output is 9.2 psi, find the inputs yielding outputs of 3 psi and 15 psi.

Section 10.5

10.16 Explain how the setpoint in Example 10.11 can be changed. Implement such a change to provide a setpoint of 90 lux. Show all new component values for a proportional band of 48%.

10.17 Show how the circuit of Problem 10.11 can be modified to provide switched reset times of 0.02, 0.04, 0.06, and 0.08 min.

10.18 Design a proportional controller for a 4- to 20-mA, ground-based input, a 0- to 9-V output, zero-error output adjustable from 0 to 100%, and K_p adjustable from 1 to 10. Design so the setpoint can be selected in the range of 4 to 20 mA.

10.19 A sensor measures pressure as 22 mV/psi. Develop signal conditioning, error detection, and a PI controller for the following specifications: 0–300-psi measurement range; setpoint adjustable from 100 to 200 psi; K_p adjustable from 1.5 to 5.0; K_I switchable between 0.8, 1.6, 2.4, and 3.2 min^{-1}; output range of 0 to 10 V.

SUPPLEMENTARY PROBLEMS

S10.1 The intensity of light on a surface must be controlled within the range of 20 to 80 mW/cm^2. The nominal setpoint is 58 mW/cm^2. Figure 10.21 shows the configuration of the system. The CdS photocells have a resistance versus intensity given by Figure 6.9, and the light source requires a 0–10-V input. Input to the error detector should be the average light intensity from the two sensors. Design signal conditioning, error detector, and an op amp PI controller with $K_p = 1.06\%/\%$ and $T_I = 1.67$ s.

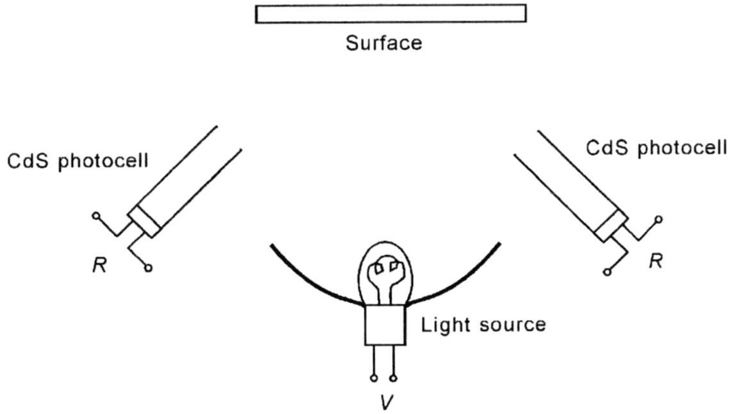

FIGURE 10.21
System for Problem S10.1.

FIGURE 10.22
Nozzle/flapper characteristic for Problem S10.2.

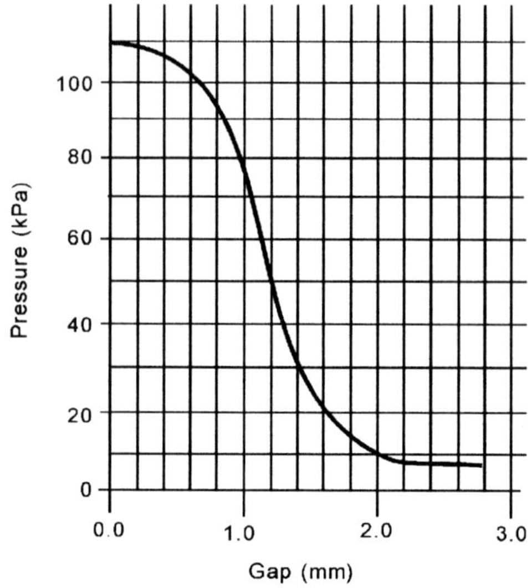

S10.2 A pneumatic proportional controller is designed as shown in Figure 10.13 such that x_1 is 5 cm and x_2 is 2.5 cm. The setpoint is 50 kPa and the effective areas of the bellows are 7.0 cm^2. The nozzle/flapper system has a pressure/displacement characteristic as shown in Figure 10.22. With no error, the output pressure is 60 kPa.

a. What nozzle/flapper gap is required to support an output of 50 kPa?
b. With no error, how much force in newtons and pounds does the input bellows exert on the flapper?
c. Suppose the input increases to 60 kPa. What force is now required by the feedback bellows? What output pressure?
d. What new gap is required? How much did the flapper move?

S10.3 A PD controller will be used to regulate the drive rate of a boring machine with a setpoint between 30 and 60 cm/min. A sensor provides the boring rate as a linear voltage of 0.07 V per cm/min. The drive motor requires 0.0 to 8.0 V. The minimum period for change is 0.01 min. The system needs a proportional gain of 3.5%/% and a derivative gain of 0.006 min. Design an op amp circuit to provide error detection and controller action.

CHAPTER

11

Computer-Based Control

INSTRUCTIONAL OBJECTIVES

In this chapter the modern implementation of control systems using computers will be studied. After reading this chapter, working through the examples, and completing the problems at the end of the chapter, you will be able to:

- Explain how basic digital systems can implement alarms and two-position controllers.
- Define the hardware characteristics of computer-based controllers.
- Describe how a computer can implement the PID mode of controller action.
- Explain the concept behind fieldbuses in control systems.
- Describe the basic characteristics of Foundation and Profibus fieldbuses.

11.1 INTRODUCTION

In Chapters 1 through 9 of this book you have learned about every aspect of a control system: sensors and measurement, error detection, final control operations, and controller action. In Chapter 10 you learned how analog electronics and pneumatic technology can be used to implement the modes of controller action presented in Chapter 9. Now you need to know that the simple truth is that nearly all modern control system implementations are carried out using computers. What does this mean? Well, the measurement and final control operations are the same as those you have already learned. The strategy for control and the modes of controller action presented in Chapter 9 are still the same. But now you will learn that the functions of the controller have been taken over by a computer. The computer inputs measurement data, determines the error, solves the controller mode equations to determine the feedback, and transmits this feedback to the final control element.

The second major revolution in control system technology comes from the technology of networks and network communication. One of the great powers of modern computing is the ability to exchange information between computers over networks: local-area networks (LAN), wide-area networks (WAN), and a worldwide network (www). This concept has been carried over into control systems by the definition of industrial "bus standards" where a bus is simply another form of computer network. In this chapter you will learn how modern methods of computer networking have led to the development of distributed control wherein sensors, computers, and final control operation signals are exchanged over a common network called the "fieldbus."

We start, however, by noting some carry-over of basic digital electronics in the implementation of control operations.

11.2 DIGITAL APPLICATIONS

There are still situations in control systems wherein basic digital electronics can provide the needed action. In some cases the use of a computer or programmable-logic-controller (PLC) is overkill for some simple need. Thus, for example, the basic two-position control needed for a simple run-off temperature control may be easier to implement with basic digital logic in lieu of a computer or PLC.

11.2.1 Alarms

One area where simple digital logic can be of practical and fiscal value is the implementation of alarms. An alarm is defined as an on/off condition wherein a warning is issued when some process variable passes a critical value. For example, we may need an alarm when a pressure exceeds some limit or when a temperature falls below some limit. The alarm may do something as simple as turning on a warning light or something as sophisticated as issuing a signal to a computer for corrective action.

Single Variable The simplest alarms are those involving only a single variable. For this type of alarm a simple implementation is to convert the measurement signal to a voltage and use a digital comparator. In general we want to use a comparator with hysteresis to avoid a staccato type of response wherein the comparator output switches between **1** and **0** repeatedly during transition. You may recall that hysteresis provides a small range of trigger within which no output transition occurs. This protects against noise on the signals and nonlinear effects in the comparator itself. The following example illustrates implementation of a single variable alarm.

EXAMPLE 11.1 Design an alarm that provides a logic high of 5 V when a liquid level exceeds 4.2 meters. The level has been linearly converted to a 0–10 volt signal for a 0–5 meter level. Hysteresis should be 0.1 volts.

Solution

The information given shows that level L and voltage output V are related by the relation $V = 2L$. Thus, for $L = 4.2$ m the output is $V_{4.2\text{m}} = 8.4$ V. Reviewing Chapter 3 we see that a hysteresis of 0.1 V means that the resistors should be $(R/R_f) = \Delta V/V = 0.1/5 = 0.02$. So if we make $R = 10\ \text{k}\Omega$ then $R_f = 500\ \text{k}\Omega$. Figure 11.1 shows the circuit.

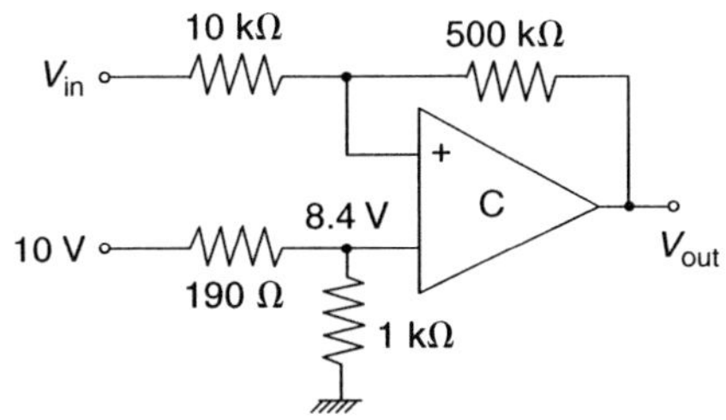

FIGURE 11.1
Circuit for Example 11.1.

Multivariable Alarms In many cases an alarm is dependent not upon the value of a single variable but rather on the combined values of one or more variables. Thus, for example, a tank with high pressure and high level may represent some unpleasant condition for which an alarm should be issued. In these cases we can use comparators combined with appropriate logic circuits to produce an alarm. In general, hysteresis should again be used to protect against noise and nonlinear effects. The following example shows how such an alarm can be constructed. (The hysteresis has not been shown on the comparators.)

EXAMPLE 11.2 In Figure 11.2, a holding tank is shown for which liquid level, inflow A, and inflow B are monitored. These measurements are converted to voltages and then, with comparators, to digital signals that are HIGH when some limit is exceeded. The flow variables FA and FB will be **0** for low flow and **1** for high flow. The level variables are such that L_2 is **1** if the level exceeds the lower limit and L_1 will be **1** if the level exceeds the upper limit. The alarm will be triggered if either of the following conditions occurs:

1. L_2 LOW and neither FA nor FB HIGH
2. L_1 HIGH and FA or FB or both HIGH

Implement this problem with digital logic circuits.

Solution

The variables FA, FB, L_1, and L_2 are already Boolean in that they have values of logic **0** or **1.** We first write Boolean equations giving an alarm output $A = \mathbf{1}$ for the given two conditions. This can be done directly as

1. $A = \overline{L_2} \cdot \overline{(FA + FB)}$
2. $A = L_1 \cdot (FA + FB)$

Now either of these conditions is provided by an OR operation:

$$A = \overline{L_2} \cdot \overline{(FA + FB)} + L_1 \cdot (FA + FB)$$

Logic gates that can be used to directly implement this equation are shown in Figure 11.3.

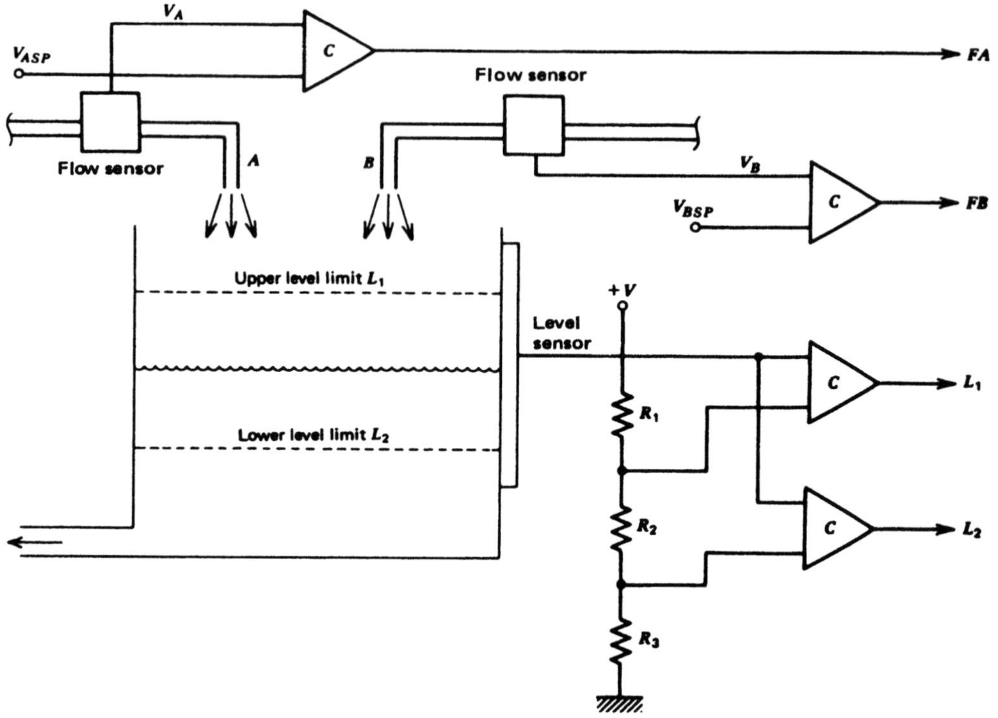

FIGURE 11.2
Holding tank level-control system for Example 11.2.

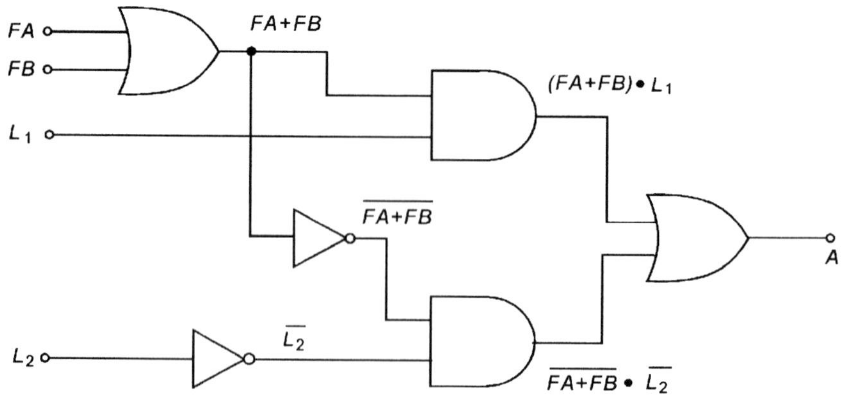

FIGURE 11.3
A logic gate solution for Example 11.2.

11.2.2 Two-Position Control

It is possible to develop a two-position controller using digital electronics methods. Such a controller differs from the simple alarm comparator just presented by the deadband that exists between ON and OFF state transitions.

The hysteresis comparator of Chapter 3 (Figure 3.7) can be used for two-position control. The only problem is that the equations given for HIGH and LOW trip points are only good if the deadband is small, which requires $R_f \gg R$. Otherwise, although the hysteresis effect is still present, the trip points and deadband depend on the resistances in a more complicated manner. This type of circuit for two-position control should be used only when the deadband is small compared to the trip voltages. It is really intended for noise suppression.

In a more general way, it is possible to develop two-position control using a combination of comparators and digital logic circuits. Figure 11.4 shows one such circuit. Two comparators and a D flip-flop (F/F) are used. One comparator determines the upper trip voltage and the other, the lower.

Operation of the circuit can be described by noting the output changes as V_{in} ranges from a low to a high value.

1. $V_{in} < V_L$; B will be LOW, and thus the output Q will be LOW because the F/F is in the clear state. A will be low but has no effect.
2. $V_{in} > V_L$ but $< V_H$; B will go HIGH, but the F/F output remains LOW because, although not in the clear state, the F/F has not been clocked to pass the D input through to the Q output. A is still low.
3. $V_{in} > V_H$; the A comparator goes HIGH. This clocks the F/F, passing the HIGH D input through to the Q output. Thus, the output goes HIGH.
4. On return, $V_{in} < V_H$ but $> V_L$; A will go LOW, but this has no effect on the F/F output, which stays in the HIGH output state.
5. $V_{in} < V_L$; B goes LOW, which places the F/F in the clear state so that the output Q goes LOW.

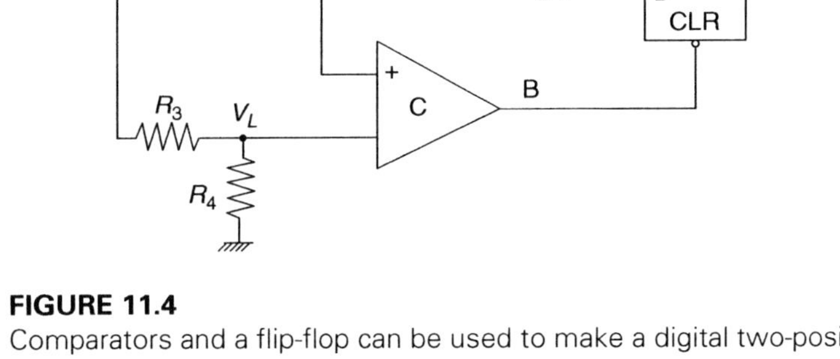

FIGURE 11.4
Comparators and a flip-flop can be used to make a digital two-position controller.

Steps 1 through 5 describe a typical two-state controller. The deadband is determined by

$$\Delta V = V_H - V_L \tag{11.1}$$

where the HIGH and LOW trip voltages can be obtained from dividers as shown.

$$V_H = R_2V_0/(R_1 + R_2) \tag{11.2}$$
$$V_L = R_4V_0/(R_3 + R_4) \tag{11.3}$$

EXAMPLE 11.3 Temperature is measured with a response of 15 mV/°C. Devise a two-position controller that turns a 115-Vac fan ON if the temperature reaches 70°C and OFF when it falls to 40°C.

Solution

Let us use the two-position controller of Figure 11.4. The trip voltages are

$$V_H = (0.015\ \text{V}/°\text{C})(70°\text{C}) = 1.05\ \text{V}$$

and

$$V_L = (0.015\ \text{V}/°\text{C})(40°\text{C}) = 0.6\ \text{V}$$

Any combination of divider resistors will do. Practically, it is desirable to keep currents in the mA range, so let us make $R_2 = R_4 = 1\ \text{k}\Omega$. Then, assuming $V_0 = 5$ V, the other resistors are easily found to be $R_1 = 3.76\ \text{k}\Omega$ and $R_3 = 7.33\ \text{k}\Omega$. A simple 5-V relay is used to switch the fan ON and OFF. The circuit is shown in Figure 11.5 using a standard TTL 7474 F/F and a type LM 319 dual comparator. The 1 kΩ resistors on the comparator outputs are necessary because the LM 319 has open-collector outputs.

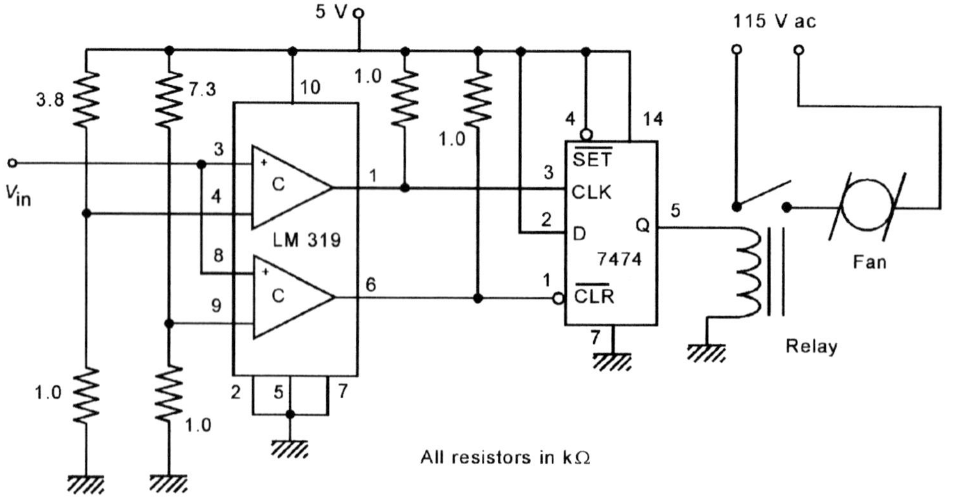

FIGURE 11.5
Solution for Example 11.3.

11.3 COMPUTER-BASED CONTROLLER

At the heart of nearly all modern control systems is a computer-based controller. Here a dedicated computer inputs measurement data from the plant, performs all the calculations for controller modes, and supplies the necessary output. Such an application is historically referred to as direct digital control (DDC). There are many hardware configurations of such a system. The controlling computer could be similar to a microprocessor-based personal computer (PC) mounted in a rack or even on a desktop. It could be a microprocessor-based computer on a small printed-circuit board mounted inside measurement equipment or even a large computer. In this section we will explore the hardware and software configurations of typical process-control computers.

11.3.1 Hardware Configurations

Any microprocessor-based computer has a standard array of devices as shown in Figure 11.6. All of the devices shown are available as small integrated circuits that can be mounted on a relatively small printed-circuit board. The ROM stands for read-only-memory, nonvolatile memory that holds the programs that the processor executes. Typically the ROM consists in part of electrically erasable memory so that programming updates can be downloaded over the communication network. The RAM stands for read-write memory and is used to hold the transient results of calculations and other results of data processing. The data I/O typically consists of ADCs and DACs as well as digital I/O channels. The final element is the network interface communication system. This provides for serial communication of the computer over a serial fieldbus or LAN.

Smart Sensors One of the most exciting developments of modern process control is to embed the controller computer directly into a sensor. Figure 11.7 shows one possible implementation of such a system. Here you see that the sensor and computer are housed directly at the site of the measurement. In this case we see that the feedback signal is delivered to the valve via the standard 4–20 mA current transmission. Operation of the flow control loop is monitored via the serial interface, which is also used to update the setpoint, controller mode gains, and other operating parameters.

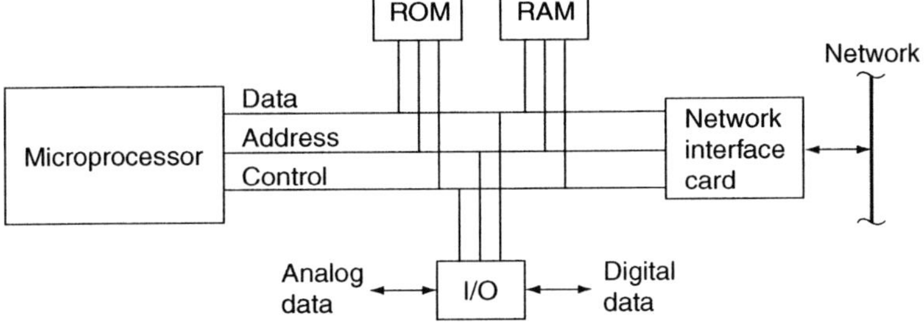

FIGURE 11.6
Basic structure of a microprocessor-based computer.

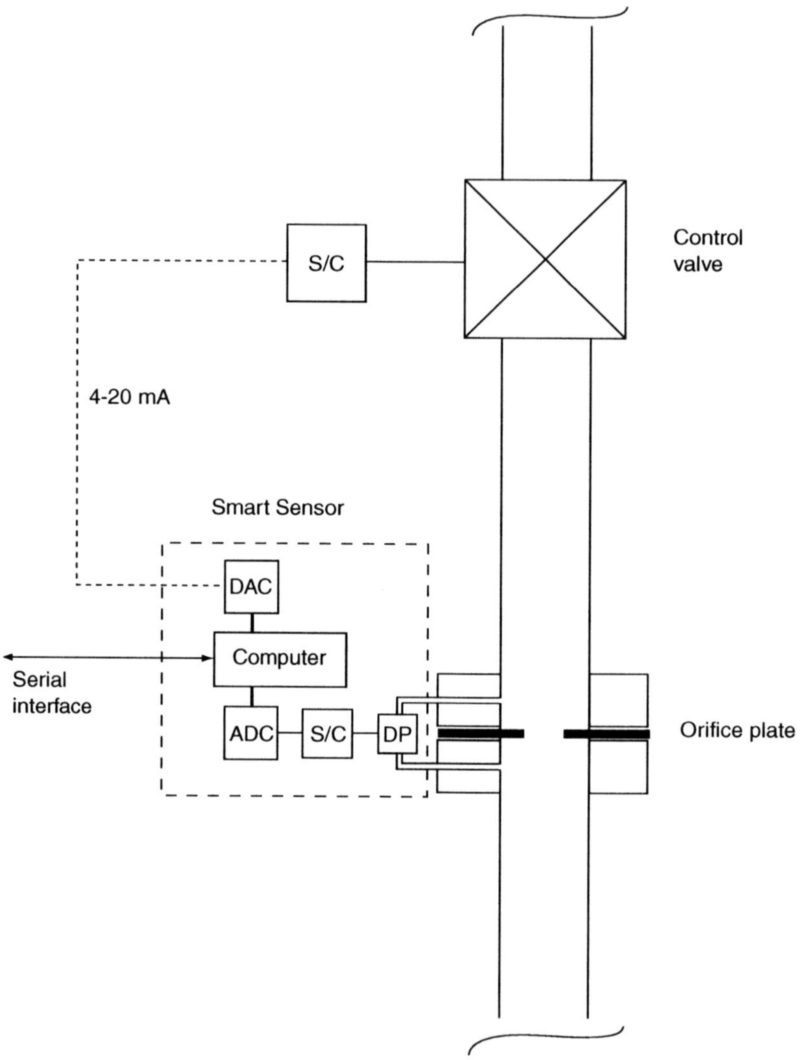

FIGURE 11.7
Smart sensors encase the computer-based controller with the sensor.

In a further refinement of the smart sensor concepts the 4–20 mA connection is eliminated. The signal-conditioning system of the valve contains a network interface circuit so that it can be connected to the serial bus. In this case the smart sensor sends feedback information to the valve via the serial bus. This is discussed further later in this chapter when fieldbuses are presented.

Multiple-Loop Controllers In some cases there may be an advantage to having a single computer operate as the controller of more than one process-control

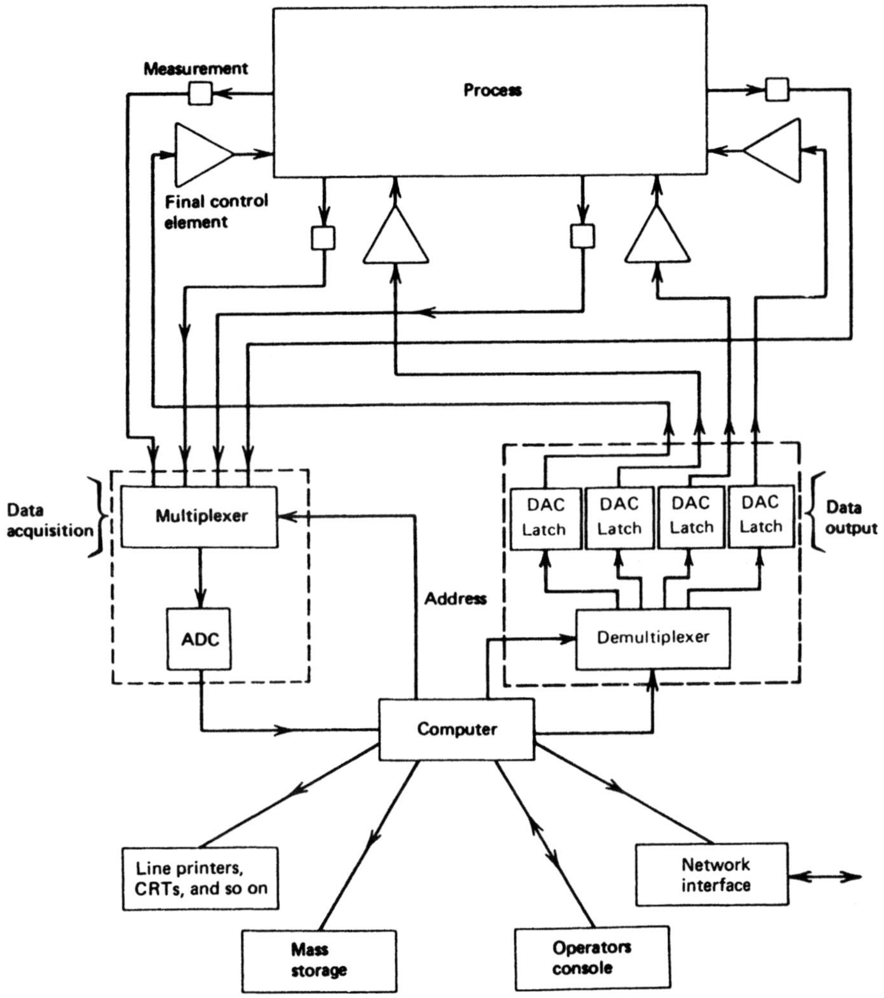

FIGURE 11.8
One computer can control several loops.

loop. This may be to account for interactions between loops, for example, or just for economy. Historically this was done because computers were relatively large and expensive investments and economics showed that multiple-loop control was called for. Furthermore, computers were (and are) fast enough that such multiple-loop control was feasible.

Figure 11.8 shows how a process could be placed under the control of a single computer. Notice the use of multiplexers and demultiplexers to allow the computer to read from various sensors and direct outputs to the correct control elements. In this case the network interface allows the computer to communicate with other computers so that operating parameters of the plant can be updated. However, even with modern fast computers control of multiple loops stretches the ability of the computer.

Consider, for example, a computer that is to control four loops as in Figure 11.8. Let's see what kind of constraint this may place on allowable variations of the controlled variables. Suppose the processor requires modestly 50,000 machine language instructions to process the data from a single channel; in other words, to input the data, linearize, determine the error, solve the control mode equations (i.e., like PID), and output the feedback to the appropriate final control element. Suppose the microprocessor instruction execution time averages about 50 ns. In this case the instructions would take about 2.5 ms. The ADC would take about 20 μs, the DAC about 5 μs, and the multiplexer about 2 μs. Therefore, these are negligible compared to the computation time. The total time for a channel would be about 2.5 ms. For the four channels this would mean a minimum time between processed samples of any one channel of about 10 ms or a sampling frequency of 100 Hz. Recall that the sampling time and maximum signal frequency are related by the (practical) limit of $f_s = 10f_{\max}$. Thus, the maximum signal frequency in this case would be 10 Hz. For many process loops this would be much too low. With the lowered price of microprocessor-based computers it is fiscally and maybe technically better to let each control loop be controlled by a single computer.

With the introduction of fieldbuses to serially carry information between computers, sensors, and feedback elements, the picture of Figure 11.8 has changed, as described in the next section.

11.3.2 Software Requirements

The use of a computer as the controller means the computer must solve the control equations introduced in Chapter 9. If computer control is implemented on a general purpose computer, the software is available as a "control package" that can be installed in the computer. In the case of smart sensors and other dedicated control computers the control equations are built into the embedded computer. External commands can be used to select the desired mode (PI, PID, etc.) and the gains for each mode.

To understand the limitations of computer solutions to control equations it will be helpful if you know how these equations are typically implemented. The following sections describe the algorithms commonly employed to implement the control equations.

Error The computer accepts an input of the value, y, of the controlled variable from an ADC or over the bus (network) from the sensor. The value will have been encoded as a binary number. In describing the algorithms we assume the measurement range of the controlled variable is known, $y_{\min}$ to $y_{\max}$. In Chapter 9 we wrote the error as percent of range,

$$e = \frac{y - y_{sp}}{y_{\max} - y_{\min}}100 \qquad \textbf{(11.4)}$$

For the purposes of algorithm description we will assume the variable has been converted from a binary encoding to its actual value as a floating-point variable (pressure, temperature, etc.) in the control program. This may entail linearization as well as scaling. In the program the error will be used as a fractional quantity rather than a percent. Thus, the error will be as in Equation (11.4) but without the 100. Furthermore, we note that the variable value, and hence error, are only available as samples taken every Δt seconds. Thus, the error will be expressed as,

$$e_i = \frac{y_i - y_{sp}}{y_{\max} - y_{\min}} \tag{11.5}$$

Again, we assume that when the binary number is brought into the computer it is passed to a floating-point processor (i.e., y_i is a base 10, floating-point number). This is typical of modern computers.

EXAMPLE 11.4

Pressure in psi is measured and converted to a voltage by a sensor according to the relationship,

$$V = 3.1[p + 10]^{1/2} - 9.8$$

The pressure range is 0 to 30 psi and the setpoint is 15 psi. This voltage is provided as input to an 8-bit unipolar ADC with a 10.00-volt reference, and the resulting binary is provided as input to a control computer. (a) Develop the equations used to find the pressure from the binary input and then the error. (b) Contrast the actual error with the computed sample error for a pressure of 17.3 psi.

Solution

(a) Let's call the sample from the ADC N_i, which is the base 10 equivalent of the binary output of the ADC. We can then find the voltage corresponding to this sample (within the ΔV of the ADC) as,

$$V_i = \frac{10}{256} N_i$$

Now the pressure sample can be determined by using the given equation relating pressure and voltage,

$$p_i = \left(\frac{V_i + 9.8}{3.1}\right)^2 - 10$$

Then, the error as a fraction of range is found from Equation (11.5),

$$e_i = \frac{p_i - 15}{30}$$

(b) To find the actual error for 17.3 psi we should use the previous equation with $p =$ 17.3 psi, $e = (17.3 - 15)/30 = 0.077$. To find the sample error we must take into account the loss in information due to the ADC. Thus, we calculate just as the computer will. The voltage of the sensor is: $V_{17.3} = 3.1(17.3 + 10)^{1/2} - 9.8 = 6.397$ V. The output of the ADC will be,

$$N_i = Int\left[\frac{6.397}{10.00}(256)\right] = 163$$

where the ADC truncated the fractional part. Thus, in the computer, the voltage will be computed as $V_i = (10/256)163 = 6.367$, and the pressure will be calculated as,

$$p_i = \left(\frac{6.367 + 9.8}{3.1}\right)^2 - 10 = 17.198$$

Therefore, the sample error will be $e_i = (17.198 - 15)/30 = 0.073$. Thus, because of truncation by the ADC there is a difference in error representation of about 0.004 or 5%.

With these assumptions about the input value and expressing the error sample as a fraction of range, let us consider the three modes of control: proportional, integral, and derivative. The equations developed will provide a fractional number (0 to 1) representing what fraction of the controlling variable range should be sent to the final control element.

Proportional Mode In Chapter 9, we saw that the proportional-mode controller action is defined by a term that is directly proportional to the error. The equation is

$$p = K_P e_p + p_0 \quad \textbf{(9.14)}$$

where K_P = proportional gain
e_p = error
p_0 = controller output with no error

The gain is expressed as percent controller output per percent error. The concept of proportional band (PB) is defined as $1/K_P$ and represents the percentage error that will cause a 100% change in controller output. This mode is easily implemented by the computer in the form of an algorithm that simply calculates Equation (9.14) directly.

The proportional mode is provided through the software by an equation that is entirely like the analog equation. Because we are expressing the error as a fraction of range, what is calculated is the fraction of the maximum output.

$$P = P0 + KP * DE \quad \textbf{(11.6)}$$

$$POUT = P * ROUT \quad \textbf{(11.7)}$$

where $P0$ = fraction of output with no error
KP = proportional gain (%/%)
P = fraction of output with error
$ROUT$ = maximum output
$POUT$ = output
DE = error from Equation (11.5)

EXAMPLE 11.5 A proportional mode has $K_P = 2.4$, input range of 255, and setpoint of 130. The output maximum is 180, and the output fraction with no error is 0.45.

a. Develop the control equations. (What is the output for no error?)
b. Find the output for an input of 124.

Solution

a. The equations are found simply from Equations (11.5), (11.6), and (11.7).

$$DE = (130 - DV)/255$$
$$P = 0.45 + 2.4 * DE$$
$$POUT = P * 180$$

When there is no error ($DE = 0$), $POUT = 0.45 * 180 = 81$.

b. For an input of 124, we get an error fraction of

$$DE = (130 - 124)/255 = 0.024$$
$$P = 0.45 + 2.4 * (+0.024) = 0.45 + 0.0576$$
$$= 0.5076$$
$$POUT = 0.5076 * 180$$
$$= 91.4$$

Figure 11.9 shows a general flowchart for the proportional mode from which software can be developed. Of course, there may be added complications such as the need for linearization and specification of the input and output software.

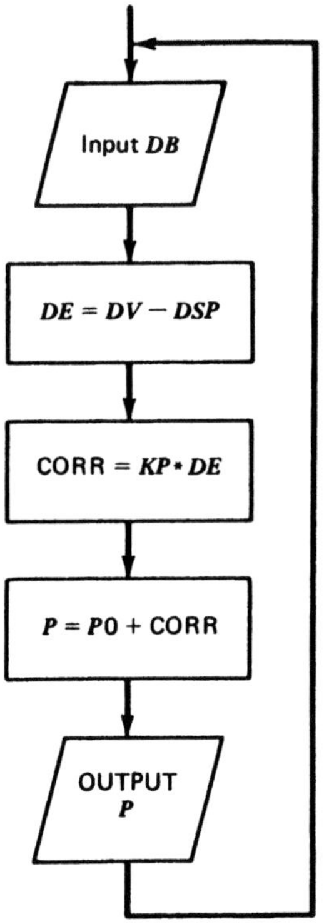

FIGURE 11.9
Flowchart for proportional mode.

Integral Mode The integral or reset mode calculates a controller output that depends on the history of the controlled variable error. In a mathematical sense, history is measured by an integral of the error

$$p = K_I \int_0^t e_p \, dt + p(0) \tag{9.17}$$

where K_I = integral gain in percent controller output per %-s error (or, more commonly, per minute)

To use this mode in computer control, we need a way of evaluating the integral of error. Many algorithms have been developed to do this, all of them only approximate, as only samples of the error in time are available. The simplest is called *rectangular,* and it is often accurate enough for use in process control. To see how this works, you should note that the integral in Equation (9.17) is merely the net area of the e_p curve from 0 to t. This is shown in Figure 11.10.

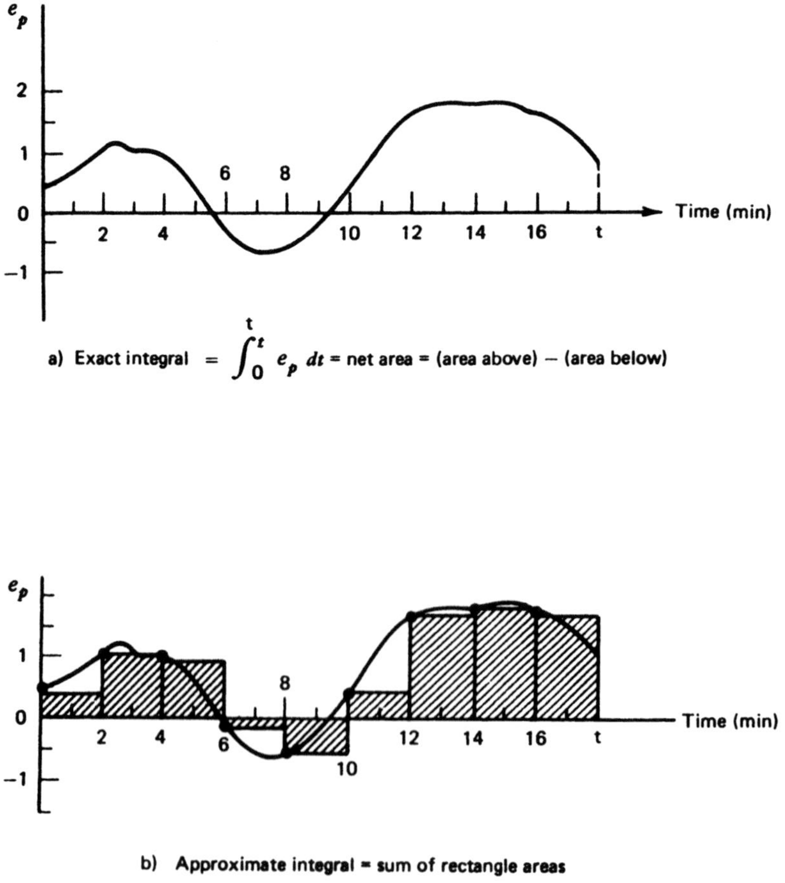

FIGURE 11.10
Finding the integral by the rectangular integration algorithm.

$$\int_0^t e_p\,dt = \text{net area} = (\text{area of } e_p > 0) - (\text{area of } e_p < 0)$$

In rectangular integration, we simply use the periodic samples of e_p to construct a series of rectangles of height equal to the sample error and of width equal to the time between samples. The integral (or area) is then approximately equal to the sum of the rectangle areas. This is shown in Figure 11.10b. In an equation, rectangular integration specifies that

$$\int_0^t e_p\,dt \approx [S + e_{pi}]\Delta t$$

where Δt = time between samples

$S = e_{p1} + e_{p2} + \ldots$ (sum of errors calculated from previous variable samples)

e_{pi} = last sample taken at time t specified in the integral

It should be clear that the smaller the time between samples, the more closely the approximate answer will approach the actual integral.

The issue of sampling time becomes important for the integral mode. If the time between samples is too large, the area will be in serious error, and control will be compromised. If the criterion developed earlier of $f_s = 10f_{\text{max}}$ is satisfied, then the integral term also will be of sufficient accuracy using the rectangular algorithm. The following example illustrates the effect of sampling time variation.

EXAMPLE 11.6

Find the approximate integral of e_p in Figure 11.10a from 0 to 14 min. Do this for the sample time shown in the figure (2 min) and again for a sample time of 1 min. What percentage change in the value of the integral results from the difference in sample time?

Solution

For the sample time of 2 min, we find the integral from the rectangular integral procedure as

$$\text{integral}_{2\text{min}} = 2(0.4 + 1.1 + 1 - 0.2 - 0.6 - 0.4 + 1.6)$$
$$\text{integral}_{2\text{min}} = 7.4\% - \text{min}$$

For a sample time of 1 min, we find the integral as

$$\text{integral}_{1\text{min}} = 1(0.4 + 0.6 + 1.1 + 1.1 + 1 + 0.4 - 0.2 - 0.6 - 0.6 - 0.4 + 0.4 + 1.1 + 1.6 + 1.7)$$
$$\text{integral}_{1\text{min}} = 7.6\% - \text{min}$$

This gives a percentage change in integral value between the two approaches as

$$\text{change} = \frac{7.6 - 7.4}{7.6} \times 100 = \mathbf{+2.6\%}$$

Implementation of this mode in software involves the following basic equations:

$$SUM = SUM + DE \tag{11.8}$$

$$PI = KI * DT * SUM \tag{11.9}$$

$$POUT = PI * ROUT$$

where SUM = a running sum of errors
KI = the integral gain
DT = time between samples
PO = fraction of maximum output

The flowchart of Figure 11.11 illustrates how such a mode can be programmed. The time-delay routine must be built in to provide the required time between samples, because time appears as part of the mode equation [Equation (11.9)] and must therefore be known. Another important point is that the units of KI and DT must be the same.

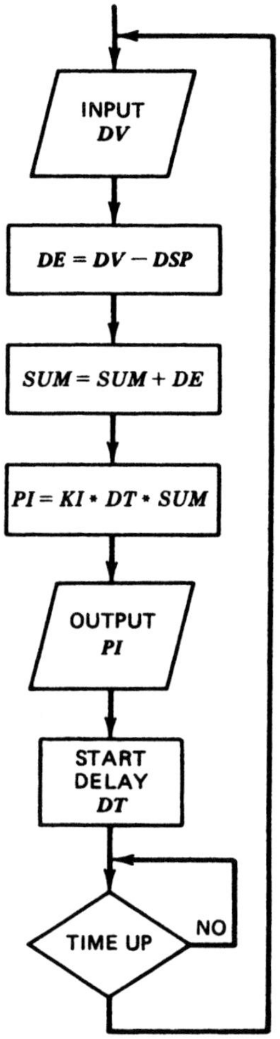

FIGURE 11.11
Flowchart for the integral mode using rectangular integration.

An alternate way of writing the integral mode equations is to use the present error sample to modify the previous output. To see this, let us substitute Equation (11.8) into Equation (11.9).

$$PI = KI * DT * (SUM + DE)$$
$$PI = KI * DT * SUM + KI * DT * DE$$

The first term in this equation is simply the *PI* from the previous sample. Therefore, we can write

$$\begin{aligned} PI &= PI0 + KI * DT * DE \\ POUT &= PI * ROUT \\ PI0 &= PI \end{aligned} \qquad \textbf{(11.10)}$$

where $PI0 = PI$ from the previous sample.

The term *DT* in the set of control equations given in Equations (11.9) and (11.10) is an important link between the software and hardware systems. This factor represents the actual time between samples in the hardware data-acquisition system. If the calculated gains and the programmed gains are to be correctly related, *DT* must be the actual time between samples. This means that *DT* cannot be arbitrarily changed without also changing the gains used in the software. Furthermore, the data-acquisition and control software must always provide the same time between samples, or errors in the equation results will occur.

This limitation is true in all of the following mode implementations using the integral or differential modes.

Derivative Mode The derivative controller mode, also called *rate,* derives a controller output that depends on the instantaneous rate of change of the error

$$p = K_D \frac{de_p}{dt} \qquad \textbf{(9.18)}$$

where K_D = derivative gain (% per %/s error)

$\frac{de_p}{dt}$ = rate of error change in percent per second (or minute)

The gain expresses the percent controller output for each percent/second change in error. This mode is implemented in computer control by calculating an approximate derivative of the error from the data samples. A derivative is defined as the rate at which a quantity is changing at an instant in time. We can calculate only the rate at which it is changing over the sample period Δt, which is therefore only an approximation. In terms of an equation, this is

$$\frac{de_{pi}}{dt} \approx \frac{e_{pi} - e_{pi-1}}{\Delta t}$$

where e_{pi} = present error sample
e_{pi-1} = previous error sample
Δt = time between samples

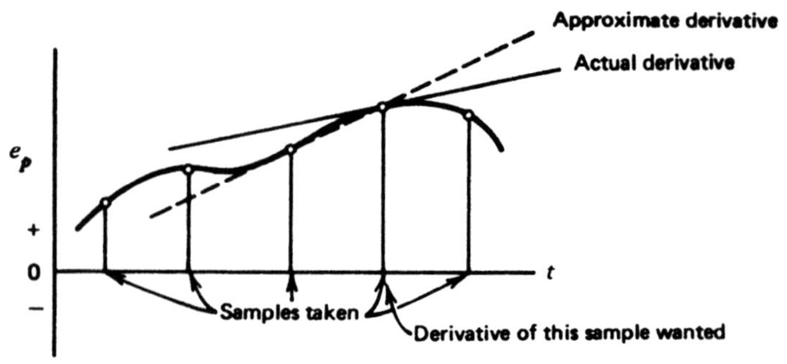

FIGURE 11.12 Approximate calculation of the derivative from sampled data.

Figure 11.12 shows that this process results in a derivative that is not the actual derivative. Notice that as the time between samples is made smaller, the error will become less.

EXAMPLE 11.7 Determine an approximate value of the derivative of e_p at a time of 12 min from Figure 11.10a, using samples every 2 min and every 1 min. Compare the results.

Solution

For 2-min samples, we get the derivative by using the sample at 10 min and at 12 min.

$$\text{dervative}_{2\text{ min}} = \frac{1.6 - 0.4}{2} = \mathbf{0.6\%/min}$$

For samples every minute, we use a sample at 11 min and at 12 min.

$$\text{derivative}_{1\text{ min}} = \frac{1.6 - 1.1}{1} = \mathbf{0.5\%/min}$$

This means there is a difference of about 17%.

The set of equations for the derivative output can be developed directly from the definitions. We find

$$DDE = DE - DEO \quad \textbf{(11.11)}$$

$$DEO = DE \quad \textbf{(11.12)}$$

$$PD = KD * DDE/DT \quad \textbf{(11.13)}$$

The flowchart for this mode is presented in Figure 11.13. As in the case of the integral mode, it is important that the units of *KD* and *DT* agree. In the following example, the PID control mode section illustrates this point.

PID Control Mode The optimum control mode is a composite of the three previous modes: proportional, integral, and derivative. With computer-based control, a composite mode is developed by simply combining the three mode equations into the computation

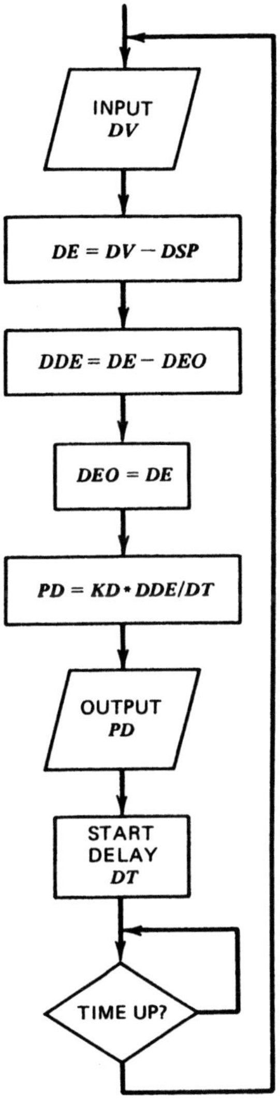

FIGURE 11.13
Flowchart of the derivative mode.

of the fractional output. According to the principles of PID control, the proportional gain should multiply all three forms. The control equations can be written

$$DDE = DE - DEO$$
$$DEO = DE$$
$$SUM = SUM + DE$$
$$PI = KP * KI * DT * SUM \tag{11.14}$$
$$PD = KP * KD * DDE/DT \tag{11.15}$$
$$P = KP * DE + PI + PD \tag{11.16}$$
$$POUT = P * ROUT$$

where all the terms have been previously defined. These equations are then programmed into the control software for determination of the required output.

An alternative expression for the PID output can be constructed by using errors to provide corrections to the current output. To develop this, let us adopt a convention that a subscript will denote a particular sample. Thus, DE_i is the ith sample, and P_i is the fractional output for that sample. The output for the P_{i-1} sample, according to Equation (11.16), can be written in the form

$$\begin{aligned} P_{i-1} = {} & KP * DE_{i-1} + KP * KI * DT[SUM + DE_{i-1}] \\ & + KP * KD * [DE_{i-1} - DE_{i_{1-2}}]/DT \end{aligned}$$

The result for P_i will be

$$\begin{aligned} P_i = {} & KP * DE_i + KP * KI * DT * [SUM + DE_{i-1} + DE_i] \\ & + KP * KD * [DE_i - DE_{i-1}]/DT \end{aligned}$$

Let us take the difference between these two expressions. This will give the correction to the previous output because of the present sample error. The result will be

$$\begin{aligned} P_i - P_{i-1} = {} & KP[DE_i - DE_{i-1}] + KP * KI * DT * DE_i \\ & + KP * KD * [DE_i - 2DE_{i-1} + DE_{i-2}]/DT \end{aligned}$$

This equation can be simplified to give

$$P_i = P_{i-1} + A * DE_i - B * DE_{i-1} + C * DE_{i-2} \qquad \textbf{(11.17)}$$

where
$$\begin{aligned} A &= KP + KP * KI * DT + KP * KD/DT \\ B &= KP + 2 * KP * KD/DT \\ C &= KP * KD/DT \end{aligned}$$

Then the result of Equation (11.17) is used to determine the output from

$$POUT = P_i * ROUT \qquad \textbf{(11.18)}$$

The following example illustrates how the controller equations are developed for a typical problem.

EXAMPLE 11.8 A digital controller is to be developed with the following specifications: $KP = 50\%/\%$, $KI = 0.5\%/(\%\text{–min})$, $KD = 0.08\%/(\%/\text{min})$, time between samples = 5 s, input range 0 to 255, setpoint = 130. The output range is 0 to 255. Set up the control equations for PID control using both approaches given in the text.

Solution

By the first approach, it is merely necessary to express the equations given in the mode description. The first set of equations is given easily by

$$\begin{aligned} DE &= (130 - DV)/(255 - 0) \\ DDE &= DE - DEO \\ DEO &= DE \\ SUM &= SUM + DE \end{aligned}$$

To compute the gains, we need to get the units the same.

$$\begin{aligned} KP * KI * DT &= (5)(0.5\ \text{min}^{-1})(5\ \text{s})(1\ \text{min}/60\ \text{s}) \\ &= 0.208 \\ KP * KD/DT &= (5)(0.08\,\text{min})/(5\ \text{s})(1\,\text{min}\,/60\ \text{s}) \\ &= 4.8 \end{aligned}$$

Thus, the remaining equations become

$$\begin{aligned} PI &= 0.208 * SUM \\ PD &= 4.8 * DDE \\ P &= 5 * DE + PI + PD \\ POUT &= P * 255 \end{aligned}$$

For the second approach, we need to evaluate the three constants of Equation (11.17).

$$\begin{aligned} A &= 5 + 0.208 + 4.8 = 10.008 \\ B &= 5 + 2 * 4.8 = 14.6 \\ C &= 4.8 \end{aligned}$$

Then Equation (11.17) can be written

$$\begin{aligned} P &= P1 + 10.008 * DE - 14.6 * DE1 + 4.8 * DE2 \\ DE2 &= DE1 \\ DE1 &= DE \end{aligned}$$

The last two equations are necessary because the next sample $DE1$ and $DE2$ contains the previous two samples. Then the previous output must be updated.

11.4 OTHER COMPUTER APPLICATIONS

Computers are used for many purposes in a process-control facility beyond acting as the controller of loops. They also serve engineering and design functions, financial analysis functions, and plant operations management functions. The following two applications support the engineering analysis and plant operations.

11.4.1 Data Logging

The efficient operation of a manufacturing process may involve the interplay of many factors, such as production rates, materials costs, and efficiencies of control. When the process requires implementation of many process-control loops, then the interaction of one stage of the system with another often can be analyzed in terms of the controlled variables of the loops. An example of this is the rate of production of one loop, expressed as a flow rate, which serves as a determining factor in the production rate of a following control system. Historically, an understanding of this type of interaction required analysis, after the fact, of strip-chart recordings taken from process parameters during a production run. Such analysis, carried out by trained personnel, may then dictate settings of operational limits of future production runs.

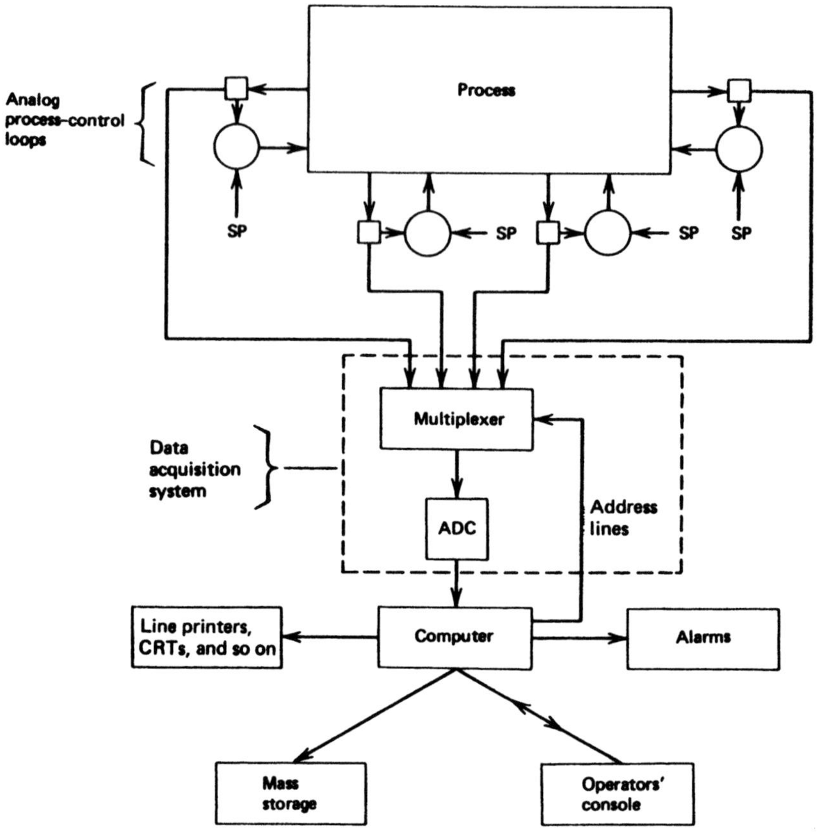

FIGURE 11.14
General features of a data-logging system using a computer.

With the development of high-speed digital computers with mass digital storage, it became possible to record such data continuously and automatically, display the data on command, and perform calculations on the data to reduce it to a form suitable for evaluation by appropriate technical individuals.

Fixed Loggers The general features of a computer data-logging system are shown in Figure 11.14. Let us assume the process is under the control of many analog process-control loops and there is provision for analog process variable measurements to be available as a commonly scaled voltage. Thus, some signal conditioning converts all measurements into a given range, often a specified voltage range, as required by a data-acquisition system. A brief accounting of the elements of the system is given later.

Data-Acquisition System (DAS) The data-acquisition system, discussed in detail in Chapter 3, is the switchyard by which the computer inputs samples of process-variable values. The concept of "samples" of these values is an important topic that will be discussed in more detail later in this chapter. The reason for concern over this point is that there are situations where the sample rate can be such that erroneous information about the variable time dependence results. The rate at which samples of a process variable can be taken depends on how long it takes for the DAS to acquire a value, how long it takes the computer to process the value, and how many other variables are to be sampled. This is illustrated in Example 11.9.

Alarms An important part of any data-logging circuit is an alarm system that monitors inputs from excursions beyond some specific limits. With scan rates of the data as high as 5000 per second, it is possible for a computer to maintain tight vigilance over variable values. Every time the computer inputs a particular variable, the value is compared to its preset limits which, if exceeded, trigger an alarm.

Computer The computer, of course, is the central element in the system. Through programming, the computer accepts inputs and performs prescribed reductions of the data through mathematical operations. The results are evaluated by further programmed tests to oversee the operation of the entire process from which the inputs are taken. Projections of future yields, evaluations of efficiency, deviation trends, and many other operations can be performed and made available to process personnel.

Peripheral Units The peripheral units are the support equipment to communicate computer operations to the outside world. These units include the *operator console* where the programs are entered and through which commands can be given to initiate specific actions, such as calculations and data outputs by the computer. The console usually has a CRT/keyboard and a typewriter unit for input and outputs. A mass-storage system, such as magnetic tape, is used to store data, such as periodically sampled inputs from the process, that can be used in later, more detailed analysis of process performance.

EXAMPLE 11.9

A data-logging system such as that shown in Figure 11.14 must monitor 12 analog loops. A small computer requires 4 μs per instruction and 100 instructions to address a multiplexer line and to read in and process the data in that line. The ADC performs the conversion in 30 μs. The multiplexer requires 20 μs to select and capture the value of an input line. Calculate the maximum sampling rate of a particular line.

Solution

The 100 instructions require a time of $(4\ \mu\text{s})(100) = 400\ \mu\text{s}$, and this must be done for 12 loops. Thus, the total instruction time is $(12)(400\ \mu\text{s}) = 4800\ \mu\text{s}$. The ADC converts in 30 μs, so that for 12 conversions we have $(12)(30\ \mu\text{s}) = 360\ \mu\text{s}$, and the total time spent in multiplexer switching is 240 μs. Adding $4800 + 360 + 240 =$ **5400 μs** as the minimum time before a particular line can be readdressed. The maximum sampling rate is the reciprocal, or **185** samples per second.

Portable Data Loggers There are many cases when data need to be logged for a period of time from a loop and no fixed logger is provided. A portable data logger can be temporarily connected to the measurement output of the loop for this purpose.

A portable data logger has many of the same features as a fixed installation, such as a DAS, a computer, or an operator's console. Of course, some portable data loggers do not use a computer; they are merely strip-chart recorders or magnetic tape recorders.

In general, the computer-based portable data loggers have some mechanism for saving the logged data for later analysis. Possible recording media are

1. Printed output
2. Digitized strip-chart recording
3. Magnetic tape
4. Magnetic floppy disks
5. Networked data communication

In network data communications, the data logger may be connected to a local area network (LAN). The data then can be transmitted over the network to another, fixed computer installation. The data can be stored on mass-storage facilities that are part of the network.

11.4.2 Supervisory Control

A natural extension of a computer data-logging system involves computer feedback on the process through automatic adjustment of loop setpoints. As various loads in a process change, it is often advantageous to alter setpoints in certain loops to increase efficiency or to maintain the operation within certain precalculated limits. In general, the choice of setpoint is a function of many other parameters in the process. In fact, a decision to alter one setpoint may necessitate the alteration of many other loop setpoints as interactive effects are taken into account. Given the number of loops, interactions, and calculations required in such decisions, it is more natural and expedient to let a computer perform these operations under program control.

Such a system is represented in Figure 11.15, where the effect is shown by the addition of a data-output system (DOS). Such a system assumes the controllers of analog loops have been designed to accept setpoint values as some properly scaled voltage. By proper switch addressing, the computer then outputs a signal through the multiplexer and DACs, representing a new setpoint to a controller connected to that output line.

It might be helpful in understanding use of this type of control if a hypothetical example is given. Study for a moment the reaction process shown in Figure 11.16. The process specifications are

1. Reactants *A* and *B* combine such that one part *A* to two parts *B* produces one part *C*.
2. Volume production of *C* varies as the square root of the *A* and *B* flow rate product.
3. The operating temperature must be linearly decreased with *C* volume production rate.
4. For stability, the reaction must occur with the pressure maintained below a critical value.

A decision is made to increase production in this operation. The first step to accomplish this is to increase the flow rate of *A* by a change in setpoint. Let us see the consequences of this in the rest of the process.

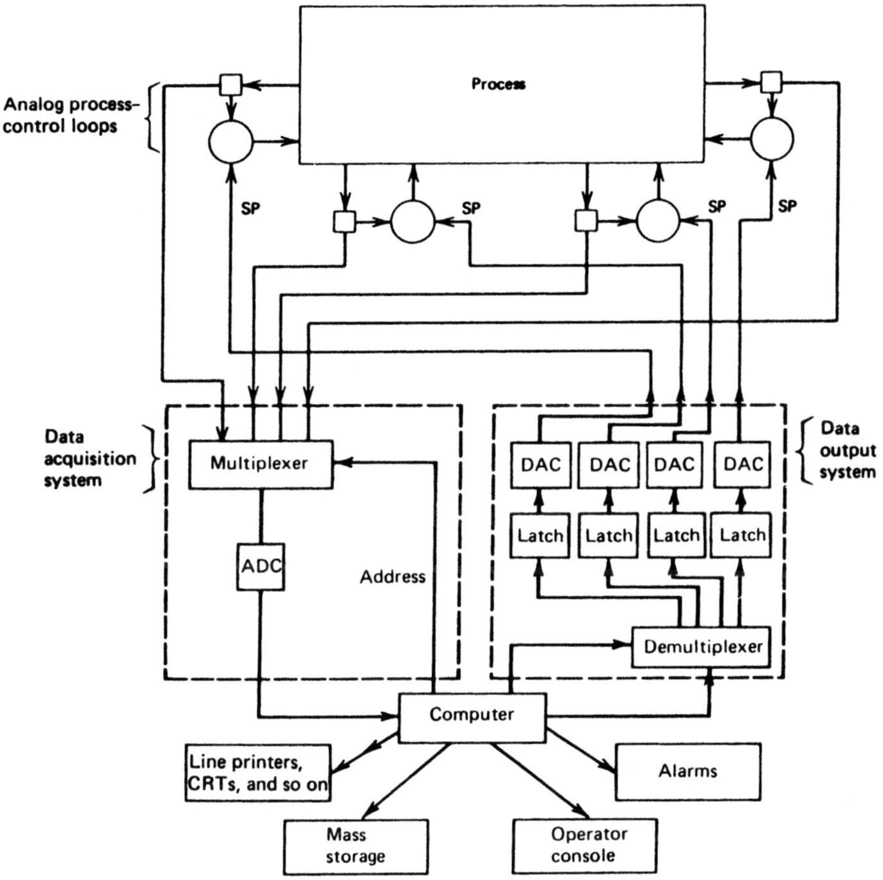

FIGURE 11.15
In computer supervisory control, the computer monitors measurements and outputs the loop setpoints.

1. The setpoint of B flow must be set to twice the A setpoint, keeping pressure below p_{max}.
2. The setpoint of C flow must be increased by the square root of the new A and B flow rates.
3. The temperature setpoint must be decreased by a proportion of the new C setpoint.

To accomplish this change in a purely analog system requires monitoring pressure constantly while the operations of the three steps are gradually performed and the new production rate is finally established. With each new setting of setpoint, we must wait until all parameters have adopted the new setpoints and wait for a safe pressure. To perform this manually requires constant human monitoring as the adjustments are made. In a supervisory control system, the computer performs these operations automatically while still performing other activities in the production.

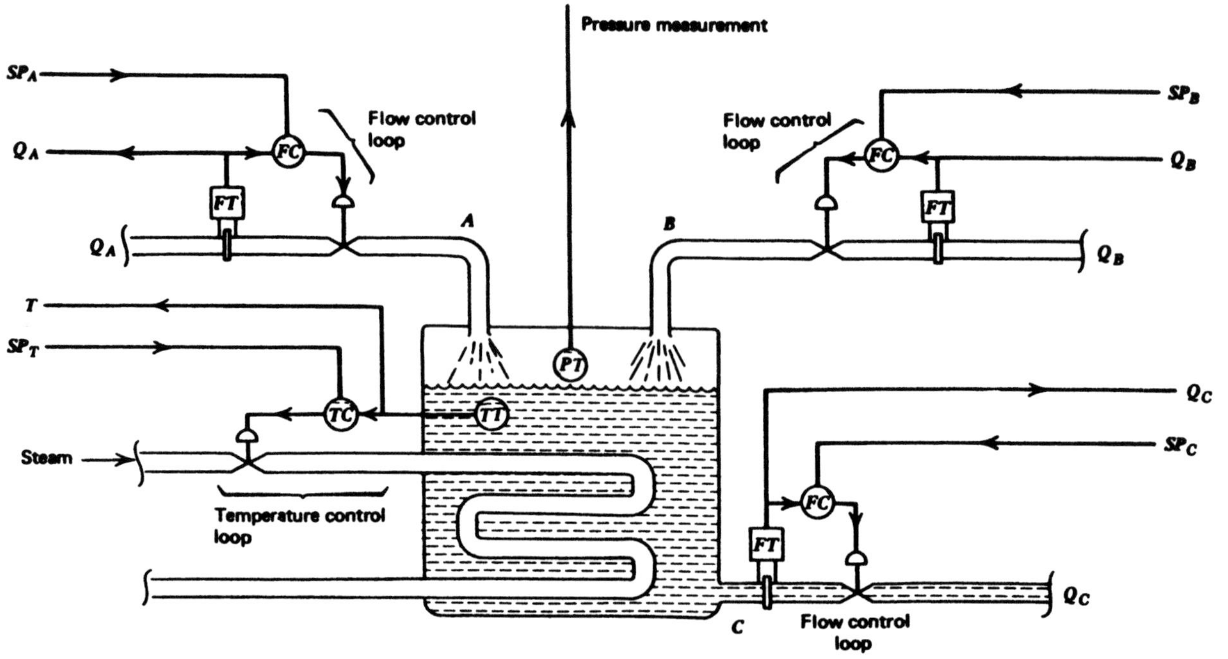

FIGURE 11.16
Computer supervisory control is ideally suited to strongly interacting control problems.

Flowchart To describe the steps a computer must go through to operate in some specified manner, we use an event flowchart. This is the same as the flow diagram used by computer programmers, and in fact some required programs can often be written (coded) directly from such a process-operation diagram. For purposes of illustration, we will use only three types of symbols to prepare a flowchart. These symbols are presented in Figure 11.17. In using such a diagram, we do not have to get lost in the details of how the input, output, operations, and decisions are made, and we can thus better design the overall solution. The next step would be to consider these details.

The event flowchart by which a computer might accomplish monitoring is shown in Figure 11.17. Remember, analog control loops are driving the control variables to the setpoint values. The boxes labeled INPUT refer to computer commands to address the input multiplexer to obtain the current values of these parameters. The OUTPUT boxes serve a similar function for the output multiplexer. The notes on the flow diagram indicate the function of each section. An important feature of a computer supervisory-control system is that it produces the desired change in operation rate in the minimum possible time. The completion of one run through the instructions in Figure 11.17 might typically require less than 100 μs for an average computer. Most of the adjustment time is spent waiting for the loops to stabilize. The instant such stabilization occurs, the next increment of setpoints is made by the computer via the controller.

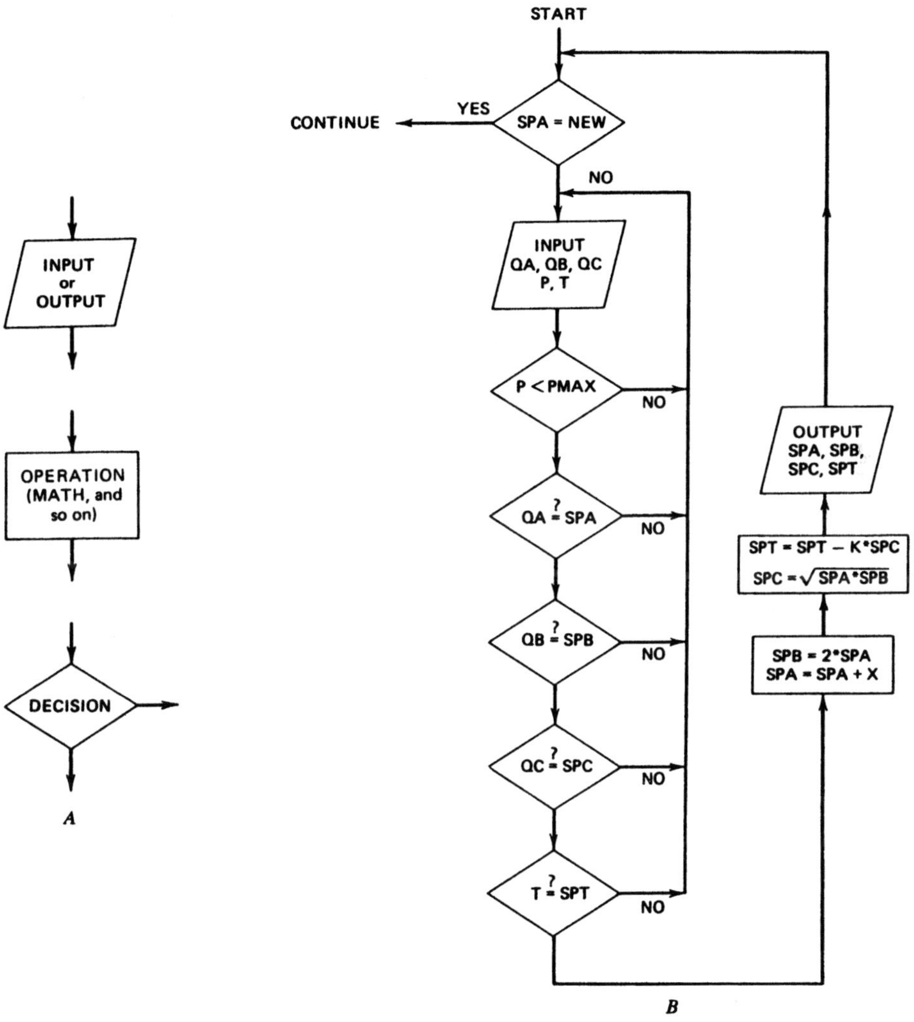

FIGURE 11.17
Flowchart symbols and an example of setpoint changes in a supervisory control system.

EXAMPLE 11.10 Develop the supervisory control flowchart of a system to increase the temperature setpoint of a pressure reaction vessel to a new value, TSPNU. The temperature setpoint (TSP) is to be increased in steps of 0.2% with a 5-s delay between increases. If the pressure (P) rises above a critical value (PCR), the TSP should be decreased by 0.1% until P falls below PCR. Then setpoint increases can begin again.

Solution

Basically, we just build up a flowchart that satisfies each specification of the description. Setpoint increases are done by the operation of 1.002 * TSP → TSP and decreases by

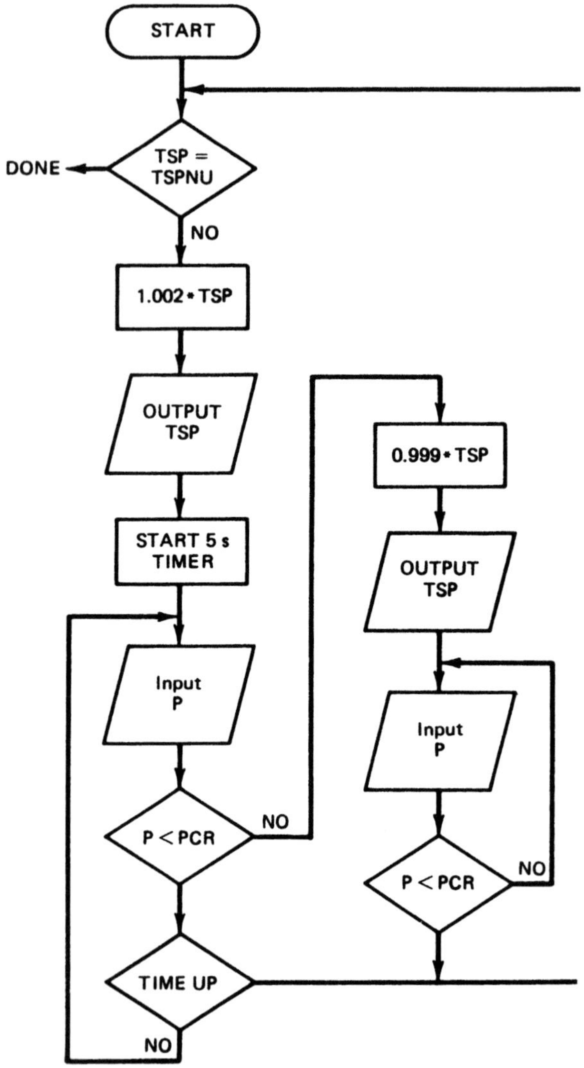

FIGURE 11.18
Flowchart solution of Example 11.5.

0.99 * TSP → TSP. It is assumed that either a hardware or software timer is available with its status available to the software. The result is shown in Figure 11.18.

11.5 CONTROL SYSTEM NETWORKS

There have been two major revolutions in process control and control systems in the last 20 years. The first was the replacement of analog controllers by digital controllers using embedded computers to perform the control function. The first part of this chapter studied this technology and showed how controller action is accomplished by software in the computer. The second revolution, which is happening even as we speak, is replacement of the standard 4- to 20-mA analog signals for communication throughout a process plant by serial digital communication via a network. For control system data communication we call the serial communication system a "fieldbus" because of the kind of network it employs. In this section the basic features of fieldbus serial communication systems will be presented.

11.5.1 Development

In the recent past a plant would typically house nearly all controllers in a building room (the control room) centrally located in the plant facility. All measurement data taken in the plant (referred to as the *field*) was communicated to the control room over pairs of wires carrying 4- to 20-mA signals, with one pair for each sensor. These were connected to the controllers. The feedback control signals to the final control element were sent back to the field in other pairs of wires carrying 4- to 20-mA signals. The 4- to 20-mA range was an *open standard* that all manufacturers of controllers, sensors, and final control elements agreed upon. It was open because there were no patents, so anyone could develop equipment to use the 4- to 20-mA communication. Equipment from different manufacturers was said to be *interoperable* because of this communication standard. The bundle of wire pairs was called a *parallel bus* carrying data from and to the field. Of course this led to a very extensive and expensive deployment of wires between the plant and the control room, as suggested in Figure 11.19. A petrochemical plant might extend over several acres so that wire runs could be very long. In the early deployment of computers as controllers, data was still carried back and forth between the control room and field over the bus of 4- to 20-mA wire pairs.

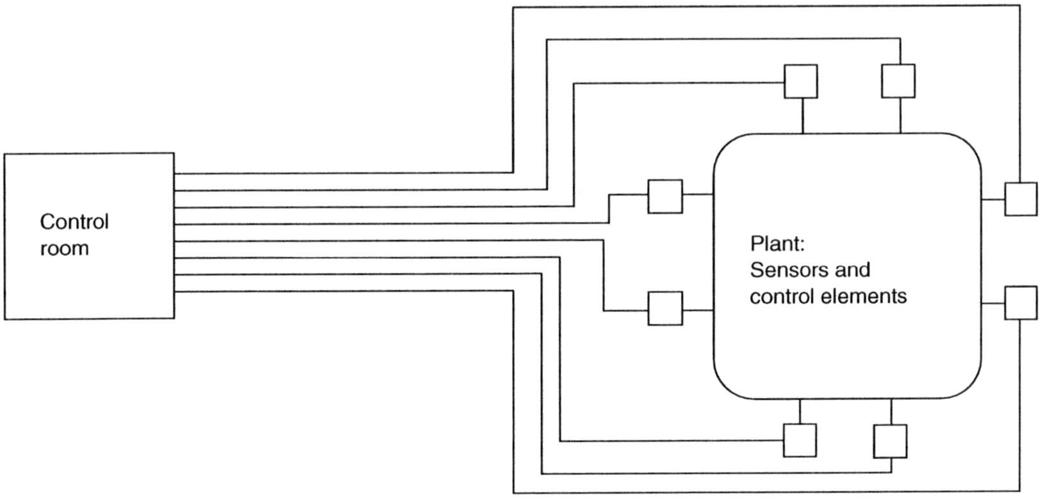

FIGURE 11.19
A myriad of wires and pipes connect the control room to the plant.

Meanwhile, in the computer industry, systems were developed to allow computers to communicate serially over a pair of wires. In this case data expressed in binary was sent serially as a stream of 1's and 0's from one computer to another. This could be done very fast, so the wires were idle most of the time. As the technology of serial communication matured, methods were developed to allow several computers to use a single pair of wires by sharing time slots on the wires so that one pair of computers communicated and then another pair, and so on, but so fast that it seemed that all computers were communicating at once over a single pair of wires. This kind of multiple-user communication required development of protocols about how a computer could gain access to time on the line for communication and how computers could be identified so the communication could be between specific computers. This eventually gave rise to a myriad of open standard technologies including the Ethernet, local-area networks (LANs), TCP/IP addressing and transmission protocols, wide-area networks (WANs), and of course the world wide web (www) as a super network. The use of open standards for protocols and communication allows many manufacturers to design and market equipment and software to further develop this networking technology.

Process Control Networks At first, as computers began to play a larger role in control systems both as controllers and for plant operation management, computer networks were established connecting the computers with digital serial technology. However, process signals were still predominantly carried by analog 4- to 20-mA current. In such systems the operations of controller computers were managed over a serial network, including changing setpoints, modes of controller action, and gains. This is often called *supervisory control.*

Eventually there was a move to eliminate the 4- to 20-mA data communication standard and let actual process data be carried from sensors to controllers and controllers to final control elements via digital serial networks. Figure 11.20 shows that there are several ways networks can be wired such as stars, rings, and another which uses straight runs of, for example, parallel wires with devices simply connected in parallel across the wires. This type of network is called a *bus.* Traditionally in a manufacturing facility the plant outside of the operations control room is referred to as the *field,* so we have a *fieldbus.* Since there existed no standard for how the process data was to be represented, many types of fieldbuses were put forward by process equipment manufacturers, many of which were patented and thus not open. Finally several open systems were developed and, at the present time, there is competition between these standards for adoption by the process industries.

Again, the advantage of an open standard is that a process plant or manufacturing facility can buy equipment (sensors, actuators, controllers, etc.) from a variety of manufacturers and connect them all together seamlessly on the network.

In this modern installation of a process plant the control room takes on a much lesser role because control has been distributed throughout the plant. Controller computers are often located in the field and communicate back to the control room process-management computers over the network. Often the controller computer is even in sensors themselves (the smart sensor introduced earlier). Final control elements have internal processors that can pick up data from the bus as sent from the controller. Figure 11.21 shows how the serial fieldbus connects field hardware and computers.

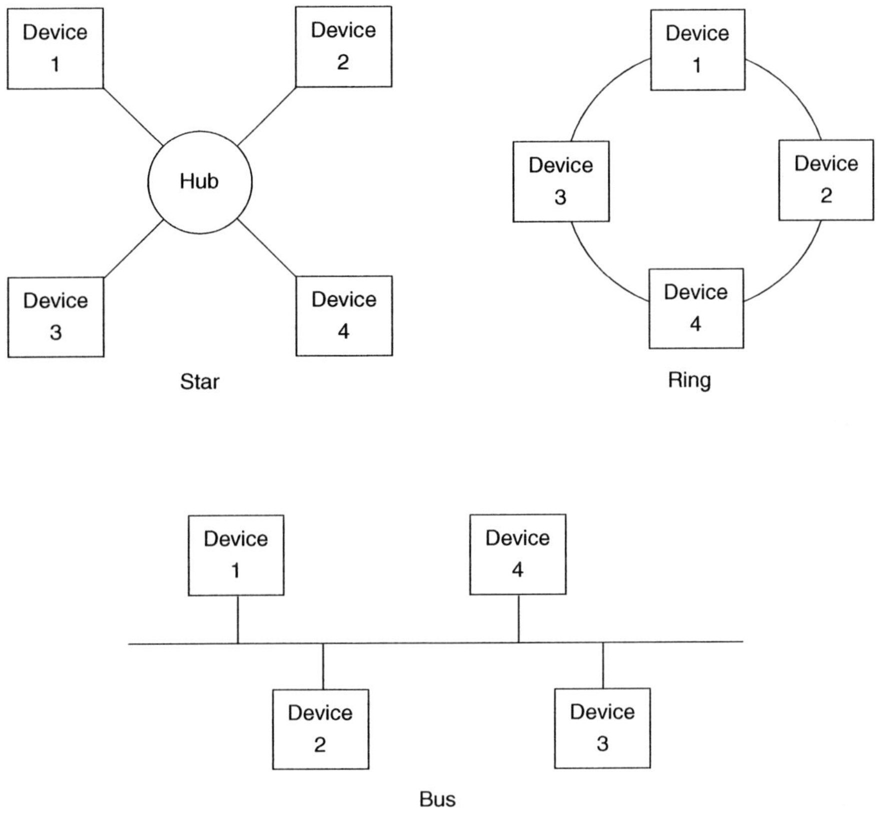

FIGURE 11.20
Networks can be wired into various configurations such as stars, rings, and a bus.

Fieldbus Operations Figure 11.22 demonstrates generically how the control process operates under this new technology. Suppose a computer in charge of level must input level data, process it, and feed back corrections to a valve. In this technology every device, including the sensor, valves, and computer, have a unique network address. Each device also has the basic intelligence to communicate over the serial bus. The control process might go as follows, with all communications over the serial bus as digital bit streams:

1. Computer 02 sends a request to level sensor 12 asking for a reading. To do this the computer forms a packet with its address (02), the sensor's address (12), and a code to request a reading.
2. The level sensor takes a reading and sends level data back to computer 02. The level sensor must have the intelligence to form a packet with the computer address, its address, and the digitally encoded level.
3. After solving the control equations, computer 02 sends a valve-setting signal to valve 24. To do this the computer forms a packet with its address, the valve address (24), and a digitally encoded valve setting.
4. Valve 24 drives the valve to the new setting and, in some cases, may send an acknowledgement back to computer 02.

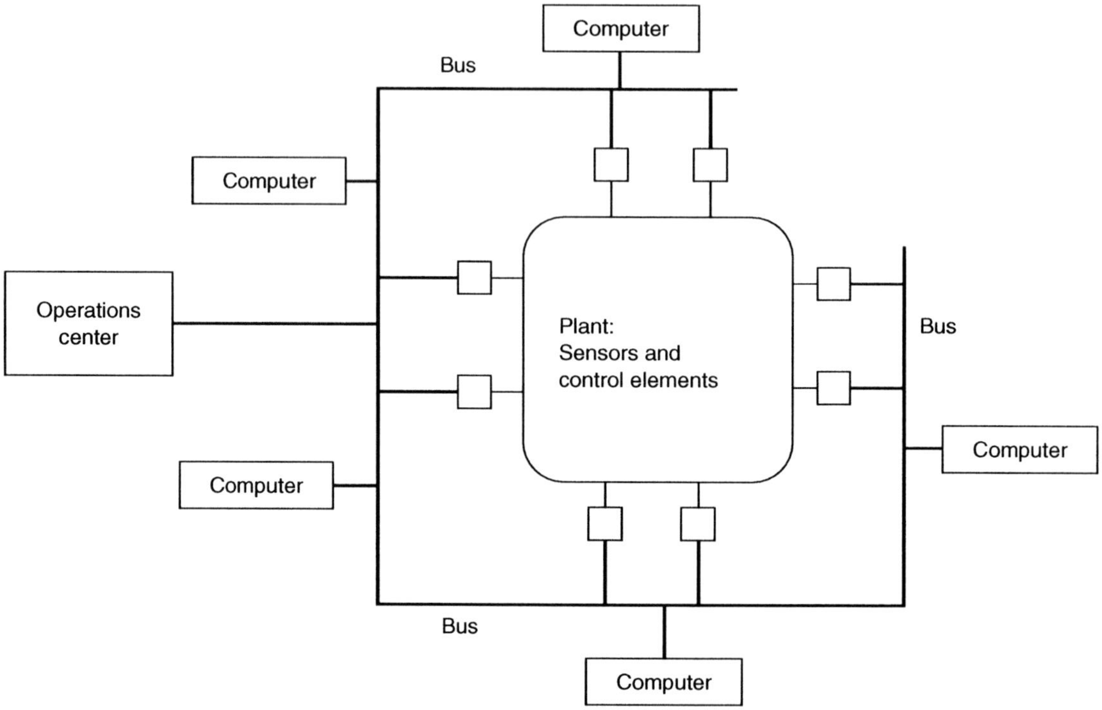

FIGURE 11.21
Using a bus eliminates the multitude of wires and pipes.

Of course computer 05, temperature sensor 75, and heater 33 all ignore this traffic because their unique addresses are not involved in the communication. However, a similar measurement, evaluation, and feedback process may be going on at virtually the same time with the temperature control system. By sharing time on the serial line at very high speeds the two loops seem to use the bus at the same time. Meanwhile the operations center computer can also be requesting readings from the sensors, so plant operations can be monitored. The operations center computer can also request reports from the control computers.

Of course one of the big advantages of this system is that the large number of wire pairs necessary with the current communication are eliminated. In terms of retrofit of existing plants one pair of the same wires used for the old 4- to 20-mA current signals can often be used for the serial digital communication.

11.5.2 General Characteristics

There are many types of fieldbuses used in the process industries, each with their own features. Certain common characteristics can be identified and used to compare the buses. In general, the set of these characteristics is called the protocol of the bus or network.

The overall commonality is that these buses carry *serial digital data.* So if you looked at the signal on a bus wire, you would see a time series of pulses representing ones and zeros being propagated across the network.

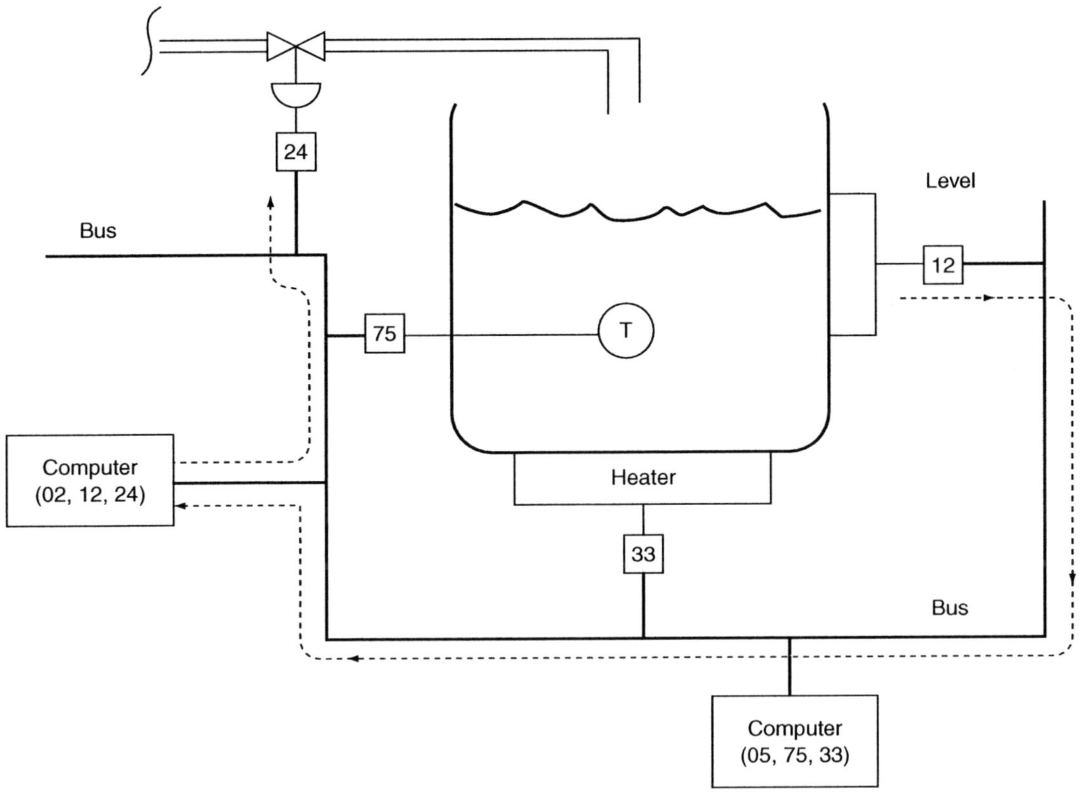

FIGURE 11.22
Unique addresses allow computers and devices to share the bus.

Addresses It stands to reason that, if many computers are to employ this bus for communication, there must be an addressing protocol so each device connected to the network can be uniquely identified. The number of bits allocated to the address determines the maximum number of devices that can be connected to the bus.

Data Packets If a computer is using the bus to send an updated signal to its final control element device, it must package the new data along with the address of the device. The final control element recognizes its own address and will therefore accept the data and perform the update. A data packet consists of the device address, often the address of the sender, the data itself, and auxiliary bits for error correction and other functions.

Figure 11.22 illustrates that if a control computer launches a data packet for its final control element (a valve) onto the bus, then only the valve addressed will accept the packet. Other devices on the bus will ignore the information.

Physical Media The actual nature of the medium that carries data varies greatly between buses. Typical media include twisted pair copper wire, coaxial cable, fiber-optic cable, and radio frequency (rf) propagation (wireless).

Speed An important characteristic of the bus is the speed by which data can be propagated across the network. The speed is typically given as the maximum number of bits per second (bps) that the bus can carry without serious degradation of the data. Speed is a function of not only the physical medium of the bus but also the number of devices connected to the bus and the complexity of the data being sent.

Cycle Time An important aspect of the bus is the time required for a data update to propagate through the system. For example, if the control computer of Figure 11.22 needs to update the valve setting, how much time lapses between when the update is launched by the computer and when the control valve receives the data? This is determined by the speed of the bus and also by how many other devices are on the bus and competing to send data.

Interoperability An important feature of a bus is the ease by which computers and devices can be added to the bus and configured to operate as part of the plant. In traditional computers, this is referred to as the "plug-and-play" feature. The manufacturer of a process-control computer can provide software that makes it compatible with a variety of bus protocols. However, manufacturers of control valves, sensors, and other control system devices generally have to build the protocol into the device. Thus, they are interoperable on a particular bus but not on another.

11.5.3 Fieldbus Types

Over the past 20 years many types of fieldbuses have been developed that are in use in the manufacturing and process industries throughout the world. Many are suitable for only restricted applications; some are proprietary (patented) and can only be used with the approval of their "owner." In this section some of the more common fieldbuses will be presented and their characteristics and protocols presented. At the time of this writing, there are two open standards that seem to have the greatest support: Foundation Fieldbus and Profibus. Although both have their custom protocols they are often being allied with the Ethernet since this technology is so well established. There is some movement in the industry to simply use an improved Ethernet throughout the plant and skip the other fieldbuses.

OSI 7-Layer Protocol The protocol of a fieldbus refers to how the bus system packages data to be transmitted. There is a kind of universal standard against which all serial communication protocols are measured. This is called the 7-layer OSI model. The first layer, called the physical layer, is associated with the actual physical equipment that provides the communication between two points in space. This could be a twisted pair of wires, coaxial cables, or a fiber-optic cable. All the other layers refer to how the actual data, called the application layer (number 7), is packaged with control information (such as the address of the sender, address of the receiver, size of the data, error correction, etc.). Figure 11.23 shows definitions of the 7-layer OSI model and a pictorial representation of how a communication packet is assembled. You can see that each layer adds its own information to the packet. The resulting packet can be quite large, such as 1000 bytes. However, it is not nec-

FIGURE 11.23
The OSI 7-layer communication protocol showing packet encapsulation by each layer.

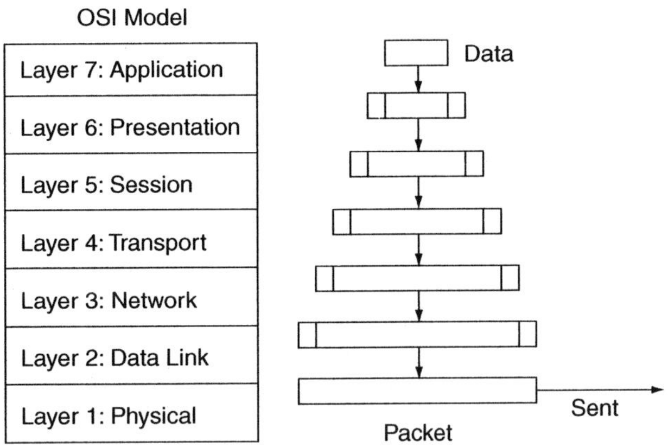

essary for a communication system to employ all seven layers; in fact, they employ only what is actually needed. Many fieldbus systems employ only three.

Fieldbus Category It is possible to approximately distinguish control system and manufacturing communication buses into one of three categories depending upon how much data the system must carry.

- *Sensor or Bit Level* This type of bus is intended for applications where single bits are the primary information carrier. This would principally include inputs with on/off conditions as from switches and alarms. Outputs would also be on/off such as lights, solenoid actuators, quick-acting valves, and so forth. Often basic PLC installations can use this type of bus.
- *Device or Byte Level* This is an intermediate application where one to several bytes (8-bits, also called an octet) are the primary information carrier. This type of bus is used when the state of some device requires more than one bit. For example, when a control computer wants to input or output the entire state of some machine, many bits may be required to specify solenoids on or off, motors on or off, valves open or closed, and so forth. This system can also be used to send and receive analog data that has been digitally encoded.
- *Fieldbus or Message Level* Finally, we come to the actual fieldbus for which the information may be hundreds of bits; when combined with the encapsulation, it may be a thousand bits. This system includes error checking, correcting, and complex addresses as well. This type of bus can be used for any level but is most suited to operating control loops over the bus and the passing of large amounts of data. Data rates are often slower than sensor or device buses because of the large packet size.

As often happens there is considerable overlap in the previous definitions. Generally the higher-complexity buses can be used for all three applications but the price is that they are generally slower because of data overhead.

Important Features A fieldbus is used to communicate control signals in a process plant among the sensors, controllers, and final control elements. As such, certain characteristics are important. Some of the important characteristics are:

1. *Physical media of transmission* This refers to layer 1 of the OSI model and defines the actual communication carrier, such as twisted wire pairs, coaxial cable, and fiber-optic cable.
2. *Number of devices* This refers to the fact that device addresses are carried in the information packet. Therefore, there will be a limit to the number of devices that can be addressed.
3. *Distance over which the bus can extend* The longer the bus, the weaker and more distorted the signals become. In some cases special repeater circuits can extend the basic bus distance.
4. *Speed of transmission* The serial data is like a square wave in the communication media where the state changes of the square wave represent bits. The bus speed refers to the frequency of the possible bit changes which is better described by the number of bits per second (bps) that the serial system can handle.
5. *Bus powered* This is a feature carried over from the 4- to 20-mA analog systems. In those systems the current conveyed the value of the input or output but also provided power to the device. So it served a dual purpose, and thus saved on wiring. Some fieldbuses have this same capability so that the data measurement

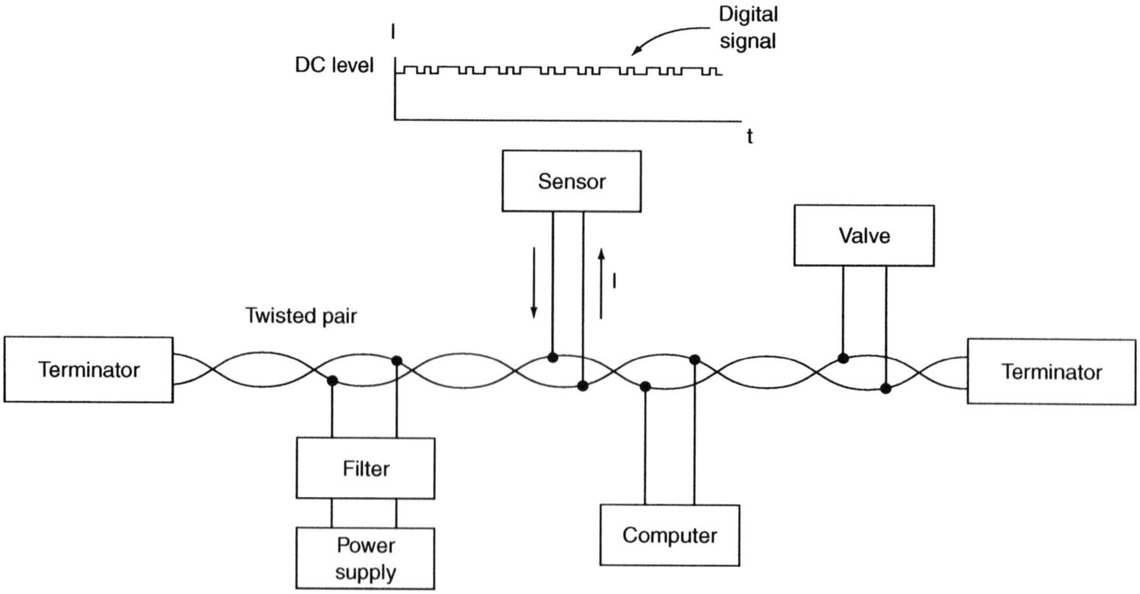

FIGURE 11.24
A fieldbus can use digitally modulated current for transmission and power.

serial bit stream rides on top of a DC current providing power to the devices on the bus. Figure 11.24 shows how a twisted pair provides power to a sensor while at the same time the sensor impresses the measurement on the current drawn as a digital modulation.

Table 11.1 shows some of the fieldbuses in use in the process industry and gives their approximate properties as previously defined. Some of the buses use the Ethernet for layer 1 but it is possible to simply use the Ethernet directly without the other protocol. The advantage to this is that Ethernet is an old and established system and its widespread use for LANs and office automation in general makes it a very familiar technology. In fact, existing information technology (IT) personnel can often then provide support for the fieldbus.

TABLE 11.1
Comparison of control system bus types

Bus	Category	Physical Media	Number of Devices	Distance	Speed	Power from Bus
ASI	Sensor	Twisted pair	31	100 m	167 kbps	Yes
Seriplex	Sensor	4-wire shielded	500	150 m	200 Mbps	No
CAN	Sensor	Twisted pair	127 nodes	25 m to 1 km (speed dependent)	10 kbps to 1 Mbps	No
DeviceNet	Device	Twisted pair	64	500 m (6 km with repeater)	125 kbps to 500 kbps	No
LONWorks	All	Twisted pair Fiber Power line	32,000 per domain	2 km @ 78 kbps	1.25 Mbps	Yes
Profibus DP/PA	Fieldbus	Twisted pair Fiber	127 nodes	100 m twisted pair 24 km fiber	DP: 500 kbps PA: 31.25 kbps	PA: Yes
Foundation (H1)	Fieldbus	Twisted pair Fiber	240/segment 65k segments	1900 m	31.25 kbps	Yes
Foundation (HSE)	Fieldbus	Twisted pair Fiber	Unlimited	100 m twisted pair 2 km fiber	100 Mbps	No
Industrial Ethernet	Fieldbus	Twisted pair Fiber Coax	Unlimited With routers	100 m twisted pair 2.5 km fiber	10 Mbps 100 Mbps	No

11.6 COMPUTER CONTROLLER EXAMPLES

The following examples illustrate some of the difficulties that arise from seemingly simple control problems implemented digitally. It is not the purpose of this chapter or book to give you all the expertise to fully design a DDC system, but merely to give you an understanding of the steps of such a design and how some of the features of the design are actually built into the system.

EXAMPLE 11.11 A proportional control system with a 50% *PB* controls conveyor speed by using weight measurement. Specifications are as follows:

1. Weight range: 40 to 90 lb.
2. Weight setpoint: 65 lb.
3. Signal conditioning already available converts 0 to 100 lb to a voltage scaled at 0.05 V/lb—that is, 0 to 5 V.
4. An 8-bit ADC converts an input of 0 to 5 V to 00H to FFH, where 5-V input just produces FFH.
5. Conveyor speed is regulated by a motor-control circuit that operates directly from the output of an 8-bit DAC. There are 256 speeds selected according to the following: 00H is OFF, FFH is full speed, and 80H is the speed corresponding to the setpoint of 75 lb—that is, zero error.

Find the digital controller equations. Express all numbers in their hex form.

Solution

The proportional-mode equations are given by Equations (11.6) and (11.7). We must define the constants that appear in these equations.

$W_{\text{min}} = 40$ lb, so

$$DMIN = (40/100)256 = 102.4 \rightarrow \text{66H}$$

$W_{\text{max}} = 90$ lb

$$DMAX = (90/100)256 = 230.4 \rightarrow \text{E6H}$$

$W_{sp} = 65$ lb

$$DSP = (65/100)256 = 166.4 \rightarrow \text{A6H}$$

The error equation becomes

$$DE = (DW - \text{A6H})/(\text{E6H} - \text{66H})$$
$$DE = (DW - \text{A6H})/\text{80H}$$

Because the zero-error output is given to be 80H, $P0 = \text{80H}$. A 50% *PB* means $K_P = 2\%/\%$. Both the error and output are given as fractions of range so that this value of K_P can be used directly. Thus

$$P = \text{80H} + 2 * DE$$

This is the fraction of full-scale output. The final output would be

$$POUT = P * FFH$$

Actually, multiplying an 8-bit number by FFH is effectively equivalent to a shift of 8 bits, so that in terms of numbers,

$$POUT = P$$

You can see from this example that the development of actual coding from the original specification involves consideration of many factors, including the encoding of numerical data. For more complicated operations, such as finding averages, linearization, integration, and finding derivatives, the math operations in binary become even more complicated. In general, routines are written in a top-down fashion, starting with a general statement of the problems and then working into the details, as shown in this example. The following example illustrates a more complicated problem. The problem takes the control algorithm only to the stage of a general flow diagram, from which the detailed flow diagram would be developed.

EXAMPLE 11.12 A DDC system, shown in Figure 11.25 will use PI control with a 40% *PB*, 0.5%/min integration time, and a 0.75-min sample time. The input is to be the average of five

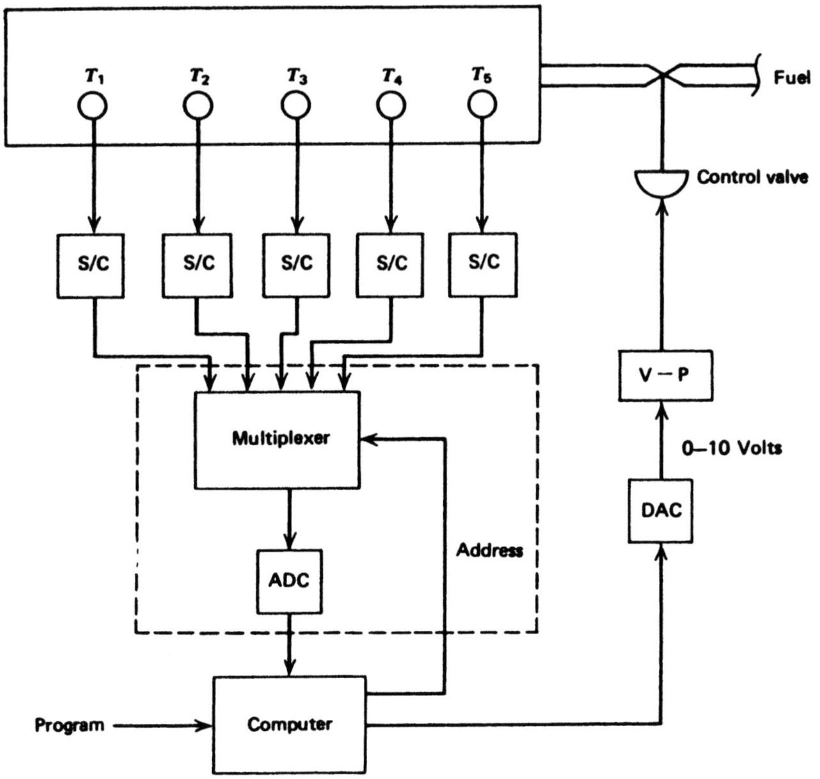

FIGURE 11.25
Process system for Example 11.12.

temperatures, and the output will be a control signal to a valve. Specifications of the system are as follows:

1. A temperature measurement system is available that provides a voltage from the temperature transducer with the following relationship:

$$V = 0.075T + 0.0004T^2 \qquad \textbf{(11.19)}$$

2. Temperature range is 0°C to 90°C with a 57°C setpoint.
3. For initial start-up, the output signal to the valve should be 5 V. The valve is driven directly from a DAC, adjusted so that 0H is 0 V and FFH is 10 V.
4. A data-acquisition system is available with five channels and an ADC that converts 0 V to 0H and 10 V to FFH.
5. The computer has floating-point hardware so that inputs are converted automatically to floating-point numbers for use in the programs. Thus, 0H input produces 0.0, and FFH produces 225.0_{10}.

Construct the general flow diagram for the control algorithm using rectangular integration.

Solution

We find the solution by working through the system from input to output in steps. For convenience, we will express the temperature error in terms of percent.

1. *Find the temperature* The input section will provide a value from 0 to 255 that is the voltage representing the temperature. We can find the voltage by dividing by 25.5. Knowing the voltage, we find the temperature by inversion of the relationship between voltage and temperature, Equation (10.9). First we write it in a quadratic form:

   $$0.0004T^2 + 0.075T - V = 0$$

   From the quadratic root equation, we find

   $$T = \frac{-0.075 \pm \sqrt{(0.075)^2 - 4(-V)(0.0004)}}{2(0.0004)}$$

   This relation reduces to

   $$T = -93.75 + \sqrt{8789 + 2500\text{ V}} \qquad \textbf{(11.20)}$$

   where we have used the positive sign because we know T must be 0 when the voltage is 0.
2. *Find the average temperature* We can find the average by simply inputting all five temperatures, by addressing the proper input channels, and then summing and dividing by 5.
3. *Find the error* The error will be found in percent, as usual, using Equation (9.3)

   $$e_p = \left(\frac{AT - 57}{90 - 0}\right) \times 100 \qquad \textbf{(11.21)}$$

   where AT = temperature average.

4. *Proportional gain* A 40% *PB* means that, when the error changes by 40%, the output changes by 100%. Because e_p is expressed in percent, we have $K_P = 100/40 = 2.5\%/\%$.
5. *Integral term* The integral gain is the 0.5%/min given in the specification of the problem. For a 1% error, the term contributes 0.5% change in proportional-mode controller output every minute. Because the samples from which the integral is constructed are taken every 0.75 min, the contribution actually will be (0.5%/min)(0.75 min/sample), or 0.375%/sample.
6. *Output* The output signal can be constructed from Equation (9.19) along with Equations (11.8) and (11.9), adjusted so that 100% corresponds to 255, so that an output of FFH is sent to the DAC when $P = 100\%$.

The final flow diagram for the controller is given in Figure 11.26 on page 554. It has been assumed here that the input channels are identified by a single integer, 1 through 5.

SUMMARY

The important features of digital and computer applications in process control are as follows:

1. Application of digital methods started with the simple application of digital logic methods to simple control problems and the use of computer installations to provide data-logging services in process control.
2. In data logging, the computer proved its value in process control by its ability to oversee the operation of a process through high-speed sampling of loop data. ADC and multiplexer systems allow rates of 5000 samples per second to be taken.
3. Supervisory control lets the computer adjust process-loop setpoints for optimum process performance even though standard analog control loops are still used for control.
4. Computer-based control eliminates the analog controller and replaces the mode implementation by programs within the computer.
5. Computer control can implement the proportional, integral, and derivative modes using approximation algorithms. Sampling time effects must be considered.
6. In distributed control, smart sensors with the control computer built-in are connected to operations management computers by networks.
7. Much of process control is now implemented using a network called a *fieldbus* to connect controllers, sensors, and final control elements.

PROBLEMS

Section 11.2

11.1 A force transducer has a transfer function of 2.2 mV/N. Design an alarm using a comparator that triggers at 1050 N.

11.2 A type-J TC with a 0°C reference monitors a process temperature. Design a two-alarm system turning on one LED at 100°C and another at 150°C.

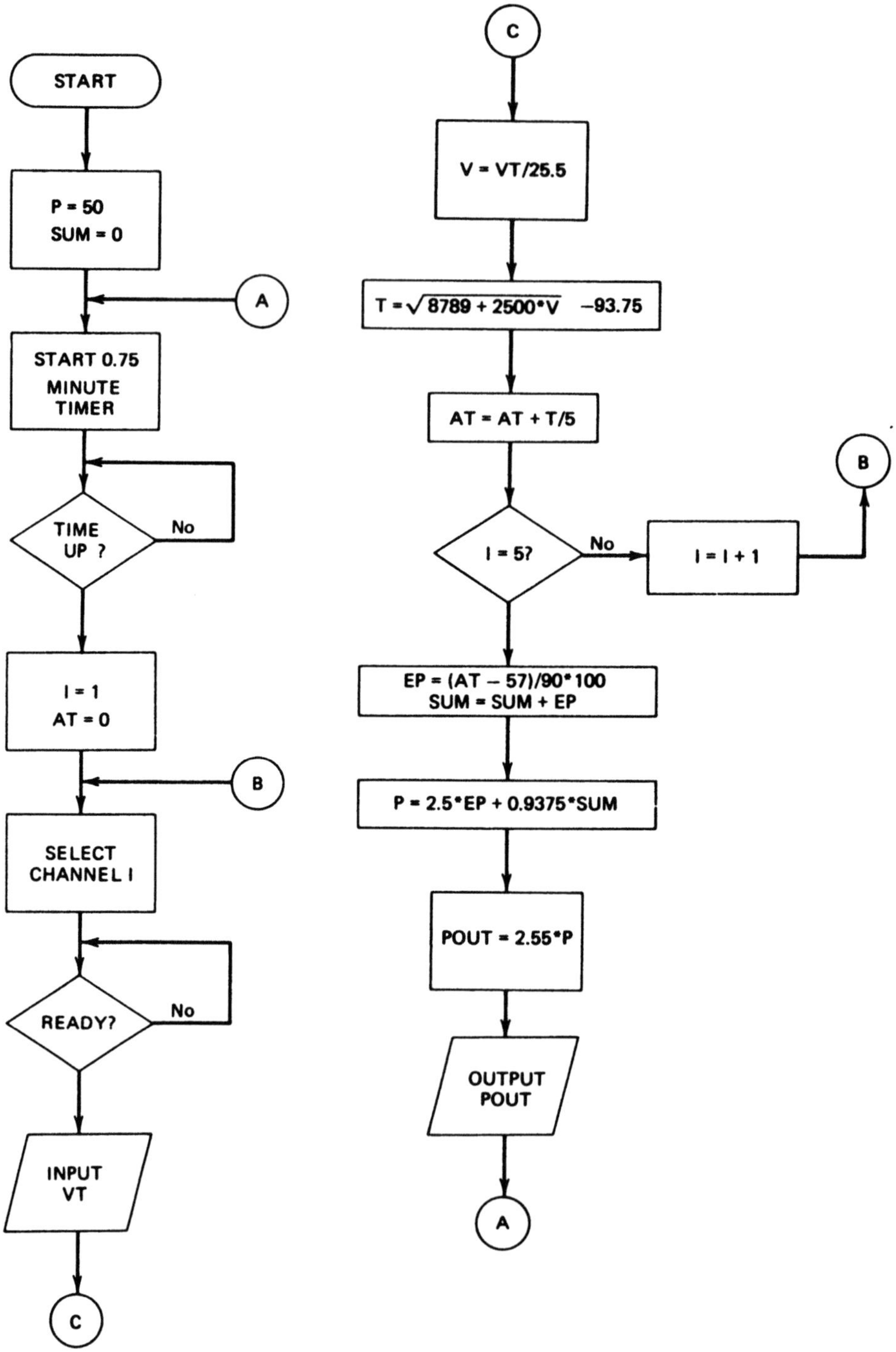

FIGURE 11.26
Flowchart solution for Example 11.12.

11.3 Design a two-position digital controller using the force transducer of Problem 11.1. The output should go high at 500 N, and the deadband should be 75 N.

11.4 Assume that a conveyor system speed *S*, load *L*, and loading rate *R* have been converted to binary signals by comparators. What Boolean equation provides an alarm, *A*, when either the load and loading rate are high when the speed is low, or the speed is high when the load is low?

11.5 Show how digital circuit elements can implement the alarm of Problem 11.4.

11.6 A multiple-variable digital control uses a reaction vessel for which binary inputs of level, temperature, and pressure drive binary outputs to the input valve and the output valve. A digital HIGH opens a valve. The following conditions must be satisfied:

Pressure	Temperature	Level	Input Valve	Output Valve
H	H	H	L	H
L	H	H	L	H
L	H	L	L	L
L	L	L	H	L
H	L	L	H	H
H	L	H	L	L

Devise a Boolean equation for this controller and the digital logic circuit.

Sections 11.3 and 11.4

11.7 A data-logging system must take samples of 40 variables at 100 samples per second each. What is the maximum signal-acquisition and processing time in microseconds?

11.8 A computer must sequentially sample 100 process parameters. It requires 14 instructions at 5.3 μs/instruction for the computer to address and process one line of data. The multiplexer switching time is 2.3 μs, and the ADC conversion time is 34 μs. Find the maximum sampling rate for a line.

11.9 Develop a block diagram similar to Figure 11.15 of a supervisory computer control system for the process illustrated in Figure 11.27. This is a firing operation for ceramic articles. The conveyor motor speed setpoint is determined by the feed rate of articles.

11.10 For the process of Figure 11.27 under supervisory control, to obtain an increase in conveyor speed by 5% requires that the preheat setpoint be increased by 3.9%, the oven setpoint by 7.2%, and the cooler by 4.4%. The speed increase must be performed in 1% steps with reacquisition of each setpoint before continuing. Prepare a flowchart showing how computer control can provide this increase.

11.11 An air-conditioner (AC) will provide control of temperature in the range 60° to 100°F. Temperature information is input to a computer from a 6-bit ADC as value N_T; that is, 00H (0_{10}) at 60°F and 3FH (63_{10}) at 100°F. The setpoint is 75°F, the zero-error output is 50%, and the proportional gain is to be 8%/%. The output goes to a 6-bit DAC for which 00H is AC off and 63H is AC maximum cooling. Specify the control equations that must be used. What is the resolution in temperature control?

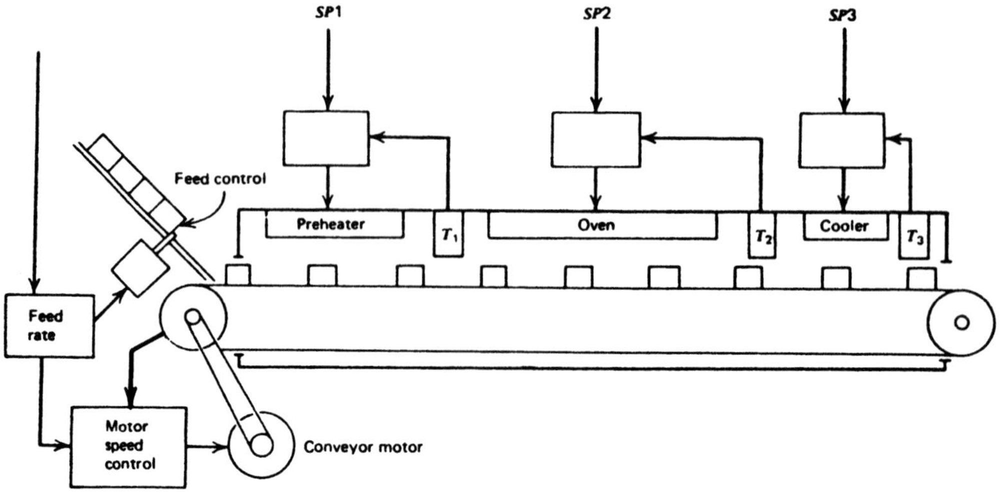

FIGURE 11.27
Process used in the problems.

11.12 The setpoint for the measurement of Figure 3.34 is 75 psi. Find the approximate integral term response at 0.8 s, assuming a 0.05-s sample time and an integral gain of 2.45 min^{-1}. The control range is 25 to 100 psi.

11.13 A process is to operate under PID with a 60% *PB*, 1.2-min integration time, and 0.05-min derivative time. If the error is available as percent of span, develop the control equations and show a flowchart of computer controller action with all constants evaluated. The sample time is 0.8 mins.

11.14 A PI controller is required for the following control problem:

a. Temperature is to be controlled from 50° to 150°C with a setpoint of 100°C.

b. A sensor provides temperature-dependent voltage as

$$V = (T - 100)/10 + 5$$

c. An 8-bit ADC is used for which 00H is 0 V and FEH to FFH occurs at 10.0 V.

d. The *PB* is 70%, and the integral gain is 1.5 min^{-1}.

e. Sample time is 0.5 min.

f. Output is through an 8-bit DAC with a 10.0-V reference.

Develop a flowchart for the control process, control equations, and a program in the language of your choice.

11.15 Measurement of position L in mm is provided in terms of voltage by the relation

$$V = (e^{0.02L} - 1)/2$$

if fed directly into an 8-bit ADC with a 5.00-V reference. Develop the control equations for the following controller specifications: $K_P = 3.7, K_I = 0.9\,\text{min}^{-1}$, 2-s sample time, setpoint of 50 mm, 8-bit output.

11.16 Add derivative action to Problem 11.15 with $K_D = 0.1$ s.

SUPPLEMENTARY PROBLEMS

S11.1 A microcomputer will be used to provide automatic temperature regulation for a shower faucet using the system shown in Figure 11.28. The user sets the desired temperature between 30°C and 50°C by a sliding knob attached to a 1-kΩ variable resistor. The user sets the water volume by manually adjusting the cold-water valve. The hot-water valve is controlled by the computer to provide the selected temperature.

The particulars are that a two-channel multiplexer is attached to an 8-bit ADC with a 5.0-V reference. The temperature sensor is a thermistor with a characteristic given by Figure 11.29 and a 5-mW/°C dissipation constant. The control valve operates from a 0.0–5.0-V signal from an 8-bit DAC. ADC input is via PORT 301,

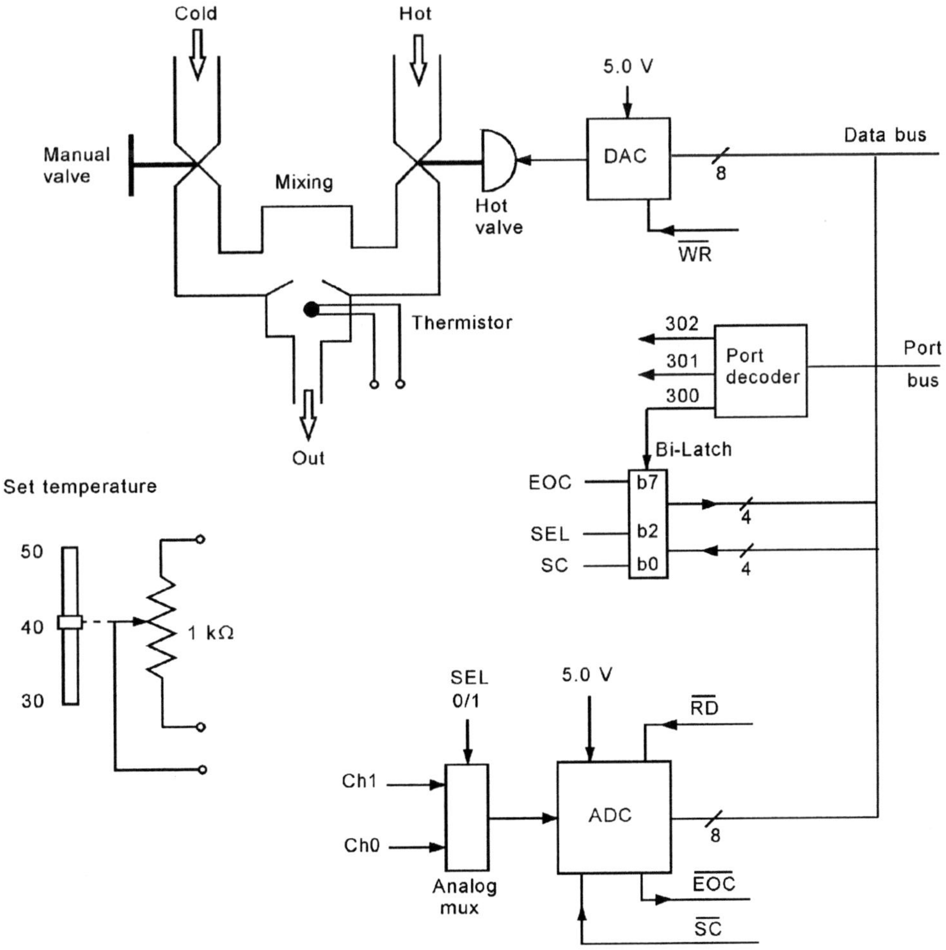

FIGURE 11.28
Temperature-control system for Problem S11.1.

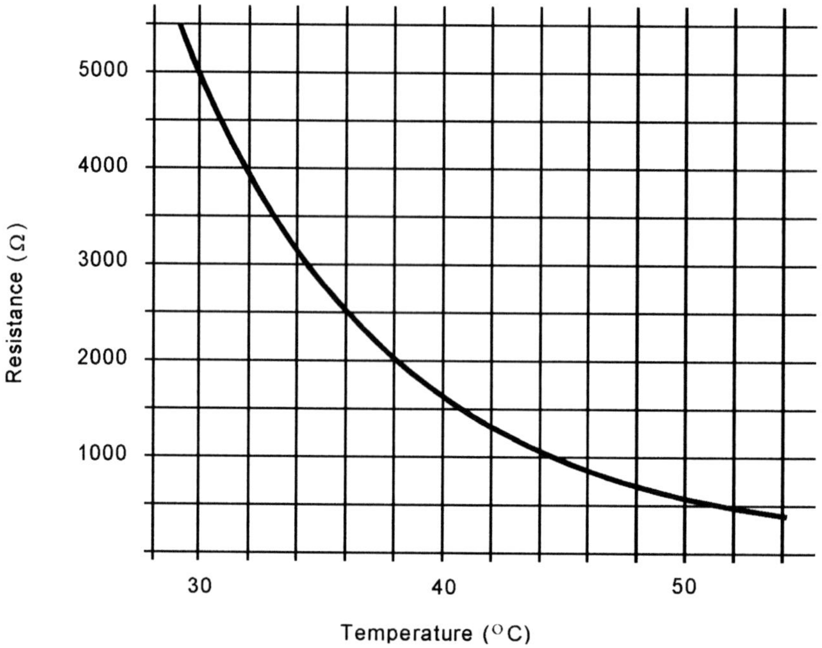

FIGURE 11.29
Thermistor characteristic for Problem S11.1.

DAC output is via PORT 302, and control signals are exchanged via a bidirectional PORT 300, as shown in Figure 11.28. The port-select lines go low when that port is addressed. It has been determined that PI control is needed with gains of $K_P = 2.5\%/\%$ and $K_I = 0.08\%/(\%\text{-min})$.

a. Design the appropriate analog hardware for getting the thermistor temperature measurement and the user temperature setting into the computer via the ADC.

b. Complete appropriate digital connections in Figure 11.28 for control interface of the ADC, DAC, and multiplexer.

c. Prepare a flowchart of the software required to take samples every second.

d. Develop the control equations for implementation of the error detector and specified PI control mode. Remember that the thermistor resistance change is nonlinear with temperature.

e. (Optional) Write a program in assembly language, C, or any other appropriate language to implement the design.

CHAPTER 12

Control-Loop Characteristics

INSTRUCTIONAL OBJECTIVES

This chapter provides process-control technologists with a general background in the practical considerations of process-control loop implementations. After reading this chapter, you should be able to

- Explain the characteristics of single-variable, compound, cascade, and multivariable control.
- Define three standard measures of quality in a control system.
- Describe the control-loop stability criteria with respect to a Bode plot.
- Describe the open-loop transient disturbance method of loop tuning.
- Describe the Ziegler-Nichols method of process-control loop tuning.
- Define phase and gain margin.
- Explain how the frequency response method can be used to tune a process-control loop.

12.1 INTRODUCTION

In the previous chapters of this book, the detailed elements of process control have been described. The instrumentation used to perform the process-control function has been presented in some detail. At this point, the reader is similar to an individual who has acquired detailed knowledge of all the elements of an airplane. Such a person, however knowledgeable about the airplane parts, is certainly *not* competent to *pilot* the aircraft! In process control, there also is a remarkable difference between knowing the elements of a process-control loop and being able to tune and operate a process-control loop installation.

In this chapter, consideration is given to the various ways of setting up a process-control loop system and tuning this system for optimum performance. A word of caution: A complete understanding of such *optimization* requires detailed mathematical study of stability theory, a substantial depth of knowledge of the process, and many years of experience. The general intent of this chapter is merely to make the reader familiar with the general concepts.

The objectives of this chapter have been carefully selected to provide the reader with as much background as can be expected without a more extensive mathematical background. The emphasis is not on the theoretical development of process-control loop characteristics but on the interpretation and employment of the results of such theory. Many volumes are devoted to studies of loop stability, optimization techniques, multiloop DDC operations theory, and a host of other topics that could never be included in this brief chapter. In short, this chapter provides the process-control technologist with a general background in the practical considerations of process-control loop implementations.

12.2 CONTROL SYSTEM CONFIGURATIONS

We consider first the types of control-loop configurations that are encountered in typical industrial applications. The control-loop concept, as outlined in the previous chapters, is a correct one, but necessarily oversimplifies many of the actual configurations used in industry. The decision of which arrangement to select is made by the process designers based on the goal of the process relative to production requirements, and the physical characteristics of operations under control.

12.2.1 Single Variable

The elementary process-control loop (which has formed the basis for most of our discussion in previous chapters) is a *single-variable* loop. The loop is designed to maintain control of a given process variable by manipulation of a controlling variable, regardless of the other process parameters.

Independent Single Variable In many process-control applications, certain regulations are required regardless of other parameters in the process. In these cases, a setpoint is established, controller action is started, and the system is left alone. Thus, in Figure 12.1, a flow-control system is used to regulate flow into a tank at a fixed rate determined by the setpoint. This system then makes adjustments in valve positions as necessary following a load change to maintain flow rate at the setpoint value.

Interactive Single Variable A second single-variable control loop, also shown in Figure 12.1, regulates the temperature of liquid in the tank by adjustment of heat input. This also is a single-variable loop that maintains the liquid temperature at the setpoint value. Under nominal conditions, the flow into the tank is held constant and the temperature is also held constant, both at their respective setpoint values. Note, however, that a change in the setpoint of the flow-control system appears as a load change to the temperature-

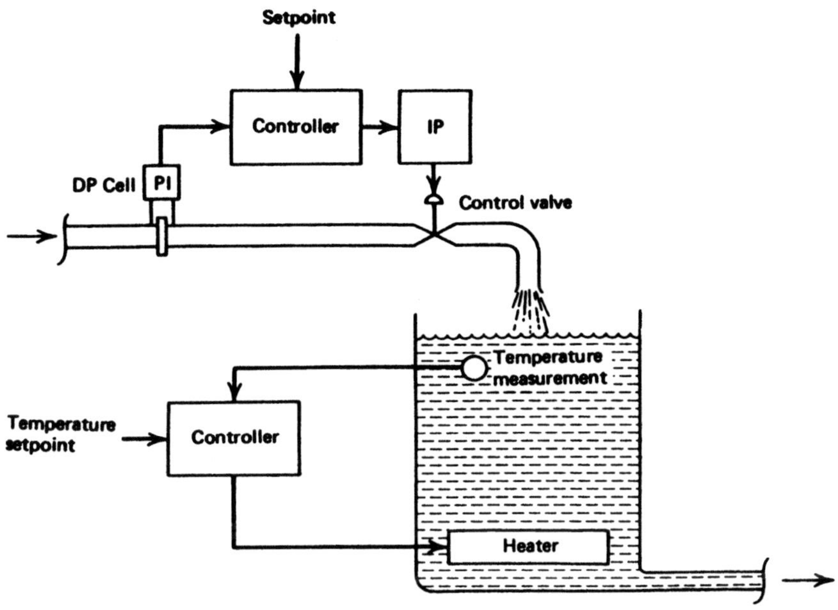

FIGURE 12.1
Two variable process-control loops that interact.

control system, because the fluid level in the tank or rate of passage through the tank must change. The temperature system now responds by resetting the heat flux to accommodate the new load and bring the temperature back to the setpoint.

We say, then, that these two loops *interact.* Almost any process where several variables are under control shows such interactive behavior. Any cycling or other instability of the flow-control loop causes cycling in the temperature system because of this interaction.

Compound Variable In some cases, a single process-control loop is used to provide control of the relationship between two or more variables. This can be accomplished by using measurements from, say, two sensors as input to the process controller. A signal-conditioning system must scale the two measurements and add them prior to input to the controller for evaluation and action. The analysis of such systems can become quite complicated.

A common example is when the ratio of two reactants must be controlled. In this case, one of the flow rates is measured but allowed to float (that is, not regulated), and the other is both measured and adjusted to provide the specified constant ratio. An example of this system is shown in Figure 12.2. The flow rate of reactant A is measured and added, with appropriate scaling, to the measurement of flow rate B. The controller reacts to the resulting input signal by adjustment of the control valve in the reactant B input line.

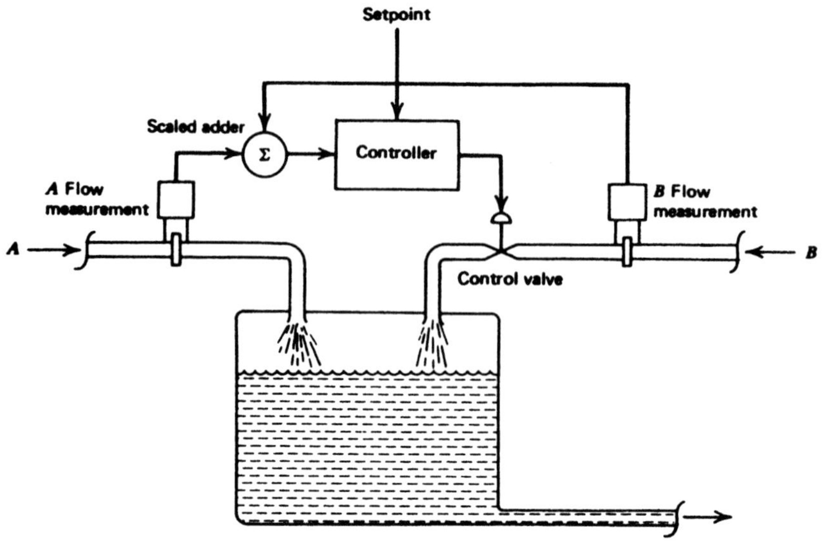

FIGURE 12.2
A compound system for which the ratio of two flow rates is controlled.

EXAMPLE 12.1 In a compound control system, the ratio between two variables is to be maintained at 3.5 to 1. If each has been converted to a 0–5-V range signal, devise a signal-conditioning system that will output a zero signal to the controller when the ratio is correct.

Solution

In this case, we use a summing amplifier (see Section 2.5.2). Then the output is related to the input by

$$V_{out} = -\frac{R_f}{R_1}V_1 - \frac{R_f}{R_2}V_2$$

If we make $V_{out} = 0$, then the voltage ratio is

$$\frac{V_1}{V_2} = -\frac{R_1}{R_2}$$

One input voltage must be negative (because we cannot use negative resistance) because the ratio of the resistance should be 3.5 to 1.

$$R_1 = 3.5R_2$$

Because the gain is unspecified, we can use $R_f = R_1$, for example. Then the circuit of Figure 12.3 accomplishes the desired function, where

$$V_{out} = -\left(\frac{3.5\ \text{k}\Omega}{3.5\ \text{k}\Omega}\right)(-V_1) - \left(\frac{3.5\ \text{k}\Omega}{1\ \text{k}\Omega}\right)V_2$$

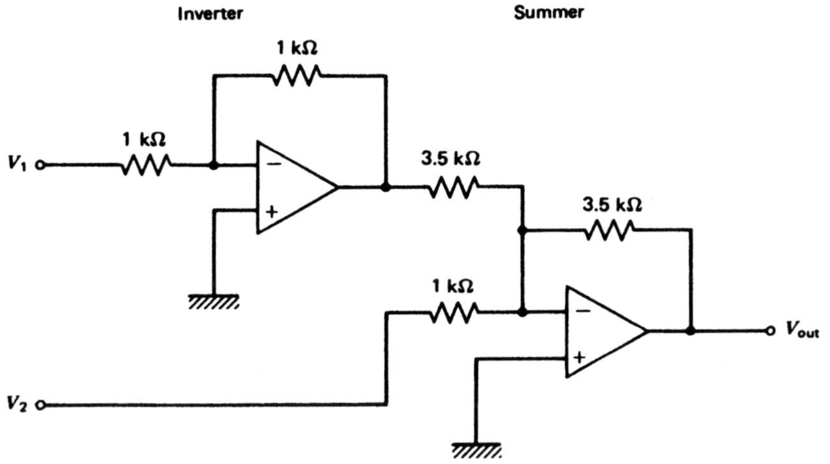

FIGURE 12.3
Circuit for Example 12.1.

or

$$V_{out} = V_1 - 3.5V_2.$$

Thus, whenever $V_1 = 3.5V_2$, the output is zero.

In some cases, the compound control system is known as a cascade, although it differs from true cascade systems, which will be described next.

12.2.2 Cascade Control

The inherent interaction that occurs between two control systems in many applications is sometimes used to provide better overall control. One method of accomplishing this is for the setpoint in one control loop to be determined by the measurement of a different variable for which the interaction exists. A block diagram of such a system is shown in Figure 12.4. Two measurements are taken from the system and each is used in its own control loop. In the outer loop, however, the controller output is the *setpoint* of the inner loop. Thus, if the outer loop controlled variable changes, the error signal that is input to the controller effects a change in setpoint of the inner loop. Even though the measured value of the inner loop has not changed, the inner loop experiences an error signal, and thus new output by virtue of the setpoint change. Cascade control generally provides better control of the outer loop variable than is accomplished through a single-variable system.

An example of a cascade control system not only shows how it works, but suggests how control is improved. Consider the problem of controlling the level of liquid in a tank through regulation of the input flow rate. A single-variable system to accomplish this is

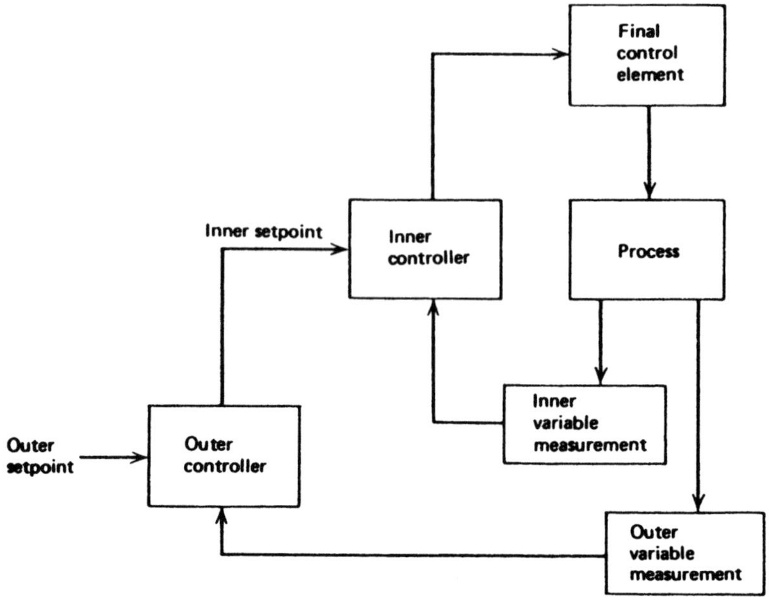

FIGURE 12.4
General features of a cascade process-control system.

shown in Figure 12.5a. A level measurement is used to adjust a flow-control valve as a final control element. The setpoint to the controller establishes the desired level. In this system, upstream load changes cause changes in flow rate that result in level changes. The level change is, however, a second-stage effect here. Consequently, the system cannot respond until the level has actually been changed by the flow rate change.

Figure 12.5b shows the same control problem solved by a cascade system. The flow loop is a single-variable system as described earlier, but the setpoint is determined by a measurement of level. Upstream load changes are never seen in the level of liquid in the tank because the flow-control system regulates such changes before they appear as substantial changes in level.

12.3 MULTIVARIABLE CONTROL SYSTEMS

One could correctly say that any reasonably complex industrial process is multivariable because many variables exist in the process and must be regulated. In general, however, many of these are either noninteracting or the interaction is not a serious problem in maintaining the desired control functions. In such cases, either single-variable controls or cascade loops suffice to effect satisfactory control of the overall process. The use of the word *multivariable* in this section refers to those processes wherein many strongly interacting variables are involved. Such a multivariable system can have such a complex interaction

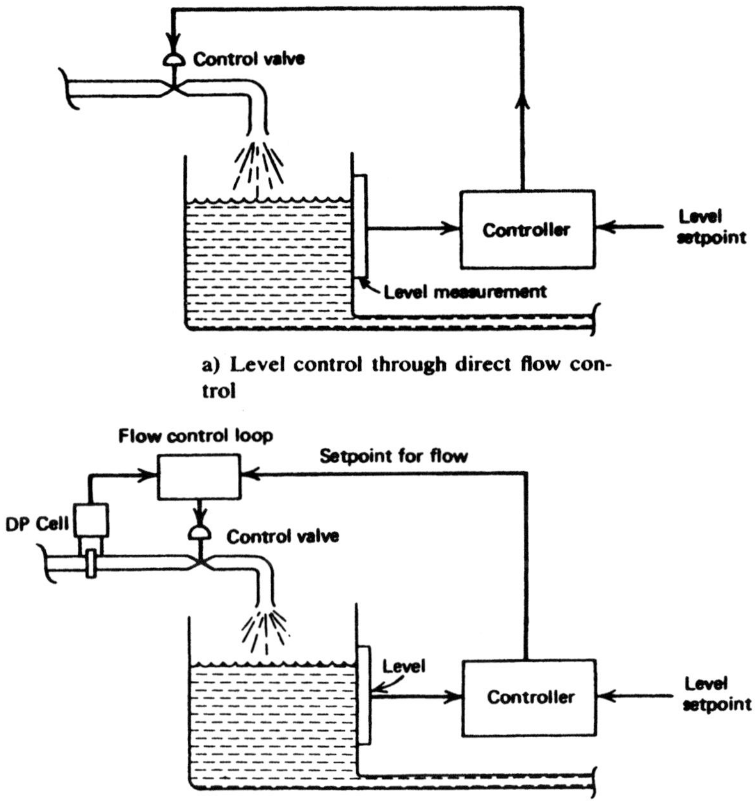

a) Level control through direct flow control

b) Level control through cascade flow control

FIGURE 12.5
Cascade control often provides better control than direct methods.

pattern that the adjustment of a single setpoint causes a profound influence on many other control loops in the process. In some cases, instabilities, cycling, or even runaway result from the indiscriminate adjustment of a few setpoints.

12.3.1 Analog Control

When analog control loops are used in multivariable systems, a carefully prepared instructional set must be provided to the process personnel regarding the procedure for adjustment of setpoints. Generally, such adjustments are carried out in small increments to avoid instabilities that may result from large changes. Let us try to give an idea of the kind of situation under consideration. Suppose we have a reaction vessel in which two reactants are mixed, react, and the product is drawn from the bottom. We are now concerned with controlling the reaction rate. It is also important, however, to keep the reaction temperature and vessel pressure below certain limits, and finally, the level is to be controlled at some nominal value. If the reactions are exothermic—that is, self-sustaining and heat-producing—then the relation among all of these parameters can be critical. If the temperature is low, then an indiscriminate increase in steam-flow setpoint could cause an unstable runaway of the reaction. Perhaps, in this case, the level and reaction flow rates must be altered as the steam-flow rate is increased to maintain control. The necessary steps are often empirically determined or from numerical solutions of complicated control equations.

12.3.2 Supervisory and Direct Digital Control

The computer is ideally suited to the type of control problem presented by the multivariable control system. The computer can make any necessary adjustments of system operating points in an incremental fashion, according to a predetermined sequence, while monitoring process parameters for interactive effects. The problem in such a system is determining the algorithm that the computer must follow to provide the control function of the setpoint-change sequence. In some cases, control equations are used. Usually, in complex interactions, these relations are not analytically known. In some cases, *self-adapting* algorithms are used, causing the computer to sequence through a set of operations and letting the result of one operation determine the next operation. As an example, if the temperature is slightly raised and the pressure rises, then drop the temperature, and so on. The computer can sequence through a myriad of such microadjustments of setpoints looking for the optimum adjustment path.

EXAMPLE 12.2

A process requires adjustment of setpoints to increase production. A particular sequence must be followed to provide the increase. SP1, SP2, and SP3 are the setpoints, P and PCR are the pressure and a critical pressure, respectively, and T and TCR are the temperature and critical temperature, respectively. Develop a flowchart that increases the setpoints as follows:

1. Increase SP1 by 1%.
2. Wait 10 s, test for pressure compared to critical.
3. If the pressure is less than critical, then
   a. decrease SP2 by $\frac{1}{2}$%
   b. increase SP3 by $\frac{3}{4}$%
   c. wait for T $<$ TCR
   d. increase SP2 by 1%
   e. go to step 1
4. If the pressure is above critical,
   a. decrease SP1 by $\frac{1}{2}$%
   b. decrease SP2 by $\frac{1}{4}$%
   c. go to step 2

Solution

To implement this with either DDC or supervisory systems requires a flowchart from which the program instructions are developed. Such a flowchart is shown in Figure 12.6.

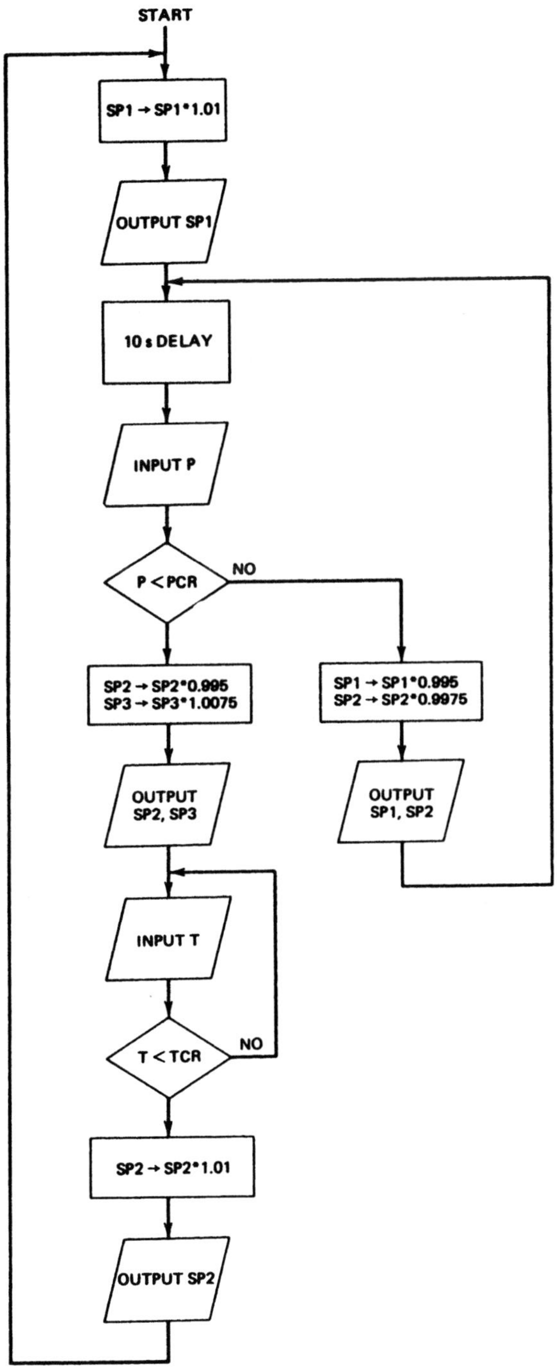

FIGURE 12.6
Flowchart for an interactive control problem.

The computer sequences through these steps in an optimum fashion according to the stipulations of the problem. If the sequence were performed manually, several operators would be required to monitor equipment and make necessary adjustments.

12.4 CONTROL SYSTEM QUALITY

When a manufacturing concept is to be implemented, the ultimate goal is to develop a product that satisfies present "design" criteria. If the product is crackers, they should have a certain color, flavor, salinity, size, and so forth. If it is gasoline, it should possess certain octane, antioxidants, viscosity, and so forth. The manufacturing process depends on the operation of a set of process-control loops to impart the desired characteristics to the product. The ultimate gauge of control system quality is, therefore, whether such control provides a product that is within specifications. The operation cannot even get started unless that level of quality is provided. We assume that in any practical situation it *is* satisfied. There are, however, other levels of quality that represent a deeper study of the process-control system, and these are the subject of the present discussion. In brief, given that a control system *can* provide a product that meets specifications, we ask how well it does perform this job, what variation in parameters exists, what percentage of rejected product occurs, and so on. To answer these questions, we must first describe measures of quality in a control system and then analyze how the loop characteristics affect these measures. It is most important to note that no absolute answers exist. What is considered good control in one manufacturing process may be unsatisfactory in another. Some of the general criteria that are applied are discussed later.

12.4.1 Definition of Quality

Let us consider for a moment one control loop in a manufacturing process. We need a control loop because the variable under control is dynamic, changing under many influences, and therefore needs regulatory operations. It is impossible for a control loop to regulate this variable to *exactly* the setpoint. Let's face it, the variable must *change* before the loop can generate a corrective action to oppose the change. Then, in considering quality, we must accept from the outset that perfect control is impossible and that some inevitable deviations of variables from the optimum values will occur. In fact, then, a definition of quality is concerned with these deviations and their interpretation in terms of the ultimate product. A process-control technologist is not necessarily in a position to evaluate how deviations of a loop variable from the setpoint actually (1) affect the specific properties of the final product, (2) cause it to be rejected, or (3) may affect the cost efficiency of the manufacturing process. To provide a communication link between the process experts and the product experts, a set of measures or criteria has been devised. These criteria provide a common language so that a product can be evaluated in terms of the dynamic characteristics of a specific loop and serve, therefore, as our measure of quality. To understand the measure, we must first define quality in terms of the process-control loop.

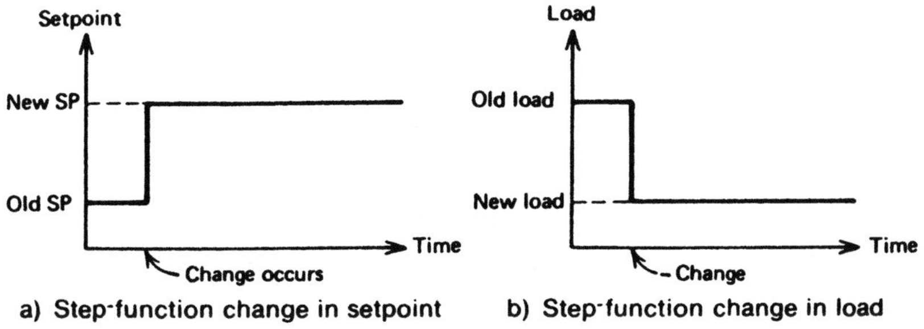

FIGURE 12.7
Loop disturbances can occur from intentional setpoint changes or changes in process load.

Loop Disturbance The process-control system is supposed to provide regulation so that disturbances in the system will cause minimum deviation of the controlled variable from the setpoint value. The quality of the control system is defined by the degree to which the deviations that result from the disturbances are minimized.

There are three basic types of disturbances that can occur in a process-control system:

1. Transient
2. Setpoint changes
3. Load changes

A transient disturbance results from a temporary change of some parameter in the system that affects the controlled variable. It is impractical to use transient disturbances to define control quality because the nature of a transient cannot be well defined. That is, the transient can vary in duration, peak amplitude, and shape. For definition of quality, we need a more regular type of disturbance. To be specific, the two other disturbance conditions are used, both of which introduce a step-function change into the loop. A step-function change in *setpoint,* as shown in Figure 12.7a, is an instantaneous change of the loop setpoint from an old value to a new value. The second possible disturbance is a step-function change in *process load,* as suggested in Figure 12.7b, that also occurs instantaneously in time. The load change can come from the sudden permanent change of any of the process parameters that constitute the process load.

To provide measures of quality, we evaluate how the system responds to either of these sudden changes.

Optimum Control The most universal definition of quality in a control system is that the system provides *optimum control*—that is, the best control possible. If anything is modified in the system, then the deviation of controlled variable from a load or setpoint change is always worse. Thus, the overall settings of the system are at an optimum. This does not mean that the control is "perfect" or even very good; it simply means that it is the best it can be.

Optimum control, and therefore control quality, can be defined in terms of the three effects resulting from a load or setpoint change:

1. Stability
2. Minimum deviation
3. Minimum duration

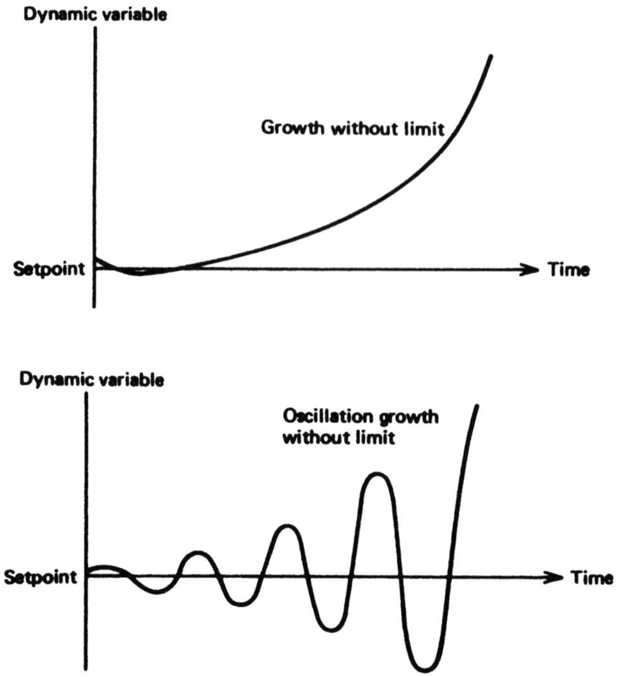

FIGURE 12.8
Instability in a process-control loop refers to the uncontrolled growth of the controlled variable.

Stability The most basic characteristic in defining process-loop quality is that it provides stable regulation of the dynamic variable. Stable regulation means that the dynamic variable does not grow without limit. In Figure 12.8, two types of unstable responses are shown. In one case, a disturbance causes the dynamic variable simply to increase without limit. In the other, the variable begins to execute growing oscillations, where the amplitude is increasing without limit. In both cases, some nonlinear breakdown (such as an explosion or other malfunction) eventually terminates the increase. We will consider stability in the next section, but for a measure of quality we assume stable operation has been achieved. A controlled variable in some process may be stable and still cyclic. This is the case, for example, in two-position control where the controlled variable oscillates between two limits under nominal load conditions. A change in load may change the period of oscillations, but the amplitude swing remains essentially the same; hence, the variable is under stable control.

Minimum Deviation If a process-control loop has been adjusted to regulate a variable at some setpoint value, then an obvious definition of quality is the extent to which a disturbance causes a deviation from that setpoint. Where a disturbance is a change in setpoint, this can be considered as any overshoot or undershoot of the variable in achieving the new setpoint. In general, we want to minimize any deviation of the dynamic variable from the setpoint value.

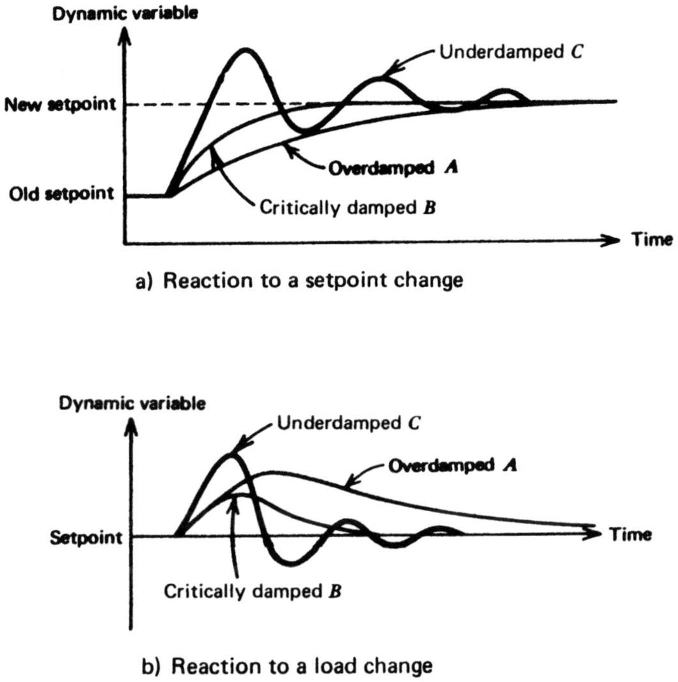

FIGURE 12.9
Tuning determines the reaction of a controlled variable to changes.

Minimum Duration If a disturbance occurs, we can conclude that some deviation will occur. Another definition of quality is the length of time before the controlled variable regains or adopts the setpoint value, or at least falls within the acceptable limits of that value.

Thus, the quality of a process-control loop is defined through an evaluation of *stability, minimum deviation,* and *minimum duration* following a disturbance of the dynamic variable.

12.4.2 Measure of Quality

In general, it is not enough simply to state that we will design or adjust the process-control loop to provide for stable, minimum-deviation, minimum-duration operation. For example, the achievement of minimum deviation may result in a less-than-minimum duration (it usually occurs). The final product also may favor a less-than-absolute minimum adjustment to provide for a faster production rate at an acceptable degradation in product specification. To accommodate such circumstances, we distinguish several measures of quality by which we can convey the degree to which we have approached the ideals.

Assume stable operation has been achieved. There are three possible responses to a disturbance that a dynamic variable in a process-control loop can execute. The specific response depends on the controller gains and lags in the process. Referring to Figure 12.9b for a load change and Figure 12.9a for a setpoint change, we have the following definitions.

Overdamped The loop is *overdamped* in case *A* of Figure 12.9; the deviation approaches the setpoint value smoothly (following a disturbance) with no oscillations. The duration is not a minimum in such a case. For that matter, the deviation itself usually is not a minimum either. Such a response is safe, however, in ensuring that no instabilities occur and that certain maximum deviations never occur.

Critically Damped Careful adjustment of the process-control loop brings about curve *B* of Figure 12.9. In this case, the duration is a minimum for a noncycling response. This is the optimum response for a condition where no overshoot or undershoot is desired in a setpoint change, or no cycling, in general, is desired.

Underdamped The natural result of further adjustments of the process-control loops is cyclic response, where the deviation executes a number of oscillations about the setpoint. This is shown in curve *C* of Figure 12.9. It is possible that this response gives minimum deviation and minimum duration in some cases. If the cycling can be tolerated, then such a response is preferred.

Two specialized measures of control are used when none of these conditions serves to define the measure of control desired in a process.

Quarter Amplitude When a process-control loop has a damped cyclic response to a disturbance, a criterion is sometimes used that is neither minimum deviation nor minimum duration. This measure of quality is found by adjusting the loop until the deviation from a disturbance is such that each deviation peak is down to one-quarter of the preceding peak, as shown in Figure 12.10. In this case, the actual magnitude of the devi-

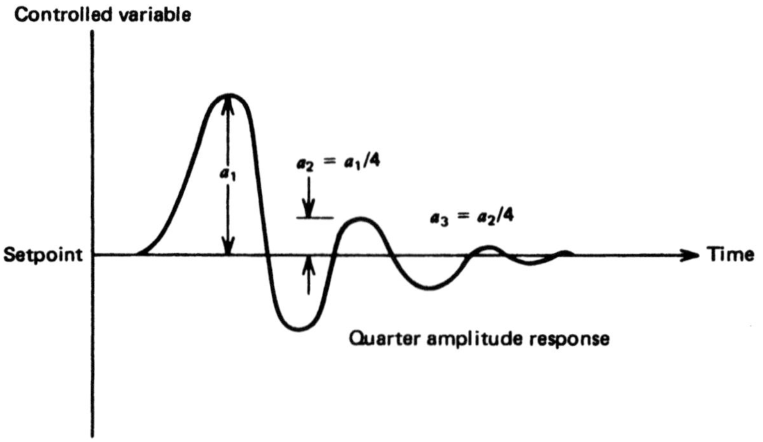

FIGURE 12.10
In one type of cyclic response, the system is adjusted to make each peak down to one-quarter of the previous peak.

ation is not included in the measure, nor is the time between each peak. In this sense, neither duration nor magnitude of the deviation is directly involved in a quarter-amplitude criterion.

EXAMPLE 12.3 Suppose the deviation following a step-function disturbance is a 4.7% error in the controlled variable. If a quarter-amplitude criterion is used to evaluate the response, find the error of the second and third peak.

Solution

The amplitude of each peak must be one-quarter of the previous peak. In this case, the second peak is 4.7%/4, or 1.18%. The third peak is an error of 1.18%/4, or 0.30%.

Minimum Area In cases of cyclic or underdamped response, the most critical element is sometimes a combination of duration and deviation, which must be minimized. Thus, if minimum deviation occurs at one loop setting and minimum duration at another, then neither is optimum. One type of optimum measure of quality in these cases is to minimize the net area of the deviation as a function of time. In Figure 12.11, this is shown as the sum of the shaded areas. Analytically, this can be expressed as

$$A = \int |r - b| dt \tag{12.1}$$

where A = area of deviation
b = measured value
r = setpoint value

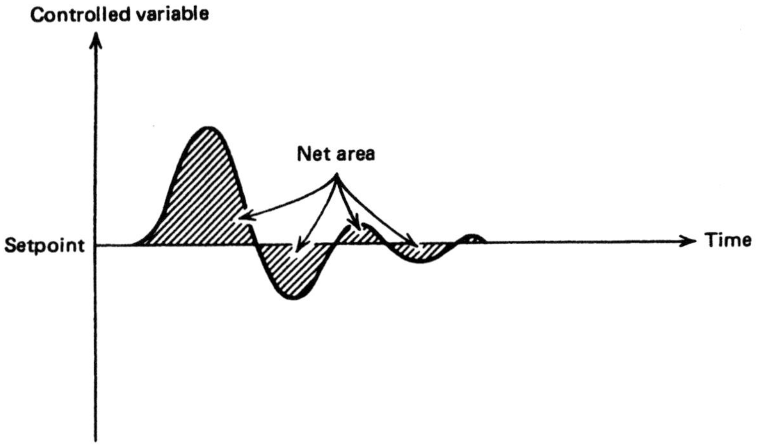

FIGURE 12.11
The minimum-area tuning response characteristics.

Another way of representing this is to use the full-scale percent error (see Section 9.3.1) to write

$$A_p = \int |e_p| dt \tag{12.2}$$

where e_p = full-scale error in %
A_p = area as %-s

By adopting this criterion of loop response, we are keeping the extent and duration of deviation, and thus product rejections, to a minimum.

EXAMPLE 12.4 Given the two response curves of Figure 12.12 for deviation versus time, following an initial disturbance, find the response preferred using the minimum-area criterion.

Solution
We find the area by application of

$$A_p = \int |e_p| dt$$

In these cases, we find the areas geometrically by finding the net area of each curve. For curve 1, we have

$$A_1 = 8.5\%\text{-min}$$

and for curve 2, we get

$$A_2 = 8\%\text{-min}$$

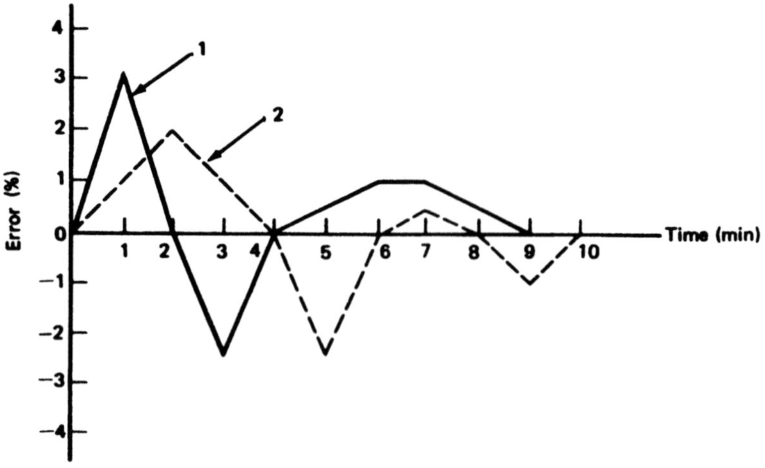

FIGURE 12.12
Error versus time for Example 12.4.

Thus, it is clear that curve 2 is preferred under the minimum-area test.

12.5 STABILITY

Earlier in this chapter, we assumed that the loop response is stable as a prerequisite for use in a practical control application. In fact, a great deal of effort is expended in the design and development of process-control loops to achieve this stability. Although a detailed treatment of stability theory in process control is beyond the intent of this book, it is of value to discuss some of the important considerations of such studies.

12.5.1 Transfer Function Frequency Dependence

The static transfer function of an element in a process-control loop tells how the output is determined from the input when the input is constant in time. The dynamic transfer function of an element tells how the output is determined from the input when the input varies in time. For the study of stability, we are interested in the particular time variation that is sinusoidal (i.e., the dynamic transfer function when the input is oscillating at some frequency, f).

Consider some element block as shown in Figure 12.13 with a transfer function, $T(\omega)$, and where the input is a sinusoidal given by

$$r = a \sin(\omega t)$$

The frequency has been expressed in terms of the angular frequency, $\omega = 2\pi f$, measured in radians/second. There are only two things that can happen, in a linear study at least: The amplitude can change, and there can be a phase shift. Thus, the output can be described as

$$c = b \sin(\omega t + \phi)$$

The ratio of the amplitudes is called the gain; gain $= b/a$, and the phase shift is called the phase lag, ϕ. In general, both the gain and amount of phase lag of an element vary with frequency. The gain decreases, and the phase lag becomes larger.

The whole issue of stability is tied up with the frequency variation of gain and phase of all elements in a control loop.

Source of Instability To see how a process-control loop can cause instability, consider the open-loop block diagram of Figure 12.14, for which an oscillating transient disturbance has been imposed at r. Notice that the feedback line has been broken at the error detector, so no actual feedback occurs. Each element of the loop has a gain and phase lag, including the process itself. The net gain is the product of all gains, and the net phase lag is the sum of all phase lags.

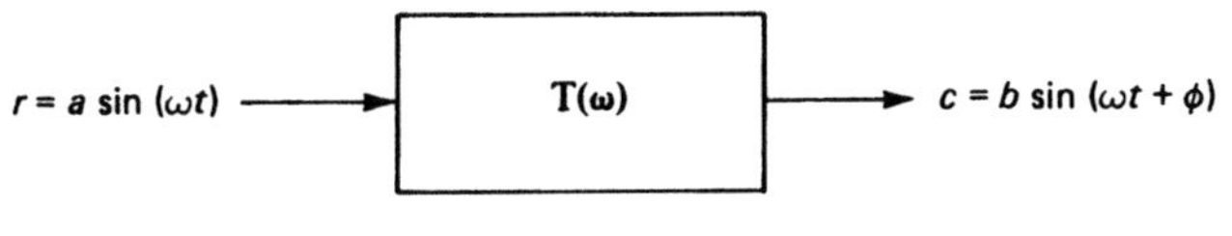

FIGURE 12.13
A transfer function changes the amplitude and phase of a sinusoidal input.

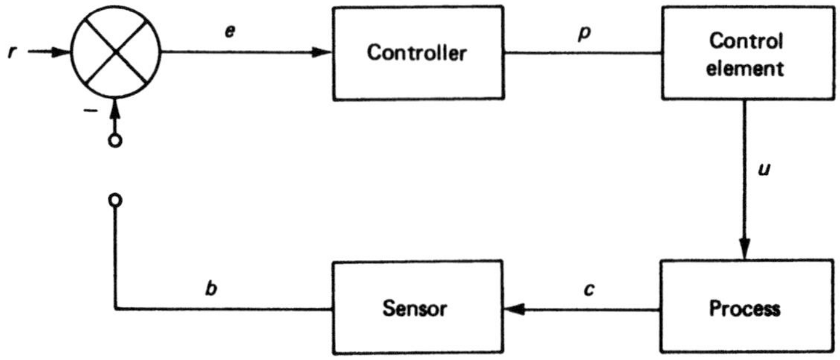

FIGURE 12.14
This control loop has been opened so that the effect of a disturbance on the loop can be traced.

In a perfect world, the feedback, *b,* would be an exact replica of the disturbance at every frequency. This would mean a loop gain of unity and no phase shift. Then the $-180°$ phase shift (lag) of the error detector would subtract the feedback from the disturbance, and it would be cancelled. In reality, there are gain variations and extra phase shifts, and both vary with the disturbance frequency.

If the gain for the disturbance frequency is greater than one and the system phase lag small, the disturbance is cancelled within a few cycles of oscillation. However, as the phase lag becomes greater with increasing frequency, the effectiveness of the feedback is reduced, and the oscillation will persist longer. This is the cyclic response to a transient discussed elsewhere in this book.

Similarly, as the gain becomes smaller with frequency, the effectiveness of the feedback to cancel error is reduced. But the control system is still working and stable.

Consider, however, the particular case of a frequency where the phase lag of the system reaches $-180°$ while the gain remains unity or greater. When combined with the error-detector phase shift, the net shift around the loop will be 360°, and so the feedback will be summed instead of subtracted. If the gain at that frequency is just unity, then the disturbance will persist forever, with constant amplitude. If the gain is greater than unity, the disturbance will grow in amplitude. This is the instability caused by the control system.

Instability Illustration Instability is caused by a condition where for some frequency the transfer function is such that feedback to the error summer actually increases the error because of the gain and phase shift. Now, if there is any frequency for which this condition exists, then oscillations will always start and grow at that frequency. To illustrate this, we have assumed that a small transient oscillation is introduced from external sources as a disturbance in *r.* Assume that at this frequency the gain is greater than 1—say, 2—and the phase shift is $-180°$—that is, a lag of 180°. Let us study the result, frozen instant by instant. The summer algebraically subtracts the feedback from the input, giving the error signal e_1 of the next instant in Figure 12.15a. In the next instant, Figure 12.15b, the transient is gone but the feedback, b_2, is the original e_1 amplified by 2 and phase-shifted by 180°. This passes through the negative summing point and becomes e_2. Figure 12.15c shows e_2

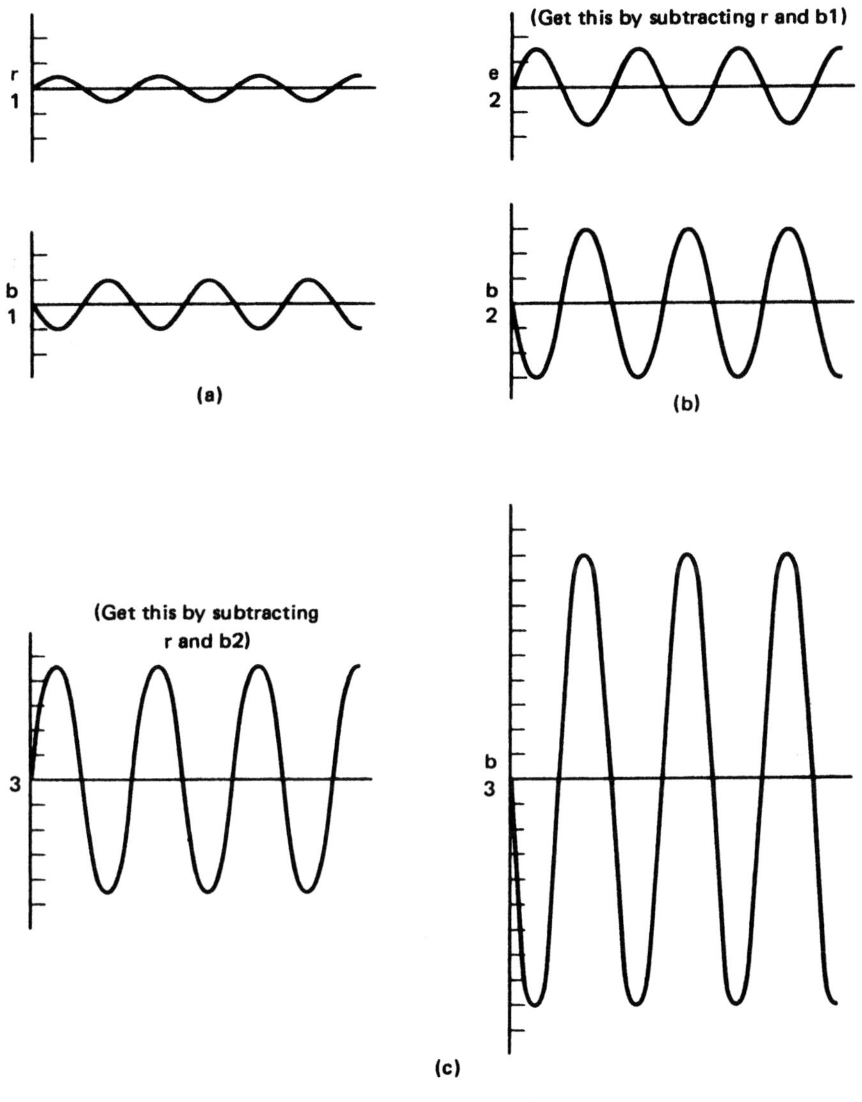

FIGURE 12.15
This figure suggests how an initial transient oscillation can grow under the appropriate feedback conditions.

amplified by 2 and phase-shifted by 180°, again becoming b_3 and then e_3. Thus, you can see that the error is actually growing! The control system is forcing the oscillating error to increase, instant by instant. Of course, in actuality, it happens smoothly, and the output would look something like Figure 12.8, for oscillating growth. If there is *any* frequency where such conditions for growth exist, the system is unstable, and something like random noise will eventually set the system into growing oscillation. When a process-control installation is designed, one has the objective of regulating the controlled variable *without* instability in the loop. Stability can be ensured by designing the controller gains so that oscillation growth is never favorable, according to certain criteria.

12.5.2 Stability Criteria

We can derive stability criteria by determining just what conditions of system gain and phase lag can lead to an enhancement of error. We are led to two conclusions:

1. The gain must be greater than 1.
2. The phase shift must be $-180°$ (lag).

Thus, if there is any frequency for which the gain is greater than one and the phase is $-180°$, the system is unstable. From this argument, we develop two ways of specifying when a system is *stable.* These rules are as follows:

- **Rule 1** A system is stable if the phase lag is more positive than $-180°$ at the frequency for which the gain is unity (one).
- **Rule 2** A system is stable if the gain is less than one (unity) at the frequency for which the phase lag is $-180°$.

The application of these rules to an actual process requires evaluation of the gain and phase shift of the system for all frequencies to see if Rules 1 and 2 are satisfied. This is easier to do if a plot of gain and phase versus frequency is used.

Bode Plot A particular type of graph, called a *Bode plot* or diagram, normally is used to plot the gain and phase of the control system versus frequency. In this case, the frequency is plotted along the abscissa (horizontal) on a log scale and expressed commonly as rad/s. Remember, if you want to know the actual frequency in Hz, you just divide the rad/s by 2π. Use of a log scale permits a large range of frequency to be displayed. The ordinate actually consists of two parts: Gain is plotted on a log scale in one part, and phase is plotted linearly as degrees in another part. Figure 12.16 shows the Bode plot with a gain/phase plotted for an example. We have labeled the gain as *open-loop gain,* since we do not include the effect of actually feeding that signal back into the system.

EXAMPLE 12.5

Determine whether the system of Figure 12.16 is stable according to the rules given earlier.

Solution

By Rule 1, we find the frequency for which the gain is unity (1). The system is stable if the phase lag is less than 180°. Unity gain occurs at 0.5 rad/s; this is $0.5/2\pi = 0.08$ Hz, or about five cycles per minute. Such frequencies are not uncommon in many industrial processes. For this frequency, the phase is about $-140°$, so the lag is less than 180°, and the system is **stable** under Rule 1. For Rule 2, we find the frequency for which the phase lag is 180° and check the gain, which should be less than one for stability. The phase lag is 180° at 0.8 rad/s, for which the gain is 0.5. Thus, the system is **stable** by Rule 2.

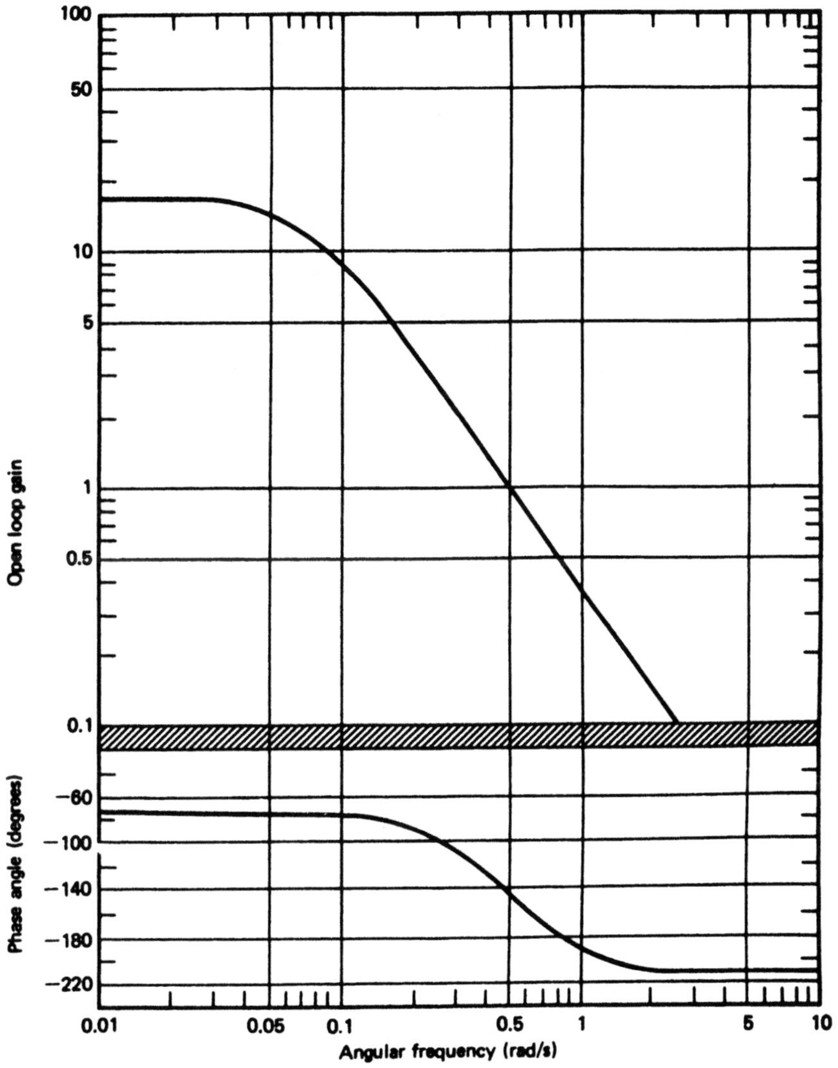

FIGURE 12.16
A Bode plot shows the open-loop gain and phase versus frequency of an applied sinusoid.

In some cases, a modified form of the Bode plot given in Figure 12.16 is used, in which the gain is plotted in dB. This gain is then plotted on a linear scale where

$$\text{gain (dB)} = 20\log_{10}(|T(\omega)|) \tag{12.3}$$

Thus, a gain of 100 gives 40 dB, unity gain is 0 dB, and a gain of less than one gives a negative dB, as 0.1 gives −20 dB. The stability rules would be revised to read 0 dB instead of unity gain.

If the transfer functions of several elements are known, the transfer functions of a series of such cascaded elements are found by a *product of the magnitudes* and a *sum of the phases,* or

$$|T(\omega)| = |T_1(\omega)| \cdot |T_2(\omega)| \cdot |T_3(\omega)| \quad \textbf{(12.4)}$$

$$\angle T(\omega) = \angle T_1(\omega) + \angle T_2(\omega) + \angle T_3(\omega) \quad \textbf{(12.5)}$$

where $T(\omega)$ = system transfer function
T_1, T_2, T_3 = element transfer function

On log-log scales, the product of Equation (12.4) becomes addition. The transfer function for an entire process-control loop therefore involves summation of the transfer functions of all the elements of the loop, *including the process itself.*

$$\log|T(\omega)| = \log|T_1(\omega)| + \log|T_2(\omega)| + \log|T_3(\omega)| \quad \textbf{(12.6)}$$

In most cases, an exact solution cannot be found using Bode plots because the *process transfer function* is too complicated or unknown. Stability can still be ensured through adjustment of the proportional gain, derivative time, and integral time. The final adjustment of these quantities *tunes* the system for stable operation.

12.6 PROCESS LOOP TUNING

The last aspect of process-control technology we consider refers to the actual start-up and adjustment of a process-control loop. We have seen how the various settings of the controller can have a profound effect on loop performance. Now, the most natural question is how to select these settings. There are, in fact, many methods for determination of the optimum mode gains, depending on the nature and complexity of the process. We consider here three common tuning methods to give a basic idea of how optimum adjustments are found. Two of the methods given are semi-empirical in that they depend on measurements made on the system to determine factors used in the adjustment formulas. The last is more analytical, in that it is based on a known transfer function of both the process and loop.

12.6.1 Open-Loop Transient Response Method

This method of finding controller settings was developed by Ziegler and Nichols and is sometimes referred to as a *process-reaction* method. The basic approach is to open the process-control loop so that no control action (feedback) occurs. This usually is done by disconnecting the controller output from the final control element. All of the process parameters are held at their nominal values. This method can be used only for systems with self-regulation.

At some time, a transient disturbance is introduced by a small, manual change of the controlling variable using the final control elements. This change should be as small as practical for making necessary measurements. The controlled variable is measured (recorded) versus time at the instant of and following the disturbance.

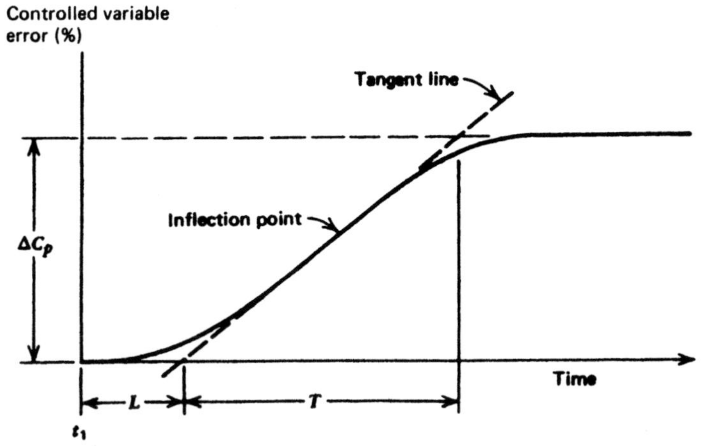

FIGURE 12.17
Process-reaction graph for loop tuning.

A typical open-loop controller response is shown in Figure 12.17, where the disturbance is applied at t_1. We have expressed the deviation as a percent of range as usual, and we assume the final control element disturbance also is expressed as a percentage change. A tangent line, shown as a dashed line, is drawn at the *inflection point* of the curve. The inflection point is defined as that point on the curve where the slope stops increasing and begins to decrease. Where the tangent line crosses the origin, we get

$$L = \text{lag time in minutes} \tag{12.7}$$

as the time from disturbance application to the tangent line intersection as shown in Figure 12.17. Also from the graph we get T, the process reaction time, and

$$N = \frac{\Delta C_p}{T} \tag{12.8}$$

where N = reaction rate in %/min
ΔC_p = variable change in %
T = process reaction time in minutes

The quantities defined by Equations (12.7) and (12.8) are used with the controlling variable change ΔP to find the controller settings. The following paragraphs give the stable control definitions for the various modes as developed by Ziegler and Nichols and corrections developed by Cohen and Coon (when the quarter-amplitude response criterion is indicated). In the latter case, a log ratio is used, defined by

$$R = \frac{NL}{\Delta C_p} \tag{12.9}$$

where R = log ratio (unitless)

Proportional Mode For the proportional mode, the proportional gain setting, K_P, is found from

$$K_P = \frac{\Delta P}{NL} \tag{12.10}$$

Corrections to the value of K_P are sometimes used to obtain the quarter-amplitude criterion of response. One given by Cohen and Coon is shown bracketed as

$$K_P = \frac{\Delta P}{NL}\left[1 + \frac{1}{3}\frac{NL}{\Delta C_P}\right] \tag{12.11}$$

Proportional-Integral Mode When the controller mode is proportional-integral, the appropriate settings for proportional gain and integration time are

$$\left.\begin{aligned} K_p &= 0.9\frac{\Delta P}{NL} \\ 1/K_I = T_I &= 3.33L \end{aligned}\right\} \tag{12.12}$$

If the quarter-amplitude criterion is used, the gain is

$$\left.\begin{aligned} K_P &= \frac{\Delta P}{NL}\left[0.9 + \frac{1}{12}R\right] \\ T_I &= \left[\frac{30 + 3R}{9 + 20R}\right]L \end{aligned}\right\} \tag{12.13}$$

Three-Mode For the three-mode controller, we find the appropriate proportional gain, integration time, and derivative time from

$$\left.\begin{aligned} K_P &= 1.2\frac{\Delta P}{NL} \\ T_I &= 2L \\ T_D &= 0.5L \end{aligned}\right\} \tag{12.14}$$

If the quarter-amplitude criterion is used, these equations are corrected by

$$\left.\begin{aligned} K_P &= \frac{\Delta P}{NL}[1.33 + R/4] \\ T_I &= \left[\frac{32 + 6R}{13 + 8R}\right]L \\ T_D &= L\left(\frac{4}{11 + 2R}\right) \end{aligned}\right\} \tag{12.15}$$

EXAMPLE 12.6 A transient disturbance test is run on a process loop. The results of a 9% controlling variable change give a process-reaction graph as shown in Figure 12.18. Find settings for three-mode action.

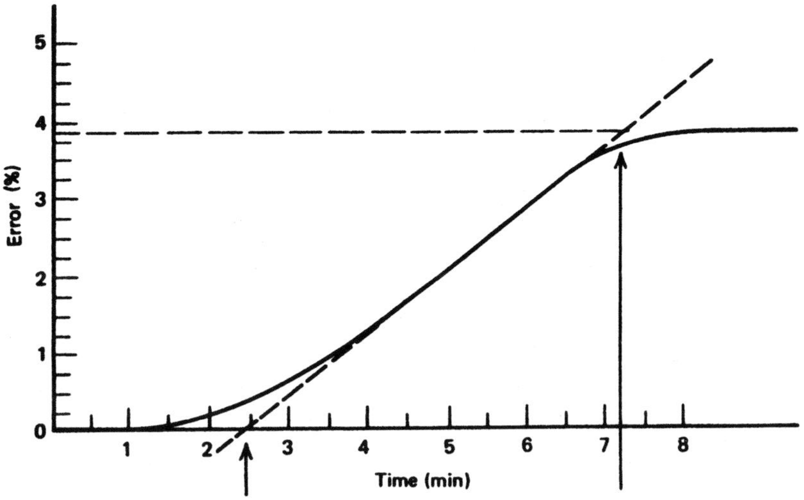

FIGURE 12.18
Process-reaction graph for Example 12.7.

Solution

By drawing the inflection point tangent on the graph, we find a lag $L = 2.4\,\text{min}$ and a process reaction time of 4.8 min. The reaction rate is

$$N = \frac{\Delta C_P}{T} = \frac{3.9\%}{4.8\,\text{min}} = \mathbf{0.8125\%/min}$$

The controller settings are found from Equation (12.14):

$$K_P = 1.2\frac{\Delta P}{NL} = 1.2\frac{9\%}{(0.8125)(2.4)}, \qquad \text{yielding } \boldsymbol{K_P} = \mathbf{5.54}$$

or a proportional band of

$$\frac{100}{K_P} = \frac{100}{5.54} = \mathbf{18\%}$$
$$T_I = 2L = (2)(2.4\,\text{min}) = \mathbf{4.8\ min}$$
$$T_D = 0.5L = (0.5)(2.4\,\text{min}) = \mathbf{1.2\ min}$$

EXAMPLE 12.7 For Example 12.6, find the three-mode settings for a quarter-amplitude response.

Solution

We use Equation (12.15) together with the lag ratio:

$$R = \frac{NL}{\Delta C_P} = \frac{(0.8125)(2.4)}{3.9}$$
$$R = \mathbf{0.50}$$

Then we get

$$K_P = \frac{\Delta P}{NL}\left(1.33 + \frac{R}{4}\right)$$

$$K_P = \frac{9}{(0.8125)(2.4)}\left(\frac{1.33 + 0.50}{4}\right)$$

$$K_P = \mathbf{6.72}$$

or a **14.9%** proportional band.

$$T_I = L\left(\frac{32 + 6R}{13 + 8R}\right)$$

$$T_I = 2.4\left(\frac{32 + 6(0.5)}{13 + 8(0.5)}\right)$$

$$T_I = \mathbf{4.94\ min}$$

$$T_D = L\left(\frac{4}{11 + 2R}\right)$$

$$T_D = (2.4)\frac{4}{11 + 2(0.5)}$$

$$T_D = \mathbf{0.80\ min}$$

12.6.2 Ziegler-Nichols Method

Ziegler and Nichols also developed another method of controller setting assignment that has come to be associated with their name. This technique, also called the *ultimate cycle method,* is based on adjusting a closed loop until steady oscillations occur. Controller settings are then based on the conditions that generate the cycling.

The particular method is accomplished through the following steps:

1. Reduce any integral and derivative actions to their minimum effect.
2. Gradually begin to increase the proportional gain while providing periodic small disturbances to the process. (These are to "jar" the system into oscillations.)
3. Note the critical gain, K_c, at which the dynamic variable just begins to exhibit steady cycling—that is, oscillations about the setpoint.
4. Note the critical period, T_c, of these oscillations measured in minutes.

This method can be used for systems without self-regulation. Now, from the critical gain and period, the settings of the controller are assigned as follows:

Proportional For the proportional mode alone, the proportional gain is

$$K_P = 0.5K_c \tag{12.16}$$

A modification of this relation is often used when the quarter-amplitude criterion is applied. In this case, the gain is simply adjusted until the dynamic response pattern to a step change in setpoint obeys the quarter-amplitude criterion. This also results in some gain less than K_c.

Proportional-Integral If proportional-integral action is used in the process-control loop, then the settings are determined from

$$\left.\begin{aligned} K_P &= 0.45K_c \\ T_I &= T_c/1.2 \end{aligned}\right\} \tag{12.17}$$

In case the quarter-amplitude criterion is desired, we make

$$T_I = T_c \tag{12.18}$$

and adjust the gain for that necessary to obtain the quarter-amplitude response.

Three-Mode The three-mode controller requires proportional gain, integral time, and derivative time. These are determined for nominal response as

$$\left.\begin{aligned} K_P &= 0.6K_c \\ T_I &= T_c/2.0 \\ T_D &= T_c/8 \end{aligned}\right\} \tag{12.19}$$

For adjustment to give quarter-amplitude response, we set

$$\left.\begin{aligned} T_I &= T_c/1.5 \\ T_D &= T_c/6 \end{aligned}\right\} \tag{12.20}$$

and adjust the proportional gain for satisfaction of the quarter-amplitude response.

EXAMPLE 12.8 In an application of the Ziegler-Nichols method, a process begins oscillation with a 30% proportional band in an 11.5-min period. Find the nominal three-mode controller settings.

Solution

First, a 30% proportional band means the gain (from Section 9.5.1) is

$$K_c = \frac{100}{PB} = \frac{100}{30} = 3.33$$

Then, from Equation (12.19), we find settings of

$$\begin{aligned} K_P &= 0.6K_c = (0.6)(3.33) \\ K_P &= 2 \\ T_I &= T_c/2 = 11.5/2 \\ T_I &= 5.75\,\text{min} \\ T_D &= T_c/8 = 11.5/8 \\ T_D &= 1.44\,\text{min} \end{aligned}$$

12.6.3 Frequency Response Methods

The frequency response method of process-controller tuning involves use of Bode plots for the process and control loops. The method is based on an application of the Bode plot stability criteria given in Section 12.5.2 and the effects that proportional gain, integral time, and derivative time have on the Bode plot.

Gain and Phase Margin The stability criteria given in Section 12.5.2 represent *limits* of stability. For example, if the gain is slightly less than 1 when the phase lag is 180°, the system is stable. But if the gain is slightly greater than 1 at 180°, the system is unstable. It would be well to design a system with a margin of safety from such limits to allow for variation in components and other unknown factors. This consideration leads to the *revised stability criteria,* or more properly, a margin of safety provided to each condition. The exact terminology is in terms of a *gain margin* and *phase margin* from the limiting values quoted. Although no standards exist, a common condition is

1. If the phase lag is less than 140° at the unity gain frequency, the system is stable. This, then, is a 40° *phase margin* from the limiting value of 180°.
2. If the gain is 5 dB below unity (or a gain of about 0.56) when the phase lag is 180°, the system is stable. This is a 5-dB *gain margin.*

EXAMPLE 12.9 Determine whether the system with a Bode plot of Figure 12.19 satisfies both the gain and phase margin conditions.

Solution

We first examine the phase at unity gain. Unity gain occurs at an angular frequency of 0.15 rad/s, and the phase is −120°. Thus, the first stability condition is satisfied with a 60° phase margin.

Now a 180° phase occurs at 0.3 rad/s angular frequency where the gain is 0.7, which is too high. Thus, the 0.56 or less gain margin is *not* satisfied, and the controller gain will have to be reduced slightly.

Tuning The operations of tuning using the frequency response method involve adjustments of the controller parameters until the stability is proved by the appropriate phase and gain margins. If the process and control elements' transfer functions are known, the correct settings can be determined analytically. If not, the Bode plot can be determined experimentally by opening the loop and providing a variable-frequency disturbance of the controlling variable. If measurements of phase and gain are made, then the Bode plot can be constructed. From this, the proper settings can be determined. The significance of the unity gain crossover in frequency is that the system can correct any disturbances of frequency less than that of the unity gain frequency. Any disturbance of higher frequency has little effect on the control system.

The tuning operation is based on the fact that the gains of each mode have a particular effect on the system Bode plot. By adjusting these gains, we can alter the Bode plot until it satisfies the gain and phase margins of safety for a stable system. Remember that on the Bode plot, the gains appear as products, and phases of the modes simply add algebraically.

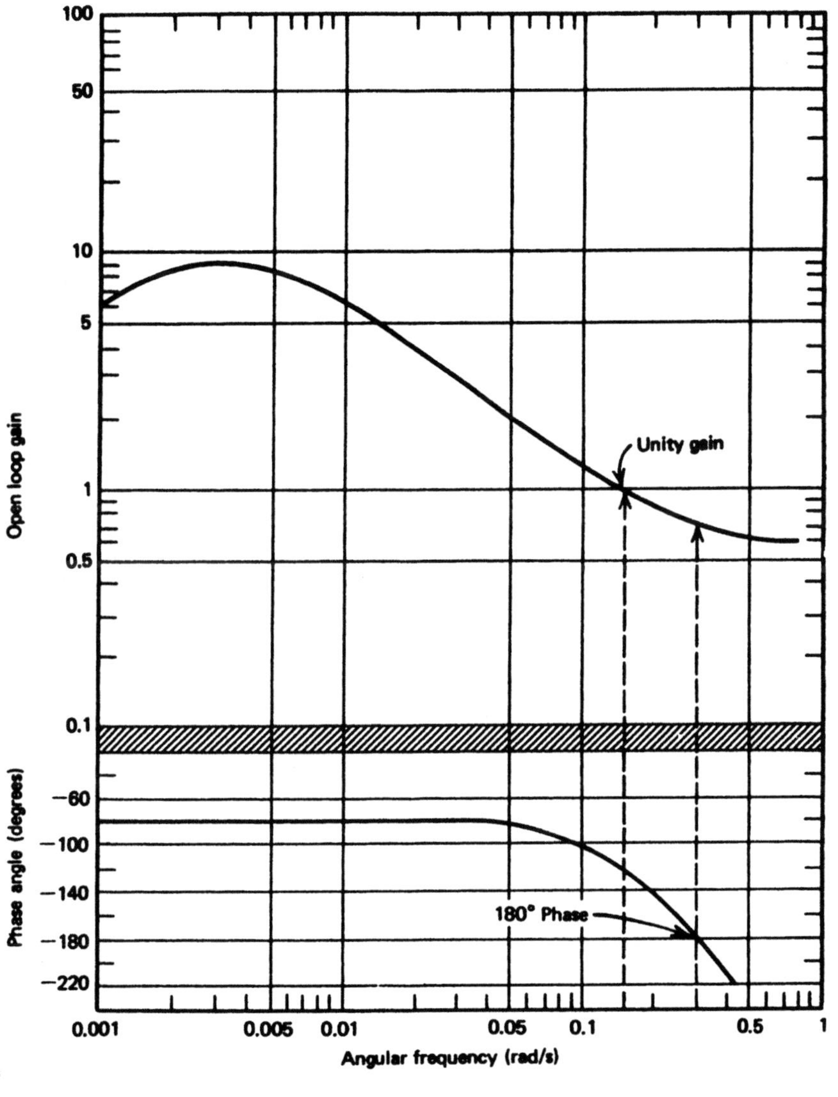

FIGURE 12.19
Bode plot for Example 12.9.

Proportional Action This mode of the controller simply multiplies the gain curve by a constant, independent of frequency, and has no phase effect at all. Thus, if a system gain curve is found for a proportional gain of 2 and we increase this to 4, then the entire gain curve is multiplied by 2. This term can be used to move the intact gain curve up (increased gain) or down (decreased gain). Generally, moving the curve up extends the range of frequency that the system can control, provided the stability margins are maintained.

Integral Action In its pure form, the integral mode contributes

$$\text{integral gain} = \frac{K_I}{\omega}$$

$$\text{integral phase} = -90°(\text{lag})$$

Thus, we see that the integral mode contributes more gain at lower frequency and less gain at higher frequency because of the inverse dependence on the frequency, ω. The phase shift is made more lagging by 90°. The integral mode gain, K_I, can increase the gain at lower frequencies. In practical form, the integral mode has some lower frequency (called the *breakpoint*) at which its effects cease, and the gain curve becomes flat while the phase lag goes to zero. This is shown in Figure 12.20, where $K_I = 1$ has been assumed.

Derivative Action Because of problems of instability at higher frequencies, the derivative mode is almost never used in its pure form. If it were, the gain and phase would be

$$\text{derivative gain} = K_D\omega$$

$$\text{derivative phase} = +90° \text{ lead}$$

Because the gain increases with frequency, it must be limited at some upper frequency. In Figure 12.20, this mode action is shown (for a mode gain $K_D = 10$) with the characteristic upper-limit frequency at which the gain becomes constant, and the phase shift is zero. In general, then, this mode can be used to increase the gain at higher frequency, but more important, it can be used to drive the phase away from the −180° (lag) shift where instability can begin.

EXAMPLE 12.10 Find the fractional decrease in proportional gain that would make the system of Example 12.9 have a 5-dB gain margin.

Solution

The system failed to satisfy the 5-dB gain margin at −180°, because the gain was 0.7 instead of 0.56 or less. Thus, if the proportional gain is reduced by a factor of 0.7/0.56, or about 0.8, then the gain margin will be satisfied. The phase margin will not be affected.

EXAMPLE 12.11 Consider the Bode plot of Figure 12.21 for some hypothetical process and controller with a proportional gain of 10. Show that the system is unstable, and find a proportional gain for satisfying gain and phase margins.

Solution

Evaluation of the curve shows that the phase lag at unity gain is 170°, which, although stable, does not satisfy the required 40° phase margin. The gain at 180° phase lag is approximately 0.8, which, although less than one, does not satisfy the stipulated gain margin. Notice that if the entire gain curve is reduced or moved down by a constant amount, as indicated, with no effect on the phase lag, then the stability criteria can be satisfied. This is

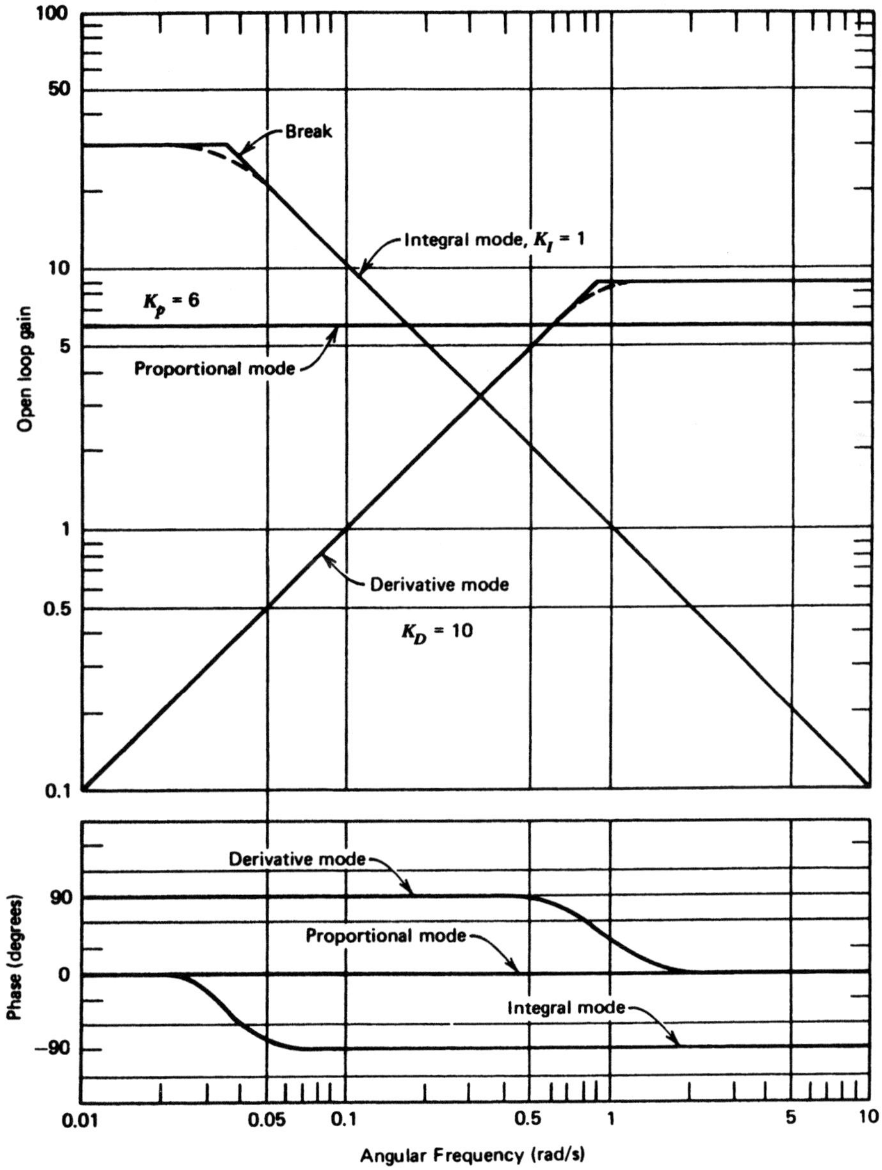

FIGURE 12.20
Bode plots showing proportional, integral, and derivative mode contributions.

done by changing the *proportional gain,* because this affects gain only. In particular, the gain must be derated by a factor of 1/2 (note difference at 5 rad/s) to achieve both margins. Thus, the new proportional gain should be 10/2, or 5.

In general, the derivative, integral, and proportional gains are adjusted until the system satisfies the stability criteria and the specified unity gain frequency.

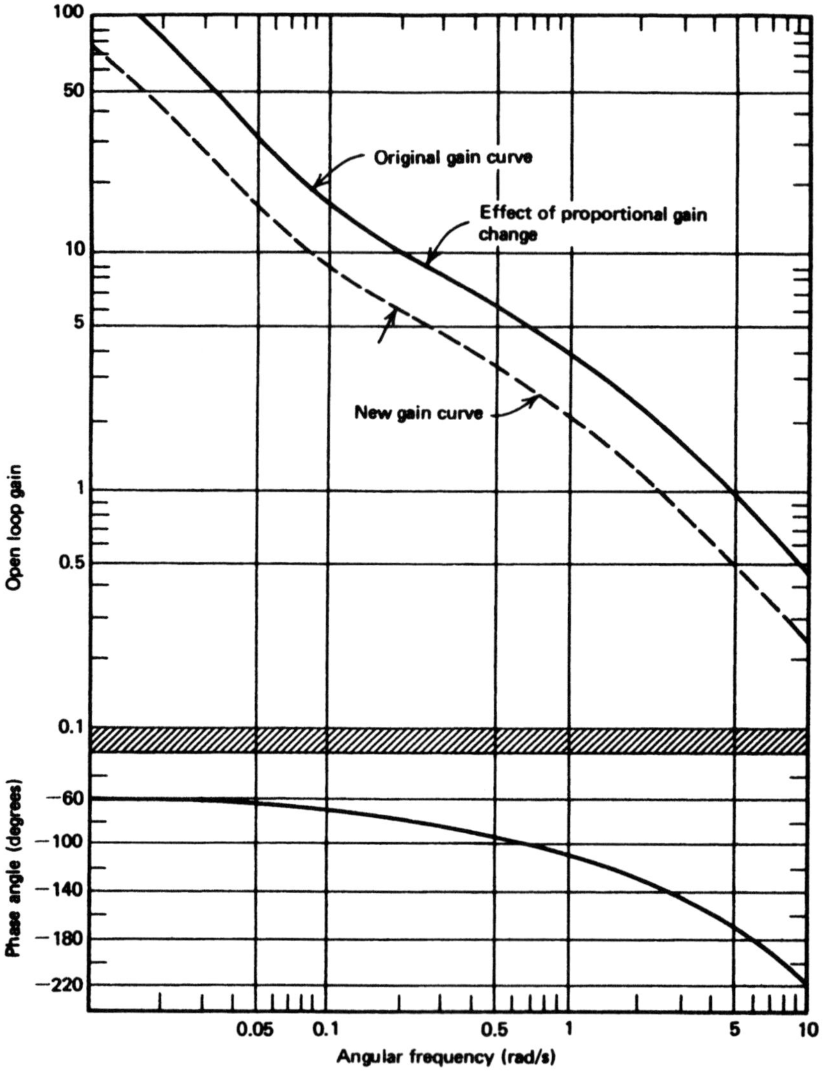

FIGURE 12.21
Bode plots for Example 12.11. The proportional mode does not affect phase.

SUMMARY

An overview of the operational characteristics of the entire process-control loop includes the following points:

1. Although the single-variable control loop is widely used in process control, we often must account for interactive effects between variables.
2. In many cases, a cascade control loop is used, where the setpoint of one loop is determined by the controller output of another loop.

3. Multivariable control systems are used where strong interactions between variables exist. In many ways, supervisory or DDC techniques are best suited for these control problems.
4. The quality of control system response is defined by characterizing the response of the system to a disturbance. The stability, deviation magnitude, and duration are important to this quality.
5. The *minimum area* and *quarter-amplitude* criteria are often used to measure quality, together with damping characteristics.
6. The stability of a process-control loop can be studied with a Bode plot, consisting of gain and phase lag versus frequency.
7. The tuning of a process-control loop consists of finding the optimum settings of controller gains for good control.
8. The open-loop transient method of tuning relies on a semi-empirical technique using the reaction of the open loop to a transient disturbance.
9. The Ziegler-Nichols method finds the conditions that generate steady oscillation and derates the settings for optimization.
10. The frequency-response method relies on the stability conditions determined from a Bode plot. Adjustments of gains provide for the phase and gain margin and frequency bandwidth.

PROBLEMS

Section 12.2

12.1 A compound control system specifies that the ratio of pressure in Pa to temperature in K be held to 0.39.

a. Diagram a control loop to provide this control.

b. Identify the function of each element in the loop necessary to accomplish the control.

12.2 Suppose that for Problem 12.1 the temperature is available at 2.75 mV/K and the pressure at 11.5 mV/Pa. Design signal conditioning that outputs zero when the ratio is correct. *Hint:* First convert each to scales of 1 mV/K and 1 mV/Pa.

12.3 Draw a diagram of a cascade control system that has an inner loop of temperature input to a flow system and an outer loop of viscosity to determine the flow controller setpoint.

Section 12.3

12.4 Draw a block diagram of an operational flow diagram to show how Problem 12.1 is implemented by DDC.

12.5 Draw a block diagram and flowchart to show how the cascade system of Problem 12.3 can be implemented by DDC.

Section 12.4

12.6 Assume we want an electronic means of finding the area of the error-time graph following a disturbance.

a. Show that the circuit of Figure 12.22 does this.

b. Find the values of R_3 and R_4 so that the output is the area of Equation 12.2 in V·s.

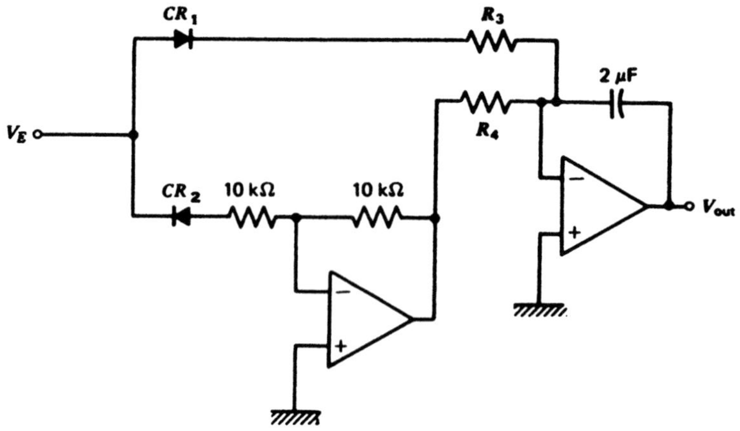

FIGURE 12.22
Circuit for Problem 12.6.

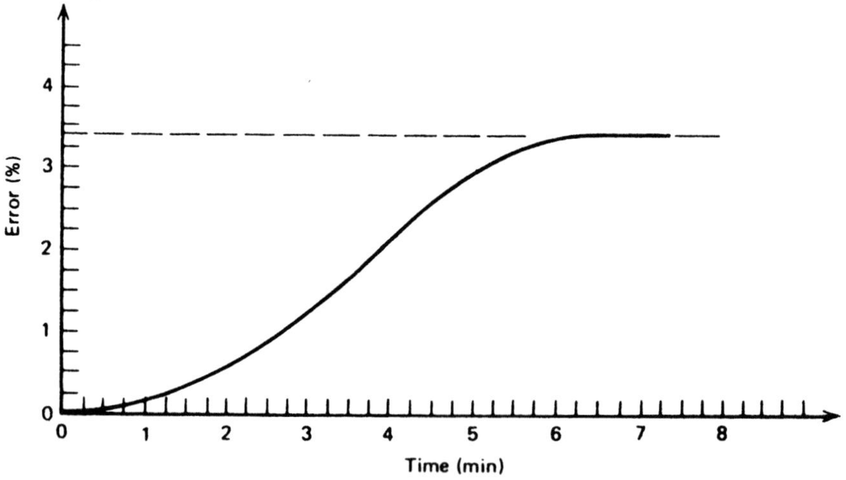

FIGURE 12.23
Process-reaction graph for Problem 12.8.

Section 12.5

12.7 A process and control system has components with transfer functions at 10 rad/s of $20\angle 0°$, $1.4 \angle -90°$, $0.05 \angle 90°$, and $3.2 \angle 85°$. Find the total system-transfer function.

Section 12.6

12.8 An open-loop transient test provides the process-reaction graph given in Figure 12.23 for a 7.5% disturbance.

a. Find the standard proportional-integral gain settings.
b. Find the three-mode quarter-amplitude settings.

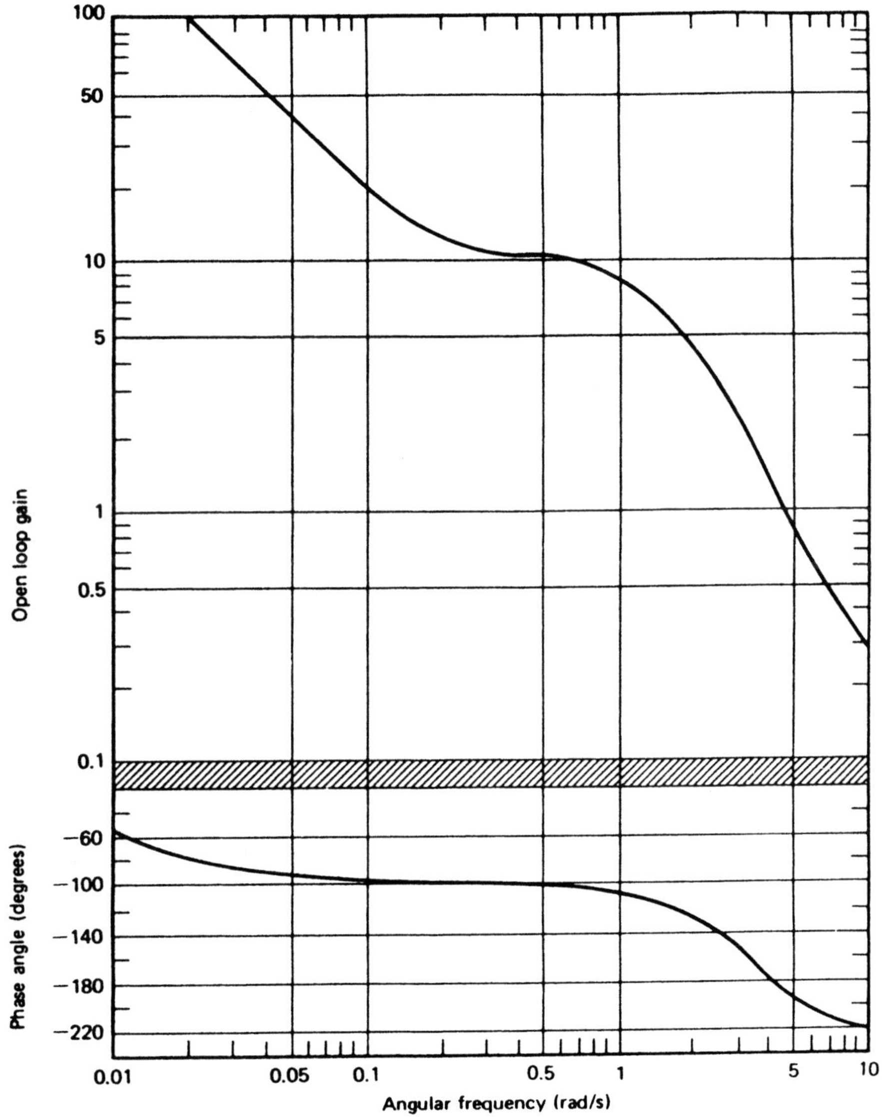

FIGURE 12.24
Bode plot for Problems 12.10 and 12.11.

12.9 In the Ziegler-Nichols method, the critical gain was found to be 4.2, and the critical period was 2.21 min. Find the standard settings for (**a**) proportional-mode control, (**b**) PI control, and (**c**) PID control.

12.10 Specify the gain and phase margins for Figure 12.24. Is the system stable?

12.11 If the nominal proportional gain for the process in Figure 12.24 was 11.5, determine the gain that will just provide a gain margin of at least 0.56 and a phase margin of at least 40°.

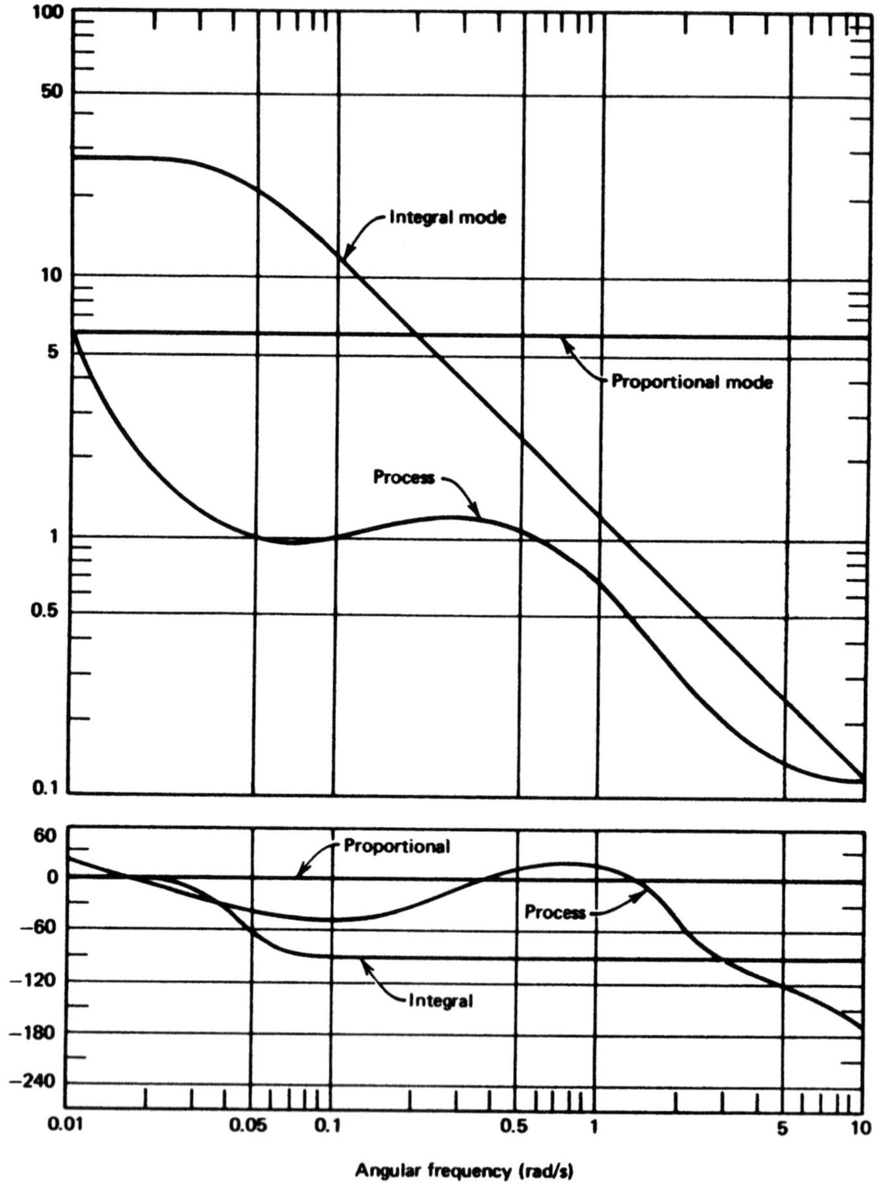

FIGURE 12.25
Bode plot for Problem 12.13.

12.12 Draw the resultant Bode plot of the combination of the proportional, integral, and derivative modes shown in Figure 12.20.

12.13 Figure 12.25 shows the Bode plots of a process, proportional mode, and integral mode of an associated controller. Find the composite transfer function Bode plot and evaluate the stability.

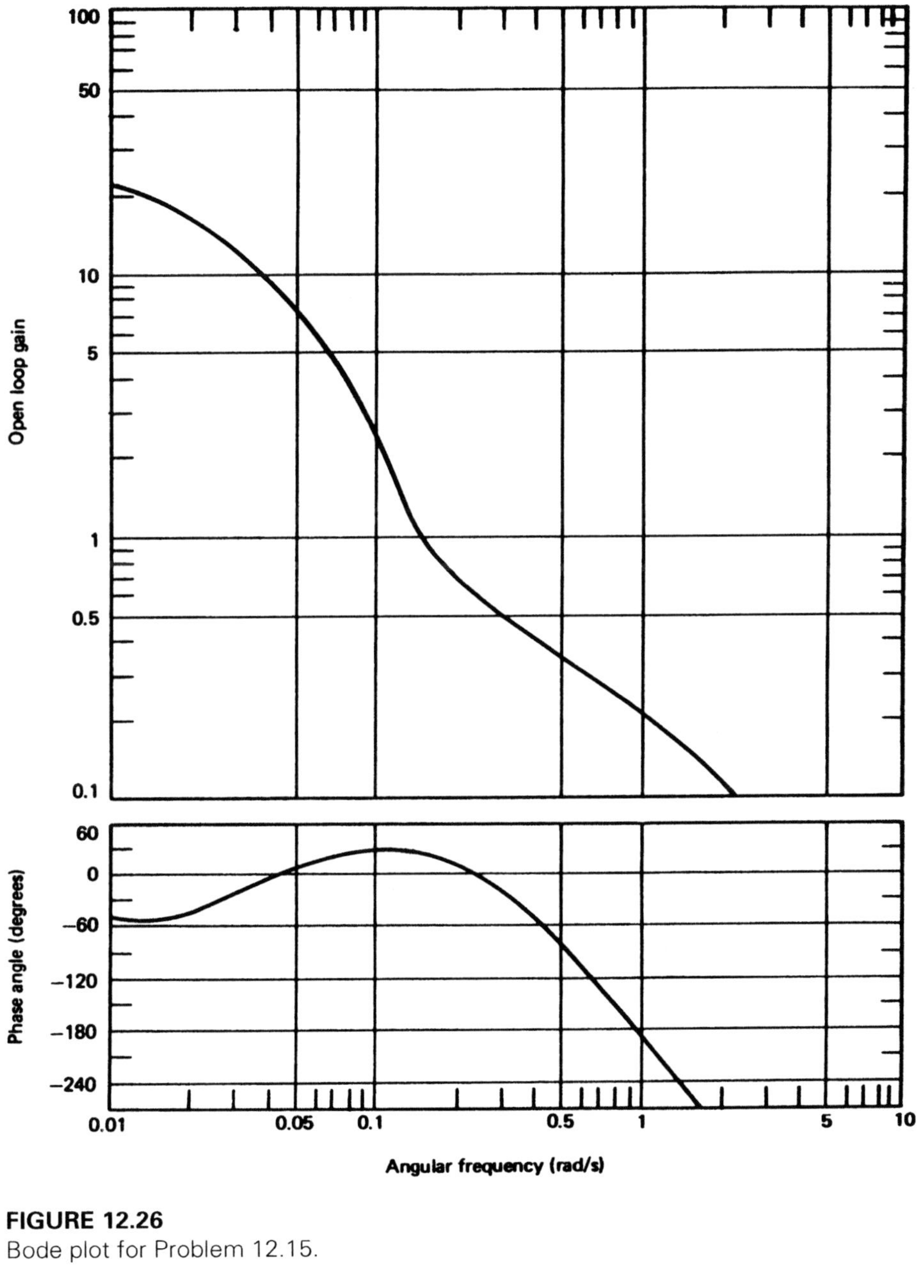

FIGURE 12.26
Bode plot for Problem 12.15.

12.14 Show how adding derivative action to the system of Problem 12.13 would change the composite Bode plot. Discuss the effect on stability of adding this action.

12.15 The composite Bode plot of an open-loop system is shown in Figure 12.26. This is for proportional action only with $K_P = 1$. How much proportional gain will drive the system into instability? The critical gain of the ultimate cycle method is

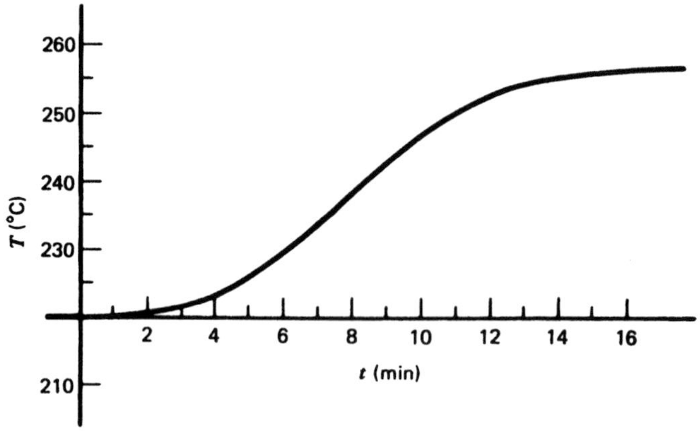

FIGURE 12.27
Process-reaction graph for Problem 12.16.

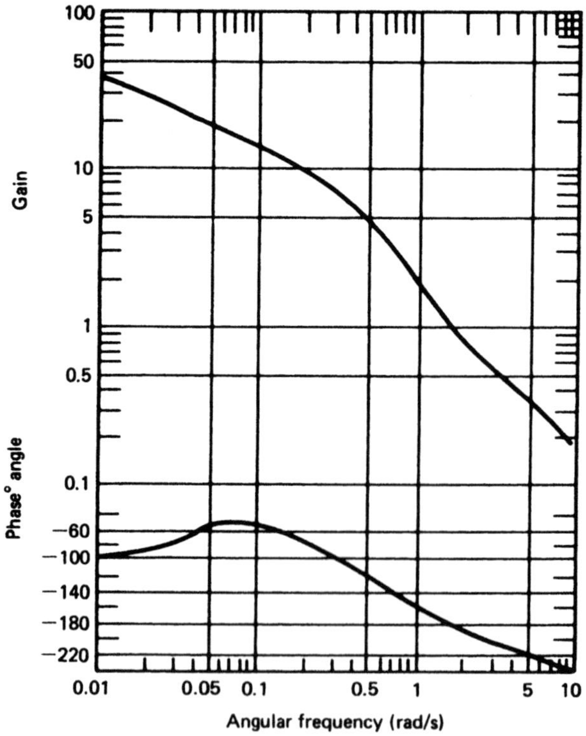

FIGURE 12.28
Bode plot for Problem 12.17.

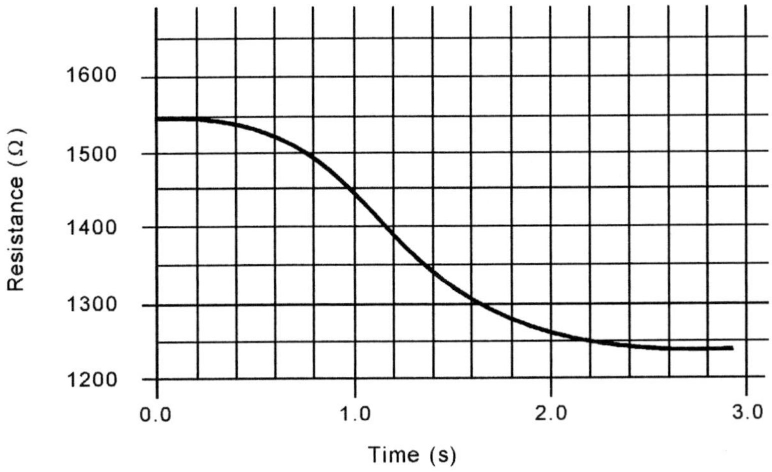

FIGURE 12.29
Process-reaction graph for Problem S12.1.

that which will drive the system to the margin of instability. What would be the proper gain setting for ultimate cycle tuning? What phase and gain margin does this provide?

12.16 A process is to be tuned by transient response to quarter amplitude using a PID controller. The system control temperature varies from 140° to 330°C with a 220°C setpoint. The output is a heater control voltage ranging from 0 to 24 V. The test is started with the system having a 14-V output. The output is increased suddenly to 16.5 V. The resulting temperature graph is shown in Figure 12.27. Find the proper PID gain.

12.17 A process Bode plot is shown in Figure 12.28. Prove that the system does not satisfy a 5-dB gain margin. The plot represents a proportional gain of 5.0. What gain will make the system satisfy a 5-dB gain margin? What is the phase margin?

SUPPLEMENTARY PROBLEMS

S12.1 The light-regulation system shown in Figure 10.27 is tuned by conducting an open-loop transient response test. For this test, the source voltage was suddenly increased by 1.5 V, and the variation in sensor resistance with time was recorded. Figure 12.29 shows the average sensor resistance in time. Determine the proper gains for a PI controller operation for both Ziegler-Nichols and quarter-amplitude response.

A.1

Units

A.1.1 SI UNITS

The system of units that is gaining acceptance throughout the world is called the Systéme International d'Unités (SI). This system is maintained by the Conférence Genérale des Poids et Measures. The basic set of these units is given in Section 1.6.1. The description of other physical quantities can be provided in terms of *derived* units that are expressible in terms of the basic set. Table A.1.1 shows the more common derived units and their descriptors.

TABLE A.1.1
Derived units

Quantity Symbol	Quantity Definition	Unit Name	Unit Symbol	Unit Definition
f	Frequency	Hertz	Hz	s^{-1}
W	Energy	Joule	J	$kg \cdot m^2/s^2$
F	Force	Newton	N	$kg \cdot m/s^2$
R	Resistance	Ohm	Ω	$kg \cdot m^2/(s^3 \cdot A^2)$
V	Voltage	Volt	V	$A \cdot \Omega$
p	Pressure	Pascal	Pa	N/m^2
ω	Angular frequency	Radians per second	rad/s	rad/s
E	Illuminance	Lux	lx	lm/m^2
Q	Charge	Coulomb	C	$A \cdot s$
L	Inductance	Henry	H	$kg \cdot m^2/(s^2 \cdot A^2)$
C	Capacity	Farad	F	$s^4 \cdot A^2/kg\text{-}m^2$
G	Conductance	Siemen	S	Ω^{-1}
Φ	Luminous flux	Lumen	lm	cd/sr
	Luminous efficacy	Lumen per watt	lm/W	lm/W
P	Power	Watt	W	J/s

A.1.2 OTHER UNITS

Many systems of units are employed in the world, although a strong effort is being made to adopt the SI units universally. Two widely employed unit systems are the English and the centimeter-gram-seconds.

English System

The English system of units originated in Great Britain. Although Great Britain converted to the metric, or SI, system of units, the United States continues to use the English system for everyday social and commercial activity. In fact, some now call this system U.S. units or U.S. customary units. A gradual conversion to the metric system is occurring in the manufacturing, science, and engineering disciplines. Table A.1.2 gives the basic units of this system and the factors for converting to SI.

Mass Note In the English system, the proper unit of mass is the *slug,* which is defined as that mass which will accelerate at 1 ft/s^2 when operated on by a force of 1 lb. This unit is not in common use, however. Instead, the English system uses the *weight* of an object to infer its mass. Weight is the force due to gravity and is therefore measured by the unit pound (lb) in the English system. This causes some confusion when converting units of mass between the English system and SI, which uses the proper mass unit, the kilogram (kg). The mass conversion factor of 0.454 kg = 1 pound (mass) takes into

TABLE A.1.2
English system of units

Quantity	English Unit	Variations	Conversion to SI
Length	Foot (ft)	inch (in.): 12 in. = 1 ft yard (yd): 1 yd = 3 ft mile (mi): 1 mi = 5280 ft	1 in. = 0.02540 m (exact)
Area	$Foot^2(ft^2)$	$inch^2$ ($in.^2$): 144 $in.^2$ = 1 ft^2 acre: 1 acre = 21,789 ft^2	10.76 ft^2 = 1 m^2
Volume	$Foot^3(ft^3)$	$inch^3$ ($in.^3$): 1728 $in.^3$ = 1 ft^3 gallon (gal): 1 gal = 0.134 ft^3 quart (qt): 4 qt = 1 gal pint (pt): 2 pt = 1 qt	35.31 ft^3 = 1 m^3 1 gal = 3.785 liters 1 gal = 0.003875 m^3
Force	Pound (lb)	ounce (oz): 16 oz = 1 lb	1 lb = 4.448 N
Mass	Slug (not used) Pound (lb) (see "Mass Note")		1 slug = 14.59 kg 1 lb (mass) = 0.454 kg
Energy	Foot-pound (ft-lb)		1 ft-lb = 1.356 J
Pressure	$Pound/inch^2$ (psi)	atmosphere (atm): 1 atm = 14.7 psi	1 psi = 6895 Pa
Power	Horsepower (hp)		1 hp = 746 W

TABLE A.1.3
CGS system of units

Quantity	CGS Unit	SI Unit Conversion
Length	centimeter (cm)	100 cm = 1 m
Mass	gram (gm)	1000 g = 1 kg
Time	second (s)	1 s = 1 s
Force	dyne	10^5 dyne = 1 N
Energy	erg	10^7 erg = 1 J
Pressure	dyne/cm^2	10 dyne/cm^2 = 1 Pa

account the nominal acceleration due to gravity to convert the English force unit to the SI mass unit:

$$1 \text{ lb (mass)} = 4.448 \text{ N}/9.8 \text{ m/s}^2 \approx 0.454 \text{ kg}$$

Centimeter-Gram-Second (CGS) System

The CGS system of units is often employed in scientific work and publications reporting the results of scientific research. The system is now defined as a subset of the SI according to the definitions of Table A.1.3.

A.1.3 STANDARD PREFIXES

The SI decimal multiple and submultiple designations are shown in Table A.1.4.

TABLE A.1.4
Prefixes

Multiple	SI Prefix	Symbol
10^{12}	tera	T
10^9	giga	G
10^6	mega	M
10^3	kilo	k
10^2	hecto	h
10	deka	da
10^{-1}	deci	d
10^{-2}	centi	c
10^{-3}	milli	m
10^{-6}	micro	μ
10^{-9}	nano	n
10^{-12}	pico	p
10^{-15}	femto	f
10^{-18}	atto	a

Note: It has become common practice for computers to express the number of bytes of memory using prefixes such as kilo-, mega-, and giga-. It is important to realize that these prefixes do not have the same meaning as the metric prefixes just defined. The reason is that they are based upon the binary, or base 2, counting system. Therefore, the following definitions should be used when dealing with computer prefixes:

$$\begin{aligned} 1 \text{ kilo (K)} &= 2^{10} = 1024 \\ 1 \text{ mega (M)} &= 2^{20} = (1024)^2 = 1{,}048{,}576 \\ 1 \text{ giga (G)} &= 2^{30} = (1024)^3 = 1{,}073{,}741{,}824 \end{aligned}$$

Therefore, a 4-GB hard disk actually has $(4)(1{,}073{,}741{,}824) = 4{,}294{,}967{,}296$ bytes.

A.2

Digital Review

A.2.1 NUMBER SYSTEMS

The base 10, or decimal, number system is normally employed in the analog description of information. In digital applications, a base 2, or binary, counting system is often used with the octal, or base 8, system for writing numbers. The hexadecimal system, or base 16, also has become popular. It is important that an individual involved in digital applications be prepared to translate numbers between these systems. This can be accomplished by the following operations and equations.

Binary

A mathematical relationship can be constructed between a base 10 number and its binary equivalent using the equation

$$N_{10} = a_n 2^n + a_{n-1}2^{n-1} + \cdots + a_3 2^3 + a_2 2^2 + a_1 2^1 + a_0 2^0 \qquad \textbf{(A.2.1)}$$

where

$$\begin{aligned} N_{10} &= \text{base 10 number} \\ a_n a_{n-1} \cdots a_3 a_2 a_1 a_0 &= \text{base 2 number} \\ n + 1 &= \text{number of digits in the binary number} \end{aligned}$$

The reverse procedure for finding the binary equivalent of a decimal number can be represented in several ways. The easiest way is to use a set of conversion tables. Lacking that, a process of successive division by 2 is employed. In any division, the remainder will always be zero or one-half, and this determines the value of that binary digit. The first division yields the value of the least significant bit (LSB), which is the a_0 term, and the last remainder gives the most significant bit (MSB), which is a_n.

EXAMPLE A.2.1 Find the binary equivalent of 259_{10}.

Solution

We use successive division by 2 as follows:

$$\begin{aligned} 259/2 &= 129 + 1/2 & a_0 &= \mathbf{1} \\ 129/2 &= 64 + 1/2 & a_1 &= \mathbf{1} \\ 64/2 &= 32 + 0/2 & a_2 &= \mathbf{0} \\ 32/2 &= 16 + 0/2 & a_3 &= \mathbf{0} \\ 16/2 &= 8 + 0/2 & a_4 &= \mathbf{0} \\ 8/2 &= 4 + 0/2 & a_5 &= \mathbf{0} \\ 4/2 &= 2 + 0/2 & a_6 &= \mathbf{0} \\ 2/2 &= 1 + 0/2 & a_7 &= \mathbf{0} \\ 1/2 &= 0 + 1/2 & a_8 &= \mathbf{1} \end{aligned}$$

Thus, the answer is that 259_{10} is $\mathbf{100000011_2}$.

Octal

Because it is frequently employed in digital work, we need to consider a number base formed of eight counting states. A principal reason for using this numbering system is that it becomes both easy and convenient to represent binary numbers in the octal system. This can be done by grouping the binary numbers in groups of three digits from the right and noting that these three digits represent the full range of octal numbers. Thus, $\mathbf{000_2}$ corresponds to 0_8, and $\mathbf{111_2}$ corresponds to 7_8. The small subscript 2 and 8 denote base 2 and base 8, respectively. Thus, 101111_2 can be grouped as $\mathbf{101_2}$ followed by $\mathbf{111_2}$ and written 57_8 in the octal system. The correspondence between octal and decimal could be accomplished through conversion first to binary. The direct method, however, is to use the equation

$$N_{10} = d_n 8^n + d_{n-1} 8^{n-1} + \cdots + d_2 8^2 + d_1 8^1 + d_0 8^0 \qquad \textbf{(A.2.2)}$$

where

$$\begin{aligned} N_{10} &= \text{decimal number} \\ d_n d_{n-1} \cdots d_1 d_0 &= \text{octal number digits} \\ n &= \text{one less than the number of digits in the octal number} \end{aligned}$$

EXAMPLE A.2.2 Find the binary and decimal equivalent of the octal number 33_8.

Solution

We can find the binary most easily by

$$\begin{aligned} 3_8 &= \mathbf{011_2} \text{ and thus} \\ 33_8 &= 011011_2 = \mathbf{11011_2} \end{aligned}$$

The decimal number could be found by either Equation (A.2.1) or (A.2.2). Using

$$N_{10} = a_n 8^n + a_{n-1} 8^{n-1} + \cdots + a_1 8^1 + a_0 8^0$$

we have $n = 1$, so

$$N_{10} = (3)(8^1) + (3)(8^0)$$
$$N_{10} = \mathbf{27_{10}}$$

To convert from decimal to octal, we can either convert first to binary and then to octal, use tables, or use the procedure of successive division where the division will be eight instead of two. As before, the least significant value is found by the first division.

EXAMPLE A.2.3 Find the decimal number 356_{10} as both octal and binary numbers.

Solution

Let us use the successive division procedure to find the octal number first and then find the binary from that.

$$\frac{356}{8} = 44 + \frac{4}{8} \quad \text{so } d_0 = 4$$

$$\frac{44}{8} = 5 + \frac{4}{8} \quad \text{so } d_1 = 4$$

and

$$\underline{d_2} = 5$$

We find, then, that 356_{10} is $\mathbf{544_8}$, and from the octal to decimal conversion of 5_8, 4_8, and 4_8, we get a binary number of $\mathbf{101100100_2}$.

Hexadecimal

The rapid development of 4-bit, 8-bit, and 16-bit microcomputers has brought about the common use of base 16 number representations. This is called the *hexadecimal,* or *hex,* system. Any binary system with 4, or a multiple of 4, bits to the word can be conveniently expressed in hex because a hex number is formed from the counting states of 4 binary numbers. Because we must go beyond 9 before the hex set is complete, it is necessary to have new symbols to complete the set. The accepted standard is to use the letters A through F for the remaining counts. Instead of a subscript 16 to denote hex, it also has become common practice to use a capital H to indicate that a number is hex. Thus, from 0_{10} through 9_{10}, we have the hex numbers 0H through 9H, and from 10_{10} through 15_{10}, we have the hex equivalents AH, BH, CH, DH, EH, and FH. Conversion of a hex number into the decimal equivalent can be made through the equation

$$N_{10} = c_n 16^n + c_{n-1}16^{n-1} + \cdots + c_2 16^2 + c_1 16^1 + c_0 16^0 \qquad \textbf{(A.2.3)}$$

where

$$N_{10} = \text{decimal number (base 10)}$$
$$c_n c_{n-1} \cdots c_1 c_0 = \text{hex number}$$
$$n = \text{one less than the number of hex digits}$$

Conversion of a decimal number to hex can be made by successive division by 16, as was done for base 2 and base 8 systems. The following examples illustrate these conversions.

EXAMPLE A.2.4

Find the decimal equivalents of 47H, 30DH, and A2FH.

Solution

We use Equation (A.2.3) directly, where for the actual evaluation, the individual hex digits are expressed in their respective base 10 equivalents. For 47H, we have $c_0 = 7$ and $c_1 = 4$, so that

$$N_{10} = 4 \times 16^1 + 7 \times 16^0 = 64 + 7$$
$$N_{10} = \mathbf{71_{10}}$$

For 30DH, we have $c_0 = \text{DH} = 13_{10}, c_1 = 0$, and $c_2 = 3$

$$N_{10} = 3 \times 16^2 + 0 \times 16^1 + 13 \times 16^0 = 3 \times 256 + 13$$
$$N_{10} = \mathbf{781_{10}}$$

For A2FH, we have $c_0 = \text{FH} = 15, c_1 = 2$, and $c_2 = \text{AH} = 10$

$$N_{10} = 10 \times 16^2 + 2 \times 16^1 + 15 \times 16^0$$
$$N_{10} = \mathbf{2607_{10}}$$

EXAMPLE A.2.5

Find the hex equivalent of binary $\mathbf{10110101_2}$ and octal 422_8.

Solution

The simplest way to convert binary to hex is to group the binary number into 4-bit sets and find the hex of each set. Thus, $\mathbf{10110101_2}$ is $\mathbf{1011_2}$ and $\mathbf{0101_2}$. But $\mathbf{1011_2}$ is 11_{10}, which is BH, and $\mathbf{0101_2}$ is 5_{10}, which is 5H. Thus, the hex is B5H. The octal number can be converted easily to binary and then to hex. Thus, 422_8 is $\mathbf{100010010_2}$, which can be grouped into three groups of 4 bits (using leading zeros) as $\mathbf{0001_2}$, $\mathbf{0001_2}$, and $\mathbf{0010_2}$. The hex is then 112H.

EXAMPLE A.2.6

Find the hex equivalents of 29_{10}, 175_{10}, and 3412_{10}.

Solution

Conversion of decimal to hex is accomplished by successive division by 16. This finds the least significant digit first.

$$29/16 = 1 + 13/16 \qquad c_0 = 13 = \text{DH}$$
$$1/16 = 0 + 1/16 \qquad c_1 = 1 = \text{1H}$$

Thus, the hex is 1DH.

For the next decimal number, we have

$$175/16 = 10 + 15/16 \qquad c_0 = 15 = \text{FH}$$
$$10/16 = 0 + 10/16 \qquad c_1 = 10 = \text{AH}$$

Thus, the answer is AFH.

The last decimal number is converted as follows:

$$3412/16 = 213 + 4/16 \qquad c_0 = 4$$
$$213/16 = 13 + 5/16 \qquad c_1 = 5$$
$$13/16 = 0 + 13/16 \qquad c_2 = 13 = \text{DH}$$

Thus, the hex number is D54H.

Negative Numbers

The representation of negative numbers can be accomplished in both binary and octal by preceding the number with a negative sign as in decimal. In many cases, however, it is desirable to let a binary number itself carry information about the sign. This is accomplished in binary by one of several artificial techniques, such as the following.

Sign and Magnitude In this method, the binary number has another term added to it. There is no convention here; in some cases, a **1** represents a negative, whereas in others a **0** represents a negative. Thus, we would write the binary number −**1011** as **11011** if our convention choice was to make the highest bit **1** for a negative number.

2s Complement In this method, an extra digit is (again) provided that is always **1** for a negative number, but the code of the number is altered. This method is popular because of the ease with which math operations can be performed. In this method, a negative number is formed by the relation

$$(-N_2) = (2^{n+1})_2 - N_2 \tag{3.3}$$

where
$(-N_2)$ = the 2s complement binary number
n = highest digit in the number
$(2^{n+1})_2$ = a binary number 2 digits larger than N_2

Thus, the number $-\mathbf{01011}_2$ becomes in 2s complement with $n = 4$

$$\mathbf{100000}_2 - \mathbf{1011}_2 = \mathbf{10101}_2 \text{ (2s complement)}$$

A simple rule for finding the 2s complement of a binary number can be stated as follows:

To find the 2s complement of a binary number, copy all zeros from LSB up to and including the first **1**, and then replace all remaining **0**s by **1**s and **1**s by **0**s. Finally, place a **1** before the MSB.

EXAMPLE A.2.7 Find the 2s complement of $\mathbf{1101010}_2$.

Solution

Using the aforementioned rule, we leave the **0** and **1** on the right alone, change all the rest, and include a final **1.**

10010110 (2s complement)

A.2.2 BOOLEAN ALGEBRA

The application of digital techniques to the solution of control problems often involves the use of complex logical operations. Boolean algebra is a mathematical method that is very useful for setting up and solving logical equations.

The (only) values of variables in the logical math of Boolean algebra are those of true or false. Thus, the variables are binary in that they can take on only two values that are assigned as **1** for a true state and **0** for a false state. As an example, if the level of liquid in a tank is the variable, we let the variable A denote this level. Then, if the level is higher than that desired, we set $A = \mathbf{1}$ (a logical true) and if lower, $A = \mathbf{0}$ (a logical false).

There are four math operations in Boolean algebra that relate how combinations of variables may operate on and be related to one another.

Equality "="

Equality between two variables is defined by the same symbol employed in traditional algebra and has the same meaning. If we write $A = B$ and know $A = 1$, then $B = 1$ also.

Complement "$\bar{A}$"

The complement operation is denoted by a bar over a variable and indicates a variable of the opposite value to the original. If we know $B = \mathbf{1}$, then $\overline{B} = \mathbf{0}$. The complement operation is sometimes referred to as a *NOT.*

AND "·"

The AND operation is denoted by a centered dot "·" between two variables. If $C = A \cdot B$, then C will be **1** if and only if $A = \mathbf{1}$ and $B = \mathbf{1}$; otherwise $C = \mathbf{0}$. In some publications, the dot will be omitted such that $A \cdot B = AB$.

OR "+"

The plus symbol is employed to define the OR operation between two variables. Thus, we note that if $C = A + B$, then C will be **1** if A or B or both are **1** and will be **0** otherwise.

These math operations may be combined into many combinations for the description of some required logical operation. Thus, we may have $C = \overline{A \cdot B}$, which would mean C was equal to the complement of A AND B or NOT A AND B. In many cases, a particular logical combination can be represented by a different logical combination but give the same result. Thus, we can show that the previous $C = \overline{A \cdot B}$ is exactly the same as the operation $\overline{A + B}$—that is, NOT A OR NOT B. In both cases, C takes on the same logic state for the same logic states of A and B. There are several theorems that aid in changing the form of Boolean math operations without resorting to a truth table in each case.

The basic theorems of Boolean algebra are separated into single-variable and multivariable categories. Table A.2.1 shows the single-variable theorems, and Table A.2.2 the multivariable theorems.

TABLE A.2.1
Single-variable theorems

$A \cdot 0 = 0$	$A + A = A$
$A + 0 = A$	$A \cdot A = A$
$A \cdot 1 = A$	$A + \overline{A} = 1$
$A + 1 = 1$	$A \cdot \overline{A} = 0$

TABLE A.2.2
Multivariable theorems

$A + B = B + A$	Commutation
$A \cdot B = B \cdot A$	
$A + (B + C) = (A + B) + C$	Association
$A \cdot (B \cdot C) = (A \cdot B) \cdot C$	
$A \cdot (B + C) = A \cdot B + A \cdot C$	Distribution
$A + B \cdot C = (A + B) \cdot (A + C)$	
$A + A \cdot B = A$	Absorption
$A \cdot (A + B) = A$	
$\overline{A \cdot B} = \overline{A} + \overline{B}$	DeMorgan
$\overline{A + B} = \overline{A} \cdot \overline{B}$	

The theorems can be of great value in reducing a logical equation into another equivalent form that may not be intuitively obvious and may be much simpler.

EXAMPLE A.2.8

Find a simpler logic expression for the equation

$$D = A \cdot B + C \cdot (A \cdot B + \overline{C})$$

Solution

Let us apply the theorems in steps to simplify this equation. First, by distribution, we can write

$$D = A \cdot B + C \cdot A \cdot B + C \cdot \overline{C}$$

But $C \cdot \overline{C} = 0$ by Table A.2.1. By applying association, we can write

$$D = A \cdot B + A \cdot B \cdot C$$

Then, from distribution again,

$$D = A \cdot B \cdot (1 + C)$$

But $1 + C = 1$, so that

$$D = A \cdot B + C \cdot (A + B + \overline{C}) = A \cdot B$$

So the rather complex expression becomes a simple AND between A and B and is independent of C.

A.2.3 DIGITAL ELECTRONIC BUILDING BLOCKS

The building blocks for digital electronic circuits are a set of black boxes that implement the basic Boolean math operations for electrical signals. These elements are referred to as *gates,* and each has a characteristic symbol that has been adopted for universal indication of the device.

AND Gate

The AND gate accepts as input two signals, A and B, and outputs a signal that is given by $C = A \cdot B$. The symbol for this element is given in Figure A.2.1a.

OR Gate

The OR gate takes as input two signals, A and B, and outputs a signal that is given by $C = A + B$. The symbol for this device is given in Figure A.2.1b.

FIGURE A.2.1
Digital gate symbols.

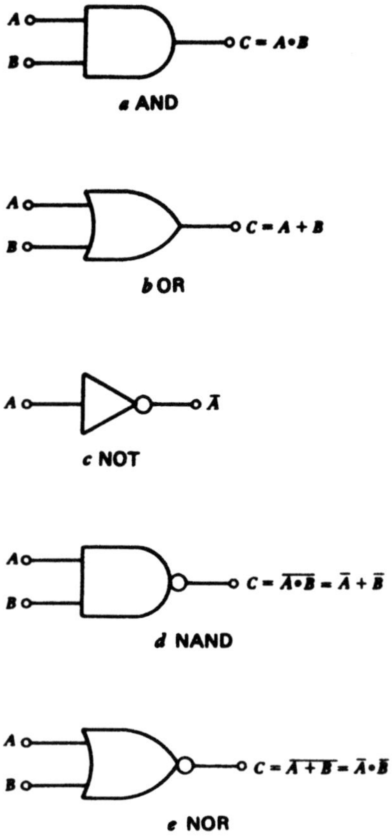

Inverter

The inverter, which is sometimes referred to as a NOT operation, accepts an input signal, A, and outputs an inverted version of this signal, $\overline{A}$. The symbol for this element is given in Figure A.2.1c. This is equivalent to providing the complement of a variable.

Two additional building blocks are frequently employed in electronic digital testing. These elements are used because of the ease with which they may be implemented using the popular circuit families.

NAND Gate

The NAND gate is essentially a device that performs the AND of two inputs and then inverts the output of that operation. Thus, if the inputs are A and B, the output will be $C = \overline{A \cdot B}$. We note from DeMorgan's theorem that this operation is identical to $\overline{A} + \overline{B}$. The symbol for this device is shown in Figure A.2.1d.

NOR Gate

As in the previous case, the NOR gate performs an OR operation between two inputs and then inverts the results. Thus, if the inputs are A and B, the output will be $C = \overline{A + B}$, which we know, again from DeMorgan's theorem, can be expressed as $C = \overline{A} \cdot \overline{B}$. The symbol for this device is shown in Figure A.2.1e. Using the five logic gates described, most of the basic logic equations developed from problems and using Boolean algebra can be implemented.

A.3

Thermocouple Tables

The following tables give the output voltage of several thermocouple (TC) types over a range of temperature in 5°C increments. In each case, the TC reference temperature is 0°C. The first-named material will be the positive terminal, as *iron-constantan;* the iron will be the positive lead when the reference temperature is lower than the measurement. The temperature is in °C, and the output is mV. Each column is in 5°C increments from the temperature of that row.

TYPE J: IRON-CONSTANTAN

	0	5	10	15	20	25	30	35	40	45
−150	−6.50	−6.66	−6.82	−6.97	−7.12	−7.27	−7.40	−7.54	−7.66	−7.78
−100	−4.63	−4.83	−5.03	−5.23	−5.42	−5.61	−5.80	−5.98	−6.16	−6.33
−50	−2.43	−2.66	−2.89	−3.12	−3.34	−3.56	−3.78	−4.00	−4.21	−4.42
−0	0.00	−0.25	−0.50	−0.75	−1.00	−1.24	−1.48	−1.72	−1.96	−2.20
+0	0.00	0.25	0.50	0.76	1.02	1.28	1.54	1.80	2.06	2.32
50	2.58	2.85	3.11	3.38	3.65	3.92	4.19	4.46	4.73	5.00
100	5.27	5.54	5.81	6.08	6.36	6.63	6.90	7.18	7.45	7.73
150	8.00	8.28	8.56	8.84	9.11	9.39	9.67	9.95	10.22	10.50
200	10.78	11.06	11.34	11.62	11.89	12.17	12.45	12.73	13.01	13.28
250	13.56	13.84	14.12	14.39	14.67	14.94	15.22	15.50	15.77	16.05
300	16.33	16.60	16.88	17.15	17.43	17.71	17.98	18.26	18.54	18.81
350	19.09	19.37	19.64	19.92	20.20	20.47	20.75	21.02	21.30	21.57
400	21.85	22.13	22.40	22.68	22.95	23.23	23.50	23.78	24.06	24.33
450	24.61	24.88	25.16	25.44	25.72	25.99	26.27	26.55	26.83	27.11
500	27.39	27.67	27.95	28.23	28.52	28.80	29.08	29.37	29.65	29.94
550	30.22	30.51	30.80	31.08	31.37	31.66	31.95	32.24	32.53	32.82
600	33.11	33.41	33.70	33.99	34.29	34.58	34.88	35.18	35.48	35.78
650	36.08	36.38	36.69	36.99	37.30	37.60	37.91	38.22	38.53	38.84
700	39.15	39.47	39.78	40.10	40.41	40.73	41.05	41.36	41.68	42.00

TYPE K: CHROMEL-ALUMEL

	0	5	10	15	20	25	30	35	40	45
−150	−4.81	−4.92	−5.03	−5.14	−5.24	−5.34	−5.43	−5.52	−5.60	−5.68
−100	−3.49	−3.64	−3.78	−3.92	−4.06	−4.19	−4.32	−4.45	−4.58	−4.70
−50	−1.86	−2.03	−2.20	−2.37	−2.54	−2.71	−2.87	−3.03	−3.19	−3.34
−0	0.00	−0.19	−0.39	−0.58	−0.77	−0.95	−1.14	−1.32	−1.50	−1.68
+0	0.00	0.20	0.40	0.60	0.80	1.00	1.20	1.40	1.61	1.81
50	2.02	2.23	2.43	2.64	2.85	3.05	3.26	3.47	3.68	3.89
100	4.10	4.31	4.51	4.72	4.92	5.13	5.33	5.53	5.73	5.93
150	6.13	6.33	6.53	6.73	6.93	7.13	7.33	7.53	7.73	7.93
200	8.13	8.33	8.54	8.74	8.94	9.14	9.34	9.54	9.75	9.95
250	10.16	10.36	10.57	10.77	10.98	11.18	11.39	11.59	11.80	12.01
300	12.21	12.42	12.63	12.83	13.04	13.25	13.46	13.67	13.88	14.09
350	14.29	14.50	14.71	14.92	15.13	15.34	15.55	15.76	15.98	16.19
400	16.40	16.61	16.82	17.03	17.24	17.46	17.67	17.88	18.09	18.30
450	18.51	18.73	18.94	19.15	19.36	19.58	19.79	20.01	20.22	20.43
500	20.65	20.86	21.07	21.28	21.50	21.71	21.92	22.14	22.35	22.56
550	22.78	22.99	23.20	23.42	23.63	23.84	24.06	24.27	24.49	24.70
600	24.91	25.12	25.34	25.55	25.76	25.98	26.19	26.40	26.61	26.82
650	27.03	27.24	27.45	27.66	27.87	28.08	28.29	28.50	28.72	28.93
700	29.14	29.35	29.56	29.77	29.97	30.18	30.39	30.60	30.81	31.02
750	31.23	31.44	31.65	31.85	32.06	32.27	32.48	32.68	32.89	33.09
800	33.30	33.50	33.71	33.91	34.12	34.32	34.53	34.73	34.93	35.14
850	35.34	35.54	35.75	35.95	36.15	36.35	36.55	36.76	39.96	37.16
900	37.36	37.56	37.76	37.96	38.16	38.36	38.56	38.76	38.95	39.15
950	39.35	39.55	39.75	39.94	40.14	40.34	40.53	40.73	40.92	41.12
1000	41.31	41.51	41.70	41.90	42.09	42.29	42.48	42.67	42.87	43.06
1050	43.25	43.44	43.63	43.83	44.02	44.21	44.40	44.59	44.78	44.97
1100	45.16	45.35	45.54	45.73	45.92	46.11	46.29	46.48	46.67	46.85
1150	47.04	47.23	47.41	47.60	47.78	47.97	48.15	48.34	48.52	48.70
1200	48.89	49.07	49.25	49.43	49.62	49.80	49.98	50.16	50.34	50.52
1250	50.69	50.87	51.05	51.23	51.41	51.58	51.76	51.94	52.11	52.29
1300	52.46	52.64	52.81	52.99	53.16	53.34	53.51	53.68	53.85	54.03
1350	54.20	54.37	54.54	54.71	54.88					

TYPE T: COPPER-CONSTANTAN

	0	5	10	15	20	25	30	35	40	45
−150	−4.603	−4.712	−4.817	−4.919	−5.018	−5.113	−5.205	−5.294	−5.379	
−100	−3.349	−3.488	−3.624	−3.757	−3.887	−4.014	−4.138	−4.259	−4.377	−4.492
−50	−1.804	−1.971	−2.135	−2.296	−2.455	−2.611	−2.764	−2.914	−3.062	−3.207
−0	0.000	−0.191	−0.380	−0.567	−0.751	−0.933	−1.112	−1.289	−1.463	−1.635
+0	0.000	0.193	0.389	0.587	0.787	0.990	1.194	1.401	1.610	1.821
50	2.035	2.250	2.467	2.687	2.908	3.132	3.357	3.584	3.813	4.044
100	4.277	4.512	4.749	4.987	5.227	5.469	5.712	5.957	6.204	6.453
150	6.703	6.954	7.208	7.462	7.719	7.987	8.236	8.497	8.759	9.023
200	9.288	9.555	9.823	10.093	10.363	10.635	10.909	11.183	11.459	11.735
250	12.015	12.294	12.575	12.857	13.140	13.425	13.710	13.997	14.285	14.573
300	14.864	15.155	15.447	15.740	16.035	16.330	16.626	16.924	17.222	17.521
350	17.821	18.123	18.425	18.727	19.032	19.337	19.642	19.949	20.257	20.565

TYPE S: PLATINUM-PLATINUM/10% RHODIUM

	0	5	10	15	20	25	30	35	40	45
+0	0.000	0.028	0.056	0.084	0.113	0.143	0.173	0.204	0.235	0.266
50	0.299	0.331	0.364	0.397	0.431	0.466	0.500	0.535	0.571	0.607
100	0.643	0.680	0.717	0.754	0.792	0.830	0.869	0.907	0.946	0.986
150	1.025	1.065	1.166	1.146	1.187	1.228	1.269	1.311	1.352	1.394
200	1.436	1.479	1.521	1.564	1.607	1.650	1.693	1.736	1.780	1.824
250	1.868	1.912	1.956	2.001	2.045	2.090	2.135	2.180	2.225	2.271
300	2.316	2.362	2.408	2.453	2.499	2.546	2.592	2.638	2.685	2.731
350	2.778	2.825	2.872	2.919	2.966	3.014	3.061	3.108	3.156	3.203
400	3.251	3.299	3.347	3.394	3.442	3.490	3.539	3.587	3.635	3.683
450	3.732	3.780	3.829	3.878	3.926	3.975	4.024	4.073	4.122	4.171
500	4.221	4.270	4.319	4.369	4.419	4.468	4.518	4.568	4.618	4.668
550	4.718	4.768	4.818	4.869	4.919	4.970	5.020	5.071	5.122	5.173
600	5.224	5.275	5.326	5.377	5.429	5.480	5.532	5.583	5.635	5.686
650	5.738	5.790	5.842	5.894	5.946	5.998	6.050	6.102	6.155	6.207
700	6.260	6.312	6.365	6.418	6.471	6.524	6.577	6.630	6.683	6.737
750	6.790	6.844	6.897	6.951	7.005	7.058	7.112	7.166	7.220	7.275
800	7.329	7.383	7.438	7.492	7.547	7.602	7.656	7.711	7.766	7.821
850	7.876	7.932	7.987	8.042	8.098	8.153	8.209	8.265	8.320	8.376
900	8.432	8.488	8.545	8.601	8.657	8.714	8.770	8.827	8.883	8.940
950	8.997	9.054	9.111	9.168	9.225	9.282	9.340	9.397	9.455	9.512
1000	9.570	9.628	9.686	9.744	9.802	9.860	9.918	9.976	10.035	10.093
1050	10.152	10.210	10.269	10.328	10.387	10.446	10.505	10.564	10.623	10.682
1100	10.741	10.801	10.860	10.919	10.979	11.038	11.098	11.157	11.217	11.277
1150	11.336	11.396	11.456	11.516	11.575	11.635	11.695	11.755	11.815	11.875
1200	11.935	11.995	12.055	12.115	12.175	12.236	12.296	12.356	12.416	12.476
1250	12.536	12.597	12.657	12.717	12.777	12.837	12.897	12.957	13.018	13.078
1300	13.138	13.198	13.258	13.318	13.378	13.438	13.498	13.558	13.618	13.678
1350	13.738	13.798	13.858	13.918	13.978	14.038	14.098	14.157	14.217	14.277
1400	14.337	14.397	14.457	14.516	14.576	14.636	14.696	14.755	14.815	14.875
1450	14.935	14.994	15.054	15.113	15.173	15.233	15.292	15.352	15.411	15.471
1500	15.530	15.590	15.649	15.709	15.768	15.827	15.887	15.946	16.006	16.065
1550	16.124	16.183	16.243	16.302	16.361	16.420	16.479	16.538	16.597	16.657
1600	16.716	16.775	16.834	16.893	16.952	17.010	17.069	17.128	17.187	17.246
1650	17.305	17.363	17.422	17.481	17.539	17.598	17.657	17.715	17.774	17.832
1700	17.891	17.949	18.008	18.066	18.124	18.183	18.241	18.299	18.358	18.416
1750	18.474	18.532	18.590	18.648						

A.4

Mechanical Review

A.4.1 MOTION

The mechanical description of motion is based on a specification of the position, velocity, and acceleration of an object relative to some reference section. If the object moves in an arbitrary manner with respect to the reference, we speak of *linear* or *rectilinear motion.* If the object is constrained to move *about* the reference, as in rotation, we speak of *angular motion.* The relationships among position, velocity, and acceleration are given in Table A.4.1. Linear position is measured in meters (m), with velocity in m/s and acceleration in $\mathrm{m/s^2}$. Angular motion is measured in radians (rad), angular velocity in rad/s, and angular acceleration in $\mathrm{rad/s^2}$.

In some cases, a special unit of measure is used for linear acceleration. This unit is the **g** that is defined as the acceleration of a freely falling body at the earth's surface. The value of **g** is given by $\mathbf{g} = 9.8\ \mathrm{m/s^2}$. In this context, an acceleration of 50 **g** in any direction would be

$$a = (50\,\mathbf{g})(9.8\ \mathrm{m/s^2})$$
$$a = 490\ \mathrm{m/s^2}$$

TABLE A.4.1
Position, velocity, and acceleration

Motion	Position	Velocity	Acceleration
Linear	x	$v = \dfrac{dx}{dt}$	$a = \dfrac{d^2x}{dt^2}$
Angular	θ	$\omega = \dfrac{d\theta}{dt}$	$\alpha = \dfrac{d^2\theta}{dt^2}$

TABLE A.4.2
Special motion cases

Motion	Linear	Angular
Constant speed position	v	ω
	$x = vt + x_0$	$\theta = \theta_0 + \omega t$
Constant acceleration speed	α	α
	$v = v_0 + \alpha t$	$\omega = \omega_0 + \alpha t$
Position	$x = x_0 + v_0 t + \frac{at^2}{2}$	$\theta = \theta_0 + \omega_0 t + \frac{\alpha t^2}{2}$

$\chi_0, \theta_0, v_0, \omega_0$ are initial positions and speed.

Special cases of motion frequently occur under constant speed or constant acceleration. In these cases, the motion can be described in terms of the speed or acceleration as shown in Table A.4.2.

A.4.2 FORCE

If an object is in uniform motion, it is found that a change in this motion (acceleration) can be effected through the action of some external agent with appropriate exchange of energy. The action that is required is defined as *force,* and the resultant energy exchange is defined as *work.* The unit of force is the newton (N), a derived quantity, equal to 1 $\text{kg}\cdot\text{m/s}^2$. The measurement of force is a significant part of process-control technology. The relationship of force to other mechanical variables can be formulated only after definition of a property of objects called *mass.*

Mass

Any object at rest or in uniform motion resists attempts to cause a change of its state. This resistance is indeed a reflection of the need for work to be done before a state change can occur. Even in a zero-gravity condition (0 **g**) in deep space, we would find it much more difficult to move a locomotive than a ball bearing. This resistance, or *inertia,* is reflected in a characteristic of an object called *mass.* Mass is a pure property, independent of the location or state of motion (neglecting relativity) of the object. Because a locomotive has much more mass than a ball bearing, it is more difficult to move. Mass is measured in the SI unit kilogram (kg).

Newton's Law

The relationship between the force applied to an object, its mass, and the resultant change in motion expressed in terms of acceleration is formulated as *Newton's law.*

$$F = ma \tag{A.4.1}$$

where F = net force in N
m = total mass in kg
a = acceleration in m/s^2

EXAMPLE A.4.1 An accelerometer indicates that a 2-kg object is accelerating at $2.2\ \text{m/s}^2$. Find the net force that acts on the object in newtons and pounds.

Solution

We can find this force by using

$$F = ma$$

$$F = 2(\text{kg})(2.2\ \text{m/s}^2) \times \frac{1\ \text{N}}{\text{kg}\cdot\text{m/s}^2}$$

$$F = \mathbf{4.4\ N}$$

Note that we have used the fact that $1\ \text{kg}\cdot\text{m/s}^2 = 1\ \text{N}$, and Appendix 1 shows that $1\ \text{lb} = 4.448\ \text{N}$, so

$$F = (4.4\ \text{N})\left(\frac{1\ \text{lb}}{4.448\ \text{N}}\right)$$

$$F = \mathbf{0.989\ lb}$$

Gravity and Weight

A fundamental law of nature states that all objects with finite mass exert a force of attraction on each other. In the case of the massive Earth and objects close to the surface, this force is (approximately)

$$F_g = mg \qquad \textbf{(A.4.2)}$$

where F_g = force of gravity in N
m = mass in kg
g = acceleration because of gravity, defined earlier ($9.8\ \text{m/s}^2$)

Equation (A.4.2) expresses the *weight* an object possesses, measured as the *force* with which the Earth attracts an object close to its surface. The proper measure of weight is also the newton, because it is a force. Scales used for weighing read weight. These are, of course, calibrated to read the force of the Earth on a mass and thus do not measure mass. A metric scale may indicate the weighed object's mass in kg in terms of F_g/g. This is a true measure of mass, because the scale has been calibrated for the Earth's surface. An English system scale may indicate a weight in pounds, and this is actually, again, the force of the Earth's attraction of the object.

Hooke's Law and Springs

An important relationship exists between force and displacement for the special case of objects defined as *springs.* A traditional spring of coiled steel wire serves as a good descriptive example. Such a spring is shown in Figure A.4.1a in the equilibrium or relaxed state. Here no forces act on the spring, and its length is x_0 as shown. In Figure A.4.1b, a force is applied to the spring. Application of this force caused the spring to extend from x_0 to a new length, x, where the system has again become stationary. The net force must be zero, and

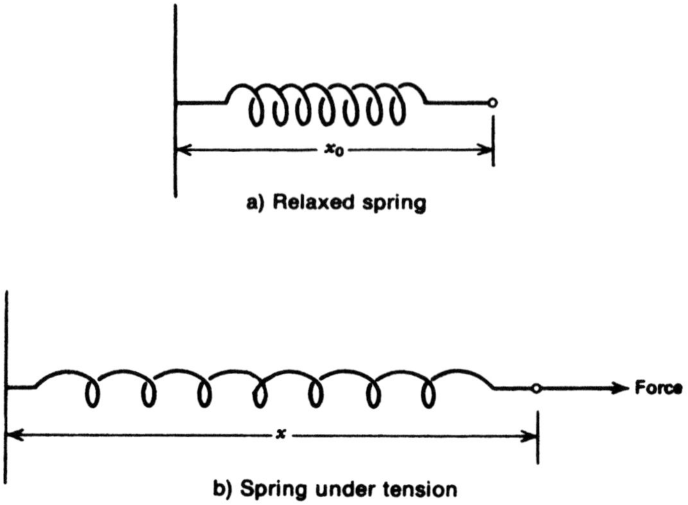

FIGURE A.4.1
A force applied to a spring extends the length from a relaxed x_0 to a new x.

hence the spring must be pulling back with an equal force, *F. Hooke's law* states that the equilibrium force of a spring under compression or extension is given by

$$F = -k\Delta x \tag{A.4.3}$$

where
F = force in N
k = spring constant in N/m
$\Delta x = x - x_0$ = change in length

The significance of the negative sign is to indicate that the spring force is *opposite* the applied force so that they balance. The spring constant k is a specification of the spring itself and is found from calibration or from a manufacturer specification sheet. The term *spring* is used loosely here. The devices described under Hooke's law could include bellows, diaphragms, coils, springs, leaf springs, or indeed any number of configurations that relate displacement to force.

EXAMPLE A.4.2 Find the extension that a spring with a spring constant of 5 N/m experiences when a force of 14 N is applied.

Solution

The equation for Hooke's law is

$$F = k\Delta x$$

$$\Delta x = \frac{14\ \text{N}}{5\ \text{N/m}}$$

$$\Delta x = \mathbf{2.8\ m}$$

A.5

P&ID Symbols

A.5.1 INTRODUCTION

Just as electronics has standard symbols to represent components in circuit schematics, process control has symbols to represent the elements of a process-control system. Instead of a schematic, we call the process-control diagram a piping and instrumentation drawing, or simply P&ID. The symbols and their meanings have been standardized in the industry through the American National Standards Institute (ANSI) and the Instrumentation, Systems and Automation (ISA) society. The resulting standard is called the Instrumentation Symbols and Identification, ANSI/ISA S5.1-1984 (R1992). This appendix presents an overview of the standard, including only the most common elements. In some cases, simplified drawings are prepared, especially in the design phase, which show only the essential features of a system. Finally, however, the fully detailed drawing is constructed, and this is the true P&ID for the plant.

A.5.2 CONNECTING LINES

The standard specifies what types of lines should be used to represent connections in the plant, including process product flow lines and the interconnections between instruments. Figure A.5.1 shows the line definitions. It is necessary to indicate on the P&ID the exact nature of these lines. For example, a dashed line represents an electric signal, but the P&ID would have to indicate the nature of the signal—for example, a 4- to 20-mA current or a 0 to 5-V voltage. In a similar fashion, it would be necessary to specify the nature of the data link—for example, 10-Mb Ethernet LAN line or serial-bit stream.

The solid line is primarily used to represent actual process product flow, often as a bold line. A plain solid line is used to denote operational control connections to the process, such as steam for heating. Notations are made on the P&ID to further explain the nature of the line.

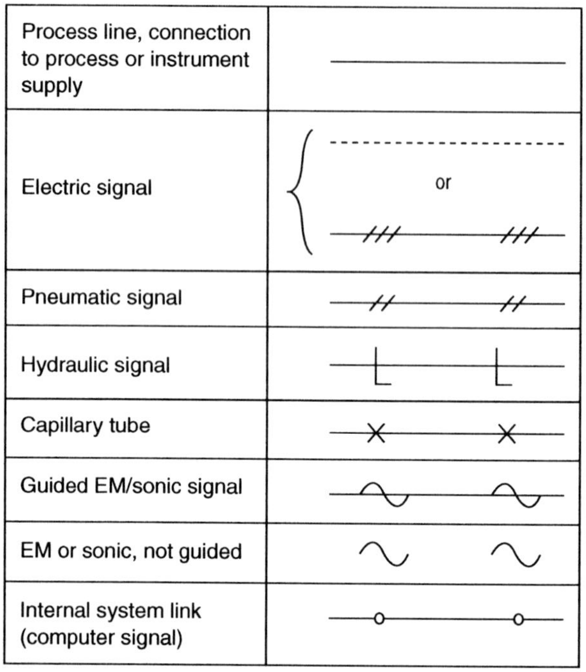

FIGURE A.5.1
P&ID signal and process lines.

A.5.3 GENERAL INSTRUMENTS OR FUNCTIONS

This aspect of the standard defines the symbols to be used for the various instrumentation needed to measure and control the process and the plant. This instrumentation includes sensors, transmitters, data converters, controllers, computers, and even programmable logic controllers (PLCs). Figure A.5.2 shows that balloons, rectangles, hexagons, and diamonds are used to denote the instrumentation. Notice that a line through a symbol means it is accessible to an operator, such as by being in a panel in the control room. No line means that the instrument is located in the field, perhaps at the control site itself, and is not accessible to the operator. A dashed line means that the device is inaccessible by virtue of being located within other equipment, such as behind the panel of a control room.

Generally, the symbol will also contain a combination of letters and numbers. The letters serve as a shorthand way of indicating the purpose of the instrument in the system. Table A.5.1 shows the assigned meanings of the letters depending upon if they are the first letter in the symbol or subsequent letters. Thus, for example, the designation FC would mean flow controller, while TT would mean temperature transmitter (sensor and transmitter). A first letter of Y is used for PLC instrumentation. Note that the letters C, D, G,

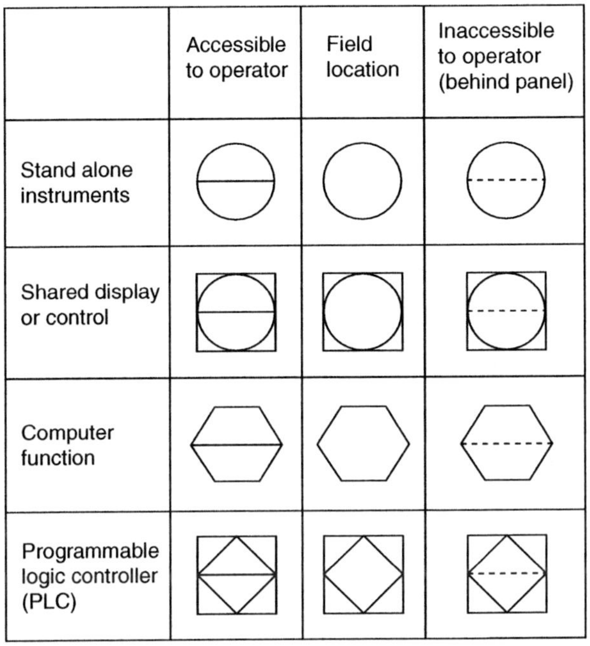

FIGURE A.5.2
Special symbols are used for instrumentation in the P&ID.

M, N, O are all at the discretion of the user to define, although M is often used for a motor. A computer may still carry the letter designation to define the control loop under its supervision.

The numbers serve to identify in which part of the overall plant the instrument operates. Assignment of these numbers is at the discretion of the user, but often the loops are numbered, as 100, 101, and so forth, and the instruments in the loop carry this number.

Figure A.5.3 shows some typical instrumentation designations. By reference to Figure A.5.2 and Table A.5.1, you can see that these symbols have the following meanings:

a. First letter P means pressure, second letter R means a recording unit, and third letter C means a controller, so this is a recording pressure controller located in loop or plant location 103. The unit is accessible to an operator and is probably in a panel of the control room.
b. The interpretation is L for level, C for control, and symbol for computer, located in the field of loop or plant location 330.
c. Y for event controller generally means a PLC, here for Z meaning for position control, not accessible to an operator and in loop or plant location 200.

TABLE A.5.1
P&ID element identification letters

	First Letter	Subsequent Letters
A	Analysis	Alarm
B	Burner, combustion	
C	Unspecified	Control, controller
D	Unspecified	
E	Voltage	Sensor, primary element
F	Flow rate	
G	Unspecified	Glass, viewing device
H	Hand	High
I	Current	Indication, readout
J	Power	
K	Time, time schedule	Control station
L	Level	Light, low
M	Unspecified	Middle, intermediate
N	Unspecified	
O	Unspecified	Orifice, restriction
P	Pressure, vacuum	Point, test point
Q	Quantity	
R	Radiation	Record, recorder
S	Speed, frequency	Switch
T	Temperature	Transmit, transmitter
U	Multivariable	Multifunction
V	Vibration, mechanical	Valve, damper, louver
X	Weight, force	Well
Y	Event, state	Relay, compute, convert
Z	Position, dimension	Driver, actuator

d. As part of a T for temperature-measurement system, the Y means a converter, converting 4 to 20 mA into 3 to 15 psi. The unit is not accessible to an operator and is part of loop or plant location 203.

Math Functions

An added element to the symbols presented in Figure A.5.2 is the employment of a rectangle to indicate a math operation. Figure A.5.4 presents a few of these. Many of these operations are now being performed by software in computer-control systems, but when they appear in a P&ID, they generally denote a hardware implementation. For example, a common example is extraction of a square root to convert pressure measurement to flow data.

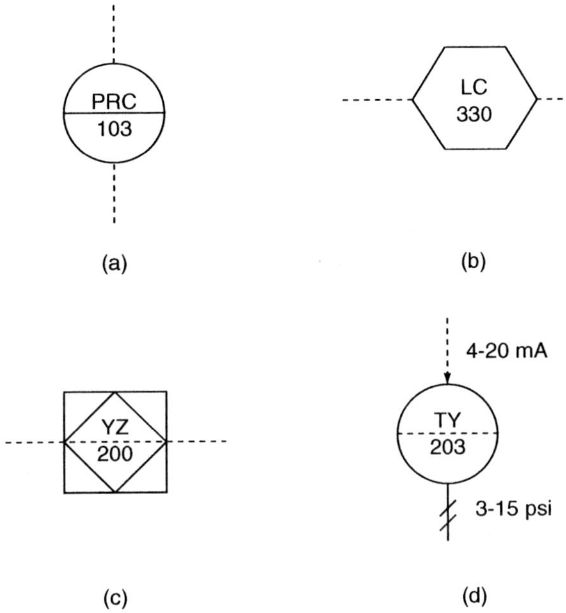

FIGURE A.5.3
Examples of the letter/number coding.

FIGURE A.5.4
The PLC includes many math operations.

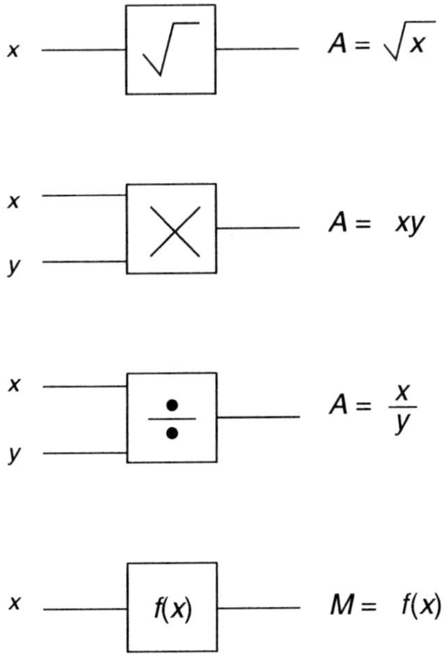

A.5.4 ACTUATORS AND PROCESS ELEMENTS

The last aspect of the P&ID are those elements that are part of the process itself. This includes control valves, actuators for control valves, conveyors, tanks, and so forth. Figure A.5.5 shows some of the more common elements that fall within this category. In many cases, the standard allows the user to pictorially represent specialized equipment in the process, such as heaters and pumps, in ways consistent with common drafting practice. Bubbles are often used to identify the element in terms of the control loop it serves.

Figure A.5.6 shows a plant P&ID with many of the symbols and connections defined in the previous sections.

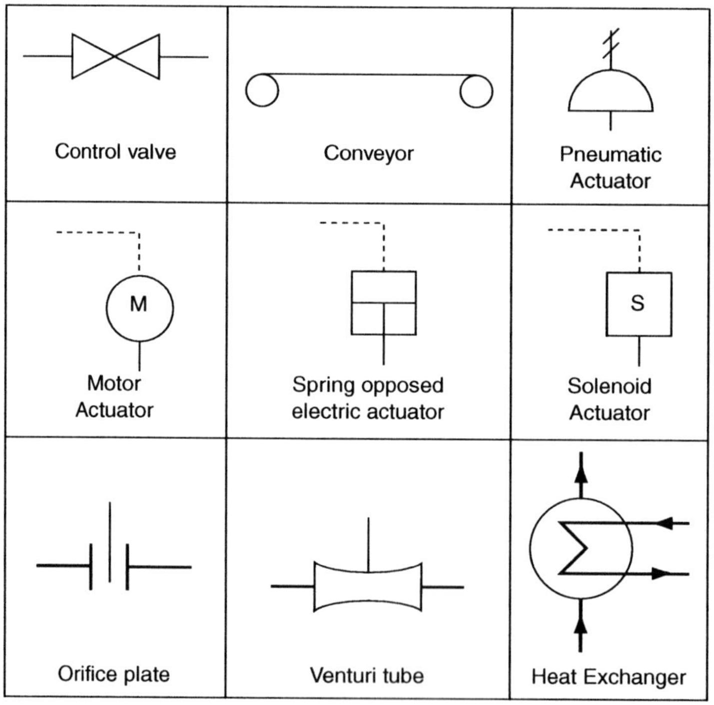

FIGURE A.5.5
Symbols for some final control elements.

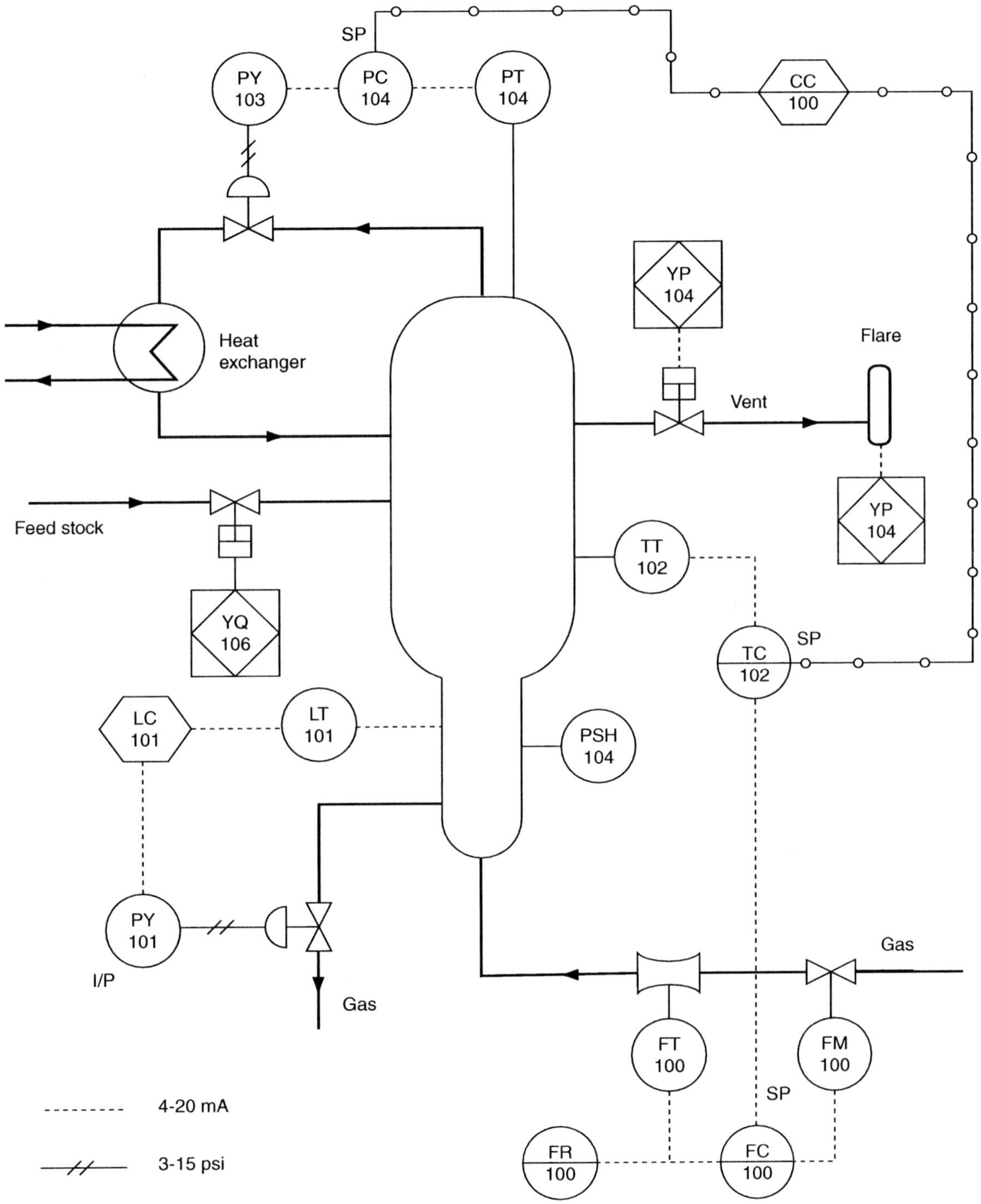

FIGURE A.5.6
Illustration of a P&ID for a chemical process.

A.6

Derivations

A.6.1 DIFFERENTIAL AMPLIFIER

In Section 2.5.4, the basic differential amplifier is shown in Figure 2.34 with a transfer function given by Equation (2.40). In order to derive this result, Figure A.6.1 shows the differential amplifier circuit but with four different resistors and with node voltages and line currents assigned. We need to find an equation for V_{out} in terms of the resistors and the input voltages. First let us apply KCL to the two nodes, *a* and *b*.

$$I_1 - I_- - I_2 = 0$$
$$I_3 - I_4 - I_+ = 0$$

The first basic design rule for op amp circuits says that the currents into the op amp input terminals must be zero, so $I_- = I_+ = 0$, and we are left with

$$I_1 - I_2 = 0$$
$$I_3 - I_4 = 0$$

FIGURE A.6.1
Op amp differential amplifier.

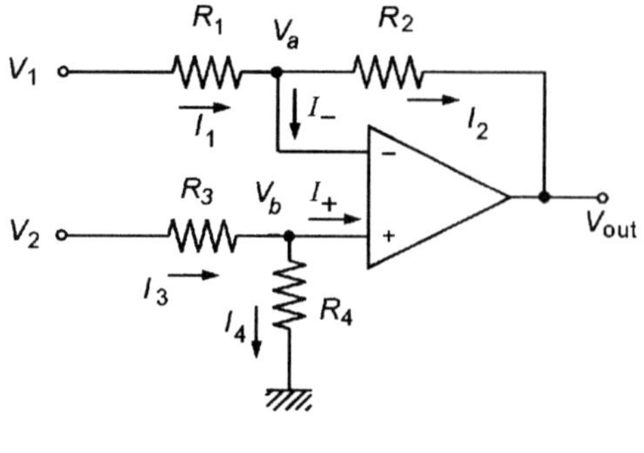

Ohm's law allows us to express these currents in terms of the given voltages:

$$\frac{(V_1 - V_a)}{R_1} - \frac{(V_a - V_{out})}{R_2} = 0$$

$$\frac{(V_2 - V_b)}{R_3} - \frac{V_b}{R_4} = 0$$

The second rule of op amp design tells us that there will be no voltage across the op amp input terminals. This means that $V_a = V_b$. So now we solve the second equation for V_b:

$$V_b = \frac{R_4}{R_3 + R_4} V_2$$

and substitute this into the first equation for V_a. The resultant equation can then be solved for V_{out} in terms of V_1 and V_2 as needed. After some algebra, we get

$$V_{out} = \frac{R_4(R_1 + R_2)}{R_1(R_3 + R_4)} V_2 - \frac{R_2}{R_1} V_1$$

In order to realize the differential amplifier, we now stipulate that $R_3 = R_1$ and $R_4 = R_2$. Substituting into the previous equation gives the familiar result

$$V_{out} = \frac{R_2}{R_1}(V_2 - V_1)$$

A.6.2 VOLTAGE-TO-CURRENT CONVERTER

Figure 2.39 presents a circuit that linearly converts input voltage into a current driving some load. The relationship between voltage and current is given in Equations (2.42) and (2.43). The transfer function of Equation (2.42) is independent of load resistance as long as op amp specifications are not exceeded. In order to derive this result, the circuit is redrawn in Figure A.6.2 with assigned node voltages and currents. Note that we have already imposed the common op amp design stipulation of no voltage across the input terminals by making V_1 appear on both op amp terminals. The load current I produces a voltage across the load resistance given by $V_L = IR_L$. So the objective will be to solve for I (or, equivalently, V_L) in terms of the input voltages and resistances. There are three unknowns, V_1, V_2, and V_L, so we need three equations. The equation for the V_2 node will involve current into or out of the op amp, which cannot be specified in any easy way. Therefore, we select the two nodes labeled with V_1 and the node labeled with V_L. Applying KCL and using the stipulation of no current into the op amp gives

$$I_1 - I_2 = 0$$
$$I_3 - I_5 - I = 0$$
$$I_5 - I_4 = 0$$

Ohm's law can now be used to express this set of equations in terms of voltages:

$$\frac{V_{in} - V_1}{R_1} - \frac{V_1 - V_2}{R_2} = 0$$

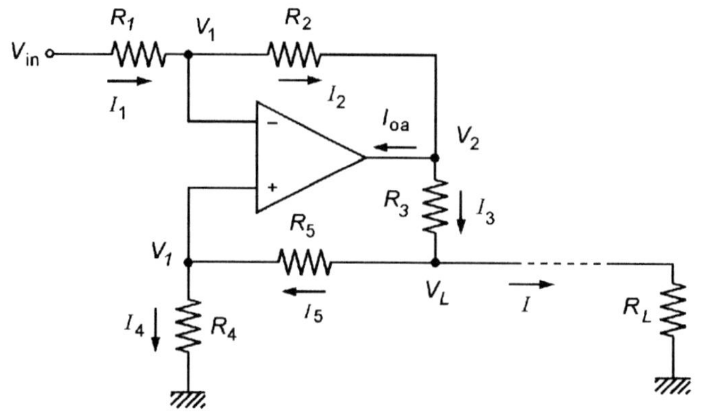

FIGURE A.6.2
Voltage-to-current converter.

$$\frac{V_3 - V_L}{R_3} - \frac{V_L - V_1}{R_5} - I = 0$$

$$\frac{V_L - V_1}{R_5} - \frac{V_1}{R_4} = 0$$

Since we really want to solve for the current, we substitute IR_L for V_L in the last two equations. The last equation of the set can be easily solved for V_1 in terms of I:

$$V_1 = \frac{R_4 R_L}{R_4 + R_5} I$$

The first equation of the set can be solved for V_2 in terms of V_{in} and V_1, but then by using the equation just found for V_1, V_2, can be expressed in terms of V_{in} and I. After some algebra, these two substitutions give

$$V_2 = -\frac{R_2}{R_1} V_{in} + \frac{R_4 R_L R_1 + R_2}{R_1 \ R_4 + R_5} I$$

These expressions for V_1 and V_2 are now substituted into the remaining equation of the set. After a lot of algebra, the result provides an expression for I in terms of the input voltage and resistances. This expression can be written

$$I\left[1 + \frac{R_2 R_4 - R_1(R_3 + R_5)}{R_1 R_3 (R_4 + R_5)} R_L\right] = -\frac{R_2}{R_1 R_3} V_{in}$$

This equation shows that I depends not only on V_{in} but also in a complicated way on the various resistors and the value of load resistance to which the circuit is connected. But, notice that the numerator of the complicated term contains a *difference*. If we select the resistors so that this difference is zero, the dependence on R_L is eliminated. Therefore, by selecting the resistors so that

$$R_2 R_4 = R_1(R_3 + R_5)$$

the equation for current becomes

$$I = -\frac{R_2}{R_1 R_3} V_{\text{in}}$$

A.6.3 PI-MODE OP AMP CONTROLLER

Figure 10.10 shows one method of using a single op amp to construct a proportional-integral mode controller. To derive the transfer function for this circuit, we first define nodes and currents as in Figure A.6.3. We stipulate that there is no current through the op amp input terminals and no voltage across the input terminals. Therefore, $V_a = 0$ and we can form the equations

$$I_1 + I_2 = 0$$
$$I_3 - I_2 = 0$$

The relationship between voltage across a capacitor and current through a capacitor is given by

$$I_C = C\frac{dV_C}{dt}$$

Combining this with Ohm's law allows the preceding current equations to be written in terms of voltage:

$$\frac{V_e}{R_1} + \frac{V_b}{R_2} = 0$$
$$C\frac{d}{dt}[V_{\text{out}} - V_b] - \frac{V_b}{R_2} = 0$$

The first equation can be easily solved for V_b in terms of V_e.

$$V_b = -\frac{R_2}{R_1} V_e$$

This is now substituted into the second equation:

$$C\frac{dV_{\text{out}}}{dt} - C\frac{d}{dt}\left(-\frac{R_2}{R_1}V_e\right) - \frac{1}{R_2}\left(-\frac{R_2}{R_1}V_e\right) = 0$$

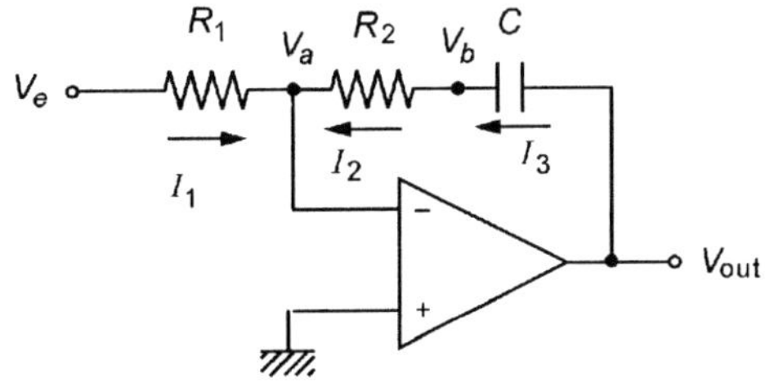

FIGURE A.6.3
Op amp proportional-integral controller.

This simplifies somewhat to

$$\frac{dV_{\text{out}}}{dt} + \frac{R_2}{R_1C}\frac{dV_e}{dt} + \frac{1}{R_1C}V_e = 0$$

In order to solve, we integrate this equation to eliminate the derivative on V_{out}. After some algebra and solving for V_{out}, the result is

$$V_{\text{out}} = -\frac{R_2}{R_1}V_e - \left(\frac{R_2}{R_1}\right)\frac{1}{R_2C}\int_0^t V_e\,dt + V(0)$$

In the second term on the right, we leave the R_2 in both places so that the relationship between component values and the proportional gain, K_P, and integral gain, K_I, can be seen explicitly. The negative signs are a result of the inverting configuration of the circuit. A unity gain inverter can then be used to convert these to positive signs, as shown in Figure 10.10.

A.6.4 PD-MODE OP AMP CONTROLLER

Figure 10.11 showed how a proportional-derivative controller could be implemented using a single op amp (other than the inverter to make the sign right). Analysis of this circuit can be performed using the circuit shown in Figure A.6.4 showing currents and nodes. As usual, the voltage across the op amp input terminals is zero, and there is no current into the op amp inputs.

Application of KCL to the two active nodes provides the equations

$$I_1 + I_2 - I_3 = 0$$
$$I_4 + I_3 = 0$$

Ohm's law and the differential relation between current and voltage for a capacitor can be used to express these equations in terms of voltage.

$$\frac{V_e - V_a}{R_3} + C\frac{d}{dt}[V_e - V_a] - \frac{V_a}{R_1} = 0$$
$$\frac{V_a}{R_1} + \frac{V_{\text{out}}}{R_2} = 0$$

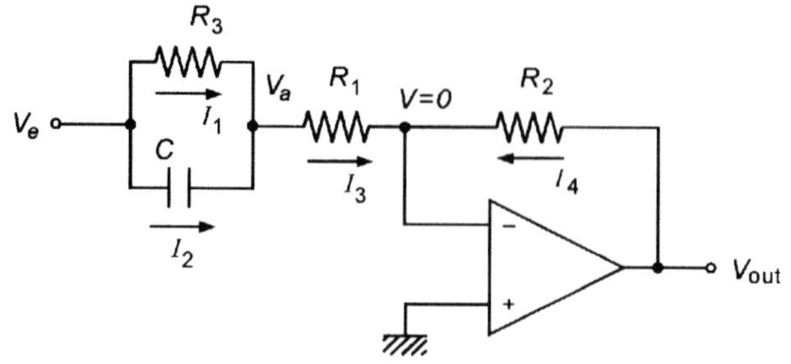

FIGURE A.6.4
Op amp proportional-derivative controller.

The second equation can be solved for V_a as

$$V_a = -\frac{R_1}{R_2} V_{\text{out}}$$

This is substituted into the first equation, and after a little algebra, the following rather complicated-looking equation results:

$$\frac{V_e}{R_3} + \frac{R_1}{R_2 R_3} V_{\text{out}} + C \frac{dV_e}{dt} + \frac{R_1}{R_2} C \frac{dV_{\text{out}}}{dt} + \frac{1}{R_1} V_{\text{out}} = 0$$

After rearranging and some more algebra, this reduces to the equation

$$V_{\text{out}} + \frac{R_1}{R_1 + R_3} R_3 C \frac{dV_{\text{out}}}{dt} = -\frac{R_2}{R_1 + R_3} V_e - \frac{R_2}{R_1 + R_3} R_3 C \frac{dV_e}{dt}$$

As was noted when the derivative-mode op amp circuit was introduced, it is necessary to clamp the response at higher frequencies. For this circuit, the criterion is found by determining the maximum frequency, f_{max}, and then forming the equation

$$2\pi f_{\text{max}} \frac{R_1}{R_1 + R_3} R_3 C = 0.1$$

A.7

Internet Resources

This appendix gives the URLs and a brief description of Web pages related to process control and instrumentation technology. These pages may be used for information and for links to many other sites that may have much more information on topics of interest. This list is not exhaustive but is intended to give the reader a start on finding more sites related to process control and instrumentation. *The reader should also know that Web pages are not static; whereas these pages are active now, they may not continue to be so in subsequent years.*

A.7.1 MAGAZINES

The following Web pages are associated with magazines related to the process industries.

1. *Sensors* www.sensorsmag.com

This site has general information about the sensor industry, including new developments, brief technical articles, and general information about process-control technology. It also has extensive links to manufacturers' Web pages, where specific information about hardware and software can be found.

2. *Control* www.controlmagazine.com

This is an actual *online* version of the hard-copy magazine. It has articles about control, measurement, and instrumentation. Back issues have been archived and are also available online. The site has a large list of links to manufacturers of process-control hardware and software and other relevant sites.

3. *Instrumentation and Control Systems* www.icsmagazine.com

This site provides very extensive links to manufacturers of process-control hardware and software and other organizations related to the process industries. Information is provided about the content of the latest issue of the hard-copy magazine.

4. *Instrumentation and Automation News* www.ianmag.com
This site contains a searchable database of products related to the process industries. A search will result in links to Web pages of companies that manufacture the product. General information about new developments in the process industries is included.

A.7.2 GENERAL INFORMATION AND ORGANIZATIONS

The following Web sites are related to professional organizations with an interest in the process-control and instrumentation industries.

1. *Instrumentation, Systems, and Automation Society (ISA)* www.isa.org
The ISA (formerly Instrument Society of America) is the primary professional organization for those working in process control and instrumentation. The organization publishes journals, provides training programs, and sponsors conferences. This Web site contains information about the ISA and its activities.

2. *Institute of Electrical and Electronics Engineers (IEEE)* www.ieee.org
The IEEE has subsections for practitioners working and studying in control systems, data acquisition, signal conditioning, and other areas related to process control and instrumentation.

3. *General online resource* www.automation.com
This site presents a vast collection of articles related to automation, process control and instrumentation. Links are also provided for seminars, job placement, upcoming events and even directories of products suppliers.

A.7.3 SIGNAL CONDITIONING

The following Web sites are for integrated circuit (IC) companies that produce hardware and software related to analog and digital signal conditioning. In many cases, in addition to providing information online about their products, the sites contain application information and even tutorials. All the other major IC companies have similar Web sites.

1. *National Semiconductor* www.national.com
In addition to downloadable specifications of their products, this site has a vast library of downloadable IC applications with design notes, schematics, and suggested component values. It includes op amps, V/F converters, comparators, ADC, DAC, and so forth.

2. *Freescale* www.freescale.com
Freescale is the former semiconductor division of Motorola. This site has information about their vast array of semiconductor products. It also has good explanations of the technologies involved and application notes.

3. *Burr-Brown* www.burr-brown.com
This site has information about the Burr-Brown line of semiconductor products and application notes.

4. *Linear Technologies* www.linear.com
This Web page has, at last count, 189 downloadable design notes on all aspects of analog and digital signal conditioning using, of course, Linear Technology ICs.

5. *Hewlett Packard* www.tmo.hp.com/tmo/iia/edcorner/English
This Web page has a great many tutorial and educational pages on analog and digital signal conditioning, filters, and many other related topics. Experiments are even described that can be conducted as an aid to learning.

6. *CyberResearch* www.cyberresearch.com/content/articles/tutorials/tutorials.aspx
CyberResearch is a shopping source for scientists and engineers. It provides access to a vast array of technical parts, equipment and supplies. The basic URL is simply www.cyberresearch.com which can be used to access the overall site. The more specific URL given above provides access to a useful array of tutorials about data acquisition and motion control.

A.7.4 SENSORS

1. *Temperature: Temperature World* www.temperatureworld.com
This is a great all-around site on temperature and temperature measurement. It contains good explanations and information about various measurement technologies. Links to many other temperature-related sites are included.

2. *Temperature: National Semiconductor* www.national.com
The site contains a line of solid-state temperature sensors, specifications, and application notes.

3. *Temperature: Alpha* www.alphasensors.com
These Web pages are devoted to the Alpha line of semiconductor temperature sensors, including thermistors. Application notes and some background on measurement technique are also included.

4. *Temperature: Pyrometry* www.wintron.com
This site contains a good review of pyrometric terms and concepts and includes descriptions of their line of infrared pyrometers.

5. *Strain gauges: BLH* www.blh.de
This is one of the founding companies of strain gauge technology. These excellent Web pages are devoted to SG concepts, application notes, and product specifications.

6. *LVDT: Macrosensors (Schaevitz)* www.macrosensors.com/primer/primer.htm
This site provides a tutorial on LVDT design and application and includes information on applications and descriptions of product lines.

7. *LVDT: Micro Strain* www.microstrain.com
This Web page presents their product line of miniature and subminiature LVDTs along with some application notes.

8. *Pressure: Entran* www.entran.com
This site is devoted to the product lines of pressure sensors, load cells, accelerometers, and other sensors. A good list of applications papers is included to illustrate how others have used these sensors.

9. *Pressure: Kavlico* www.kavlico.com
This manufacturer of pressure and displacement sensors presents Web pages with product specifications and some application notes.

10. *GlobalSpec* www.designinfo.com

This site functions as an aid to the selection of equipment for engineers and technologists. It includes many types of sensors. In addition it has excellent descriptions of the technology associated with sensors and measurement.

A.7.5 PROGRAMMABLE LOGIC CONTROLLERS

1. *PLC Tutor* www.plcs.net

This Web site is an excellent tutorial on the construction of ladder diagrams and the application of programmable logic controllers. Animation and graphics are used to clearly show how ladder-rung activations are related to the applications.

A.7.6 CONTROLLERS AND CONTROL SYSTEMS

The following Web pages are for companies and organizations associated with analog and digital controllers, control systems in general, and the tuning of control loops.

1. *Honeywell* www.honeywell.com

This Web site includes descriptions of a great variety of hardware and software associated with process control and instrumentation. Sensors and actuators are included, as are control valves and the controllers themselves. There is not a lot of application information, but a great deal of product data are included.

2. *Omega Engineering* www.omega.com

This is a vast Web site for a multifaceted worldwide company associated with the process industries. It includes many downloadable files with application notes, design aids, and product information.

3. *Foxboro* www.foxboro.com

This is a good general-purpose Web site for a company that manufactures digital and analog controllers as well as many other process-control-related products. Catalogs can be downloaded.

4. *Valves: K Controls* www.k-controls.co.uk

This manufacturer of control valves and actuators has a free, downloadable guide on valve and actuator selection and application.

5. *Tuning: Expertune* www.expertune.com/tutor.html

This company, which produces software for tuning control systems, presents an excellent tutorial on control system tuning. Graphs and animation are used to enhance the educational presentation.

6. *Tutorial on control system* chem.engr.utc.edu

This Web site is quite complicated but presents an excellent set of online experiments on control loops, as well as their operation and tuning. The user can select gains and observe the results.

7. *DesignInfo* www.designinfo.com

This Web site contains a vast database of process control and instrumentation vendors and their products. You can specify the type and characteristics of hardware you need, and the online search engine will provide a list of possible vendors and products. Some application guides and notes are also included.

REFERENCES

Bateson, *Introduction to Control Systems Technology,* 7th Edition, Prentice Hall, 2002.
Bucek, *Control Systems,* Prentice Hall, 1989.
Carr, *Elements of Electronic Instrumentation and Measurement,* 3rd Edition, Prentice Hall, 1996.
Cassell, *Microcomputers and Modern Control Engineering,* Reston, 1983.
Hunter, *Automated Process Control Systems,* 2nd Edition, Prentice Hall, 1987.
Jacob, *Industrial Control Electronics,* Prentice Hall, 1988.
Kilian, *Modern Control Technology,* West, 1996.
Otter, *Programmable Logic Controllers,* Prentice Hall, 1988.
Tompkins and Webster, *Interfacing Sensors to the IBM PC,* Prentice Hall, 1988.

GLOSSARY

Accelerometer A sensor that measures the acceleration of the object to which it is attached.

Actuator A part of the final control element that translates the control signal into action of the final control device in the process.

ADC An analog-to-digital converter that converts an analog input of electric voltage or current into a proportional digital signal.

Alarm In process control, an indicator that some process variable has exceeded preset limits.

Bellows A pressure sensor that converts pressure into a nearly linear displacement.

Binary A number representation system of base 2. This is the working system of digital computers.

Bode plot A graph of transfer function versus frequency where the gain (often in decibels) and phase (in degrees) are plotted against the frequency on a log scale.

Bourdon tube A pressure sensor that converts pressure to a displacement. The device is essentially a coiled, flattened tube that tends to straighten when pressure is applied.

Cascade control A control system composed of two loops where the setpoint of one loop (the inner loop) is the output of the controller of the other loop (the outer loop).

Controlled variable The process variable regulated by the process-control loop.

Controller The element in a process-control loop that evaluates error of the controlled variable and initiates corrective action by a signal to the controlling variable.

Controlling variable The process variable changed by the final control element under command of the controller to effect regulation of the controlled variable.

Cyclic A condition of either steady-state or transient oscillation of a signal about the nominal value.

DAC A digital-to-analog converter that converts a digital signal, often from a computer, into a proportional analog voltage or current.

DAS A data-acquisition system that interfaces many analog signals, called channels, to a computer. All switches, controls, and the ADC are included in the system.

DDC Direct digital control, where a computer performs all the functions of error detection and controller action.

Derivative-control mode A controller mode in which controller output is directly proportional to the rate of change of controlled variable error.

DP cell A pressure sensor that responds to the difference in pressure between two sources, most often used to measure flow by the pressure difference across a restriction in the flow line.

Dynamic variable Process variable that can change from moment to moment because of unspecified or unknown sources.

Error The algebraic difference between the measured value of a variable and the ideal value.

Floating-control mode A controller mode in which an error in the controlled variable causes the output of the controller to change at a constant rate. The error must exceed preset limits before controller change starts.

Foreground/background A control system that uses two computers, one performing the control functions and the other used for data logging, off-line evaluation of performance, financial operations, and so on. Either computer is able to perform the control functions.

Frequency response method A method of tuning a process-control loop for optimum operation by proper selection of controller settings. This method is based on a study of the frequency response of the open process-control loop.

Gas thermometer A temperature sensor that converts temperature to pressure of gas in a closed system. The relation between temperature and pressure is based on the gas laws at constant volume.

Hardware When used in the context of computers, hardware refers to the physical equipment associated with the computer (e.g., ICs, printed circuit boards, cables, and so on).

Hex A number of the representation system of base 16. The hex number system is useful in cases where computer words are composed of multiples of 4 bits (i.e., 4-bit words, 8-bit words, 16-bit words, and so on).

Hysteresis The tendency of an instrument to give a different output for a given input, depending on whether the input results from an increase or decrease from the previous value.

Integral-control mode A controller mode in which the controller output increases at a rate proportional to the controlled variable error. Thus, the controller output is the integral of the error over time with a gain factor called the integral gain.

Ionization gauge A pressure sensor based on conduction of electric current through ionized gas of the system whose pressure is to be measured, useful only for very low pressures (e.g., below 10^{-3} atm).

I/P converter A device that linearly converts electric current into gas pressure (e.g., 4 to 20 mA into 3 to 15 psi).

LASER Stands for light amplification by stimulated emission of radiation. It is a source of EM radiation generally in the IR, visible, or UV bands, and is characterized by small divergence, coherence, monochromaticity, and high colimation.

Linearity The closeness to which the curve relating two variables approximates a straight line. It is usually expressed as the maximum deviation between the actual curve and the best-fit straight line.

Load The process load is a term to denote the nominal values of all variables in a process that affect the controlled variable.

Load cell A sensor for the measurement of force or weight. Action is based on strain gauges mounted within the cell on a force beam.

LVDT A linear variable differential transformer that measures displacement by conversion to a linearly proportional voltage.

Microcomputer A computer based on the use of a microprocessor integrated circuit. The entire computer often fits on a small, printed-circuit board and works with a data word of 4, 8, or 16 bits.

Microprocessor A large-scale integrated circuit that has all the functions of a computer, except memory and input/output systems. The IC thus includes the instruction set, ALU, registers, and control functions.

Multiplexer This device allows selection of one of many input channels of analog data under computer control. The device is often an integral part of a DAS.

Nozzle A particular type of restriction used in flow systems to facilitate flow measurement by pressure drop across a restriction.

Nozzle/flapper A fundamental part of pneumatic signal processing and pneumatic control operations. Basically, the device converts a displacement of the flapper to a pressure signal.

Octal A number representation system of base 8. The octal system is most useful for computer systems with digital words of multiples of three bits (i.e., 3 bits, 6 bits, 9 bits, 12 bits, and so on).

Orifice plate A type of restriction used in flow systems to facilitate measurement of flow by the pressure drop across a restriction.

P&ID Stands for piping and instrumentation drawing, which is the primary schematic drawing used for laying out a process-control installation.

PD The designation of a controller operating in the proportional-derivative mode combination.

Photoconductive cell A sensor that converts the intensity of EM radiation, usually in the IR or visible bands, into a change of cell resistance.

Photodiode A sensor in which a *pn* diode junction is exposed to variations of light intensity, causing variations in the diode reverse-bias current.

Photovoltaic cell A sensor that converts the intensity of EM radiation, usually in the IR or visible bands, into a voltage.

PI The designation of a controller operating in the proportional-integral mode combination.

P/I A pressure-to-current converter, linearly converts a signal pressure range into a signal current range (e.g., 3-15 psi into 4-20 mA).

PID The designation of a controller operating in the proportional-integral-derivative mode combination. The PID is also called a three-mode controller.

Pirani gauge A pressure sensor based on measurement of the resistance of a heated wire as a function of wire temperature that is a function of the gas pressure; useful primarily for pressures less than one atmosphere.

Pneumatic Systems that employ gas, usually air, as the carrier of information and the medium to process and evaluate information.

Process Any system composed of dynamic variables, usually involved in manufacturing and production operations.

Process-reaction method A method of determining optimum controller settings when tuning a process-control loop. The method is based on the reaction of the open loop to an imposed disturbance.

Proportional band (PB) The change in input of a proportional-controller mode required to produce a full-scale change in output. Thus, if a 10% change in error causes a 100% change in controller output, the PB is 10.

Proportional-control mode A controller mode in which the controller output is directly proportional to the controlled variable error.

Pyrometer A temperature sensor that measures temperature by the EM radiation emitted by an object, which is a function of the temperature.

Quarter amplitude A process-control tuning criterion where the amplitude of the deviation (error) of the controlled variable, following a disturbance, is cyclic, so that the amplitude of each peak is one-quarter of the previous peak.

Range The region between the limits within which a variable is to be measured. Thus, "a temperature is to be measured in the range of 20°C to 250°C" defines the range. (See also *Span*.)

Rate action Another name for the derivative-control mode.

Reset action Another name for the integral-control mode.

Resolution The minimum detectable change of some variable in a measurement system.

Rotameter A flow-measurement system that is based on the proportionality of the rise of a float in a tapered tube, arranged vertically, placed in the flow system.

RTD A temperature sensor that provides temperature information as the change in resistance of a metal wire element, often platinum, as a function of temperature.

Self-regulation The property of some variables in a process to adopt a stable value under given load conditions without regulation via a process-control loop.

Sensitivity The ratio of the change in output magnitude to the change of input magnitude, under steady-state conditions, for a measurement device.

Sensor A transducer that converts a physical variable such as pressure, temperature, flow, and so forth, into an analog quantity, often electrical in nature.

Setpoint The desired value of a controlled variable in a process-control loop.

Software When used in the context of computers, software refers to the programs that provide the instructions to the computer on operations and calculations to be performed.

Span The algebraic difference between the upper-range value and the lower-range value. Thus, a temperature in the range of 20°C to 250°C has a span of 230°C. (See also *Range*.)

Strain gauge A sensor that converts information about the deformation of solid objects, called the strain, into a change of resistance.

Supervisory control A process-control installation for which a computer oversees the operation of the control systems and provides the setpoint for the process-control loops, which are themselves still analog.

Thermistor A temperature sensor constructed from semiconductor material and for which the temperature is converted into a resistance, usually with a negative slope and highly nonlinear.

Thermocouple A temperature sensor in which a voltage is produced nearly linearly with the difference in temperature to be measured and a known reference.

Three-mode controller Another name for a PID controller.

Time constant A number characterizing the time required for the output of a device to reach approximately 63% of the final value following a step change of its input.

Transducer A device that converts variation of one variable into variation of another variable (for example, current change into voltage change).

Transfer function The response of an element of a process-control loop that specifies how the output of the device is determined by the input.

Transmitter In process control, a device that converts a variable into a form suitable for transmission of information to another location (e.g., resistance changed to current that is propagated on wires to a control installation).

Ultimate cycle method See *Ziegler-Nichols method*.

Vapor pressure thermometer A temperature sensor for which the pressure of vapor in a closed system of gas and liquid is a function of temperature.

Venturi A type of restriction used in flow systems to facilitate measurement of flow by the pressure drop across a restriction.

Ziegler-Nichols method A method of determining optimum controller settings when tuning a process-control loop (also called the ultimate cycle method). It is based on finding the proportional gain that causes instability in a closed loop.

SOLUTIONS TO THE ODD-NUMBERED PROBLEMS

Following are the solutions to most of the odd-numbered problems at the end of each chapter. In many cases, more than one answer to a design question is correct, so the given answer is to be taken as typical.

Chapter 1

1.5 The area is estimated by rectangular areas; curve a is found to give the least area.

1.7 First peak error is 22.5°C, so the quarter amplitudes should be 5.6 and 1.8. Actual values are 7 and 2, so quarter amplitude is approximately satisfied.

1.9 1110_2

1.11 19°C, 19.3°C, 20.7°C, and 21°C

1.15 101379 Pa

1.17 $p = 9.6$ psi, $L = 5.9$ m

1.19 $R = 397 \pm 7.5\ \Omega$

1.21 Worse case $= \pm 4.2\%$, $RMS = \pm 2.5\%$

1.23 $T = 199°\text{C}$

1.25 See Figure S.1.

1.27 $\tau = 12$ s

1.29 $t = 1.99$ s

1.31 $\tau_{\text{ave}} = 0.25$ s

1.33 $Q_{\text{ave}} = 10.4$ gal/min, $\sigma = 1.24$ gal/min

Supplementary Problems

S1.1 See Figure S.2.

S1.3 Temperature and flow are self-regulating, level is not.

Figure S.1

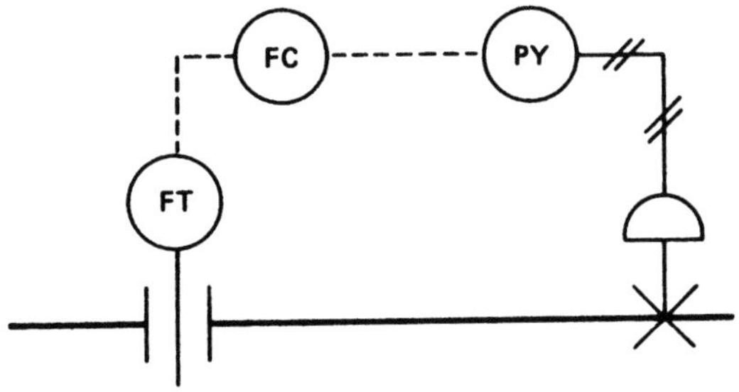

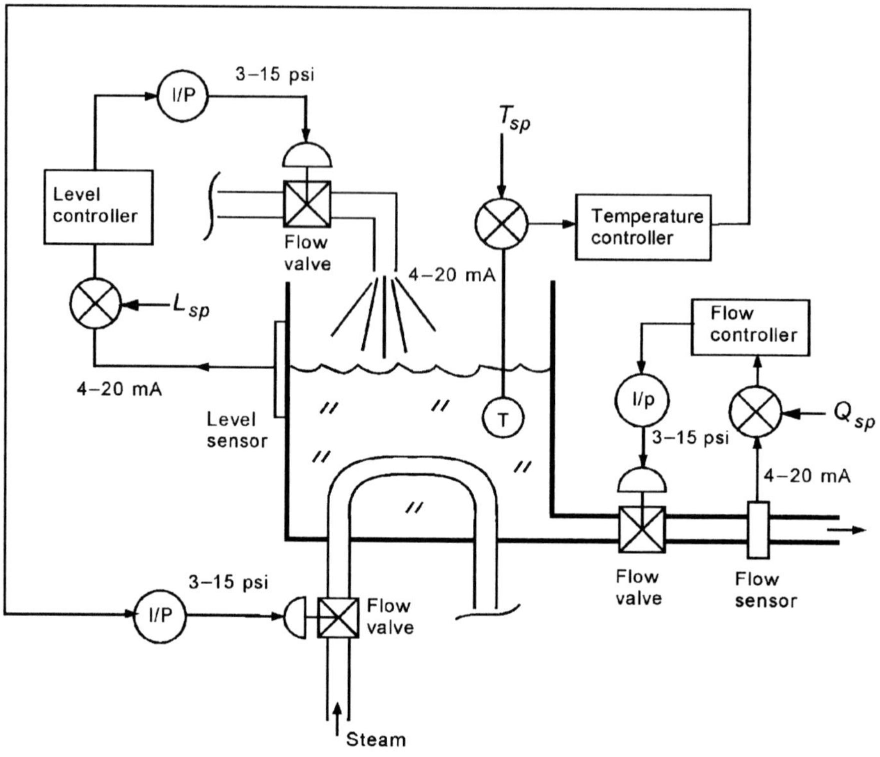

Figure S.2

S1.5 **a.** $K = 3.75$
b. L (low) $= 1.1$ m
c. period $= 3.23$ min

S1.7 **a.** Use known first-order time-response equation since input is a step.
b. final pressure $= 1035$ psi
c. $V(1\text{ s}) = 1.76$ V
d. $V_f = 1.12 + \dfrac{V(1) - 1.12}{1 - e^{-0.5}}$; then $p = \left(\dfrac{Vf}{0.05}\right)^2 - 500$

Chapter 2

2.3 4.90 to 1.67 V, 49.98 to 27.78 mW

2.7 $R_4 = 2790\ \Omega$

2.9 $R_4 = 1285\ \Omega$

2.11 $R_3 = 600\ \Omega, R_3 = 600.64\ \Omega$ for 0.25 mA

2.13 $C_4 = 0.21\ \mu\text{F}$

2.15 $f_c = 15$ kHz, $R = 1060\ \Omega, C = 0.01\mu\text{F}$, down 0.04% at 400 Hz

2.17 $f_c = 12$ kHz, $R = 1326\ \Omega, C = 0.01\ \mu\text{F}$, down 7% at 30 kHz

2.19 See Figure S.3.

2.21 Referring to Figure 2.20, $R_H = 1300\ \Omega, C_H = 0.05\ \mu\text{F}, R_L = 65\ \text{k}\Omega$, and $C_L = 50$ pF.

2.23 See Figure S.4.

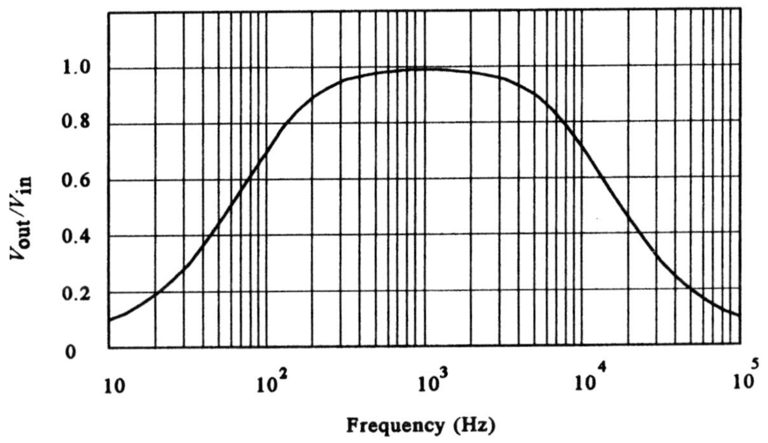

Figure S.3

Figure S.4

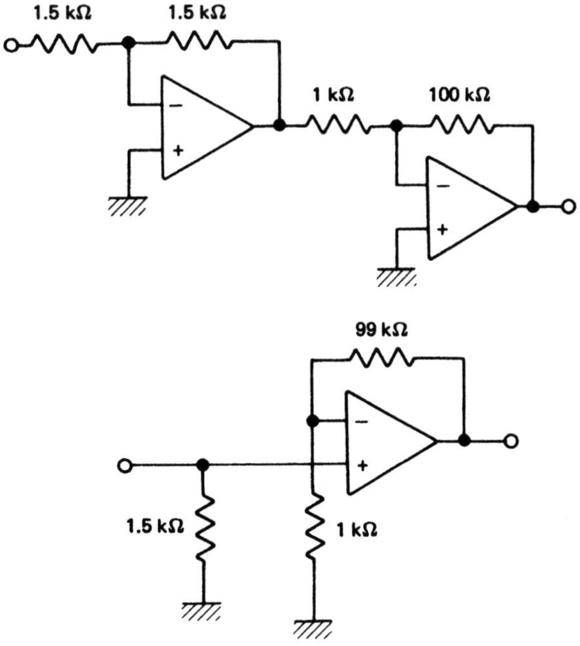

2.25 See Figure S.5.

2.27 CMRR = 5002, CMR = 74 dB

2.29 See Figure S.6.

2.31 See Figure S.7.

2.33 See Figure S.8 for the circuit and Figure S.9 for a plot of V_{out} vs. R_4. Nonlinearity is slight; least-squares-curve-fit straight line shows a 1% FS max deviation.

2.35 See Figure S.10.

2.37 See Figure S.11.

2.39 See Figure S.12.

Figure S.5

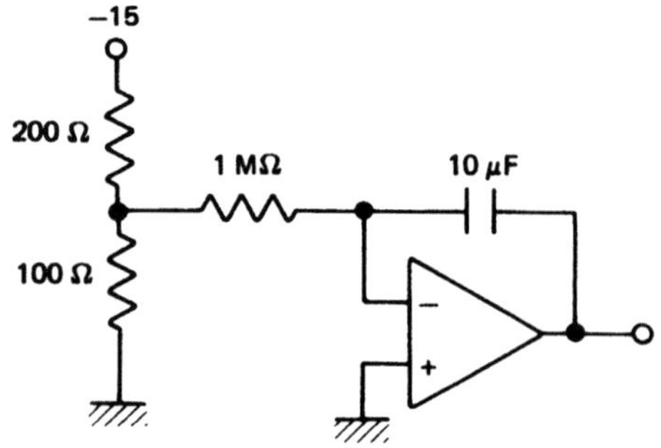

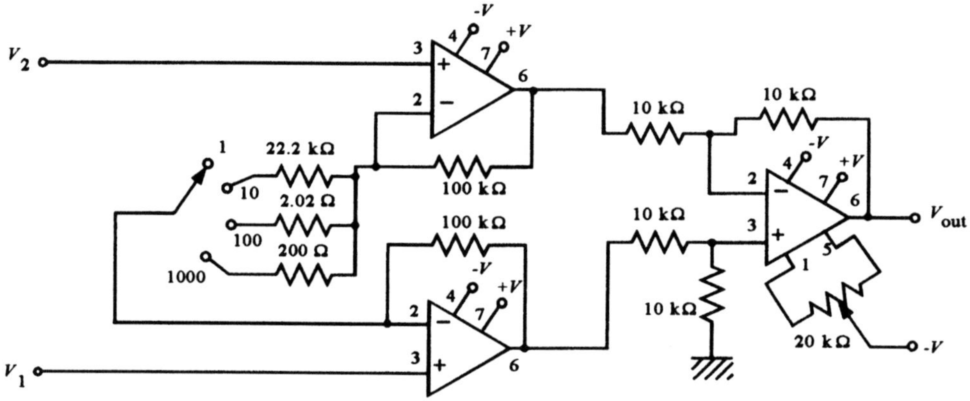

Figure S.6

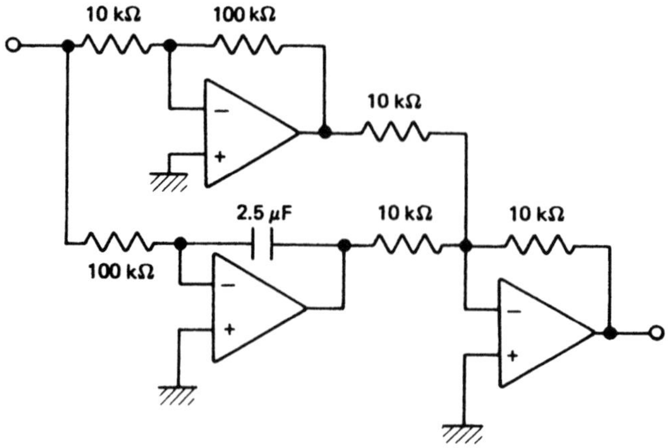

Figure S.7

Figure S.8

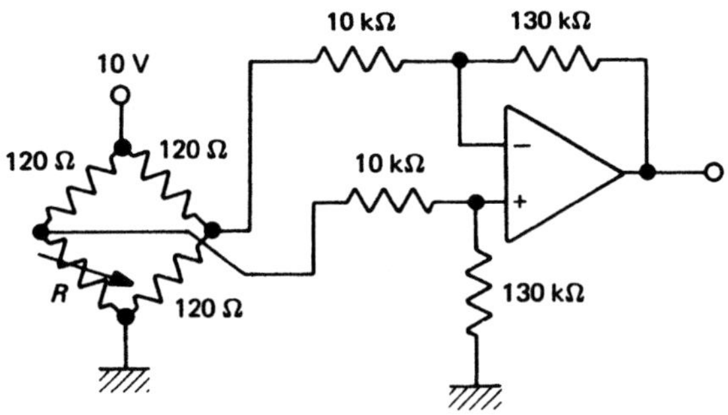

Figure S.9

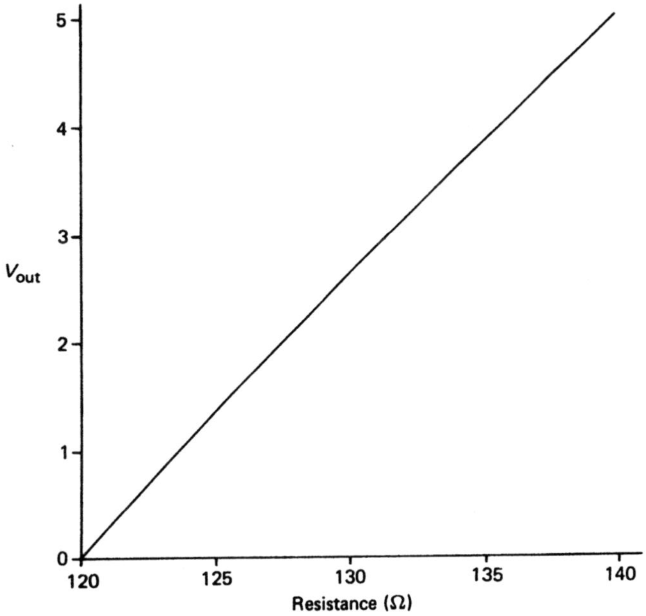

Figure S.10

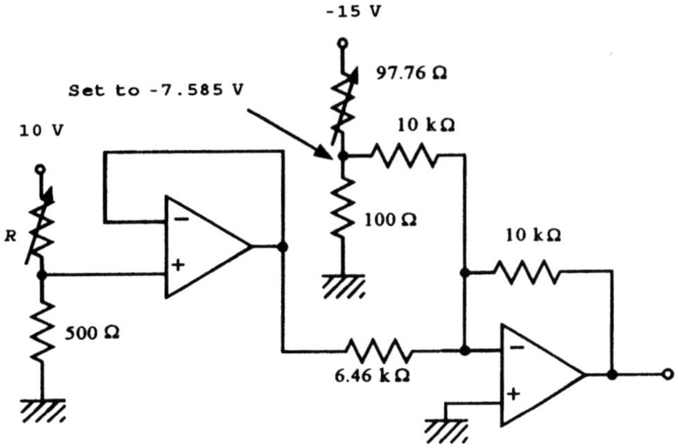

Figure S.11

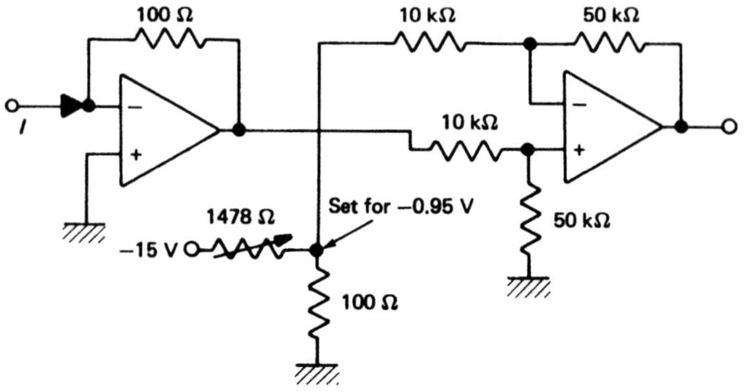

Figure S.12

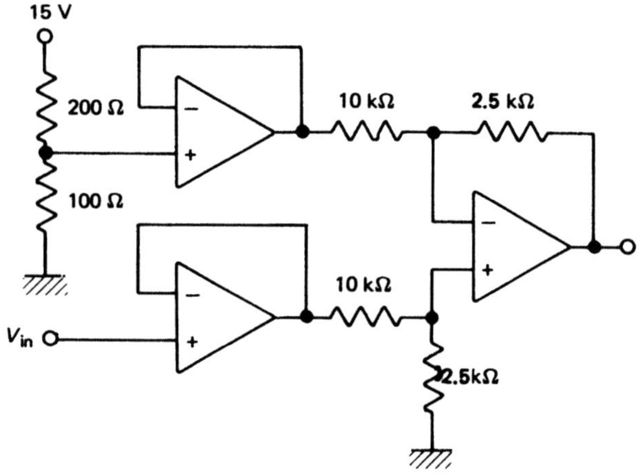

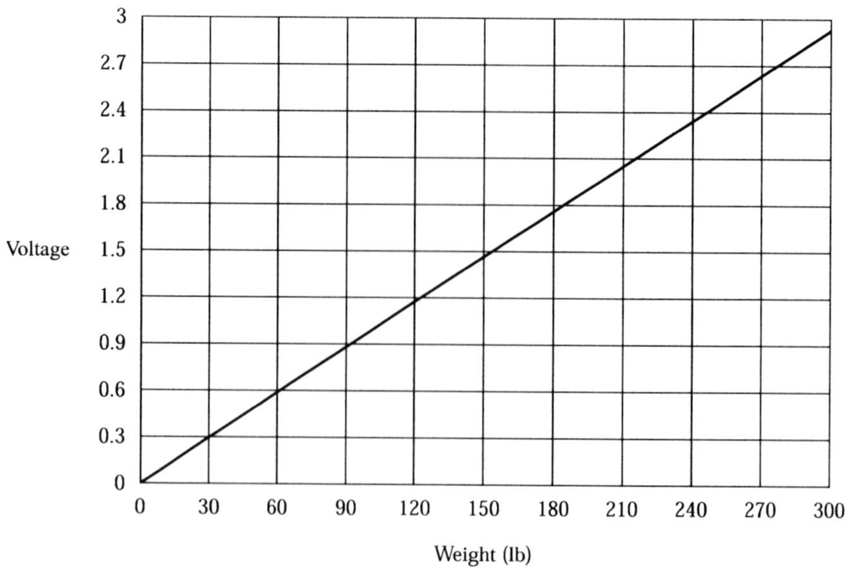

Figure S.13

Supplementary Problems

S2.1 **a.** 98.51 Ω

b. 117.99 to 122.17 Ω

c. 0.0829 V

S2.3 **a.** 1.5 V or 1-1b error, nonlinear circuit response

b. Error at 0 lb is 0 lb, error at 299 is 2.95, so readout error is −4 lb.

c. See Figure S.13.

S2.5 See Figure S.14.

S2.7 See Figure S.15.

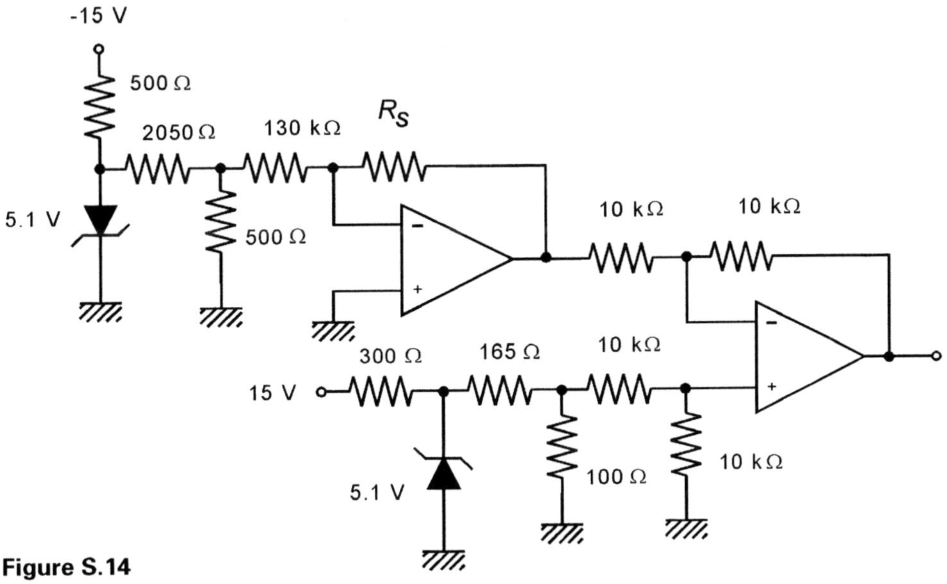

Figure S.14

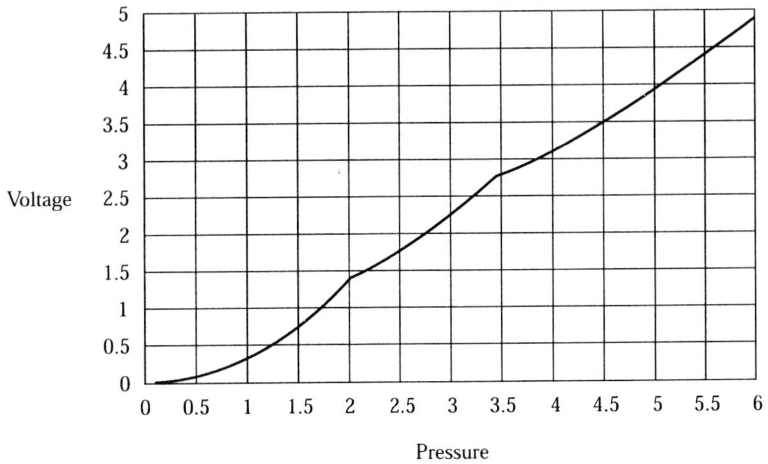

Figure S.15

Chapter 3

3.1 **a.** Ah, 12_8, 10_{10}

b. 3Bh, 73_8, 59_{10}

c. 16h, 26_8, 22_{10}

3.3 **a.** 5h, 10101_2, 25_8

b. 276h, 1001110101_2, 1165_8

c. 1ABh, 110101011_2, 653_8

3.5 **a.** 10101_2

b. 101010100_2

3.9 See Figure S.16.

3.11 See Figure S.17.

3.13 Refer to Figure 3.10. $R = 2.56$ kΩ, $R_f = 100$ kΩ

3.15 See Figure S.18.

3.17 $V_{\text{ref}} = 8.53$ V

3.19 **a.** 4A6H: −2.095V, 0B2H: −4.895V, D5DH: 3.352 V

b. F96H, 0.0026%

c. Bits 8, 9, 10, 11, and 12 are lost to noise.

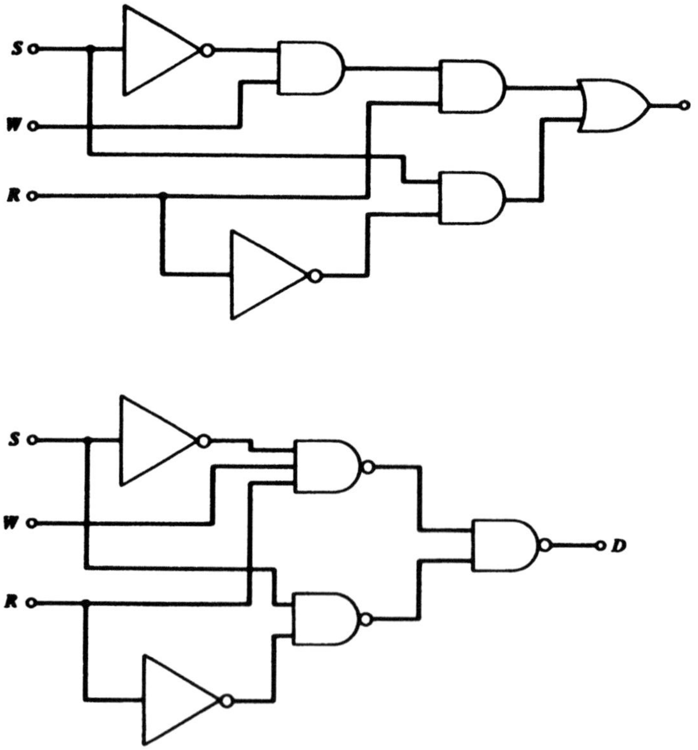

Figure S.16

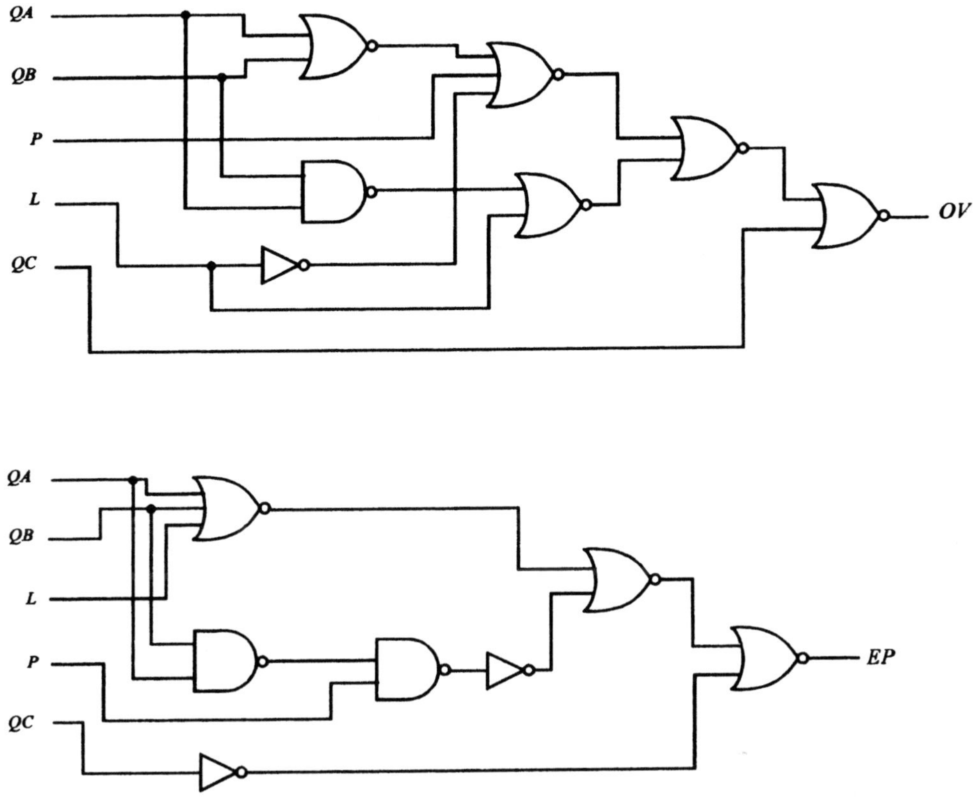

Figure S.17

3.21 Output = 61 H or 01100001_2; input could be in the range of 3.789 to 3.828 V; input for 10110111_2 could be 7.148 to 7.187 V.

3.23 For −4.3 V, 47H; for −0.66 V, 1BCH; for 2.4 V, 2F5H; for 4.8 V, 3EBH. For 30BH out, input is 2.607 to 2.617 V.

3.25 $f_{max} = 51.8$ Hz

3.27 **a.** $f_{sample} < 24.5$ kHz

b. Droop is −42 V/s, which exceeds ADC limit of 35 V/s, so bit-12 significance is lost.

c. Sampling low-pass critical frequency is 279 kHz, so at f_{sample} the output is 0.996 of input.

3.29 $T_c = 27.65$ ms, count for 4.6 kHz is 127.

3.31 See Figure S.19: Resolution: 0.3125°C, 0.3867 psi, 0.2343 gal/min.

3.33 $\tau_s = 3.9$ ms

3.35 **a.** 6Ch for 3.4 V and D6h for 6.7 V

b. 5.71875 to 5.75 V

3.37 See Figure S.20.

3.39 28.6 kHz, 3.6 kHz

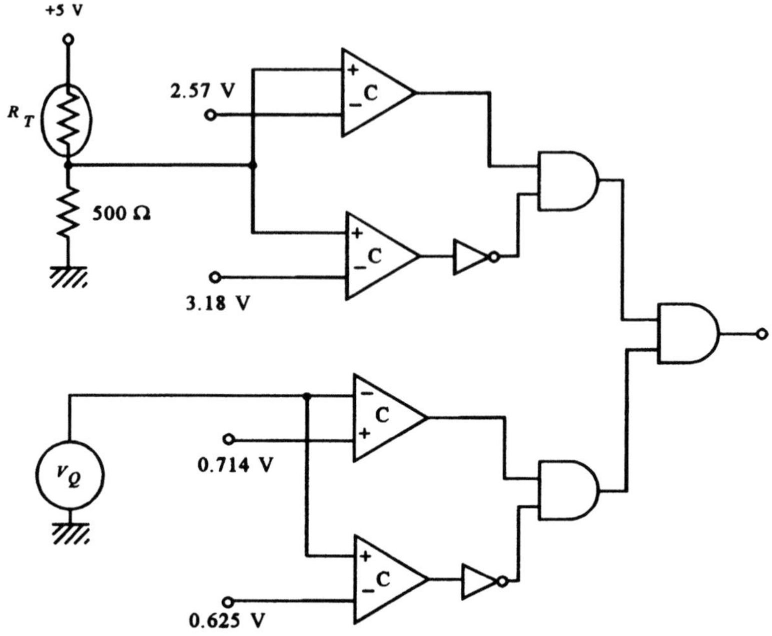

Figure S.18

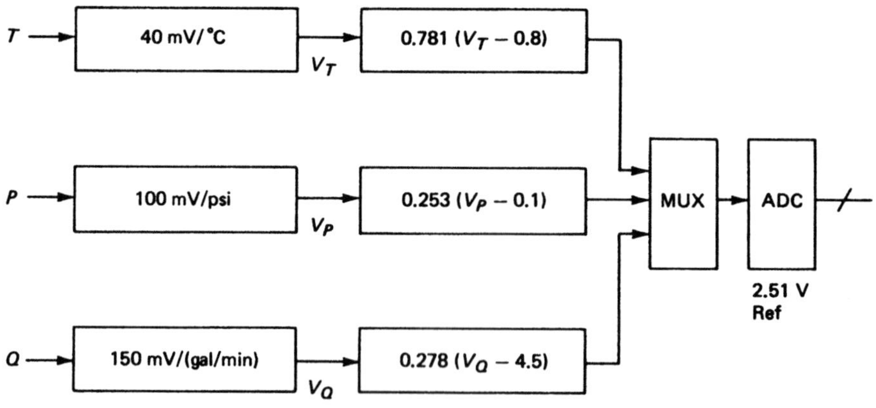

Figure S.19

3.41 10 bits

3.43 See Figure S.21.

3.45 Equation is $\Delta 1/1 = -0.9756\Delta V/(5 + \Delta V)$. This is used directly in the higher-level language, as

STRAIN = −0.9756 * DELTA V/(5 + DELTA V)
MICROS = 1000000 * STRAIN

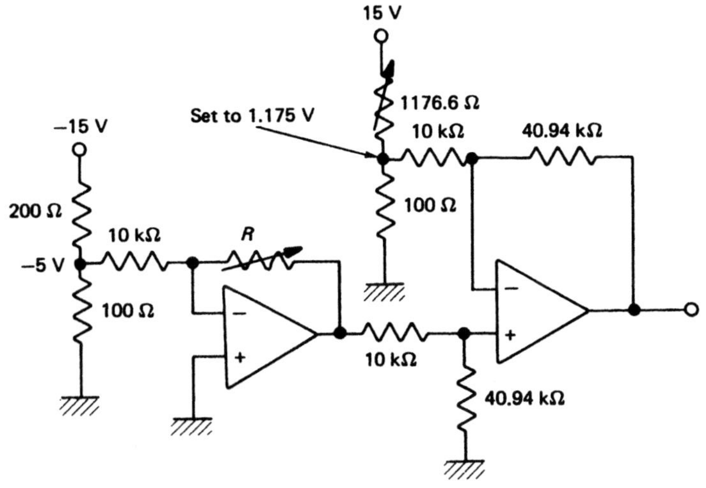

Figure S.20

Supplementary Problems

S3.1 See Figure S.22.

S3.3 **a.** 14°C is 0, 19°C is 160 (A0H), and 22°C is 255 (FFH).

b. 100% is 5 V and 255 (FFH), 70% is 3.5 V and 179 (B3H), 38% is 1.9 V and 97 (61H).

c. See Figure S.23.

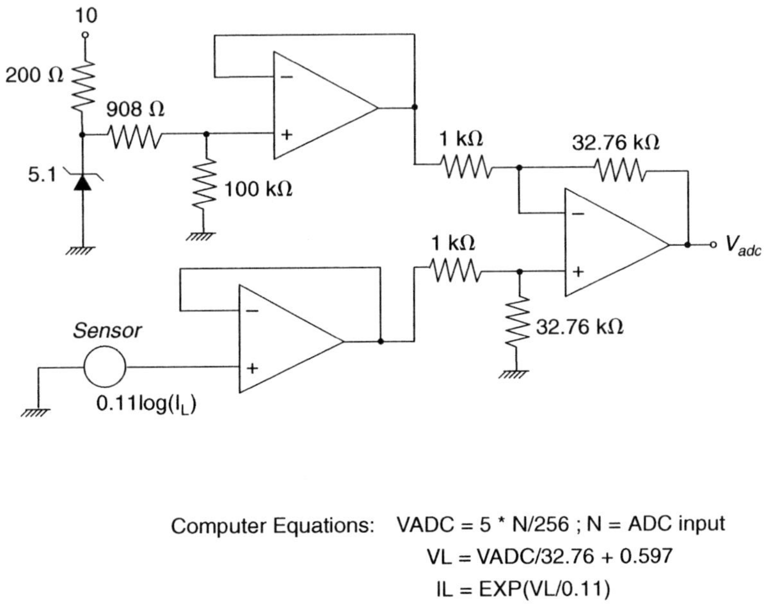

Figure S.21

Figure S.22

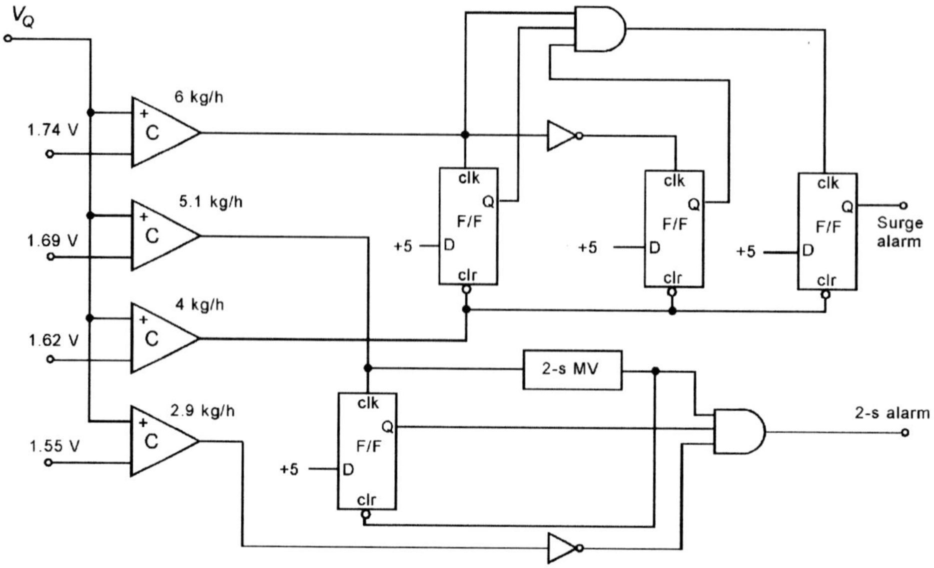

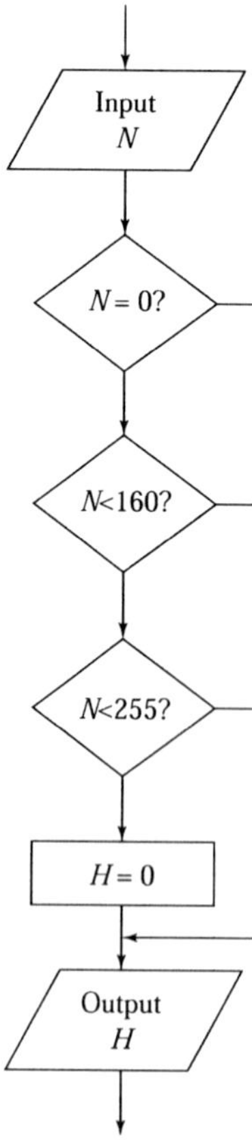

Figure S.23

Chapter 4

4.1 251.7 K, −21.4°C, −6.5°F

4.3 423.1 K, 302°F

4.5 3114 m/s, 10217 ft/s

4.7 $R = 108.1\ \Omega$

4.9 $R(T) = 589.48[1 + 0.0018(T - 115)]$; $R(T) = 589.48[1 + 0.0018(T - 115) - 0.00000015(T - 115)^2]$. Linear has −0.89% error, quadratic has a +0.016% error.

4.11 Resolution required is 263 mV.

4.13 See Figure S.24, nonlinear, 0.4°C heating.

4.15 1225.25°C

4.17 **a.** 12.21 mV

b. 9980 Ω

4.19 0.22 mm

4.21 64 psi to 420 psi

4.23 See Figure S.25.

4.25 See Figure S.26.

4.27 See Figure S.27.

4.29 See Figure S.28.

Supplementary Problems

S4.1 **a.** 398 thermocouples, so we use 400 with 50 evenly distributed on each wing.

b. The total internal resistance is 20 Ω, so maximum power is delivered into a 20-Ω load; 453 mW at 300°C and 846 mW at 400°C.

c. For 10°C ambient and 300°C heat, the voltage rises to 5.28 V.

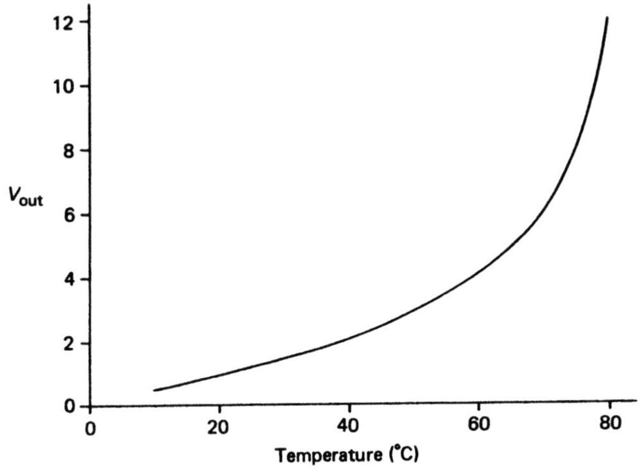

Figure S.24

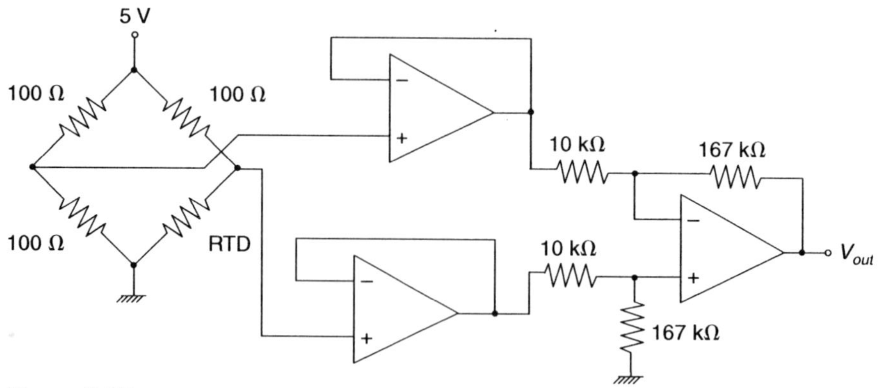

Figure S.25

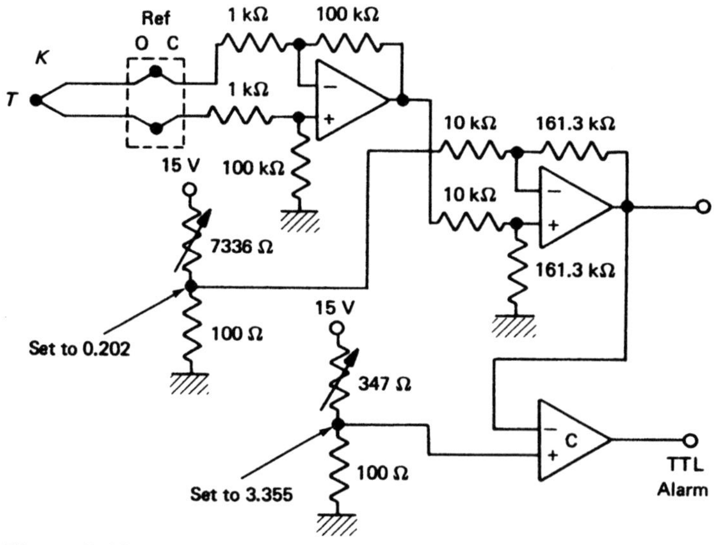

Figure S.26

S4.3 **a.** See Figure S.29 for the circuit. The table follows:

Speed (mph)	ΔT(°C)	$\Delta R(\Omega)$	Bridge ΔV (V)	V_{out}(V)
0	0	0	0	0.0
10	−3.0	−72	0.112	2.44
20	−4.5	−108	0.166	3.62
30	−5.5	−132	0.201	4.38
40	−6.3	−151.2	0.228	4.97
50	−7.1	−170.4	0.255	5.56
60	−7.7	−184.4	0.275	6.00

Figure S.27

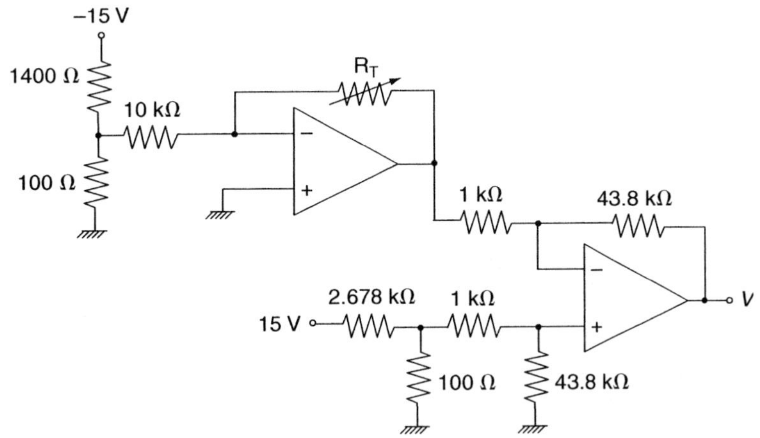

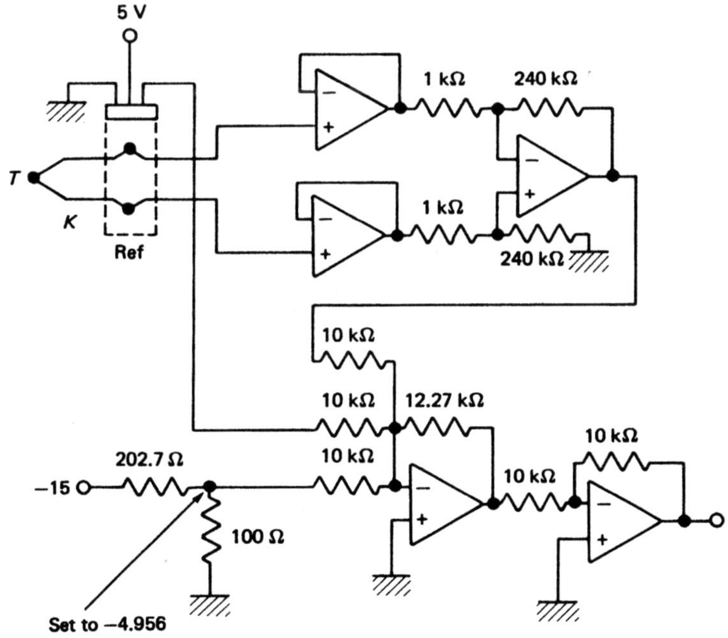

Figure S.28

b. See Figure S.30.

c. Greatest error is at a wind speed of 10 mph, where the voltage indicates 2.44 V or 24.4 mph for a 146% error. The error is due to nonlinearities of the wind/temperature, thermistor resistance versus temperature, and bridge voltage versus resistance. All in all, it is not a very good sensor.

Chapter 5

5.1 0.6 mm, 250 Ω

5.3 862.75 pF to 897.96 pF, or approximately ±17.6 pF

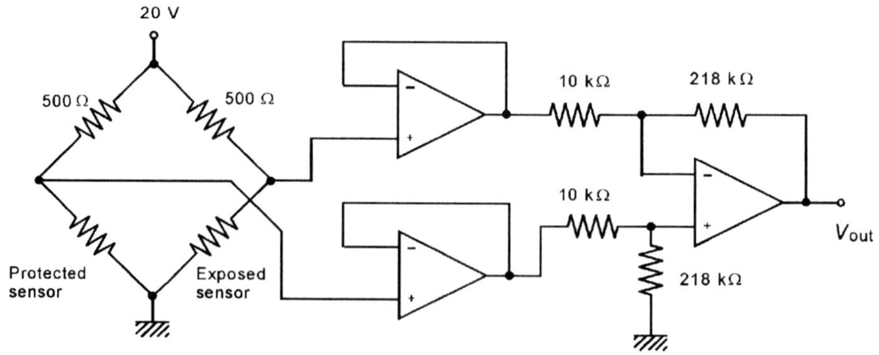

Figure S.29

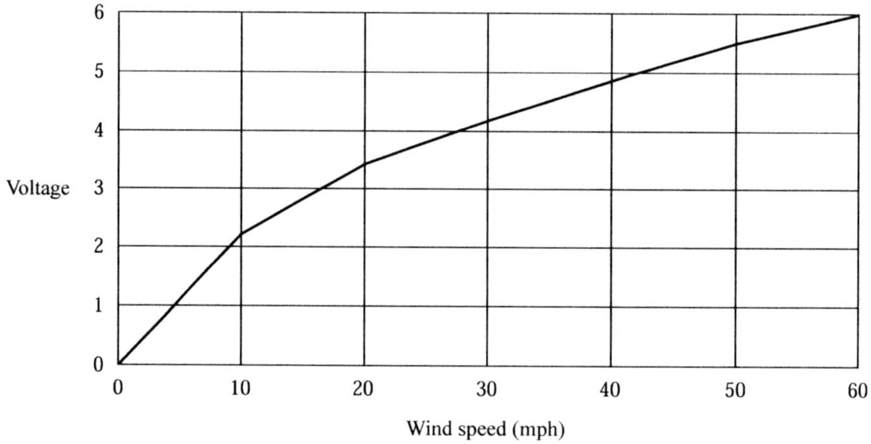

Figure S.30

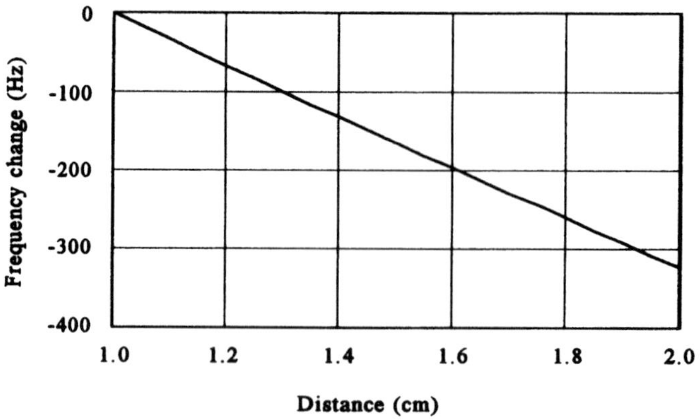

Figure S.31

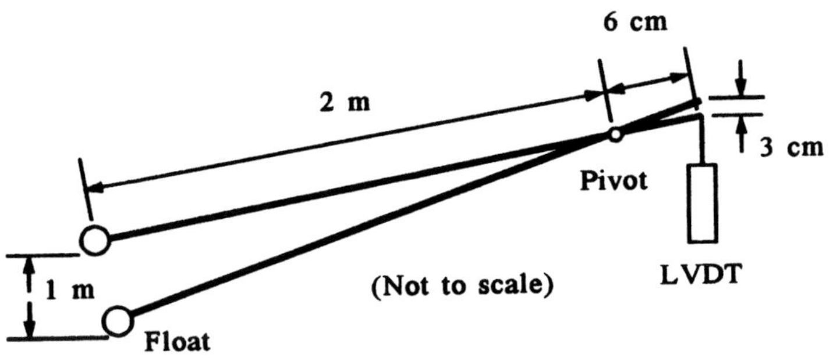

Figure S.32

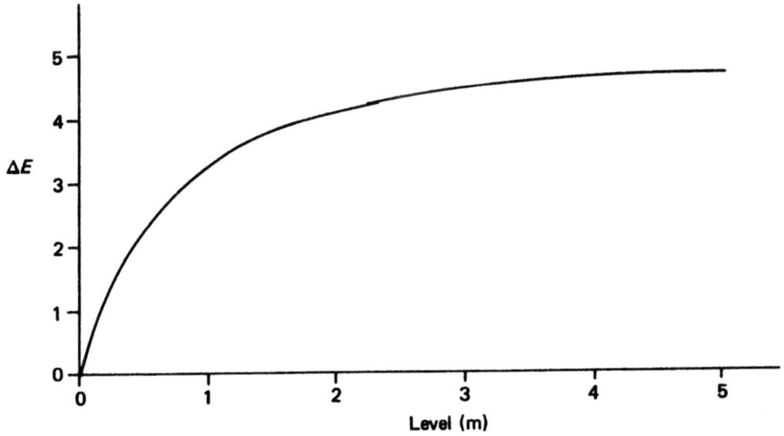

Figure S.33

5.5 9.961 to 9.637 kHz, see Figure S.31 linear.

5.7 See Figure S.32; resolution is 0.97 mm.

5.9 Refer to Figure 2.10; $R_1 = R_3 = 1\ \text{k}\Omega, R_2 = R_4 = 6.25\ \text{k}\Omega, C_3 = 0.02\ \mu\text{F}$. See Figure S.33 for plot.

5.13 $0.037\ \Omega$

5.15 .05% error

5.17 See Figures S.34 and S.35 for circuit and plot; linear; 4.88 lb/LSB.

5.19 1047 rad/s

5.21 $2.18\ \text{m/s}^2$

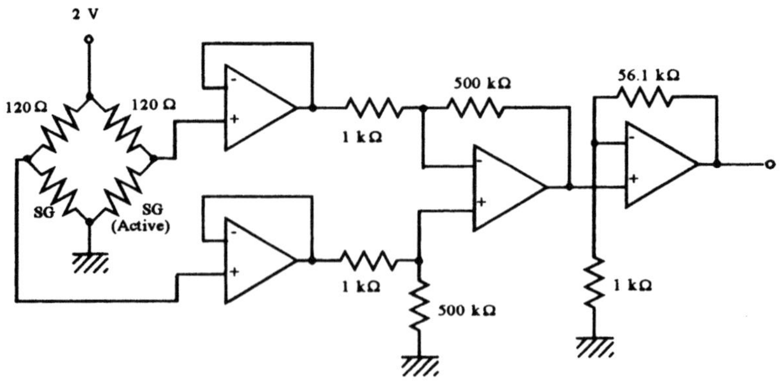

Figure S.34

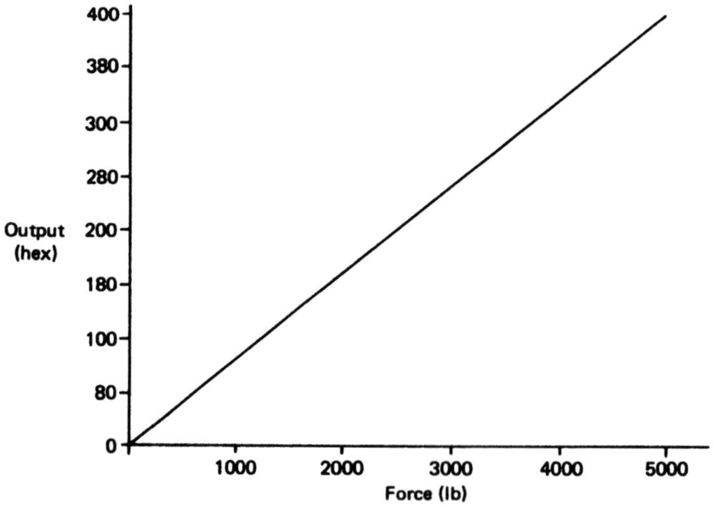

Figure S.35

5.23 2.58 gs

5.25 $V_{out} = [0.0000646\ \mathrm{V/(m/s^2)}]a, a_{max} = 96\ \mathrm{m/s^2}$, 11 Hz

5.27 1.2 mm

5.29 Use bridge with 120Ω in each arm, 5-V source. See Figure S.36 for the plot.

5.31 10.34 MPa, 102 atm

5.33 Max is 3.16 m/s; see Figure S.37, nonlinear.

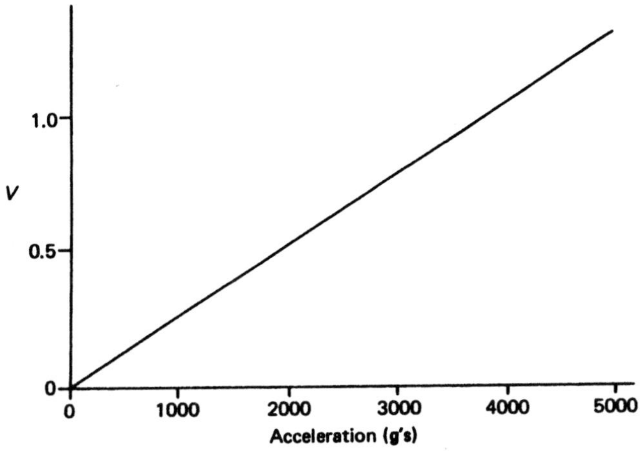

Figure S.36

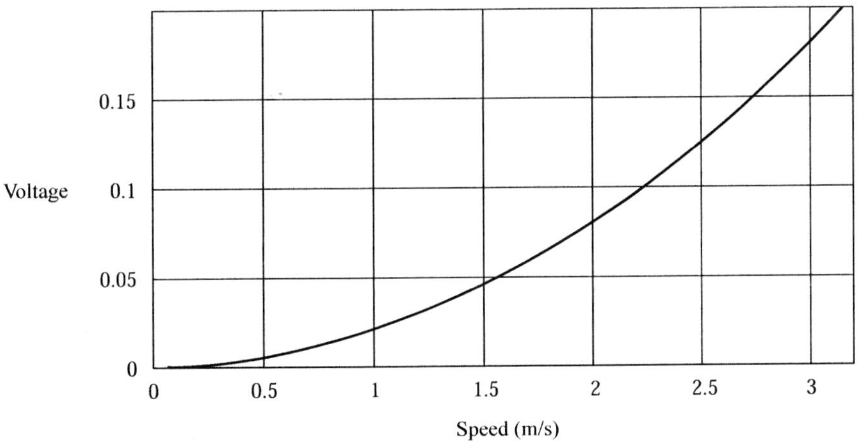

Figure S.37

5.35 197.7 m/hr

5.37 See Figure S.38.

Supplementary Problems

S5.1 **a.** Level at 9 m takes 13.33 ms, level at 1 m takes 66.67 ms.

b. Preset value is 110_{10} or $\mathbf{1101110_2}$, clock frequency is 1500 Hz.

c. See Figure S.39.

S5.3 **a.** See Figure S.40.

b. See Figure S.41.

S5.5 Use a bridge like Figure 2.5 with $R_1 = R_2 = 80\ \Omega$, R_3 and R_4 sensors, and a 24-V supply. See Figure S.42 for the plot.

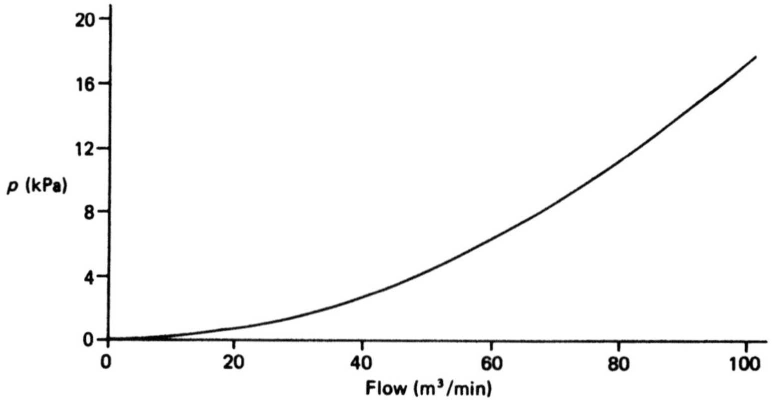

Figure S.38

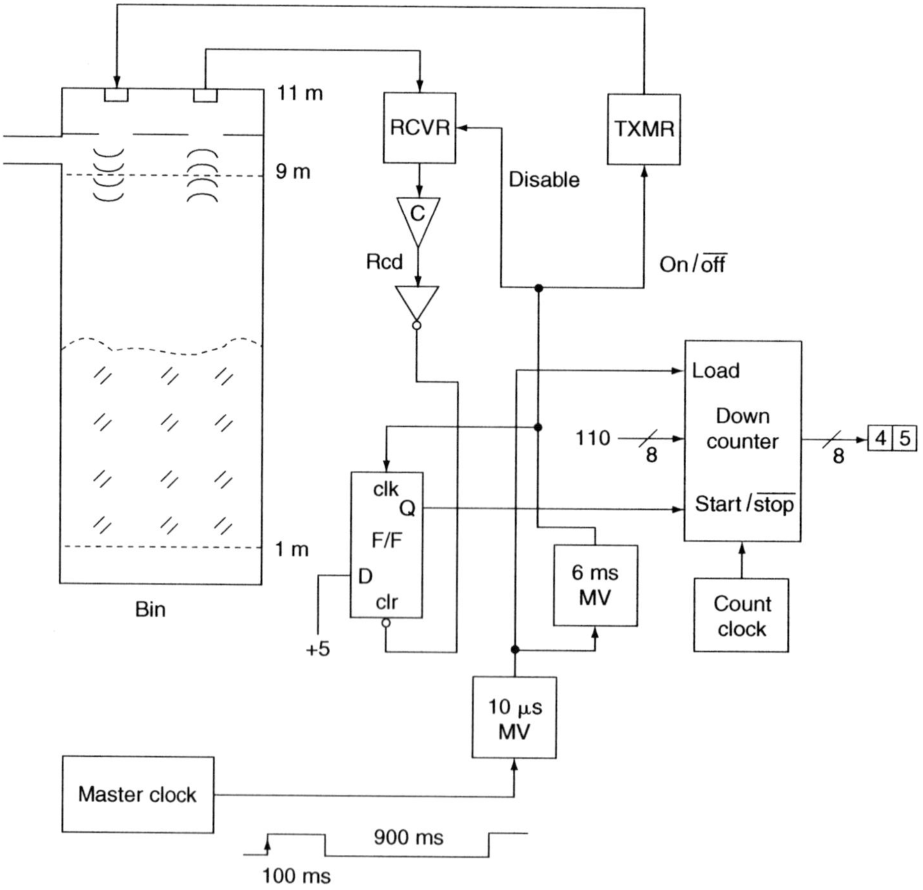

Figure S.39

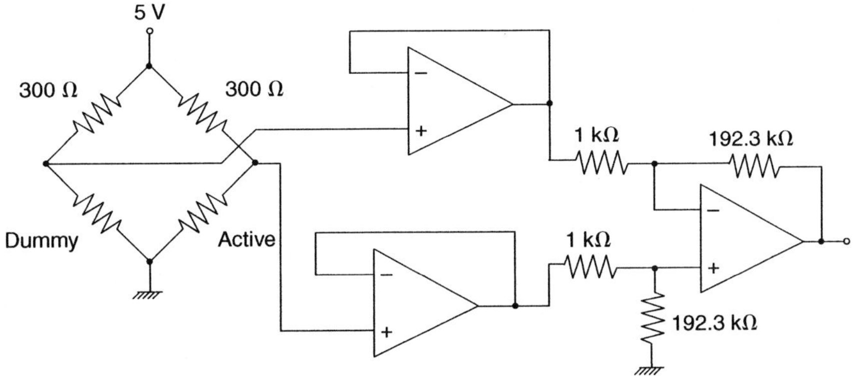

Figure S.40

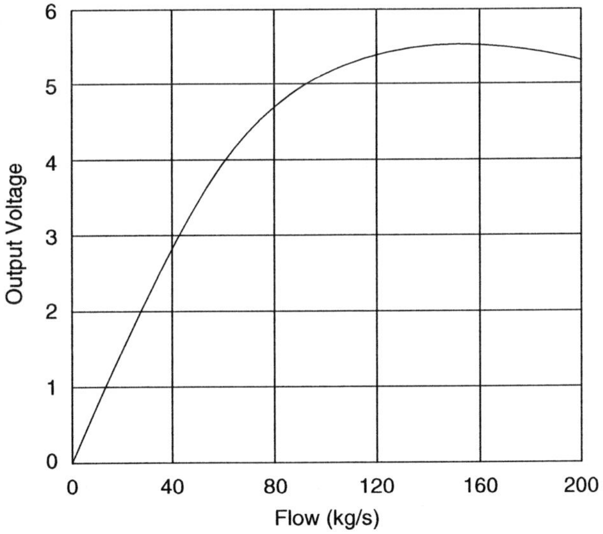

Figure S.41

Chapter 6

6.1 10^{10} Hz, microwave

6.3 0.33 μs without liquid, 0.57 μs with liquid

6.5 5.64 m

6.7 23-km diameter, 2.4 $\mu W/m^2$

6.9 See Figure S.43; 124.8 kΩ at 20 ms.

6.11 See Figure S.44.

6.13 See Figure S.45; $V_{max} = 0.8$ V

6.15 See Figure S.46.

6.17 See Figure S.47.

6.19 12.7 MW/m^2

6.21 0.36 m

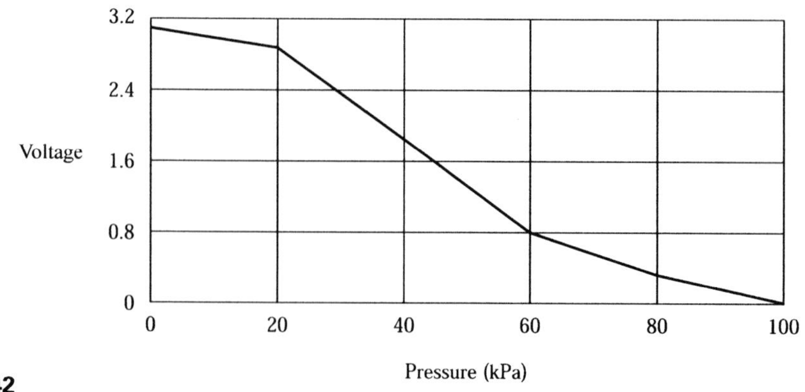

Figure S.42

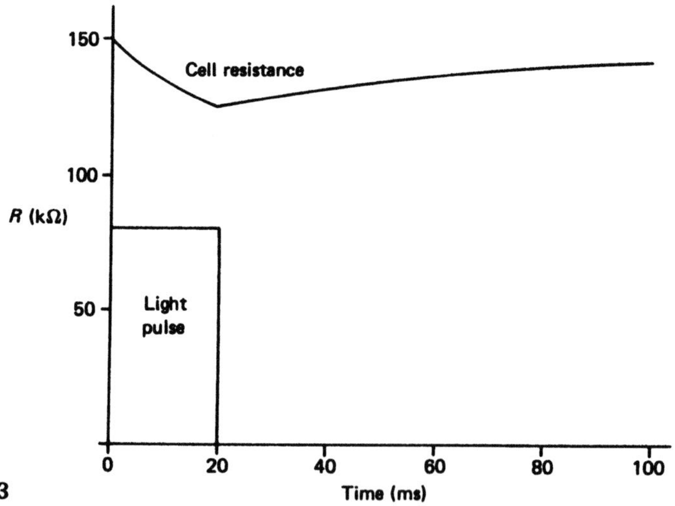

Figure S.43

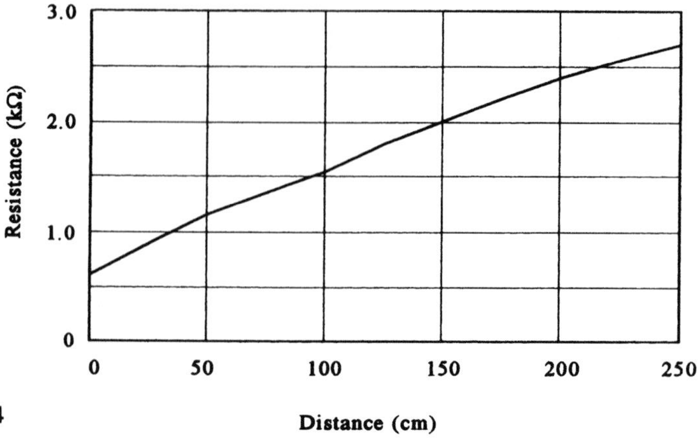

Figure S.44

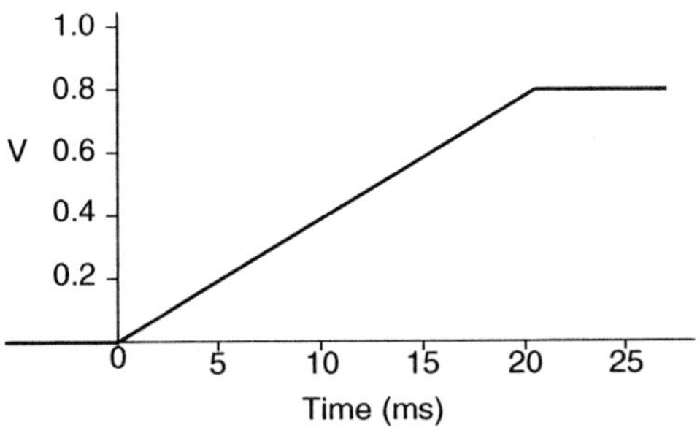

Figure S.45

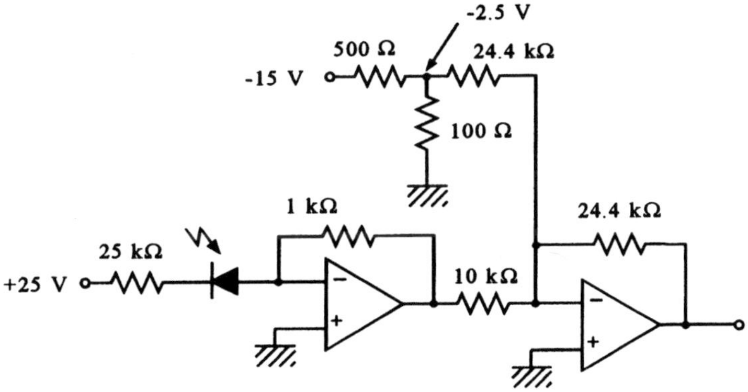

Figure S.46

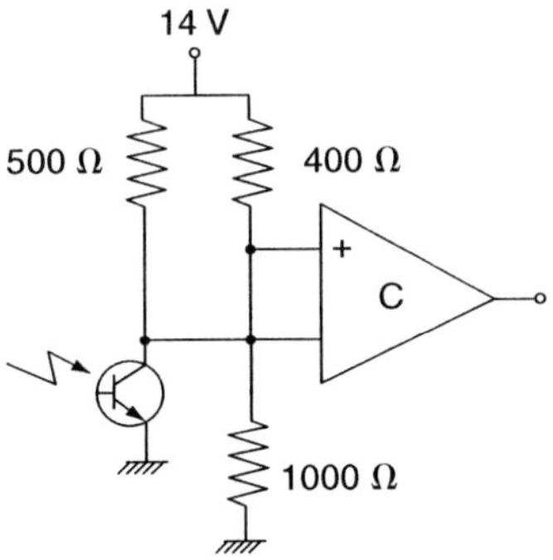

Figure S.47

Supplementary Problems

S6.1 **a.** See Figure S.48.

b. See Figure S.49, nonlinear.

S6.3 See Figure S.50 for the circuit. See Figure S.51 for the plot, linear from −0.5 to +0.5 cm, slightly nonlinear at each extreme.

Chapter 7

7.1 $V_{\text{out}} = 1250I + 15$

7.3 min: 01010_2, max: 10101_2, 93.75 rpm/LSB

7.5 R ranges from 19.6 to 372 kΩ. <P>=1.25 W. See Figure S.52.

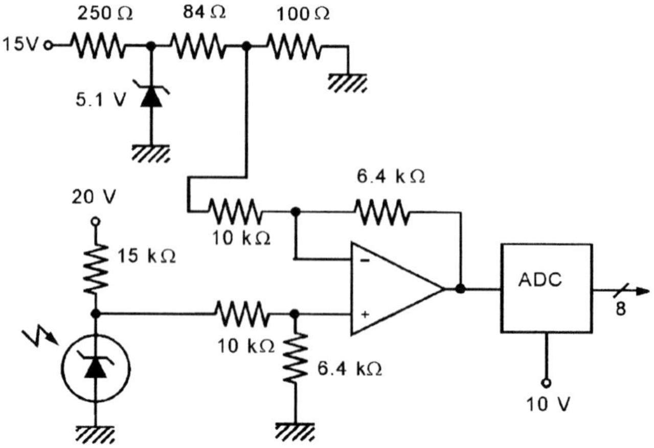

Figure S.48

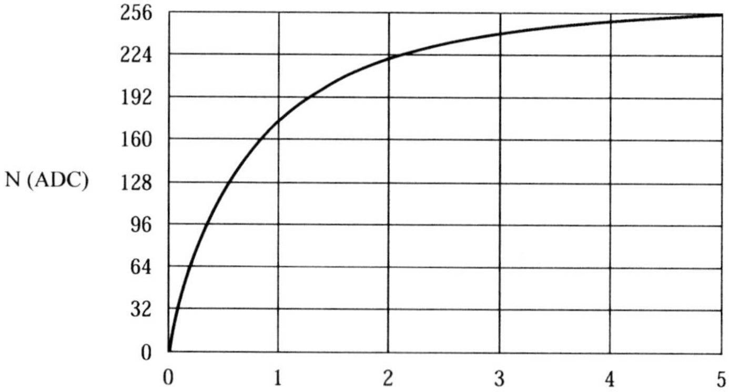

Figure S.49

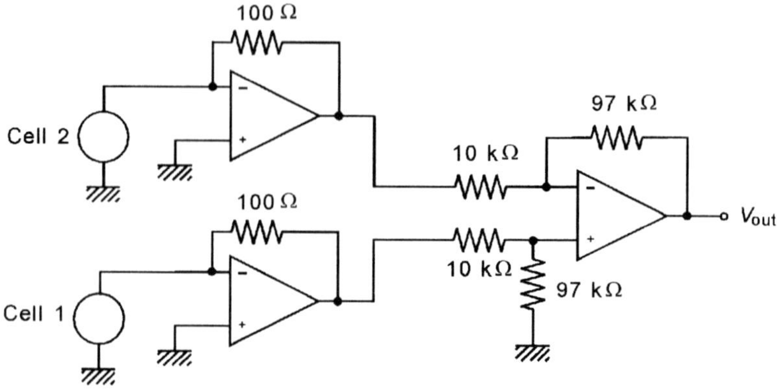

Figure S.50

Figure S.51

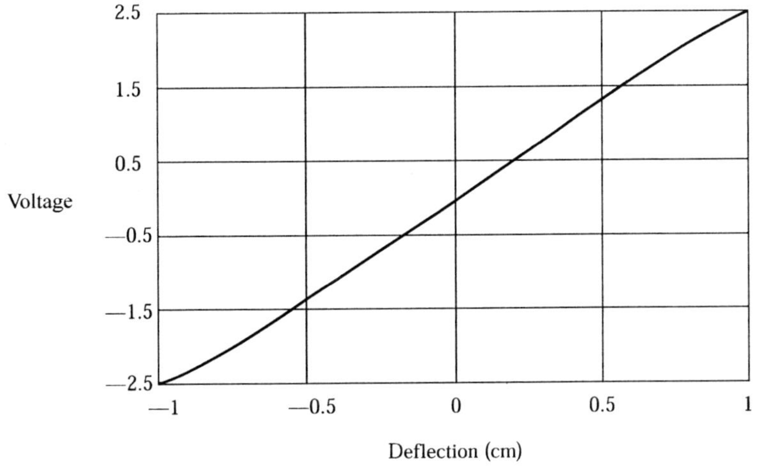

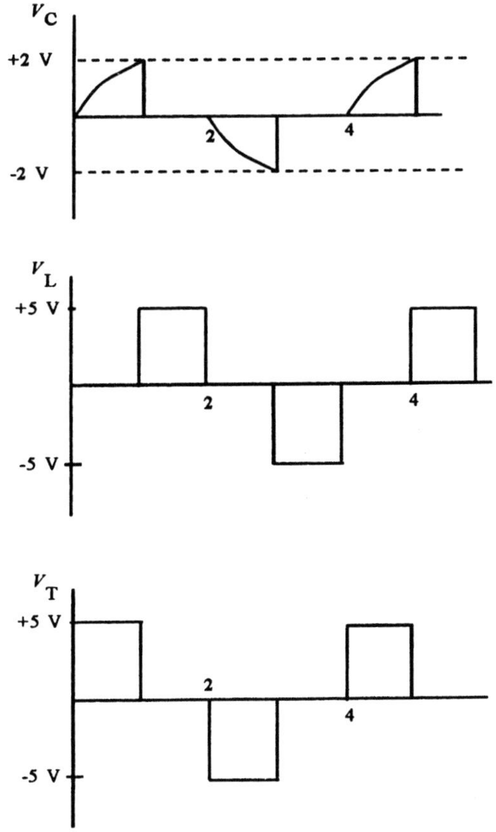

Figure S.52

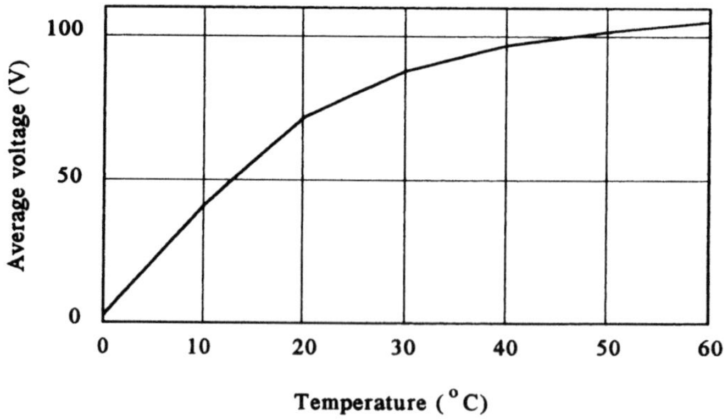

Figure S.53

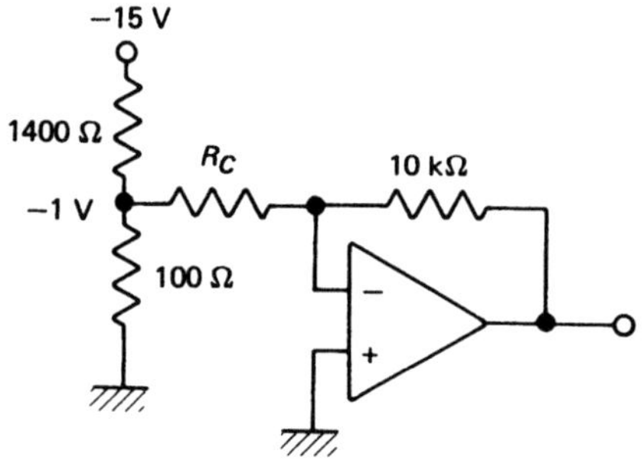

Figure S.54

7.7 $R = 875\ \Omega, C = 1.275\ \mu\text{F}$. See Figure S.53.

7.9 Distances are 8 cm and 5 cm. $F = 3.75$ N.

7.11 2500 rpm

7.13 12.25 N

7.15 See Figure S.54.

7.17 3.33 in^2.

7.19 $Q(2/3) = 31.5\ \text{m}^3/\text{h}, Q(4/5) = 50\ \text{m}^3/\text{h}$

7.21 $Q_{\text{min}} = 11.9\ \text{m}^3/\text{h}$, $Q_{\text{max}} = 356\ \text{m}^3/\text{h}$, 3.13 cm for a flow of 100 m^3/hr

7.23 **a.** 8.33 psi or 44.4%

b. 9.57 psi or 54.7%

Supplementary Problems

S7.1 Use the V/F converter of Figure 3.26 with $R_S = 10\ \text{k}\Omega, R_L = 10\ \text{k}\Omega, R_t = 19.9\ \text{k}\Omega$, and $C_t = 0.1\ \mu\text{F}$. By Equation 3.27, the frequency is $f = 240V_{\text{out}}$. See Figure S.55 for conversion of current to V_{out}. V_{out} varies from 1 .0 to 2.0V as I varies from 4 to 20 mA.

S7.3 See Figure S.56.

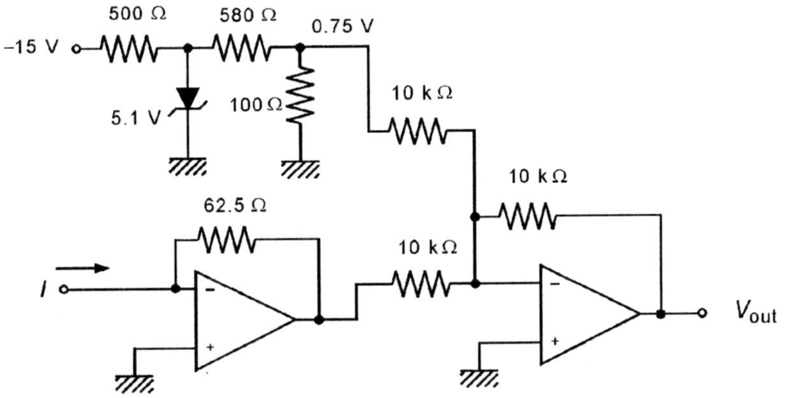

Figure S.55

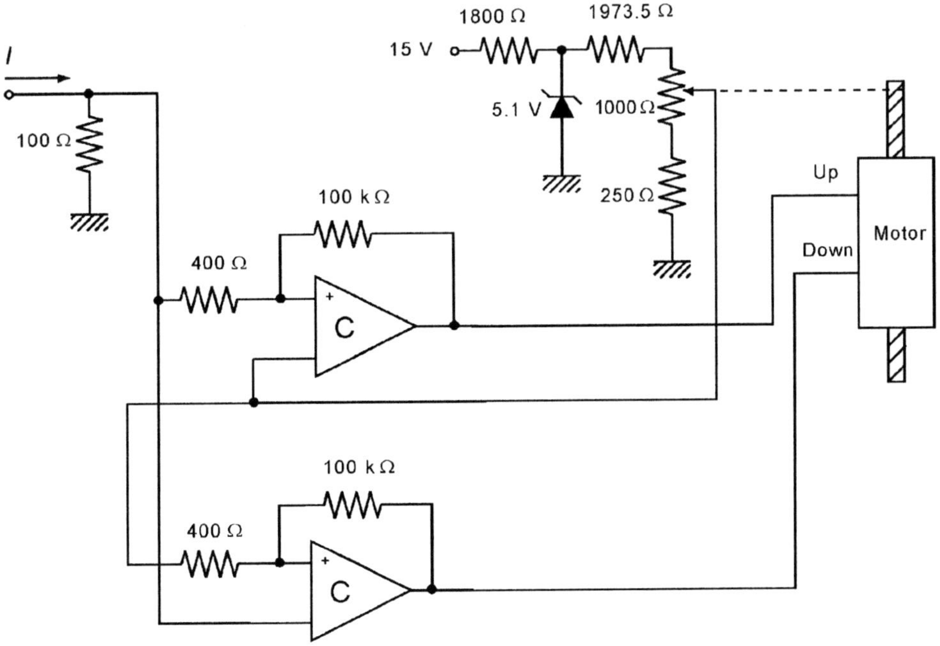

Figure S.56

Chapter 8

8.1 One possible solution is as follows:

Condition	Generator	Fan	Light	Door SW	PWR SW	Timer
1. Open door	OFF	OFF	ON	ON	OFF	OFF
2. Place food	OFF	OFF	ON	ON	OFF	OFF
3. Close door	OFF	OFF	OFF	OFF	OFF	OFF
4. Set timer	OFF	OFF	OFF	OFF	OFF	OFF
5. Power on	ON	ON	OFF	OFF	ON	ON
6. Open door	OFF	OFF	ON	ON	ON	OFF
7. Close door	ON	ON	OFF	OFF	ON	ON
8. Time up	OFF	OFF	OFF	OFF	OFF	OFF

8.3 See Figure S.57.

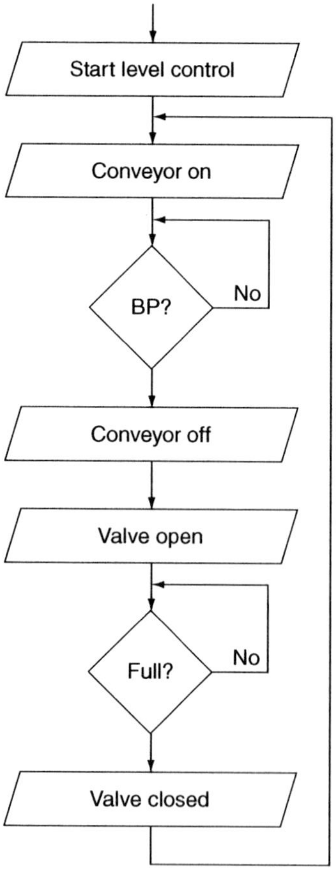

Figure S.57

8.5 Assume a state is given by

$$(BP)(BF)(R)(LC)(V)(M)$$

State solution is

State		Output	
1. XX0000	→	000	Idle
2. XX1000	→	001	Conveyor on
3. 0X1000	→	001	Conveyor on
4. 0X1001	→	001	Waiting for BP
5. 1X1001	→	110	Conveyor off, start level and fill
6. 101110	→	110	Waiting for BF
7. 111110	→	000	All off
8. XXX000	Go to 1		

8.7 See Figure S.58.

8.9 See Figure S.59.

8.11 See Figure S.60.

8.13 See Figure S.61.

8.15 See Figure S.62.

Supplementary Problems

S8.1 See Figure S.63.

S8.3 See Figure S.64.

Figure S.58

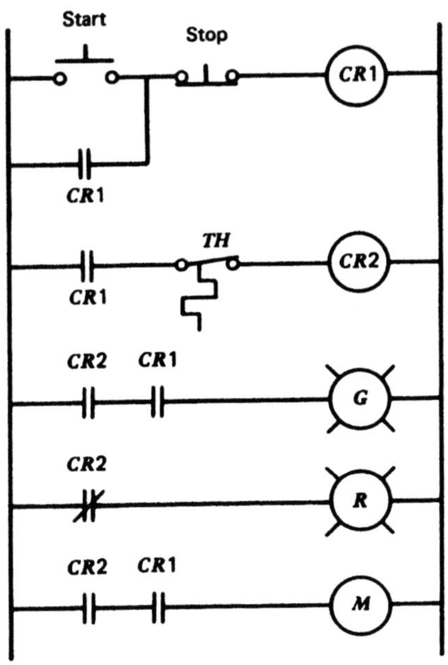

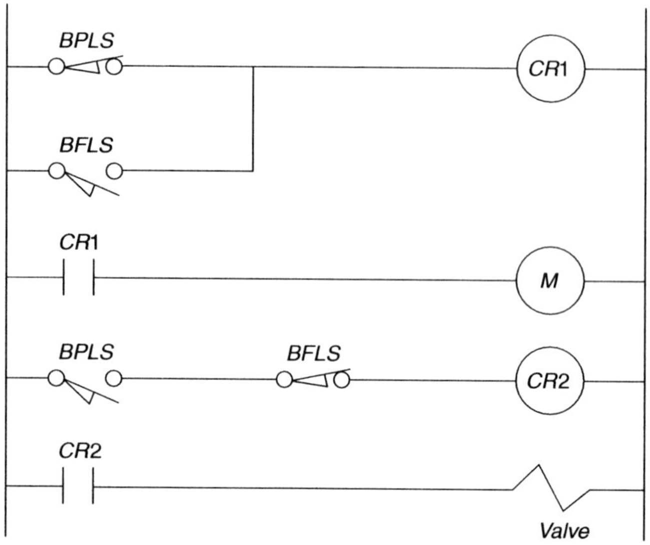

Figure S.59

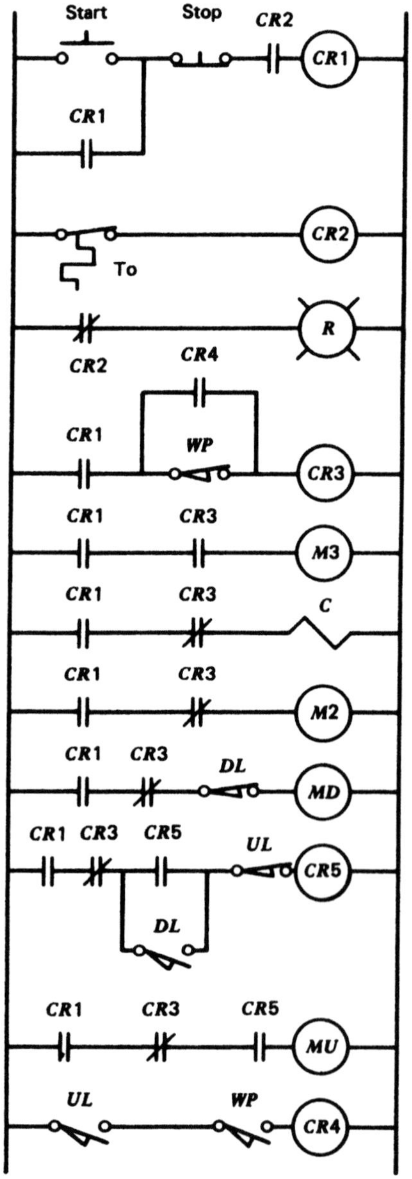

Figure S.60

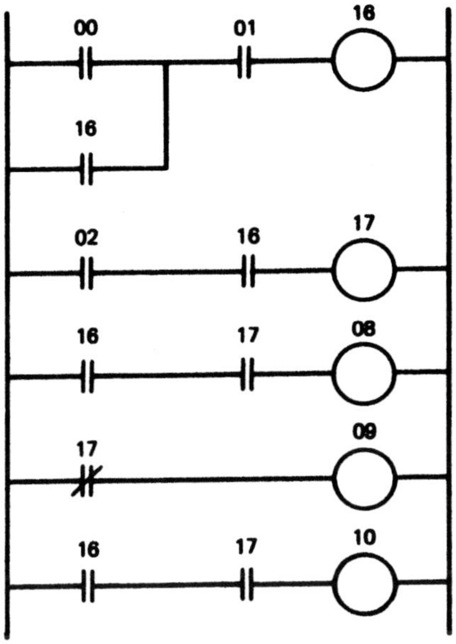

Figure S.61

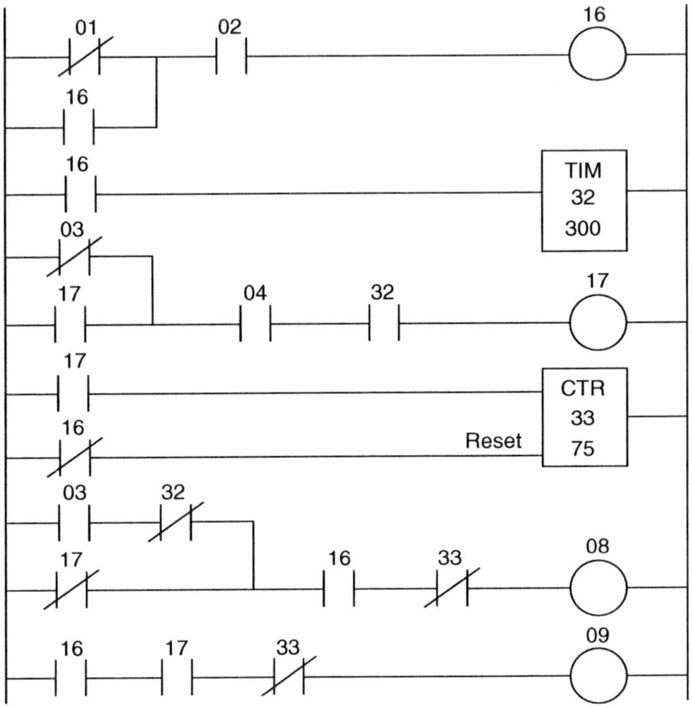

Inputs:
01 = ON switch (NC)
02 = OFF switch (NC)
03 = Right limit switch (NC)
04 = Left limit switch (NC)

Outputs:
08 = CW motor
09 = CCW motor

Timer tick = 0.1 second

Figure S.62

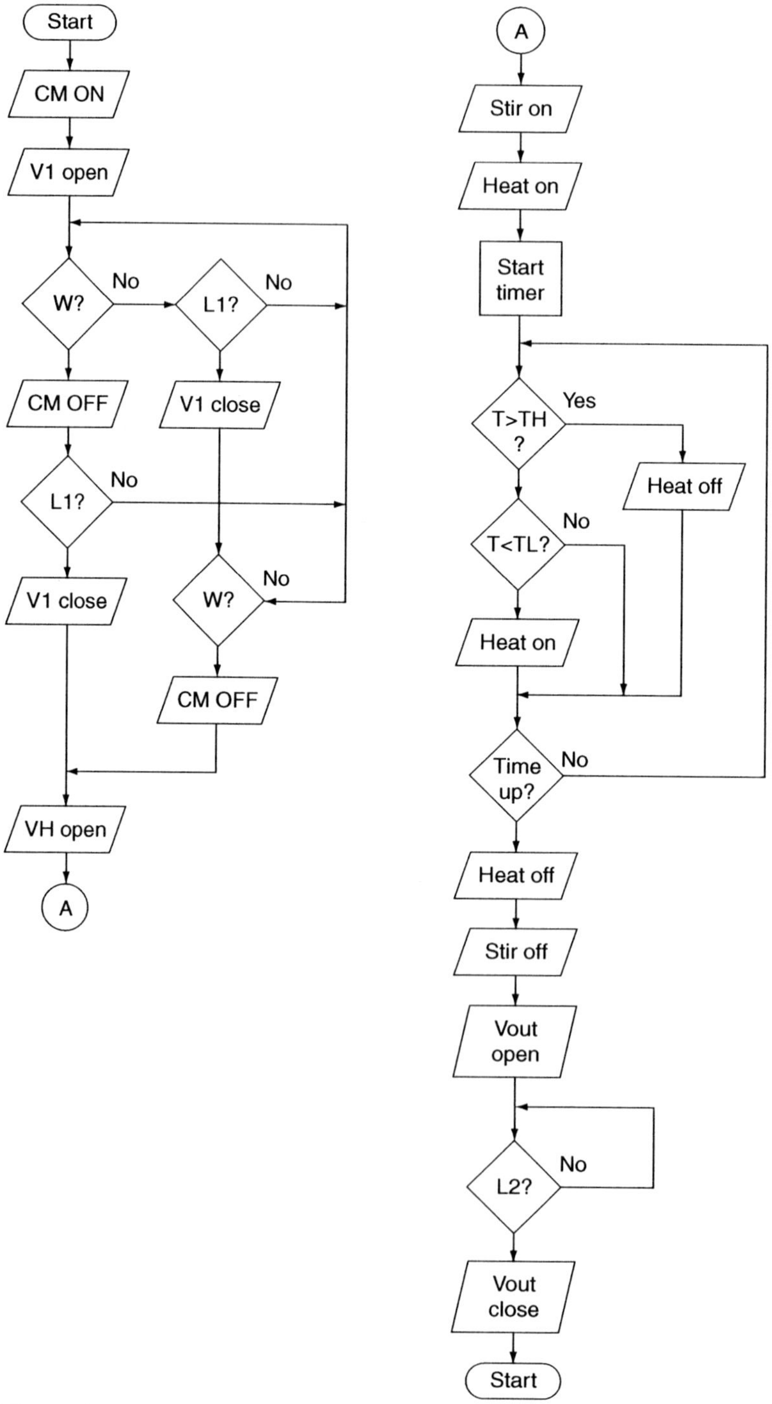
Start
CM ON
V1 open
W?
No
L1?
No
CM OFF
V1 close
L1?
No
V1 close
W?
No
CM OFF
VH open
A
A
Stir on
Heat on
Start timer
T>TH ?
Yes
Heat off
T<TL?
No
Heat on
Time up?
No
Heat off
Stir off
Vout open
L2?
No
Vout close
Start

Figure S.63

W = 00	CM = 08	Timer = 32
L1 = 01	V1 = 09	
L2 = 02	VH = 10	
TL = 03	SM = 11	
TH = 04	VOUT = 13	

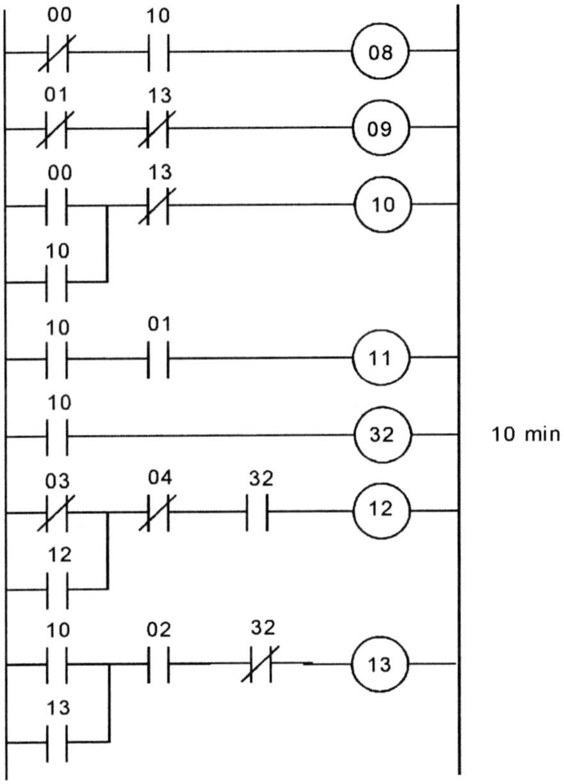

Figure S.64

Chapter 9

9.1 Q_A, T_0, T_A, Q_B

9.3 $e_p = 13.75\%$

9.5 Period = 39.3 min; see Figure S.65.

9.7 See Figure S.66.

9.9 **a.** 83.75%

b. 0%

c. $28.75\% - 12.5t\%$

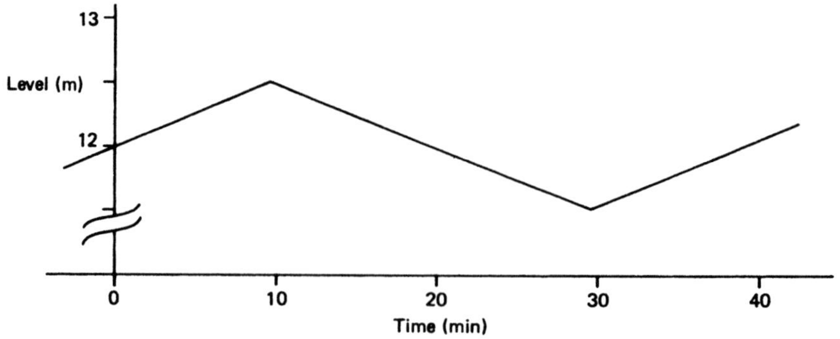

Figure S.65

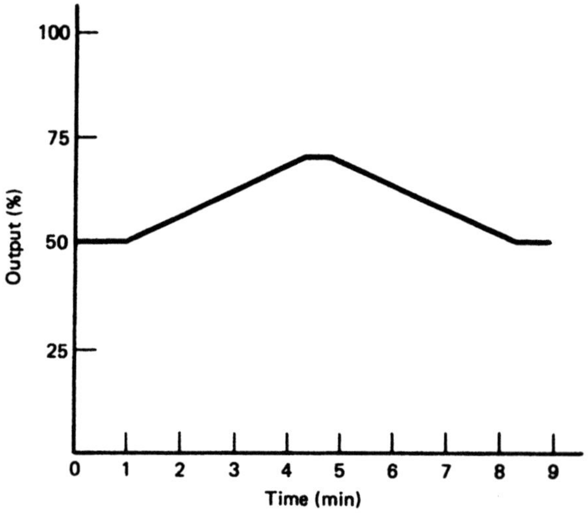

Figure S.66

9.11 $P_D = 0.352 \cos(0.04t)$

9.13 See Figure S.67.

9.15 See Figure S.68.

9.17 0.72 min

9.19 $p(t) = 13.5 \sin(\pi t) - 30[\cos(\pi t) - 1]$; phase shift $= 65.8°$ lag

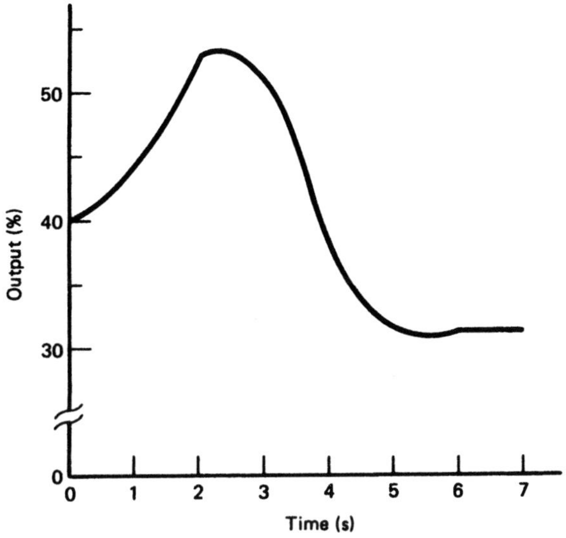

Figure S.67

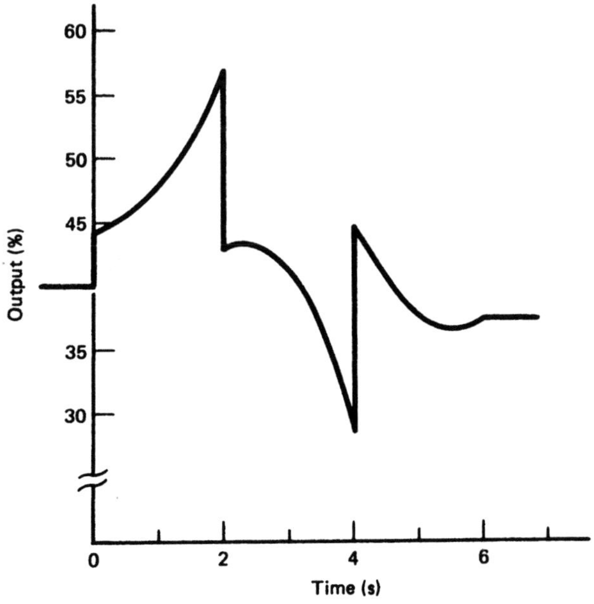

Figure S.68

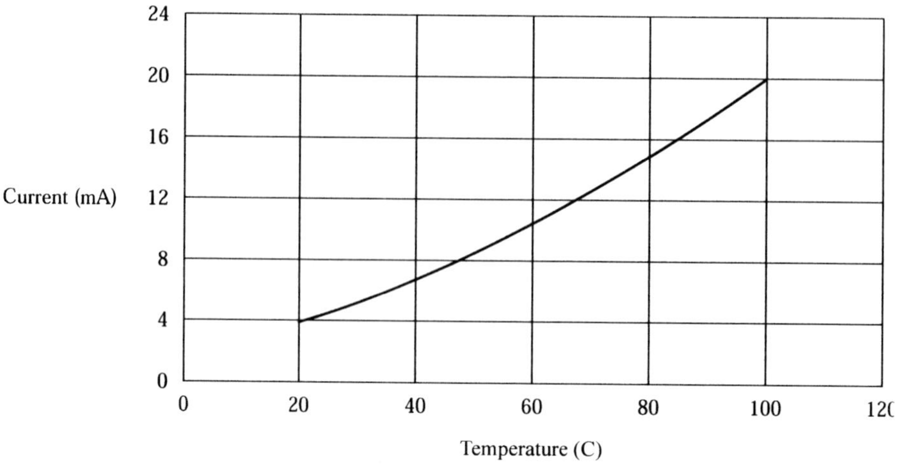

Figure S.69

Supplementary Problems

S9.1 **a.** See Figure S.69.

b. 0.738 mA and 1.215 mA or 5.27% and 7.59%.

c. $e_p(t) = 5.27t$ and $e_p(t) = 7.59t$.

S9.3 See Figure S.70 (dc amplitude is not included). Amplitude increases without limit at low and high frequency. Phase shift changes from −90° to +90°.

Chapter 10

10.1 **a.** 1.08 V

b. −1.1 V to 0.5 V

10.3 Use Figure 10.5 with $R_1 = 2.2\ \text{k}\Omega$, $R_2 = 10\ \text{k}\Omega$, and V_{sp} set by a divider of the +15-V supply to 5.4 V.

10.5 Use Figure 10.5 with V_{sp} selected from a divider of the 15-V supply to be 1.356 V, $R_2 = 100\ \text{k}\Omega$, and $R_1 = 1.1\ \text{k}\Omega$.

10.7 See Figure S.71. Offset error = 4.5%.

10.9 $G_I = 0.343\ \text{s}^{-1}, R = 292\ \text{k}\Omega, C = 10\ \mu\text{F}$

10.11 In Figure 10.10, $R_1 = 100\ \text{k}\Omega$, $R_2 = 300\ \text{k}\Omega$, and $C = 2.5\ \mu\text{F}$.

10.13 Use Figure 10.9 with $R_1 = 10\ \text{k}\Omega, R_2 = 62.5\ \text{k}\Omega, C_D = 10\ \mu\text{F}, R_D = 937.5\ \text{k}\Omega$, $C_I = 1\ \mu\text{F}, R_I = 576\ \text{k}\Omega, R_3 \ll 75\ \text{k}\Omega$.

10.15 8.33 psi and 5.57 psi

10.17 Replace single capacitor with a four-position switch with a capacitor for each required reset time.

10.19 See Figure S.72.

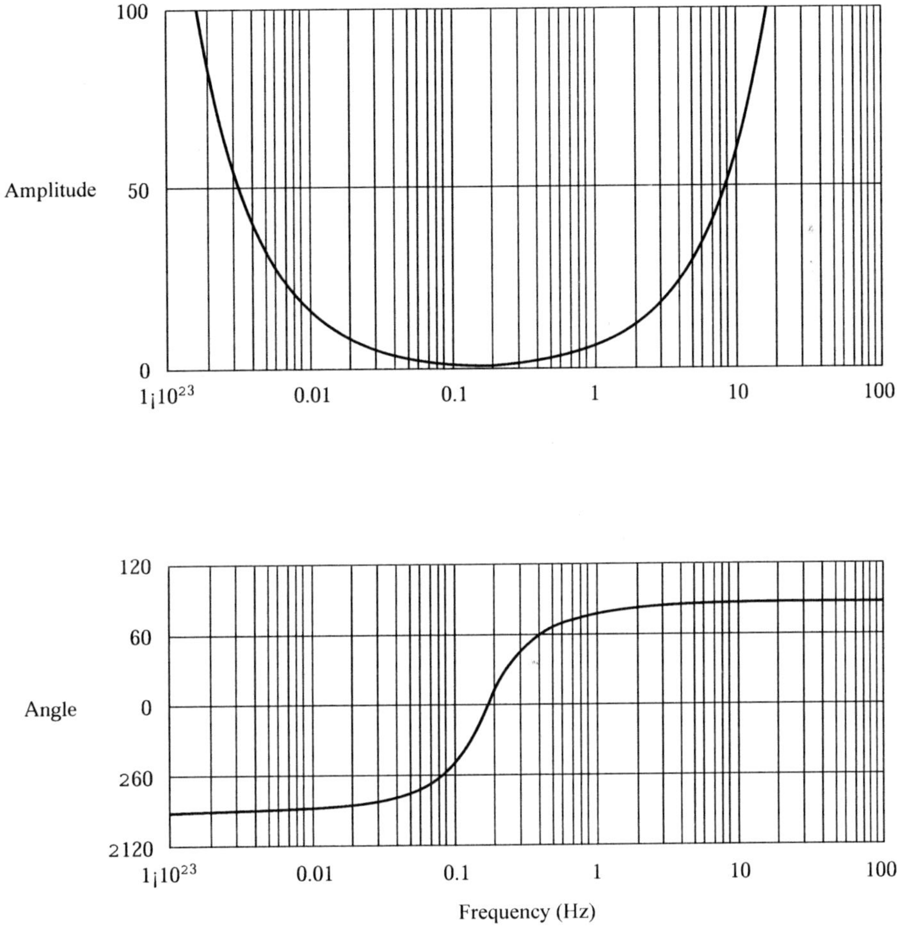

Figure S.70

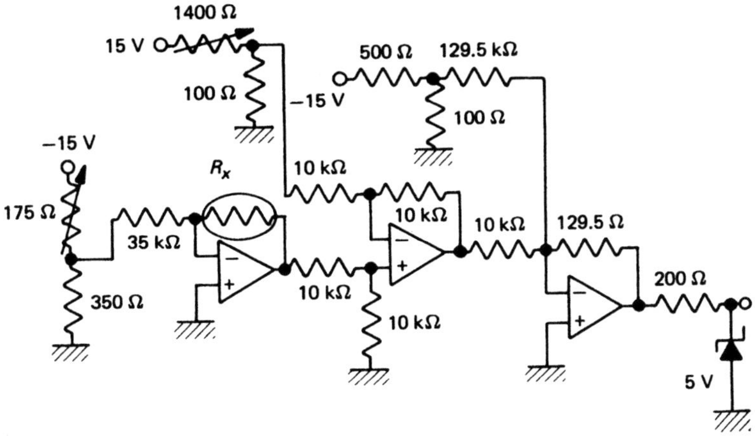

Figure S.71

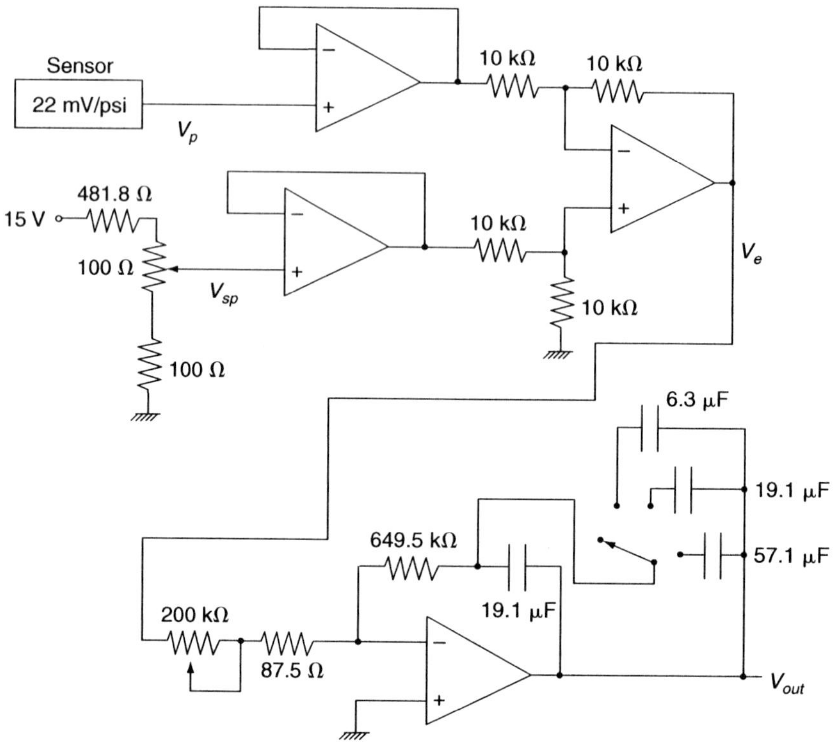

Figure S.72

Supplementary Problems

S10.1 See Figure S.73.

S10.3 Error detector is shown in Figure S.74. Controller is Figure 10.11 with $C = 10\ \mu F$, $R_1 = 962\ \Omega$, $R_2 = 1.89\ M\Omega$, and $R_3 = 137\ k\Omega$.

Chapter 11

11.1 Use a comparator with a trigger level of 2.31 V.

11.3 Use the circuit of Figure 11.1 with dividers to produce 1.1 V and 0.935 V.

11.5 See Figure S.75.

11.7 250 μs

11.9 See Figure S.76.

11.11 Error is $E = (24 - N_T)/63$, controller output is $P = 8 * E + 32$. Resolution is 0.625°F.

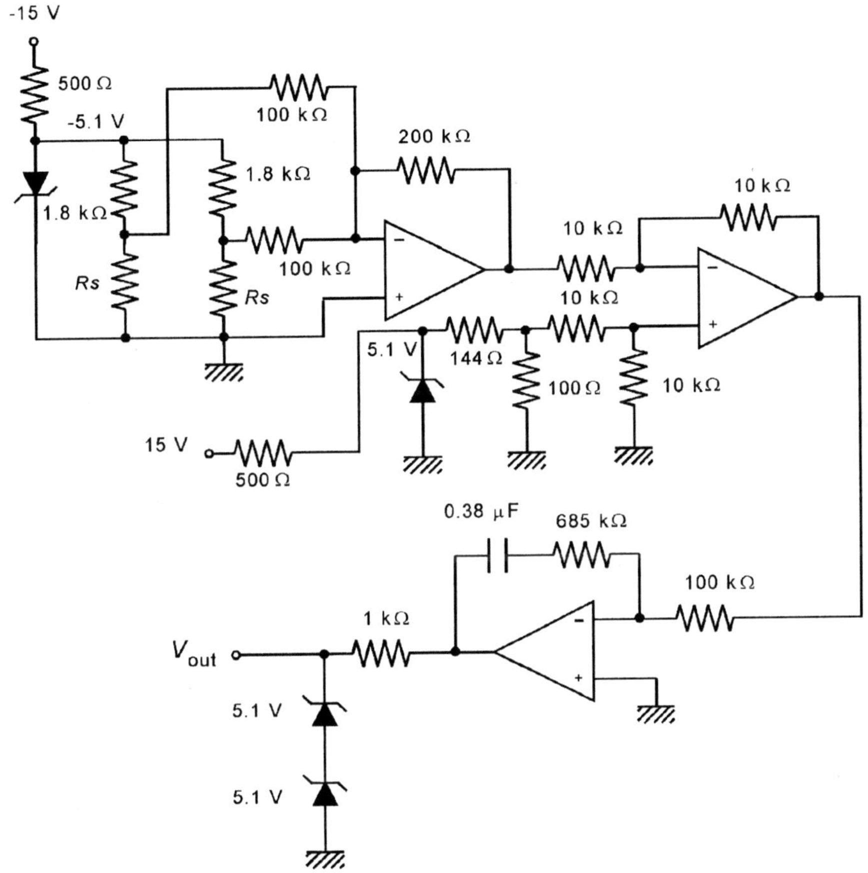

Figure S.73

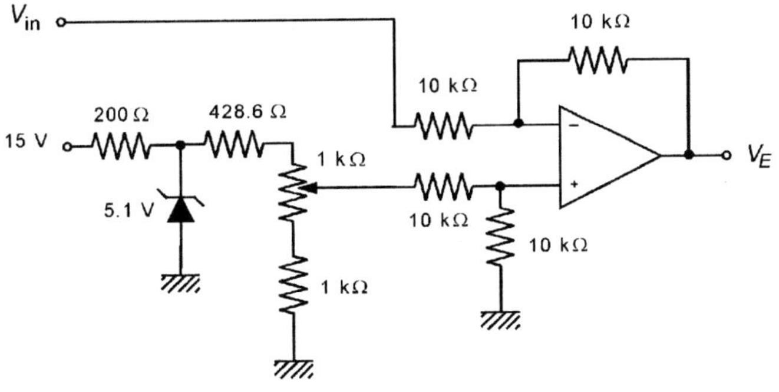

Figure S.74

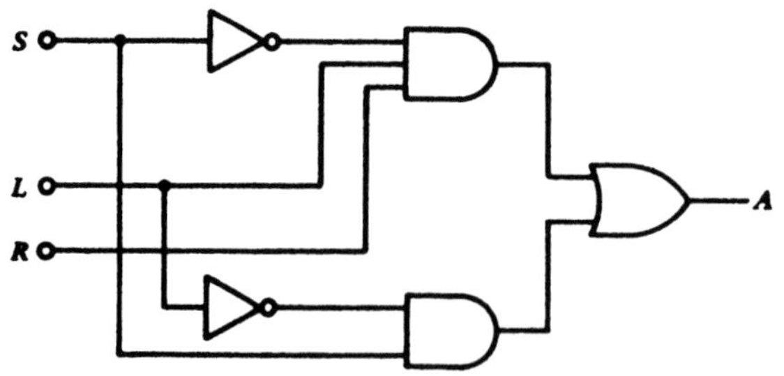

Figure S.75

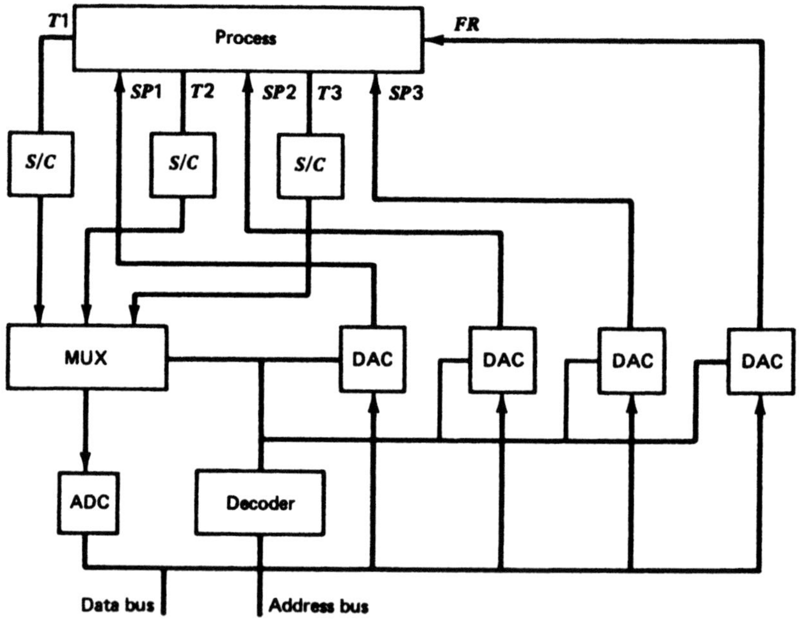

Figure S.76

11.13 See Figure S.77.

11.15 The appropriate equations, in program format, are as follows:

$$\begin{aligned}
\text{Assuming } N &= \text{input from ADC}\\
V &= N * 5/256\\
L &= 50 * \text{LOG}(2 * V + 1)\\
E &= 100 * (50 - L)/119.7; \text{Error}\\
SUM &= SUM + E\\
P &= 3.7 * E + 0.111 * SUM\\
OUT &= P * 255
\end{aligned}$$

Figure S.77

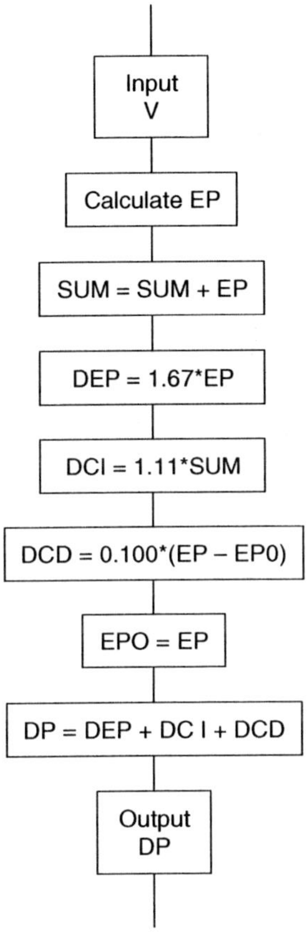

or, alternatively,

$$
\begin{aligned}
P &= P_0 + 3.811 * E - 3.7 * E0 \\
P0 &= P \\
E0 &= E \\
OUT &= P * 25
\end{aligned}
$$

Supplementary Problems

S11.1 **a.** See Figure S.78.

b. See Figure S.78.

c. See Figure S.79.

d. **1.** Set Temperature: Input VTSP $(0 - 255)$. Use in equation:

$$\mathrm{TSP} = 30 + 20 * \mathrm{VTSP}/255$$

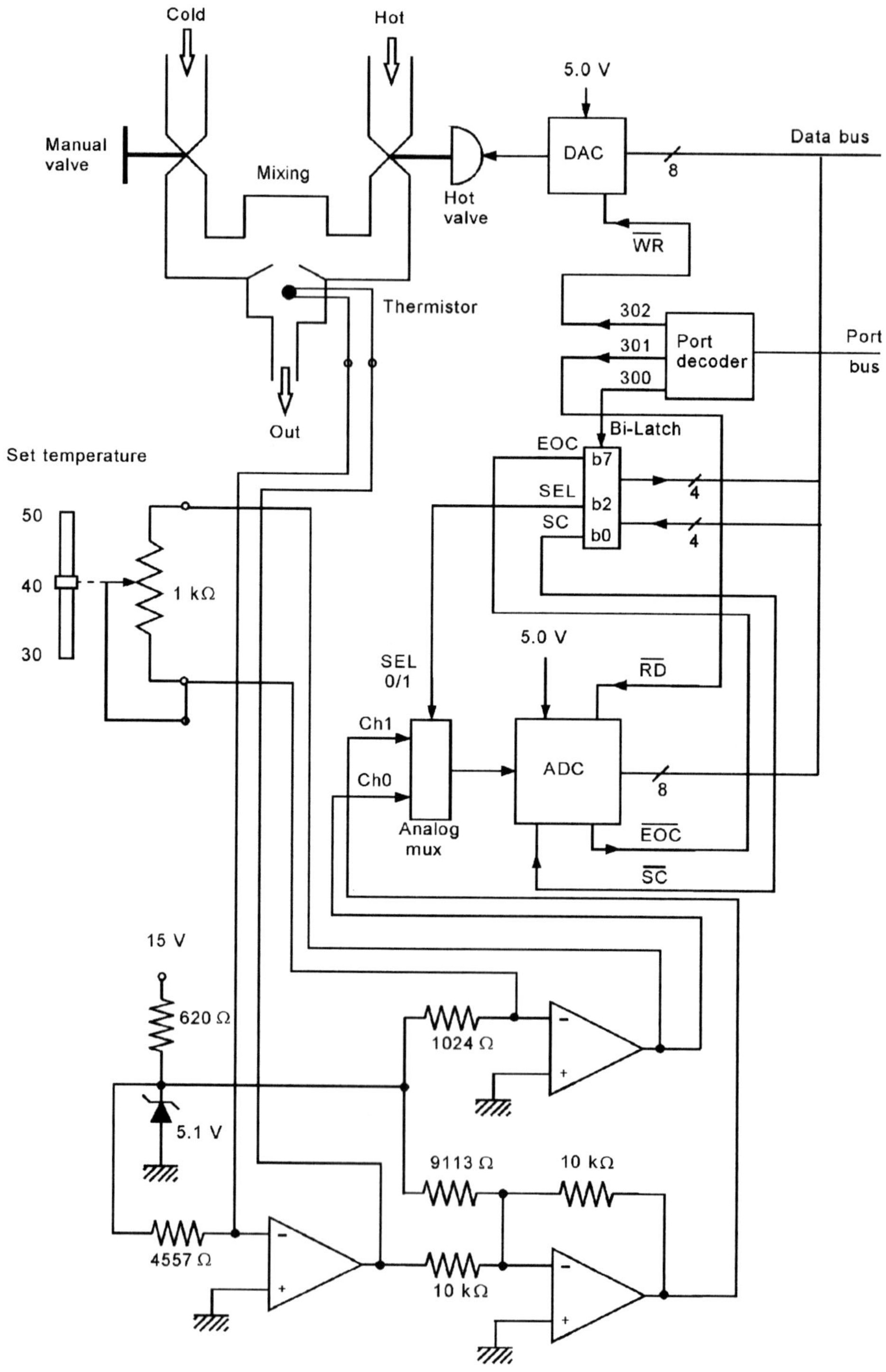
Cold
Hot
5.0 V
Manual
valve
Mixing
Hot
valve
DAC
Data bus
8
WR
Thermistor
302
301
300
Port
decoder
Port
bus
Out
Bi-Latch
EOC
b7
SEL
b2
SC
b0
4
4
Set temperature
50
40
30
1 kΩ
5.0 V
SEL
0/1
RD
Ch1
Ch0
Analog
mux
ADC
8
EOC
SC
15 V
620 Ω
1024 Ω
5.1 V
9113 Ω
10 kΩ
4557 Ω
10 kΩ

Figure S.78

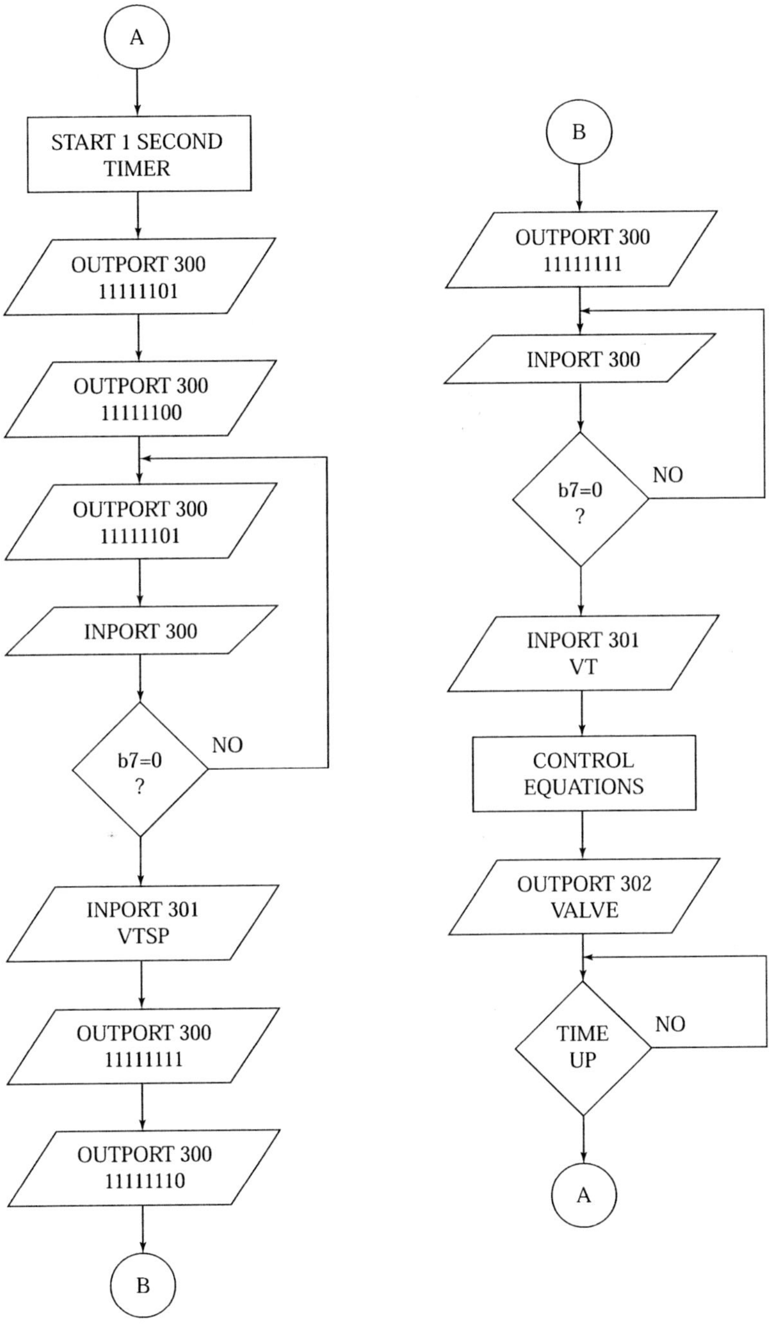
A
START 1 SECOND TIMER
OUTPORT 300 11111101
OUTPORT 300 11111100
OUTPORT 300 11111101
INPORT 300
b7=0 ?
NO
INPORT 301 VTSP
OUTPORT 300 11111111
OUTPORT 300 11111110
B
B
OUTPORT 300 11111111
INPORT 300
b7=0 ?
NO
INPORT 301 VT
CONTROL EQUATIONS
OUTPORT 302 VALVE
TIME UP
NO
A

Figure S.79

S11.2. Input Temperature: Input VT $(0 - 255)$. Use table-lookup and linear interpolation. Try 8-point table.

VT	0	32	64	96	128	160	192	224	255
T	30	31	32	33.5	35.2	37.2	40	43.8	50

Interpolation: $T = TL + (TH - TL) * (VT - VL)/32$

Error: $E = (TSP - T)/20$ (fraction of range)

S11.3. Control equations (use Equation (11.17) with KD = 0):

$$\begin{aligned} VALVE &= VALVE0 + 8.9 * E - 6.7 * E1 \\ VALVE0 &= VALVE \\ E1 &= E \end{aligned}$$

Chapter 12

12.1 See Figure S.80.

12.3 See Figure S.81.

12.5 See Figure S.82.

12.7 4.48; 85°

12.9 **a.** $K_P = 2.1$

b. $K_P = 1.89, T_I = 1.84\,\text{min}$

c. $K_P = 2.52, T_I = 1.11\,\text{min}, T_D = 0.28\,\text{min}$

12.11 $K_P = 3.29$

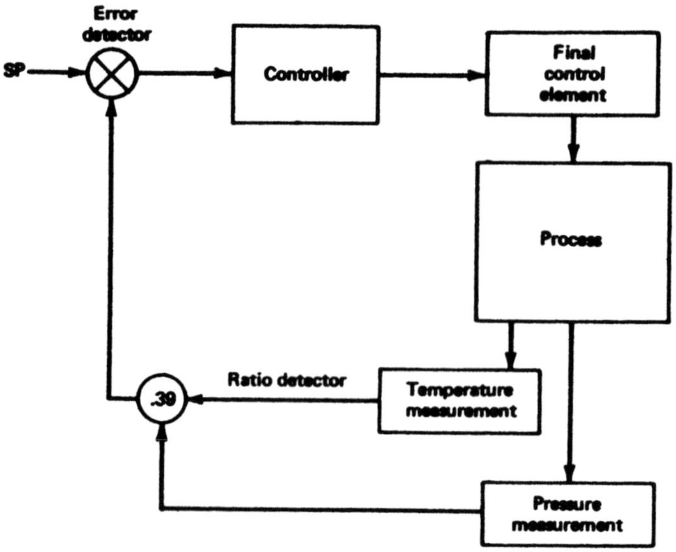

Figure S.80

Figure S.81

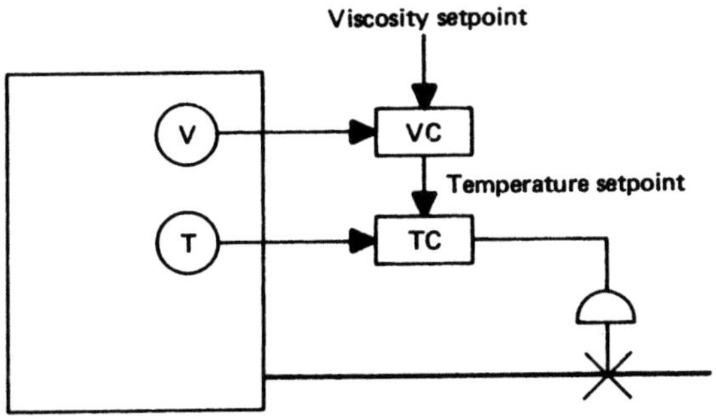

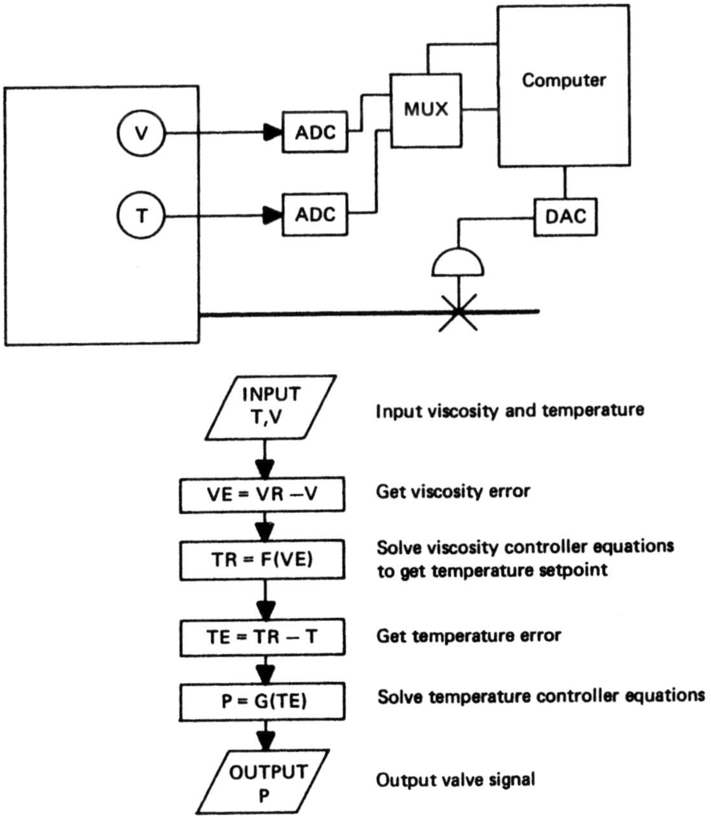

Figure S.82

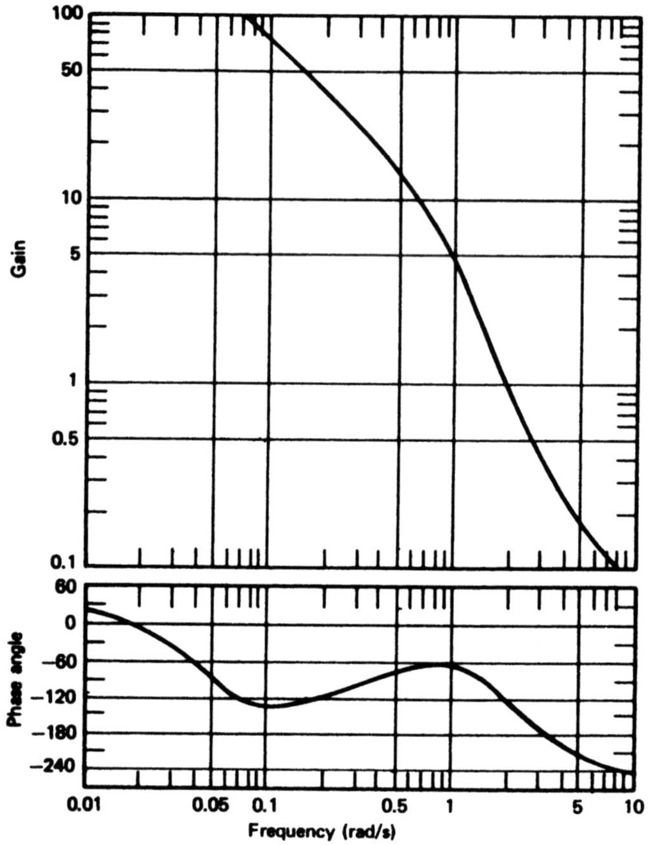

Figure S.83

12.13 See Figure S.83. Stable.

12.15 Instability for $K_P = 4.55$; ultimate cycle gives $K_P = 2.3$, gain margin of 0.5, or about 6 dB, and a phase margin of 140°.

12.17 **a.** At unity gain, the phase lag is about 180° and therefore unstable. At 180°, the gain is about 1.0, greater than 0.56, and therefore unstable.

b. Reduce gain to about 2.8.

Supplementary Problems

S12.1 $K_P = 0.57$ and $T_I = 2.16$ s; quarter amplitude: $K_P = 0.6$ and $T_I = 0.96$

INDEX